# Environmental Pollution

Volume 30

**Series Editor**

J. Trevors
GUELPH, ON, Canada

The Environmental Pollution book series includes current, comprehensive texts on critical national and global environmental issues useful to scientists in academia, industry and government from diverse disciplines. These include water, air, and soil pollution, organic and inorganic pollution, risk assessment, human and environmental health, environmental biotechnology, global ecology, mathematics and computing as related to environmental pollution, environmental modelling, environmental chemistry and physics, biology, toxicology, conservation and biodiversity, agricultural sciences, pesticides, environmental engineering, bioremediation/biorestoration, and environmental economics. Environmental problems and solutions are complex and interrelated. Complex problems often require complex solutions. The linkage of many disciplines can result in new approaches to old and new environmental problems as well as pollution prevention. This knowledge will assist in understanding, maintaining and improving the biosphere in which we live. Proposals for this book series can be sent the Series Editor:

Jack Trevors at jtrevors@uoguelph.ca
or the Publishing Editor
Zachary Romano at Zachary.Romano@springer.com

More information about this series at http://www.springer.com/series/5929

M. A. Rodrigo • E. V. Dos Santos
Editors

# Electrochemically Assisted Remediation of Contaminated Soils

## Fundamentals, Technologies, Combined Processes and Pre-Pilot and Scale-Up Applications

Springer

*Editors*
M. A. Rodrigo
Department of Chemical Engineering
Faculty of Chemical Sciences & Technologies
University of Castilla La Mancha
Campus Universitario
Ciudad Real, Spain

E. V. Dos Santos
School of Science and Technology
Federal University of Rio Grande Do Norte
Natal, RN, Brazil

ISSN 1566-0745 ISSN 2215-1702 (electronic)
Environmental Pollution
ISBN 978-3-030-68142-5 ISBN 978-3-030-68140-1 (eBook)
https://doi.org/10.1007/978-3-030-68140-1



This Springer imprint is published by the registered company Springer Nature Switzerland AG
The registered company address is: Gewerbestrasse 11, 6330 Cham, Switzerland

# Preface

It gives us immense pleasure to introduce the book *Electrochemically Assisted Remediation of Contaminated Soils: Fundamentals, Technologies, Combined Processes and Pre-Pilot and Scale-Up Applications* based on an overview of current developments in remediation technologies, related to the application of electrochemistry, for cleaning up the contaminated soil with organic and inorganic compounds and their fundamentals.

The treatment of polluted soils is a matter of major relevance. Among the great variety of technologies that can help to remediate polluted sites, electrochemically assisted processes have gained great relevance in recent years, because of the promising results obtained in research studies: hundreds of papers have been published in the last few decades providing a very favorable view on the expected impact of these technologies in the near future.

These processes are driven by the application of electric fields among electrodes placed in the polluted soil and, unfortunately, they are not simple. In fact, the application of the electric fields triggers a plethora of processes of different sort including (1) physical processes like soil heating, which can help to mobilize volatile and semivolatile pollutants; (2) chemical processes, like the precipitation or the redissolution or the ionic exchange of pollutants; (3) electrokinetic processes like the electro-osmotic transport of water, the electromigration of ions, and the electrophoresis of charged micelles; and (4) electrochemical processes, like water oxidation and reduction on the surfaces of the electrodes. These processes influence importantly on the concentration of the pollutants contained in soil and sometimes they can be synergetic: others, antagonistic. However, always experimental observations have a clear explanation based on the fundamentals of the processes involved. In fact, this interaction of the electric fields with the pollutants contained in the soil can be modified with the many inputs that affect all these processes and which include not only the characteristics of soil but also many others such as electrodes placement, addition of soil flushing fluids, and use of permeable reactive barriers. The advances in this interdisciplinary area have recently encouraged a close collaboration between chemists, electrochemists, engineers, and other scientists, particularly in the applicability of these technologies for many industrial processes.

With this background, this book gives an overview of current developments in remediation technologies, related to the application of electrochemistry, for cleaning up the contaminated soil with organic and inorganic compounds and their fundamentals. It has been written by many internationally recognized experts and comprises 21 chapters describing the characteristics and theoretical fundaments of the different electrochemically assisted soil remediation processes. The book is aimed to be used for students at the university level and professionals in the industry.

The book is organized into three different parts. The first part, with four chapters (Chapters "Physicochemical and Hydrodynamic Aspects of Soil," "Fundamental of Electrokinetic Processes," "Fundamental of Reactive and Thermal Processes in Electrochemically Assisted Soil Remediation," and "Conceptual and Mathematical Modeling of the Transport of Pollutants in Soil by Electric Fields"), tries to shed light on the fundamentals of the electrochemically assisted processes by describing the complexity of the soil and the most important aspects of the thermal, electrokinetic, and chemical processes that occur in soil when electric fields are applied. It starts with the description of soil characteristics, which are essential to understand the performance of the different electrochemically based technologies and to develop processes that provide higher efficiency in removing organic and inorganic compounds. This part also discusses the primary processes resulting from soil remediation under application of electric fields, helping to understand how operating conditions can affect the mobility or destruction of pollutants.

The second part of the book, with 12 chapters (Chapters "Treatment of Soil Washing Solutions by Electrochemical Advanced Oxidation," "Electrokinetic Soil Flushing," "Electrokinetic Remediation of Soil Polluted with Inorganic Ionic Species," "Fenton Processes for Remediation of Polluted Soils," "Coupling of Anodic Oxidation and Soil Remediation Processes," "Persulfate in Remediation of Soil and Groundwater Contaminated by Organic Compounds," "Electro-phytoremediation of Cropland and Mine Tailings Polluted by Mercury, Using $IrO_2$-$Ta_2O_5$/Ti electrodes, *Lavandula vera*, and *Solanum tuberosum*," "Electrobioremediation of Polluted Soils," "Soil and Groundwater Remediation with Zero-Valent Iron Nanoparticles," "Adsorption and Ion Exchange Permeable Reactive Barriers," "Electrochemically Assisted Thermal-Based Technologies for Soil Remediation," and "Electrochemically Assisted Dewatering"), deals with practical applications of technologies related to the separation of organic and inorganic compounds from soil or to their destruction. From soil washing to soil flushing passing through phytoremediation and bioremediation processes, the most relevant technologies are described. Most of the technologies are still at a research stage but results are as promising as to recommend implementation in the near future. Special emphasis is given to the expected characteristics of soil after the treatment. Permeable reactive barriers are also well described because of their outstanding performance. Experts in the field of electrochemistry and engineering participate discussing about the laboratory applications as well as the pilot plants that have been developed to understand the processes associate soil remediation.

Finally, the third part focuses on the real application of the technology and includes not only processes that are really being applied but also some important considerations that have to be taken into account in the design of full-scale processes. It has five chapters (Chapters "Fundamentals of the Scale-Up of the Electrochemically Assisted Soil Remediation Processes," "Electrochemical Technologies for Petroleum Contaminated Soils," "Treatment of Gaseous Effluents Produced During Electrochemically Assisted Soil Remediation Processes," "Solar-Powered Electrokinetic Remediation for Treatment to Soil Polluted with Organic Compounds," and "Electrokinetic Processes: Directions for Future Research and Constraints") and describes how the scale-up of these technologies has to be faced, real applications of the technology, and how generated gases should be processed. Also, it discusses the use of solar photovoltaic panels as an energy source for powering electrochemical systems, in order to decrease both the investment and maintenance costs of electrokinetic remediation processes. Finally, it concludes with a view on the future research and constraints of the technology.

Natal, Brazil E. V. Dos Santos
Ciudad Real, Spain M. A. Rodrigo

# Contents

# Part I
# Fundamentals

# Physicochemical and Hydrodynamic Aspects of Soil

**Ángel Yustres, Rubén López-Vizcaíno, Virginia Cabrera, and Vicente Navarro**

## 1 Introduction

Soil is a particulate material which is usually of natural origin. Its porous nature allows the movement of fluids within its interior. Solid soil particles are mostly of a mineral origin and, to a greater or lesser extent, chemically interact with the pore fluids, which are usually constituted by water. Due to the numerous thermal, hydraulic, mechanical, electrical, and even biological phenomena associated with the interaction between mineral particles and pore fluids, soil is a complex medium that is difficult to model and understand.

This chapter reviews the basic principles that govern the behavior of soils and influence the electrokinetic remediation processes of contaminated soils.

## 2 Origin of Soils

Soils are formed in a variety of ways, but the fundamental origin is the alteration of lithospheric rocks. The disintegration of the minerals that compose a rock gives rise to the particulate matter of soil. The type of soil that is generated depends on the different mechanisms responsible for decomposing the rocks and the subsequent process of aggregation of the mineral particles.

Residual soils are formed by the weathering (physical–chemical attack of environmental agents) of parent rock that is exposed on the earth's surface. The resulting

Á. Yustres (✉) · R. López-Vizcaíno · V. Cabrera · V. Navarro
Geoenvironmental Group, Civil Engineering Department, University of Castilla-La Mancha, Ciudad Real, Spain
e-mail: angel.Yustres@uclm.es

 
M. A. Rodrigo, E. V. Dos Santos (eds.), *Electrochemically Assisted Remediation of Contaminated Soils*, Environmental Pollution 30,
https://doi.org/10.1007/978-3-030-68140-1_1

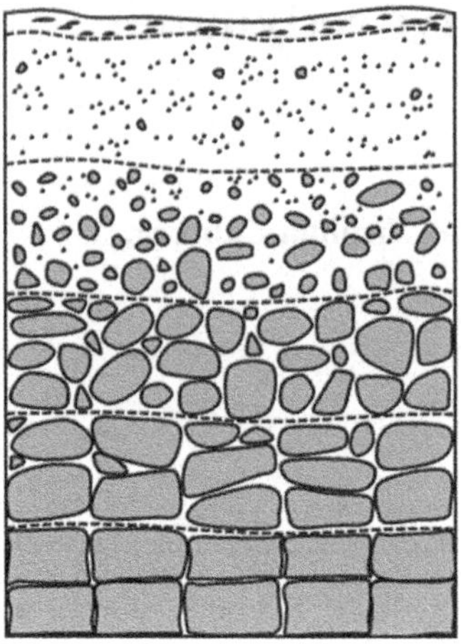


**Fig. 1** Typical vertical profile of a residual soil

mineral particles or rock relicts are not transported, so they remain at the site where they were formed, eventually covering the underlying rock material.

Residual soils can be found in any region of the globe, but thicker soils are found in very humid climates [1]. Climatic, topographical, and, of course, lithological factors [2, 3] have a decisive influence on the genesis of residual soils. In general, residual soils present a high degree of heterogeneity because they are not classified according to particle size or reorganized by mechanical effects. Some rock relicts persist in their internal structure and are more abundant at lower depths (Fig. 1).

Colluvial soils are formed by the accumulation of weathered material in situ that is transported to low points of the terrain due to gravity. Different types of phenomena called landslide processes [4] are involved in their formation. Although they are very diverse in nature (from slow moving creep processes to sudden debris flow processes), residual soils have low or no internal organization and are very heterogeneous materials as a result [5, 6].

Soils of sedimentary origin (or transported soils [5]) come from the erosion, transport, and sedimentation of weathered material. Given that this type of soil transport involves a fluid (wind or water), the velocity of the fluid determines the size of the particles that can be mobilized [5, 7]. This factor confers a greater degree of homogeneity to sedimentary soils, at least on a small scale. However, at larger scales (tens to thousands of meters), the continuous variations in the velocity of the eroding and transporting agent generate significant spatial variability [8]. The transport process over long distances causes progressive wear of the particles, leading to rounded or subrounded grains. Another common characteristic of these soils is the existence of discontinuities due to stratification. Deposition usually occurs in subhorizontal layers and in different events, which favors the presence of bedding planes [2].

Soils that are formed by the action of water are called alluvial soils. They have a wide spatial distribution and are generally associated with large depressions of the terrain through which rivers flow. Given that these environments are ideal for human settlements, alluvial soils often receive discharges of pollution. However, given that the flows of the river courses in these areas tend to be slow, the predominant soils are fine-grained (mainly clay and silts), with low permeabilities and for which electrokinetic remediation techniques are particularly recommended [9].

Soils formed by aeolian processes are characteristic of arid or semiarid zones, where wind action is the main erosive agent. These are well-sorted soils compared to alluvial soils. Sandy sediments derived from dune formation are the most recognizable, but the finest grained aeolian soils called loess are very important and are quite common in cold and arid areas [10]. Typically, aeolian soils have a very porous and metastable internal structure that can suddenly collapse in the presence of water [11, 12].

There are also soils for which humans are primarily responsible for their genesis. The main agent of erosion, transport, and sedimentation is construction machinery. In this manner, human beings determine the size of particles, their location, and their degree of compaction. These artificial soils are increasingly being used in urban infrastructure [13]. Common examples are large landfills for airport construction [14, 15], artificial islands, and land reclamation [16], all of which have considerable environmental impacts.

Based on the multiple factors that affect soil formation (climatology, vegetation, topography, fluvial, and/or aeolian regime, lithology, etc.) and the alteration factors that appear once a soil has developed (precipitation or dissolution of salts, compaction, etc.), each soil will be different because the combination of factors is almost infinite. The genetic conditioning of a soil is the influence that all factors have on its genesis, development over time, and final behavior [5].

## 3 Distribution of the Phases of a Soil

To be able to successfully address any soil remediation action, an abstraction and simplification of the associated processes should be performed. Due to their genesis and their temporal evolution, soils show very high internal and spatial variabilities. It is impossible to approach the study of a soil from an exhaustive point of view. Therefore, tools should be provided to obtain a qualitative and integrated description that allows rationally grouping, classifying, and organizing soils into types with homogeneous behaviors.

### *3.1 Phases of a Soil: Phase Diagram*

The fluids inside the pores usually consist of water (with dissolved salts) and air (with a certain amount of water vapor). If we examine a soil sample point by point, the amount of water, air, or solid particles can change very abruptly. Thus, it is necessary to establish a series of hypotheses to rationally analyze the distribution of soil phases.

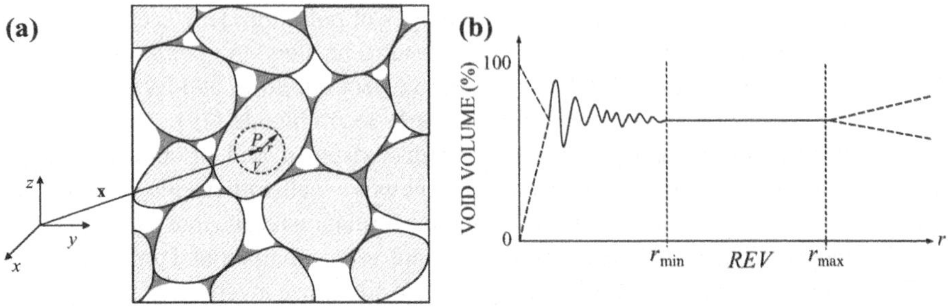


**Fig. 2** (**a**) Sampling point *P* and its domain volume *V*. (**b**) Representation of average soil properties stabilization for representative elementary volume

If we take point *P* (Fig. 2a) and its immediate domain, as determined by a characteristic radius *r* that defines the volume *V*, the average properties of the domain highly depend on the position of the point. If *P* is on a mineral particle, 100% of volume *V* will be occupied by solid particles. If *P* is in a void, 100% of volume *V* may be occupied by air or water. As the value of *r* increases, there is a greater probability that the volume consist of a mixture of solid particles, liquid, and gases. As *V* continues to increase, the proportion of the volumes occupied by each of the phases is likely to stabilize around an average value that can be considered representative of the entire soil. Figure 2b shows how from $r_{min}$ the percentage of voids in the total soil volume remains constant until $r_{max}$ is reached. For small values of *r* the percentage fluctuates, in the most extreme case being 100% or 0% as discussed above. Between $r_{min}$ and $r_{max}$ lies the representative elementary volume (REV). The representative elementary volume [17] is the characteristic volume of a soil for which the average properties remain constant, even with small changes in that volume.

Based on this approach, the soil can be conceptualized as a multiphase mixture [18], with a series of index properties relating the volumes and masses of each phase. The variation of these average index properties is assumed to be continuous in space, and all these continua are superimposed at each point. Despite its particulate nature, the soil is assumed to be a continuous medium whenever it is possible to define a representative elemental volume.

Some of the most important and widely used index properties in soil characterization are:

1. Porosity ($\varphi$) is defined by the following expression:

$$\varphi = \frac{V_V}{V_T}. \tag{1}$$

   where $V_V$ is the void volume, and $V_T$ is the total volume of the soil (void volume plus volume of solids).

2. Void ratio. The void ratio ($e$) is given by the following expression:

$$e = \frac{V_V}{V_S}. \tag{2}$$

where $V_S$ is the volume of solids contained in the soil.
Porosity and void ratio give us an insight into the amount of space available for the movement of fluids and the internal organization of the soil, so they are fundamental magnitudes for their characterization.
3. Gravimetric water content. The gravimetric water or moisture content ($w$) of a soil is given by the following expression:

$$w = \frac{M_W}{M_S} \approx \frac{M_L}{M_S} \tag{3}$$

where $M_S$ is the mass of solid particles, and $M_W$ and $M_L$ are the water and liquid masses, respectively, which are assumed to be equal.
4. Degree of saturation. The degree of saturation is defined as the volume occupied by the liquid phase ($V_L$), which is assumed to be equal to the water volume ($V_W$), versus the volume of voids in a soil ($V_V$) and is given by the following expression:

$$Sr = \frac{V_W}{V_V} \tag{4}$$

The water content and the degree of saturation provide information on the importance of the liquid phase in the soil, since both index properties determine the thermodynamic state of the water and the type of movement it has within the porous medium.
5. Density of solid particles and specific gravity. The density of solid particles ($\rho_s$) is given by the following ratio:

$$\rho_S = \frac{M_S}{V_S} \tag{5}$$

where $V_S$ is the volume occupied by the solid particles. The use of specific gravity ($G$) to characterize the minerals that make up a soil is quite common in the literature on soil mechanics. This variable is given by the following expression:

$$G = \frac{\rho_S}{\rho_W} \tag{6}$$

where $\rho_W$ is the density of water. The value of $G$ is dependent on the predominant mineralogy, although the most common minerals [19] have values between 2.6 and 2.8 [5].
6. Dry density. The dry density ($\rho_d$) is defined as:

$$\rho_d = \frac{M_S}{V_T} \tag{7}$$

where, as previously stated, $M_S$ is the mass of solid particles, and $V_T$ is the total volume of the soil.

7. Bulk density. The bulk density ($\rho_b$) of a soil is defined by the following expression:

$$\rho_b = \frac{M_T}{V_T} \tag{8}$$

where, as previously stated, $M_T$ is the total mass (mass of solids, liquid, and gas, although the latter is usually disregarded), and $V_T$ is the total volume of the soil.

8. Saturated density. The saturated density ($\rho_{sat}$) is defined as the bulk density when the soil is saturated, i.e., all pores are occupied by water.

The dry density and the saturated density are particular cases of the bulk density when the degrees of saturation are 0 and 1, respectively.

## *3.2 Experimental Determination and Soil Phase Relations*

The determination of phase indices is usually not cumbersome and only requires very basic laboratory equipment. Furthermore, only a few indices need to be determined and the rest can be calculated from them. For these reasons phase distribution characterization in a soil can be performed easily and inexpensively in almost any laboratory.

The three most common characterization tests are:

(a) Determination of the water content by mass [20] using oven drying. This test consists of measuring the mass of a soil sample before and after being placed in an oven with a temperature between 105 and 115 °C. The difference in mass between the two measurements returns the amount of water contained in the soil, and the last weighing provides the value of the mass of solid particles. With both masses, the gravimetric water content can be calculated using Eq. (3).

(b) Determination of the specific gravity of soil solids using a water pycnometer [21]. This test can be somewhat more cumbersome, but, as previously specified, the density of solid particles presents a very low range of variation [5]. In many cases, in the absence of data, a value of 2700 kg/m$^3$ is used [22].

(c) Determination of the density (unit weight) of a soil specimen [23]. In this case, it is crucial to precisely determine the value of the total volume of the soil sample, which should remain unaltered after extraction. For this purpose, sampling tubes and paraffin wax are used for the samples [24].

From these values, the rest of the indices can be calculated. For this purpose, the use of the unit phase diagram is of special interest. Figure 3a shows a phase diagram of a soil for which the volume (to the left) and mass (to the right) are schematically represented. It is assumed that the mass of the gas phase is equal to zero. Dividing all the terms in the diagram in Fig. 3a by $V_S$ (volume of solid particles) yields the

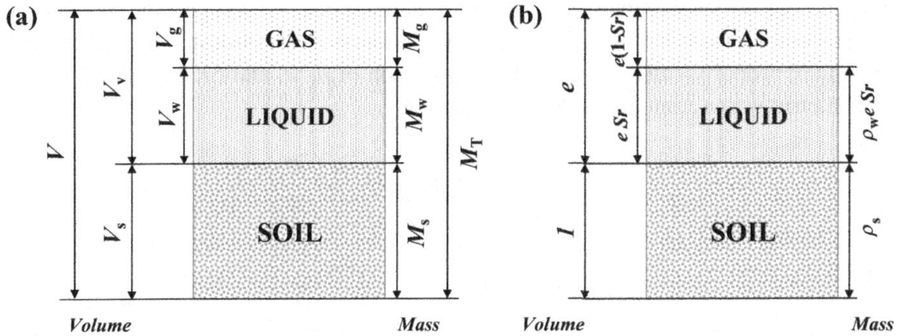


**Fig. 3** (**a**) Soil phase diagram. (**b**) Unit diagram. All the masses and volumes of (**a**) are divided by $V_S$

**Table 1** Soil phase relationships

| Index property | Expressions in terms of $\rho_s$, $e$, and $Sr$ | Expressions in terms of $\rho_s$, $w$, and $\rho_b$ |
|---|---|---|
| $e$ | $e$ | $\frac{\rho_S + w\rho_S}{\rho_b} - 1$ |
| $\varphi$ | $\frac{e}{1+e}$ | $1 - \frac{\rho_b}{\rho_S + w\rho_S}$ |
| $w$ | $\frac{\rho_W \, e \, Sr}{\rho_S}$ | $w$ |
| $Sr$ | $Sr$ | $\frac{w\rho_S \, \rho_b}{\rho_w(\rho_S + w\rho_S - \rho_b)}$ |
| $\rho_S$ | $\rho_S$ | $\rho_S$ |
| $\rho_d$ | $\frac{\rho_S}{1+e}$ | $\frac{\rho_b}{1+w}$ |
| $\rho_b$ | $\frac{\rho_S + \rho_W \, e \, Sr}{1+e}$ | $\rho_b$ |
| $\rho_{sat}$ ($Sr = 1$) | $\frac{\rho_S + \rho_W \, e}{1+e}$ | $\frac{\rho_b\left(\rho_S + \rho_W\left(\frac{\rho_S + w\rho_S}{\rho_b} - 1\right)\right)}{\rho_S + w\rho_S}$ |

diagram in Fig. 3b, which is called the unit diagram because the volume of the solid particles takes the value of 1. From this diagram and the definitions of Eqs. (1)–(8), any index property can be obtained as a function of $\rho_s$, $e$, and $Sr$. For example, the gravimetric water content can be calculated as:

$$w = \frac{M_W}{M_S} = \frac{\rho_W V_W}{\rho_S V_S} = \frac{\rho_W \frac{V_W}{V_V} V_V}{\rho_S V_S} = \frac{\rho_W \frac{V_W}{V_V} \frac{V_V}{V_S}}{\rho_S \frac{V_S}{V_S}} = \frac{\rho_W Sr\, e}{\rho_S} \tag{9}$$

where the expression can be obtained directly by dividing the terms on the right of the unit diagram that correspond to the mass of the liquid and solid phases. Thus, all indices can be obtained through the relationships shown in Table 1. These relationships can also be obtained as a function of $\rho_s$, $w$, and $\rho_b$ (Table 1), which can be useful in the laboratory when calculating the distribution of phases of a soil.

Soils are characterized by all these index properties. Even if two soils are mineralogically and texturally identical, their behavior under same external actions will be completely different if their void ratio and degree of saturation are not equal. It is common in many studies (both in laboratory tests and in situ) to ignore the initial soil phase distribution and to compare the remediation efficiency with completely different initial conditions.

## 4 The Solid Phase

### *4.1 Soil Mineralogy*

The mineralogy of soils varies greatly, depending on the parent rock and the physicochemical alterations subsequent to their formation. However, due to the nature of the composition of the terrestrial lithosphere, silicates, clay minerals, metal oxides, oxyhydroxides, hydroxides, carbonates, and sulfates predominate [19].

The reactivity of all these minerals generates a very significant range of pore waters that condition soil decontamination processes. Highly soluble minerals such as gypsum, calcite, or other carbonates, which can be important pH regulators in the electrokinetic treatment of contaminated soils [25, 26], are of special interest.

### *4.2 Soil Texture*

The first approach for characterizing the soil texture is the granulometric analysis of the particles that compose the soil. Although the basic characterization tests for the analysis of the particle size distribution or gradation are sieving [27] and

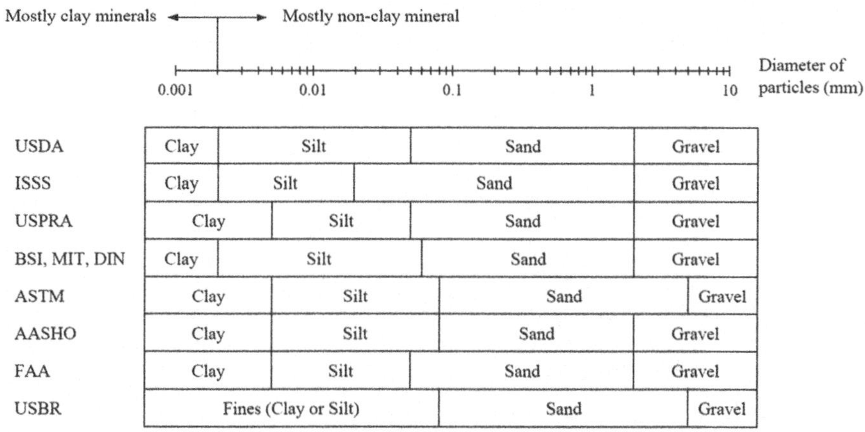


USDA (US Department of Agriculture), ISSS (International Soil Science Society), USPRA (US Public Roads Administration), BSI (British Standards Institute), MIT (Massachusetts Institute of Technology), DIN (German Standards), ASTM (American Society for Testing and Materials), AASHO (American Association of State Highway Officials), FAA (Federal Aviation Agency) and USBR (US Bureau of Reclamation)

**Fig. 4** Characteristic sizes of soil particles according to different classification systems

sedimentation [28], there are several techniques to determine the size distribution of solid particles, such as laser diffractometry [29], which is increasingly common. The predominant particle size marks the behavior of the soils, both mechanically [5] and hydraulically [30, 31]. The particle size variation in a soil can range immensely, from tens of centimeters to a micrometer in size. Although there is a convention regarding the characteristic sizes of each type of particle (see Fig. 4), there are various classifications, mainly from the perspective of agronomy (such as the USDA classification [32]) and civil engineering (such as the Unified Soil Classification System [33] and the AASHTO system [34]).

Traditionally, soil has been divided into two types of particles according to their size: a fine fraction, which is less than 0.074 mm in diameter, and a coarse fraction, which is above 0.074 mm in diameter. The division is marked by the difficulty of sieving below that size, but it is also conditioned by the presence of clay minerals that have a quite different behavior from the rest of the solid particles and play a fundamental role in soils of low or very low permeability.

Clay minerals are formed by weathering in an advanced phase of different materials of siliceous origin. They are formed by two types of planar mineral structures. The first structure is a phyllosilicate sheet that is formed by $SiO_4^{4-}$ tetrahedra that share, by a covalent bond, the oxygen atoms in their bases (Fig. 5a). The second sheet structure is formed by metal hydroxides, generally $Al_2(OH)_6$ (also called gibbsite) and $Mg_3(OH)_6$ (also called brucite). Spatially, both hydroxides are in the form of octahedra combined by covalent bonding and share hydroxyl groups in the vertices (Fig. 5b). Because tetrahedral and octahedral sheets are not electrically neutral, they combine to form covalent bonds between the free apices of the tetrahedrons and the

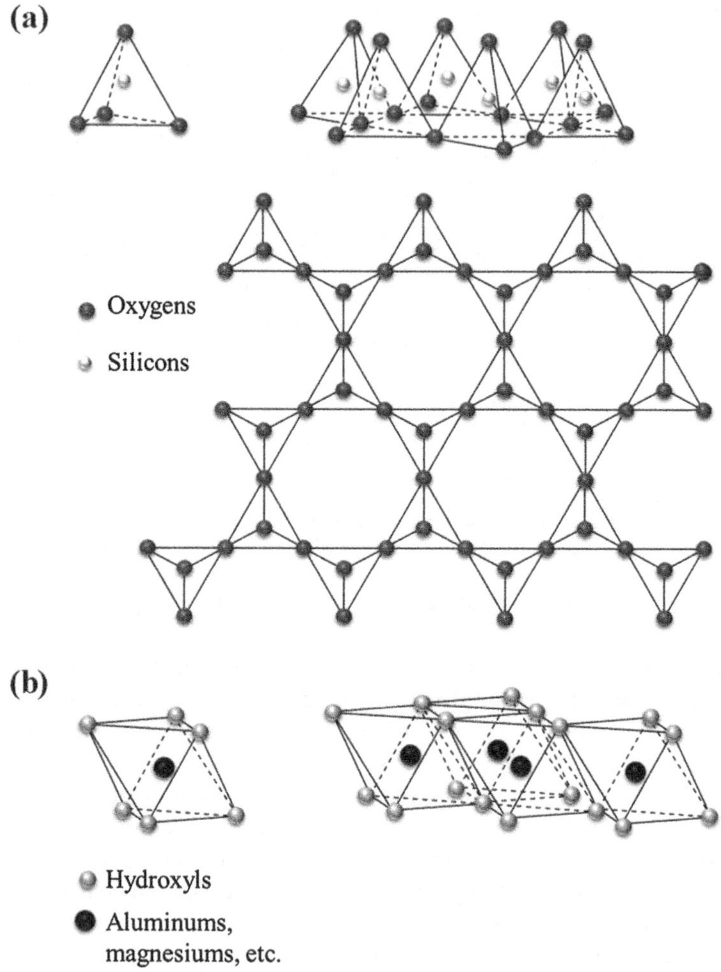


**Fig. 5** (**a**) Structural diagram of phyllosilicates and (**b**) structural diagram of metal hydroxides

oxygen of the hydroxyl groups of the octahedra, giving rise to different clay minerals. The combination of one layer of phyllosilicates (tetrahedra) and one layer of hydroxides (octahedra) gives rise to a clay mineral type 1:1 (see Fig. 6a). In contrast, the sandwiching of an octahedral layer by two tetrahedral layers gives rise to a mineral type 2:1 (see Fig. 6b) or TOT (tetrahedron-octahedron-tetrahedron). These elementary pieces, in turn, have charge deficits and affinities for other like-layers, which give rise to a wide variety of minerals depending on the mode of combination [5, 35]. Most clay minerals are found in nature not as simple 1:1 or 2:1 sheets but as stacks of layers.

Regarding complexity, the phenomenon of isomorphic substitution must also be considered. During the formation of the clay layers, once the basic structure is formed, the silicon of the tetrahedra or the metallic cations of the octahedra can be displaced by elements with a lower charge without structural changes. Substitution

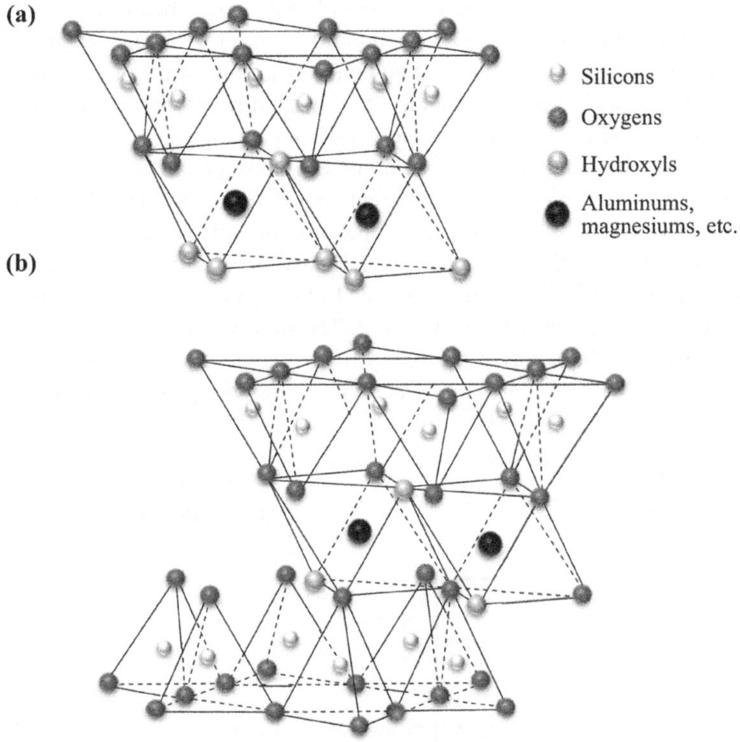


**Fig. 6** (**a**) Structural diagram of a 1:1 clay sheet and (**b**) structural diagram of a 2:1 clay sheet

involves a loss of positive charges and, therefore, results in a net negative charge. The layers of clay become highly electrically charged planar structures, which give rise to behavioral phenomena and trends that are very different from those of soils formed only by large inert mineral particles [5].

From the point of view of the electrokinetic remediation of soils, isomorphic substitution is responsible for many of the phenomena observed during treatment. Firstly, the clay sheets are able to attract water molecules that are electrical dipoles so that the mineral particles are hydrated and form a near crystalline structure. In addition, mineral surfaces are also able to attract a large amount of hydrated cations to compensate for negative net charges [35]. Both types of water molecules are virtually immobilized and are referred to as the water adsorbed on the surface of the clay particles. This water does not flow under conventional hydraulic gradients [36] and is responsible for the low permeability of this type of soils.

However, this property of clay soils makes them especially suited for electrokinetic remediation. When an external electric field is applied, the ions of the hydrating water travel toward the electrodes due to electromigration, which causes the entire liquid phase to flow. In clay soils, there are more cations than anions due to the excess negative charge of the clay particles, and more water is pulled toward the cathode than the anode; this phenomenon is called electroosmosis [9, 37, 38]. Some authors have confirmed [39] that if cationic surfactants are used, the electroosmotic flow is reversed. Due to the positive charge of hydrophilic heads, more anions are required in the solution to attain electroneutrality, and the balance between anions and cations begins to equilibrate or even tend toward an excess of negative charges in the pore water with a subsequent electroosmotic flow toward the anode.

The study of the electrostatic properties of clay particles requires sophisticated equipment, which is unusual in most common commercial laboratories. For this reason, in the classification of soils, indirect tests that give an idea of the ability of a soil to retain water and solutes in its internal structure are used. The most common tests are the Atterberg limits [40, 41], which have been widely correlated with the macroscopic mechanical properties of soils [42–44], including the swelling capacity and the hydraulic conductivity [45, 46]. The methylene blue value [47–49] has also been used to characterize the behavior of clayey soils [50, 51]. These tests allow the classification of the fine fraction according to its nature whether it is

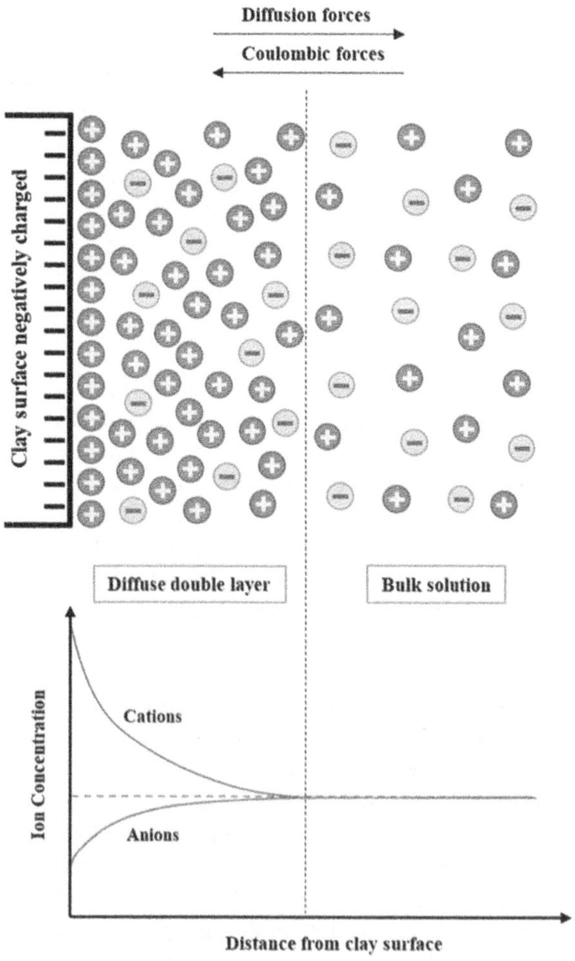


**Fig. 7** Diffuse double layer in clays

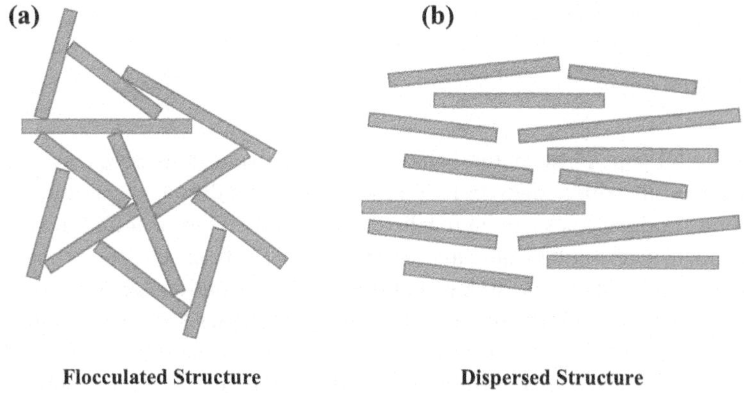


**Fig. 8** Idealized representation of a clay with (**a**) flocculated structure and (**b**) dispersed structure

predominantly silty (with particles with a diameter of between 74 and 2 μm) or clayey (with a diameter of less than 2 μm). For this, the plasticity is studied, which is given fundamentally by the surface area and amount of charge of a soil, as measured through the cation exchange capacity (CEC) [19, 52].

As previously mentioned, the negative electric charge of clay particles results in an attraction of the cations toward their surface and a repulsion of the anions. This attraction alters the state of the pore water around minerals (Fig. 7), forming a diffuse double layer [52, 53] that consists of the mineral particle itself and its surrounding, which is affected by the electric field. The equilibrium between the electrostatic repulsive forces of clay surfaces and the attractive forces (London-van der Waals forces) cause the internal structure to be dispersed or flocculated [5] (see Fig. 8), with highly differentiated macroscopic behavior [5, 52].

Another additional effect of the nature of clays is the adsorption of contaminants on mineral surfaces. The electrostatic adsorption mechanism is very important, but it is not the only one [54]; many mechanisms can be superimposed for the same soil depending on the mineralogical composition, organic matter content, type of pollutant, and geochemical conditions of the pore water. For this reason, the adsorption of pollutants is usually described empirically through the use of sorption isotherms [19, 54, 55]. In many cases, the desorption of pollutants at high concentrations requires the use of anionic surfactants [56] that can alter the direction of water movement during electrokinetic soil remediation [39].

# 5 The Liquid Phase

## 5.1 Water

In virtually all soils in the natural state, water is present in its pores, even in small amounts. In soils with a high void ratio in saturated conditions, a greater volume of water than solid particles can be reached. Liquid water can be found in two forms. The first form is adsorbed water, as described above, which is strongly bound to the surface of the clay particles. It does not flow with conventional hydraulic gradients and can be considered immobile for all practical purposes [36]. Some authors have suggested that the molecular structure and thermodynamic properties of adsorbed water change [5, 57] and are different from the liquid-free water. In addition, it can remain in contact with the clay particles even if the soil is desiccated in an oven to determine the water content [20].

The second type of water is called free, capillary, or gravitational water and can flow with hydraulic gradients [36]. Due to the small size of the pores in which it is found, surface tension is of fundamental importance in its behavior. In partially saturated soils (Fig. 9), water adheres to solid particles to form curved surfaces called menisci. In the menisci, a balance of forces (Fig. 10) is established between the surface tension of the water and the forces exerted by the pressures of the liquid and gas. In the arrangement of Fig. 10, the equilibrium is given by the following expression:

$$P_{\mathrm{G}} - P_{\mathrm{L}} = \frac{2\sigma}{R} \tag{10}$$

where $\sigma$ is the surface tension, $P_{\mathrm{G}}$ and $P_{\mathrm{L}}$ are the pressures of the gas phase and liquid phase, respectively, and $R$ is the radius of curvature of the meniscus. In a three-dimensional arrangement, the menisci are warped surfaces in which two radii of curvature ($R_1$ and $R_2$) can be defined. Thus,

$$P_{\mathrm{G}} - P_{\mathrm{L}} = \sigma\left(\frac{1}{R_1} + \frac{1}{R_2}\right) \tag{11}$$

The difference between the gas pressure ($P_{\mathrm{G}}$) and the liquid pressure ($P_{\mathrm{L}}$) shown in Eqs. (10) and (11) is called capillary suction ($s$). The greater the value of $s$, the smaller the radius of the meniscus, the closer the warped surface is to the solid particles, and the water content of the soil is, consequently, also lower. The function that relates the degree of saturation ($Sr$) and the capillary suction is called the soil water retention curve (SWRC). The SWRC can present hysteretic behavior (Fig. 11) with a main wetting path and a main drying path in which the hysteresis cycles are inscribed. However, especially for modeling purposes, these curves are simplified using the central tendency for both wetting and drying. There are many formulations for this type of curve [58–65], although their parameterization is usually simple (no more than four parameters). The SWRC is also called the characteristic curve because

**Fig. 9** Water in unsaturated soils

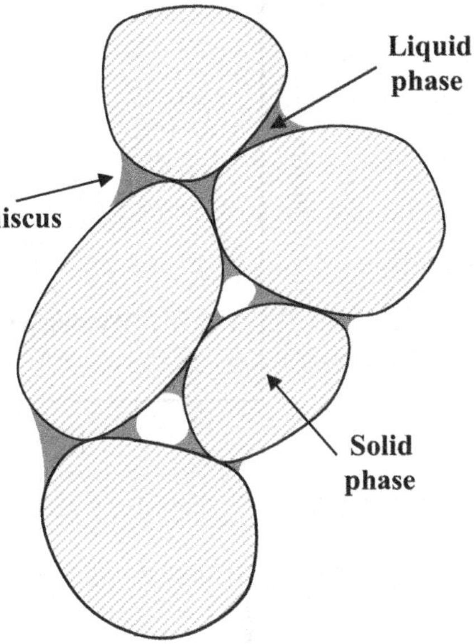


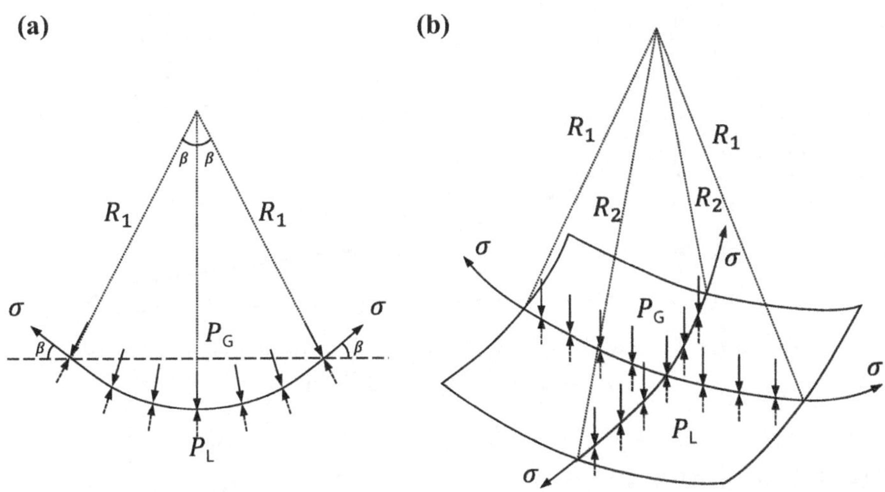


**Fig. 10** Force balance of the menisci in (**a**) a two-dimensional configuration and (**b**) a three-dimensional configuration

it is representative of the type of soil [66]. As seen in Fig. 12, the curves for clay and sand are very different. The first type of soil has a more extended curve, with very high degrees of saturation, even for high suctions. However, the curve for a granular-type material, such as coarse sand, is much more step-like, with the soil changing from almost complete saturation to desaturation with only a small change in suction. The value of the suction in which the jump occurs for coarse granular materials is called the air entry pressure. Although in theory it also exists in clays and reaches much higher values, an abrupt jump in the value of *Sr* is generally not observed.

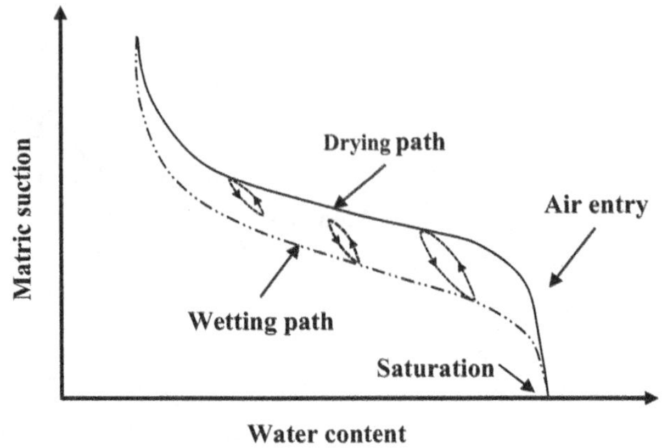


**Fig. 11** Hysteretic behavior of the soil water retention curve

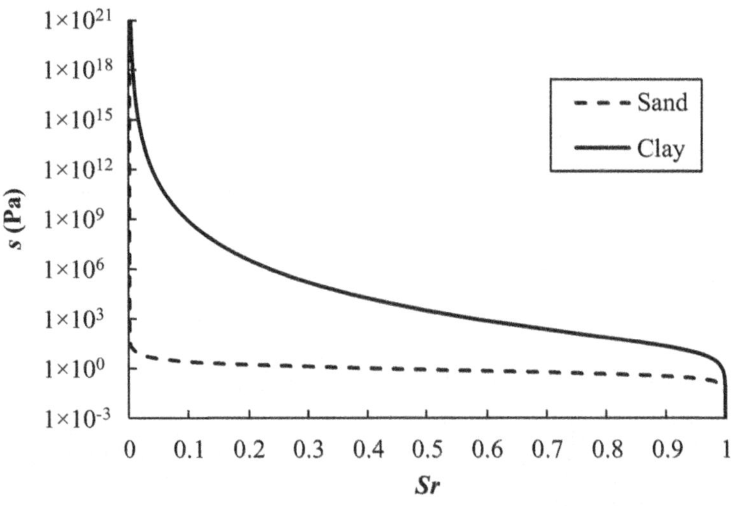


**Fig. 12** Typical retention curves of a clayey soil and a sandy soil

If it is assumed that $P_G$ is equal to or close to the value of atmospheric pressure ($P_{atm}$), which may be true in soils with a low degree of saturation, and $P_{atm} = 0$ (reference pressure level), $P_L$ must be negative; therefore, its internal energy is lower than the water found in liquid form on the surface of the earth. For this reason, extraction for the subsequent treatment of pore water in a partially saturated soil is very complex and energy demanding.

### 5.2 *Solutes in the Liquid Phase*

Due to the chemical interaction of water with the gases present in the pores and with the minerals of the solid phase, the concentrations of the solutes in pore water are higher than in rainwater. Due to the composition of the lithosphere, the major ions in the water include $Ca^{2+}$, $Mg^{2+}$, $Na^{+}$, and $K^{+}$ as cations and $HCO_3^{-}$, $SO_4^{2-}$, and $Cl^{-}$ as anions [67, 68]. Other minor ions or trace elements can be present, and some of them can be considered as pollutants [69].

An effect of dissolved salts is that they alter the partial pressure of water vapor in equilibrium with the liquid phase. This variation induces an increase in suction called osmotic suction ($s_O$). Osmotic suction can be calculated as:

$$s_O = RT\rho_W \sum_{i=1}^{j} a_i \tag{12}$$

where $R$ is the ideal gas constant (8.314463 J $K^{-1}$ $mol^{-1}$), T is the temperature, and $a_i$ is the activity of the $i$th species in the pore water solution.

Osmotic suction and capillary suction make up the value of the total suction ($s_{TOT}$), which will influence the mechanical behavior of the soil.

### 5.3 *Electrical Conductivity*

Another important effect that solutes have on pore water is the variation of the electrical conductivity of the soil. If this is taken as a parallel system with two conductivities, that of the solid phase and that of the liquid phase [70], the increase in the concentration of solutes substantially increases the total conductivity. There are several approaches to estimate water conductivity based on the its chemical composition. A good compilation and comparison of the best-known approaches can be found in Tarantino et al. [71].

## 6 The Gas Phase

The gas phase in a soil consists mainly of air (a mixture of nitrogen, oxygen, and other minor constituents) that is enriched by other gases due to interactions with minerals and the pore water. For example, the content of $CO_2$ or methane will be strongly related to biological activity and pH of the water. However, the gas most dependent on the physicochemical state of the pore water is water vapor. If a thermodynamic equilibrium between water vapor and free water is assumed, then the psychrometric equation or Kelvin's equation is fulfilled. This equation is given by the expression [72]:

$$P_V = P_V^o \exp\left(\frac{\text{WMM}}{RT\rho_W} s_{TOT}\right) \tag{13}$$

where $P_V$ is the partial vapor pressure, $P_V^o$ is the saturation vapor pressure (equilibrium partial vapor pressure for a planar surface of water) that is only temperature-dependent, WMM is the water molecular mass, $\rho_W$ is the density of pure water, and $s_{TOT}$ is the total suction, which is equal to

$$s_{TOT} = s + s_O \tag{14}$$

where $s$ is the capillary suction, and $s_O$ is the osmotic suction. Given that $P_V / P_V^o$ is the relative humidity, if we are able to measure this ratio in the soil pores, the total suction can be determined. Subtracting the osmotic suction from the total suction yields the capillary suction. From the SWRC of the soil, the degree of saturation and water content can then be obtained. This is the operating principle of psychrometers, which are suitable for determining the suction in soils with low degrees of saturation. For other conditions (low suction and greater degree of saturation), tensiometers [73] are more appropriate and are capable of determining capillary suction directly. There are also other methodologies to obtain the soil water content based on the measurement of thermal and electrical conductivities [74] or on the dielectric properties of the soil [73]. Therefore, there are a variety of techniques to measure the suction or water content, both in the laboratory and on site.

## 7 Coupled Hydromechanical Behavior of Soils

The cornerstone of soil mechanics is the principle of effective stress [73–75]. In the formulation of this principle, for a saturated soil, the effective stress ($\boldsymbol{\sigma}'$), which is responsible for the soil strains, is given by:

$$\sigma' = \sigma - P_L \mathbf{I} \tag{15}$$

where $\boldsymbol{\sigma}$ is the tensor of total stress to which the soil is subjected, $P_L$ is the pressure of the liquid phase, and $\mathbf{I}$ is the identity tensor. For unsaturated soils, the formulation of the effective stress is slightly different [76].

$$\sigma' = \bar{\sigma} - \chi \cdot s \cdot \mathbf{I} \tag{16}$$

$$\bar{\sigma} = \sigma - P_G \cdot \mathbf{I} \tag{17}$$

where $\chi$ is the Bishop factor [77], which is usually formulated as a function of the degree of soil saturation and $\bar{\sigma}$ is the net stress tensor. Consequently, there is a coupling between mechanical and hydraulic behaviors in soils. This interdependence can generate unexpected phenomena [78] when applying electrokinetic remediation techniques that generate significant changes in the distribution of water pressure and the degree of soil saturation [79].

## 8 Flows in Soils

The high microscopic and macroscopic variability of soils, the changes in the environmental variables, and the human actions cause flows of a diverse physicochemical nature. The gradients of chemical potential experienced by all soil components involve flows of matter and energy in all phases and, consequently, in all species.

Any flow $\mathbf{q}_i$ in a soil can be formulated generically by the following expression [80]:

$$\mathbf{q}_i = \sum_{j=1}^{n} \mathbf{L}_{ij} \nabla X_j \tag{18}$$

where $\mathbf{L}_{ij}$ is the coupling coefficient tensor that links the flow $\mathbf{q}_i$ and the gradient of the driving variable $X_j$, and $n$ denotes the total number of driving variables and types of flow considered.

In most of the conceptual and numerical models that have been used to simulate electrokinetic remediation treatments, movements in the solid or gaseous phase have not been considered. However, they have been used in other applications, such as the analysis of engineered barriers for the storage of radioactive waste [81, 82].

The flows that are most commonly considered in electrokinetic remediation processes include the water mass, which is practically equal to that of the entire liquid phase, the mass of chemical species, heat and electric charge. The most intuitive part of the description of the flows is to consider only those in which $i = j$, that is, those in which the flow and driving force are of the same nature. Thus, the mass flow of water is given by Darcy's Law.

$$\mathbf{q}_{\mathrm{W}}^{\mathrm{h}} = -\mathbf{K}^{\mathrm{h}} \nabla h \tag{19}$$

where $\mathbf{q}_{\mathrm{W}}^{\mathrm{h}}$ is the hydraulic water flux, $\mathbf{K}^{\mathrm{h}}$ is the hydraulic conductivity tensor, and $\nabla h$ is the hydraulic gradient, being $h$ the water head, it is defined as:

$$h = z + \frac{P_{\mathrm{L}}}{\gamma_{\mathrm{W}}} \tag{20}$$

where $z$ is the vertical coordinate, $P_{\mathrm{L}}$ is the liquid pressure, and $\gamma_{\mathrm{W}}$ is the specific weight. An additional term to the water head, which corresponds to kinetic energy of the porewater, is not included because the velocity of water movement in the soil is very low.

The flow of solutes is driven by gradients in the concentration of chemical species and is given by Fick's Law.

$$\mathbf{j}_i = -\mathbf{D}_i \nabla C_i \tag{21}$$

where $\mathbf{j}_i$ is the diffusive flux of chemical species $i$, $\mathbf{D}_i$ is the diffusion/dispersion tensor, and $\nabla C_i$ is gradient of the concentration of the species considered.

In the case of conductive heat flux ($\mathbf{q}^t$), the formulation using Fourier's law is similar to the previous ones and is given by the expression:

$$\mathbf{q}^t = -\mathbf{K}^t \ \nabla T \tag{22}$$

where $\mathbf{K}^t$ is the thermal conductivity tensor and $\nabla T$ is the temperature gradient in the soil.

Finally, the electrical current ($\mathbf{I}^e$) is given by Ohm's law can be obtained by as:

$$\mathbf{I}^e = -\sigma^e \ \nabla E \tag{23}$$

where $\boldsymbol{\sigma}^e$ is electrical conductivity tensor, and $\nabla E$ is the electric potential gradient.

In certain cases, it is necessary to include in the conceptual model coupled flows ($i \neq j$ in Eq. 18) for which the nature of the flow and that of the driving force are different. In the case of electrokinetic remediation treatments, the most important coupled flows to consider are the electroosmotic flow and the electromigration. The first is given by the following expression:

$$\mathbf{q}^{eo} = -\mathbf{K}^{eo} \ \nabla E \tag{24}$$

where $\mathbf{K}^{eo}$ is the electroosmotic conductivity (or permeability). In the same way, charged species experience electromigration due to the presence of an electric field. This chemical flux can be expressed as:

$$\mathbf{q}_i^{em} = -\mathbf{U}_i \ \nabla E \tag{25}$$

where $\mathbf{q}_i^{em}$ is the electromigratory flow, and $\mathbf{U}_i$ is the ionic electromobility tensor of chemical species $i$.

Many other coupled flows can occur (see Table 2 adapted from [5]), and it can be shown [82] that due to Onsager reciprocal relations, the coupling coefficients must be symmetric; thus, according to Eq. (18),

$$\mathbf{L}_{ij} = \mathbf{L}_{ji} \tag{26}$$

The hydraulic, electroosmotic, thermal, and electrical conductivities, diffusion/dispersion, and electromobility ion coefficients are phenomenological and are

**Table 2** Coupled flows in soils

| Flow | Driving force (∇) Chemical concentration | Hydraulic head | Temperature | Electrical current/potential |
|---|---|---|---|---|
| Chemical species | Fickian diffusion | Streaming current | Soret effect | Electrophoresis |
| Fluid | Chemo-osmosis | Darcyan conduction | Themo-osmosis | Electroosmosis |
| Heat | Dufour effect | Isothermal heat transfer | Fourier conduction | Ohmic conduction |
| Electric current | Diffusion and membrane potentials | Streaming potential | Thompson effect | Peltier effect |

Adapted from [5]

therefore dependent on other variables. Although tensor is used in a general way in Eqs. (19)–(25), if the medium is isotropic, these coefficients are taken as constants so that

$$\mathbf{L}_{ij} = l_{ji}\mathbf{I} \tag{27}$$

where $\mathbf{L}_{ij}$ is the coupling tensor, $l_{ij}$ is the scalar coupling coefficient, and $\mathbf{I}$ is the identity tensor.

The presence of all these coupled transport phenomena, together with the chemical reactivity of the mineral and liquid phases, makes any chemical–physical process of remediation in soils extremely complex. Therefore, from the point of view of modeling and understanding the phenomena, many of these mutually coupled equations are disregarded based on their lower relative importance. However, such an evaluation must be careful when there are changes in scale or changes in soil or environmental conditions. For example, the influence of thermal effects due to ohmic heating of the electrodes is usually disregarded in the analysis of electrokinetic remediation processes. Nevertheless, it has been shown that it is relevant when the process is scaled up [83, 84].

## 9 Conclusions

Because of their special multiphasic and multicomponent nature, soils are highly complex porous media. Their character as natural materials makes them highly heterogeneous at all scales. In addition, multiple physical–chemical phenomena take place within soils when the initial equilibrium conditions are altered, leading to unexpected or irregular behaviors in many cases.

From the point of view of electrokinetic treatments, empirical approaches have often been used for understanding the main behavioral trends; however, these methods fail to understand all the intrinsic complexity of soils. To achieve this objective, it is necessary to completely characterize the distribution of the soil phases, the mineralogical composition, the geochemistry of the pore water, and the coefficients that govern the different flows of matter and energy in soil. Clearly, this characterization is an extremely complex task, but it should not be forgotten that an oversimplification of conceptual approaches can omit relevant phenomena in the behavior of soil.

**Acknowledgments** The authors acknowledge funding support from: (1) the Ministerio de Economía, Industria y Competitividad from Spanish Government and the European Union through the project [BIA2017-89287-R (AEI/FEDER, UE)] and (2) the Ministerio de Ciencia, Innovación y Universidades from Spanish Government through the Postdoctoral Grant [IJC-2018-035212] awarded to Dr. López-Vizcaíno.

## References

1. I.M.R. Duarte, C.M.G. Rodrigues, Residual Soils, in *Encyclopedia of Engineering Geology*, ed. by P. T. Bobrowsky, B. Marker, (Springer International Publishing, Cham, 2018), pp. 751–752
2. L.D. Wesley, *Geotechnical Engineering in Residual Soils* (Wiley, Hoboken, 2010)
3. B.B.K. Huat, D.G. Toll, A. Prasad, *Handbook of Tropical Residual Soils Engineering* (CRC Press, Boca Raton, 2012)
4. L.M. Highland, P. Bobrowsky, *The Landslide Handbook—A Guide to Understanding Landslides* (Geological Survey Circular 1325, Reston, 2008)
5. J.K. Mitchell, K. Soga, *Fundamentals of Soil Behavior* (John Wiley & Sons, Inc., Hoboken, 2005)
6. T. Zádorová, V. Penížek, Formation, morphology and classification of colluvial soils: a review. Eur. J. Soil Sci. **69**, 577–591 (2018)
7. R.P.C. Morgan, *Soil Erosion and Conservation*, 3rd edn. (Blackwell Science, Hoboken, 2005)
8. R.E. Hunt, *Geotechnical Engineering Investigation Handbook*, 2nd edn. (Taylor and Francis Group, Abingdon, 2005)
9. A.T. Yeung, S. Datla, Fundamental formulation of electrokinetic extraction of contaminants from soil. Can. Geotech. J. **32**, 569–583 (1995)
10. A.S. Goudie, *Arid and Semi-Arid Geomorphology* (Cambridge University Press, New York, 2013)
11. A. Assadi-Langroudi, S. Ng'ambi, I. Smalley, Loess as a collapsible soil: some basic particle packing aspects. Quat. Int. **469**, 20–29 (2018)
12. S.L. Houston, W.N. Houston, C.E. Zapata, C. Lawrence, Geotechnical engineering practice for collapsible soils. Geotech. Geol. Eng. **19**, 333–355 (2001)
13. M. Bendixen, J. Best, C. Hackney, L.L. Iversen, Time is running out for sand. Nature **571**, 29–31 (2019)
14. G. Mesri, J.R. Funk, Settlement of the Kansai International Airport Islands. J. Geotech. Geoenviron. **141** (2015)
15. The construction record of Kansai International Airport. World Dredg. Min. Constr. **37**, 6–30 (2001)
16. S.Y. Chee, A.G. Othman, Y.K. Sim, A.N. Mat Adam, L.B. Firth, Land reclamation and artificial islands: walking the tightrope between development and conservation. Glob. Ecol. Conserv. **12**, 80–95 (2017)
17. J. Bear, *Dynamics of fluids in porous media* (Dover, New York, 1972)
18. M. Hassanizadeh, W.G. Gray, General conservation equations for multi-phase systems: 1. Averaging procedure. Adv. Water Resour. **2**, 131–144 (1979)
19. G. Sposito, *The Chemistry of Soils* (Oxford University Press, Cary, 2008)

20. ASTM, ASTM D2216-19. Standard Test Methods for Laboratory Determination of Water (Moisture) Content of Soil and Rock by Mass (2019)
21. ASTM, ASTM D854-14. Standard Test Methods for Specific Gravity of Soil Solids by Water Pycnometer (2014)
22. Skaven-Haug, S.V.: Volumetric relations in soil materials. Proceedings of the Fourth International Peat Congress, Espoo (1972)
23. ASTM, ASTM D7263-09. Standard Test Methods for Laboratory Determination of Density (Unit Weight) of Soil Specimens (2018)
24. ASTM, ASTM D4531-15. Standard Test Methods for Bulk and Dry Density of Peat and Peat Products (2015)
25. R. López-Vizcaíno, E.V. dos Santos, A. Yustres, M.A. Rodrigo, V. Navarro, C.A. Martínez-Huitle, Calcite buffer effects in electrokinetic remediation of clopyralid-polluted soils. Sep. Purif. Technol. **212**, 376–387 (2019)
26. T. Grundl, C. Reese, Laboratory study of electrokinetic effects in complex natural sediments. J. Hazard. Mater. **55**, 187–201 (1997)
27. ASTM, ASTM D6913/D6913M—17. Standard Test Methods for Particle-Size Distribution (Gradation) of Soils Using Sieve Analysis (2017)
28. ASTM, ASTM D7928-17. Standard Test Method for Particle-Size Distribution (Gradation) of Fine-Grained Soils Using the Sedimentation (Hydrometer) Analysis (2017)
29. A. Bieganowski, M. Ryżak, A. Sochan, G. Barna, H. Hernádi, M. Beczek, C. Polakowski, A. Makó, Chapter 5: Laser Diffractometry in the Measurements of Soil and Sediment Particle Size Distribution, in *Advances in Agronomy*, ed. by D. L. Sparks, (Academic Press, London, 2018), pp. 215–279
30. A. Scheuermann, A. Bieberstein, Determination of the Soil Water Retention Curve and the Unsaturated Hydraulic Conductivity from the Particle Size Distribution, in *Experimental Unsaturated Soil Mechanics*, ed. by T. Schanz, (Springer Berlin Heidelberg, Berlin, Heidelberg, 2007), pp. 421–433
31. M.D. Fredlund, G.W. Wilson, D.G. Fredlund, Use of the grain-size distribution for estimation of the soil-water characteristic curve. Can. Geotech. J. **39**, 1103–1117 (2002)
32. Soil Science Division Staff, Soil Survey Manual, in *USDA Handbook 18*, ed. by C. Ditzler, K. Scheffe, H. C. Monger, (Government Printing Office, Washington, 2017), p. 603
33. ASTM, ASTM D2487-17. Standard Practice for Classification of Soils for Engineering Purposes (Unified Soil Classification System), (2017)
34. ASTM, ASTM D3282-15. Standard Practice for Classification of Soils and Soil-Aggregate Mixtures for Highway Construction Purposes (2015)
35. F. Bergaya, G. Lagaly, *Handbook of Clay Science* (Elsevier, London, 2013)
36. T.A. Hueckel, Water–mineral interaction in hygromechanics of clays exposed to environmental loads: a mixture-theory approach. Can. Geotech. J. **29**, 1071–1086 (1992)
37. A.T. Yeung, C.-n. Hsu, R.M. Menon, Physicochemical soil-contaminant interactions during electrokinetic extraction. J. Hazard. Mater. **55**, 221–237 (1997)
38. D.H. Gray, J.K. Mitchell, Fundamental aspects of electro-osmosis in soils. J. Soil Mech. Found. Div. **93**, 209–236 (1967)
39. R. López-Vizcaíno, C. Sáez, P. Cañizares, V. Navarro, M.A. Rodrigo, Influence of the type of surfactant on the mobility of flushing fluids for electro-remediation processes. Sep. Sci Techn. **46**, 2148–2156 (2011)
40. K.H. Head, *Manual of Soil Laboratory Testing. Volume 1: Soil Classification and Compaction Tests* (Whittles Publishing, Dunbeath, 2006)
41. ASTM, ASTM D4318—17e1. Standard Test Methods for Liquid Limit, Plastic Limit, and Plasticity Index of Soils (2017)
42. B. Voight, Correlation between Atterberg plasticity limits and residual shear strength of natural soils. Geotechnique **23**, 265–267 (1973)
43. L.D. Wesley, Residual strength of clays and correlations using Atterberg limits. Geotechnique **53**, 669–672 (2003)

44. M.A. Kanji, The relationship between drained friction angles and Atterberg limits of natural soils. Geotechnique **24**, 671–674 (1974)
45. J.M. Lee, C.D. Shackelford, C.H. Benson, H.Y. Jo, T.B. Edil, Correlating index properties and hydraulic conductivity of geosynthetic clay liners. J. Geotech. Geoenviron. **131**, 1319–1329 (2005)
46. A. Bouazza, S. Jefferis, T. Vangpaisal, Investigation of the effects and degree of calcium exchange on the Atterberg limits and swelling of geosynthetic clay liners when subjected to wet-dry cycles. Geotext. Geomembr. **25**, 170–185 (2007)
47. E. Çokça, Relationship between methylene blue value, initial soil suction and swell percent of expansive soils. Turkish J. Eng. Environ. Sci. **26**, 521–529 (2002)
48. J. Zhang, J. Peng, Y. Chen, J. Li, F. Li, Estimation of soil-water characteristic curve for cohesive soils with methylene blue value. Adv. Civil Eng. (2018)
49. M. Schaeffner, Introduction of methylene blue value of a soil into soil classification of recommendation for road earthworks. Bull. Liaison Lab. Ponts Chaussees, 9–16 (1989)
50. J.C. Santamarina, K.A. Klein, Y.H. Wang, E. Prencke, Specific surface: determination and relevance. Can. Geotech. J. **39**, 233–241 (2002)
51. Y. Yukselen, A. Kaya, Suitability of the methylene blue test for surface area, cation exchange capacity and swell potential determination of clayey soils. Eng. Geol. **102**, 38–45 (2008)
52. D. Hillel, *Introduction to Environmental Soil Physics* (Elsevier Science & Technology, Burlington, 2003)
53. D.G. Strawn, H.L. Bohn, G.A. O'Connor, *Soil Chemistry* (John Wiley & Sons, Incorporated, Somerset, 2015)
54. R.N. Yong, M. Nakano, R. Pusch, *Environmental Soil Properties and Behaviour* (CRC Press LLC, Baton Rouge, 2012)
55. N.M. Nagy, J. Konya, A.T. Hubbard, *Interfacial Chemistry of Rocks and Soils* (CRC Press LLC, Baton Rouge, 2009)
56. S.D. Haigh, A review of the interaction of surfactants with organic contaminants in soil. Sci. Total Environ. **185**, 161–170 (1996)
57. R.T. Martin, in *Adsorbed Water on Clay: A Review*, ed. By E. Ingerson. Clays Clay Miner. (1962), pp. 28–70
58. R.H. Brooks, A.T. Corey, *Hydraulic Properties of Porous Media* (Colorado State University, Fort Collins, 1964)
59. G.S. Campbell, A simple method for determining unsaturated conductivity from moisture retention data. Soil Sci. **117**, 311–314 (1974)
60. D.G. Fredlund, X. Anqing, Equations for the soil-water characteristic curve. Can. Geotech. J. **31**, 521–532 (1994)
61. W.R. Gardner, Some steady-state solutions of the unsaturated moisture flow equation with application to evaporation from a water table. Soil Sci. **85**, 228–232 (1958)
62. P.H. Groenevelt, C.D. Grant, A new model for the soil-water retention curve that solves the problem of residual water contents. Eur. J. Soil Sci. **55**, 479–485 (2004)
63. K. Kosugi, Lognormal distribution model for unsaturated soil hydraulic properties. Water Resour. Res. **32**, 2697–2703 (1996)
64. M. Kutílek, Soil hydraulic properties as related to soil structure. Soil Till. Res. **79**, 175–184 (2004)
65. M.T. van Genuchten, A closed-form equation for predicting the hydraulic conductivity of unsaturated soils. Soil Sci. Soc. Am. J. **44**, 892–898 (1980)
66. Meyer, P.D., Rockhold, M.L., Gee, G.W.: Uncertainty Analyses of Infiltration and Subsurface Flow and Transport for SDMP Sites, Pacific Northwest National Laboratory. U.S. Nuclear Regulatory Commission., (NUREG/CR-6565, PNNL-11705) NRC Job Code W6503 (1997)
67. Hem, J.D.: Study and Interpretation of the Chemical Characteristics of Natural Water. Water Supply Paper, U.S. Geological Survey, Reston, VA (1985), p. 263
68. Bartos, T.T., Ogle, K.M.: Water Quality and Environmental Isotopic Analyses of Ground-Water Samples Collected from the Wasatch and Fort Union Formations in Areas of Coalbed

Methane Development—Implications to Recharge and Ground-Water Flow, Eastern Powder River Basin, Wyoming. Water-Resources Investigations Report 02-4045, USGS, Cheyenne, Wyoming (2002)
69. J.D. Rhoades, N.A. Manteghi, P.J. Shouse, W.J. Alves, Soil electrical conductivity and soil salinity: new formulations and calibrations. Soil Sci. Soc. Am. J. **53**, 433–439 (1989)
70. R.B. McCleskey, D.K. Nordstrom, J.N. Ryan, Comparison of electrical conductivity calculation methods for natural waters. Limnol. Oceanogr.-Meth. **10**, 952–967 (2012)
71. A. Tarantino, A.M. Ridley, D.G. Toll, Field Measurement of Suction, Water Content, and Water Permeability, in *Laboratory and Field Testing of Unsaturated Soils*, ed. by A. Tarantino, E. Romero, Y.-J. Cui, (Springer Netherlands, Dordrecht, 2009), pp. 139–170
72. R. Bulut, E.C. Leong, Indirect Measurement of Suction, in *Laboratory and Field Testing of Unsaturated Soils*, ed. by A. Tarantino, E. Romero, Y.-J. Cui, (Springer Netherlands, Dordrecht, 2009), pp. 21–32
73. K. Terzaghi, *Erdbaumechanik* (Deuticke, Vienna, 1925)
74. A.W. Skempton, in *Terzaghi's Discovery of Effective Stress*. From Theory to Practice in Soil Mechanics (1960), pp. 42–53
75. R. de Boer, W. Ehlers, The development of the concept of effective stresses. Acta Mech. **83**, 77–92 (1990)
76. E.E. Alonso, J.M. Pereira, J. Vaunat, S. Olivella, A microstructurally based effective stress for unsaturated soils. Géotechnique **60**, 913–925 (2010)
77. A.W. Bishop, The principle of effective stress. Teknisk Ukeblad. **106**, 859–863 (1959)
78. R. López-Vizcaíno, V. Navarro, M.J. León, C. Risco, M.A. Rodrigo, C. Sáez, P. Cañizares, Scale-up on electrokinetic remediation: engineering and technological parameters. J. Hazard. Mater. **315**, 135–143 (2016)
79. R. López-Vizcaíno, A. Yustres, M.J. León, C. Saez, P. Cañizares, M.A. Rodrigo, V. Navarro, Multiphysics implementation of electrokinetic remediation models for natural soils and porewaters. Electrochim. Acta **225**, 93–104 (2017)
80. A.T. Yeung, J.K. Mitchell, Coupled fluid, electrical and chemical flows in soil. Geotechnique **43**, 121–134 (1993)
81. V. Navarro, E.E. Alonso, Modeling swelling soils for disposal barriers. Comput. Geotech. **27**, 19–43 (2000)
82. D.W. Pollock, Simulation of fluid flow and energy transport processes associated with high-level radioactive waste disposal in unsaturated alluvium. Water Resour. Res. **22**, 765–775 (1986)
83. S. Barba, R. López-Vizcaíno, C. Saez, J. Villaseñor, P. Cañizares, V. Navarro, M.A. Rodrigo, Electro-bioremediation at the prototype scale: what it should be learned for the scale-up. Chem. Eng. **334**, 2030–2038 (2018)
84. R. López-Vizcaíno, C. Risco, J. Isidro, S. Rodrigo, C. Saez, P. Cañizares, V. Navarro, M.A. Rodrigo, Scale-up of the electrokinetic fence technology for the removal of pesticides. Part II: Does size matter for removal of herbicides? Chemosphere **166**, 549–555 (2017)

# Fundamental of Electrokinetic Processes

**M. A. Rodrigo**

## 1 Introduction

From a soil treatment perspective, application of electric fields in soil throughout a set of electrodes is a very important topic because it generates the phenomena in which most of the electrochemically assisted soil remediation processes are based [1–3]. In this context, although sometimes the remediation of soil using electrochemical technology is improperly known as "electrokinetic soil remediation processes," really electrokinetic processes are only a limited number of the phenomena that can take place in soil during the remediation with the application of electric fields, all of them characterized by being related to the transport of species throughout the soil. In other scientific disciplines, the definition of electrokinetics is wider and, hence, these phenomena involved all processes that related mass transport and electric field and that develop in heterogeneous fluids or in porous bodies filled with fluid. They include the most important as [4]: (1) transport under the influence of an electric field, such as the motion of liquid (electroosmosis) or particles (electrophoresis), (2) transport under the influence of a chemical potential gradient such as the motion of liquid (capillary osmosis) or particles (diffusiophoresis), (3) electric field generated by the transport of colloids or fluids such as the generated by the sedimentation of colloids (sedimentation potential), their transport under the influence of ultrasounds (colloid vibration current), or the pass of fluid through a porous body (streaming potential).

From these processes, only three are important in soil remediation [5]: those that involved the transport of ions, charged colloids, or pore fluid under the influence of an electric field. With this, electrokinetic refers to all transport processes that occur

M. A. Rodrigo (✉)
Department of Chemical Engineering, Faculty of Chemical Sciences & Technologies, University of Castilla-La Mancha, Campus Universitario, Ciudad Real, Spain
e-mail: manuel.rodrigo@uclm.es



M. A. Rodrigo, E. V. Dos Santos (eds.), *Electrochemically Assisted Remediation of Contaminated Soils*, Environmental Pollution 30,
https://doi.org/10.1007/978-3-030-68140-1_2

in a soil or, by extension, in any kind of mixture of solid and liquid (such as sludge), when an electric field (*V*/cm) is applied between one/several anode/s and cathode/s placed in that system. Electrodes can be placed in the soil, either directly or they can be positioned inside an electrolyte solution in direct contact with the soil (electrolyte wells). Typically, the rate of these electrokinetic processes depends linearly on the electric field exerted, following Eq. (1), where $E_x$ is the electric field in the direction of the movement. This rate also depends linearly on the charge of the species and on the properties of the liquid and/or soil particles which made the soil, as it will be explained later on.

$$v_{ek} = k_{ek} E_x \tag{1}$$

Driving force of electrokinetic processes is the electric field applied between the electrodes, and the key mechanism that explains the processes occurring in the system is the electrostatic attraction/repulsion forces (coulombic forces) between charged species that may be transported and the electrodes. Hence, it is important to take into account that electrokinetic phenomena are not a chemical but a set of physical processes. However, simultaneously with these transport processes, with the application of an electric field to soil, it is expected that many other processes occur, including physical (heating, evaporation, changes in physical properties such as viscosity, etc.), chemical (ion exchange, dissolution of precipitates, precipitation of salts, etc.), and electrochemical processes (water oxidation and reduction, deposition of metals, oxidation of chloride, etc.). All these processes, which will be described in other chapters of this book, may have an impact on the species transported. In addition, there are significant interactions among all these processes, which influence on the observations made in the system. This means that in a real system, explanation of the effects observed in the transport of species when an electric field is applied is not always simple and many of these processes have to be recalled in order to reach a clear understanding of the observations [6].

Thus, in applying an electric field between electrodes, the key process to be considered is the set of electrolytic reactions developed on their surfaces. The most important processes, although not the only processes occurring, are the oxidation of water on the anodic surface to produce oxygen and protons (Eq. 2) and the reduction of water (that produced hydrogen and hydroxyl anions) (Eq. 3) or, sometimes, the electrodeposition of metals on the cathodic surface.

$$2H_2O \leftrightarrow O_2 + 4H^+ + 4e^- \tag{2}$$

$$2H_2O + 2e^- \leftrightarrow H_2 + 2OH^- \tag{3}$$

From a soil remediation viewpoint, the formation of gases produced in these electrolytic reactions is not a relevant aspect, at least not as important as the production of protons and hydroxyl cations. Bubbles may have a small impact, although just in the nearness of the electrodes, because both gases are easily dissipated into the atmosphere in a real treatment. In fact, the primary consequence of the oxidation of water is the formation of an "acidic front," which really is a gradient of protons concentration, which has its maximum on the surface of the anodes. This front displaces towards the cathode/s mostly by electromigration (and also by electroosmosis), and it contributes to the release of the pollutants that were fixed in the soil either by dissolution of the precipitates in which they are contained and/or by ionic exchange. Water reduction produced on the cathode generates a basic front in the opposite direction to the acidic front (gradient in the concentration of hydroxyl anions), with exactly the opposite effects on the pollutants. Both fronts can be modified by adding suitable reactants to the soil and they also interact in intermediate position between the anodes and the cathodes. Acidic and basic fronts may also affect the *z*-potential of the soil particles (because of the interaction of protons and hydroxyl radicals with the surface of the soil particles) and they can affect importantly the magnitude of the electroosmotic flux.

Electrical heating increases the soil temperature, generating a temperature gradient, with maximum temperatures in the nearness of the electrodes [7, 8]. This temperature increase is caused by ohmic losses that are generated by large ionic resistances of the soil and that are higher in soils with low ionic conductivities or in soil treatment with large interelectrode distances. This rise in temperature is caused by the Joule–Thomson effect, and obviously, it is proportional to the intensity of the current that flows externally to the soil (between the anodes and cathodes) and the soil resistances and can be modeled using Ohm's law (Eq. 4). The main effect related to electrokinetic treatments is that this increase in temperature may affect the transport properties of volatile and semi-volatile organic pollutants and thus, it can have influence on their desorption and mobility. Hence, these physical processes favor transport of species and, up to a certain point, they could be partially classified as electrokinetic, although the controlling mechanism is not mass transport but heat transmission.

$$W = I^2 \cdot R \tag{4}$$

Three processes are considered as pure electrokinetic, and each of them implies the transport of a different species: electroosmosis, which is responsible for the transport of water; electromigration, which is responsible for the transport of ions; and electrophoresis, which explains the transport of charged particles. These processes can be suitably combined to ensure the removal of many inorganic and organic contaminants from soils by setting optimum configurations and operation conditions in a soil remediation process such as promoting specific processes such as electrochemical soil flushing (Figs. 1, 2, 3, 4, 5, and 6).

## 2 Electromigration

The simplest electrokinetic processes correspond to those in which charge species are transported. As it is well-known, water contained in soil may contain ions, noncharged molecules, and suspended species of larger size such as colloids or

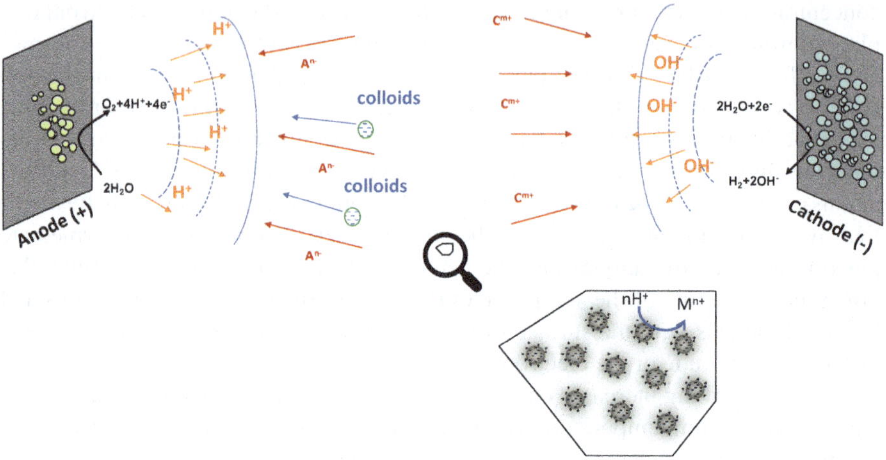


**Fig. 1** Electrochemical effects of applying an electric field: acidic and basic fronts

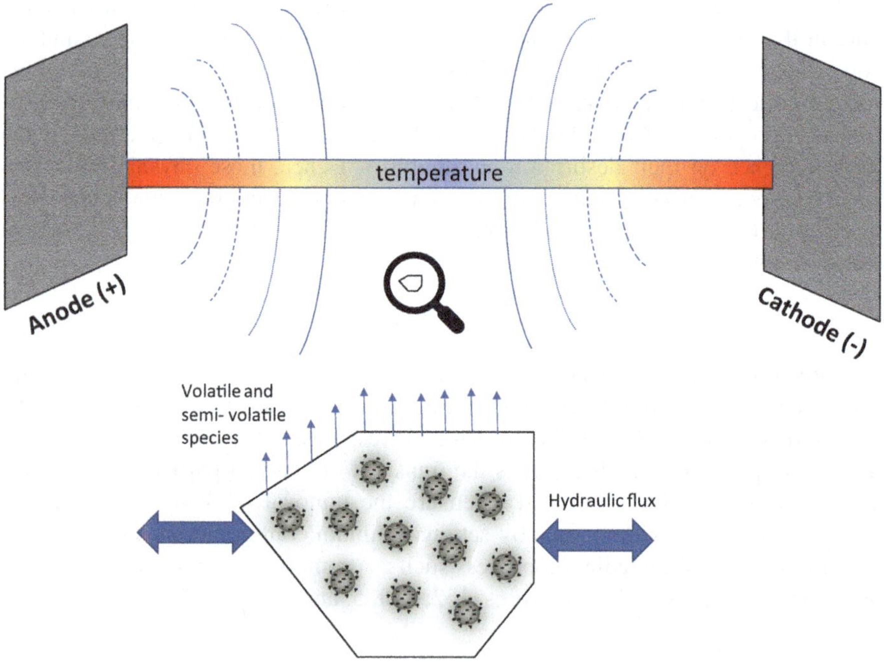


**Fig. 2** Thermal processes undergone by soil with the application of an electric field

microdrops. From this group, non-charge molecules are not expected to be transported because of the application of an electric field, except for the dragging that they may undergo into the liquid in movement (either by electroosmotic or by hydraulic fluxes). There are no charges involved, so there are no physical

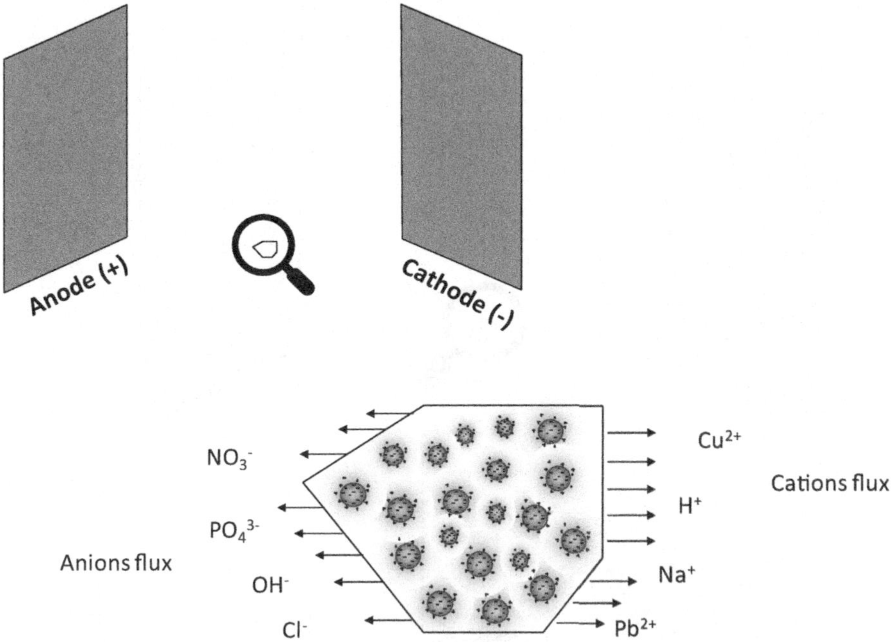


**Fig. 3** Transport of ions by electromigration

electrostatic attraction or repulsion forces acting on them when an electric field is applied.

Opposite ions consist of charged groups of atoms and, because of their net electric charge, they can be easily attracted by electrodes of different polarity and repulsed by electrodes of the same charge. Thus, anions (negatively charge ions) are attracted to the anode/s and repulsed by the cathode/s, while cations (positively charged ions) are attracted to the cathode/s and repulsed by the anode/s within the electric field generated in the soil. This electrokinetic process is known as electromigration. To fully understand, it has to be taken into account that the speciation of ions may depend on (1) the pH and p$K_a$/p$K_b$ of the corresponding acid/base and (2) the presence of other counterions, which may form insoluble salts. This speciation motivates that ions do not always transport as expected. In addition, it is necessary to understand that electromigration does not mean that any part of the soil can be charged positively or negatively from the macroscopic point of view. This is explained by taking into account that this transport of charged molecules from/to electrodes is balanced with the production of other charged species formed electrochemically on the surface of the electrodes, being the most important the protons produced on the anode surfaces and the hydroxyl ions produced on the surface of the cathodes. Alternatively, the consumption of ionic species on the surface of the electrodes (in case of the formation of chlorine from chloride anion or metal from metal ions) can also help to balance charges.

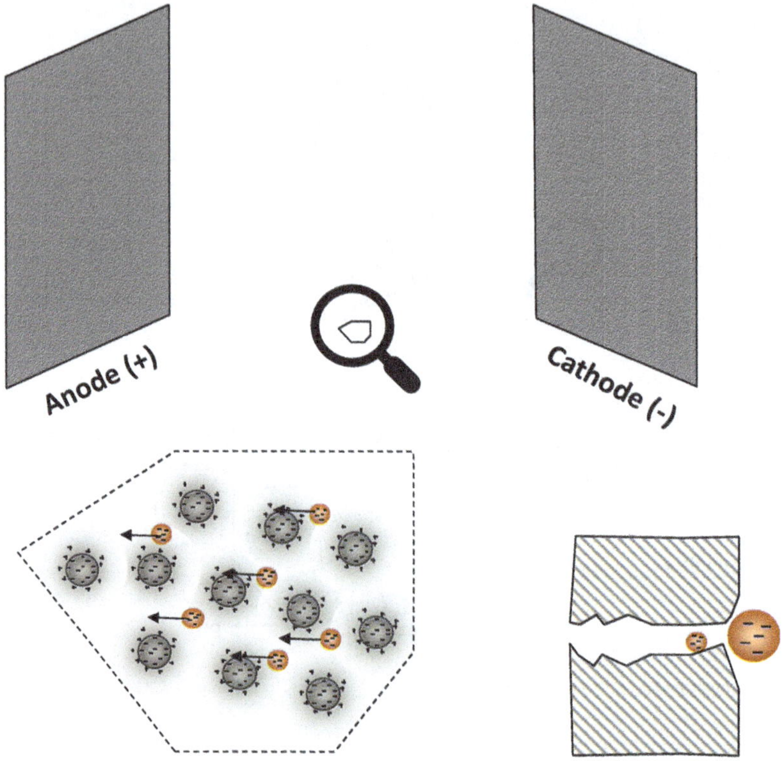


**Fig. 4** Transport of colloids by electrophoresis

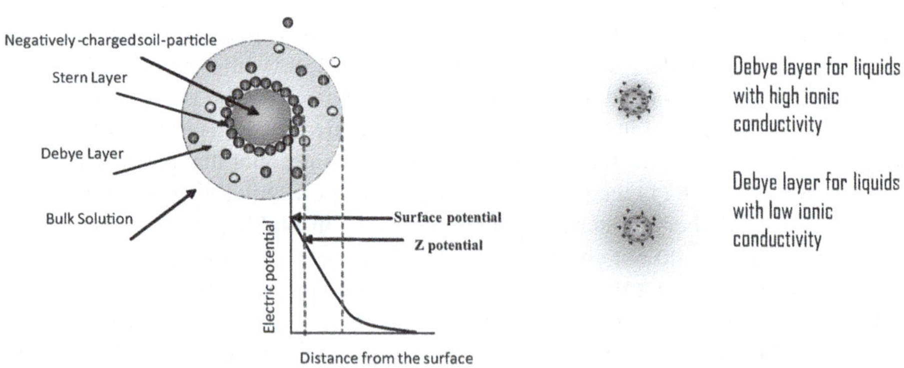


**Fig. 5** The Debye layer and the effect of the ionic conductivity on its thickness

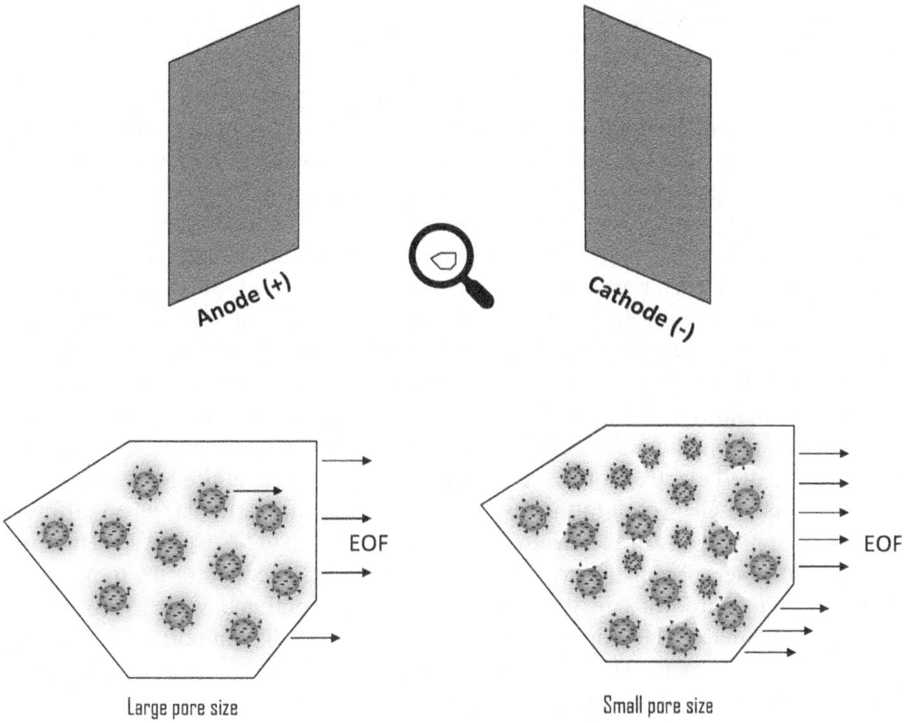


**Fig. 6** Electroosmotic flux. Influence of pore size

The electromigration rate of an ion depends (Eq. 5) on the dielectric constant ($D$), the electric charge of the ion ($z$), the temperature ($T$), and the electric field ($E_x$).

$$v_{\mathrm{em}} = \frac{z \cdot F \cdot D}{R \cdot T} E_x \tag{5}$$

When the ionic species to be transported is a weak acid or base, the mobility by electromigration can be strongly affected by the pH, because it may transform the ion into a non-charged molecule, which is no longer affected by the electric field applied to the soil. This is the case of weak organic acids such as clopyralid. In addition, when there is the possibility of formation of insoluble salts by combination of different ionic species, consequences of electromigration are also affected because the insoluble salt is fixed in the pores and electromigration does no longer exist for these ions.

## 3 Electrophoresis

Other species that can be contained in polluted soil are micelles, which in a first approach can be understood as particles of colloidal dimensions in equilibrium with the water contained in soil and which consists of aggregates of surfactants in which the hydrophilic part is in contact with the surrounding water and the hydrophobic part is in direct contact with a nonpolar compound. The size of these species is much higher than the molecular level of ions but despite of that, often, they are susceptible to undergo transport when the pore size of the soil is higher than the size of the particles. The mechanisms that activate this transport are the same as in electromigration: the electrostatic attractive/repulsion forces between their electric charges and the electric charge of the electrodes. However, the size of the particle becomes a major challenge and, here, what is found is not transport of a simple group of atoms but the movement of a much more complex and larger structure. The process is called electrophoresis and although its overall magnitude is less important than that of electromigration, its relevance is huge, in particular when different full-scale applications of the electrokinetic processes are to be explained. Thus, the transport of microorganisms in electro-bioremediation processes or the transport of hydrocarbon microdrops in the remediation of soil polluted with hydrocarbons using surfactants need to understand this process [9–11]. It is important to note that not only the size of the species to be transported is larger but also the number of charges contained in the surface of these larger particles.

The rate of the electrophoresis of a particle depends on the *z*-potential of the particle ($\zeta_{\text{particle}}$), viscosity of the fluid ($\eta$), and electric field ($E_x$) according to Eq. (6).

$$v_{\text{ep}} = \frac{D \cdot \xi_{\text{particle}}}{\eta} E_x \tag{6}$$

Sometimes, the *z*-potential of most colloids depends strongly on the pH, because surface charge depends on the protonation of functional groups (such as carboxylic groups). This is particularly relevant with microorganisms, but it can also be applied to many colloids negatively charged (such as the extensively used SDS). This points out the relevance of the basic and acid fronts, which may promote or disfavor, respectively, the mobility of pollutants depending on the affectation of soil by extreme pHs. This is particularly important in the nearness of the anodes. Likewise, temperature increases generated by ohmic heating may have a very important impact, because they can contribute to a decrease in the viscosity and, hence, to a promotion in the transport of colloids. As the highest temperatures are found in the proximities of the electrodes, this mobility will be much higher in this region that in zones far from them. The last important point to be considered in the understanding of this process is the comparison of the pore size of soil with the size of colloids. For low-permeability soils (in which other electrokinetic processes, such as the electro-osmosis, are favored), the mobilization of colloids can be prevented by the trapping of particles. For this reason, these electrophoretic processes are more relevant in soils with larger pore size.

## 4 Electroosmosis

Transport of water is the third process to be considered when describing the electrokinetic phenomena, although it is not the less important. The phenomenon is well known since the beginning of the nineteenth century. Electroosmosis can be defined as the motion of the pore water in the soil under the action of an electrical field. Thus, it is the mass flux of pore fluid. Typically, the fluid transported is the pore water itself, although sometimes an aqueous solution may be added to promote the motion of pollutants and, frequently, pore water that has already been treated is simply fed back to the process. The liquid typically flows from the anode to the cathode but under certain conditions the flow direction can be reversed.

Electroosmosis is not a simple process [12]. In fact, it is the result of the combined effect of two different phenomena that may happen in heterogeneous fluids or in porous bodies filled with fluid: (1) the accumulation of a net electrical charge on the surface of a solid that is in contact with an electrolyte solution and (2) the accumulation of a thin counterion layer of the liquid surrounding the solid surface.

It is important to take into account that the electrolytic fluid is electrically neutral beyond this thin layer (which is known as the electrical double layer or the Debye layer). However, the Debye layer in contact with the solid surface has a net charge and is therefore attracted by charges of the opposite sign and repelled by charges of the same sign. Therefore, this portion of the fluid can move within the electrical field that is generated between the electrodes of an electrochemical cell. Obviously, this transport does not depend on pressure gradients and, because of that, it may cause pore water flow in low permeability soils. In addition, it can be used to promote the flow of water added to flush the soil.

The mineral particles that make up soil are usually negatively charged. This is because of the release of protons or the exchange of aluminum or silicon atoms in the mineral structure by monovalent cations in the soil particles. In general, the total electrical charge increases as the specific surface of the soil mineral increases. This means that surface charge density increases in the sequence: sand < silt < kaolinite < montmorillonite. Because the soil particles are negatively charged, typically, the water in the Debye layer is positively charged. For this reason, the electroosmotic flux in an electroremediation treatment normally flows from the anode to the cathode. However, by changing the pH the surface charge can be reversed and this motivates that the electroosmotic flux can develop in the opposite direction.

The thickness of the Debye layer depends on the ionic concentration. The higher the ionic concentration, the smaller is this diffuse layer and the more restricted is the electroosmotic pore fluid flux, which will be more confined to the nearness of the capillary. Hence, as a rule, the higher the ionic conductivity, the lower the electroosmotic flux. Maximum electroosmotic flux can be obtained with very low pore fluid conductivity (below 100 $\mu S\ cm^{-1}$). In addition, it is important to remark that the low ground pore size enhances rather than inhibits this fluid flow. This result is obtained because under these conditions, the higher the amount of liquid in contact with the solid, the higher is the volume of liquid in the Debye layer, which can consequently be mobilized by the action of an electrical field. This result is important, because small pore sizes inhibit the hydraulic flux (i.e., via higher pressure losses). In addition, it explains why an electroosmotic flux can be generated in soils whose small pore size do not enable appreciable hydraulic fluxes to develop. Typical electroosmotic fluxes of $10^{-4}$ cm$^3$ cm$^{-2}$ s$^{-1}$ under electric gradients of 1 V cm$^{-1}$ can be obtained in low activity clays, with low electrolyte concentration and high water content [13].

Electroosmotic flux depends on the dielectric constant of the fluid ($D$), porosity of the soil ($n$), zeta potential ($\zeta$), viscosity of the fluid ($\eta$), and electric field ($E_x$) according to Eq. (7) (Helmholtz–Smoluchowski model).

$$v_{eo} = \frac{n \cdot D \cdot \xi}{\eta} E_x \tag{7}$$

Changes in the $z$-potential of soil during the application of electric field can be produced by the acidic and the basic fronts. As explained in the introduction, the protons contained in the first (together with other cations released) and the hydroxyl anions contained in the second (and anions released) may interact with the surface of the soil particles modifying the value of the $z$-potential. Moreover, the addition of reagents may have the same impact and in certain situations the electroosmotic flux may be reverted and flow from cathode to anode instead of the normal direction. To prevent this extreme influence of the pH fronts, the use of reagents such as citric acid is proposed and applied with very promising results [14].

In understanding electrokinetic processes, the direction of each of the flows has to be taken into account (as shown in Fig. 7). Thus, transport processes can be in the same or in opposite directions and in addition colloids can be trapped in the soil

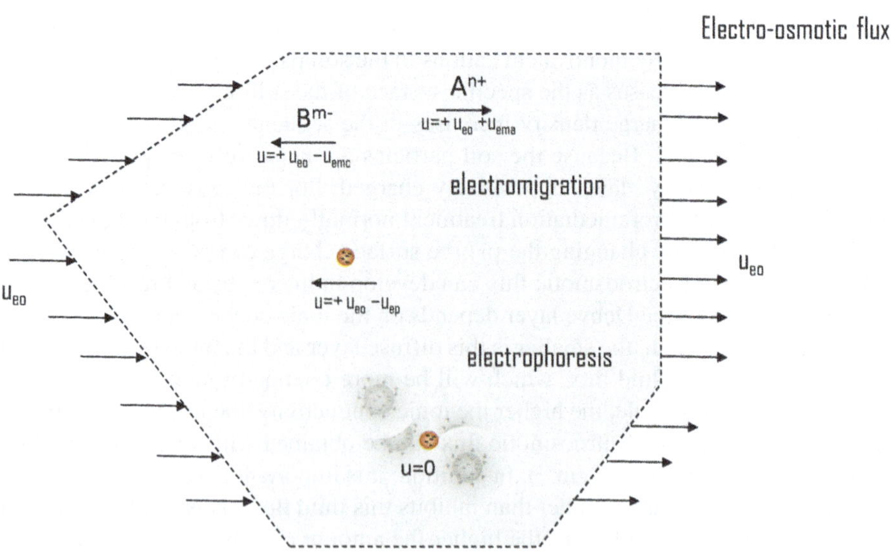


**Fig. 7** Direction of the transport of species during electrokinetic processes

structure. All these processes are of extreme significance to understand what really happens in soil when an electric field is applied.

## 5 Applications of Electrokinetic Processes

This section aims to be only a brief introduction to the very different type of processes that can be promoted when an electric field is applied. Throughout the book, all these processes will be described in a more detailed way, but we have considered important to introduce them right here for showing the great importance of the processes.

The first important treatment based on electrokinetic processes is the removal of metal ions, and it can be easily arranged simply by placing electrodes in the soil and preventing the negative effects of the basic front formed by the reduction of water on the surface of the cathode, which can produce the immobilization of the ions by precipitation [7, 15]. The key transport phenomenon is the electromigration, and pH plays a very important role. The prevention of the effect of the basic front can be carried out in different ways, being the most important the addition of acids such as citric acid in order to neutralize the hydroxyl ions formed. This technology can be applied in soils of very different permeability because for its success the promotion of the electroosmotic fluxes is not necessary. When the deposition of the metal on the cathode is obtained, the valorization of the pollution is carried out to the maximum advantage.

A little bit more complex is the electrokinetic soil flushing, which is not based on the electromigration but on the electroosmosis [12, 16]. In this case, water or a more complex soil washing fluid is added typically near the cathodes and it is collected in wells near the anodes. During the transport, it drags pollutants without necessity of excavating the soil. This technique is particularly efficient in soils with low solubility, because in those soils the electroosmotic flux is more important. The pH is also a very important factor in these processes because it can influence the magnitude of the flux, as it affects the size of the Debye layer. It is a variety of the pump and treat process especially suited for soils with low hydraulic conductivity [17, 18].

Reactive barriers are the natural evolution of the EKSF processes [19–23]. Once it is known that electroosmosis can produce the transport of water dragging part of the pollution from soil, the insertion of a barrier which interacts with these pollutants seems to be a very easy continuing step. Many types of barriers can be used, including biological (electro-bioremediation processes) and vegetables (electro-phytoremediation processes), GAC, ZVI, etc. [24, 25]. In this case, water is not aimed to be collected, so a good practice is to reverse polarity periodically because it may help control the effects of the acidic and basic fronts without requiring the addition of any reagents.

# 6 Conclusions

The application of an electric field in soil promotes the transport of different species, from pore water to ions and charged colloids. These processes can be explained by the electrostatic repulsion or attraction forces between the electric charges of the species and the net charges generated on the surface of the electrodes, although many other side processes have to be considered in order to understand the full processes occurring in soil. When these side processes are not limiting, the rate of this transport depends linearly on the electric field applied and on the properties of soil and transported species. While electroosmosis is promoted in fine-grained soils with low hydraulic conductivities, electrophoresis performs better in soil with large pore size. Many direct and enhanced soil remediation technologies can be proposed based on the understanding of the electrokinetic processes.

**Acknowledgments** Financial support from the Spanish Ministry of Economy, Industry and Competitiveness and European Union through project CTM2016-76197-R (AEI/FEDER, UE) is gratefully acknowledged.

# References

1. A.T. Yeung, Y.-Y. Gu, A review on techniques to enhance electrochemical remediation of contaminated soils. J. Hazard. Mater. **195**, 11–29 (2011)
2. A.T. Yeung, Contaminant extractability by electrokinetics. Environ. Eng. Sci. **23**, 202–224 (2006)
3. A.N. Alshawabkeh, A.T. Yeung, M.R. Bricka, Practical aspects of in-situ electrokinetic extraction. J. Environ. Eng.-ASCE **125**, 27–35 (1999)
4. J. Lyklema, Fundamentals of Interface and Colloid Science: Soft Colloids. **5**, Elsevier, Amsterdam, Netherlands (2005)
5. K.R. Reddy, A. Urbanek, A.P. Khodadoust, Electroosmotic dewatering of dredged sediments: bench-scale investigation. J. Environ. Manag. **78**, 200–208 (2006)
6. M.A. Rodrigo, N. Oturan, M.A. Oturan, Electrochemically assisted remediation of pesticides in soils and water: a review. Chem. Rev. **114**, 8720–8745 (2014)
7. R. Lopez-Vizcaino, C. Risco, J. Isidro, S. Rodrigo, C. Saez, P. Canizares, V. Navarro, M.A. Rodrigo, Scale-up of the electrokinetic fence technology for the removal of pesticides. Part I: Some notes about the transport of inorganic species. Chemosphere **166**, 540–548 (2017)
8. R. Lopez-Vizcaino, C. Risco, J. Isidro, S. Rodrigo, C. Saez, P. Canizares, V. Navarro, M.A. Rodrigo, Scale-up of the electrokinetic fence technology for the removal of pesticides. Part II: Does size matter for removal of herbicides? Chemosphere **166**, 549–555 (2017)
9. S. Barba, R. Lopez-Vizcaino, C. Saez, J. Villasenor, P. Canizares, V. Navarro, M.A. Rodrigo, Electro-bioremediation at the prototype scale: what it should be learned for the scale-up. Chem. Eng. J. **334**, 2030–2038 (2018)
10. E. Mena, S. Barba, C. Saez, V. Navarro, J. Villasenor, M.A. Rodrigo, P. Canizares, Prescale-Up of Electro-Bioremediation Processes, in *Geo-Chicago 2016: Sustainable Waste Management and Remediation*, ed. by N. Yesiller, D. Zekkos, A. Farid, A. De, K. R. Reddy, (Geotechnical Special Publication, Reston, 2016), pp. 264–273

11. B.S. Ramadan, G.L. Sari, R.T. Rosmalina, A.J. Effendi, Hadrah, An overview of electrokinetic soil flushing and its effect on bioremediation of hydrocarbon contaminated soil. J. Environ. Manag. **218**, 309–321 (2018)
12. R. Lopez-Vizcaino, C. Saez, E. Mena, J. Villasenor, P. Canizares, M.A. Rodrigo, Electro-osmotic fluxes in multi-well electro-remediation processes. J Environ Sci Health A Tox Hazard Subst Environ Eng **46**, 1549–1557 (2011)
13. Y.B. Acar, R.J. Gale, A.N. Alshawabkeh, R.E. Marks, S. Puppala, M. Bricka, R. Parker, Electrokinetic remediation—basics and technology status. J. Hazard. Mater. **40**, 117–137 (1995)
14. C. Cameselle, Enhancement of electro-osmotic flow during the electrokinetic treatment of a contaminated soil. Electrochim. Acta **181**, 31–38 (2015)
15. P.R. Buchireddy, R.M. Bricka, D.B. Gent, Electrokinetic remediation of wood preservative contaminated soil containing copper, chromium, and arsenic. J. Hazard. Mater. **162**, 490–497 (2009)
16. R. Lopez Vizcaino, A. Yustres, L. Asensio, C. Saez, P. Canizares, M.A. Rodrigo, V. Navarro, Enhanced electrokinetic remediation of polluted soils by anolyte pH conditioning. Chemosphere **199**, 477–485 (2018)
17. R. Lopez-Vizcaino, V. Navarro, J. Alonso, A. Yustres, P. Canizares, M.A. Rodrigo, C. Saez, Geotechnical behaviour of low-permeability soils in surfactant-enhanced electrokinetic remediation. J Environ Sci Health A Tox Hazard Subst Environ Eng **51**, 44–51 (2016)
18. R. Lopez-Vizcaino, V. Navarro, M.J. Leon, C. Risco, M.A. Rodrigo, C. Saez, P. Canizares, Scale-up on electrokinetic remediation: engineering and technological parameters. J. Hazard. Mater. **315**, 135–143 (2016)
19. E. Mena, C. Ruiz, J. Villasenor, M.A. Rodrigo, P. Canizares, Biological permeable reactive barriers coupled with electrokinetic soil flushing for the treatment of diesel-polluted clay soil. J. Hazard. Mater. **283**, 131–139 (2015)
20. C. Weng, Coupled Electrokinetics-Permeable Reactive Barriers, in *Electrochemical Remediation Technologies for Polluted Soils, Sediments and Groundwater*, ed. by K. R. Reddy, C. Cameselle, (John Wiley & Sons, Inc., Hoboken, 2009), pp. 483–503
21. H.I. Chung, M. Lee, A new method for remedial treatment of contaminated clayey soils by electrokinetics coupled with permeable reactive barriers. Electrochim. Acta **52**, 3427–3431 (2007)
22. A.D. Henderson, A.H. Demond, Long-term performance of zero-valent iron permeable reactive barriers: a critical review. Environ. Eng. Sci. **24**, 401–423 (2007)
23. P.L. Palmer, *Permeable Treatment Barriers, In Situ Treatment Technology*, 2nd edn. (Lewis Publishers, Boca Raton, 2001), pp. 459–482
24. J. Vidal, C. Saez, P. Canizares, V. Navarro, R. Salazar, M.A. Rodrigo, ZVI—reactive barriers for the remediation of soils polluted with clopyralid: are they really worth? Chem. Eng. J. **350**, 100–107 (2018)
25. C. Ruiz, E. Mena, P. Canizares, J. Villasenor, M.A. Rodrigo, Removal of 2,4,6-trichlorophenol from spiked clay soils by electrokinetic soil flushing assisted with granular activated carbon permeable reactive barrier. Ind. Eng. Chem. Res. **53**, 840–846 (2014)

# Fundamental of Reactive and Thermal Processes in Electrochemically Assisted Soil Remediation

**Salvador Cotillas**

## 1 Introduction

Increasing soil pollution has led to the development of novel efficient and environmentally friendly treatment technologies. Soil pollution can come from different sources such as petrochemical industries, fertilizers, or pharmaceuticals. Hence, it is necessary to evaluate the nature of contaminants before the application of a given technology [1–3]. Likewise, the chemical and geological soil characteristics, soil vegetal cover, the contaminated area, and the climate are factors that should be considered for the design of a proper technology. In this context, electrokinetic soil remediation has become as a promising technology for the removal of different pollutants in the soils [4, 5]. Specifically, it has been successfully tested for the treatment of the soils polluted with hydrocarbons, polar and nonpolar herbicides, or heavy metals [6, 7]. This technology is highly recommended for the remediation of soils with low hydraulic permeability, and, for this reason, this parameter should be estimated before carrying out an electrokinetic soil remediation process.

Hydraulic soil permeability can be defined as the capacity of the soil for allowing the pass of a fluid without an alteration of the soil structure. It is calculated by the permeability coefficient which means the water transport velocity in the soil for a hydraulic gradient. This coefficient can be easily calculated according to Darcy's law (Eq. 1), where $k$ is the permeability coefficient (m/s), $Q$ is the flowrate ($m^3/s$), $i$ is the hydraulic gradient (m/m) and $A$ is the soil section ($m^2$).

$$k = \frac{Q}{i \bullet A} \tag{1}$$

S. Cotillas (✉)
Chemical Engineering Department, Technical School of Industrial Engineering, University of Castilla-La Mancha, Avenida de España S/N, Albacete, Spain
e-mail: salvador.cotillas@uclm.es



M. A. Rodrigo, E. V. Dos Santos (eds.), *Electrochemically Assisted Remediation of Contaminated Soils*, Environmental Pollution 30,
https://doi.org/10.1007/978-3-030-68140-1_3

The application of electrokinetic soil remediation technique involves different processes that can simultaneously occur when applying an electric potential to a set of electrodes placed into the soil: electromigration, electrophoresis, and electroosmosis [5]. Electromigration consists of the transport of ions from water retained in soil by the action of the electric field. Anions (negative charge) are moved to the anode whereas cations (positive charge) are moved to the cathode. Electrophoresis is based on the transport of charged particles contained in soil or added for the treatment by the action of the same electric field. Electroosmosis is the phenomenon that promotes the electrolyte transport throughout the soil. This electrolyte can be natural groundwater, or a synthetic solution added to enhance the transport of pollutants in the soil. Usually, the electrolyte transport follows the anode–cathode direction [8]. To ensure the development of these processes, the soil must present a minimum water content [9–11]. A detailed description of these processes is reported in Chap. 2. In addition to these transport processes, other reactive and thermal processes take place during electrokinetic soil remediation, specifically, the electrolysis, directly over the electrodes or mediated in the electrolyte solution, and soil electrical heating [12]. The most important electrochemical processes are water oxidation (Eq. 2) and reduction (Eq. 3) and metal electrodeposition over the cathode (Eq. 4).

$$2H_2O \rightarrow O_2(g) + 4H^+ + 4e^- \tag{2}$$

$$2H_2O + 2e^- \rightarrow H_2 + 2OH^- \tag{3}$$

$$Me^{n+} + ne^- \rightarrow Me \tag{4}$$

From a soil remediation viewpoint, water oxidation produces large amounts of protons, generating an acid front that is moved to the cathode due to electromigration process. This can help to release the pollutants retained in the soil by dissolution of precipitates or ion exchange [13, 14]. On the contrary, the production of hydroxyl ions is promoted by water reduction over the cathode surface and a basic front is moved to the anode. Both fronts can be modified by the addition of chemicals. Regarding thermal processes, electrical heating promotes an increase in the soil temperature, being more remarkable in the vicinity of the electrodes [15]. This is due to the ohmic drop generated by soil ionic resistances (low ionic conductivity) and is proportional to the current intensity that externally passed to the soil between anodes and cathodes and the soil resistance.

## 2 Reactive Processes in Electrokinetic Soil Remediation: Electrolysis

An electrolytic process consists of an electrical energy conversion to chemical energy by applying a current intensity that promotes a chemical reaction. The extension of these electrochemical processes is directly proportional to the current intensity that circulates between anode and cathode in an electrochemical cell. The oxidation and reduction rate (*r*) can be easily calculated according to Faraday's law for electrolysis (Eq. 5), where *I* is the current intensity between the electrodes, *n* is the number of electrodes changed in the electrode reaction and *F* is the Faraday's constant (96,485 C $mol^{-1}$).

$$r = \frac{I}{n \cdot F} \quad (5)$$

The application of high electric potentials between the electrodes placed into the soil only give rise to low electrical current intensity due to the large ohmic drop produced by the low ionic soil conductivity. This leads to a limited development of the electrolytic processes, and most of the electrical energy supplied is released as heat. The main reaction that takes place over the anode surface is the water oxidation to form oxygen and protons (Eq. 2), favoring the acid front which is moved to the cathode by electromigration. Thus, the release of pollutants retained in the soil by dissolution of precipitates or ion exchange is favored (Fig. 1). It is important to highlight that the soil usually has more affinity with protons than with heavy metals and, hence, acid pH conditions can significantly contribute to the release of these pollutants, allowing the subsequent treatment [16]. Proton concentration is higher

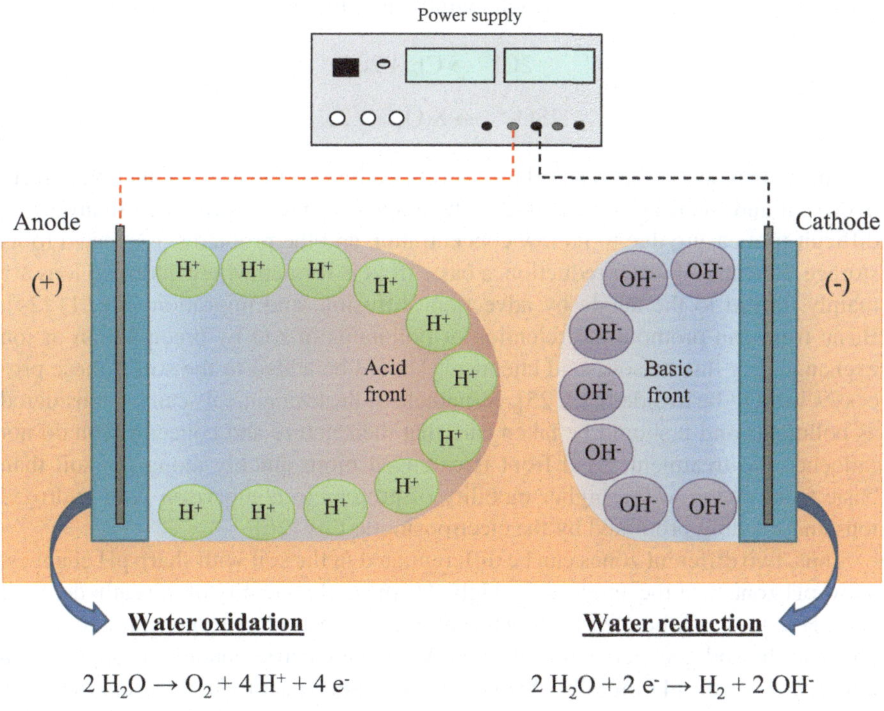


**Fig. 1** Electrolysis in electrokinetic soil remediation processes

near the anode and will decrease as it moves on to the cathode, depending on the buffering capacity of the soil [17]. The transport of protons takes place by advection (drag by the electroosmotic or hydraulic flow), diffusion, and migration, the last one being the main mechanism [18]. The generation of oxygen gas does not have a significant influence on the removal of pollutants from soil. However, its presence can cause several operating problems because the oxygen accumulation near the anode decreases the electrical conductivity and, therefore, the treatment costs. On the other hand, other electrochemical reactions can take place over the anode surface which directly compete with water oxidation [19]. These reactions will depend on the composition and concentration of the electrolyte. Specifically, the production of free chlorine can occur from the oxidation of chlorides contained in the electrolyte solution (Eq. 6). This species could play a key role during electrokinetic remediation of soils polluted with organics since it can attack these compounds, favoring their degradation. Likewise, the generation of persulfate could also take place from the electrooxidation of the electrolyte containing sulfates (Eq. 7). This anionic species is more important because it can also move on to the cathode by electromigration, enhancing the remediation process [20]. The occurrence of these oxidizing species is mainly influenced by the nature of the electrode [21, 22]. Mixed metal oxides (MMO) over Ti-support or carbon-based electrodes such as boron-doped diamond (BDD) favors the production of large amounts of these species. However, the main electrodes used in electrokinetic soil remediation processes are graphite, carbon cloth, or titanium [23] which mainly favor the production of chlorine.

$$2Cl^- \rightarrow Cl_2 + 2e^- \tag{6}$$

$$2SO_4^{2-} \rightarrow S_2O_8^{2-} + 2e^- \tag{7}$$

The main reaction that takes place over the cathode is the water reduction to form hydrogen and hydroxyl ions (Eq. 3). The amount of electrogenerated hydrogen is difficult to measure due to its low concentration and the problems associated to its storage. Because of water reduction, a basic front is formed where hydroxyl ions are mainly moved to the anode by advection, diffusion, and migration (Fig. 1) [24]. Basic front can promote the retention of pollutants in soil by precipitation or ion exchange. For this reason, acid chemicals should be added to the soil if these processes want to be avoided [13, 25]. Nonetheless, these chemicals can be considered as pollutants and it should be taken care that their nature and concentration do not influence the treatment. Acid front is extended more quickly along the soil than basic front owing to the higher mobility of protons in comparison with hydroxyl ions and the drag promoted by the electroosmotic flux [26].

Thus, two different zones can be differentiated in the soil with sharp pH changes: a low pH zone near the anode and a high pH area in the vicinity of the cathode. Real soil pH values will be clearly influenced by the transport of protons and hydroxyl ions and the soil geochemical properties. When an electric potential is applied to a polluted soil, soil pH initially changes (in time and space) which favors the mass transfer from one phase to others, such as dissolved or solid/precipitated. This promotes chemical speciation changes, being sorption/desorption, precipitation/dissolution, and oxidation/reduction the most important reactions [5]. Sorption is the fixing of pollutants in the soil from the liquid and it involves two different processes: adsorption and ion exchange. It depends on the pollutant nature, pore fluid characteristics, and the soil type. Desorption is the reverse process and allows to release the pollutants retained in the soil. Both sorption and desorption are influenced by pH changes caused by the migration of protons and hydroxyl ions. Precipitation and dissolution of pollutants retained in the soil can also affect the electrokinetic remediation process. Proton migration acidifies the soil and favors the dissolution of metal hydroxides. However, this process can be hampered by a high soil buffering capacity which will depend on its nature [17]. The formation of hydroxyl ions over the cathode surface (Eq. 3) leads to an increase in the electrode solution pH and the area near the cathode. Under these operating conditions, heavy metals precipitate, avoiding the transport of pollutants [27]. Furthermore, metal electrodeposition can take place over the cathode surface (Eq. 4). This process can be developed if metals are not fixed in the soil by basic front, and, hence, the addition of acid chemicals to avoid metal precipitation or the fixation by ion exchange is required [28]. This is an important reaction because it allows to recover metal ions from the soil in a more valuable form, as metallic elements.

Oxidation and reduction reactions are very important when treating soils polluted with metallic ions such as chromium because it mainly exits in two different states (III and VI). Cr(III) is found as cationic hydroxide which will migrate to the cathode during the electrokinetic process whereas Cr(VI) will migrate to the anode because it exists as oxyanion. The soil composition influences the metal valence, particularly by the presence of organic matter and Fe(II) (reducing agents) or Mn(IV) (oxidizing agent). For this reason, it is necessary to know the valence of metals retained in the soils as well as the possible redox reactions that could take place during the treatment. This knowledge will allow to evaluate the transport of pollutants in soil.

## 3 Thermal Processes in Electrokinetic Soil Remediation: Electrical Heating

The application of a potential difference to the electrodes placed in a polluted soil generates an electric current (electrons flow) between the anodes and cathodes, as has been previously commented. Current flow is proportional to the electrochemical reactions that take place over the electrodes surface (mainly water oxidation/reduction and metals electrodeposition), since one species must lose an electron on the anode (oxidation process) and another one must earn an electron on the cathode (reduction process) to guarantee the electrons flux and the electroneutrality condition.

However, the electrical power supplied to the system not only promotes the electrons flux (and consequently the oxidation/reduction reactions) but also a significant fraction is dissipated as heat. In this context, it is important to bear in mind that the electric potential applied to an electrochemical cell (potential difference between the anode/s and the cathode/s) is the sum of different contributions which are described below.

Anodic cell potential and overpotential. These can be simply defined as the contributions required to carry out the electrode reaction from the thermodynamic and kinetic viewpoint, respectively.

Cathodic cell potential and overpotential. These contributions are similar to previously described but referred to the cathodic process. Both anodic and cathodic processes simultaneously occur in the electrochemical cell, and, therefore, the sum of anodic and cathodic contributions is required for developing the electrochemical process.

Overcoming ionic and electrical resistances potential. when an electrical current is circulated through a conductive material, this is heated by Joule-Thompson effect. This heating is proportional to the resistance of the conductive material and is responsible of heat release from any electrical equipment (from a television to an electrochemical cell). In the case of electrokinetic soil remediation processes, there are different resistance types: electrodes, conductive elements, and current suppliers (electrochemical cell external elements). On the other hand, there is the soil ionic resistance which will depend on its hydration and the presence of conductive species in the water contained therein. All these resistances are needed to close the ionic circuit and to guarantee the interest processes from electrokinetic soil remediation viewpoint for developing electrochemically assisted soil remediation processes, this last contribution is the most important and, hence, a significant fraction of electric potential ($V$) will be used to overcome this resistance. Heat released ($W$) can be calculated from the estimation of electric resistance (Ohms) associated to this process, according to Eq. (8) which involves the electric power and Ohm's law.

$$W = I_{\text{cell}} \cdot V_{\text{soil}} = I^2_{\text{cell}} \cdot R_{\text{soil}} \tag{8}$$

$I_{\text{cell}}$ is the current generated when an electric potential ($V_{\text{soil}}$) is applied to a set of electrodes placed in a polluted soil and $R_{\text{soil}}$ is the soil electric resistance during the process. The amount of heat generated promotes an increase in the temperature which can be estimated through a energy balance, assuming a macroscopic description of the system. This means that it is assumed uniform matter and energy distributions into the soil (Eq. 9).

$$I^2_{\text{cell}} \cdot R_{\text{soil}} = m_{\text{soil}} \cdot C_{\text{p}} \cdot \Delta T \tag{9}$$

where $m_{\text{soil}}$ is the amount of treated soil, $C_{\text{p}}$ is the specific heat capacity, and $\Delta T$ is the temperature increase during the process.

However, temperature distribution is more complex since heat release is higher near the electrodes (Fig. 2). This promotes the release of significant amounts of heat

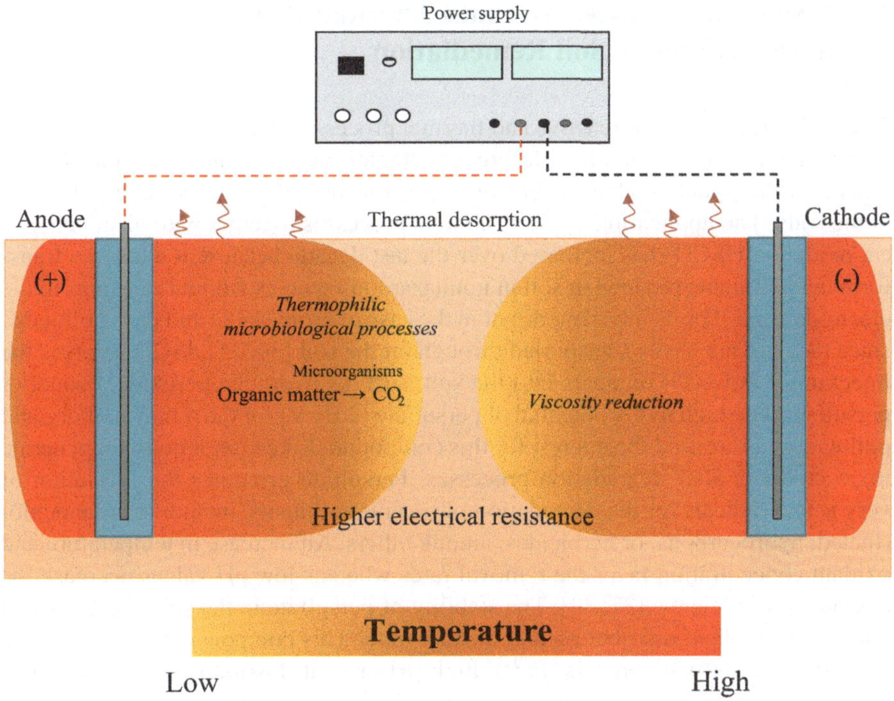


**Fig. 2** Electrical heating in electrokinetic soil remediation processes

to the system limits (the surrounding air) and will depend on the soil nature (texture and composition). Therefore, an accuracy energy balance, including maximum or multiple temperature gradient estimations, should be carried out for the correct temperature distribution calculation. A hydraulic gradient promotes an isothermal heat transfer whereas an electrical gradient favors the Peltier effect. Likewise, a thermal gradient influences the thermal conduction (Fourier's law) and a chemical gradient promotes the Dufour effect [8, 29]. Furthermore, a thermal gradient also promotes the thermo-osmosis on the fluid, the seebeck effect on the electric current, and soret effect on the ions [8].

Temperature increase has important effects on organic pollutants. Specifically, an increase in temperature leads to a decrease in the viscosity of hydrocarbons polluted fluids, favoring a higher movement of fluids in the soil. On the other hand, the higher temperatures achieved, which can even be close to the water boiling point, favor the volatilization of low-molecular-weight pollutants. This can be used as a complementary or alternative treatment to conventional soil vapor extraction (SVE). Finally, another important point is the effect of temperature on biological processes. In this context, higher values favor thermophilic microorganisms' growth which can also contribute to the removal of organic pollutants retained in the soil, improving the electrokinetic soil remediation efficiency.

## 4 Development of Reactive and Thermal Processes in Electrokinetic Soil Remediation

Due to the advantages of reactive and thermal processes that naturally occur during electrokinetic soil remediation, different strategies based on in situ chemical reactions or heat application have been developed for the removal of several pollutants in the soils. The application of electrokinetic processes together with in situ chemical oxidation (ISCO) has increased over the last decade because it allows to transform the pollutants retained in soil in nonhazardous species without transport to the electrode zone. This is more important in the case of more stable and inert pollutants since they are not easily transported throughout the soil [30, 31]. ISCO involves the injection of powerful oxidants into the soil, such as hydrogen peroxide, ozone, or persulfate. The high redox potential of persulfate (2.01 V) for the removal of organic pollutants has aroused the interest for this compound. Likewise, it does not generate toxic products after degradation processes. Persulfate promotes the formation of very active radicals for the degradation of several pollutants, including organochlorinated, hydrocarbons, or herbicides, among others. An increase in temperature and oxidant concentration favor the removal rates whereas low pH values decrease the degradation efficiency [32, 33]. The stability of persulfate in the soil is good that it allows an excellent distribution and contact time. This compound can be activated by hydrogen peroxide, metals ($Fe^{2+}$), high pH or heat, favoring the production of sulfate ($SO_4^{-}\cdot$, 2.6 V), and hydroxyl (OH, 2.7 V) radicals which are more reactive than persulfate [34]. Electrokinetic processes help to add persulfate into the soil. This compound should be introduced in the cathode chamber and then moved by electromigration. Likewise, it can also be introduced in the anode chamber but, in this case, it will be moved by electroosmosis. Yukselen-Aksoy and Reddy [31] reported the remediation of polychlorobiphenyls (PCBs) polluted soils by enhanced electrokinetic with persulfate activated by heat and high pH. They achieved a maximum PCBs removal of 77.9% when treating kaolin soil with 30% Na-persulfate activated by heat. Fan et al. [35] studied different activation methods for persulfate in electrokinetic remediation of PCBs polluted soil: ZVI, citric acid chelated $Fe^{2+}$, iron electrode, alkaline pH, and peroxide. They concluded that the most efficient method for activating persulfate in electrokinetic remediation was alkaline pH, reaching a 40% removal of PCBs in soil. Likewise, they also demonstrated that electroosmosis was the most efficient electrokinetic process for persulfate distribution in the soil and the removal of PCBs.

Another efficient oxidant for in situ applications is permanganate which can also be introduced by electromigration. The main advantages of this species are its high oxidation potential (1.7 V), excellent stability, strong ability to remove several organic pollutants, high efficiency at different pH values, ease to deliver and identify for its color, and low cost [30]. This oxidant has been used for the removal of organochlorinated pollutants from soil [36–38].

Finally, Fenton reaction has been widely used as in situ technology for the removal of organic pollutants during electrokinetic soil remediation owing to its high efficiency, low cost, and it is environmentally friendly. This is based on the production of large amounts of hydroxyl radicals as oxidizing agent using hydrogen peroxide and $Fe^{2+}$ as catalyst (Eq. 10) [39].

$$H_2O_2 + Fe^{2+} \rightarrow Fe^{3+} + OH^- + \cdot OH \tag{10}$$

Occurrence of hydroxyl radicals in soil guarantees a strong oxidation conditions in the whole system. The main parameters that affect Fenton reaction are the amount of hydrogen peroxide, catalyst type, pH solution, ratio $H_2O_2$:$Fe^{2+}$, and temperature [40]. Yang and Liu [41] reported the application of Fenton reaction during the electrokinetic process for the remediation of trichloroethylene polluted soil. They studied the influence of soil and catalyst types, amount of catalyst, and electrode type. The removal of phenanthrene by electrokinetic-Fenton process was described by Kim et al. [42]. In this study, it is concluded that the introduction of large amounts of hydrogen peroxide in the anode chamber promoted the formation of intermediate anions, decreasing the removal efficiency. Alcántara et al. [43] also reported the electrokinetic-Fenton remediation of phenanthrene polluted soils. Kaolinite was polluted with iron (catalyst), and 10% hydrogen peroxide was added to anodic and cathodic chambers. Soil pH was maintained constant in acid values (around 3.5). Results reported a 99% removal when applying 3 V $cm^{-1}$ in 14 days. More recently, Paixao et al. [44] evaluated the remediation of soils polluted with petroleum using the electrokinetic-Fenton process. In this study, they used iron as electrode material to take advantage of the electrochemical reactions during electrokinetic process. Thus, the iron source for the Fenton reaction comes from the electrodissolution of the anode (Eq. 11), and the addition of catalyst is not required. An 89% removal of total petroleum hydrocarbons (TPHs) was achieved when citric acid was added to control the pH conditions, avoiding the generation of a secondary effluent with higher organic content.

$$Fe \rightarrow Fe^{2+} + 2e^- \tag{11}$$

Regarding the development of thermal processes coupled to electrokinetic remediation, this is very limited due to the high energy requirements. Electrical heating caused during electrokinetic soil remediation contributes to the removal of pollutants retained in soil by desorption and vapor extraction. The application of electrokinetic remediation processes at prototype scale for the removal of organics has allowed to verify the high temperatures reached which also favor the volatilization of pollutants and the soil water evaporation [15, 45]. Specifically, López-Vizcaíno et al. [46] found that the temperature increased three times in the soil and four times in the wells during the remediation of soils polluted with the herbicide 2,4-D. To avoid the soil water evaporation, capillary barriers are usually included on the soil surface. Based on these results and although there are not relevant thermal/heat treatments integrated with electrokinetic, single thermal processes have been proposed for the removal of petroleum hydrocarbon from soil [47, 48]. A brief description of the most relevant thermal/heat treatment methods is detailed below.

Thermal desorption. It is a physical treatment based on pollutant volatilization and desorption from soil by direct or indirect heat application under vacuum or with a carrier gas that allows to separate the pollutants retained in soil [49]. Different mechanism can be used to achieve thermal desorption such as thermal cracking, pyrolytic reactions, oxidation, or incineration. It depends on the temperature and the oxygen distribution, and it is highly recommended for the removal of volatile and semi-volatile pollutants [50]. The process takes place at temperatures within the range 100–500 °C, and soil can reach values up to 800–900 °C. Liu et al. [51] reported the thermal desorption of PCBs polluted soil with calcium hydroxide, applying temperatures between 300 and 600 °C. A 94% removal was attained at 600 °C.

Smoldering combustion. It is a flameless combustion that generates a self-sustaining wave of exothermic combustion from fuel and oxygen. Pollutants and oxygen are transformed to carbon dioxide, water, and heat. Thus, the process occurs without fuel or external energy after ignition up to complete remediation [52]. Pollutants reaction promotes a heat generation and a temperature increase which are used to preheat and to start a higher pollutants combustion. Thus, a combustion front is spread through the polluted zone in the presence of air [53]. Heat and temperature generated during smoldering combustion achieve values within the range 600–1100 °C. Pironi et al. [54] studied the smoldering combustion of soil polluted with nonaqueous phase liquids, reaching a 99.5% removal for crude oil and 99.9% for coal tar.

Incineration. This process promotes the total removal of pollutants by a high temperature combustion within the range 870–1200 °C. It can be carried out in liquid injection systems, rotatory kilns, fluidized bed incinerators, circulating bed combustors, and infrared combustion systems [47]. The oxygen level should be maintained around 10% to burn volatile organic compounds. To ensure a safe incineration, the oxygen level and the soil loading must be considered together with the pollutant lower explosion limit [55]. After incineration, soil is rehydrated for controlling the dust before using in any applications. The main advantage of this technology is that it guarantees a complete removal of hazardous pollutants. It is a high-cost technology but highly efficient for destruction of several contaminants. The treatment of fuel oil polluted soil by incineration has been reported in the literature of Bucala et al. [56]. In this study, the removal efficiency was 100% in 0.7 s. When applying 1000 °C/s. They concluded that it is necessary to apply temperatures higher than 500 °C to attain a remarkable chemical transformation.

Pyrolysis. It consists of thermal heating or cracking of organics polluted soil under anoxic conditions or inert atmosphere at temperature values between 400 and 1200 °C. Hydrocarbons are removed by thermal desorption when the boiling temperature is achieved during pyrolysis of hydrocarbons polluted soil. However, char can be formed by sequential aromatic condensation reactions from free radicals and broken chemical bonds if the temperature is within the range 300–500 °C [57]. This process is similar to incineration and thermal desorption, but it works under inert atmosphere or anoxic conditions. Indirect electrical heating is used to avoid the presence of oxygen in the treatment. Vidonish et al. [58] reported the pyrolysis of soils polluted with heavy hydrocarbons. They concluded that it is possible to attain a 99% removal of total petroleum hydrocarbons, maintaining the nutrients and soil properties that are lost during incineration.

Vitrification. This process is based on heating at higher temperatures, ranging between 1600 and 2000 °C, to convert organic pollutants retained in soil into vitreous compounds [59]. Most of the pollutants are immobilized and degraded by pyrolysis and they are volatilized. However, remaining pollutants are transformed to chemical stable, inert, glass-like, and crystalline products [60]. Three different processes are developed to carry out vitrification which require large amounts of energy to treat one ton of polluted soil [61]: Electrical heating, thermal process, and plasma arc process. The remediation of hexavalent chromium polluted soil by vitrification has been reported by Ballesteros et al. [62]. Results showed the formation of glassy products with silicate composition which present environmental stability, high mechanical resistance, and chemical stability. Toxic characteristics leaching test was performed to confirm these properties. It consists of evaluating the possible leach of toxic compounds from vitrified samples.

## 5 Conclusion

The application of electrokinetic soil remediation promotes different physical and chemical processes that help to remove the pollutants. Water electrolysis occurs over the electrodes surface, favoring the production of large amounts of protons and hydroxyl ions which can be transported throughout the soil by electromigration. This generates different pH zones in soil, and these species, mainly protons, contribute to the removal of pollutants by dissolution of precipitates and ion exchange. Furthermore, the application of an electric potential leads to an increase in soil temperature owing to the ohmic drop generated by soil ionic resistances. This is more significant near the electrodes where it is possible to achieve the water boiling point. Thus, the volatilization of low-molecular-weight pollutants can be carried out by soil vapor extraction. Finally, different combined electrokinetic technologies have been developed based on the advantages of reactive processes that allow to increase the removal efficiencies.

**Acknowledgments** Financial support from the Spanish Ministry of Science, Innovation and Universities through "Juan de la Cierva-Incorporación" postdoctoral grant (IJC2018-036241-I) is gratefully acknowledged.

## References

1. C.N. Mulligan, R.N. Yong, B.F. Gibbs, Remediation technologies for metal-contaminated soils and groundwater: an evaluation. Eng. Geol. **60**(1–4), 193–207 (2001). https://doi.org/10.1016/S0013-7952(00)00101-0
2. C.N. Mulligan, R.N. Yong, B.F. Gibbs, Surfactant-enhanced remediation of contaminated soil: a review. Eng. Geol. **60**(1–4), 371–380 (2001). https://doi.org/10.1016/S0013-7952(00)00117-4
3. S.C. Wilson, K.C. Jones, Bioremediation of soil contaminated with polynuclear aromatic hydrocarbons (PAHs): A review. Environ. Pollut. **81**(3), 229–249 (1993). https://doi.org/10.1016/0269-7491(93)90206-4

4. J. Virkutyte, M. Sillanpää, P. Latostenmaa, Electrokinetic soil remediation – critical overview. Sci. Total Environ. **289**(1–3), 97–121 (2002). https://doi.org/10.1016/S0048-9697(01)01027-0
5. K.R. Reddy, C. Cameselle, *Electrochemical Remediation Technologies for Polluted Soils, Sediments and Groundwater. Electrochemical Remediation Technologies for Polluted Soils, Sediments and Groundwater* (John Wiley and Sons, Hoboken, 2009). https://doi.org/10.1002/9780470523650
6. M.A. Rodrigo, N. Oturan, M.A. Oturan, Electrochemically assisted remediation of pesticides in soils and water: a review. Chem. Rev. **114**(17), 8720–8745 (2014). https://doi.org/10.1021/cr500077e
7. R.E. Saichek, K.R. Reddy, Electrokinetically enhanced remediation of hydrophobic organic compounds in soils: a review. Crit. Rev. Environ. Sci. Technol. **35**(2), 115–192 (2005). https://doi.org/10.1080/10643380590900237
8. A.T. Yeung, Chapters, Electrokinetic flow processes in porous media and their applications. Adv. Porous Media **2** (1994)
9. R. López-Vizcaíno, C. Sáez, E. Mena, J. Villaseñor, P. Cañizares, M.A. Rodrigo, Electro-osmotic fluxes in multi-well electro-remediation processes. J. Environ. Sci. Health A Tox. Hazard. Subst. Environ. Eng. **46**(13), 1549–1557 (2011). https://doi.org/10.1080/10934529.2011.609458
10. C. Cameselle, K.R. Reddy, Development and enhancement of electro-osmotic flow for the removal of contaminants from soils. Electrochim. Acta **86**, 10–22 (2012). https://doi.org/10.1016/j.electacta.2012.06.121
11. Á. Yustres, R. López-Vizcaíno, C. Sáez, P. Cañizares, M.A. Rodrigo, V. Navarro, Water transport in electrokinetic remediation of unsaturated kaolinite. Experimental and numerical study. Sep. Purif. Technol. **192**, 196–204 (2018). https://doi.org/10.1016/j.seppur.2017.10.009
12. R.E. Saichek, K.R. Reddy, Effect of pH control at the anode for the electrokinetic removal of phenanthrene from kaolin soil. Chemosphere **51**(4), 273–287 (2003). https://doi.org/10.1016/S0045-6535(02)00849-4
13. M. Villen-Guzman, J.M. Paz-Garcia, J.M. Rodriguez-Maroto, C. Gomez-Lahoz, F. Garcia-Herruzo, Acid enhanced electrokinetic remediation of a contaminated soil using constant current density: strong vs. weak acid. Separ. Sci. Tech. **49**(10), 1461–1468 (2014). https://doi.org/10.1080/01496395.2014.898306
14. Y.B. Acar, R.J. Gale, A.N. Alshawabkeh, R.E. Marks, S. Puppala, M. Bricka, R. Parker, Electrokinetic remediation: basics and technology status. J. Hazard. Mater. **40**(2), 117–137 (1995). https://doi.org/10.1016/0304-3894(94)00066-P
15. R. López-Vizcaíno, V. Navarro, M.J. León, C. Risco, M.A. Rodrigo, C. Sáez, P. Cañizares, Scale-up on electrokinetic remediation: engineering and technological parameters. J. Hazard. Mater. **315**, 135–143 (2016). https://doi.org/10.1016/j.jhazmat.2016.05.012
16. R. Lageman, R.L. Clarke, W. Pool, Electro-reclamation, a versatile soil remediation solution. Eng. Geol. **77**(3), 191–201 (2005). https://doi.org/10.1016/j.enggeo.2004.07.010
17. R. López-Vizcaíno, E.V. dos Santos, A. Yustres, M.A. Rodrigo, V. Navarro, C.A. Martínez-Huitle, Calcite buffer effects in electrokinetic remediation of clopyralid-polluted soils. Sep. Purif. Technol. **212**, 376–387 (2019). https://doi.org/10.1016/j.seppur.2018.11.034
18. S.K. Puppala, A.N. Alshawabkeh, Y.B. Acar, R.J. Gale, M. Bricka, Enhanced electrokinetic remediation of high sorption capacity soil. J. Hazard. Mater. **55**(1), 203–220 (1997). https://doi.org/10.1016/S0304-3894(97)00011-3
19. J.M. Paz-García, B. Johannesson, L.M. Ottosen, A.N. Alshawabkeh, A.B. Ribeiro, J.M. Rodríguez-Maroto, Modeling of electrokinetic desalination of bricks. Electrochim. Acta **86**, 213–222 (2012). https://doi.org/10.1016/j.electacta.2012.05.132
20. P. Isosaari, R. Piskonen, P. Ojala, S. Voipio, K. Eilola, E. Lehmus, M. Itävaara, Integration of electrokinetics and chemical oxidation for the remediation of creosote-contaminated clay. J. Hazard. Mater. **144**(1), 538–548 (2007). https://doi.org/10.1016/j.jhazmat.2006.10.068
21. F. Sopaj, M.A. Rodrigo, N. Oturan, F.I. Podvorica, J. Pinson, M.A. Oturan, Influence of the anode materials on the electrochemical oxidation efficiency. Application to oxidative

degradation of the pharmaceutical amoxicillin. Chem. Eng. J. **262**, 286–294 (2015). https://doi.org/10.1016/j.cej.2014.09.100
22. P. Cañizares, C. Sáez, A. Sánchez-Carretero, M.A. Rodrigo, Synthesis of novel oxidants by electrochemical technology. J. Appl. Electrochem. **39**(11), 2143–2149 (2009). https://doi.org/10.1007/s10800-009-9792-7
23. E. Méndez, M. Pérez, O. Romero, E.D. Beltrán, S. Castro, J.L. Corona, A. Corona, M.C. Cuevas, E. Bustos, Effects of electrode material on the efficiency of hydrocarbon removal by an electrokinetic remediation process. Electrochim. Acta **86**, 148–156 (2012). https://doi.org/10.1016/j.electacta.2012.04.042
24. J.M. Paz-Garcia, K. Baek, I.D. Alshawabkeh, A.N. Alshawabkeh, A generalized model for transport of contaminants in soil by electric fields. J. Environ. Sci. Health A **47**(2), 308–318 (2012). https://doi.org/10.1080/10934529.2012.640911
25. A.T. Yeung, Y.-Y. Gu, A review on techniques to enhance electrochemical remediation of contaminated soils. J. Hazard. Mater. **195**, 11–29 (2011). https://doi.org/10.1016/j.jhazmat.2011.08.047
26. R. López Vizcaíno, A. Yustres, L. Asensio, C. Saez, P. Cañizares, M.A. Rodrigo, V. Navarro, Enhanced electrokinetic remediation of polluted soils by anolyte pH conditioning. Chemosphere **199**, 477–485 (2018). https://doi.org/10.1016/j.chemosphere.2018.02.038
27. L.M. Ottosen, H.K. Hansen, A.B. Ribeiro, A. Villumsen, Removal of Cu, Pb and Zn in an applied electric field in calcareous and non-calcareous soils. J. Hazard. Mater. **85**(3), 291–299 (2001). https://doi.org/10.1016/S0304-3894(01)00231-X
28. D.-M. Zhou, C.-F. Deng, L. Cang, Electrokinetic remediation of a Cu contaminated red soil by conditioning catholyte pH with different enhancing chemical reagents. Chemosphere **56**(3), 265–273 (2004). https://doi.org/10.1016/j.chemosphere.2004.02.033
29. A.T. Yeung, J.K. Mitchell, Coupled fluid, electrical and chemical flows in soil. Geotechnique **43**(1), 121–134 (1993). https://doi.org/10.1680/geot.1993.43.1.121
30. L. Ren, H. Lu, L. He, Y. Zhang, Enhanced electrokinetic technologies with oxidization–reduction for organically-contaminated soil remediation. Chem. Eng. J. **247**, 111–124 (2014). https://doi.org/10.1016/j.cej.2014.02.107
31. Y. Yukselen-Aksoy, K.R. Reddy, Effect of soil composition on electrokinetically enhanced persulfate oxidation of polychlorobiphenyls. Electrochim. Acta **86**, 164–169 (2012). https://doi.org/10.1016/j.electacta.2012.03.049
32. C. Liang, Z.-S. Wang, C.J. Bruell, Influence of pH on persulfate oxidation of TCE at ambient temperatures. Chemosphere **66**(1), 106–113 (2007). https://doi.org/10.1016/j.chemosphere.2006.05.026
33. X. Xie, Y. Zhang, W. Huang, S. Huang, Degradation kinetics and mechanism of aniline by heat-assisted persulfate oxidation. J. Environ. Sci. **24**(5), 821–826 (2012). https://doi.org/10.1016/S1001-0742(11)60844-9
34. C. Cameselle, S. Gouveia, Electrokinetic remediation for the removal of organic contaminants in soils. Curr. Opin. Electrochem. **11**, 41–47 (2018). https://doi.org/10.1016/j.coelec.2018.07.005
35. G. Fan, L. Cang, H.I. Gomes, D. Zhou, Electrokinetic delivery of persulfate to remediate PCBs polluted soils: effect of different activation methods. Chemosphere **144**, 138–147 (2016). https://doi.org/10.1016/j.chemosphere.2015.08.074
36. M.Z. Wu, D.A. Reynolds, A. Fourie, D.G. Thomas, Optimal field approaches for electrokinetic in situ oxidation remediation. Ground Water Monit. Remed. **33**(1), 62–74 (2013). https://doi.org/10.1111/j.1745-6592.2012.01410.x
37. A.I.A. Chowdhury, J.I. Gerhard, D. Reynolds, B.E. Sleep, D.M. O'Carroll, Electrokinetic-enhanced permanganate delivery and remediation of contaminated low permeability porous media. Water Res. **113**, 215–222 (2017). https://doi.org/10.1016/j.watres.2017.02.005
38. P. Thepsithar, E.P.L. Roberts, Removal of phenol from contaminated kaolin using electrokinetically enhanced in situ chemical oxidation. Environ. Sci. Technol. **40**(19), 6098–6103 (2006). https://doi.org/10.1021/es060883f

39. S. H-w, Y. Q-s, Influence of pyrene combination state in soils on its treatment efficiency by Fenton oxidation. J. Environ. Manag. **88**(3), 556–563 (2008). https://doi.org/10.1016/j.jenvman.2007.03.031
40. M. Pazos, O. Iglesias, J. Gómez, E. Rosales, M.A. Sanromán, Remediation of contaminated marine sediment using electrokinetic–Fenton technology. J. Ind. Eng. Chem. **19**(3), 932–937 (2013). https://doi.org/10.1016/j.jiec.2012.11.010
41. G.C.C. Yang, C.-Y. Liu, Remediation of TCE contaminated soils by in situ EK-Fenton process. J. Hazard. Mater. **85**(3), 317–331 (2001). https://doi.org/10.1016/S0304-3894(01)00288-6
42. S.-S. Kim, J.-H. Kim, S.-J. Han, Application of the electrokinetic-Fenton process for the remediation of kaolinite contaminated with phenanthrene. J. Hazard. Mater. **118**(1), 121–131 (2005). https://doi.org/10.1016/j.jhazmat.2004.10.005
43. T. Alcántara, M. Pazos, S. Gouveia, C. Cameselle, M.A. Sanromán, Remediation of phenanthrene from contaminated kaolinite by electroremediation-Fenton technology. J. Environ. Sci. Health A **43**(8), 901–906 (2008). https://doi.org/10.1080/10934520801974418
44. I.C. Paixão, R. López-Vizcaíno, A.M.S. Solano, C.A. Martínez-Huitle, V. Navarro, M.A. Rodrigo, E.V. dos Santos, Electrokinetic-Fenton for the remediation low hydraulic conductivity soil contaminated with petroleum. Chemosphere **248**, 126029 (2020). https://doi.org/10.1016/j.chemosphere.2020.126029
45. S. Barba, R. López-Vizcaíno, C. Saez, J. Villaseñor, P. Cañizares, V. Navarro, M.A. Rodrigo, Electro-bioremediation at the prototype scale: what it should be learned for the scale-up. Chem. Eng. J. **334**, 2030–2038 (2018). https://doi.org/10.1016/j.cej.2017.11.172
46. R. López-Vizcaíno, C. Risco, J. Isidro, S. Rodrigo, C. Saez, P. Cañizares, V. Navarro, M.A. Rodrigo, Scale-up of the electrokinetic fence technology for the removal of pesticides. Part II: Does size matter for removal of herbicides? Chemosphere **166**, 549–555 (2017). https://doi.org/10.1016/j.chemosphere.2016.09.114
47. I.C. Ossai, A. Ahmed, A. Hassan, F.S. Hamid, Remediation of soil and water contaminated with petroleum hydrocarbon: a review. Environ. Technol. Innov. **17**, 100526 (2020). https://doi.org/10.1016/j.eti.2019.100526
48. J.E. Vidonish, K. Zygourakis, C.A. Masiello, G. Sabadell, P.J.J. Alvarez, Thermal treatment of hydrocarbon-impacted soils: a review of technology innovation for sustainable remediation. Engineering **2**(4), 426–437 (2016). https://doi.org/10.1016/J.ENG.2016.04.005
49. F. Kastanek, P. Topka, K. Soukup, Y. Maleterova, K. Demnerova, P. Kastanek, O.J.J.C.P. Solcova, Remediation of contaminated soils by thermal desorption; effect of benzoyl peroxide addition. J. Clean. Prod. **125**, 309–313 (2016)
50. C. Zhao, Y. Dong, Y. Feng, Y. Li, Y. Dong, Thermal desorption for remediation of contaminated soil: a review. Chemosphere **221**, 841–855 (2019). https://doi.org/10.1016/j.chemosphere.2019.01.079
51. J. Liu, H. Zhang, Z. Yao, X. Li, J. Tang, Thermal desorption of PCBs contaminated soil with calcium hydroxide in a rotary kiln. Chemosphere **220**, 1041–1046 (2019). https://doi.org/10.1016/j.chemosphere.2019.01.031
52. G.C. Scholes, J.I. Gerhard, G.P. Grant, D.W. Major, J.E. Vidumsky, C. Switzer, J.L. Torero, Smoldering remediation of coal-tar-contaminated soil: pilot field tests of STAR. Environ. Sci. Technol. **49**(24), 14334–14342 (2015). https://doi.org/10.1021/acs.est.5b03177
53. G.P. Grant, D. Major, G.C. Scholes, J. Horst, S. Hill, M.R. Klemmer, J.N.J.R.J. Couch, Smoldering combustion (STAR) for the treatment of contaminated soils: examining limitations and defining success. Remediat. J. **26**(3), 27–51 (2016)
54. P. Pironi, C. Switzer, J.I. Gerhard, G. Rein, J.L. Torero, Self-sustaining smoldering combustion for NAPL remediation: laboratory evaluation of process sensitivity to key parameters. Environ. Sci. Technol. **45**(7), 2980–2986 (2011). https://doi.org/10.1021/es102969z
55. E.K. Nyer, *In Situ Treatment Technology* (CRC Press, Boca Raton, 2000)
56. V. Bucala, H. Saito, J.B. Howard, W.A. Peters, Thermal treatment of fuel oil-contaminated soils under rapid heating conditions. Environ. Sci. Technol. **28**(11), 1801–1807 (1994). https://doi.org/10.1021/es00060a008

57. R.S. Baker, M. Kuhlman, in *A Description of the Mechanisms of In-Situ Thermal Destruction (ISTD) Reactions*. Current Practices in Oxidation and Reduction Technologies for Soil and Groundwater, and Presented at the 2nd International Conference on Oxidation and Reduction Technologies for Soil and Groundwater, ORTs-2, Toronto, Ontario, Canada (2002)
58. J.E. Vidonish, K. Zygourakis, C.A. Masiello, X. Gao, J. Mathieu, P.J.J. Alvarez, Pyrolytic treatment and fertility enhancement of soils contaminated with heavy hydrocarbons. Environ. Sci. Technol. **50**(5), 2498–2506 (2016). https://doi.org/10.1021/acs.est.5b02620
59. R.M. Abousnina, A. Manalo, J. Shiau, W. Lokuge, An overview on oil contaminated sand and its engineering applications. Int. J. GEOMATE **10**(1), 1615–1622 (2016)
60. F.I. Khan, T. Husain, R. Hejazi, An overview and analysis of site remediation technologies. J. Environ. Manag. **71**(2), 95–122 (2004). https://doi.org/10.1016/j.jenvman.2004.02.003
61. A.T. Yeung, Remediation Technologies for Contaminated Sites, in *Advances in Environmental Geotechnics*, ed. by Y. Chen, L. Zhan, X. Tang, (Springer Berlin Heidelberg, Berlin, Heidelberg, 2010), pp. 328–369
62. S. Ballesteros, J.M. Rincón, B. Rincón-Mora, M.M. Jordán, Vitrification of urban soil contamination by hexavalent chromium. J. Geochem. Explor. **174**, 132–139 (2017). https://doi.org/10.1016/j.gexplo.2016.07.011

# Conceptual and Mathematical Modeling of the Transport of Pollutants in Soil by Electric Fields

**Rubén López-Vizcaíno, Ángel Yustres, Virginia Cabrera, and Vicente Navarro**

## 1 Introduction

As discussed in previous chapters, electrokinetic soil remediation (EKR) is a soil decontamination technique aimed at containing, mobilizing, and/or extracting the pollutant species present in each contamination scenario. EKR is based on the application of a low electrical current field between pairs of electrodes (anode/cathode) located in the contaminated soil. As a result, an electrical potential gradient is generated, which acts as a driving force for different electrokinetic transport processes. These transport processes allow for (i) the controlled and/or directed movement of ionic (charged) species towards the electrodes of opposite charge by means of the electromigration process, (ii) the generation of flow of water in soils with a high clay content, in which the Darcy flow is limited due to its low hydraulic permeability, by means of the electroosmosis process, and (iii) the mobilization of charged particles (micelle-type) through the electrophoresis process. In addition to these electrokinetic transport processes, there may also be a water flow driven by a hydraulic gradient (the magnitude of this transport mechanism will be determined by the hydraulic characteristics of the soil to be treated) and a Fickian diffusive flux generated by the concentration gradients of the species present in the evaluated domain. Taking into account all of these transport mechanisms, the mass balance equation for species "i" can be defined by the following general expression applying Fick's second law in the Nernst–Planck equation:

$$\frac{\partial C_i}{\partial t} = \nabla\left(-D_i \nabla C_i\right) + \nabla\left(-u_i z_i F C_i \nabla E\right) + \left(u_i z_i F \nabla E + v\right) \nabla C_i \tag{1}$$

R. López-Vizcaíno (✉) · Á. Yustres · V. Cabrera · V. Navarro
Geoenvironmental Group, Civil Engineering Department,
University of Castilla-La Mancha, Ciudad Real, Spain
e-mail: Ruben.LopezVizcaino@uclm.es



M. A. Rodrigo, E. V. Dos Santos (eds.), *Electrochemically Assisted Remediation of Contaminated Soils*, Environmental Pollution 30,
https://doi.org/10.1007/978-3-030-68140-1_4

where $C_i$ is the concentration, $D_i$ is the diffusion coefficient, $u_i$ is the ion mobility, and $z_i$ is the electrical charge of species "i," respectively. $F$ is the Faraday constant, $E$ is the electrical potential, and $v$ is the advective velocity (the sum of the electro-osmotic and hydraulic contributions). The first term corresponds to diffusion transport, the second term corresponds to the electromigration process, and the third term is the advection contribution to the total mass transport of species "i."

In an EKR process, other electrochemical processes (the reduction–oxidation reactions of the species present in the system) also occur simultaneously with all of these transport mechanisms. Among these reactions, the most remarkable is the water electrolysis reaction. The half-reactions produced in this process are:

Water oxidation at the anode:

$$2H_2O \rightleftarrows O_2 + 4H^+ + 4e^- \tag{2}$$

Water reduction at the cathode:

$$2H_2O + 2e^- \rightleftarrows H_2 + 2OH^- \tag{3}$$

Ionic species are generated in this redox process; protons are generated at the anode and hydroxyl ions at the cathode, which produce a marked pH gradient of the porewater between the two electrodes. Due to their electrical charge, these species can be transported across the soil, mainly by the electromigration process (the protons are attracted by the negative charge of the cathode and the hydroxyl ions are attracted by the positive charge of the cathode). This means that the pH gradient is not static, but rather is quite the opposite as the ions move between the pair of electrodes (anode–cathode). These changes in pH, both temporal and spatial, have a strong influence on a multitude of physical–chemical processes that can take place in the soil such as: (i) the dissolution of metals in acidic environments and complexation under high pH values, (ii) the precipitation/dissolution of minerals present in the soil, and (iii) the adsorption–desorption processes of species in the solid soil skeleton, among others. In addition to these processes, it is important to emphasize the high effect of the pH on the chemical speciation of the species present in the porewater (contaminants and species present in the natural environment), since the pH defines the ionic state (anion, cation, or neutral) of a species in the soil. This will indirectly affect the main transport mechanism by which each species could be mobilized.

All of these electrokinetic, physicochemical, and electrochemical phenomena occur simultaneously. This coupling makes the comprehension of the EKR process and the interpretation of the obtained experimental results very complex. Therefore, it is necessary to develop mathematical models to improve the study of this technology. The goal of these models is to provide the researcher with a numerical tool focused on facilitating the analysis of all of the processes that take place as well as the interactions generated between them and to predict the behavior of the system in the face of changes in the operational variables. With this information, it will be easier to design more efficient experimental campaigns and consequently obtain useful results that can be utilized in future scale-up of the technology prior to its implementation in the field. Therefore, the scientific community has dedicated great efforts to this issue. In the following section, a review of the conceptual models defined to reproduce a soil decontamination process using electrokinetic techniques is presented.

## 2 State of the Art

One of the first works carried out within the framework of electrokinetic transport modeling in porous media was by Lewis and Garner [1] who formulated a two-dimensional mathematical model of coupled hydrodynamic and electroosmotic flows in porous media. However, it was not until the early 1990s that more work began to be developed.

From Northeastern University in Boston (Massachusetts, U.S.A), Yeung [2] proposed a transport model for water, electricity, and ionic species in a clayed soil under hydraulic, electrical, and chemical gradients. The formulation was conducted under the formalism of nonequilibrium thermodynamics. Next, the model was validated by the simulation of migration of sodium and chloride ions in compacted clay under hydraulic, electrical, and chemical coupled gradients. The numerical results were compared with the experimental data obtained in a laboratory-scale experiment under isothermal conditions, obtaining satisfactory fitting [3]. In 1995, Yeung and Datla [4] developed the code NEUTRAL using a finite difference method that considered electroosmosis and electromigration transport. In this case, the model was validated with experimental data from a test conducted by Acar et al. [5] using an air-floated kaolinite clay produced by the Thiele Kaolin Company in Wrens, Georgia. The comparison between the numerical and experimental data of the spatial distribution of the pH at different times was very acceptable. NEUTRAL is one of the first models to take pH as a control variable, which is a very relevant consideration. However, NEUTRAL code, for simplification, does not consider several aspects such as: a nonuniform electrical potential distribution, the influence of pH on sorption–desorption processes, complexation reactions, and the precipitation of minerals, among others.

In 1991, Corapcioglu presented a theoretical formulation focused on simulating the electrical, chemical, and hydraulic-osmotic flows through soil under the application of an electric field [6]. A distinction of this work is that the conceptual model is based on the macroscopic conservation of mass equations and the principle of continuity, in contrast to the previous models, which were developed based on the irreversible thermodynamics of coupled flows. Another important aspect is that the model is valid for compressible soils and assumes an elastic solid with a one-dimensional displacement field.

Alshawabkeh and Acar [7] proposed a generalized theoretical model for the electrokinetic treatment of polluted soils applied to one dimension. This model contemplates water flux (hydraulic and electroosmotic), chemical species transport (Fick diffusion and electromigrations), and electrical current flux connected with the flux of charged species. Similar to Corapcioglu, they used an approach based on the conservation equations in a continuum domain. In addition, the procedure to calculate the chemical speciation was presented. These researchers highlighted the importance of the pH distribution in the efficiency of the removal of pollutants. For this reason, the model was applied to simulate the pH evolution across the sample at different observation times for the experimental tests conducted by Acar et al. [5], obtaining promising results. In 1996, Alshawabkeh and Acar [8] presented an improved multicomponent model (lead, protons, hydroxyl, and nitrate ions) that included sorption and precipitation/dissolution processes at instantaneous equilibrium. The more important assumptions were: (i) the soil is a homogeneous and saturated media and, consequently, the hydraulic conductivity and electroosmotic permeability are constant, (ii) the fluxes are linearly proportional with the gradients of each driving force, (iii) the porous system is composed of soil particles with negative charge, and the fluid surrounding the soil has a high concentration of cations and free water with dissolved species, and (iv) kinetic reactions are not considered for any phenomena. The model was applied to simulate a pilot-scale study of the electrokinetic extraction of lead from a spiked kaolinite/sand mixture [9]. The numerical results obtained for pH, lead concentration, and liquid pressure presented a good fit with the experimental data. In addition, the predictions obtained allowed for an improved understanding of coupled electrochemical and mechanical phenomena in the electroosmotic consolidation of soils. This aspect was studied extensively by Alshawabkeh et al. [10], where the authors investigated the direct influence of the electrolysis, transport, and geochemical changes on the mechanical phenomena such as shear strength, consolidation, and swelling of the soil. Two different cases were studied: a conventional EKR process to remove metal species, and an electrobioremediation process with pH conditioning (by lactic acid) in the cathodic reservoir. In the first case, the ionic strength distribution induced a negative pore pressure and consolidation was observed. In the other case, the behavior of the ionic strength distribution was the opposite, and the soil swelled.

On the other hand, Jacobs et al. [11] studied the efficiency of the EKR treatment applied to remove zinc from a clayed soil by a laboratory experiment A one-dimensional model was developed to improve the interpretation of the experimental results. The complexation, adsorption, and precipitation processes of the zinc were considered. The influence of the pH in the chemical speciation of zinc was identified with the numerical model. This model was extended to carry out a two-dimensional analysis using a Galerkin finite element formulation with added isotropic diffusivity [12]. The model was used to simulate an EKR process applied to kaolin polluted with phenol and it was validated, assuming there were no charged species in the system because the p*K*a of phenol is higher than pH in most of the soil. This simulation demonstrated the effect of the electroosmosis process in the advective transport of phenol. Two other simulations were conducted to verify the effects of the electromigration process and the pH (acid/base chemistry).

Choi et al. [13] proposed a mathematical model for electrokinetic remediation for the removal of heavy metals. The transport phenomena considered were electromigration and diffusion, and the advective process was ignored. The model was used to simulate a hypothetical one-dimensional EKR test where the selected pollutant was cadmium. Two different boundary conditions were evaluated: (i) no recirculation of electrolytes in the electrode reservoirs and (ii) recirculation of the HCl solution to maintain the pH in electrode reservoirs. In the same way, Wilson et al. (1995) developed a one-dimensional model based on the finite differences method and implemented it in TurboBASIC to simulate the electrokinetic treatment of saturated porous media contaminated with an ionic salt. One hypothetical case was analyzed: an enhanced EKR process for the removal of cadmium by the addition of an acid solution in the cathodic compartment.

Yu and Neretnieks (1996) presented a one-dimensional model with the compartment concept (the "flexible" finite difference model) for transport (hydraulic advection, diffusion, ionic electromigration, and electroosmosis). Redox and precipitation–dissolution processes were defined by kinetics equations, the complexation reactions were defined by chemical equilibrium, and the sorption–desorption phenomenon was determined by a linear isotherm. This model simulated three real cases: the removal of copper from sand [14], electrokinetic soil remediation with cathode rinsing [15], and electrokinetic soil remediation with ion-exchange membranes [16]. In addition, Yu and Neretnieks (1997) simulated a synthetic case of soil polluted with $^{137}C$ and $^{9}Sr$ [17].

In 2002, Mattson et al. [18] developed a numerical model with important improvements compared to the previous models. This model is valid for three-dimensional domains and is formulated to simulate the electromigration phenomenal of nonreactive ions in saturated/unsaturated soils. The authors assumed that the flow lines of the current under an electric gradient through soil are equivalent to the flow lines of water molecules under a hydraulic gradient. In this way, they were able to calculate the electric field with MODFLOW [19] using a block-centered finite difference approach. When the electrical potential field was known, the ion migration velocities were calculated in an intermediate software (LKMT), which was used to transfer the information to the transport software, MT3D [20]. MT3D is used to solve the mass transport balance using the upstream finite difference method. First, the model was validated using a one-dimensional experimental test of EKR on a soil spiked with a dye (as a model pollutant). In the next paper of Mattson et al. [21], an analysis of a real field EKR test was conducted using the proposed 3D model. The experimental test was based on the electrokinetic transport of acetate through an unsaturated heterogeneous soil for 6 months.

Park et al. [22] presented a mathematical model able to simulate a one-dimensional and isothermal EKR process. The model has these implicit simplifications: (i) it considers a saturated soil with negative superficial charge, (ii) the fluxes of the electric current, water, and chemical species depend linearly on the gradients of the driving force for each transport phenomena, (iii) the chemical speciation is conducted under instantaneous equilibrium, and (iv) the electric current is transported in the porewater, therefore the soil can be assumed to be an "inert" material with respect to this charge transport. The model was used to carry out a numerical analysis of the influence of pH in the efficiency of phenol removal from kaolinite, conducted by Kim et al. [23] and to determine the best pH conditioning method. The geochemical model implemented is limited, since only phenol, protons, and hydroxyl, sodium, and sulfate ions were contemplated. In another study, to determine the influence of pH on the development of electrokinetic processes, Dangla et al. [24] presented a more specific model focused on predicting the electroosmotic flow between parallel planes of saturated kaolinite.

A new two-dimensional model to simulate the EKR of soil polluted with heavy metals was developed by Vereda-Alonso et al. [25]. This model considers the soils as a set of compartments distributed in a Cartesian grid connected by resistors. Only the electromigration phenomenon was considered. The model was validated with an EKR test focused on the removal of copper from a synthetic clayed soil composed of kaolin.

Ribeiro et al. [26] presented a one-dimensional model that simulates the behavior observed in a real EKR test used to remove atrazine. This work is very interesting because it presents the transport of organic compounds by advection caused by electroosmosis, and, in addition, the model can reproduce a "reverse" electroosmotic flow from the cathode to the anode due to the characteristics of tested soil.

In 2007, Mascia et al. [27] carried out an EKR test at the laboratory scale to decontaminate kaolin with polluted cadmium. The catholyte pH was controlled to avoid the precipitation of cadmium salts. This test was used to validate the proposed mathematical model, which was used to analyze the interactions between the cadmium species and surface of the low-permeability soil.

Al-Hamdan and Reddy [28] studied the effects of a complex geochemical system in the performance of an EKR process. For this purpose, the authors proposed a one-dimensional transport model able to simulate the movement and the speciation of a complex system of heavy metals (chromium, nickel, and cadmium) under the application of an electric field. The model, EKGEOCHEM, was implemented in FORTRAN 77, and it is composed of two submodules: (i) the transport module that contemplates electroosmosis and hydraulic advection, diffusion, and electromigration and (ii) the chemical module (CHEMSPC) that includes the chemical speciation, precipitation–dissolution reactions, and adsorption–desorption process.

In 2011, Paz-García et al. [29] developed an EKR model based on the Nernst–Planck–Poisson equations using a finite element method. Synthetic cases of the electrokinetic desalination of a porous solid were analyzed: first, only the diffusion of ions was simulated, and then three EKR cases were conducted (one with constant voltage difference and two with constant electric current). The model was used to study the electrokinetic remediation of polluted soil with lead [30] and the electrodesalination of yellow bricks polluted with chloride salts [31] and red bricks with sodium chloride, nitrate, and sulfate salts [32]. A new version of the model was used to study the electrokinetic remediation of kaolinite polluted with cadmium and lead [33].

Miao and Pan [34] presented a 2D model based on the finite element method verified with experimental data for the electrokinetic transport of sodium and chloride. The model was used to predict the theoretical transport of radioactive species under electrical potential gradients.

López-Vizcaíno et al. [35] developed the model Multiphysics for ElectroKinetic Remediation (M4EKR). An important feature of M4EKR is that both the transport phenomena and chemical speciation (monolithic approach) are fully implemented in COMSOL Multiphysics. In this work, the authors presented a study of the influence of the geochemical system considering the behavior of the EKR process. For this purpose, two geochemical models were implemented: (i) a simplified system that contemplated the products of the electrolysis water reactions and sodium chloride and (ii) a geochemical system corresponding to a realistic composition of natural porewater, composed of 34 species and 24 chemical equilibrium reactions. This model will be described in detail during the following sections where an application case will be explored. The flexibility provided by the multiphysics platform made it possible to carry out different technological studies with the aim of improving the knowledge of EKR process such as: (i) a study of the advective transport of water in an unsaturated kaolinitic soil under the application of different electric potential gradients [36], (ii) an evaluation of the effect of a polarity reversal strategy in the EKR treatment of a soil polluted with the pesticide 2,4-D [37], (iii) an analysis of the anolyte pH conditioning strategy applied to remove organic compounds with negative charge [38], and (iv) a study of the calcite pH buffer effects in the EKR treatment of a soil polluted with clopyralid [39].

Masi et al. [40] presented a one-dimensional mathematical model for multicomponent reactive transport and geochemical reactions to predict an EKR treatment of harbor sediments polluted with lead, zinc, and nickel. The model utilized two different software: the transport phenomena was implemented in COMSOL and the chemical processes were solved in PHREEQC. The coupling of these steps was carried out in a MATLAB environment using PhreeqcRM (an operator splitting approach). An extended geochemical model was implemented with 21 species and 3 solid compounds. The model considers (i) the chemical speciation in the liquid phase (equilibrium), (ii) the precipitation–dissolution of the solid phases, (iii) the buffer effect on the sediments assuming the complexation of the protons on the surface of the solid, and (iv) the adsorption of the species onto the solid matrix. The results obtained were very good compared with the data of a real test. In addition, the model was used to optimize a field-scale treatment of sediments [41]. This model was improved [42] by including kinetic reactions for the dissolution of calcite and could simulate the experimental data from Villen-Guzman et al. [43].

Using the same implementation strategy (COMSOL + PHREEQC), Sprocati et al. [44] developed a multidimensional model based on the Nernst–Planck–Poisson formulation. Different cases were used to verify the model: (i) a 3D cylinder EKR test, (ii) 2D migration across concrete, and (iii) a 2D EKR test. The model was validated with two real experiments: (i) EKR enhanced with the delivery of permanganate [45] and (ii) an EKR-Bio test [46] where the substrate (nitrate) was mobilized electrokinetically. In another work, Sprocati et al. [47] used the model to analyze the columbic interactions between charged species.

Hojabri et al. [48] presented a one-dimensional model focused on simulating the electrochemical treatment of groundwater. Krcmar et al. [49] proposed a two-dimensional model to predict the behavior of an EKR treatment of sediment polluted with nickel using a hexagonal electrode configuration. Pérez-Corona et al. [50] simulated an EKR process of sodium-bentonite polluted with phenanthrene with the Nernst–Planck approach.

After this review of the state of the art in the modeling of electrokinetic remediation processes, a clear evolution can be verified by the increase of processes considered by these tools and, therefore, their complexity. Currently, there are quite complete models that are very useful in regard to interpreting the results obtained experimentally; however, there are fields in which there is a great potential for improvement. Mainly, the areas where work has not yet been done extensively are: (i) thermal effects, commonly referred to as "electrical heating" produced in the vicinity of electrodes and all associated processes such as the volatilization of pollutants, drying of soil, (ii) the evaluation of solid media with double porosity, and (iii) chemomechanical coupling, i.e., effects of the porewater composition on the solid media deformability. Efforts should be made regarding these issues in the future with the aim of upgrading the numerical tools currently available.

## 3 Numerical Evaluation of an EKR Process

In this section, a numerical evaluation of a general EKR process is presented. Different high-quality generic models are available to simulate an EKR treatment, as described in the section "State of the Art" [3, 8, 12, 17, 18, 29, 30, 35, 40, 44, 50]. The M4EKR model, developed by the authors of this chapter (López-Vizcaíno et al. [35]), has been selected for the inspection exercise. The selection of this model is due solely to the great confidence, control, and knowledge that the authors have with their own numerical tool rather than with other models. First, the general formulation implemented in M4EKR is described in detail. Next, the simulation of an EKR process of a natural soil polluted by a polar pesticide, 3,6-dichloro-2-pyridinecarboxylic acid (Clopyralid), was carried out to demonstrate the scope of the model selected as an example of a numerical tool.

### *3.1 Conceptual Model*

The conceptual model implemented in this M4EKR version defines the one-dimensional transport of porewater and dissolved species in unsaturated and saturated soil where an electric potential gradient is imposed. The model involves the following transport phenomena: advection (hydraulic and electroosmotic), diffusion, and electromigration. Isothermal conditions were assumed (298.15 K). The following simplifications were considered: (i) the gas transport was ignored by maintaining a constant gas pressure at a value defined by the atmospheric pressure, (ii) the soil was defined as a solid media with a single and constant porosity, and (iii) the sorption of the pesticide onto the soil particles was not evaluated.

### 3.1.1 Water Mass Balance

The water transport was defined by the general mass balance equation, assuming that the sink/source term is disregarded:

$$\frac{\partial m_{\mathrm{w}}}{\partial t} + \nabla \cdot \mathbf{I}_{\mathrm{w}} = 0 \tag{4}$$

where $\nabla$ is the divergence operator, $\mathbf{I}_{\mathrm{w}}$ is the water mass flow (kg m$^{-2}$ s$^{-1}$), and $m_{\mathrm{w}}$ is the mass of water per unit total volume (kg m$^{-3}$), which was calculated by the equation:

$$m_{\mathrm{w}} = \varphi Sr \rho_{\mathrm{w}} \tag{5}$$

where $\varphi$ is the soil porosity, $\rho_{\mathrm{w}}$ is the porewater density, and $Sr$ is the degree of saturation of the soil calculated by the van Genuchten approach [51]. The total water mass flow was calculated by adding the contributions of the hydraulic ($\mathbf{I}_{\mathrm{w}}^{\mathrm{h}}$) and electroosmotic ($\mathbf{I}_{\mathrm{w}}^{\mathrm{eo}}$) flows:

$$\mathbf{I}_{\mathrm{w}} = \mathbf{I}_{\mathrm{w}}^{\mathrm{h}} + \mathbf{I}_{\mathrm{w}}^{\mathrm{eo}} = \rho_{\mathrm{w}} \left( \mathbf{q}_{\mathrm{w}}^{\mathrm{h}} + \mathbf{q}_{\mathrm{w}}^{\mathrm{eo}} \right) \tag{6}$$

where $\mathbf{q}_{\mathrm{w}}^{\mathrm{h}}$ and $\mathbf{q}_{\mathrm{w}}^{\mathrm{eo}}$ are the hydraulic and electroosmotic volumetric flows (m s$^{-1}$), calculated using Darcy's law and the semi-empirical Helmholtz-Smoluchowski formulation, respectively [52]. A Brooks and Corey-type power function with an exponent of 3 was selected to determine the relative permeability [53] and, consequently, to obtain both saturated permeabilities (hydraulic and electroosmotic).

Many EKR models are simplified by assuming a constant $Sr$ throughout the tests, so it is not necessary to solve the water mass balance. However, in cases where the electroosmotic permeability is much higher than the hydraulic permeability, partial drainage in the soil may occur and therefore $Sr = 1$ is not fulfilled. These phenomena can be simulated with advanced EKR models applied to unsaturated soils [18, 30, 31, 33, 36].

### 3.1.2 Chemical Species Mass Balance

When the geochemical system included in the EKR model involves an important number of species and exhibits several chemical reactions, it is common to differentiate between two types of chemical species [28, 31]: (i) components or master species and (ii) secondary species obtained as products of the chemical reactions where the components act as reactants [54]. Considering this approach, to obtain the temporal and spatial distribution of any species, it was necessary to solve the mass balance defined by the equation:

$$\frac{\partial m_{\mathrm{m}}}{\partial t} + \nabla \cdot \mathbf{I}_{\mathrm{m}} = R_{\mathrm{m}} \tag{7}$$

where $m_{\mathrm{m}}$ is the total mass of component "m" per unit total volume (mol m$^{-3}$), $\mathbf{I}_{\mathrm{m}}$ is the total molar flux of component "m" (mol m$^{-2}$ s$^{-1}$), and $R_{\mathrm{m}}$ is the sink/source term of component "m" (mol m$^{-3}$ s$^{-1}$). However, it was only necessary to solve for the total mass balance for M-2 components because one component was estimated by fulfilling the electroneutrality condition (in this case, the chloride), and the water mass was solved by Eq. (4). The total mass of component "m" was defined as:

$$m_{\mathrm{m}} = \varphi S r \rho_{\mathrm{w}} C_{\mathrm{m}} \tag{8}$$

where $C_{\mathrm{m}}$ is the total concentration of component "m" expressed in (mol kg$_{\mathrm{w}}^{-1}$). The total molar flux of component "m" was defined as the sum of the flow associated with the transport phenomena considered in the model: advection (hydraulic, $\mathbf{I}_{\mathrm{m}}^{\mathrm{h}}$, and electroosmotic, $\mathbf{I}_{\mathrm{m}}^{\mathrm{eo}}$), diffusion, $\mathbf{I}_{\mathrm{m}}^{\mathrm{dif}}$, and electromigration, $\mathbf{I}_{\mathrm{m}}^{\mathrm{em}}$, according to the Nernst–Planck equation. This formulation is extensively applied in other advanced EKR models [29, 40, 44, 50]:

$$\mathbf{I}_{\mathrm{m}} = \mathbf{I}_{\mathrm{m}}^{\mathrm{h}} + \mathbf{I}_{\mathrm{m}}^{\mathrm{eo}} + \mathbf{I}_{\mathrm{m}}^{\mathrm{dif}} + \mathbf{I}_{\mathrm{m}}^{\mathrm{em}} \tag{9}$$

The molar flux of component "m" for each transport phenomena was defined as:

$$\mathbf{I}_{\mathrm{m}}^{\mathrm{h}} = +\mathbf{I}_{\mathrm{w}}^{\mathrm{h}} C_{\mathrm{m}} \tag{10}$$

$$\mathbf{I}_{\mathrm{w}}^{\mathrm{eo}} = +\mathbf{I}_{\mathrm{w}}^{\mathrm{eo}} C_{\mathrm{m}} \tag{11}$$

$$\mathbf{I}_{\mathrm{m}}^{\mathrm{dif}} = -\rho_{\mathrm{w}} \sum_{i=1}^{N} D_{\mathrm{i}} \alpha_{\mathrm{i}}^{\mathrm{m}} \nabla c_{\mathrm{i}} \tag{12}$$

$$\mathbf{I}_{\mathrm{m}}^{\mathrm{em}} = \rho_{\mathrm{w}} \sum_{i=1}^{N} \alpha_{\mathrm{i}}^{\mathrm{m}} c_{\mathrm{i}} \left(-u_{\mathrm{i}} \nabla E\right) \tag{13}$$

where $\alpha_i^m$ is the molar contribution to the total concentration of component "m" of species "i" (see Sect. 3.1.4), and $c_{\mathrm{i}}$ is the molal concentration of the secondary species "i," calculated as the product of the activity coefficient, $\gamma_{\mathrm{i}}$, and the activity, $a_{\mathrm{i}}$, of species "i." The effective diffusion–dispersion coefficient $D_{\mathrm{i}}$ for one-dimensional flow was estimated via [50]:

$$D_{\mathrm{i}} = D_{\mathrm{i}}^{\mathrm{e}} + \delta_{\mathrm{i}}^{\mathrm{L}} \left| \mathbf{q}_{\mathrm{i}}^{\mathrm{em}} + \mathbf{q}_{\mathrm{w}}^{\mathrm{h}} + \mathbf{q}_{\mathrm{w}}^{\mathrm{eo}} \right|. \tag{14}$$

where $\partial_{\mathrm{i}}^{\mathrm{L}}$ is the longitudinal dispersivity coefficient (m). On the other hand, the effective diffusion coefficient, $D_{\mathrm{i}}^{\mathrm{e}}$, and the effective ionic mobility of the secondary species "i" were calculated by the expressions [18]:

$$D_{\mathrm{i}}^{\mathrm{e}} = \varphi S r \tau D_{\mathrm{i}}^{\mathrm{o}} \tag{15}$$

$$u_i = \varphi S r \tau \gamma_i u_i^o \tag{16}$$

where $\tau$ is the tortuosity of the medium. The ionic mobility, $u_i^o$, and the diffusion coefficient, $D_i^o$, referencing both parameters at infinite dilution in water are related by the Nernst–Einstein equation [55]:

$$u_i^o = \frac{D_i^o z_i F}{RT} \tag{17}$$

where $R$ (J mol$^{-1}$ K$^{-1}$) is the ideal gas constant and $T$ is the absolute temperature (K).

The source/sink term, $R_m$, was considered null in the M-2 component total mass balances because no component was generated or produced.

### 3.1.3 Electric Charge Balance

The electric charge balance was defined with the equation:

$$\nabla \cdot \mathbf{i} = 0 \tag{18}$$

To obtain Eq. (18), two assumptions were adopted: (i) the electroneutrality condition was fulfilled in the whole domain and (ii) no charge accumulation capacity was developed during the EKR process. The total current density, **i,** (A m$^{-2}$) was calculated by Ohm's law [12], adopting the empirical formulation proposed by Rhoades [56] to obtain the apparent electrical conductivity of the soil and the Appelo approach [57] to calculate the electrical conductivity of the porewaters.

### 3.1.4 Chemical Speciation

A classical stoichiometric formulation based on the resolution of a system of mass balance and mass action equations [58] was implemented to determine the chemical speciation, as some other numerical tools [31]. This approach was based on the solution of a system of M nonlinear equations of the type:

$$C_m - C_m^{cal} = 0 \tag{19}$$

where the concentrations $C_m$ are the state variables of the component's mass balances, Eq. (7), and $C_m^{cal}$ was calculated via the expression:

$$C_m^{cal} = \sum_{i=1}^{N} \alpha_i^m c_i \tag{20}$$

The molar contribution of species "i" to component "m," $\alpha_i^m$, coincides with the stoichiometric coefficient of component "m" in the chemical reaction to produce species "i." The sign of this coefficient depends on the role of component "m" in the chemical reaction (positive if it is a reactant and negative if it is a product) [35].

The generic chemical equilibrium equation was used to calculate the activity of each species "i":

$$\log(a_i) = \log(K_i^{eq}) + \sum_{m=1}^{M}(\alpha_i^m \log(a_m)) \tag{21}$$

where $K_i^{eq}$ is the chemical equilibrium constant of the reaction that defines the production of species "i." The activity coefficients were estimated using the WATEQ Debye Hückel formulation [59]. Other advanced models have contemplated the use of kinetic laws to define the mineral precipitation–dissolution reactions [39, 42] or other chemical equilibria to define the surface complexation process [40]. The version of M4EKR model used to conduct the inspection test presented in the following section does not consider these phenomena.

In addition, to calculate the water activity, $a_{H_2O}$, the concentration of free chlorine ions, $c_{Cl^-}$ , and the ionic strength, $_{IS}$, and it was necessary to solve three additional algebraic equations:

$$\log(a_{H_2O}) - \log(a_{H_2O}^{cal}) = 0 \tag{22}$$

$$c_{Cl^-} - c_{Cl^-}^{cal} = 0 \tag{23}$$

$$IS - IS^{cal} = 0. \tag{24}$$

where $a_{H_2O}^{cal}$ was calculated using the Garrels and Christ formulation [60], $a_{Cl^-}^{cal}$ was obtained by applying the electroneutrality condition, and $IS^{cal}$ was calculated by the ionic strength definition:

$$a_{H_2O}^{cal} = 1 - 0.017 \sum_{i=1}^{N\text{-}H_2O} c_i \tag{25}$$

$$c_{Cl^-}^{cal} = -\frac{1}{z_{Cl^-}} \sum_{i=1}^{N-Cl^-} z_i c_i \tag{26}$$

$$IS^{cal} = \frac{1}{2}\sum_{i=1}^{N} z_i^2 c_i \qquad ((27)$$

### 3.1.5 Mass Balance in Electrolyte Compartments

Generally, two different strategies can be chosen to position the electrodes in an EKR process: (i) inserting the electrodes into the soil directly or (ii) placing the electrodes into the electrolyte compartments. The M4EKR model uses the second option, allowing for the simulation of added washing/enhanced fluids and the removal of the collected pollutants. For this purpose, the mass balance of each component "m" in these reservoirs was formulated by the following ordinary differential equation:

$$\frac{\partial M_{\mathrm{m}}^{*}}{\partial t} = M_{\mathrm{m}}^{\mathrm{in},*} - M_{\mathrm{m}}^{\mathrm{out},*} + R_{\mathrm{m}}^{*} \tag{28}$$

where the * superscript indicates the kind of reservoir (*A* for the anode or *C* for the cathode), $M_{\mathrm{j}}$ corresponds to the total mass of component "m" in the reservoir, and $M_{\mathrm{j}}$ determines the input (superscript "in") or output (superscript "out") of the total mass flow of component "m." These variables are defined in detail in the literature by Skibsted et al. [33]. It is important to highlight that, considering the experimental setup (see Sect. 3.3), the extraction of the pollutants from the system is produced by the overflow in the cathodic reservoir. The sink/source flow term of component "m," $R_{\mathrm{m}}$, depends on the electrochemical reactions considered in the system, in this case, the electrolysis of water, which was calculated via Faraday's law:

$$R_{\mathrm{m}}^{*} = \frac{\varphi_{\mathrm{m}}^{*}}{F} \int_{\partial\Omega} (-\mathbf{i} \cdot \mathbf{n}) \mathrm{d}S \tag{29}$$

where *S* is the active electrode surface, $\mathbf{n}$ is the vector normal to the interface soil/electrolyte reservoir, and $\varphi_{\mathrm{m}}^{*}$ is the ratio between the number of moles of component "m" generated electrochemically and the number of moles of electrons exchanged in the redox half-reaction (taking into account the water electrolysis reaction, $\varphi_{\mathrm{m}}^{*} = 1$).

This formulation is not necessary when the numerical model assumes that the electrodes are placed directly into the soil.

## 3.2 Numerical Model

Matlab, multiphysics codes such as COMSOL, or transport modules coupled to Phreeqc (using operator splitting schemes) are the more frequent platforms to implement EKR numerical models. The M4EKR model was fully implemented in the multiphysics platform COMSOL [61]. This software solves partial differential equations based on the finite element method with Lagrange multipliers. The versatility and adaptability of this kind of program provides freedom to the developers to define the system of differential and algebraic equations. One distinction of the M4EKR code is that it includes the classical stoichiometric formulation to solve the chemical speciation problem in a manner that couples it with the transport phenomena (the monolithic approach). Thus, the model solves for 39 state variables in a coupled manner at each time step by regarding 10 partial differential equations (mass balances of water and components in the soil), 18 ordinary differential equations (mass balances in the electrolyte wells), and 11 algebraic equations (chemical speciation). It is important to highlight that the solution of the 18 ordinary

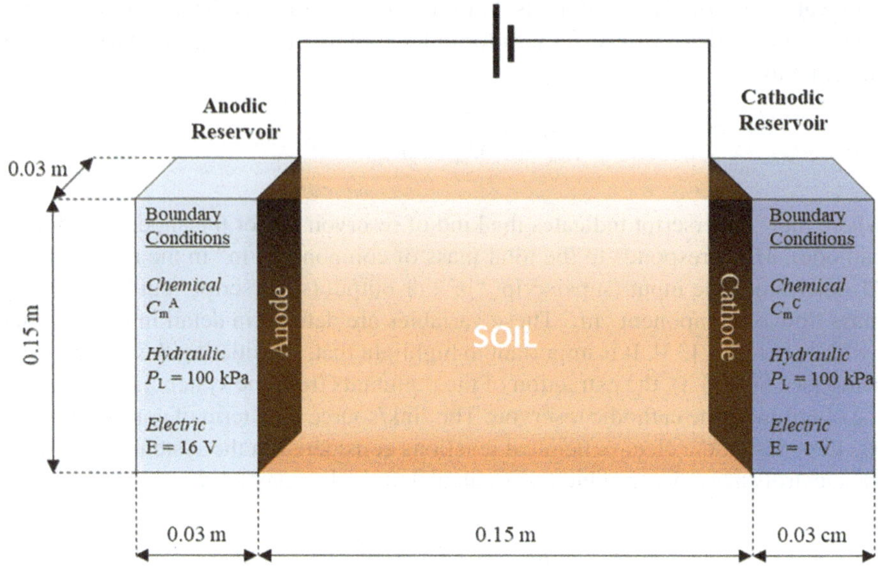


**Fig. 1** Scheme of the modeled system, including dimensions and boundary conditions

differential equations is used as boundary conditions for the 9 partial differential equation (mass balance of chemical species). This coupling is straightforward due to the great flexibility of the multiphysics platform. If another electrode configuration is selected only the boundary conditions of the mass balances equations must be changed.

Finally, for better understanding of the study, a one-dimensional configuration is adopted (Fig. 1). However, the model could be extended to a 2D or 3D configuration easily by the formulation of the different components of the flux vectors. In addition, the model could be used to analyze field-scale applications by changing the domain size in COMSOL and defining the more appropriate discretization.

## 3.3 Description of the Simulation Case

An EKR test at a laboratory scale of a soil polluted with clopyralid (the initial target pollution was 20 mg/kg$_{drysoil}$) was carried out. The soil selected is a low-permeability material taken from a clay quarry in Mora (Spain), and it is classified as a low plasticity clay (Unified Soil Classification System) or silty loam (texture classification of USDA). A bulk density of 1890 kg m$^{-3}$ and a gravimetric water content of 32.8% were fixed to reproduce a compaction state representative of the soil in the natural location. Table 1 shows the soil hydraulic parameters that were taken into account in the simulation.

**Table 1** Soil hydraulic parameters

| Parameters (units) | Values | Description |
|---|---|---|
| $\alpha_{VG}$ ($kPa^{-1}$) | 0.0147 | Parameter of the van Genuchten retention curve |
| $n_{VG}$ (−) | 1.2593 | Parameter of the van Genuchten retention curve |
| $m_{VG}$ (−) | 0.2059 | Parameter of the van Genuchten retention curve |
| $\phi$ (−) | 0.4681 | Porosity |
| $K^{h}_{sat}$ ($m\ s^{-1}$) | $2.03 \times 10^{-10}$ | Saturated hydraulic permeability |
| $K^{eo}_{sat}$ ($m^2\ V^{-1}\ s^{-1}$) | $2.4 \times 10^{-9}$ | Saturated electroosmotic permeability |
| $\rho_s$ ($kg\ m^{-3}$) | 2681.5 | Soil particle density |
| $\delta^{L}_{i}$ (m) | 0.01 | Longitudinal dispersivity of species "i" |
| $\tau$ (−) | 1.00 | Tortuosity |

Figure 1 shows a scheme of the modeled system with its dimensions and the applied boundary conditions. The electrokinetic reactor is a prismatic cell composed of three different compartments: one central compartment where the polluted soil is located (675 $cm^3$) and two external electrolyte reservoirs where the anode and cathode are placed (135 $cm^3$, constant volume during the simulation process). The electrodes have a prismatic geometry with an electrochemically active surface area equal to the area of the electrolyte compartment/soil interface.

Tables 2 and 3 show the M master species and the I secondary species of the geochemical system implemented in the model, respectively. The chemical reactions, equilibrium constants [62], hardcore diameter, and diffusion coefficient at infinite dilution in water of each species are also presented.

The initial chemical conditions are presented in Table 4. The compositions of the electrolyte extracted from a real sample of carbonate groundwater and porewater are the same as the amount of clopyralid corresponding to the selected target pollution. Initially, the soil is fully saturated. Figure 1 shows the boundary conditions applied in the simulation. The liquid pressure is constant and equal to the atmospheric pressure in both electrolyte reservoirs ($P_L$ = 100 kPa). The chemical composition at the boundaries is determined by the concentration in the electrolyte reservoirs. A constant electric potential gradient of 1 V $cm^{-1}$ is applied between the anode and cathode.

## 3.4 Simulation Results

A simulation of five days was carried out to explore the results obtained with a general EKR model. The analysis of the EKR process was divided into three studies related to the (i) hydraulic, (ii) chemical, and (iii) electric behaviors.

**Table 2** Thermodynamic properties of the master species in the modeled geochemical system

| Master species | Reactions | log $K_i^{eq}$ (25 °C) | Hardcore diameter (Å) | $D_i^o$ ($m^2s^{-1}$) |
|---|---|---|---|---|
| $Cl^-$ | $Cl^-$ | 0 | 3.6 | $2.03 \times 10^{-9}$ |
| $H_2O$ | $H_2O$ | 0 | 3.4 | $5.27 \times 10^{-9}$ |
| $H^+$ | $H^+$ | 0 | 4.1 | $9.31 \times 10^{-9}$ |
| $CO_3^{-2}$ | $CO_3^{-2}$ | 0 | 4.7 | $9.55 \times 10^{-10}$ |
| $SO_4^{-2}$ | $SO_4^{-2}$ | 0 | 4.7 | $1.07 \times 10^{-9}$ |
| $NH_4^+$ | $NH_4^+$ | 0 | 4.1 | $1.98 \times 10^{-9}$ |
| $Ca^{+2}$ | $Ca^{+2}$ | 0 | 5.7 | $7.93 \times 10^{-10}$ |
| $Mg^{+2}$ | $Mg^{+2}$ | 0 | 5.7 | $7.05 \times 10^{-10}$ |
| $Na^+$ | $Na^+$ | 0 | 4.1 | $1.33 \times 10^{-9}$ |
| $K^+$ | $K^+$ | 0 | 4.1 | $1.96 \times 10^{-9}$ |
| $Clopy^-$ | 2,4-$D^-$ | 0 | 3.4 | $6.50 \times 10^{-10}$ |

**Table 3** Thermodynamic properties of the secondary species in the modeled geochemical system

| Secondary species | Reactions | log $K_i^{eq}$ (25 °C) | Hard core diameter (Å) | $D_i^o$ ($m^2s^{-1}$) |
|---|---|---|---|---|
| $OH^-$ | $H_2O \leftrightarrow OH^- + H^+$ | −14 | 3.6 | $5.27 \times 10^{-9}$ |
| $HCO_3^-$ | $CO_3^{-2} + H^+ \leftrightarrow HCO_3^-$ | 10.33 | 3.6 | $1.18 \times 10^{-9}$ |
| $H_2CO_3$[a] | $CO_3^{-2} + 2H^+ \leftrightarrow H_2CO_3$[a] | 16.68 | 3.4 | $1.92 \times 10^{-9}$ |
| $HSO_4^-$ | $SO_4^{-2} + H^+ \leftrightarrow HSO_4^-$ | 1.98 | 3.6 | $1.33 \times 10^{-9}$ |
| $NH_3$ | $NH_4^+ \leftrightarrow NH_3 + H^+$ | −9.23 | 3.4 | $2.00 \times 10^{-9}$ |
| $CaHCO_3^+$ | $Ca^{+2} + CO_3^{-2} + H^+ \leftrightarrow CaHCO_3^+$ | 11.43 | 4.1 | $5.06 \times 10^{-10}$ |
| $CaCO_3$ | $Ca^{+2} + CO_3^{-2} \leftrightarrow CaCO_3$ | 3.22 | 3.4 | $4.46 \times 10^{-10}$ |
| $CaSO_4$ | $Ca^{+2} + SO_4^{-2} \leftrightarrow CaSO_4$ | 2.31 | 3.4 | $4.71 \times 10^{-10}$ |
| $Ca(OH)^+$ | $Ca^{+2} + H_2O \leftrightarrow Ca(OH)^+ + H^+$ | −12.78 | 4.1 | $2.13 \times 10^{-10}$[a] |
| $MgHCO_3^+$ | $Mg^{+2} + H^+ + CO_3^{-2} \leftrightarrow MgHCO_3^+$ | 11.37 | 4.1 | $4.78 \times 10^{-10}$ |
| $MgCO_3$ | $Mg^{+2} + CO_3^{-2} \leftrightarrow MgCO_3$ | 2.98 | 3.4 | $4.21 \times 10^{-10}$ |
| $MgSO_4$ | $Mg^{+2} + SO_4^{-2} \leftrightarrow MgSO_4$ | 2.23 | 3.4 | $4.45 \times 10^{-10}$ |
| $Mg(OH)^+$ | $Mg^{+2} + H_2O \leftrightarrow Mg(OH)^+ + H^+$ | −11.68 | 4.1 | $2.13 \times 10^{-10}$[a] |
| $NaHCO_3$ | $Na^+ + HCO_3^- \leftrightarrow NaHCO_3$ | 10.08 | 3.4 | $6.73 \times 10^{-10}$ |
| $NaCO_3^-$ | $Na^+ + CO_3^{-2} \leftrightarrow NaCO_3^-$ | 1.27 | 3.6 | $5.85 \times 10^{-10}$ |
| $NaSO_4^-$ | $Na^+ + SO_4^{-2} \leftrightarrow NaSO_4^-$ | 0.94 | 3.6 | $6.18 \times 10^{-10}$ |
| $NaOH$ | $Na^+ + H_2O \leftrightarrow NaOH + H^+$ | −14.75 | 3.4 | $1.89 \times 10^{-10}$[a] |
| $KSO_4^-$ | $K^+ + SO_4^{-2} \leftrightarrow KSO_4^-$ | 0.88 | 3.6 | $7.46 \times 10^{-10}$ |
| $KOH$ | $K^+ + H_2O \leftrightarrow KOH + H^+$ | −14.46 | 3.4 | $1.89 \times 10^{-10}$[a] |
| $MgCl^+$ | $Mg^{+2} + Cl^- \leftrightarrow MgCl^+$ | 0.35 | 4.1 | $2.13 \times 10^{-10}$[a] |
| $CaCl^+$ | $Ca^{+2} + Cl^- \leftrightarrow CaCl^+$ | −0.29 | 4.1 | $2.13 \times 10^{-10}$[a] |
| $KCl$ | $K^+ + Cl^- \leftrightarrow KCl$ | −0.5 | 3.4 | $1.89 \times 10^{-10}$[a] |
| $NaCl$ | $Na^+ + Cl^- \leftrightarrow NaCl$ | −0.5 | 3.4 | $1.89 \times 10^{-10}$[a] |
| $CaCl_2$ | $Ca^{+2} + 2Cl^- \leftrightarrow CaCl_2$ | −0.64 | 3.4 | $7.54 \times 10^{-10}$ |
| Clopy | $Clopy^- + H^+ \leftrightarrow Clopy$ | 2.32 | 3.4 | $8.20 \times 10^{-10}$ |

[a]Obtained using Pikal's model [63]

**Table 4** Initial conditions

| Molal | Total concentration | | | |
|---|---|---|---|---|
| | Porewater | Electrolyte | Molal | Porewater chemical speciation |
| $C_{H^+}$ | $1.98 \times 10^{-3}$ | $1.68 \times 10^{-3}$ | $a_{H^+}$ | $1.27 \times 10^{-7}$ |
| $C_{CO_3^{-2}}$ | $1.65 \times 10^{-3}$ | $1.65 \times 10^{-3}$ | $a_{CO_3^{-2}}$ | $4.39 \times 10^{-7}$ |
| $C_{SO_4^{-2}}$ | $5.12 \times 10^{-4}$ | $5.12 \times 10^{-4}$ | $a_{SO_4^{-2}}$ | $3.17 \times 10^{-4}$ |
| $C_{NH_4^+}$ | $5.54 \times 10^{-6}$ | $5.54 \times 10^{-6}$ | $a_{NH_4^+}$ | $5.08 \times 10^{-6}$ |
| $C_{Ca^{+2}}$ | $1.11 \times 10^{-3}$ | $1.11 \times 10^{-3}$ | $a_{Ca^{+2}}$ | $7.61 \times 10^{-4}$ |
| $C_{Mg^{+2}}$ | $5.31 \times 10^{-4}$ | $5.31 \times 10^{-4}$ | $a_{Mg^{+2}}$ | $3.66 \times 10^{-4}$ |
| $C_{Na^+}$ | $9.53 \times 10^{-4}$ | $9.53 \times 10^{-4}$ | $a_{Na^+}$ | $8.73 \times 10^{-4}$ |
| $C_{K^+}$ | $1.28 \times 10^{-4}$ | $1.28 \times 10^{-4}$ | $a_{K^+}$ | $1.17 \times 10^{-4}$ |
| $C_{Clopy^-}$ | $3.19 \times 10^{-4}$ | 0 | $a_{Clopy^-}$ | $2.93 \times 10^{-4}$ |
| pH | 6.89 | 6.89 | $a_{H_2O}$ | $9.99 \times 10^{-1}$ |
| | | | $c_{Cl^-}$ | $1.71 \times 10^{-3}$ |
| | | | *IS* | $6.21 \times 10^{-3}$ |

### 3.4.1 Analysis of the Hydraulic Behavior

From a hydraulic point of view, the model results can improve the comprehension of the transport of water in unsaturated porous media by the analysis of hydraulic and electroosmotic fluxes. Figure 2 shows the spatial distribution of the pore liquid pressure and the degree of saturation. At 1.25 h, a drop in the pore liquid pressure in the zone near the anodic compartment and an incremental increase in the opposite soil zone can be noted. This distribution is related to the expected behavior when the water is moved to the cathodic region. At day one of simulation, the decrease in the liquid pressure is more pronounced, producing the soil desaturation observed in experimental tests [64, 65]. As the simulation progresses, it is observed that the soil becomes unsaturated. The spatial distribution of the degree of saturation allows this behavior to be verified, and this behavior can produce cracking and other geotechnical problems such as those observed by Lopez-Vizcaíno et al. [66]. Analyzing the water flux at the anode–soil and soil–cathode interfaces (Fig. 3a) can verify the different trends that take place in the EKR process. Until the first day, the advective flux that produced both hydraulic and electric gradients is towards the cathodic region; however, when the simulation time increases, the direction of the hydraulic flux at the soil–cathode interface is inverted due to the unsaturated conditions observed in this zone. However, the degree of saturation of this soil region is not increased. This situation is often produced when a clayed soil has a hydraulic permeability lower than its electroosmotic permeability, and therefore the electroosmotic flow extracted in the cathodic reservoir is higher than the inbound hydraulic flow from the anodic reservoir (Fig. 3b).

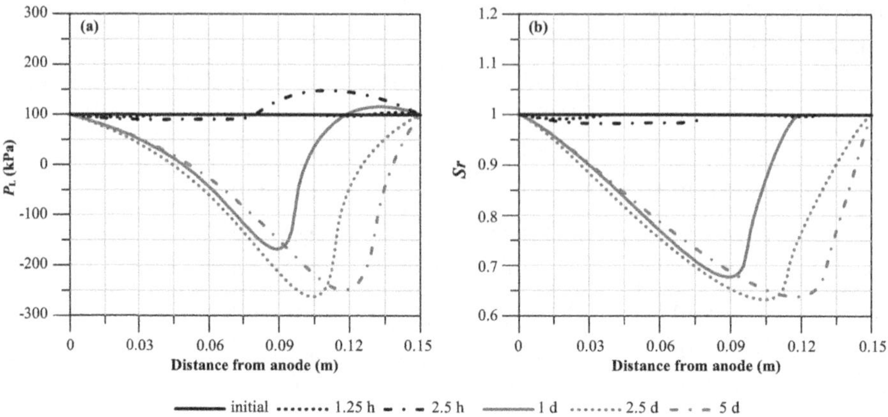


**Fig. 2** Spatial distribution of the (a) pore liquid pressure and (b) degree of saturation

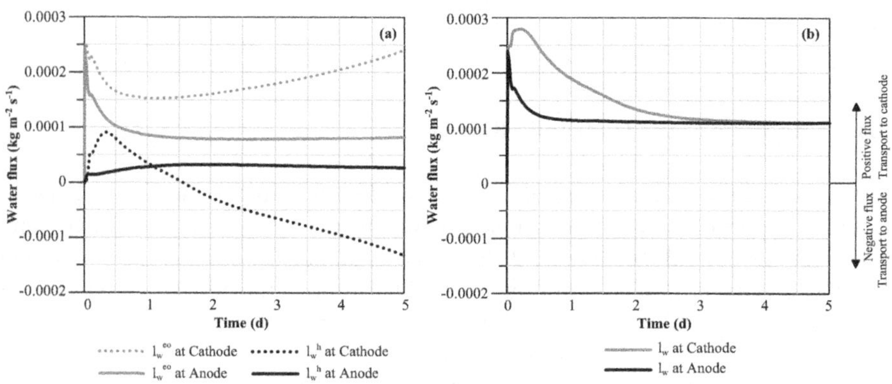


**Fig. 3** Time evolution of the water flux evaluated at the anode–soil and soil–cathode interfaces. (**a**) Individual water flux based on the transport mechanism and (**b**) the total water flux

### 3.4.2 Analysis of Chemical Behavior

From a chemical point of view, the model provides valuable information on spatial and temporal distributions of the concentration of chemical species and/or contaminants present in the soil. From a theoretical point of view, these results can be very useful to understand the interaction of the multiple physical–chemical phenomena that take place within the soil. From a functional point of view, these results can help to define the duration of the test and other operation variables.

Figure 4 shows the time evolution of the total concentration of the three representative master species in both electrolyte compartments. As expected, taking into account the process of electromigration of charged species, an increase in the concentration of negatively charged species is observed in the anodic reservoir, and consequently, the presence of these species decreases in the cathodic reservoir (Fig. 4b, c). The opposite behavior is observed for positively charge species (Fig. 4a).

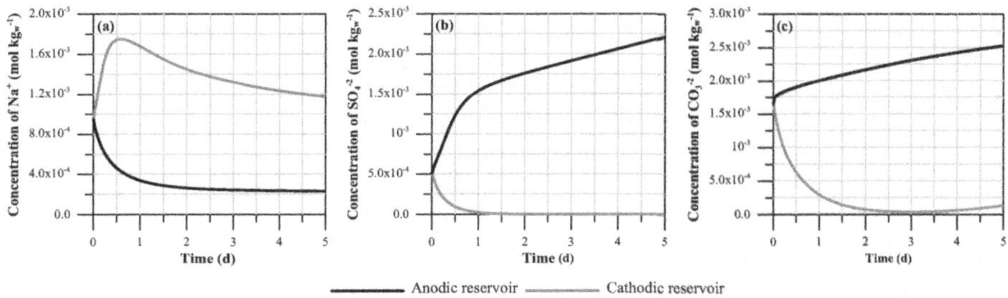


**Fig. 4** Time evolution of the total concentration of the master species in the anodic and cathodic reservoirs. (**a**) Sodium, (**b**) sulfates, and (**c**) carbonates

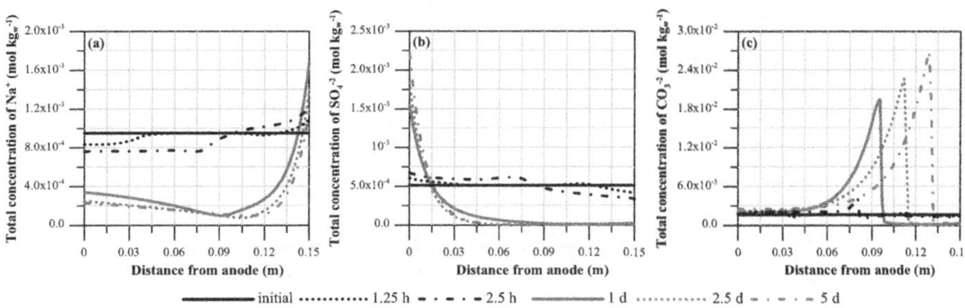


**Fig. 5** Spatial distribution of total concentration. (**a**) Sodium, (**b**) sulfates, and (**c**) carbonates

Figure 5 shows the spatial distribution of these master species at different times. The observed behavior is consistent with the expected behavior for sodium and sulfate, where the concentrations increased in the soil regions near the cathode and the anode, respectively. However, the carbonate master species presents an uncharacteristic spatial distribution where a significant accumulation is observed in the central zone of the soil and not at the zone near the anodic reservoir, as would be predicted for a species with a negative charge. To try to understand this trend, the individual spatial distributions of the carbonate species concentration and the pH are illustrated in Figs. 6 and 7, respectively.

The chemical speciation of the carbonate system is strongly related to the pH distribution of the porewater (Fig. 7) and vice versa due to the buffering capacity of these species. Considering a closed system [67], $HCO_3^-$ is present when the pH is between 6.3 and 10.25 (Fig. 6b), $CO_3^{-2}$ is predominant at higher pH values and $H_2CO_3^*$ (assumed to be a nonvolatile acid that involves the mass of the $H_2CO_3$ acid and the dissolved $CO_2$) is present at lower pH values. Taking into account this limitation and the electric charge of the carbonate species, it is possible to clarify the carbonate master species distribution. $HCO_3^-$ and $CO_3^{-2}$ are mobilized to the anode by electromigration; however, the $H_2CO_3^*$ species is not charged and therefore can only be transported to the cathodic reservoir by advection. This fact results in an intersection of the flows located at the pH front, which could explain the carbonate accumulation at this point. In this context, it is important to remark that the M4EKR does not consider, in this simulation, mineral precipitation–dissolution processes.

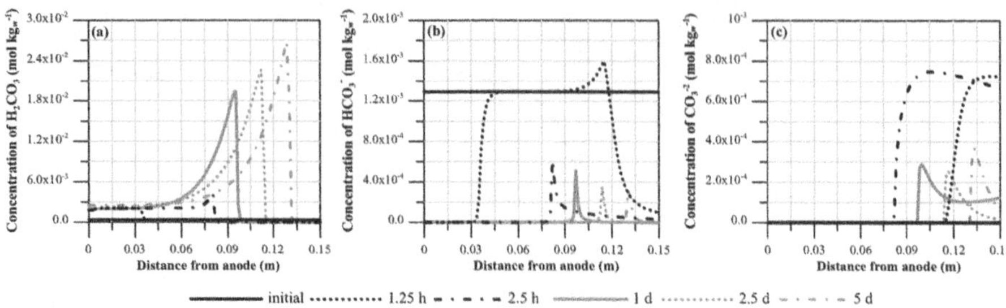


**Fig. 6** Spatial distribution of the carbonate species concentration. (**a**) $H_2CO_3$, (**b**) $HCO_3^-$, and (**c**) $CO_3^{-2}$

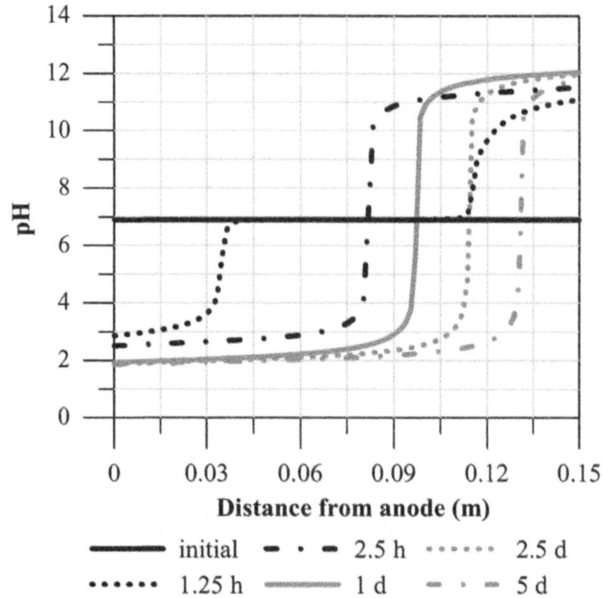


**Fig. 7** Spatial distribution of the pH

These reactions, especially those with calcite, could disrupt the system of the carbonates and therefore the pH trends, as can be seen in the simulations carried out by Masi et al. [42].

On the other hand, it can be noted that the pH distribution presents a conventional trend [8, 28]. The acidification and basification processes due to the water electrolysis reactions are identified in the zones near the anode and cathode, respectively. In addition, the movement of the pH fronts is apparent, where the acid front is displaced faster due to the higher ionic mobility of protons.

As previously discussed, the pH influences the chemical speciation of the species present in porewater, including the speciation of pollutants [28, 68]. Figure 8 presents the spatial distribution of the concentration of the clopyralid species (acid and

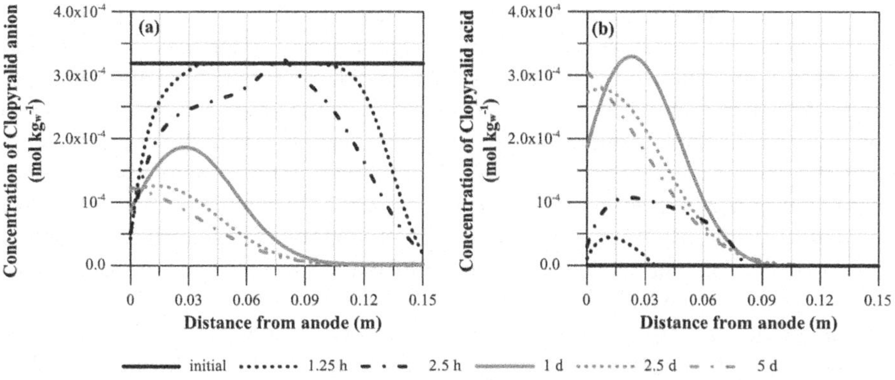


**Fig. 8** Spatial distribution of the concentration of the clopyralid species. (**a**) Anion and (**b**) acid species

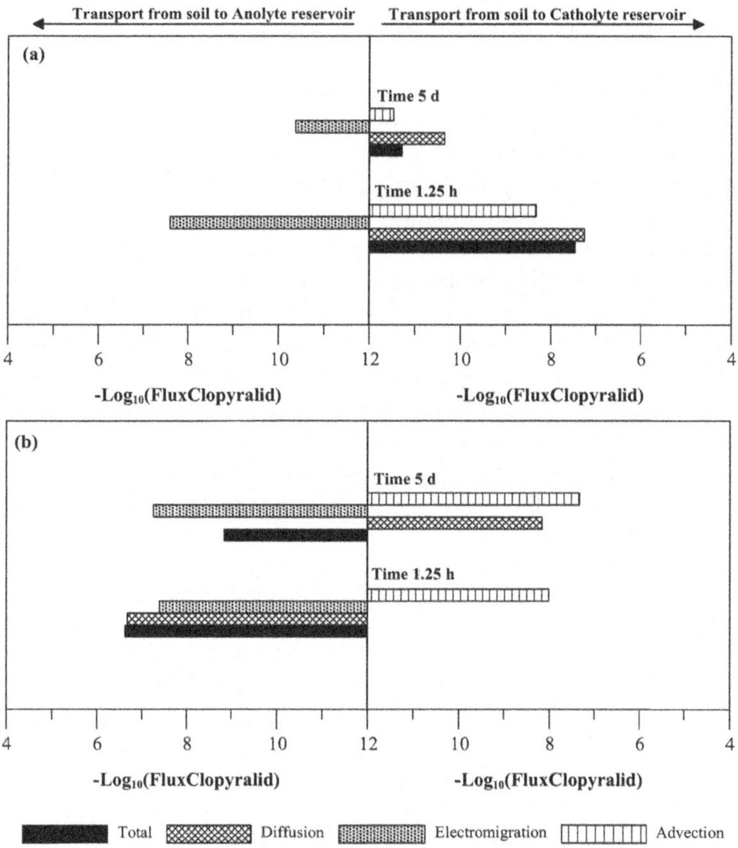


**Fig. 9** Total mass fluxes for clopyralid evaluated at (**a**) the soil–cathode and (**b**) the anode–soil interfaces. Evaluated times: 1.25 h and 5 days

anion species), and Fig. 9 shows the fluxes associated with the transport phenomena independently estimated at both soil–electrolyte reservoir interfaces. At the beginning of the simulation, all of the clopyralid pesticide is present as an anion species. At the first time point (1.25 h), the concentration of the anion species decreases in the soil near the anolyte and catholyte reservoirs, mainly due to diffusion transport (Fig. 9), since the concentration gradient is important in these moments (the initial clopyralid concentration in the electrolyte is zero). In addition, considering the movement of the pH front, the acid species of clopyralid is generated in the soil region between the anolyte reservoir and pH front. As the pH front moves towards the compartment, the acid soil region is larger. This fact disrupts the chemical equilibrium of the clopyralid acid species and its concentration increases. Consequently, the anion species concentration is progressively lower, due to the chemical transformation discussed above and the accumulation of the species in the anolyte compartment (Fig. 9).

### 3.4.3 Analysis of Electrical Behavior

Figure 10 shows the spatial distribution of the electric potential, the electrical conductivity of the porewater, $\sigma_a$, and the ionic strength.

Initially, the system is equilibrated, thus the electrical potential gradient is linear. However, when the EKR begins, the transport mechanisms produce a movement of the chemical species present in the porewater and of the electrogenerated protons and hydroxyl ions, and consequently there is a rearrangement of the ionic strength in the domain (Fig. 10c). These changes in the ionic strength are associated with the trends of the electrical conductivity of porewater illustrated in Fig. 10b, which is high near the anolyte and catholyte reservoirs and lower in the middle of the domain. Considering this distribution, it is easy to explain the changes in the behavior of the electrical potential. It can be noted how high values for the electrical conductivity produce a less pronounced electrical potential gradient (the $E$ slope decreases near the anode and cathode), and therefore, the more accentuated electrical potential gradient is observed in the center of the soil where the electrical conductivity is lower.

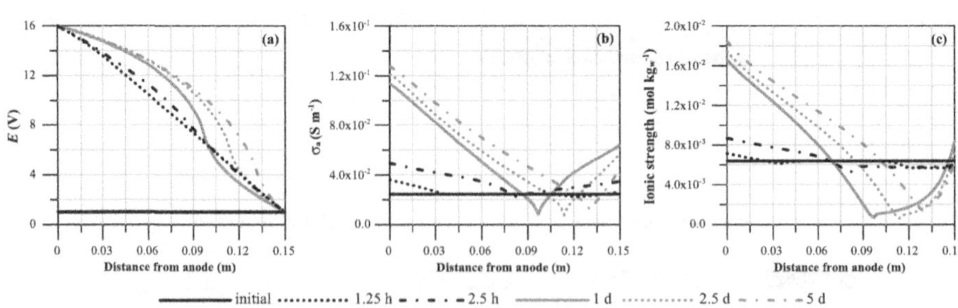


**Fig. 10** Spatial distribution of (**a**) the electric potential, (**b**) the electrical conductivity of the porewater, and (**c**) the ionic strength

## 4 Conclusions

This chapter is focused on the study of the mathematical modeling of treatments of contaminated soils using electrokinetic techniques. The requirement for the use of numerical tools in the theoretical study and the interpretation of experimental tests of electrokinetic remediation processes has become evident.

A detailed description of the main mathematical models of soil EKR processes was presented. It can be concluded that the models have improved. The first models only contemplated electroosmosis transport; the subsequent models began to introduce chemical species transport by diffusion and electromigration mechanisms but used a very limited geochemical system (simple salts and a few metals). Next, the simulation of pH and its influence on chemical speciation was a key aspect in developing a more extended geochemical model and considering other physicochemical processes such as precipitation–dissolution, adsorption, and complexation. Generally, most of the models developed are one-dimensional, although some 2D or even 3D models are already used to analysis field-scale treatments. From the point of view of the calculation methods, there has also been an evolution associated with technological development. This has resulted in the most recent models being implemented in more powerful calculation platforms and therefore allowing for more complex simulations. In addition, the potential areas of improvement of the developed numerical tools were highlighted, the most remarkable of which were the study of the thermal changes of the system, the conceptual model used to define the soil (dual-porosity media), and the interaction between the physical–chemical processes that take place and the mechanical processes in the soil itself.

Finally, a theoretical study of the electrokinetic remediation process of a natural soil contaminated with a polar pesticide, 3,6-dichloro-2-pyridinecarboxylic acid, was conducted with a generalist EKR model developed by the authors. The hydraulic behavior was analyzed, verifying the desaturation of soil and the evolution of the water flows (hydraulic and electroosmotic). Moreover, the transport of the chemical species was evaluated by analyzing the individual contribution of each transport phenomena and their coupling. In addition, the influence of pH on the chemical speciation was illustrated, and the buffer effect of the carbonate system was explained. All of this demonstrates the great usefulness of these types of numerical tools, which can be used both to enhance the general compression of the EKR process and to predict the behavior of the system in the presence of changes in operational variables.

**Acknowledgments** The authors acknowledge funding support from: (i) the Ministerio de Economía, Industria y Competitividad from Spanish Government and the European Union through the project [BIA2017-89287-R (AEI/FEDER, UE)] and (ii) the Ministerio de Ciencia, Innovación y Universidades from Spanish Government through the Postdoctoral Grant [IJC-2018-035212] awarded to Dr. López-Vizcaíno.

## References

1. R.W. Lewis, R.W. Garner, Finite element solution of coupled electrokinetic and hydrodynamic flow in porous media. Int. J. Numer. Meth. Eng. **5**, 41–55 (1972)
2. A.T. Yeung, Coupled flow equations for water, electricity and ionic contaminants through clayey soils under hydraulic, electrical and chemical gradients. J. Non-Equil. Thermody. **15**, 247–268 (1990)
3. A.T. Yeung, J.K. Mitchell, Coupled fluid, electrical and chemical flows in soil. Geotechnique **43**, 121–134 (1993)
4. A.T. Yeung, S. Datla, Fundamental formulation of electrokinetic extraction of contaminants from soil. Can. Geotech. J. **32**, 569–583 (1995)
5. Y.B. Acar, R.J. Gale, J. Hamed, G. Putnam, Acid/base distributions in electrokinetic soil processing. Transp. Res. Rec. **1288**, 23–34 (1991)
6. M.Y. Corapcioglu, Formulation of electro-chemico-osmotic processes in soils. Transport Porous Med. **6**, 435–444 (1991)
7. A.N. Alshawabkeh, Y.B. Acar, Removal of contaminants from soils by electrokinetics: A theoretical treatise. J. Environ. Sci. Heal. A. **27**, 1835–1861 (1992)
8. A.N. Alshawabkeh, Y.B. Acar, Electrokinetic remediation. II: Theoretical model. J. Geotech. Eng. **122**, 186–196 (1996)
9. Y.B. Acar, A.N. Alshawabkeh, Electrokinetic remediation. I: Pilot-scale tests with lead-spiked kaolinite. J. Geotech. Eng. **122**, 173–185 (1996)
10. A.N. Alshawabkeh, T.C. Sheahan, X. Wu, Coupling of electrochemical and mechanical processes in soils under DC fields. Mech. Mater. **36**, 453–465 (2004)
11. R.A. Jacobs, M.Z. Sengun, R.E. Hicks, R.E. Probstein, Model and experiments on soil remediation by electric fields. J. Geotech. Eng. **29**, 1933–1955 (1994)
12. R.A. Jacobs, R.F. Probstein, Two-dimensional modeling of electroremediation. AIChE J. **42**, 1685–1696 (1996)
13. Y.S. Choi, R. Lui, A mathematical model for the electrokinetic remediation of contaminated soil. J. Hazard. Mater. **44**, 61–75 (1995)
14. Z. Li, J.-W. Yu, I. Neretnieks, A new approach to electrokinetic remediation of soils polluted by heavy metals. J. Contam. Hydrol. **22**, 241–253 (1996)
15. R.E. Hicks, S. Tondorf, Electrorestoration of metal contaminated soils. Environ. Sci. Technol. **28**, 2203–2210 (1994)
16. J.R.B. Jensen, V. Kueeš, M. Kubal, Electrokinetic remediation of soils polluted with heavy metals. Removal of zinc and copper using a new concept. Environ. Technol. **15**, 1077–1082 (1994)
17. J.-W. Yu, I. Neretnieks, Theoretical evaluation of a technique for electrokinetic decontamination of soils. J. Contam. Hydrol. **26**, 291–299 (1997)
18. E.D. Mattson, R.S. Bowman, E.R. Lindgren, Electrokinetic ion transport through unsaturated soil: 1. Theory, model development, and testing. J. Contam. Hydrol. **54**, 99–120 (2002)
19. M.G. McDonald, A.W. Harbaugh, *A modular three-dimensional finite-difference ground-water flow model* (U.S. Geological Survey Techniques of Water-Resources Investigations, 1988)
20. C. Zheng, P. Wang, MT3DMS: A Modular Three-Dimensional Multispecies Transport Model for Simulation of Advection, Dispersion, and Chemical Reactions of Contaminants in Groundwater Systems; Documentation and User's Guide. (1999)
21. E.D. Mattson, R.S. Bowman, E.R. Lindgren, Electrokinetic ion transport through unsaturated soil: 2. Application to a heterogeneous field site. J. Contam. Hydrol. **54**, 121–140 (2002)
22. J.-S. Park, S.-O. Kim, K.-W. Kim, B.-R. Kim, S.-H. Moon, Numerical analysis for electrokinetic soil processing enhanced by chemical conditioning of the electrode reservoirs. J. Hazard. Mater. **99**(1), 71–88 (2003)
23. S.O. Kim, S.H. Moon, K.W. Kim, Enhanced electrokinetic soil remediation for removal of organic contaminants. Environ. Technol. (U.K.) **21**(417–426) (2000)
24. P. Dangla, T. Fen-Chong, F. Gaulard, Modelling of pH-dependent electro-osmotic flows. CR Mécanique. **332**, 915–920 (2004)

25. C. Vereda-Alonso, J.M. Rodríguez-Maroto, R.A. García-Delgado, C. Gómez-Lahoz, F. García-Herruzo, Two-dimensional model for soil electrokinetic remediation of heavy metals: Application to a copper spiked kaolin. Chemosphere **54**, 895–903 (2004)
26. A.B. Ribeiro, J.M. Rodríguez-Maroto, E.P. Mateus, H. Gomes, Removal of organic contaminants from soils by an electrokinetic process: The case of atrazine. Experimental and modeling. Chemosphere **59**(9), 1229–1239 (2005)
27. M. Mascia, S. Palmas, A.M. Polcaro, A. Vacca, A. Muntoni, Experimental study and mathematical model on remediation of cd spiked kaolinite by electrokinetics. Electrochim. Acta **52**, 3360–3365 (2007)
28. A.Z. Al-Hamdan, K.R. Reddy, Electrokinetic remediation modeling incorporating geochemical effects. J. Geotech. Geoenviron. Eng. **134**, 91–105 (2008)
29. J.M. Paz-García, B. Johannesson, L.M. Ottosen, A.B. Ribeiro, J.M. Rodríguez-Maroto, Modeling of electrokinetic processes by finite element integration of the Nernst-Planck-Poisson system of equations. Sep. Purif. Technol. **79**, 183–192 (2011)
30. J.M. Paz-Garcia, K. Baek, I.D. Alshawabkeh, A.N. Alshawabkeh, A generalized model for transport of contaminants in soil by electric fields. J. Environ. Sci. Heal. A. **47**, 308–318 (2012)
31. J.M. Paz-García, B. Johannesson, L.M. Ottosen, A.N. Alshawabkeh, A.B. Ribeiro, J.M. Rodríguez-Maroto, Modeling of electrokinetic desalination of bricks. Electrochim. Acta **86**, 213–222 (2012)
32. G. Skibsted, L.M. Ottosen, P.E. Jensen, J.M. Paz-Garcia, Electrochemical desalination of bricks – Experimental and modeling. Electrochim. Acta **181**, 24–30 (2015)
33. M. Rezaee, G. Asadollahfardi, C. Gomez-Lahoz, M. Villen-Guzman, J.M. Paz-Garcia, Modeling of electrokinetic remediation of cd- and Pb-contaminated kaolinite. J. Hazard. Mater. **366**, 630–635 (2019)
34. T. Miao, T. Pan, A Multiphysics model for evaluating Electrokinetic remediation of nuclear waste-contaminated soils. Water Air Soil Poll. **226**, 77 (2015)
35. R. López-Vizcaíno, A. Yustres, M.J. León, C. Saez, P. Cañizares, M.A. Rodrigo, V. Navarro, Multiphysics implementation of Electrokinetic remediation models for natural soils and Porewaters. Electrochim. Acta **225**, 93–104 (2017)
36. Á. Yustres, R. López-Vizcaíno, C. Sáez, P. Cañizares, M.A. Rodrigo, V. Navarro, Water transport in electrokinetic remediation of unsaturated kaolinite. Experimental and numerical study. Sep. Purif. Technol. **192**, 196–204 (2018)
37. R. López-Vizcaíno, A. Yustres, C. Sáez, P. Cañizares, M.A. Rodrigo, V. Navarro, Effect of polarity reversal on the enhanced electrokinetic remediation of 2,4-D-polluted soils: a numerical study. Electrochim. Acta **258**, 414 (2017)
38. R. López-Vizcaíno, A. Yustres, L. Asensio, C. Saez, P. Cañizares, M.A. Rodrigo, V. Navarro, Enhanced electrokinetic remediation of polluted soils by anolyte pH conditioning. Chemosphere **199**, 477–485 (2018)
39. R. López-Vizcaíno, E.V. dos Santos, A. Yustres, M.A. Rodrigo, V. Navarro, C.A. Martínez-Huitle, Calcite buffer effects in electrokinetic remediation of clopyralid-polluted soils. Sep. Purif. Technol. **212**, 376–387 (2019)
40. M. Masi, A. Ceccarini, R. Iannelli, Multispecies reactive transport modelling of electrokinetic remediation of harbour sediments. J. Hazar. Mater. **326**, 187–196 (2017)
41. M. Masi, A. Ceccarini, R. Iannelli, Model-based optimization of field-scale electrokinetic treatment of dredged sediments. Chem. Eng. J. **328**, 87–97 (2017)
42. M. Masi, J.M. Paz-Garcia, C. Gomez-Lahoz, M. Villen-Guzman, A. Ceccarini, R. Iannelli, Modeling of electrokinetic remediation combining local chemical equilibrium and chemical reaction kinetics. J. Hazard. Mater. **371**, 728–733 (2019)
43. M. Villen-Guzman, J.M. Paz-Garcia, J.M. Rodriguez-Maroto, F. Garcia-Herruzo, G. Amaya-Santos, C. Gomez-Lahoz, C. Vereda-Alonso, Scaling-up the acid-enhanced electrokinetic remediation of a real contaminated soil. Electrochim. Acta **181**, 139–145 (2015)

44. R. Sprocati, M. Masi, M. Muniruzzaman, M. Rolle, Modeling electrokinetic transport and biogeochemical reactions in porous media: a multidimensional Nernst–Planck–Poisson approach with PHREEQC coupling. Adv. Water Resour. **127**, 134–147 (2019)
45. D. Hodges, A. Fourie, D. Thomas, D. Reynolds, Overcoming permanganate stalling during electromigration. J. Environ. Eng. **139**, 677–684 (2013)
46. R.T. Gill, S.F. Thornton, M.J. Harbottle, J.W. Smith, Effect of physical heterogeneity on the electromigration of nitrate in layered granular porous media. Electrochim. Acta **199**, 59–69 (2016)
47. R. Sprocati, M. Rolle, Charge interactions, reaction kinetics and dimensionality effects on electrokinetic remediation: a model-based analysis. J. Contam. Hydrol. **229**, 103567 (2020)
48. S. Hojabri, L. Rajic, A.N. Alshawabkeh, Transient reactive transport model for physico-chemical transformation by electrochemical reactive barriers. J. Hazard. Mater. **358**, 171–177 (2018)
49. D. Krcmar, N. Varga, M. Prica, L. Cveticanin, M. Zukovic, B. Dalmacija, Z. Corba, Application of hexagonal two dimensional electrokinetic system on the nickel contaminated sediment and modelling the transport behavior of nickel during electrokinetic treatment. Sep. Purif. Technol. **192**, 253–261 (2018)
50. M. Pérez-Corona, J.M. Rodríguez-Maroto, C. Gómez-Lahoz, E. Bustos, Electroremediation of sodium bentonite contaminated with phenanthrene and its modeling with a Nernst-Planck equation. J. Appl. Electrochem. **48**, 1373–1380 (2018)
51. M.T. van Genuchten, Closed-form equation for predicting the hydraulic conductivity of unsaturated soils. Soil Sci. Soc. Am. J. **44**, 892–898 (1980)
52. J.K. Mitchell, K. Soga, *Fundamentals of soil behavior* (Wiley, New York, 2005)
53. R.H. Brooks, A.T. Corey, *Hydraulic Properties of Porous Media* (Colorado State University, Fort Collins, 1964)
54. J. Bear, *Dynamics of Fluids in Porous Media* (Dover, New York, 1972)
55. J.O.M. Bockris, A.K. Reddy, *Modern Electrochemistry 2B: Electrodics in Chemistry, Engineering, Biology and Environmental Science* (Springer Science & Business Media, USA, 2000)
56. J.D. Rhoades, N.A. Manteghi, P.J. Shouse, W.J. Alves, Soil electrical conductivity and soil salinity: New formulations and calibrations. Soil Sci. Soc. Am. J. **53**, 433–439 (1989)
57. C. Appelo, Specific Conductance: how to calculate, to use, and the pitfalls. URL: http://www.hydrochemistry.eu/exmpls/sc.html (letzter Zugriff: 27 August 2013). (2010)
58. A.M.M. Leal, M.J. Blunt, T.C. LaForce, Efficient chemical equilibrium calculations for geochemical speciation and reactive transport modelling. Geochim. Cosmochim. Acta **131**, 301–322 (2014)
59. D. L. Parkhurst, User's guide to PHREEQC—A computer program for speciation, reaction-path, advective-transport, and inverse geochemical calculations. User's Guide to PHREEQC (Version 2)—A Computer Program for Speciation, Batch-Reaction, One-Dimensional Transport, and Inverse Geochemical Calculations. (1995)
60. R.M. Garrels, C.L. Christ, *Solutions, Minerals, and Equilibria* (Harper & Row, New York, 1965)
61. COMSOL: COMSOL Multiphysics Reference Manual Version: COMSOL 5.1., (2015)
62. E. Giffaut, M. Grivé, P. Blanc, P. Vieillard, E. Colàs, H. Gailhanou, S. Gaboreau, N. Marty, B. Madé, L. Duro, Andra thermodynamic database for performance assessment: ThermoChimie. Appl. Geochem. **49**, 225–236 (2014)
63. M.J. Pikal, Ion-pair formation and the theory of mutual diffusion in a binary electrolyte. J. Phys. Chem. **75**, 663–675 (1971)
64. S. Wieczorek, H. Weigand, M. Schmid, C. Marb, Electrokinetic remediation of an electroplating site: design and scale-up for an in-situ application in the unsaturated zone. Eng. Geol. **77**, 203–215 (2005)
65. J. Yuan, M.A. Hicks, Influence of gas generation in electro-osmosis consolidation. Int. J. Numer. Anal. Meth. **40**, 1570–1593 (2016)

66. R. Lopez-Vizcaino, V. Navarro, J. Alonso, A. Yustres, P. Cañizares, M.A. Rodrigo, C. Saez, Geotechnical behaviour of low-permeability soils in surfactant-enhanced electrokinetic remediation. J. Environ. Sci. Heal. A **51**, 44–51 (2016)
67. W. Stumm, J.J. Morgan, *Aquatic Chemistry: Chemical Equilibria and Rates in Natural Waters* (Wiley, New York, 1996)
68. A.Z. Al-Hamdan, K.R. Reddy, Geochemical assessment of metal transport in glacial till during electrokinetic remediation. Environ. Monit. Assess. **139**, 137–149 (2008)

# Part II
# Technologies

# Treatment of Soil Washing Solutions by Electrochemical Advanced Oxidation

**Emmanuel Mousset, Clément Trellu, and Mehmet A. Oturan**

## 1 Introduction

Contaminated soil is not only an environmental issue but also a sanitary challenge. The various physico-chemical properties of the pollutants lead to equilibrium partition with the different environmental compartments: solid/gas partition (soil/air), solid/liquid partition (soil/water), and solid/solid partition (soil/soil)). For instance, the volatile organic compounds (VOCs) (trichloroethene, benzene, toluene, ethylbenzene, xylene known as BTEX, etc.) can be mainly found in equilibrium between soil and atmosphere, while more water soluble compounds such as hydrophilic pesticides (cyromazine, diquat, fenuron, 2,4-dichlorophenoxyacetic acid, etc.) are present in soil but can be lixiviated into the groundwater with rainwater infiltration [1, 2]. Contrastingly, the hydrophobic organic compounds (HOCs) such as heavy polycyclic aromatic hydrocarbons (PAHs) tend to remain in soil, mainly bound to soil organic matter (SOM) [3]. The VOCs can constitute a direct health risk when polluted soils are located in urban area. Pollution of groundwater rules out its use as resource for potable water unless a complementary treatment is implemented with additional costs. Moreover, the pollution remaining in soil needs to be treated before any other planning such as building.

Thus, the strategy of soil remediation needs to be adapted to the kind of pollution. When it is suitable, the in situ techniques can constitute good options to avoid excavation and soil transport. However, in narrow area and/or to ensure a more homogeneous depollution, the excavation before treatment can be a better alternative. Among ex situ treatment techniques, soil washing (SW) is a technology specifically employed to

E. Mousset
LRGP, CNRS/Université de Lorraine, Nancy Cedex, France

C. Trellu · M. A. Oturan (✉)
Université Gustave Eiffel, Champs-sur-Marne, France
e-mail: mehmet.oturan@univ-eiffel.fr

 
M. A. Rodrigo, E. V. Dos Santos (eds.), *Electrochemically Assisted Remediation of Contaminated Soils*, Environmental Pollution 30,
https://doi.org/10.1007/978-3-030-68140-1_5

remove HOCs by extraction of the contaminants from soil to liquid matrix at optimal soil/liquid ratio, contact time, and mixing rate [4, 5]. Most of the time, the extraction is enhanced with the addition of the relevant extracting agents in water. A good washing agent should be able to barely adsorb in soil, be biodegradable, have a high extraction power, and allow reuse for successive washing steps. Several kinds of agents have been proposed for HOCs extraction such as organic cosolvents, synthetic surfactants, biosurfactants, and cyclodextrins [3, 4, 6, 7]. They rely on their amphiphilic properties to increase the water solubility of HOCs.

The main limit of SW technology is that the pollution is only transferred and the SW solutions need to be further treated. Due to the biorecalcitrant properties of SW solutions, biological processes are not applicable and advanced physico-chemical treatments have been therefore developed. Electrochemical technologies are one of them and offer the advantages (1) to be modular, (2) to have a low footprint area, and (3) to avoid the addition of chemicals since the electrodes are prone to direct electrooxidation/reduction and/or indirect oxidation/reduction reactions with pollutants [8]. According to the kind of electrode material employed, electrochemical advanced oxidation processes (EAOPs) can occur by involving strong oxidizing agents such as hydroxyl radicals ($^{\bullet}OH$) ($E°(H_2O/^{\bullet}OH) = 2.8$ V/SHE) in the degradation and mineralization mechanisms [9–11]. This radical has a very short lifetime ($10^{-9}$ s) and is very reactive thanks to its four action modes [12–14]: (1) hydrogen atom abstraction (dehydrogenation), (2) electrophilic addition to an unsaturated bond (hydroxylation), (3) electron transfer (redox reactions), and (4) ipso-substitution of the halogen atom on perhalogenocarbon compounds. $^{\bullet}OH$s are therefore useful to oxidize HOCs' molecular structures that are present in SW solutions.

In this chapter, the different kinds of electrochemically assisted SW technologies, mainly the EAOPs involving homogeneous and heterogeneous catalysis are presented. The direct electrooxidation/reduction processes are deliberately not covered, since only few articles deal with such process for SW solution treatment [15]. They consist of the direct electron transfer with the target compounds to oxidize through oxidation reaction or to reduce through reduction reaction at the surface of electrode materials. This mechanism is involved only for electroactive species, which is rarely the case of pollutants from SW solutions. It therefore makes the removal efficiency lower than in indirect oxidation/reduction implemented in EAOPs.

The possible combinations of catalytic processes are then discussed in terms of enhancement of efficiency, transfer competitions, and reactor design as important engineering aspects for upscaling considerations.

# 2 EAOPs Involving Homogeneous Catalysis

## *2.1 Mechanisms of Electro-Fenton Process*

The EAOPs that involve homogeneous catalysis are mainly based on the Fenton's reaction (Eq. 1) occurred in the bulk solution through hydrogen peroxide ($H_2O_2$) electrogeneration by the two-electron oxygen reduction reaction (Eq. 2) and the ferrous ion ($Fe^{2+}$) regeneration from ferric ion ($Fe^{3+}$) reduction (Eq. 3) at the carbon cathode [9]:

$$Fe^{2+} + H_2O_2 \rightarrow Fe^{3+} + HO^- + {}^{\bullet}OH \quad (1)$$

$$O_2 + 2H^+ + 2e^- \rightarrow H_2O_2 \quad (2)$$

$$Fe^{3+} + e^- \rightarrow Fe^{2+} \quad (3)$$

This so-called electro-Fenton (EF) process has been applied in the treatment of SW solutions as represented in Table 1 and as detailed in a previous book chapter [16, 17].

The nature of cathode material is primordial in this process. Carbon-based materials have been preferred for their flexibility, cheapness, and efficiency to promote $H_2O_2$ electrogeneration [9]. Three-dimensional materials such as carbon felt, graphite felt, or carbon sponge have a great specific surface area and can therefore increase the number of $O_2$ active sites which rise the rates of $H_2O_2$ electrogeneration and $Fe^{2+}$ regeneration [18, 19]. However, the presence of HOCs in SW solution such as humic substances from SOM increases their adsorption to the cathode due to the hydrophobic properties of such carbonaceous materials [20]. Still, rebound effects of the total organic carbon (TOC) were noticed in solution due to the degradation of HOCs into more hydrophilic intermediates with $^{\bullet}OH$ attack.

## *2.2 Influence of Operating Parameters and Matrix Effects*

The efficiency of the treatment depends on the influence of operating parameters such as the current density and the catalyst concentration. There exists an optimal condition with both factors as shown in Fig. 1a, b, respectively, which is due to the increase of the rate of wasting reactions when the parameter value increases. At higher current density values, $H_2O_2$ is decomposed at the cathode (Eq. 4), at the anode (Eqs. 5 and 6), and in a lesser extent by thermal decomposition in bulk solution (Eq. 7):

$$H_2O_2 + 2H^+ + 2e^- \rightarrow 2H_2O \quad (4)$$

$$H_2O_2 \rightarrow HO_2^{\bullet} + H^+ + e^- \quad (5)$$

**Table 1** EAOPs involving homogeneous catalysis proposed in literature for the treatment of SW solutions

| SW/SF solution characteristics[a] | EF characteristics[b] | Efficiency[c] | Reference |
|---|---|---|---|
| Synthetic SW solution containing beta-cyclodextrin (BCD) (1.13 g $L^{-1}$) and TNT (45 mg $L^{-1}$) | Cathode: carbon felt (60 $cm^2$)<br>Anode: Pt grid (3 cm diameter, 4.5 cm height)<br>Undivided cell<br>$J$ = 1.0–4.2 mA $cm^{-2}$<br>Addition of $Na_2SO_4$ (50 mM) and $Fe^{2+}$ (0.2 mM); acidification to pH 3<br>$V$ = 150 mL | Total mineralization at 20 Ah $L^{-1}$ | [21] |
| Synthetic SW solution containing HPCD (9 g $L^{-1}$) and phenanthrene (16 mg $L^{-1}$) | Cathode: carbon felt (150 $cm^2$)<br>Anode: Pt grid (3 cm diameter, 5 cm height) or DSA (40 $cm^2$)<br>Undivided cell<br>$J$ = 3.3–13.3 mA $cm^{-2}$<br>Addition of $Na_2SO_4$ (150 mM) and $Fe^{2+}$ (0.05–10 mM); acidification to pH 3<br>$V$ = 400 mL | Better mineralization with DSA anode<br>Around 7% of mineralization with DSA at 20 Ah $L^{-1}$ | [22] |
| Synthetic SW solution containing tween 80 (0.75 g $L^{-1}$) or HPCD (10 g $L^{-1}$) and phenanthrene (17 mg $L^{-1}$) | Cathode: carbon felt (150 $cm^2$)<br>Anode: Pt grid (3 cm diameter, 5 cm height),<br>Undivided cell<br>$J$ = 3.3–13.3 mA $cm^{-2}$<br>Addition of $Na_2SO_4$ (150 mM) and $Fe^{2+}$ (0.5 mM); acidification to pH 3<br>$V$ = 400 mL | Around 6% of mineralization with HPCD and 85% of mineralization with Tween 80 at 20 Ah $L^{-1}$ | [23] |
| Synthetic and real SW solution from spiked uncontaminated soil, containing HPCD (6.25 g $L^{-1}$) and PCP (205 mg $L^{-1}$) | Cathode: Carbon felt (10 $cm^2$)<br>Anode: Pt sheet (1 $cm^2$)<br>Undivided cell<br>$J$ = 4.0–20.0 mA $cm^{-2}$<br>Addition of $Na_2SO_4$ (150 mM) and $Fe^{2+}$ (0.2 mM); acidification to pH 3<br>$V$ = 125 mL | Around 91% of COD removal at 18.8 Ah $L^{-1}$ | [24] |

(continued)

**Table 1** (continued)

| SW/SF solution characteristics[a] | EF characteristics[b] | Efficiency[c] | Reference |
|---|---|---|---|
| Real SW solution from spiked uncontaminated soil, with Lissamine green B (dye) (1.7–3.5 g $kg^{-1}$) or phenanthrene (430 mg $kg^{-1}$) | Cathode: graphite (1.27 $cm^2$) or stainless steel (3.14 $cm^2$)<br>Anode: Graphite (1.27 $cm^2$) or stainless steel (3.14 $cm^2$)<br>Undivided cell<br>Cell potential: 5 V<br>Addition of $Na_2SO_4$ (100 mM) and $Fe^{2+}$ (0.2 mM); pH 3<br>$V = 150$ mL | 90% of decoloration in 3 h of electrolysis and 50% of phenanthrene removal after 25 h of electrolysis | [25] |
| Real SW solution from historically contaminated soil, containing Tween 80 (7.5 g $L^{-1}$) or HPCD (7.5 g $L^{-1}$) and 16 PAHs (1090 mg $kg^{-1}$) | Cathode: carbon felt (150 $cm^2$)<br>Anode: Pt grid (3 cm diameter, 5 cm height)<br>Undivided cell<br>$J = 6.7$ mA $cm^{-2}$<br>Addition of $Na_2SO_4$ (150 mM) and no $Fe^{2+}$ added; no pH adjustment<br>$V = 400$ mL | Around 3% of mineralization with HPCD and 19% of mineralization with Tween 80 at 20 Ah $L^{-1}$ | [26] |

[a] *BCD* beta-cyclodextrin, *TNT* trinitrotoluene, *HPCD* hydroxypropyl-beta-cyclodextrin (assume a molar weight of 1250 g $mol^{-1}$), *PCP* pentachlorophenol, *PAHs* polycyclic aromatic hydrocarbons
[b] DSA: dimensionally stable anode; *Pt* platinum, *J* current density, *V* volume of treated solution
[c] COD: chemical oxygen demand

$$HO_2^{\bullet} \rightarrow O_2 + H^+ + e^- \tag{6}$$

$$2H_2O_2 \rightarrow O_2 + 2H_2O \tag{7}$$

Hydrogen ($H_2$) reduction reaction is another parasitic equation that can occur as well (Eq. 8):

$$2H^+ + 2e^- \rightarrow H_{2(g)} \tag{8}$$

Similarly, at higher concentration of $Fe^{2+}$, its reaction with $^{\bullet}OH$ (Eq. 9) is becoming more important, which decreases the amount of $^{\bullet}OH$ available for organic pollutants degradation:

$$Fe^{2+} + {}^{\bullet}OH \rightarrow Fe^{3+} + HO^- \tag{9}$$

The advantage of SW combination with EF is the possibility to take the advantage of the presence of iron ions in SW solutions as a source of catalyst allowing the Fenton reaction occurs without any external addition of iron salts [26, 27]. The amount extracted depends on the concentration and availability of iron in the

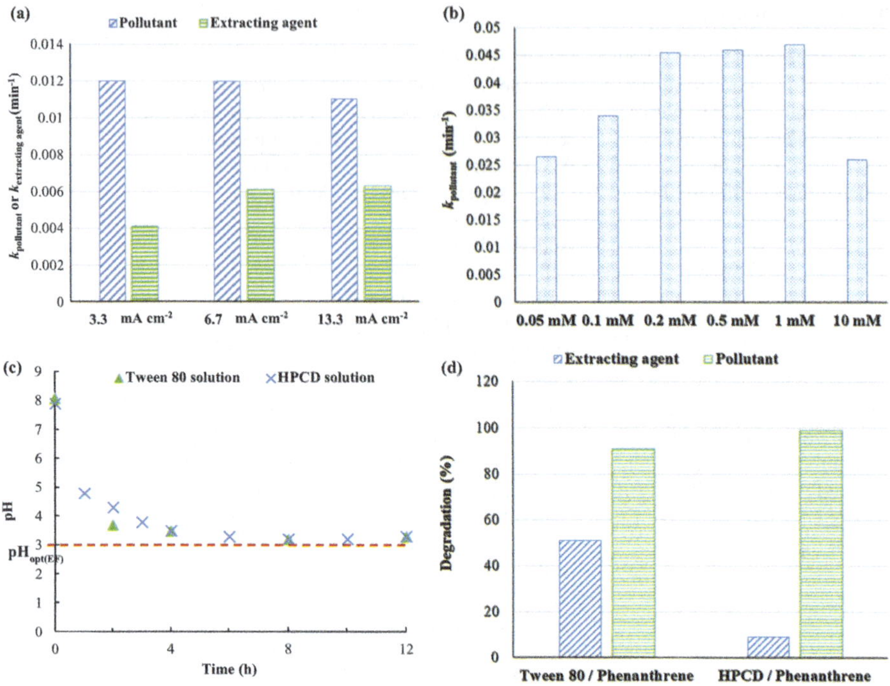


**Fig. 1** (**a**) Influence of current density on pseudo-first-order kinetics rate constant of phenanthrene and HPCD degradation during EF treatment of SW solution (Adapted with permission from [22, 23]. Copyright 2014 Elsevier); (**b**) Influence of ferrous ion concentration on pseudo-first-order kinetics rate constant of phenanthrene degradation during EF treatment of SW solution (Adapted with permission from [22, 23]. Copyright 2014 Elsevier); (**c**) Evolution of solution pH during EF treatment of SW solutions containing either HPCD or Tween 80 as washing agent; (**d**) Influence of nature of extracting agent (HPCD (10 g $L^{-1}$) or Tween 80 (0.75 g $L^{-1}$)) on pollutant (PHE (17 mg $L^{-1}$)) degradation. (Reprinted with permission from [22]. Copyright 2014 Elsevier)

studied soil. Still, 0.02–0.06 mM of dissolved iron was quantified in the SW solution coming from historically contaminated soils [26, 27]. This amount was sufficient to ensure EF process runs appropriately.

The influence of SW solution characteristics is another key factor in the combination with EF process. The pH of solution is well-known to be of primary importance when implementing Fenton reaction. The iron(III) species precipitate with $HO^-$ to form ferric hydroxide sludge ($Fe(OH)_3$) at pH higher than 4. That is why a pH around 3 has been found optimal for homogeneous EF process [9, 26]. An interesting feature is depicted in Fig. 1c, in which the pH of real SW solution decreases with electrolysis time from an initial pH around 8 until a pH around 3 [16]. This could be attributed to the formation of carboxylic acids as oxidation by-products whose pKa values vary between 3 and 4. In this condition, the pH of the solution needs to be neutralized after the treatment.

The kind of extracting agent plays a role not only in the extraction efficiency but also in the oxidation mechanisms. It has been previously shown that the iron species could be complexed with cyclodextrin according to their functional group bonded on the external ring [4, 22]. In the presence of HOCs that are trapped into the cyclodextrin having a toroidal shape, a ternary complex is formed between the pollutant, the cyclodextrin such as hydroxypropyl-beta-cyclodextrin (HPCD), and $Fe^{2+}$ as catalyst source for Fenton reaction ($Fe^{2+}$-cyclodextrin-pollutant (Eq. 10)).

$$Fe^{2+} : HPCD : HOC + {}^{\bullet}OH \rightarrow Fe^{2+} : HPCD : HOC(OH)^{\bullet} \quad (10)$$

Thanks to the formation of $^{\bullet}OH$ close to the iron species; the pollutant should be preferentially degraded as compared to the cyclodextrin as displayed in Fig. 1d. This selectivity of oxidation seems less obvious with surfactant such as Tween 80 due to the micelle structure that entraps the HOCs inside it (Fig. 1d) [22]. In addition, the extracting agent's properties could favor or not their recycling ability during the oxidative stress, which is another interesting feature considering their costs.

The matrix effect is a decisive aspect when considering the upscaling phase. As a first approach, experiments with synthetic SW solutions should depict higher removal efficiency of HOCs against experiments with real SW solutions. Real solutions contain inorganic ions and mostly organic matter (OM) that could decrease the oxidation efficiency by trapping the HOCs and by reacting very quickly with $^{\bullet}OH$ [20, 28]. Interestingly, the mineralization efficiencies were very similar during the EF treatment of synthetic and real SW solutions containing either Tween 80 or HPCD as extracting agent [23]. In such solutions, the proportion of TOC content coming from OM and HOCs is negligible (4–5% of TOC) as compared to the content in extracting agent (95–96% of TOC). Still, the impact of matrix has to be considered, since the amount of OM and HOCs extracted is dependent on the age, amount, and kind of organic pollution as well as on the nature of soil and the kind of extracting agent employed [3, 26].

## 3 EAOPs Involving Heterogeneous Catalysis

The use of EAOPs involving heterogeneous catalysis for the treatment of SW solutions has been the focus of several studies during the 5 last years (Table 2). The objective is to generate large amounts of highly oxidizing species at the anode surface. Particularly, the so-called advanced anodic oxidation (AO) process is based on the formation of physisorbed hydroxyl radicals ($^{\bullet}OH$) from water oxidation (Eq. 11) at the surface of an appropriate anode (M) with high overpotential for oxygen evolution reaction [8, 29].

$$M + H_2O \rightarrow M({}^{\bullet}OH) + H^+ + e^- \quad (11)$$

**Table 2** EAOPs involving heterogeneous catalysis proposed in literature for the treatment of SW solutions

| Contaminated soil | Extracting agent | Organic load of the SW solution (mg $L^{-1}$) | Anode/cathode | Current density (mA $cm^{-2}$) | Effectiveness | Reference |
|---|---|---|---|---|---|---|
| Kaolinite with no organic content spiked with phenanthrene | SDS, ABDMA, tween 80 | COD 17,000–20,000 | BDD or DSA/SS[a] | 30 | Between 15 and 75 ah $L^{-1}$ for 60% removal of COD with BDD anode, according to nature of surfactant. Very low mineralization with DSA anode | [41] |
| Kaolinite spiked with atrazine | SDS | – | BDD/SS | – | Total mineralization at 40 Ah $L^{-1}$ | [42] |
| Kaolinite soil spiked with atrazine | SDS | TOC 50–700 | BDD/SS | 30 | More than 95% TOC removal at 25 Ah $L^{-1}$ | [43] |
| Synthetic SW solution with phenanthrene as soil pollutant | Tween 80 | TOC 770 | BDD/SS | 42 | 70% removal of COD at 15 Ah $L^{-1}$ | [33] |
| Kaolinite soil spiked with oxyfluorfen | SDS | TOC 50–2000 | BDD/SS | 30 | More than 95% removal of turbidity at 25 Ah $L^{-1}$ | [39] |
| Clay soil with low organic content spiked with lindane | SDS | TOC 3000–8000 | BDD/SS | 50 | Total degradation of lindane between 15 and 80 Ah $L^{-1}$ | [34] |
| Clay soil spiked with pendimethalin | SDS | TOC 7000–10,000 | BDD/SS | 30 | 80% removal of pendimethalin at 20 Ah $L^{-1}$ | [44] |
| Clay soil spiked with clopyralid | Only water | TOC 30–40 | BDD/BDD | 10–100 | Total degradation of clopyralid between 10 and 20 Ah $L^{-1}$ Total mineralization at 50 ah $L^{-1}$ | [35] |
| Sand fraction of a real PAHs[b] contaminated soil | Tween 80 | TOC 3000–4000 | BDD/SS | 2.1 | 80% removal of PAHs at 3.5 Ah $L^{-1}$ | [40] |
| Clay soil with low organic content spiked with clopyralid | Only water | TOC 7–20 | BDD/SS | 10 | 80% removal of clopyralid between 4 and 6 ah $L^{-1}$ | [45] |
| Kaolinite spiked with atrazine | SDS | TOC 40–670 | BDD/SS | 30 | Total mineralization between 15 and 25 Ah $L^{-1}$ | [46] |
| Clay soil spiked with Lontrel® (herbicide containing 72% w/w of clopyralid) | Only water | TOC 65–70 | BDD or Ru-MMO[c] or Ir-MMO or CF[d]/SS | 5–15 | Total degradation of clopyralid at 14 Ah $L^{-1}$ with BDD anode | [47] |

[a]*SS* stainless steel
[b]*PAHs* polycyclic aromatic hydrocarbons
[c]*MMO* mixed metal oxide
[D]*CF* Carbon felt

### 3.1 Electrode Materials

Boron-doped diamond (BDD) is the most widely anode material for AO because of its corrosion/mechanical resistance, inertness surface, and high oxygen evolution potential (2.3 V/SHE) [29]. Thus, AO using BDD anode is able to achieve high mineralization rate of a large range of organic pollutants [20, 30–32]. These properties are also particularly suitable for the treatment of SW solutions, which are complex effluents containing large amounts of poorly biodegradable organic compounds. This is why almost all studies used BDD anode for the treatment of SW solution by heterogeneous catalysis (Table 2).

During AO, the cathode is the inert electrode. Stainless steel is the most widely used material because of several advantages such as high conductivity, low cost, robustness, and mechanical resistance. No specific catalytic property is expected from the cathode material when performing AO as a stand-alone process.

### 3.2 Technical Advantages Over Homogeneous Catalysis

By comparison to homogeneous catalysis based on EF process, AO presents several technical advantages. First, it is possible to avoid the use of carbonaceous materials as cathode. Thus, low adsorption of hydrophobic organic compounds is usually observed, which reduces issues of electrode fouling and improves process robustness. Secondly, AO does not require $O_2$ supply, which represents an additional cost and often causes important troubles for process operation because of important foaming. Thirdly, AO does not require pH adjustment to 3 and no control because of its low influence on AO efficiency [32]. However, pH can naturally decrease during the treatment because of the formation of carboxylic acids as degradation by-products [33]. Evolution of pH also depends on buffer capacity of the SW solution, which depends on the nature and amount of salts extracted during the washing step [34]. Finally, a key advantage of AO is also to use electrons as sole reagent. Even the addition of electrolyte can be avoided, thanks to salts extracted from the soil [35].

### 3.3 Degradation and Mineralization Effectiveness

As reported in Table 2, AO has been mainly applied to the treatment of SW solutions containing either PAHs or pesticides as soil pollutants. Tween 80 (nonionic surfactant) [39] and sodium dodecyl sulfate (SDS, anionic surfactant) [32, 38, 40–43] have been the most widely used extracting agents. Some studies also focused on highly soluble organic pollutants (e.g., clopyralid) for which no extracting agents were used for extraction from the contaminated soil [44]. AO allows for achieving very high mineralization rate (>90% of TOC or COD removal). However, high current density (in the range 30–50 mA $cm^{-2}$) and electrical charge (in the range 15–50 Ah $L^{-1}$, sometimes more) were always required, thus resulting in high energy consumption (e.g., 127 kWh $m^{-3}$ for 70% removal of 2.7 g $L^{-1}$ of COD) [33]. From Table 2, it is important to highlight that high degradation rate of soil pollutants also always required high electrical charge. Degradation kinetics were only slightly faster than mineralization kinetics. This phenomenon is generally observed for AO and attributed to various reasons:

1. By comparison to homogeneous catalysis, heterogeneous catalysis such as AO usually results in faster mineralization kinetics but slower degradation kinetics because of mass transport issues [36–38].
2. Micelles formed by surfactants act as a protective environment for soil pollutants, thus resulting in lower availability of pollutants for reaction with oxidizing species [22, 39, 40].
3. Extracting agents often account for almost the total organic load of the SW solution, therefore they strongly act as scavengers for oxidizing species [22, 39, 40].

## 3.4 Electrooxidation Mechanisms

The effectiveness of AO relies on two main mechanisms. Organic compounds can react either at the surface of the BDD anode with physisorbed $^{\bullet}OH$ or in the bulk by mediated oxidation with electrochemically generated oxidizing species such as $S_2O_8^{2-}$, $SO_4^{\bullet-}$, and active chlorine species [39, 40, 43]. The formation of micelles is an important specificity of SW solution, which plays a crucial role in electrooxidation mechanisms (Fig. 2). For example, at low concentration of SDS, it was reported that the size of particles mobilized during the washing of lindane-spiked clay soil was in the range 5–20 μm [34]. By comparison, at high concentration of SDS, the formation of micelles and aggregates caused an increase of mean particle size in the range 100–1000 μm [34, 42, 43]. The size of micelles/aggregates depends on soil characteristics, surfactant concentration, and liquid/solid ratio during the washing step [39, 43]. This phenomenon has a crucial influence on heterogeneous catalysis since large size micelles cannot react with physisorbed $^{\bullet}OH$ generated in a thin layer of few nanometers close to the anode surface [40, 48]; they can only be degraded through mediated oxidation in the bulk. Interestingly, SDS degradation was reported to release large amount of $SO_4^{2-}$ ions and, therefore, to further promote mediated oxidation in the bulk during the treatment [39]. Thus, mediated oxidation allows the degradation of surfactant micelles and results in the decrease of mean particle size (Fig. 3). Then, both physisorbed $^{\bullet}OH$ at the anode surface and mediated oxidation in the bulk are involved for complete mineralization of the effluent [39, 43].

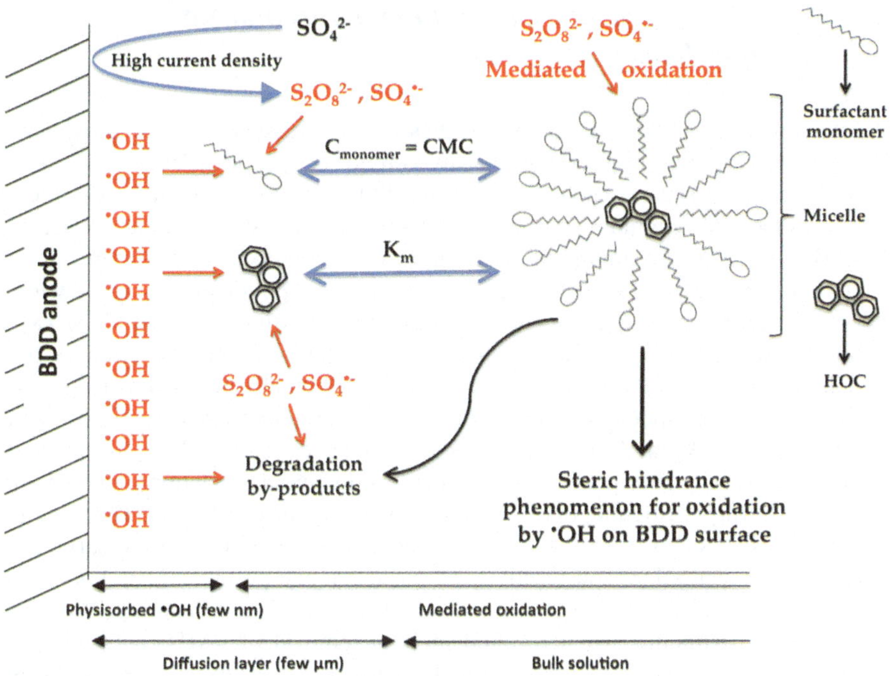


**Fig. 2** Schematic view of main degradation mechanisms during the treatment of soil washing solutions by AO using BDD anode. (Adapted from [40])

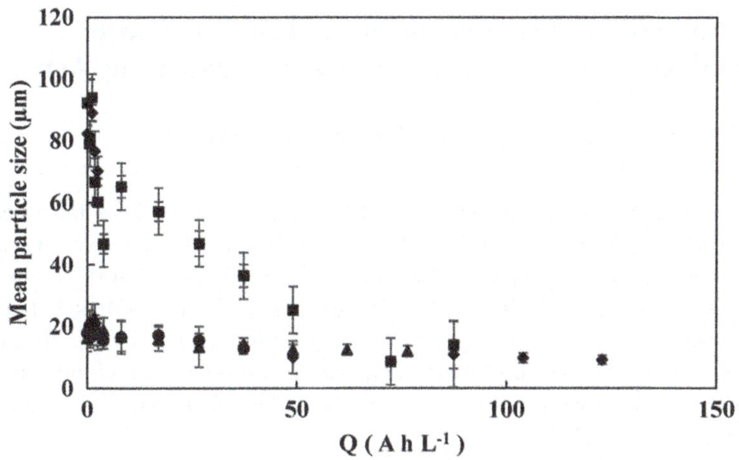


**Fig. 3** Changes in the mean particle size (and limits 10–90%) observed during the electrolysis of soil washing effluents carried out at 0.20 (■), 0.15 (◆), 0.10 (●), and 0.05 (▲) mg SDS g$^{-1}$ soil. (Reprinted with permission from [34])

### 3.5 Selective Degradation of Soil Pollutants and SW Solution Reuse

As explained above, degradation of micelles strongly relies on mediated oxidation in the bulk. However, high current density is required for electrochemical generation of oxidizing species such as $S_2O_8^{2-}$ and $SO_4^{\bullet-}$ [8, 29]. Thus, it is possible to strongly hinder the degradation of micelles by performing AO at low current density [40]. Besides, it is also important to take into consideration that there is an equilibrium between free extra-micellar organic pollutants and pollutants entrapped within micelles, according to the micellar/aqueous phase partitioning coefficient ($K_m$). While the degradation of compounds entrapped within micelles is strongly hindered, free extra-micellar organic pollutants can still be effectively degraded at the anode surface. Entrapped pollutants can then be continuously released in the solution according to $K_m$, and subsequently be oxidized at the anode [40]. This phenomenon explains why it is possible to achieve a selective degradation of target soil pollutants during the treatment of SW solution at low current density. The selectivity of the process depends on $K_m$, critical micellar concentration of the surfactant, surfactant concentration, and reactivity of surfactant and pollutant [40]. For example, AO during 23 h at 2.1 mA $cm^{-2}$ of a real SW solution containing PAHs and Tween 80 as extracting agent resulted in 80% removal of PAHs (phenanthrene, anthracene, fluoranthene, pyrene), while the extraction capacity of the treated solution was maintained. In fact, the solution was then reused for another SW step with only 5% lower PAHs removal from soil than the fresh SW solution [40].

## 4 Combination of Homogeneous and Heterogeneous Catalysis: Efficiency, Modeling, and Engineering Aspects

### 4.1 Efficiency of Combined Catalytic Processes

EAOPs involving both homogeneous and heterogeneous catalysis have been also investigated for SW treatment. Particularly, AO using BDD has been combined with various other processes such as electro-Fenton [22, 23, 26, 27, 45], sonolysis [46, 49–51], or photolysis [44, 46, 49, 51–53]. By comparison to AO as a stand-alone process, such approach allows for providing an additional source of $^{\bullet}OH$ and other oxidizing species in the bulk. Moreover, the activation of persulfate into sulfate radicals is also further promoted, thus increasing the effectiveness of mediated oxidation in the bulk [44].

## 4.2 Charge Transfer vs Mass Transfer Competition

Electrolysis is composed of three consecutive steps: (1) mass transfer of the target compounds from the electrolyte towards the electrode, (2) adsorption of the compounds at the electrode surface and charge transfer during the electrochemical reaction, and (3) mass transfer of the target compounds from the electrode towards the electrolyte. The kinetics of the global reaction therefore depends on the rates of mass transfer vs charge transfer. Based on this competition, the importance in oxidation mechanism of homogeneous catalysis occurring in the bulk of electrochemical reactor and heterogeneous catalysis occurring at the surface of electrode can vary. It has been studied with SW solution containing Tween 80 as representative washing agent and at different initial COD (1.61, 12.1, and 23.3 g-$O_2$ $L^{-1}$) [54].

A mathematical model has been proposed on the basis of that developed by Comninellis' team for AO processes in early 2000s [30, 55], and adapted for combination between EF and AO with BDD, noted as EF-BDD and AO-BDD in this subsection. Three oxidation mechanisms have been considered: (1) advanced electrooxidation ($^{\bullet}OH$) at anode surface, (2) advanced oxidation from ($^{\bullet}OH$) generated via Fenton reaction in bulk solution, and (3) mediated oxidation ($SO_4^{\bullet-}$, $S_2O_8^{2-}$, $O_3$, etc.) at anode surface and in bulk solution. The model has been plotted in Fig. 4a against the experimental data of COD removal as a function of electrolysis time in a stirred batch reactor [54]. The critical time ($t_{cr}$) has been identified as the time at which the global reaction order of the model changes corresponding to the change from charge transfer control to the mass transfer control. The $t_{cr}$ value is lower in EF-BDD than in AO-BDD. There are two production sites of $^{\bullet}OH$ in EF-BDD as compared to AO-BDD, which makes faster the COD removal with former one. An interesting feature in this model is that the mass transfer coefficient is varying with electrolysis time due to the viscosity of solution that is decreasing when the surfactant is progressively degraded, which makes more complicated the value of global reaction order [54]. In the aim at comparing the influence of the three different mechanisms of oxidation in EF-BDD cell, the evolution of their respective characteristic times, representing the velocity of the reaction that is occurring, as function of electrolysis time has been plotted in Fig. 4b [54]. It highlights that AO mechanism becomes predominant after several hours of treatment due to the decrease of mass transfer resistance as a function of time as well as due to electrooxidation at BDD anode which become more and more important. In addition, a new expression of global instantaneous current efficiency ($ICE_{global}$) has been proposed in order to take into account the three oxidation mechanisms which could even fit with $ICE_{global}$ values higher than 1 (Fig. 4c) [54]. Both experimental and theoretical ICE values are based on the fact that COD is removed only due to electrooxidation at anode, which could explain the values higher than 1 in EF-BDD experiments.

The knowledge of the importance of the different oxidation mechanisms that occur in both processes allows identifying which of the heterogeneous or homogeneous oxidation is favored and therefore which mechanism must be improved. Reactor design studies become meaningful in this step as discussed in Sect. 4.3.

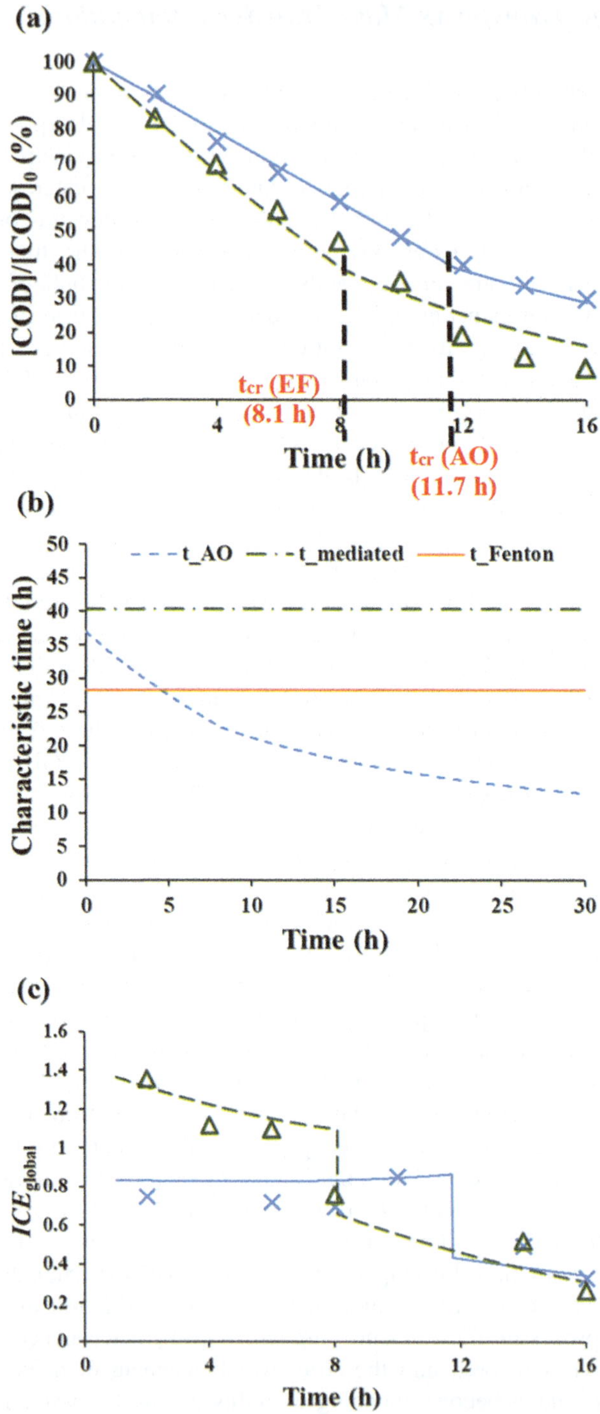


**Fig. 4** COD removal as a function of electrolysis time (**a**); competition between the different characteristic times of the processes involved (t_AO: advanced electro-oxidation, t_mediated:

## 4.3 Reactor Design: Towards Transfer Intensification

The selection of anode and cathode materials by taking into account the matrix effect according to the electrolyte to treat is an important preliminary achievement as shown in the previous sections. The reactor design is another important step to optimize the transfer and therefore the rate of (electro)chemical reactions occurring within the process. An optimized design would lead to less reagent consumption and a decrease of energy requirement, while the shape of electrode materials needs to fit with the design, which is primordial when considering the upscale of the process to reduce the costs.

There are different categories of electrochemical reactor according to the electrical connection, i.e., either monopolar system in which all electrodes are connected two by two or bipolar arrangement in which only the ends of electrode stack are connected and therefore polarizing each electrode within the stack [56]. In SW solution treatment application, the monopolar system has been preferred with lower cell potential—which is safer and requires less energy—as well as better homogeneity of current distribution [57].

The compartment separation is another criterion of electrochemical cell design. Undivided cell can be distinguished from the divided cell in which a separator is added between cathodic and anodic compartments [56]. Electrochemical treatment of SW solutions has been mainly performed in undivided cell [16]. In EF treatment, undivided cell allows for catalyst cycle of iron species.

The mode of operation can be either in batch or in continuous flow, and two kinds of reactor configuration can be proposed, i.e., stirred tank reactor or plug-flow reactor. The later configuration is assimilated to flow-by or flow-through reactors in electrochemical engineering [56]. Stirred tank reactor in batch mode has been proposed in literature to treat SW solutions at laboratory scale (Fig. 5a) [22, 40, 58]. However, there are mass transfer limitations in such reactors. Therefore, flow-by reactors involving interelectrode distances in millimeter to centimeter range have been proposed (Fig. 5b) [59]. To better maximize the transfer phenomenon, flow-through cells in which the effluent is flushed through the electrodes have also been developed (Fig. 5c) [60–62]. Still, the conductivity of solution required is quite high, which means that a supporting electrolyte needs to be added unless the initial SW solution conductivity is high enough [35]. The excess of the electrolyte has to be removed before discharging the treated effluent to avoid the contamination of water bodies. The implementation of very short interelectrodes distance in the

**Fig. 4** (continued) mediated oxidation, t_Fenton: Fenton oxidation) as a function of the treatment time of EF-BDD experiments (**b**); global instantaneous current efficiency ($ICE_{global}$) evolution (**c**). Condition: initial COD concentration at 12.1 g-$O_2$ $L^{-1}$ in a stirred batch reactor. AO-BDD (—/×) and EF-BDD (– – –/Δ); model (—/– – –) and experimental data (markers (×/Δ)). (Adapted with permission from [54]. Copyright 2019 Elsevier)

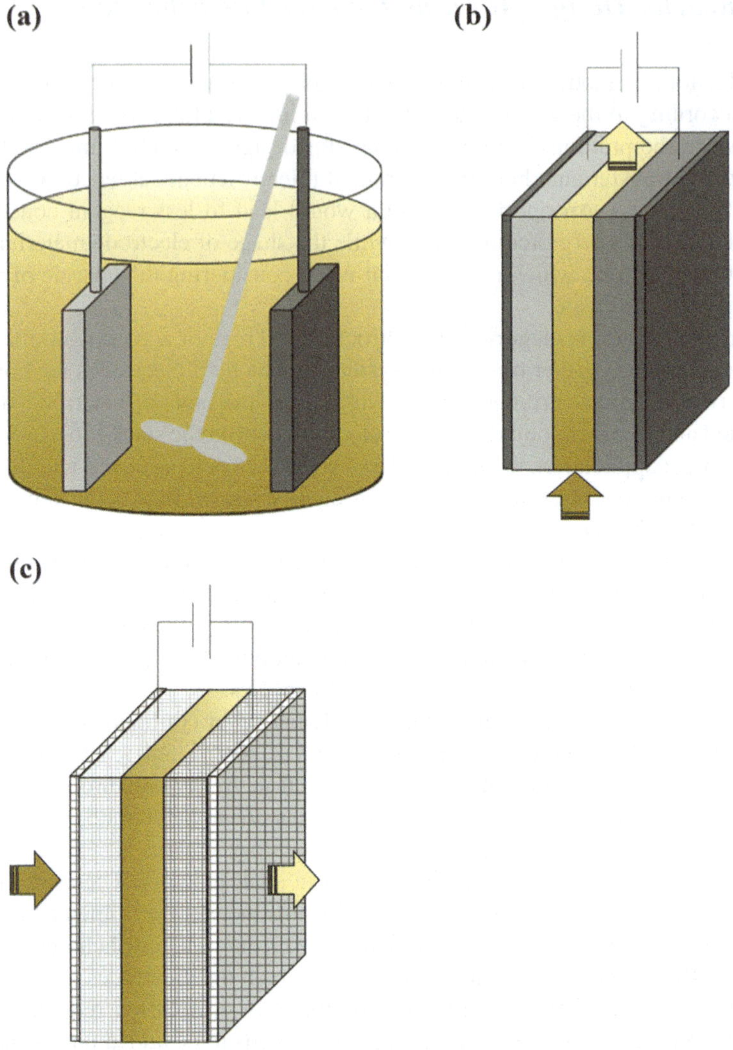


**Fig. 5** Different reactor configurations proposed during the electrochemical treatment of SW solution: (**a**) stirred batch reactor, (**b**) flow-by reactor, (**c**) flow-through reactor

micrometer range has been proposed in microfluidic device [63]. It allows reducing the cell resistance and therefore minimizing the conductivity requirement [64]. Another advantage is the reduction of energy consumption since the cell potential is diminished in this design. The efficiency between a flow-by cell with standard inter-electrodes distance (3000 μm) and a microfluidic flow-through cell (400 μm inter-electrode gap) has been compared [60]. The energy consumed at a similar removal percentage was reduced by six times in the microfluidic reactor in a low-conductive media (1 mS $cm^{-1}$), which represents a promising cost-effective approach.

## 5 Concluding Remarks

The main electrochemical technologies applied to SW solution treatments have been reviewed. The influence of main operating parameters and solution characteristics as well as the process efficiency has been reported. The transfer intensification via modeling studies and reactor design optimization are also presented.

The impact of treated SW solution on ecotoxicity and biodegradability is another critical issue to be taken into account before discharging the effluent. Their evolutions mainly depend on the kind of extracting agent, the matrix composition, current density, and electrolysis time [23, 33]. The trends often show a decrease of inhibition or increase of biodegradability after a lag phase at the beginning of the electrolysis [23, 26]. This is attributed to the first organic intermediates such as hydroxylated aromatic compounds that are known to be toxic [26]. Then the aromatic ring is broken to form carboxylic acids that are easily biodegradable [33].

The energy requirement represents the most important part of operating costs of EAOPs technology. Therefore, efforts to reduce it are important. It is also meaningful to consider the specific energy—expressed in kWh per kg of pollutant removed—for process comparison and for sizing calculation. The faradic yield—corresponding to the ratio of energy devoted to degrade the pollutant with the total applied energy—needs also to be optimized to reduce the energy consumed. Since the extracting agent represents an important part of the organic matter, these agents strongly impact on the faradic yield and therefore on the energy. In order to reduce it, some authors proposed the selective oxidation at low current density in anodic oxidation systems [40], while the combination of adsorption with EAOPs as a way to selectively degrade the contaminants instead of extracting agents has been also suggested [65]. The next step would be to bring electrochemical-assisted soil washing technologies closer to industrial applications with further engineering studies for hybrid complementary process developments.

## References

1. C.N. Mulligan, R.N. Yong, B.F. Gibbs, Surfactant-enhanced remediation of contaminated soil: a review. Eng. Geol. **60**, 371–380 (2001)
2. T.E. Ochsner, B.M. Stephens, W.C. Koskinen, R.S. Kookana, Sorption of a hydrophilic pesticide: effects of soil water content. Soil Sci. Soc. Am. J. **70**, 1991–1997 (2006)
3. C. Trellu, E. Mousset, Y. Pechaud, D. Huguenot, E.D. van Hullebusch, G. Esposito, M.A. Oturan, Removal of hydrophobic organic pollutants from soil washing/flushing solutions: a critical review. J. Hazard. Mater. **306**, 149–174 (2016a)

4. E. Mousset, M.A. Oturan, E.D. Van Hullebusch, G. Guibaud, G. Esposito, Soil washing/flushing treatments of organic pollutants enhanced by cyclodextrins and integrated treatments: state of the art. Crit. Rev. Environ. Sci. Technol. **44**, 705–795 (2014a)
5. T.B. Boving, X. Wang, M.L. Brusseau, Cyclodextrin-enhanced solubilization and removal of residual-phase chlorinated solvents from porous media. Environ. Sci. Technol. **33**, 764–770 (1999)
6. S. Paria, Surfactant-enhanced remediation of organic contaminated soil and water. Adv. Colloid Interface Sci. **138**, 24–58 (2008)
7. X. Mao, R. Jiang, W. Xiao, J. Yu, Use of surfactants for the remediation of contaminated soils: a review. J. Hazard. Mater. **285**, 419–435 (2015)
8. C.A. Martínez-Huitle, M.A. Rodrigo, I. Sirés, O. Scialdone, Single and coupled electrochemical processes and reactors for the abatement of organic water pollutants: a critical review. Chem. Rev. **115**, 13362–13407 (2015)
9. E. Brillas, I. Sirés, M.A. Oturan, Electro-Fenton process and related electrochemical technologies based on Fenton's reaction chemistry. Chem. Rev. **109**, 6570–6631 (2009)
10. I. Sirés, E. Brillas, M.A. Oturan, M.A. Rodrigo, M. Panizza, Electrochemical advanced oxidation processes: today and tomorrow. a review. Environ. Sci. Pollut. Res. **21**, 8336–8367 (2014)
11. M.A. Rodrigo, N. Oturan, M.A. Oturan, Electrochemically assisted remediation of pesticides in soils and water: a review. Chem. Rev. **114**, 8720–8745 (2014)
12. E. Mousset, N. Oturan, M.A. Oturan, An unprecedented route of OH radical reactivity evidenced by an electrocatalytical process: IPSO-substitution with perhalogenocarbon compounds. Appl. Catal. B Environ. **226**, 135–146 (2018a)
13. L.M. Dorfman, G.E. Adams, Reactivity of the hydroxyl radical in aqueous solutions. Natl. Bur. Stand (1973)
14. G.V. Buxton, C.L. Greenstock, W.P. Helman, A.B. Ross, Critical review of rate constants for reactions of hydrated electrons, hydrogen atoms and hydroxyl radicals ($\bullet OH/\bullet O^-$) in aqueous solution. J. Phys. Chem. Ref. Data **17**, 513–886 (1988)
15. J. Gómez, M.T. Alcántara, M. Pazos, M.A. Sanromán, Remediation of polluted soil by a two-stage treatment system: desorption of phenanthrene in soil and electrochemical treatment to recover the extraction agent. J. Hazard. Mater. **173**, 794–798 (2010)
16. E. Mousset, C. Trellu, N. Oturan, M.A. Rodrigo, M.A. Oturan, Soil Remediation by Electro-Fenton Process, in *Electro-Fenton process: New Trends and Scale-Up*, ed. by M. Zhou, M. A. Oturan, I. Sires, (Springer, New York, 2018b)
17. N. Oturan, M.A. Oturan, Electro-Fenton Process: Background, New Developments, and Applications, in *Electrochemical Water and Wastewater Treatment*, ed. by C. A. Martinez-Huitle, M. A. Rodrigo, O. Scialdone, (Elsevier, Oxford, 2018)
18. I. Sirés, J.A. Garrido, R.M. Rodríguez, E. Brillas, N. Oturan, M.A. Oturan, Catalytic behavior of the $Fe^{3+}/Fe^{2+}$ system in the electro-Fenton degradation of the antimicrobial chlorophene. Appl. Catal. B Environ. **72**, 382–394 (2007)
19. E. Mousset, Z. Wang, J. Hammaker, O. Lefebvre, Physico-chemical properties of pristine graphene and its performance as electrode material for electro-Fenton treatment of wastewater. Electrochim. Acta **214**, 217–230 (2016a)
20. C. Trellu, Y. Péchaud, N. Oturan, E. Mousset, D. Huguenot, E.D. van Hullebusch, G. Esposito, M.A. Oturan, Comparative study on the removal of humic acids from drinking water by anodic oxidation and electro-Fenton processes: mineralization efficiency and modelling. Appl. Catal. B Environ. **194**, 32–41 (2016b)
21. M. Murati, N. Oturan, E.D. Van Hullebusch, M.A. Oturan, Electro-Fenton treatment of TNT in aqueous media in presence of cyclodextrin. Application to Ex-situ treatment of contaminated soil. J. Adv. Oxid. Technol. **12**, 29–36 (2009)
22. E. Mousset, N. Oturan, E.D. van Hullebusch, G. Guibaud, G. Esposito, M.A. Oturan, Influence of solubilizing agents (cyclodextrin or surfactant) on phenanthrene degradation by electro-Fenton process – study of soil washing recycling possibilities and environmental impact. Water Res. **48**, 306–316 (2014b)

23. E. Mousset, N. Oturan, E.D. Van Hullebusch, G. Guibaud, G. Esposito, M.A. Oturan, Treatment of synthetic soil washing solutions containing phenanthrene and cyclodextrin by electro-oxidation. Influence of anode materials on toxicity removal and biodegradability enhancement. Appl. Catal. B Environ. **160–161**, 666–675 (2014c)
24. K. Hanna, S. Chiron, M.A. Oturan, Coupling enhanced water solubilization with cyclodextrin to indirect electrochemical treatment for pentachlorophenol contaminated soil remediation. Water Res. **39**, 2763–2773 (2005)
25. E. Rosales, M. Pazos, M.A. Longo, M.A. Sanromán, Influence of operational parameters on electro-Fenton degradation of organic pollutants from soil. J. Environ. Sci. Health A **44**, 1104–1110 (2009)
26. E. Mousset, D. Huguenot, E.D. Van Hullebusch, N. Oturan, G. Guibaud, G. Esposito, M.A. Oturan, Impact of electrochemical treatment of soil washing solution on PAH degradation efficiency and soil respirometry. Environ. Pollut. **211**, 354–362 (2016b)
27. D. Huguenot, E. Mousset, E.D. van Hullebusch, M.A. Oturan, Combination of surfactant enhanced soil washing and electro-Fenton process for the treatment of soils contaminated by petroleum hydrocarbons. J. Environ. Manag. **153**, 40–47 (2015)
28. P. Westerhoff, G. Aiken, G. Amy, J. Debroux, Relationships between the structure of natural organic matter and its reactivity towards molecular ozone and hydroxyl radicals. Water Res. **33**, 2265–2276 (1999)
29. M. Panizza, G. Cerisola, Direct and mediated anodic oxidation of organic pollutants. Chem. Rev. **109**, 6541–6569 (2009)
30. M. Panizza, P.A. Michaud, G. Cerisola, C.H. Comninellis, Anodic oxidation of 2-naphthol at boron-doped diamond electrodes. J. Electroanal. Chem. **507**, 206–214 (2001a)
31. E. Brillas, I. Sirés, C. Arias, P.L. Cabot, F. Centellas, R.M. Rodríguez, J.A. Garrido, Mineralization of paracetamol in aqueous medium by anodic oxidation with a boron-doped diamond electrode. Chemosphere **58**, 399–406 (2005)
32. A. Özcan, Y. Sahin, A.S. Koparal, M.A. Oturan, Propham mineralization in aqueous medium by anodic oxidation using boron-doped diamond anode: Influence of experimental parameters on degradation kinetics and mineralization efficiency. Water Res. **42**, 2889–2898 (2008)
33. C. Trellu, O. Ganzenko, S. Papirio, Y. Pechaud, N. Oturan, D. Huguenot, E.D. van Hullebusch, G. Esposito, M.A. Oturan, Combination of anodic oxidation and biological treatment for the removal of phenanthrene and Tween 80 from soil washing solution. Chem. Eng. J. **306**, 588–596 (2016c)
34. M. Muñoz-Morales, M. Braojos, C. Sáez, P. Cañizares, M.A. Rodrigo, Remediation of soils polluted with lindane using surfactant-aided soil washing and electrochemical oxidation. J. Hazard. Mater. **339**, 232–238 (2017)
35. S. Cotillas, L. Cañizares, M. Muñoz, C. Sáez, P. Cañizares, M.A. Rodrigo, Is it really important the addition of salts for the electrolysis of soil washing effluents? Electrochim. Acta **246**, 372–379 (2017)
36. E. Brillas, M. Angel Banos, M. Skoumal, P. Lluis Cabot, J. Antonio Garrido, R. Maria Rodriguez, Degradation of the herbicide 2,4-DP by anodic oxidation, electro-Fenton and photoelectro-Fenton using platinum and boron-doped diamond anodes. Chemosphere **68**, 199–209 (2007)
37. N. Oturan, E. Brillas, M.A. Oturan, Unprecedented total mineralization of atrazine and cyanuric acid by anodic oxidation and electro-Fenton with a boron-doped diamond anode. Environ. Chem. Lett. **10**, 165–170 (2012)
38. H. Zazou, N. Oturan, M. Sönmez-Çelebi, M. Hamdani, M.A. Oturan, Mineralization of chlorobenzene in aqueous medium by anodic oxidation and electro-Fenton processes using Pt or BDD anode and carbon felt cathode. J. Electroanal. Chem. **774**, 22–30 (2016)
39. E.V. Santos, C. Sáez, C.A. Martínez-Huitle, P. Cañizares, M.A. Rodrigo, Removal of oxyfluorfen from ex-situ soil washing fluids using electrolysis with diamond anodes. J. Environ. Manag. **171**, 260–266 (2016)

40. C. Trellu, N. Oturan, Y. Pechaud, E.D. van Hullebusch, G. Esposito, M.A. Oturan, Anodic oxidation of surfactants and organic compounds entrapped in micelles – selective degradation mechanisms and soil washing solution reuse. Water Res. **118**, 1–11 (2017)
41. C. Saez, R. Lopez-Vizcaino, P. Canizares, M.A. Rodrigo, Conductive-diamond electrochemical oxidation of surfactant-aided soil-washing effluents. Ind. Eng. Chem. Res. **49**, 9631–9635 (2010)
42. E.V.D. Santos, C. Sáez, C.A. Martínez-Huitle, P. Cañizares, M.A. Rodrigo, The role of particle size on the conductive diamond electrochemical oxidation of soil-washing effluent polluted with atrazine. Electrochem. Commun. **55**, 26–29 (2015a)
43. E.V. Santos, C. Sáez, C.A. Martínez-Huitle, P. Cañizares, M.A. Rodrigo, Combined soil washing and CDEO for the removal of atrazine from soils. J. Hazard. Mater. **300**, 129–134 (2015b)
44. P.T. Almazán-Sánchez, S. Cotillas, C. Sáez, M.J. Solache-Ríos, V. Martínez-Miranda, P. Cañizares, I. Linares-Hernández, M.A. Rodrigo, Removal of pendimethalin from soil washing effluents using electrolytic and electro-irradiated technologies based on diamond anodes. Appl. Catal. B Environ. **213**, 190–197 (2017)
45. M. Rodríguez, M. Muñoz-Morales, J.F. Perez, C. Saez, P. Cañizares, C.E. Barrera-Díaz, M.A. Rodrigo, Toward the development of efficient electro-Fenton reactors for soil washing wastes through microfluidic cells. Ind. Eng. Chem. Res. **57**, 10709–10717 (2018)
46. E.V. Santos, C. Sáez, P. Cañizares, M.A. Rodrigo, C.A. Martínez-Huitle, Coupling photo and Sono technologies with BDD anodic oxidation for treating soil-washing effluent polluted with atrazine. J. Electrochem. Soc. **165**, 262–267 (2018)
47. M.B. Carboneras, P. Cañizares, M.A. Rodrigo, J. Villaseñor, F.J. Fernandez-Morales, Improving biodegradability of soil washing effluents using anodic oxidation. Bioresour. Technol. **252**, 1–6 (2018)
48. A. Kapałka, G. Fóti, C. Comninellis, The importance of electrode material in environmental electrochemistry. Formation and reactivity of free hydroxyl radicals on boron-doped diamond electrodes. Electrochim. Acta **54**, 2018–2023 (2009)
49. E.V. Santos, C. Sáez, P. Cañizares, D.R. Silva, C.A. Martínez-Huitle, M.A. Rodrigo, Treatment of ex-situ soil-washing fluids polluted with petroleum by anodic oxidation, photolysis, sonolysis and combined approaches. Chem. Eng. J. **310**, 581–588 (2017a)
50. E.V. Santos, C. Sáez, P. Cañizares, C.A. Martínez-Huitle, M.A. Rodrigo, Treating soil-washing fluids polluted with oxyfluorfen by sono-electrolysis with diamond anodes. Ultrason. Sonochem. **34**, 115–122 (2017b)
51. K. Chair, A. Bedoui, N. Bensalah, C. Sáez, F.J. Fernández-Morales, S. Cotillas, P. Cañizares, M.A. Rodrigo, Treatment of soil-washing effluents polluted with herbicide oxyfluorfen by combined biosorption-electrolysis. Ind. Eng. Chem. Res. **56**, 1903–1910 (2017a)
52. K. Chair, A. Bedoui, N. Bensalah, F.J. Fernández-Morales, C. Sáez, P. Cañizares, M.A. Rodrigo, Combining bioadsorption and photoelectrochemical oxidation for the treatment of soil-washing effluents polluted with herbicide 2,4-D. J. Chem. Technol. Biotechnol. **92**, 83–89 (2017b)
53. E.V. Santos, C. Sáez, P. Cañizares, C.A. Martínez-Huitle, M.A. Rodrigo, UV assisted electrochemical technologies for the removal of oxyfluorfen from soil washing wastes. Chem. Eng. J. **318**, 2–9 (2017c)
54. E. Mousset, Y. Pechaud, N. Oturan, M.A. Oturan, Charge transfer/mass transport competition in advanced hybrid electrocatalytic wastewater treatment: development of a new current efficiency relation. Appl. Catal. B Environ. **240**, 102–111 (2019)
55. M. Panizza, P.A. Michaud, G. Cerisola, C.H. Comninellis, Electrochemical treatment of wastewaters containing organic pollutants on boron-doped diamond electrodes: Prediction of specific energy consumption and required electrode area. Electrochem. Commun. **3**, 336–339 (2001b)
56. S. Bebelis, K. Bouzek, A. Cornell, M.G.S. Ferreira, G.H. Kelsall, F. Lapicque, C. Ponce de León, M.A. Rodrigo, F.C. Walsh, Highlights during the development of electrochemical engineering. Chem. Eng. Res. Des. **91**, 1998–2020 (2013)

57. F.L. Souza, C. Saéz, M.R.V. Lanza, P. Cañizares, M.A. Rodrigo, Is it worth the use of bipolar electrodes in electrolytic wastewater treatment processes? Chem. Eng. J. **264**, 310–315 (2015)
58. L.H. Tran, P. Drogui, G. Mercier, J.F. Blais, Comparison between Fenton oxidation process and electrochemical oxidation for PAH removal from an amphoteric surfactant solution. J. Appl. Electrochem. **40**, 1493–1510 (2010)
59. C. Sáez, R. López-Vizcaíno, P. Canizares, M.A. Rodrigo, Conductive-diamond electrochemical oxidation of surfactant-aided soil-washing effluents. Ind. Eng. Chem. Res. **49**, 9631–9635 (2010)
60. J.F. Pérez, J. Llanos, C. Sáez, C. López, P. Cañizares, M.A. Rodrigo, A microfluidic flow-through electrochemical reactor for wastewater treatment: a proof-of-concept. Electrochem. Commun. **82**, 85–88 (2017)
61. C. Trellu, C. Coetsier, J.C. Rouch, R. Esmilaire, M. Rivallin, M. Cretin, C. Causserand, Mineralization of organic pollutants by anodic oxidation using reactive electrochemical membrane synthesized from carbothermal reduction of $TiO_2$. Water Res. **131**, 310–319 (2018a)
62. C. Trellu, B.P. Chaplin, C. Coetsier, R. Esmilaire, S. Cerneaux, C. Causserand, M. Cretin, Electro-oxidation of organic pollutants by reactive electrochemical membranes. Chemosphere **208**, 159–175 (2018b)
63. O. Scialdone, C. Guarisco, A. Galia, G. Filardo, G. Silvestri, C. Amatore, C. Sella, L. Thouin, Anodic abatement of organic pollutants in water in micro reactors. J. Electroanal. Chem. **638**, 293–296 (2010)
64. E. Mousset, M. Puce, M.N. Pons, Advanced electro-oxidation with boron-doped diamond for acetaminophen removal from real wastewater in a microfluidic reactor – Kinetics and mass transfer studies. ChemElectroChem **6**, 2908–2916 (2019)
65. S. Cotillas, C. Sáez, P. Cañizares, I. Cretescu, M.A. Rodrigo, Removal of 2,4-D herbicide in soils using a combined process based on washing and adsorption electrochemically assisted. Sep. Purif. Technol. **194**, 19–25 (2018)

# Electrokinetic Soil Flushing

**Claudio Cameselle, Susana Gouveia, and Adrián Cabo**

## 1 Introduction

Human activities have led to contamination of air, waterbodies, soil, and groundwater. The contamination of soil and groundwater is mainly associated with mining activities, industry, the use of chemicals in agriculture and the lack of proper management of municipal, industrial, and hazardous waste. The contamination of soil has received less attention than the contamination of other media; however, it is a serious problem that affects ecosystems, public health, and the economic activities associated with the use of soil (agriculture, cattle rising, residential, recreational areas, etc.). The European Union was paying attention to the problem of soil contamination, and it has issued regulations to protect the soil and to restore the contaminated sites. The European regulation stressed the need to adopt measures to prevent, limit, and reduce the impact of the human activities in soil [1, 2]. Moreover, it is necessary to develop feasible technologies for the remediation of contaminated sites.

The remediation of contaminated sites requires the application of physical, chemical, and/or biological processes to separate, remove, degrade, or eliminate the contaminants. Since late 1980s and early 1990s of the twentieth century, various innovative soil remediation technologies were developed and tested [3]. Despite the research and development during about 30 years, there is not still a reliable technology for the remediation of contaminated sites. This is probably due to the complex geochemical interactions among soil components and contaminants. In this context, the electrochemical remediation of contaminated soils has been proposed as a new technology with the capacity of removing organic and inorganic contaminants, even in low permeability soils [4]. The studies at laboratory and field scale have proved

C. Cameselle (✉) · S. Gouveia · A. Cabo
BiotecnIA, Department of Chemical Engineering, University of Vigo, Vigo, Spain
e-mail: claudio@uvigo.es; gouveia@uvigo.es; acabo@uvigo.es



M. A. Rodrigo, E. V. Dos Santos (eds.), *Electrochemically Assisted Remediation of Contaminated Soils*, Environmental Pollution 30,
https://doi.org/10.1007/978-3-030-68140-1_6

that electrokinetics is a practical technology for the remediation of contaminated soils, sediments, and sludge, especially for the removal of inorganic contaminants, such as heavy metals, metalloids, and inorganic anions. The remediation of contaminated sites with organic pollutants is more complex due to the hydrophobicity of most of the common organics. However, the operation of electrokinetics in the appropriate conditions with the use of facilitating agents may result in an effective removal and degradation of organic contaminants. The objective of this chapter is to present the scientific and technical bases of electrokinetic remediation and to give an overview of the capacity of this technology for the remediation of organic contaminants.

## 2 Basis of Electrokinetic Remediation

Electrokinetic remediation is an in situ technology specially designed and developed for the restoration of contaminated soils, sediments, and sludge. The electrokinetic process relies on the application of a low-intensity DC electric field directly to the soil to be remediated. The electric current induces the mobilization of the contaminants and their transportation toward the main electrodes, anode and cathode. The electrodes are commonly installed in situ in a well filled with a processing fluid, typically water with chemicals that favor the removal of contaminants. The processing fluid is pumped out of the well and treated to remove or eliminate the contaminants. The processing fluid is recycled back to the well. Figure 1 depicts the in situ application of electrokinetics in a contaminated site [4].

The electrokinetic treatment of a contaminated soil is basically a separation process. The electric field induces the mobilization and transportation of the contaminants by two main transport mechanisms: electromigration and electroosmosis. Electromigration is the transportation of ions toward the electrode of opposite charge. Cations, such as $Cu^{2+}$, $Pb^{2+}$, $Na^{+}$, etc. will be transported toward the cathode (the negative electrode) and anions, such as $CrO_4^{2-}$, $F^-$, $SO_4^{2-}$, etc. will be transported toward the anode (the positive electrode). Electroosmosis is the net flux of water through the soil induced by the electric field. The electroosmotic flow is the result of the interaction of ions in the interstitial fluid and the charged solid surface of the soil particles. These particles are usually negatively charged, and as a result, the electroosmotic flow goes from anode (−) to cathode (+). All the soluble contaminants in the interstitial fluid (water) can be removed from soil by electroosmosis, including ionic and non-ionic contaminants. The electrokinetic studies have proved that electromigration is the main transport mechanism for ionic contaminants (heavy metals, inorganic anions, etc.), whereas electroosmosis is more effective removing non-ionic contaminants (organic compounds) from the soil [4].

The application of the electric current to a soil specimen also induces chemical reactions upon the main electrodes but also in the mass of soil. These reactions include solubilization, precipitation, neutralization, and redox reaction. The main reaction is the electrolysis of water upon the electrodes: oxidation of water in the

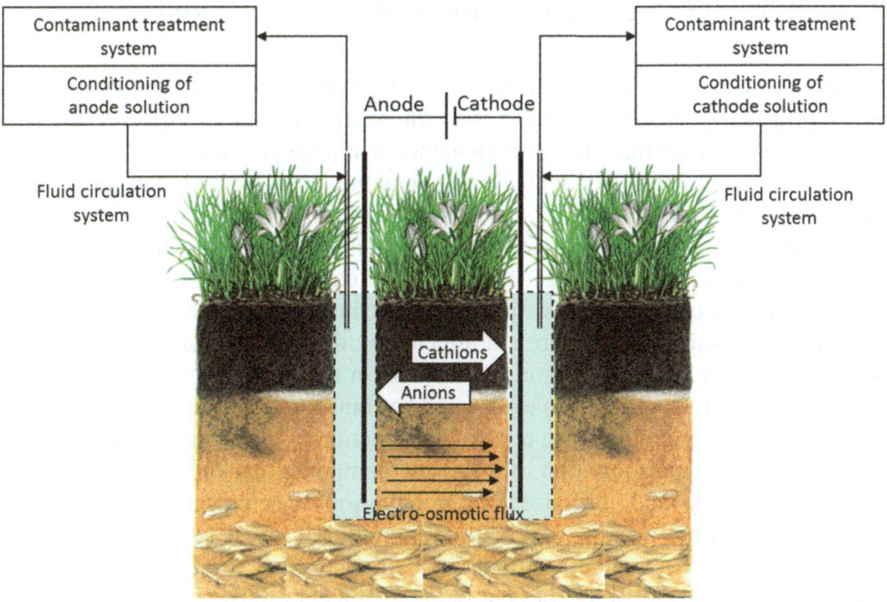


**Fig. 1** Schematic representation of an application of the electrokinetic remediation in a contaminated site

anode and reduction of water in the cathode (Eqs. 1 and 2). These reactions have an enormous influence in the solubilization and speciation of the contaminants because the hydronium ions generated at the anode and the hydroxyl ions generated in the cathode are transported through the soil modifying the soil pH. Any change in the pH of the soil and interstitial fluid affects the desorption of contaminants, the precipitation of metals, the ionization of organics, and in general, the speciation of the contaminants. The electrolysis of water is inevitable in a water–soil system, but the operation conditions can be adjusted to favor the pH conditions in the soil that enhance the mobilization of the target contaminants [4]. As an example, Ricart et al. [5] favored the acidification of soil specimen to enhance the mobilization of Mn from soil as $Mn^{2+}$ and its transportation toward the cathode. On the contrary, Ottosen et al. [6] used ammonia to increase the pH of the soil specimen and mobilize the contaminant copper as $[Cu(NH_3)_4]^{2-}$. These studies demonstrate that the understanding of the geochemistry of soils and contaminants is of utmost importance in electrokinetic remediation.

$$2H_2O \rightarrow 4e^- + 4H^+_{(aq)} + O_{2(gas)} \qquad E^0 = -1.229V \tag{1}$$

$$4H_2O + 4e^- \rightarrow 2H_{2(gas)} + 4OH^-_{(aq)} \qquad E^0 = -0.828V \tag{2}$$

## 3 Removal of Organic Contaminants by Electrokinetics

Electrokinetic remediation was initially tested for the restoration of soils contaminated with heavy metals and other ionic inorganic contaminants; however, later studies have proved that the electrokinetic treatment can also be satisfactorily applied to soils contaminated with organic contaminants. The main transport mechanism for ionic contaminants is electromigration, especially at high contaminant concentrations. At low concentrations, electroosmosis may also play a significant role in the removal of ionic contaminants. In the case of organic contaminants, the main transport mechanism is electroosmosis because most of the organic contaminants are non-ionic, and they are not affected by the presence of an electric field. Some organic molecules are ionic or ionizable in the soil under the electrokinetic treatment conditions. These molecules can be transported through the soil porous matrix by electromigration. The relative contribution of electromigration and electroosmosis to the transportation of a specific compound depends on the chemical nature of the compound, its concentration in the soil, the characteristics of soil and water content. It has been proved that ion migration if about 10–300 times higher than electroosmosis. In the case of non-ionic organics, they can be removed by soil flushing in high permeability soil, but in low permeable soils, the hydraulic flow is negligible. It is in this case where electroosmosis plays a prominent role in the transportation of non-ionic contaminants. The operating condition of the electrokinetic treatment must be adjusted to favor and maintain a high electroosmotic flow in the soil. However, electroosmosis is very much affected by the physicochemical conditions of soil, ionic concentration in the interstitial fluid, and pH. The acidification of soil in the electrokinetic treatment operated at constant electric potential provokes the drop in the soil zeta potential and the decrease of electroosmotic flow.

The capacity of electrokinetic remediation to remove organic contaminants has been tested with a variety of organic compounds of environmental concern. These compounds include hydrophobic and toxic organics such as polycyclic aromatic hydrocarbons (PAHs), polychlorinated organic compounds (PCBs), pesticides, herbicides, and energetic compounds. Considering that the most toxic, persistent, and dangerous organic contaminants are non-ionic or non-ionizable, electromigration does not play a role in the removal of these compounds, and only electroosmosis is able to remove non-ionic organic contaminants. However, the mass transport by electroosmosis implies the solubilization of organics in the interstitial fluid, water. The solubility in water of most of organic contaminants of concern is very low or even completely insoluble. The removal of organics by electroosmosis requires the addition of facilitating agents that enhance the solubility of the organic contaminants in water. This facilitating agents includes surfactants, co-solvents, cyclodextrins, and others. Other approaches to achieve an effective removal of elimination of organics imply the combination of electrokinetics with other remediation technologies such as chemical oxidation, chemical reduction, permeable reactive barriers, or electrolytic reactive barriers [4].

# 4 Electrokinetic Removal of Soluble Organics

The dye reactive black 5 is an organic compound commonly used in the textile industry and dyeing processes. This compound is toxic for living organisms, and it can be found in the discharge effluents of the textile industry. Once released to the environment, this compound can be adsorbed and retained in soils and sediments. The chemical structure of the reactive black 5 (Fig. 2) shows a complex molecule of an azo dye with four aromatic rings and four sulfonic groups. The complex structure of the azo dye explains why this molecule is difficult to degrade, as well as its negative effect in the environment that may last for long time. The sulfonic groups confers to the molecule the necessary polar characteristics to explain its solubility in water. Overall, reactive black is a toxic compound, difficult to biodegrade, and it is soluble in water, which increases its bioavailability and toxicity.

The removal of reactive black 5 from soils and sediments by electrokinetics is quite challenging because it is necessary to desorb the molecule from the solid matrix (soil or sediment) and then transport the molecule by electromigration or electroosmosis to be accumulated in the electrode chambers. The desorption of reactive black 5 from soil can be achieved using potassium sulfate in the processing fluid in anode and cathode. Potassium sulfate is transported into the soil specimen by electromigration and electroosmosis. The ionic exchange of $K^+$ with reactive black molecules allows for desorption of the dye from the solid matrix. Once the molecule of reactive black 5 was in solution in the interstitial fluid, it can be transported by electroosmosis and electromigration toward the electrode chambers. The pH of the soil (and the interstitial fluid) is also important in the speciation of the molecule and its transportation by electrokinetics. As it is shown in Fig. 2, reactive black 5 has four sulfonic groups that confer to the molecule the characteristics of a weak acid. In neutral or acid environment, reactive black 5 remains as a neutral molecule, but in alkaline environment, the molecule is ionized forming an anion with four negative charges. The different speciation of reactive black 5 in acid or alkaline pH can be used to favor its transportation by electromigration or electroosmosis. When the pH is alkaline, the anion of reactive black 5 can be transported by electromigration toward the anode. In neutral acid environment, reactive black 5

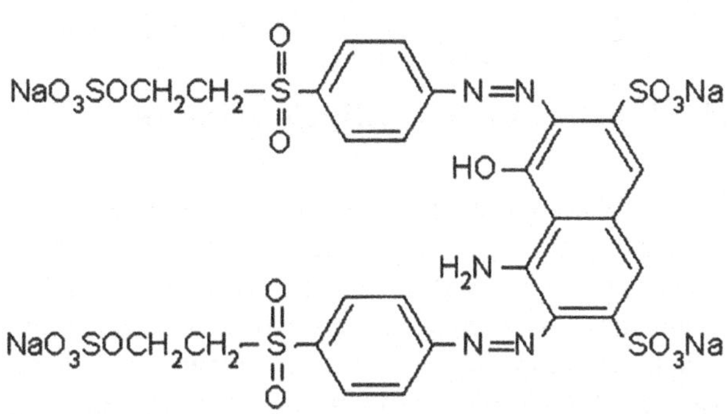


**Fig. 2** Chemical structure of reactive black 5

remains as neutral molecule, and only electroosmosis could be effective in its removal from soil.

The pH of the soil specimen can be modified by the electrokinetic process. The electrolysis of water tends to acidify the pH close to the anode and alkalinize the soil close to cathode. If the reduction of water in the cathode is suppressed by the controlled addition of an acid, the soil is acidified by the hydronium ions electrogenerated in the anode. Conversely, the addition of sodium hydroxide in the anode suppress the formation of $H^+$ ions, and the hydroxyl ions generated in the cathode electro-migrates through the soil specimen increasing the pH. In the study for the electrokinetic removal of reactive black 5 [7] very different results were obtained depending on the soil pH. The direct application of the electric field with no pH control in the electrode chambers resulted in no significant removal of reactive black 5 from soil. The dye remained adsorbed to the soil. In a second experiment, the $OH^-$ ions in the cathode were neutralized by the controlled addition of sulfuric acid. The acid pH in the soil kept the reactive black 5 as a neutral molecule. No transportation or removal of reactive black 5 was observed in these conditions. The molecule is too big to be transported through the porous matrix of kaolinite by electroosmosis, and it probably remained adsorbed to the soil in acidic pH conditions. However, the electrokinetic treatment with the neutralization of the acid environment with the addition of NaOH in the anode resulted in a complete removal of reactive black 5 from soil. In these conditions, the soil pH was alkaline due to the electromigration of $OH^-$ generated in the cathode. The alkaline environment in the soil favored the desorption of reactive black 5 (in the presence of potassium sulfate). However, the most important effect of the alkaline pH is the ionization of the molecule of reactive black 5. At pH higher than 7, reactive black 5 forms an anion with four negative charges than can be electromigrated toward the cathode. Figure 3 shows the electromigration of reactive black 5 toward the anode. The transportation of the dye is evident in the picture sequence. After 5 days of treatment, all the dye was accumulated in the anode chamber. Then, reactive black 5 was degraded by anodic oxidation. This study combined electrokinetic transport and electrochemical oxidation to remove reactive black from soil and its degradation in the anodic solution. This is a good example of the strategic use of chemistry and electrokinetic transport to achieve a complete remediation of the contaminated soil and the degradation of the contaminant simultaneously in the same experimental setup.

## 5 Electrokinetics with Co-solvents

The main limitation of the electrokinetic remediation of soils with organic contaminants is the low solubility of the contaminants in water. Hydrocarbons, trichloroethylene, pesticides, and energetic compounds usually show very low solubility in water that makes difficult the removal of these compounds in contaminated soils. The transport mechanisms in electrokinetics are electromigration and electroosmosis, and both mechanisms require that the contaminants are in solution in the

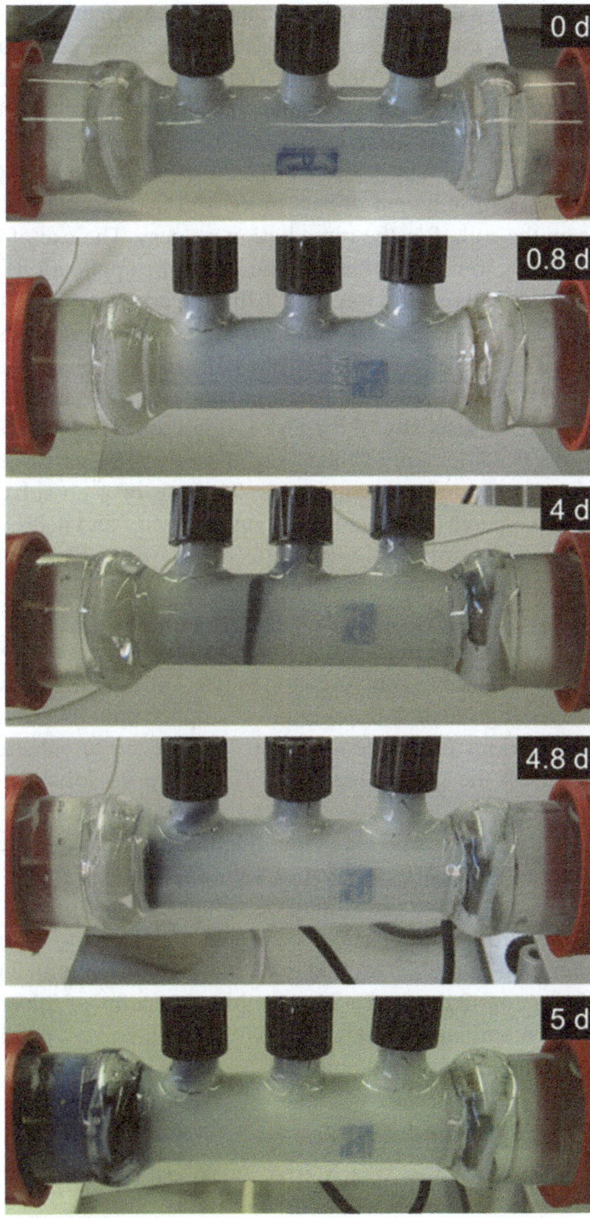


**Fig. 3** Electrokinetic remediation of kaolin specimen contaminated with reactive black 5

interstitial fluid. The fluid in the pores of soil is always water in in situ applications. As a result, the remediation of hydrophobic organics by electrokinetics needs the addition of facilitating agents that increase the solubility of the target contaminant in the interstitial fluid. One possibility to enhance the solubility of organics is the use of a co-solvent in the processing fluid.

The co-solvent for the enhanced electrokinetic remediation of organics in soil need to be carefully selected based on technical, environmental, and legal aspects. The co-solvent has to show an important solubility toward the target contaminant to assure fast and effective desorption and solubilization of the contaminants. Moreover, the co-solvent has to be miscible with water since water is always present in soil in in situ applications. The co-solvent must be safe for the environment, with minor environmental impact in the soil and groundwater, and it must be recovered from the soil after the remediation process. The use of co-solvents with water also provokes some technical limitations in the application of electrokinetics. The solubility of salts in the interstitial fluid decreases due to the presence of organic co-solvents, so the electric conductivity of the soil decreases too with the subsequent impact in the electrokinetic transport of the contaminants in soil. The co-solvent also affects the interaction between the soil surface and the interstitial fluid and may change the viscosity of the processing fluid. These two aspects have a major impact in the development of the electroosmotic flow, which is the main transport mechanism for organic contaminants. The use of a co-solvent may have two opposite effects: increases the contaminant solubility and decreases electroosmosis, and the combined result of the two effects may even be negative for the removal of the contaminants.

Various studies in the literature have tested the use of co-solvents for the removal of phenanthrene, an aromatic polycyclic hydrocarbon widely studied as a persistent and hydrophobic contaminant [8–10]. The electrokinetic treatment using water as a processing resulted in no removal of phenanthrene, independently of the pH of the soil. Various organic co-solvents were tested to enhance the solubility of phenanthrene in the processing fluid. The tested co-solvents include ethanol, n-butanol, n-butylamine, tetrahydrofuran, or acetone. The mobilization and removal of phenanthrene from soil were evident with the use of co-solvents, specially n-butylamine that resulted in the removal of 43% of phenanthrene in 127 days from a model soil specimen. The removal may be enhanced avoiding the acidification of soil due to the electrolysis of water in the anode, or enhancing the electroosmotic flow with the operation of higher voltage gradient (2 V/cm) or with a periodic voltage application (5 days on: 2 days off). The combined benefits of the enhanced electroosmosis and the contaminant solubilization with the co-solvent resulted in the effective removal of phenanthrene from soil [11].

## 6 Enhancing Solubility with Surfactants

The solubility of hydrophobic organics in soil remediation may be enhanced with the use of surfactants in the processing fluid in the electrode wells or chambers. The electrokinetic transport phenomena introduce the surfactants in the soil solubilizing the organic contaminants. Then, the solubilized organics can be transported out of the soil by electroosmosis. The combination of the electroosmotic flow and the solubilization of organics with surfactants is a practical approach for the electrokinetic removal of organic contaminants.

Surfactants are a group of chemical compounds with the capacity to modify the surface tension of water. The interest of using surfactants in soil remediation is their capacity to decrease the interfacial tension of water, increasing the solubility in water of hydrophobic organic contaminants. The molecules of surfactants include a hydrophilic group in one end of the molecule and a hydrophobic group in the opposite end. Surfactants are soluble in water due to the activity of the hydrophilic group. At the same time, the hydrophobic group assures the interaction with the organic contaminants in soil. Surfactants in water tend to form spherical structures called micelles. The micelles are formed with the surfactant molecules oriented with the hydrophilic group to the external part of the sphere and the hydrophobic group oriented to the inner part of the sphere. The inner space in the sphere is a hydrophobic environment very appropriate for the solubilization of the organic contaminants. Surfactants only form micelles when the concentration of surfactant in the interstitial fluid reach a specific critical micelle concentration (CMC). It means that the dose of surfactant to the soil needs to be adjusted to reach that specific CMC; lower concentrations are not appropriate for the effective solubilization of the organic contaminants [10].

Surfactant compounds can be classified into four groups considering the electric charge in the molecule. The four groups are neutral, cationic, anionic and zwitterionic. The latter are molecules that include positive and negative charges at the same time in the chemical structure of the molecule. In general, cationic surfactants are not effective in soil remediation application due to the electronegativity of the soil particles. The cationic surfactants tend to interact with soil particles, lowering the mass transportation and, therefore, their efficiency in contaminant solubilization. The most common surfactant in soil remediation are anionic and cationic molecules. Zwitterionic surfactants were also tested in soil remediation. The most important factor in the selection of a surfactant for an in situ application, apart from the contaminant solubilization capacity, is the toxicity for the soil microflora. This is the reason why the most interesting surfactants for soil remediation are natural compounds or biosurfactants. These compounds shows minor environmental impact, and they are easy to degrade after the remediation process [12].

Some examples of electrokinetic studies of contaminated soils with hydrophobic organics includes compounds such as phenanthrene, DDT, diesel hydrocarbons, dinitrotoluene, hexachlorobenzene, and others. These compounds were mobilized with the use of various surfactants: Sodium dodecyl sulfate (SDS), Brij 35, Tween 80, Igepal CA-720, Tergitol, and others. The removal efficiency of the target contaminants in electrokinetic applications with water as processing fluid was negligible. The contaminants remained adsorbed to the soil particles and were not mobilized in the testing conditions. However, the use of surfactants increases the removal efficiency over 80% at lab scale tests with various model and real soils [13, 14]. Phenanthrene was the target contaminant in a study by Reddy and Saichek [9]. Phenanthrene is a low solubility organic compound classified as "acute toxicity, category 4" in the globally harmonized system. In the unenhanced electrokinetic treatment using water as processing fluid, an important electroosmotic flow was registered. Despite the large electroosmotic flow, there was no removal of phenanthrene. Three surfactants were used to improve the solubility of phenanthrene in the interstitial fluid; Tween 80, Witconol, and Igepal CA-720. The addition of the surfactants in the interstitial fluid resulted in a significant decreasing of the electroosmotic flow. This decreasing was due to the increasing viscosity of the interstitial fluid, the different interaction with the soil particle surface and the lower electric conductivity. However, the removal of phenanthrene in the cathodic solution clearly increased despite the lower electroosmotic flow. This was due to the important solubilization of the phenanthrene in the interstitial fluid. Other variables that affect the solubilization and removal of phenanthrene are the soil pH, the ionic strength of the interstitial fluid, and the geochemical characteristics of soil. These variables mainly affect the development of the electroosmotic flow. The acidification of soil by the $H^+$ ions electrogenerated at the anode tend to decrease and even suppress the electroosmotic flow. The acidification of soil is only important in soils with low buffering capacity. In any case, the use of a buffering solution in the anode chamber or the controlled addition of NaOH in the anode may help to avoid the acidification of soil [10, 11]. Overall, the effective removal of phenanthrene may be achieved by the combination of a surfactant and maintaining a high electroosmotic flow. Various strategies were tested to increase the electroosmotic flow: the operation at higher voltage gradients (2 V/cm) or use a periodic application of the voltage gradient, with 5 days on and 2 days off. The off time was used to let reactions to occur in the soil sample. The periodic voltage application resulted in about 90% of the phenanthrene removed on the cathode solution [11].

## 7 Selective Complexation with Cyclodextrins

Cyclodextrins are a family of compounds with a special structure in the form of a truncated cone. They are composed of glucose units, forming three different cyclodextrins namely α-cyclodextrin (6 units of glucose), β-cyclodextrin (7 units of glucose), and γ-cyclodextrin (8 units of glucose). These compounds are soluble in water due to the hydrophilic interaction of the –OH groups on both the ends of the cone. At the same time, the inner cavity of the cyclodextrin molecule shows a hydrophobic behavior that is very appropriate to allocate non-polar and hydrophobic organic contaminants. The size of the inner cavity of cyclodextrin depends on the number of glucose units that form the molecule. The inner cavity of the α-cyclodextrin is 0.45–0.53 nm, β-cyclodextrin is 0.60–0.65 nm, and γ-cyclodextrin is 0.75–0.85 nm. The cyclodextrins show a selective behavior based on the size of the organic molecule that can be allocated in the inner cavity. The cyclodextrin used in each application can be selected based on the target contaminant and some selective solubilization and removal can be observed based on the size of the contaminant molecules.

Various studies have used cyclodextrins in soil remediation applications for the removal of contaminants such as phenanthrene [15], dinitrotoluene [16], the herbicide atrazine [17], and other contaminants [18] in kaolinite-spiked specimens and real soil samples. The use of cyclodextrins resulted in better removal of the target contaminants compared with the unenhanced electrokinetic treatment. However, the removal results with cyclodextrins are usually lower than that with surfactants, or other remediation technologies (zero iron nanoparticles, in situ chemical oxidation, etc.). The activity of cyclodextrins can be enhanced with their combination with other removal enhancing options. As an example, the combined use of cyclodextrins and ultrasounds [19] and cyclodextrins and chemical oxidation with hydrogen peroxide [20] was tested with a significant improvement in the remediation results of hexachlorobenzene and phenanthrene in kaolinite model soils. An additional aspect to be considered in the use of cyclodextrins is the high cost of these compounds and the difficulty in their recovery from soil in in situ applications. Overall, the limited capacity for organics removal, compared to other remediating alternatives, and the high cost of the compound, result in not good perspectives in contaminated site applications.

## 8 Polycyclic Aromatic Hydrocarbons

Polycyclic aromatic hydrocarbons (PAH) have received a lot of attention in soil remediation studies. These compounds show a structure based on the condensation of aromatic rings. The most common PAHs studied were phenanthrene and anthracene. These compounds are insoluble and tend to absorb in the soil. Due to these characteristics, PAH are not be removed by electrokinetics in an enhanced test using water as processing fluid. The strategies to improve the electrokinetic removal include the use of facilitating agents that increase the solubility of PAHs in the processing fluid. The most common facilitating agents are surfactants, biosurfactants, co-solvents, and cyclodextrins. As a result, PAHs can be removed from soil based on the combined effect of the solubilization by the facilitating agents and the transportation toward the cathode by electroosmosis. However, the use of facilitating agents (surfactants, co-solvents, etc.) also affects the physicochemical properties of the processing fluid and its interaction with the soil particle surface. These changes in the chemistry of soil have an enormous effect in the development and maintenance of the electroosmotic flow, and hence the removal results will be affected. The pH in the electrode chambers and in the soil must be monitored to avoid the soil acidification that tend to suppress the electroosmotic flow. The periodic voltage application or the use of higher potential gradients may help in the development of the electroosmotic flow. Overall, the combined effect of the electroosmotic flow and surfactants as solubilizing agent results in very effective remediation of soils contaminated with PAH [21].

## 9 Chlorinated Aliphatic Hydrocarbons, Chlorophenols, and Chlorobenzenes

This group of organic contaminants is characterized for its toxicity toward aquatic organisms and soil microflora. The most common representative of this group is the trichloroethylene (TCE); other common soil contaminants of this group are chlorinated aliphatic hydrocarbons: pentachlorethylene (PCE), trichloroacetate (TCA), and trichlorethylene (TCE); chlorophenols: pentachlorophenol; and chlorobenzenes: PCB (polychloro biphenyls). TCE is found as a soil contaminant due to the lack of proper management of TCE wastes. This compound is relatively more soluble in water (1.280 g/L of TCE) than other components of this group. This means that contaminant TCE in soil can be mobilized to contaminate ground water and surrounding areas. The removal of chlorinated organics from soil is difficult due to the tendency to adsorb in the soil. However, some components of this group can be dissociated, so electromigration also plays a role in the transportation during the electrokinetic treatment. The removal of chlorinated organics from soil by electrokinetics requires the use of solubilizing agents due to the low solubility of these compounds in water. The solubilizing agents are surfactants and co-solvents. Yuan and Weng [22] reach a complete removal of ethylbenzene with the use of SDS (sodium dodecylsulfate) as a surfactant in the anodic solution. The authors claimed that the surfactant aided electrokinetic treatment was cost-effective, and it can be considered as a suitable method for large-scale applications.

## 10 Herbicides and Pesticides

The continuous use of pesticides and herbicides in agriculture resulted in the contamination of many agricultural fields. Common contaminants found in agricultural soils are the pesticides: DDT, aldrin, dieldrin, and endrin; and the herbicides: atrazine, molinate, and bentazone. These compounds tend to adsorb to the soil. Therefore, it is not possible to remove these contaminants from soil by an unenhanced electrokinetic treatment using water as processing fluid. It is not possible to remove many of those pesticides by electromigration because they form neutral species, and electroosmosis is ineffective due to the negligible solubility in water. The typical approach to remove these contaminants imply the use of surfactants to desorb the contaminants from soil and the subsequent removal by electroosmosis. Suanon et al. [23] reported the effective removal of organochlorine pesticides from a historically contaminated soil. These authors claimed to remove 50% of DDT and 77% of hexachlorobenzene in 15 days using the surfactant Triton X-100 in the anodic solution. The electrokinetic removal of the herbicides, molinate and bentazone, from soil, was studied by Ribeiro et al. [24]. These two herbicides were removed from soil by a combination of electromigration and electroosmosis. Molinate was concentrated in the cathodic solution. Both transport mechanisms, electromigration and electroosmosis, contributed to molinate transportation. Conversely, bentazone was accumulated in the anode and in the cathode depending on the electrokinetic conditions. At high electric density conditions, the transport of bentazone was faster than electroosmosis and bentazone was accumulated in the anode chamber. At low electric density conditions, the transport by electroosmosis was more effective and bentazone was accumulated in the cathode compartment. These results proved the importance of the geochemical interactions of soil-interstitial fluid-contaminant and their effect on the transportation of the contaminants out of the soil.

## 11 Nitroaromatic Compounds

Manufacture and use of ammunition are the causes of the contamination of soil with a specific group of contaminants called energy compounds. In this group, the most relevant contaminating substances are nitroaromatic compounds (TNT, DNT, and RDX). These compounds show a good affinity for organic matter and clay minerals. Therefore, they tend to remain adsorbed to the soil. Moreover, these compounds show non-polar molecules and low solubility in water. Removal of these compounds from soil by electrokinetics requires the use of solvents or surfactants to enhance the solubility in the interstitial fluid. Kessler et al. [25] showed that removal of DNT can be enhanced with cyclodextrins (CD) and cyclodextrin derivatives such as carboxymethyl-β-CD, amino-β-CD, and hydroxypropyl-β-CD. Further research is required in this field to find suitable and effective solubilizing agents for a complete removal of energetic compounds in contaminated soils.

## 12 Mixtures of Heavy Metals and Organic Pollutants

Contaminated sites often contain a mixture of contaminants including heavy metals and organics. The remediation of these sites by electrokinetics is challenging due to the different physicochemical characteristics of the contaminants and the different behavior under electrokinetic test conditions [26]. In general, heavy metals and other inorganic contaminants form ions that can be transported by electromigration. Conversely, organic contaminants are usually neutral species and show very low solubility in water, and their transportation is mainly by electroosmosis. The pH conditions in soil play a key role in the solubilization and removal of heavy metals. Cationic metals ($Cd^{2+}$, $Cu^{2+}$, $Pb^{2+}$, etc.) can be solubilized at acidic pH favoring the advance of the acid front from the anode and neutralizing the alkaline environment of the cathode with the controlled addition of an acid. The penetration of $OH^-$ ions in the soil from the cathode may also be avoided with cationic exchange membranes. The use of complexing agents in the processing fluid is an alternative option to keep the metals in solution and avoid their precipitation in the alkaline environment close to the cathode. Anionic metals ($CrO_4^-$, etc.) can be mobilized at alkaline pH due to the reduced adsorption of the metal anions to the soil. On the other hand, organic contaminants require the addition of solubilizing agents to increase the solubility in water. The operating conditions of the electrokinetic treatment must be adjusted to favor the electroosmotic flow. The combined effect of the solubilizing agents and high electroosmotic flow results in the effective removal of the organic contaminants. However, the electroosmotic flow may be largely affected by the pH changes, especially the acidification in soil. When the pH of soil decreases and reaches acidic values, the electroosmotic flow sharply decreases and even reverses due to the change of the soil surface charge from negative to positive. As a result, the simultaneous removal of both heavy metals and organics is not always possible [27].

Various methods and processes have been developed for the sequential removal of organics and heavy metals from soils. Elektorowicz and Hakimipour [28, 29] developed the so-called SEKRIOP process that uses surfactants to dissolve hydrocarbons and EDTA to mobilize metals in the electrokinetic treatment of contaminated soils. This technology uses cationic exchange membranes in the cathode to retain free cationic metals transported by electromigration. The anionic complexes of heavy metals with EDTA are retained in anionic exchange membranes is the anode. This method resulted in very good removal ratios of heavy metals and hydrocarbons when used in model soils at lab scale. However, the SEKRIOP technology did not show such a good performance when processing actual soil specimens from contaminated sites with heavy metals and hydrocarbons. The research was then focused in improving the removal ratio, the treatment time, and the adequate management of the wastes from the electrode solutions [30].

## 13 Coupled Electrokinetic-Chemical Oxidation/Reduction

The coupled technology electrokinetic remediation with chemical oxidation/reduction is an interesting alternative to the treatment of contaminated soils with organic contaminants. The main limitation of organic contaminants is the low solubility in the processing fluid (water). That is the reason for the difficulty in removing organics by electrokinetics. The remediation by in situ chemical oxidation does not require the mobilization and transportation of the organic contaminants. In this technology, the contaminants are degraded in the soil by the action of chemical oxidants. The contaminants are transformed in smaller molecules, typically less toxic and harmful than the original contaminants. Eventually, the organic contaminants are completely oxidized to carbon dioxide and water. The limitation of chemical oxidation is the effective delivery of the oxidant in to the soil, especially in low permeability soils. It is exactly here where the combination of electrokinetics and chemical oxidation may result in a synergistic effect to achieve a fast remediation. In the coupled technology electrokinetic chemical oxidation, the electrokinetic transport phenomena are used to introduce into the soil the oxidants. Depending on the chemical nature of the oxidants, these reagents are dissolved in the anodic solution or the cathodic solution, and they are transported into the soil by electromigration or electroosmosis. As an example, permanganate and persulfate can be transported into the soil by electromigration from the cathodic solution, whereas hydrogen peroxide has to be added to the anodic solution and transported by electroosmosis into the soil because hydrogen peroxide is a neutral molecule. Other chemical reactants can be used for the degradation of the organic contaminants by a reductive chemical process to obtain less toxic organics. For example, the reductive dechlorination of organochloride pesticides or chlorinated solvents is based on the removal of chloride ions from the organic molecule, resulting in much less toxic products than can be degraded by monitored natural attenuation [31].

Yukselen-Aksoy and Reddy [32] tested the electrokinetic delivery of persulfate in a contaminated soil with PCB. Sodium persulfate is one of the common oxidizing agents used in environmental applications. The standard reduction potential of persulfate is 2.7 V, which assures the oxidation of organic contaminants. In this study, persulfate was added to the anodic solution, and it was transported into the soil by a combination of electromigration and electroosmosis. Persulfate needs to be activated to be able to degrade the organic contaminants. The activation in electrokinetic can be done by temperature or pH. Persulfate requires a minimum of 45 °C to be activated or a pH below 4. Those conditions can be achieved in electrokinetics adjusting the electric field intensity and favoring the advance of the acid from electrogenerated in the cathode. These authors reported that about 78% of PCB in the kaolinite soil specimen was degraded by persulfate activated by temperature and pH.

Oonnittan et al. [20, 33] tested the in situ chemical oxidation of hexachlorobenzene (HCB) contaminated soils. These authors used hydrogen peroxide as chemical reagent added to the anolyte solution. Hydrogen peroxide was transported into the soil by electroosmosis and attacked the organic contaminant in a Fenton-like process where the iron content in the soil was sufficient to activate the $H_2O_2$ for the generation of hydroxyl radicals ($^{\cdot}OH$). The Fenton reagent can be deactivated at alkaline pH, so it is important to maintain the pH of the soil below 7 to assure an effective reaction of the hydrogen peroxide over the HCB. At alkaline pH, $H_2O_2$ decomposes in water and oxygen and does not form $^{\cdot}OH$ radicals. In acid conditions, about 60% of HCB was eliminated from the soil in 10 days. The removal efficiency may be improved increasing the treatment time and controlling the soil pH in the optimum range for Fenton reaction (slightly acidic pH).

## 14 Nanoparticle Transport by Electrokinetics

The use of zero valent iron in environmental applications grew very fast due to its high capacity to reduce and degrade a variety of contaminants. One of the most interesting application of zero valent iron is the catalysis of reductive dechlorination of organic compounds such as trichloroethylene, pentachlorophenol, hexachlorobenzene, and others. The use of zero valent iron in the form of nanoparticles largely increases its activity and its application in environmental remediation. Reddy and Karri [34] tested the use of iron nanoparticles with electrokinetics to remove pentachlorophenol. In the coupled technology, the electrokinetic process is used as a driving force to introduce the nanoparticles in soil. The transportation of the nanoparticles is carried out by electroosmosis. The nanoparticles can be added to the anodic solution, but the oxidative environment in the anode may affect the stability of the nanoparticles. To avoid such effect, the addition of the nanoparticles is commonly done in the soil or in an additional chamber separated from the anode. In the study by Reddy and Karri [34], the advance of the nanoparticles through the soil specimen was followed by analyzing the increase in iron concentration. This method does not assure that the zero valent iron was transported as nanoparticles. The concentration of pentachlorophenol decreases at the end of the treatment by about 50% in the soil specimen. A complete removal of pentachlorophenol was found in the cathode side due to the combined effect of nanoparticles and reductive dechlorination. In order to improve the remediation results, it is necessary to favor the transportation of nanoparticles through the soil. The main limitation for an effective transportation of iron nanoparticles is their tendency to interact among them, to aggregate, and to settle very fast. Moreover, the high reactivity of iron nanoparticles results in their premature oxidation. Cameselle et al. [35] studied the zeta potential of iron nanoparticles and the influence of groundwater in the electrokinetic behavior. These authors have proposed the use of dispersants in the addition of iron nanoparticles to soil in order to avoid premature aggregation and settlement. Cameselle et al. [35] concluded that aluminum lactate showed good properties in the dispersion of iron nanoparticles in large-scale applications for the removal of organochloride contaminants. Other microscale and nanoscale particles, including bimetallic particles with copper and iron, or palladium and iron, were tested in the dechlorination of organic contaminants. Zheng et al. [36] proved the good activity of Cu/Fe nanoparticles in the complete removal of hexachlorobenzene, whereas the Pd/Fe nanoparticles showed only 60% removal in the study by Wan et al. [37].

## 15 Coupled Electrokinetic-Permeable Reactive Barriers

Permeable reactive barriers (PRB) have been satisfactorily used for the remediation of contaminant plumes in groundwater. PRB are a passive remediation system installed in the path of groundwater. Basically, a PRB is a trench in the path of groundwater filled with a reactive material. The contaminants in the groundwater are absorbed or react with the filling material of the PRB. The reactive filling material must be carefully selected to remove the target contaminants. In the case of organic contaminants, hydrocarbons and organochlorides, zero iron nanoparticles are preferred for their capacity to degrade organics and to decrease the toxicity of the contaminants by reductive dechlorination. Activated carbon can be used for the removal of organics and heavy metals. Precipitations reagents, calcium carbonate, can be used for the neutralization and precipitation of heavy metals in the groundwater. The PRB have to be designed to receive all the flow of groundwater, avoiding bypass. A PRB with a hydraulic conductivity much higher than the surrounding soil assures the flow of groundwater through the PRB. This technology is, in general, practical and affective in the removal of contaminants. Moreover, the operation costs and maintenance are minimal. A well-designed PRB may operate for a year with minor supervision.

A modification of the PRB is the electrokinetic barriers, which consist of a series of electrodes around a contaminated area or in front of the advance of contaminated groundwater. The electrodes are arranged in rows perpendicular to the direction of the groundwater flow. The depth of the electrodes must be at least coincident with the depth of the contaminated area. The electric current is established within the soil by the alternation of anodes and cathodes, which are connected to independent hydraulic circuits to adjust the most suitable conditions for the anolyte and the catholyte (pH, addition of solubilizing or complexing agents). Periodically, the contaminants in the electrolytes are removed by different techniques such as adsorption and ion exchange.

## *15.1 Electrokinetic Bio-barriers*

Electrokinetic bio-barriers are basically electrokinetic barriers designed to contain pollutants and promote their biodegradation, both in soils and in groundwater. This technology consists of the installation of a row of anodes and cathodes perpendicular to the direction of the groundwater flow. A series of wells sandwiched between anodes and cathodes are also drilled and used to inject the nutrient solutions such as nitrogen, phosphorous, and substances capable of supplying oxygen to the medium. The chemical species that make up the nutrients are electrically charged and can therefore be dispersed through the soil homogeneously by electromigration. The organic pollutants present in the soil and transported by the groundwater are degraded at the level of the bio-barrier and downstream, thanks to the microbial activity favored by the supply of nutrients and oxygen.

## *15.2 Reactive Electrolytic Barriers*

The electrolytic reactive barriers consist of two rows of electrodes (anodes and cathodes) very close to each other with a permeable filler material between both rows of electrodes. The barrier is installed in a trench perpendicular to the direction of the groundwater flow so that it intercepts the advance of contaminants carried by the groundwater (similar to reactive permeable barriers). A low electrical potential is applied to the electrodes that induces oxidation conditions at the anodes and reduction conditions at the cathodes. This system allows the transformation or degradation of pollutants by redox reactions into new products that are less toxic or dangerous for the environment. A wide range of redox contaminants such as arsenic, chlorinated hydrocarbons (TCE, TCA), and energy compounds (TNT and RDX), including mixtures of contaminants (difficult to treat with other technologies), can be treated with electrolytic barriers. This approach offers several advantages including: (1) the effective degradation of contaminants and reaction intermediates through oxidation and sequential reduction, (2) controlling the formation of contaminant precipitates through periodic inversion of the electrode potentials, (3), the contribution of chemical products is not necessary for the transformation; (4) simple operation, and (5) low operating cost. This technology has shown very good results in the treatment of soils and groundwater contaminated with chromium or TCE.

## 16 Bioelectroremediation

Electrokinetics can be combined with bioremediation to achieve a synergistic effect in the remediation of soil contaminated with organics. Bioremediation uses the capacity of the soil microflora to degrade the organic contaminants in situ. The main limitation of the biological degradation in the bioavailability of the contaminants. Electrokinetics may be used to mobilize and increase the availability of the contaminants. The electric field favors the desorption of the contaminants to be dissolved in the interstitial fluid, and transport the contaminants out of small pores where the microorganisms cannot enter. Moreover, electromigration and electroosmosis can be used for the supply of nutrients (ammonium, phosphate, etc.) and oxygen (e.g., oxygen in the form of $H_2O_2$) to the subsoil in in situ applications. An interesting effect of the coupled technology electrokinetic bioremediation is the transport of bacteria. The electric field may transport the bacteria, even in short distances, increasing the probability to access the contaminants, i.e., the electrokinetic transport of bacteria and contaminants increases the bioavailability of contaminants [38, 39]. Overall, the electric field is a simple and effective way to increase the bioremediation activity.

The electrokinetic biofence (EBF) technology developed by Lageman and Pool [40] is another way to combine electrokinetics and biodegradation. In the EBF, a series of alternating anodes and cathodes are installed perpendicular to the contaminated water flow. The electrode wells are filled with a nutrient solution (ammonium nitrate, potassium phosphate, etc.) that is dispersed in the subsoil by the electrokinetic transport. The increasing concentration of nutrients in the contaminated groundwater favors the biodegradation of the contaminants. This technology was applied for the remediation of a contaminated soil with organochlorine solvents. After 2 years of operation, the decreasing of the chlorine index (the amount of compounds with chlorine in the molecules) was observed. The operation requires low maintenance and supervision. The electricity for the electrokinetic process was provided by solar panels.

## 17 Electric Amendment of Phytoremediation

Phytoremediation is a benign and sustainable technology for the remediation of contaminated soils with heavy metals, inorganic contaminants, and biodegradable organic contaminants. Phytoremediation uses green plants to remove and/or degrade contaminants in the rhizosphere, the layer of soil occupied by the roots. The benefits of phytoremediation are the capability of removing organic and inorganic contaminants, minimum maintenance, and operational costs, which is visually pleasing and improves the quality of soil during the remediation, unlike other remediation technologies that seriously damage the quality of soil. The main limitations of the phytoremediation are the bioavailability of contaminants, the slow growth of the plants, the remediation limited to the layer of soil occupied by the roots, and the contaminant concentration has to be low or moderate, because plants will not survive in highly contaminated soils. The coupled technology electrokinetics phytoremediation was proposed to avoid in part the limitations of phytoremediation. The application of an electric field around a growing plant shows various positive effects in the plant and in the remediation process. The electric field mobilizes the nutrients that are transported toward the roots. Selected nutrient solutions can be added to the electrode wells and transported toward the roots. Similarly, the contaminants in soil can be mobilized by the electric field and transported toward the roots, where the contaminants are accumulated and degraded. The contaminants can be transported from soil zones out of the rhizosphere. Overall, the simultaneous application of electrokinetics to phytoremediation enhances the plant growing, increases de bioavailability of contaminants, and extends the remediation further than the rhizosphere.

The research results in electro-phytoremediation have proved that low or moderate electric gradients (below 2 DCV/cm) are beneficial for the plant and the remediation process. High-intensity electric field may provoke damage in soil microflora and plants. The damage is associated with pH changes due to the electrolysis of water upon the electrodes (acid pH on the anode side and alkaline pH on the cathode side). Rapid mobilization and transportation of contaminants toward the roots, reaching concentrations that may be toxic for the plant, is another limitation associated with high-intensity electric fields. These limitations may be avoided using alternate current instead of direct current. More research is still needed to determine the real benefits of electro-phytoremediation over the phytoremediation itself. The research must focus on the physiological changes induced by the electric current and how those changes in plant physiology may contribute to a better and faster soil remediation [41, 42].

## 18 Future Perspectives

The design of an electrokinetic application for the remediation of soils contaminated with organics must consider the scientific knowledge accumulated in the last three decades. In general, removal of organics requires the addition of facilitating agents (surfactants) to enhance the solubility of contaminants and, at the same time, keep a high electroosmotic flow. The combined effect of surfactants and electroosmotic flow results in the effective removal of organics. However, the surfactant-enhanced electrokinetics is costly, requires long treatment time, and generates wastes that require further treatment. As alternative, the combination of electrokinetics with chemical oxidation, bioremediation, or phytoremediation shows better perspectives. Chemical oxidation can be used in soils with toxic contaminants at high concentrations, whereas biological technologies can be used in low to moderate contaminated sites. Biological technologies are preferred because they do not damage the soil properties and do not require expensive chemicals, and the implementation and operational costs are relatively low.

## References


1. S. Paleari, Is the European Union protecting soil? A critical analysis of community environmental policy and law. Land Use Policy **64**, 163–173 (2017)
2. B. Vanheusden, Recent developments in European policy regarding brownfield remediation. Environ. Pract. **11**(4), 256–262 (2009)
3. H.D. Sharma, K.R. Reddy, *Geoenvironmental Engineering: Site Remediation, Waste Containment, and Emerging Waste Management Technologies* (Wiley, Hoboken, 2004)
4. K.R. Reddy, C. Cameselle, *Electrochemical Remediation Technologies for Polluted Soils, Sediments and Groundwater* (Wiley, Hoboken, 2009)
5. M.T. Ricart, C. Cameselle, T. Lucas, J.M. Lema, Manganese removal from spiked kaolinitic soil and sludge by electromigration. Sep. Sci. Technol. **34**(16), 3227–3241 (1999)
6. L.M. Ottosen, H.K. Hansen, G. Bech-Nielsen, A. Villumsen, Electrodialytic remediation of an arsenic and copper polluted soil-continuous addition of ammonia during the process. Environ. Technol. **21**(12), 1421–1428 (2000)
7. M. Pazos, M.T. Ricart, M.A. Sanromán, C. Cameselle, Enhanced electrokinetic remediation of polluted kaolinite with an azo dye. Electrochim. Acta **52**(10), 3393–3398 (2007)
8. A. Li, K.A. Cheung, K.R. Reddy, Cosolvent-enhanced electrokinetic remediation of soils contaminated with phenanthrene. J. Environ. Eng. **126**(6), 527–533 (2000)
9. K.R. Reddy, R.E. Saichek, Effect of soil type on electrokinetic removal of phenanthrene using surfactants and cosolvents. J. Environ. Eng. **129**(4), 336–346 (2003)
10. R.E. Saichek, K.R. Reddy, Electrokinetically enhanced remediation of hydrophobic organic compounds in soils: a review. Crit. Rev. Environ. Sci. Technol. **35**(2), 115–192 (2005)
11. R.E. Saichek, K.R. Reddy, Effect of pH control at the anode for the electrokinetic removal of phenanthrene from kaolin soil. Chemosphere **51**(4), 273–287 (2003)
12. S.N. Seyed Razavi, A. Khodadadi, H. Ganjidoust, Treatment of soil contaminated with crude-oil using biosurfactants. J. Environ. Stud. **37**(60), 107–116 (2012)
13. H.I. Gomes, C. Dias-Ferreira, A.B. Ribeiro, Electrokinetic remediation of organochlorines in soil: enhancement techniques and integration with other remediation technologies. Chemosphere **87**(10), 1077–1090 (2012)
14. A.T. Yeung, Y. Gu, A review on techniques to enhance electrochemical remediation of contaminated soils. J. Hazard. Mater. **195**, 11–29 (2011)
15. K. Maturi, K.R. Reddy, Simultaneous removal of organic compounds and heavy metals from soils by electrokinetic remediation with a modified cyclodextrin. Chemosphere **63**(6), 1022–1031 (2006)

16. A.P. Khodadoust, K.R. Reddy, O. Narla, Cyclodextrin-enhanced electrokinetic remediation of soils contaminated with 2,4-dinitrotoluene. J. Environ. Eng. **132**(9), 1043–1050 (2006)
17. G. Wang, W. Xu, X. Wang, L. Huang, Glycine-β-cyclodextrin-enhanced electrokinetic removal of atrazine from contaminated soils. Environ. Eng. Sci. **29**(6), 406–411 (2012)
18. K.R. Reddy, P.R. Ala, S. Sharma, S.N. Kumar, Enhanced electrokinetic remediation of contaminated manufactured gas plant soil. Eng. Geol. **85**(1–2), 132–146 (2006)
19. T.D. Pham, R.A. Shrestha, M. Sillanpää, Removal of hexachlorobenzene and phenanthrene from clayey soil by surfactant- and ultrasound-assisted electrokinetics. J. Environ. Eng. **136**(7), 739–742 (2010)
20. A. Oonnittan, R.A. Shrestha, M. Sillanpää, Effect of cyclodextrin on the remediation of hexachlorobenzene in soil by electrokinetic Fenton process. Sep. Purif. Technol. **64**(3), 314–320 (2009)
21. M.O. Boulakradeche, D.E. Akretche, C. Cameselle, N. Hamidi, Enhanced electrokinetic remediation of hydrophobic organics contaminated soils by the combination of non-ionic and ionic surfactants. Electrochim. Acta **174**, 1057–1066 (2015)
22. C. Yuan, C.H. Weng, Remediating ethylbenzene-contaminated clayey soil by a surfactant-aided electrokinetic (SAEK) process. Chemosphere **57**(3), 225–232 (2004)
23. F. Suanon, L. Tang, H. Sheng, Y. Fu, L. Xiang, Z. Wang, et al., Organochlorine pesticides contaminated soil decontamination using TritonX-100-enhanced advanced oxidation under electrokinetic remediation. J. Hazard. Mater. **122**, 388 (2020)
24. A.B. Ribeiro, E.P. Mateus, J.M. Rodríguez-Maroto, Removal of organic contaminants from soils by an electrokinetic process: the case of molinate and bentazone. Experimental and modeling. Sep. Purif. Technol. **79**(2), 193–203 (2011)
25. D.A. Kessler, C.P. Marsh, S.W. Morefield, *Electrokinetic Removal of Energetic Compounds* (Wiley, Hoboken, 2009)
26. M.T. Ricart, M. Pazos, S. Gouveia, C. Cameselle, M.A. Sanroman, Removal of organic pollutants and heavy metals in soils by electrokinetic remediation. J. Environ. Sci. Health A **43**(8), 871–875 (2008)
27. K.R. Reddy, C. Cameselle, P. Ala, Integrated electrokinetic-soil flushing to remove mixed organic and metal contaminants. J. Appl. Electrochem. **40**(6), 1269–1279 (2010)
28. M. Elektorowicz, M. Hakimipour, Electrical field applied to the simultaneous removal of organic and inorganic contaminants from clayey soil, in *18th Eastern Research Symposium on Water Quality, October, Montreal, Canada: CAWQ* (2001)
29. M. Elektorowicz, M. Hakimipour, Practical consideration for electrokinetic remediation of contaminated soil, in *CSCE 8th Environmental and Sustainable Eng. Specialty Conference and Offshore Engineering, Moncton* (2003), pp. 689–698
30. K.R. Reddy, K. Maturi, C. Cameselle, Sequential electrokinetic remediation of mixed contaminants in low permeability soils. J. Environ. Eng. **135**(10), 989–998 (2009)
31. S. Grandel, A. Dahmke, Monitored natural attenuation of chlorinated solvents: assessment of potential and limitations. Biodegradation **15**(6), 371–386 (2004)
32. Y. Yukselen-Aksoy, K.R. Reddy, Effect of soil composition on electrokinetically enhanced persulfate oxidation of polychlorobiphenyls. Electrochim. Acta **86**(24), 164–169 (2012)
33. A. Oonnittan, R.A. Shrestha, M. Sillanpää, Remediation of hexachlorobenzene in soil by enhanced electrokinetic Fenton process. J. Environ. Sci. Health A Tox. Hazard. Subst. Environ. Eng. **43**(8), 894–900 (2008)
34. K.R. Reddy, M.R. Karri, Effect of electric potential on nanoiron particles delivery for pentachlorophenol remediation in low permeability soil. in *Proceedings of the 17th International Conference on Soil Mechanics and Geotechnical Engineering: The Academia and Practice of Geotechnical Engineering* (2009)
35. C. Cameselle, K.R. Reddy, K. Darko-Kagya, A. Khodadoust, Effect of dispersant on transport of nanoscale iron particles in soils: zeta potential measurements and column experiments. J. Environ. Eng. **139**(1), 23–33 (2013)

36. Z. Zheng, S. Yuan, Y. Liu, X. Lu, J. Wan, X. Wu, et al., Reductive dechlorination of hexachlorobenzene by Cu/Fe bimetal in the presence of nonionic surfactant. J. Hazard. Mater. **170**(2–3), 895–901 (2009)
37. J. Wan, Z. Li, X. Lu, S. Yuan, Remediation of a hexachlorobenzene-contaminated soil by surfactant-enhanced electrokinetics coupled with microscale Pd/Fe PRB. J. Hazard. Mater. **184**(1–3), 184–190 (2010)
38. S.T. Lohner, A. Tiehm, S.A. Jackman, P. Carter, Coupled electrokinetic-bioremediation: applied aspects, in *Electrochemical Remediation Technologies for Polluted Soils, Sediments and Groundwater*, ed. by K. R. Reddy, C. Cameselle, (Wiley, Hoboken, 2009), pp. 389–416
39. L.Y. Wick, Coupling electrokinetics to the bioremediation of organic contaminants: principles and fundamental interactions, in *Electrochemical Remediation Technologies for Polluted Soils, Sediments and Groundwater*, ed. by K. R. Reddy, C. Cameselle, (Wiley, Hoboken, 2009), pp. 369–387
40. R. Lageman, W. Pool, Electrokinetic biofences, in *Electrochemical Remediation Technologies for Polluted Soils, Sediments and Groundwater*, ed. by K. R. Reddy, C. Cameselle, (Wiley, Hoboken, 2009), pp. 357–366
41. C. Cameselle, S. Gouveia, Phytoremediation of mixed contaminated soil enhanced with electric current. J. Hazard. Mater. **361**, 95–102 (2019)
42. C. Cameselle, R.A. Chirakkara, K.R. Reddy, Electrokinetic-enhanced phytoremediation of soils: status and opportunities. Chemosphere **93**(4), 626–636 (2013)

# Electrokinetic Remediation of Soil Polluted with Inorganic Ionic Species

**Marco Vocciante and Sergio Ferro**

## 1 Introduction

Since the late nineteenth century, with the introduction of electricity, chemicals, and petroleum-based components, the various industrial, agricultural, and chemical activities have generated countless cases of environmental contamination, with the release of large quantities of inorganic and organic pollutants. Since then, the anthropogenic impact has affected the planet at ever-deeper levels, through the reduction of nutrient cycling, the water confinement, the supply of physically and chemically stable compounds, and the lack of support for biodiversity.

Estimates reveal that around 20 million hectares worldwide are contaminated with heavy metals [1] and that over five million sites have been polluted due to improper waste disposal practices and accidental spills [2]. In 2007, the European Environmental Agency identified 80,000 contaminated sites [3] and 2.5 million *potentially contaminated* sites (PCS), 38% of which being the result of improper landfilling of municipal solid waste [4]. The annual cost of remediation has been estimated at around six billion Euro [4]. In 2004, the United States' Environmental Protection Agency stated that by 2030 approximately 294,000 contaminated sites are to be treated, at a cost of over 200 billion US$; more recently, the estimates have been updated: the number of sites has been increased to over 500,000 [5], with an expected reclamation cost of approximately 650 billion US$ [6]. Australia has over 50,000 confirmed contaminated sites and 160,000 PCS [6–8], with an estimated annual remediation cost that exceeds 3 billion US$, while Japan has over 500,000

M. Vocciante (✉)
Dipartimento di Chimica e Chimica Industriale, Università degli Studi di Genova, Genova, Italy

S. Ferro
Ecas4 Australia Pty Ltd, Mile End South, SA, Australia



M. A. Rodrigo, E. V. Dos Santos (eds.), *Electrochemically Assisted Remediation of Contaminated Soils*, Environmental Pollution 30,
https://doi.org/10.1007/978-3-030-68140-1_7

PCS, which require investments for remediation that were estimated in 2010 to be around 3 billion US$ a year [6].

Although the above problems have been documented in the past 70 years, less than a tenth of the contaminated sites have so far been remediated [9], due to the complexity and specificity of each problem and the high costs of the interventions [9–11]. Several different physicochemical or biological approaches have been suggested for the remediation of contaminated water and soils [12–14]; however, selecting a suitable technology is often a difficult yet crucial step for the successful remediation of a contaminated site [15, 16]. Among the various methods investigated, electrokinetic remediation has proved particularly interesting and efficient, owing to its possibility of being applied in situ [17–21] as well as its more environmentally friendly character, compared to other methods [22].

Figure 1 shows the trend in the number of articles related to ElectroKinetic Remediation Technology (EKRT) for soil remediation, since 1993 (data source: Web of Science Core Collection, Clarivate Analytics). The growing interest of academic and industrial communities in the EKRT is reflected in a high number of publications; unfortunately, not all sources are included in the analysis (for example, books and patent applications are excluded). A similar research performed with SciFinder® produced 1746 items (instead of 1038), including 303 patents. Interestingly, some items deal with "generic" contaminants (e.g., heavy metals: 352; polycyclic aromatic hydrocarbons: 73), while some others focus on a specific target (lead: 297; copper: 274; cadmium: 270; chromium: 195; zinc: 181; nickel: 120; arsenic: 106; phenanthrene: 88; mercury: 62; trichloroethylene: 36; petroleum hydrocarbons: 31).

The purpose of this chapter is to present the state of the art in the electrokinetic remediation of soils polluted by inorganic ionic compounds. The fundamentals of

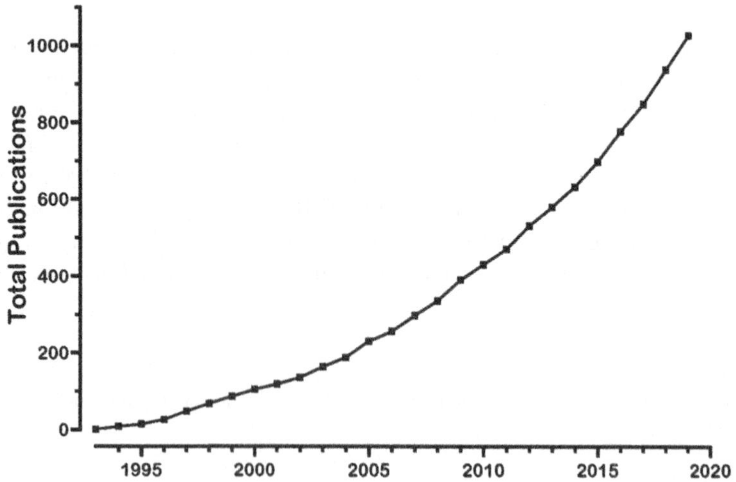


**Fig. 1** Total publications dealing with an electrokinetic (or electrochemical) remediation. (Data source: Web of Science Core Collection, Clarivate Analytics)

the technology will be briefly mentioned, together with the phenomena that occur in the soil, which are the basis of its effectiveness. As regards the aspects relating to the modeling of the technological approach, as well as its applicability to organic contaminants or mixtures, only a few words will be provided, since these topics are discussed more widely in dedicated chapters.

## 2 Fundamentals of Electrokinetic Remediation

EKRT is a very effective approach (applicable both in situ and ex situ) for the removal of pollutants present in an earthy matrix, as it allows a high efficiency even if applied in soils with low permeability [23–27]. During an EKRT process, an electric potential gradient is applied to induce a low electric current across a portion of the contaminated soil to be treated using electrodes suitably located in the subsurface [28]. The applied electric field can easily reach contaminants embedded deep in the subsurface, which other technologies are unable to reach [29, 30]. The electrodes are generally inserted in suitably constructed wells containing an electrolytic solution, and a low potential gradient is then applied between them (Fig. 2a). Depending on the intensity of the resulting electric current and the characteristics of the system (salt content, humidity, soil composition, etc.), different physical, chemical, and electrochemical processes are induced, which can cause substantial migration of the species through the soil and towards the electrode wells, from where they can eventually be removed [31]. Once in the electrode well, their extraction from the solution can take place by electroplating, adsorption, precipitation or coprecipitation on the electrode [32], or simply by removing the contaminated electrolyte solution (which can therefore be treated and reused [33]).

The most relevant phenomena induced in a soil system by the applied electric field are electroosmosis (i.e., the displacement of the solution naturally present in the soil) and the (electro)migration of species that have an electric charge (Fig. 2b), in addition to electrolysis and electrophoresis [34]. Moreover, since the soil acts as an ohmic resistance, an increase in the temperature of the soil is normally observed [35].

Electrophoresis refers to the transport of charged particles of colloidal size within a stationary fluid, due to the application of an electric gradient. Compared to ionic migration and electroosmosis, mass transport by electrophoresis is negligible in low permeability soil systems [36].

Electrolysis is a reactive process that occurs on anodic and cathodic surfaces due to the application of the electric field, which normally takes place at the expense of water (decomposition reactions). During an electrokinetic remediation, the electrolytic reactions at the electrodes generate gaseous oxygen and protons ($H^+$) at the anode (due to the oxidation of water) and gaseous hydrogen and hydroxyl ions ($OH^-$) at the cathode (due to the reduction of water), as schematized by Eqs. (1) and (2).

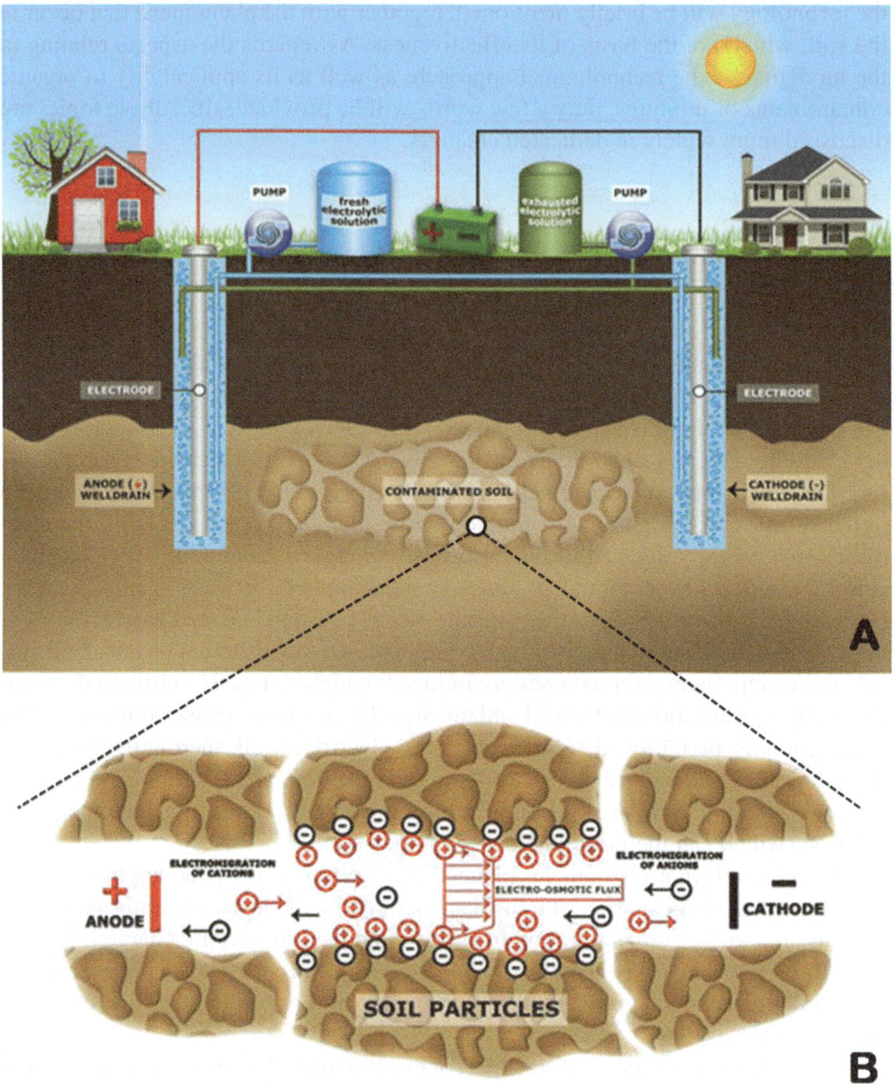


**Fig. 2** (**a**) Schematic of a typical EKRT installation; (**b**) detail of the main mechanisms occurring during an EKRT remediation. (Adapted from [21]. Copyright 2015, Elsevier B.V.)

Anode (water oxidation, pH 0):

$$2H_2O \rightarrow O_2 + 4H^+ + 4e \rightarrow E^\circ = 1.229\,\text{V} \quad (1)$$

Cathode (water reduction, pH 14):

$$4H_2O + 4e \rightarrow 2H_2 + 4OH^- \rightarrow E^\circ = -0.828\,\text{V} \quad (2)$$

The anodic oxidation of water generates an acidic front that moves towards the cathodes under the action of the electric field (mainly by electromigration). As a secondary yet useful effect, the acidic front can facilitate the release of pollutants fixed in the soil through dissolution and ion-exchange reactions [37]. On the other hand, the reduction of water at the cathodes produces an alkaline front that is then dragged towards the anodes [38]. Further reactions may occur at the electrodes, depending on the nature and concentration of the available species, as indicated by Eqs. (3) and (4) for a generic metallic specie $M^{n+}$.

$$\text{Anode}(\text{oxidation}): aM^{n+} + bH_2O \rightarrow M_aO_b + 2bH^+ + (2b - \text{an})e \quad (3)$$

$$\text{Cathode}(\text{reduction}): M^{n+} + ne \rightarrow M \quad (4)$$

The reactions that occur at the electrodes also depend on their composition and electrocatalytic properties, which in turn decide the value of the electrode potentials. Anodic materials, generally inert to chemical reactions and with a high electrical conductivity, include activated titanium (i.e., titanium coated with noble metals), graphite, or, more rarely, sacrificial electrodes such as iron and aluminum; cathode materials are generally chosen from those resistant to corrosion in an alkaline environment and include aluminum and iron [39, 40].

Electromigration is the movement of the ions present in the soil solution under the action of the electric field generated between the electrodes: the anions move towards the anodes, while the cations move towards the cathodes. The transport of the $H^+$ and $OH^-$ ions generated by the electrolytic reactions (i.e., the movement of the acid and alkaline fronts) is also mainly due to electromigration, since diffusion provides a less important contribution [19]. Consequently, during an electrokinetic remediation, changes in pH can influence soil chemistry with consequent different chemical reactions, mainly precipitation or dissolution of salts and minerals in the soil. As for the ionic species originally present in the soil solution, the formation of poorly soluble compounds near the cathodic region can cause clogging of the soil pores, with loss of hydraulic permeability, while in the anodic region there is normally no formation of deposits, due to its low pH. In general, electromigration contributes to the transport of metallic species, polar organic molecules, ionic micelles, and colloidal electrolytes [41].

The transport of water during an electrokinetic remediation occurs mainly due to the electroosmotic flow, that is, under the action of the electric field. The electroosmotic flow can involve the groundwater itself or an aqueous solution that has been added to improve the removal of contaminants. As known, soil permeability influences the transport of water in the soil [42]. The parameter that regulates the electroosmotic flow in a soil mass is the electroosmotic permeability coefficient, $k_e$, which is empirically defined as the ratio between the rate of the volume of water flowing through a unit cross-sectional area of soil ($Q_e$, m/s) and the intensity of the electric field ($E$, V/m). Consequently, $k_e$ is expressed in $m^2/V$ s and depends on the effective bulk electrical conductivity of the soil. Both the hydraulic gradient and electric gradient contribute to the transport of water, but the former is much less effective than the latter due to the very low hydraulic conductivity ($10^{-8}$–$10^{-11}$ m/s) that characterizes clay soils. The electroosmotic flow is directly proportional to the applied electrical gradient (V/cm) and, for systems with pores that are large relative to the size of the electric double layer that surrounds soil particles, is practically independent of the porosity of the soil/sediment (unlike the hydraulic conductivity) [43]. This makes this remediation approach ideal for the removal of non-ionic contaminants from soils and sediments with low hydraulic permeability.

The establishment of a pH gradient in the soil, due to the electrolytic reactions taking place at the electrodes, also affects the electroosmotic flow [44]. Near the anode, if the pH value is lower than the pH of zero charge (PZC) of the soil, the surfaces of the particles are positively charged; the opposite occurs near the cathode, as the alkaline pH value can increase the negative charge on the soil particles. Since the surface of the clay particles typically has a negative charge (due to the dissociation of the ionogenic groups at the normal pH of the soil), within the electric double layer the soil solution presents an excess of positive charge that can be dragged towards the cathodes under the effects of the electric field, resulting in a water movement (electroosmotic flow).

Diffusion is another important transport mechanism, which however normally has a minor influence on the movement of pollutants during an EK treatment.

Numerous research studies have been developed to understand the processes and make the most of the phenomena that occur when an adequate electric field is induced [21, 31, 45, 46].

## 3 Evolution of the Technology and Its Applications

Although the first investigations on electroosmotic flow date back to 1807, when the German scientist Ferdinand Friedrich Reuss began his research in Moscow, his discoveries were practically ignored by the scientific community (probably because the results were published in Russia, and in French [28]). About 10 years later, the British amateur chemist Robert Porrett Jr. independently rediscovered electroosmosis [47], but it was only several decades later that the electrokinetic approach was applied, initially as a consolidation process for fine soils [48] and subsequently to recover heavy metals [49, 50].

In first investigations [23, 34], the removal of contaminants was attempted simply by inducing an electric current through electrodes directly inserted into the contaminated soil, but the remediation efficiency was low. To increase the remediation efficiency, numerous improvements have been proposed, including:

- the optimization of the pH of electrolytic solutions [51];
- the use of an ion-exchange membrane to prevent the migration of protons (acidity) and hydroxyl ions (alkalinity) from the electrodes in the soil [51–53];
- the increase of the mobility of pollutants by adding complexing agents [54] and surfactants [55]; and,
- the optimization of the effective volume, varying the arrangement of the electrodes according to the nature of the site and the target contaminants [39, 40].

Since electromigration generally provides a greater impact than electroosmosis, for many years EK remediation has focused mainly on charged species such as heavy metals [56–58]. Acidic pH enhances metal solubilization and their transport towards the cathode reservoirs, whereas complexing agents convert soil-bound metals into soluble complexes thus improving their removal [21, 59, 60]. However, the chemical transformations that occur during treatment can modify the mobility/accessibility of these substances and possibly make them more hazardous for living organisms; for this reason, toxicity tests are increasingly being taken into account to prove the effectiveness of a remediation process [61].

More recently, research has focused on the use of EKR technology for the removal of hazardous organic substances from the soil [62] or marine and river sediments [63]. A plethora of technologies have been reported, ranging from simple electrokinetic soil flushing for soils with low hydraulic conductivity [64] to the use of permeable reactive barriers loaded with granular activated carbon (GAC), zero-valent iron (ZVI) [65–67], or even microorganisms [68], to be used to retain or transform the organic species mobilized by the applied electric fields.

The main challenge of EK technology is the conversion of low solubility pollutants into mobile forms, in order to extract them. Therefore, enhancing agents, added to the processing fluids, are necessary to obtain an effective removal of every type of contaminants.

During an electrokinetic soil flushing, surfactants are normally introduced into the processing fluid to allow the formation of micelles (charged particles) with the species target of the remediation. These micelles are then transported across the soil under the effect of the electric field that is applied to promote the removal of organic and inorganic compounds [23, 32], making electrophoresis a significant contribution to the remediation process of relatively permeable soils. Only few studies have treated fine-grained sediments [69, 70]. López-Vizcaíno and coworkers [71] studied the mobility of different solutions during a lab-scale electrokinetic soil flushing. Comparing the electroosmotic flows obtained by working with different surfactants (cationic, anionic, and nonionic), they observed significant changes depending on the nature of the washing fluid and the voltage gradient applied between the electrodes. In particular, ionic surfactants led to low electroosmotic flow rates, while nonionic surfactants contributed more efficiently to the displacement of the target contaminant. When contamination was caused by complex organic substances such as polycyclic aromatic hydrocarbons (PAHs) and polychlorobiphenyls (PCBs), surfactants allow their better desorption from sediment particles and solubilization in the aqueous pore fluid [70, 72–74]. For nonionic organic compounds, electroosmosis is the main transport mechanism.

The ever-growing attention towards environmental sustainability is now catalyzing the interest in novel (bio)remediation technologies, which involve a minimal use of chemicals (and external energy) and consequently imply a lower environmental impact [75]. In this context, electrobioremediation technologies have recently attracted considerable attention, in particular following the discovery of many microorganisms capable of degrading environmental contaminants, including PAHs, using electrodes as terminal electron acceptors virtually inexhaustible in their metabolism [76–87]. Furthermore, bioelectrochemical systems allow manipulating the redox potential of the contaminated matrix, thus establishing in situ conditions that favor the biodegradation of contaminants [88–91]. For example, dimensionally stable anodes (DSA) buried within a contaminated sediment have been successfully used to optimize oxygen generation within the sediment via low voltage seawater electrolysis, and in doing so accelerating (up to three times compared to non-electrified controls) the biodegradation of crude oil hydrocarbons [92]. The coupling of electrokinetics with bioremediation and phytoremediation could represent the most sustainable approach, since the energy requirements are very low, the addition of chemicals is not necessary, and the physicochemical and 'biological' (e.g. the fertility) characteristics of the soil at the end of the treatment are improved compared to the initial situation. However, appropriate operating conditions must be selected, in order to ensure the survival and development of microorganisms and/or plants [93]. Extreme pH values and high temperatures can be produced during the process, which are two of the most critical parameters for keeping microorganisms active [94, 95]. Since most of the processes induced by the electric field have a negative effect on the viability of microorganisms, the simultaneous optimization of both, the electrokinetic and biological processes, can be very challenging [88]. The EK treatment can increase the bioavailability of organic pollutants by facilitating the contact between microbes and nutrients and/or pollutants, and the weak electric current may also directly stimulate microbial activity [96, 97] or degrade some of the pollutants through an electrolytic reaction [98]. In addition, the application of an EK treatment can improve the growth and respiration of plants (which in turn can facilitate the removal of metals) and facilitate the diffusion of rhizosphere microorganisms (with possible enhancement of organic contaminants biodegradation) [99].

According to Lemström [100], plants exposed to an alternating current become greener and produce more biomass than those grown under the same conditions but without an electric field; as already commented, the EK mobilization of nutrients (and contaminants) can make them more available for the absorption of plant [101]. O'Connor and collaborators have reported on the influence of an alternating electric field on ryegrass phytoremediation of a soil contaminated by heavy metal [102], concluding that the electric field contributes to the transport of metals, which are then accumulated by the ryegrass.

The elimination of organic and/or inorganic contaminants through the combined use of plants and an electric field applied through the soil to be treated is an effective approach [103], able to control the transport of contaminants to the rhizosphere as well as to prevent the establishment of strong acid or alkaline fronts in the soil [104].

Low to moderate voltage gradients (0.67 V/cm) induce small changes in the physical and chemical properties of the soil, which do not compromise the survival of the plants [105]. The electric field-supported phytoremediation was faster and more effective than that with plants alone. Electro-phytoextraction tests were conducted in soils contaminated by heavy metals (Cd, Co, Cr, Cu, Pb, and Zn), selecting two species of plants (*Brassica rapa* L. subsp. *Rapa* and *Lolium perenne* L.) already adapted to the climatic and soil conditions. The electric field improved the growth of plants, mainly of *L. perenne*, and increased the phytoremediation of the metals. Mixed crops of the two plant species have shown interesting results for large-scale applications.

The biological activity of plants also improves in the presence of the electric field: the application of a direct or alternating electric current favors germination and biomass production. This finding, first described by Lemström [100], has been confirmed in recent works, which have shown how ryegrass, sunflower, and oat plants grow faster in the presence of a low-intensity electric field [101, 106]. As already mentioned, the electric field would make nutrients more bioavailable, thanks to their mobilization and transport to the rhizosphere. Other authors instead argue that the benefits are related to the influence of the electric field on enzymatic reactions, water activity, and membrane transport [107–109].

The application of an electric field can also cause negative effects: for example, O'Connor et al. [102] reported growth inhibition and death of plants located near the electrodes. The negative effects of electricity would be related to changes in soil pH associated with water electrolysis and phytotoxicity due to the increasing bioavailability of metals.

### *3.1 Contamination by Complex Organic Compounds*

In order to focus on the treatment of soils contaminated by inorganic species in the next section, the electrokinetic remediation of soils contaminated by complex organic compounds will be briefly discussed discussed below. However, the reader is invited to refer to the dedicated chapter(s).

Among the many organic contaminants, particular attention was paid to PAHs [24], pesticides [45], organochlorinated species [110], and the mixture of hydrocarbons found in crude oil, i.e., petroleum hydrocarbons (PHs) [111]. As already mentioned, phytoremediation has been considered for the removal of organic pollutants in combination with an electrokinetic approach. Although both technologies are efficient in removing organic pollutants from the soil, a significant intensification can be obtained from their association: the optimization of mass transport [112,

113] allows improving the spatial scale and the speed of phytoremediation [101, 114–116].

The electrokinetic removal of simpler organic compounds has shown that volatile and/or soluble organic contaminants are more easily removed; however, it is necessary to understand the phenomena that influence the efficiency of the process.

In this regard, it should be remembered that most of the research has been carried out on a laboratory scale with a few hundred grams of soil, since this size allows for easier characterization of the processes that occur in the experimental configuration of the system. However, as already commented, the conclusions drawn in a laboratory or bench-scale (where the electrokinetic processes control the rate) are poorly reproduced or cannot be easily extrapolated to the full scale (where the ohmic and heating effects can dominate). According to Vidal et al. [117], bench-scale plants have to be used in order to draw sound conclusions in an experimental study; on the other hand, scale-up is critical, as it affects not only the speed of the treatment [118, 119] but also the efficiency and selectivity reported.

## *3.2 Petroleum*

The electrokinetic removal of petroleum implies the displacement and subsequent removal of PAHs and PHs, which unfortunately are poorly soluble in water (petroleum compounds are hydrophobic) and therefore also in the soil solution. To improve their removal, the use of surfactants, bio-surfactants, cosolvents, and cyclodextrins has been attempted [24, 46]. These species are not only able to act as solubilizing agents but also to modify the surface characteristics of soil particles, as well as the properties of the pore fluid, influencing its dielectric constant, pH, and viscosity [120].

The removal of diesel from spiked kaolin was investigated by Mena et al. [46] by coupling an electrokinetic soil flushing (enhanced by the addition of sodium dodecyl sulfate, SDS) with bioremediation, i.e., by implementing a biological permeable reactive barrier between the electrodes. They observed that both nutrients and SDS were efficiently transported through the soil by a combination of electromigration and electroosmotic processes; diesel was also displaced, but not through a purely electrokinetic approach. For efficient bioremediation, it is important to monitor the pH and nutrients supplied to the bio-barrier, since extreme pH values may kill the microorganisms, while the lack of nutrients limits their growth and hence the remediation.

For the removal of petroleum contaminants from soil, an in situ chemical oxidation (ISCO) integrated with EKRT has recently been investigated [121]. The role of ISCO has been tested considering two different electrolytic solutions, namely NaCl and $NaNO_3$, as well as different values for the electric field. By using a chloride-containing solution, the produced active chlorine contributed to the reclamation of sand spiked with diesel (10 g $kg^{-1}$). In particular, a higher concentration of NaCl (from 10 to 40 g $L^{-1}$) allowed a greater production of active chlorine (from 12.5 to 30 mM) and therefore to increase the efficiency of diesel removal (from 33% to 44% for 1 V $cm^{-1}$ and from 43% to 67% for 2 V $cm^{-1}$, respectively, using an NaCl concentration of 20 g $L^{-1}$).

Great attention has been devoted also to Fenton and electro-Fenton processes, owing to their capability to operate at room temperature and the fact that reagents are readily available. Both approaches are based on the use of $H_2O_2$ that, in presence of iron(II), generates hydroxyl radicals according to Eq. (5).

$$Fe^{2+} + H_2O_2 + H^+ \rightarrow Fe^{3+} + \bullet OH + H_2O \quad (5)$$

Seo et al. [122] explored an electrokinetic approach coupled with Fenton oxidation (thanks to the iron(II) naturally present in the soil) for the simultaneous removal of PHs, PAHs, and zinc from a contaminated soil. They showed that the removal of PHs and zinc can be significantly improved by adding 20 mM SDS in 10–20% hydrogen peroxide (the concentration depends on soil properties) at the anode and 20 mM SDS and 20 mM NaOH at the cathode. Moreover, they concluded that the SDS and NaOH introduced in the cathode reservoir help dissolving the PAHs by forming micelles that are then transported towards the anode.

In the case of soils where the iron(II) concentration is not sufficient to sustain a Fenton reaction, the catalyst can be introduced, e.g., using iron electrodes; this approach has recently been investigated by Paixão et al. [123] for the remediation of a soil contaminated with TPHs. To avoid alkalinization due to water reduction at the cathode and, consequently, the precipitation of iron hydroxide, citric acid was added to the cathode well. The purely electrokinetic approach allowed removing about 27% of the hydrocarbons in 15 days, while an 89% removal was obtained with the EK approach coupled to Fenton. Interestingly, most of the contaminants were removed by the ISCO (Fenton reaction), not because of the EK mobilization and subsequent removal from reservoirs, which means that no process fluid treatment is required.

### 3.3 *Pesticides and Herbicides*

The production and use of pesticides has increased in recent years, drawing attention to their various chemical properties and to their hazardousness. The methods for treating this heterogeneous class of contaminants have been studied in several research groups [45, 124–126].

Li et al. [127] examined the performance of a PRB filled with reactive particles of Pd/Fe, installed between the electrodes, with the aim of obtaining a dechlorination of pentachloro-phenol (PCP), during its electrokinetic displacement. Due to the different pH values that can be found in the soil, between anode and cathode, PCP species can be present in unprotonated or protonated form, obviously characterized by different transport behaviors. The position of the PRB, with respect to the anode, proved to be important: with the reactive barrier installed at one-third of the distance from the cathode, the PCP originally present between the anode, and the PRB moved towards the cathode (and through the PRB), while that present between the PRB and the cathode accumulated near the PRB due to the dissociation caused by the alkaline pH. By positioning the PRB halfway from the cathode and adjusting the pH of the PRB by periodic additions of acetic acid, it was possible to reduce the contamination homogeneously, removing 49% of the PCP. A 22.9% was recovered as phenol, mainly from the catholyte, suggesting that the transport of PCP is mainly due to electromigration, while that of phenol is obtained mainly through electroosmosis.

An electrochemically assisted soil washing was investigated for the remediation of soils contaminated by pesticides [45], considering four different bench-scale pilot tests containing soils spiked with 2,4-dichlorophenoxyacetic acid (2,4D), oxyfluorfen, chlorsulfuron, and atrazine, respectively. The efficiency of the electrokinetic soil washing has been shown to depend on the chemical characteristics of pesticides; moreover, suitable surfactants must be added to allow the displacement of nonpolar pesticides. Over 80% of the pollution could be removed (up to 95% in the case of 2,4D, while the adsorption of chlorsulfuron to the soil limits the effectiveness of the treatment); polar pesticides were moved towards the anode by electromigration, while electroosmosis dragged them to the cathode, although to a lesser extent.

The applicability of an electrokinetic fence technology for the removal of 2,4D and oxyfluorfen from a 32-$m^3$ portion of polluted soil was investigated by López-Vizcaíno et al. [128]. While removal in a laboratory test was mainly due to electrokinetic phenomena, pesticide elimination through volatilization was observed in the case of a large-scale test, due to ohmic electrical effects. The investigation pointed out that small-scale test results cannot be extrapolated to large-scale applications, as the control mechanisms can be very different, due to their significantly different relevance.

## 4 Electrokinetic Removal of Heavy Metals

Soil contamination by heavy metals represents a worldwide problem for human health and the environment. These contaminants are inadvertently introduced to the soil as byproducts of various anthropogenic activities, which include metals extraction through smelting, waste disposal, and agrochemical applications. Nowadays, estimates reveal that over 20 million hectares of land are contaminated by heavy metals (Cd, Hg, Co, Cu, Pb, Ni, Cr, and others), at soil concentrations higher than the normal geochemical baseline or regulatory levels [18].

Starting from the late 1980s, the EKRT approach has been experimented for the removal of heavy metals from soils [23, 129, 130]. Under the effects of a suitable electric field, normally resulting in a low-intensity direct current (DC), heavy metals can be accumulated at the electrodes and subsequently removed by precipitation, electroplating, as well as by treating the solutions extracted from the soil, through a suitable decontamination approach (e.g., ion exchange) [33]. Metal contaminants may exist in soil in mobile states (as species already dissolved or sorbed on colloidal particulates suspended in soil solution) or immobile states (as sorbed species on soil particles or solid precipitates) [95]; however, only metals that exist in mobile states can be extracted from soil by a pure electrokinetic approach. During an electrokinetic extraction, a low pH environment can be easily generated through the electrolytic decomposition of fluids at the anode.

The low pH environment enhances the desorption and removal of heavy metals from soil by electroosmosis and electromigration [131, 132]. In fact, the acidic front that develops from the anodic side of the soil dominates the chemistry across the specimen, with the exception of some small sections close to the cathode, since the advance of the alkaline front is slower than the acidic one. When the acid front reaches the cathode, the remediation action is complete [133].

Unfortunately, it is difficult to generate an acidic environment by the electrokinetic remediation process alone in soils of high acid buffer capacity. The cations sorbed on soil particle surfaces, salts of weak acids (such as carbonates, bicarbonates, and silicates), and amphoteric substances (such as amino acids) in soil can increase the acid buffer capacity of soil. In addition, when an acidic environment is generated, the kinetic dissolution of species such as calcium and magnesium carbonates may further complicate the situation and should be considered in order to understand the mechanism of metal dissolution [134]. Then, the transport competition between the target heavy metals and the released buffering species (e.g., $Ca^{2+}$ and $Mg^{2+}$ from dissolved $CaCO_3$ and $MgCO_3$) greatly influences the metal mobilization efficiency and thus the energy cost of the technology [135]. Another possible obstacle to the effective mobilization of target contaminants (for example, heavy metals) is represented by the salinity of the soil, which can further hinder the mobility, extraction, and transfer of ionic species of an EK treatment [136]. To improve removal efficiency, particularly in soils with a high pH buffer capacity, the use of improvement agents is often essential as they can solubilize target contaminants and keep them in a mobile state.

Similarly, the EKRT approach has been shown to improve the soil washing results through the electrically induced mobilization of metal species present in the soil. Several research groups have observed that suitable extraction agents can help eliminating the pollutants from the soil by promoting ion exchange, the dissolution of precipitates, or simply by dragging pollutants through complexation reactions [24, 33, 137]. As a result, the contamination is transferred from the soil to the pore water, turning the problem of decontamination of the soil into a wastewater treatment problem.

Giannis et al. [138] studied the removal of Cd from soils by an enhanced EKRT using either an anionic surfactant (SDS) or acid conditions plus humic acid. When the SDS was used as a washing solution for the soil, over 90% of the contamination was removed in a very short time. In contrast, when SDS was added to the electrode wells, no improvement was achieved because the surfactant was unable to migrate into the soil specimen. Under the acidic conditions obtained by adding acetic acid in the cathode compartments, the supply of humic acid to the soil proved to be poorly effective, since the weak complexes that form between the organic substrate and Cd are not stable at low pH values.

Ng et al. [139] tested the electrokinetic approach in soils contaminated by lead and chromium, using different electrolytic solutions, namely 0.01 M $NaNO_3$, 0.1 M citric acid, and 0.1 M EDTA. Based on the obtained results, only EDTA provided satisfactory results thanks to the formation of water-soluble anionic complexes with both heavy metals. In presence of $NaNO_3$ or citric acid, the electromigration of Pb is poor, and the same applies to Cr, especially at low pH, due to the adsorption of Cr(VI) to the soil particles and the possible formation of Cr(III).

Different organic substances, such as ethylenediaminetetraacetic acid, nitrilotriacetic acid, hydroxy carboxylates like citric acid, and organophosphonates have been widely investigated as enhancement agents in electrokinetic remediation due to their strong chelating ability and availability [140–144]. Kim et al. [56] reported efficiencies of 90% for the removal of Pb and Cd (from a soil contaminated with Cd, Cu and Pb) by adding 50 mmol/L of citric acid. Lower removals (between 15 and 23%) have been obtained by Gu et al. [141] using 0.1 M organophosphonates (nitrilotrimethylene triphosphonate, NTMP, and ethylene-dinitrilo-tetramethylene phosphonate, EDTMP) or EDTA to help mobilizing the sorbed cadmium ions, in 5 days of EK treatment under a constant voltage gradient of about 1.0 V $cm^{-1}$. Unfortunately, most enhancement agents are suspected to induce secondary pollution in the environment [145–147] and their use (especially when considering commercial formulations) significantly increases the cost of soil remediation.

Many innovative techniques have been developed in the past few decades to improve EK removal efficiency. An electrochemical system with six anodes positioned hexagonally around a central cathode made it possible to increase the production of alkali at the latter and remove heavy metals [148]; however, to avoid clogging of the pores near the electrode and the direct precipitation of species on the cathode, the latter was periodically replaced. After 480 h, the average removal efficiency was 89.9% for Cd and 69.3% for Pb [148].

The EK approach has also been used in combination with other remediation techniques, such as permeable reactive barriers and microorganisms, in order to increase the efficiency of the treatment and to treat soils and groundwater polluted by complex contaminants.

Hu [149] studied the use of a PRB enhanced with EK transport for the treatment of wastewater contaminated with COD and $Pb^{2+}$. Under the effect of the electric field, contaminants were forced to migrate from the anode to the cathode, passing through a PRB located near the cathode. The removal efficiencies of COD and Pb were 57.0% and 81.6% in the anode zone, but reached higher values in the cathode zone, respectively 90.4% and 94.6%.

Fu et al. [150] proposed to use permeable reactive composite electrodes, in order to add the effectiveness of the PRB and improve the EK treatment. In particular, a mixture of zeolite and Fe (ZVI) would allow controlling the pH and absorbing heavy metals. This combination produced removal efficiencies for lead, cadmium, nickel, and copper of 49.4%, 47.1%, 39.2%, and 36.7%, respectively.

Also, Shariatmadari et al. [151] studied the effect of coupling a PRB with EK, in this case for the removal of Cr(VI) from a clayey soil. A reactive barrier of iron nanoparticles and sand was inserted at a distance of 2 cm from the anode. Although chromium is one of the most difficult heavy metals to be removed from contaminated soils, according to the results obtained the reduction of Cr(VI) and total removal efficiency of Cr increased by 88% and 19%, respectively.

In a similar way, according to Yan et al. [152], the insertion of an activated carbon barrier loaded with iron(II) in soils contaminated with Cr(VI) improved the removal efficiency of the latter thanks to its adsorption and reduction. The maximum removal efficiency (80.2%) was achieved with an AC-Fe ratio of 5%.

In another study, García et al. [153] devised a lab-scale EK process combined with a PRB containing anion exchange resins, with the aim of treating clayey soil contaminated with nitrates. Using a 1.2 V $cm^{-1}$ potential gradient, a 90% removal efficiency was achieved in less than a week.

Adikesavan and Rajasekar [154] reported that a bioleaching-enhanced EK remediation (BEER) improved the efficiency of zinc removal (93.08% in 72 h, compared to 56.6% in 96 h with EK alone). At the anode, the optimization of the pH stimulated the growth of acidophilic bacteria (*Serratia marcescens*), which in turn were able to increase the solubility of heavy metals, thus promoting their electromigration.

Silva et al. [155] investigated the applicability of an EK remediation of $Pb^{2+}$-polluted soil by testing different electrolytes ($NaNO_3$, citric acid, EDTA) and the effects of using direct currents (DC) as well as a periodically reversed polarity (RP). The results obtained showed that both RP and citric acid improved the mobilization of $Pb^{2+}$ towards the electrode reservoirs, minimizing the otherwise pH-dependent precipitation of ionic species. In contrast, when $NaNO_3$ and EDTA were used as electrolytes and a DC polarization was used, the development of a significant pH gradient eventually caused the precipitation and immobilization of the target contaminant. Finally, germination of *Helianthus annuus* seeds was used to assess the phytotoxicity of the treated soil: a reduced germination (40–70%) was obtained from soils treated with DC and $NaNO_3$ or EDTA as cathodic solutions, due to the altered soil pH, while a much higher germination (65–90%) was obtained using EK and $NaNO_3$ or citric acid as cathodic solutions.

In light of the results obtained using citric acid, Gu et al. [156] studied the use of citric acid-containing industrial wastewater (CAIW) as an enhancement agent in the EK extraction of cadmium from a natural clay soil. CAIW contains large quantities of citric acid and acetic acid, which have proven effective in improving the extraction of heavy metals from soils [157–160]. In addition, the buffering capacity of CAIW can prevent the increase of the pH at the cathode during the EK treatment.

Karaca et al. [161] studied the EK treatment of a real sediment taken from a mine pond in Turkey, aiming at the removal of As, Al, Fe, and Mn. Using deionized water as the electrolyte and applying an electrical potential gradient of 1 V $cm^{-1}$ for 18 days, with no pH control, Al and Mn were removed, while Fe and As were only mobilized. The authors believe that better results could be obtained by neutralizing the alkaline environment at the cathode, using facilitating agents, and increasing the treatment time.

In another study, Chilean copper mine tailings were taken into consideration, focusing on the effects of initial acidity (1 M, 1.5 M, and 2 M $H_2SO_4$ added to 1.5 kg of soil, to obtain a 20% humidity) and intensity of the applied electric field (1 and 2 V $cm^{-1}$) on the EK remediation of Mn and Zn [162]. After up to 7 days of treatment, the maximum removals were around 32% for Mn and 18% for Zn. As expected, a lower initial pH made it possible to obtain greater solubility of the metals and increase their migration under the effect of the applied electric field.

A different approach has been investigated by Wang et al. [163], who used an auxiliary electrode to improve the performance of a conventional EK remediation of a soil contaminated with Cr(VI). Only a limited number of papers has focused on the structure and configuration of electrode system used for the remediation intervention [164–167]. These usually require energizing multiple electrodes, while any auxiliary electrode do not contribute to the removal of contaminants. However, according to Wang et al. [163], the auxiliary electrode (a polyaniline composite supported on a nonwoven fabric) has not only the function to adsorb the target species but also to act as an intermediate electrode (polarized by the electric field) that strengthen the process by improving the electromigration. Polyaniline is a widely studied conductive polymer, which has also been used as a promising material in the treatment of wastewater, thanks to its adsorbent character; however, its application in soil remediation is not common. In the above example, its insertion allowed improving the Cr(VI) removal by up to 20%.

In another contribution from the same research group [168], the auxiliary electrode was made of polypyrrole (PPy) on linen fabric; although PPy has already been used as an adsorbent to remove heavy metal ions, organic dyes, and so on (see original paper for references), its involvement in soil electrokinetic remediation is original. In the presence of the auxiliary electrodes, the distribution of the electric field was more homogeneous and the concentration of Cr(VI) around them decreased significantly, allowing local removal efficiency close to 92%.

Whatever the approach applied for the remediation, every electrochemical process requires electricity to work; if the power demand is not too high, green sources such as photovoltaic solar panels or wind turbines can be exploited. However, to avoid the reversibility of transport processes [169, 170], with consequent loss in efficiency overnight or when no wind is available, the use of energy storage devices is highly suggested. In particular, the use of redox flow batteries, which can store important amounts of energy and help regulating the supply of renewable energy, is a good option [171]. The use of solar energy and batteries has been tested by Hussein and Alatabe [133] in the EK remediation of soils contaminated with lead. After 7 days of treatment, with a potential gradient of about 1.2 V $cm^{-1}$, overall removal efficiencies of 90.7%, 63.3%, and 42.8% were achieved for sandy, sandy loam, and silty loam soils, respectively. In the absence of the batteries, the removal efficiencies were lower (53.7%, 31.2%, and 23.1%, respectively), since the night interruptions resulted in a considerable waste of time.

Table 1 presents some data relating to the electrokinetic treatment of soils polluted with heavy metals, providing additional information on the experimental setup and/or the results obtained.

**Table 1** Electrokinetic removal of heavy metals from soils

| Soil sample size | Pollutant/type of soil/ concentration | Technical solution | Observations—best results | Ref. |
|---|---|---|---|---|
| 0.7–1.2 kg | Cd (spiked)/real soil/150 mg $kg^{-1}$ Initial moisture content: 33% w/w (saturated soil) | Use of enhancing agents (0.001–0.01 M SDS, 0.8–3.2 g $kg^{-1}$ humic acid) | Removals after 7, 13, and 18 days at 2 V $cm^{-1}$: 81%, 90%, and 94%, respectively | [138] |
| 1.7 L | Cd, Pb, Cu (spiked)/ real soil/145, 455, and 1005 mg $kg^{-1}$ Initial moisture content: 33% w/w (saturated soil) | Use of enhancing agents (0.001–0.1 M NTA, DTPA, EGTA) | Removals after 23 days at 1.3 V $cm^{-1}$: 68–96% with NTA, 59–99% with DTPA, and 36–82% with EGTA | [137] |
| 0.125 kg | Cr(VI) (spiked)/real soil/100 mg $kg^{-1}$ Initial moisture content: 45% (saturated soil) | EK enhanced with PBR (iron nanoparticles and sand) inserted near the anode | Removal after 24 h at 2 V $cm^{-1}$: 14.8% without the PRB, 42% with the PRB and pH control | [151] |
| 1.5 kg | As, Cr, Cu, Ni, Pb, Zn, and various PAHs/real sediment/17, 45, 89, 30, 93, and 714 mg $kg^{-1}$ (see paper for PAHs' details) Initial moisture content: 40% w/w (saturated soil) | Use of enhancing agents (0.2 M EDTA, 3% w/w Tween 80) | Best removals, obtained after 23 days at 2 V $cm^{-1}$: 84% As, 62% Cr, 28–44% for Cd, Cu, Ni, Pb, and Zn | [69] |
| 0.05 kg | Pb, Cu, Zn/real soil/3529, 209, and 78 mg $kg^{-1}$ Initial moisture content: 15–17% | Bio-electrokinetic treatment, enhanced with acid or EDTA | Best removals after 20 days at 2 mA $cm^{-2}$: 92% Pb, 64% Cu, and 45% Zn, using EDTA BioEK | [172] |
| 1.2 kg | As, Cd, Cr, Co, Cu, Hg, Ni, Pb, Zn/real soil/3.8, 8.4, 19.5, 4.3, 263, 9.3, 18.6, 1160, and 1100 mg $kg^{-1}$ Initial moisture content: 30% | Use of enhancing agents (0.2 M EDTA first, then 5% Igepal CA-720) | Removals after 120 h at 1 V $cm^{-1}$: 80% Pb, 60% Zn, and 32% Cu | [173] |
| 0.32 L | Cu, Ni, Pb, Zn/ real sediment/183, 18, 85, and 270 mg $kg^{-1}$ Initial moisture content: most likely saturated soil | Use of enhancing agents (0.1 M EDTA, citric acid, $HNO_3$, HCl) | Best removals after 15 days at 1 V $cm^{-1}$: 68% Cu, 71% Ni, 65% Pb, 62% Zn, using HCl | [56] |

(continued)

**Table 1** (continued)

| Soil sample size | Pollutant/type of soil/ concentration | Technical solution | Observations—best results | Ref. |
|---|---|---|---|---|
| 0.32 L | As, Cu, Pb/real soil/72, 273, and 299 mg kg$^{-1}$ Initial moisture content: 30% | Soil pretreated with 0.05 M $HNO_3$, NaOH, or tap water before EK. 0.1 M NaOH and 0.1 M $HNO_3$ used in anode and cathode compartments | Best removal after 30 days at 1 V cm$^{-1}$: 60% Cu, 75% Pb, with acidic pretreatment and catholyte conditioning. A higher amount of As was extracted from soil after EK | [174] |
| ~0.25 L | Pb, Cd, Ni, Cu (spiked)/real soil/500, 395, 371, and 465 mg kg$^{-1}$ Initial moisture content: N/A | Permeable reactive composite electrodes (mixture of zeolite and Fe) | Removals after 15 days at 1.5 V cm$^{-1}$: 49.4% Pb, 47.1% Cd, 39.2% Ni, 36.7% Cu | [150] |
| N/A | Pb (spiked), COD/real soil/92 mg kg$^{-1}$ Initial moisture content: N/A | PRB (ash + zeolite and AC + Fe) enhanced with EK | Removals after 31 h at 0.35 V cm$^{-1}$: 57.8% Pb and 40% COD near the anode, 92.6% Pb and 90.4% COD near the cathode | [149] |
| ~125 L | Cd, Pb (spiked)/real soil/109 mg kg$^{-1}$ Initial moisture content: N/A | PRB (zeolite + Fe) around the cathode | Removals after 480 h at 0.75 V cm$^{-1}$: 90% Cd, 69% Pb | [148] |
| 0.1 kg | Pb and Cr(VI) (spiked)/real sandy soil/402 and 798 mg kg$^{-1}$ Initial moisture content: N/A (saturated soil) | Use of enhancing agents (0.01 M $NaNO_3$, 0.1 M citric acid, or EDTA) | Removals after 24 h at 1 V cm$^{-1}$: N/A; contaminants were simply mobilized (EDTA > $NaNO_3$ > Citric acid) | [139] |
| 0.5 kg | As/real soil/87 mg kg$^{-1}$ Initial moisture content: 23% | Soil pre-washed with water or NaOH | Removal after 28 days at 1 V cm$^{-1}$: 13–48% depending on pretreatment | [175] |
| 250 L | As, Cd, Cu, Hg, Pb, Tl, Zn/real soil/247, 265, 1074, 38, 2478, 16, and 75,274 mg kg$^{-1}$ Initial moisture content: 45% (saturated soil) | Use of enhancing agents (0.02 M KI in 0.02 M $HNO_3$) | Two current densities were tested (1.6 and 16.6 A/m$^2$); results have been reported in terms of removal rates (mg day$^{-1}$ m$^{-3}$) | [33] |

(continued)

**Table 1** (continued)

| Soil sample size | Pollutant/type of soil/ concentration | Technical solution | Observations—best results | Ref. |
|---|---|---|---|---|
| N/A (≤0.5 L) | Zn (spiked)/dried tannery sludge/400 mg $kg^{-1}$<br>Initial moisture content: N/A | Bioleaching-enhanced EK remediation (BEER) | Removals: 93% in 72 h, compared to 56.6% in 96 h with EK alone | [154] |
| 0.2 kg | Cr(VI), Pb, Cd (phenanthrene, anthracene)/real soil/400, 500, 50 (200, 100) mg $kg^{-1}$<br>Initial moisture content: 41% | Phytoremediation enhanced with electric field | Removals after 45 days at 1 V $cm^{-1}$: <20% for the metals, >80% for PAHs, in presence of an alternating electric field | [93] |
| ~3.5 kg | Cd (spiked)/real soil/285 mg $kg^{-1}$<br>Initial moisture content: 31–34% (saturated soil) | Use of enhancing agents (0.1 M NTMP or EDTMA or EDTA) | Removals after 5 days at 1.0 V $cm^{-1}$: 22.4–22.8% with phosphonates, 15.1% with EDTA | [141] |
| ~4 kg | Pb/real soil/1000 mg $kg^{-1}$<br>Initial moisture content: 15% | Polarity reversed periodically<br>Use of enhancing agents (0.1 M $NaNO_3$ or EDTA or citric acid) | Removals after 14 days at 1 V $cm^{-1}$: 69% with $NaNO_3$; 72% with $NaNO_3$ + EDTA; 87% with $NaNO_3$ + citric acid | [155] |
| 0.48 L | Cr(VI)/real soil/1172 mg $kg^{-1}$<br>Initial moisture content: N/A | AC-Fe (1–11%) | Removal after 10 days at 1 V $cm^{-1}$: 80%, with an AC-Fe ratio of 5% | [152] |
| ~1.2 L | Pb (spiked)/real soil/1500 mg $kg^{-1}$<br>Initial moisture content: 30% w/w | Use of solar energy and batteries in different kinds of soil | Removals after 7 days at 1.2 V $cm^{-1}$: 90.7%, 63.3%, and 42.8% for sandy, sandy loam, and silty loam soils respectively (compared to 53.7%, 31.2%, and 23.1%, respectively for an intermittent approach based on solar alone) | [133] |
| ~8.4 kg | Cr(VI) (spiked)/ sand/500 mg $kg^{-1}$<br>Initial moisture content: 30% w/w | Polyaniline auxiliary electrodes | Removals improved by 5–20% (after 120 h at 1 V $cm^{-1}$) | [163] |
| ~2.6 kg | Cr(VI) (spiked)/quartz sand/500 mg $kg^{-1}$<br>Initial moisture content: 34.6% w/w | Polypyrrole auxiliary electrodes | Removal improved by ~10% (after 72 h at 1 V $cm^{-1}$) | [168] |

## 4.1 Electrokinetic Treatment of Complex Contaminations

Both fundamental and applied studies (in laboratory) often focus on a single contaminant, having an organic or inorganic nature [176–178], as this simplified approach allows understanding the processes and operational variables, opening the way to the designing of systems for the treatment of complex contaminated sites [179].

In November 2015, the worst environmental disaster occurred in the history of Brazil, when the breakdown of a dam containing mineral sludge discharged 43 million cubic meters of material, contaminating with heavy metals the Gualaxo do Norte River, in the south-eastern state of Minas Gerais [179–181]. In other cases, sites prove to be contaminated by a combination of heavy metals and organic compounds [182], consequently requiring the application of synergistic approaches for the effective removal of the different contaminants. Such complex investigations have been evaluated in a limited number of studies.

Wang et al. [183] investigated the applicability of an electrokinetic remediation to treat kaolin contaminated with heavy metals (Cu and Pb) and organic compounds (*p*-xylene and phenanthrene), testing different current densities, various fluids for the cathodic chamber, treatment duration, and reactor size. They observed a high removal efficiency for *p*-xylene and phenanthrene (around 67% and 93%, respectively), but lower efficiencies for Cu and Pb (62% and 35%), during tests with a duration of 6 days. The low removal of Pb was attributed to precipitation in the cathodic section of the soil, with a consequent reduction in the electromigration of ionic species.

The removal of neutral contaminants by electroosmosis and electrophoresis was investigated by Ammami et al. [184], who evaluated different process fluids (nitric acid, citric acid, SDS, Tween 20) taken singly or as a mixture, in order to remove five heavy metals (Cd, Pb, Cr, Cu, and Zn) and five PAHs from a model-aged sediment (a mixture of kaolinite, silt, and sand). Nitric acid showed the highest removal efficiencies (76.8–99.9% in the case of metals, 70.3–89.7% for PAHs) in a single run (10–14 days under a constant voltage gradient of 1 V $cm^{-1}$). The more environmentally friendly mixture of Tween 20 and citric acid allowed the simultaneous removal of metals (up to 90.8%) and PAHs (up to 61.6%) from the fine-grained sediment, while the SDS mixed with citric acid yielded lower results (up to a removal of 65% for metals and 41% for PAHs).

The removal of PAHs and heavy metals from a polluted soil was also examined by Reddy et al. [173]. Their integrated use of hydraulic flushing and EKRT, using EDTA and a surfactant (5% Igepal CA-720), proved successful. The best results were obtained by using 0.2 M EDTA for soil flushing in two phases (without and with a voltage gradient of 1 V $cm^{-1}$), followed by a washing with Igepal, again in two stages: heavy metals were removed thanks to the formation of soluble species, with efficiencies ranging from 30% (for Cu) to 80% (for Pb; about 60% for Zn). During the subsequent flushing with Igepal, no heavy metals were removed, but the treatment was effective against PAHs (minus 40% of phenanthrene, minus 30% of pyrene, and minus 20% of benzo[a]pyrene). The use of the voltage gradient retarded the PAHs removal, while improving the displacement of the heavy metals.

Pazos et al. [185] investigated an electrokinetic process coupled with Fenton for the treatment of dredged marine sediment contaminated with heavy metals and TPHs. Electrokinetic treatments exploited EDTA and Tween 80 as process fluids to improve the removal of inorganic and organic contaminants, respectively. In addition, a Fenton oxidation combined with EDTA for the in situ degradation of TPHs and metal solubilization was examined. Approximately 90% of TPH and 55–60% of metals were removed after 30 days of treatment with EK-Fenton-EDTA. EDTA increased the solubility of metal species, as it forms soluble complexes in a wide range of pH. By exploiting the presence of iron in the sediment, the simple addition of $H_2O_2$ allowed a Fenton reaction that proved to be effective in degrading the TPHs and, consequently, in improving the mobilization of heavy metals.

Cameselle and Gouveia [93] investigated the phytoremediation enhanced with electric field for the remediation of soils contaminated with PAHs and heavy metals. *Brassica rapa* was selected among 14 plant species for its fast germination and growth even in soil purposely polluted with anthracene, phenanthrene, Cr(VI), $Pb^{2+}$, and $Cd^{2+}$. Although phytoremediation combined with EK did not provide better results than a simple electrokinetic approach, the removal of PAHs was significantly improved with the coupled technology, allowing removals higher than 80%. Interestingly, the best results were obtained using an alternating electric field, which avoids inducing physico-chemical changes in the soil and transporting or concentrating contaminants. The biological activity of the plants and soil microflora thus preserved leads to an optimal metabolism of organic contaminants.

## 4.2 Mathematical Modeling of Electrochemical Remediation

The coupling between the various transport processes and phenomena involved (including electrolysis of water and pH modifications) makes the design, analysis, and implementation of an electrokinetic process in a contaminated soil a very complex task. Since the electrokinetic technology is still under development, most of the information related to its mechanisms has inevitably been obtained through small-scale studies, in which the use of hermetic cylinders or prismatic cells as electrokinetic reactors is common [21, 186–188]. Investigations conducted on a lab-scale allow isolating the soil from the surrounding environment, considerably reducing the variables that need to be controlled during the treatment and allowing a detailed analysis of the various electrokinetic processes that occur in the soil. In this way, a suitable mathematical description of the system is also possible.

Increasingly, realistic mathematical models have been proposed, which allow to reliably characterize the interactions among the various physical, chemical, and electrochemical processes that occur simultaneously, thus facilitating the scalability of the technology.

Starting with very simple representations, and attempting progressive extensions and generalizations, several models have been developed:

- 1D models for decontamination of synthetic soil (kaolin) enriched with acetic acid and phenol [189, 190], with inclusion of coupling phenomena, adsorption, and precipitation/dissolution processes [191] and allowing the estimate of pH profiles generated in the soil contaminated with many metals [192].
- 2D models [193] focused on the electrokinetic removal of metals [194, 195] and organic pollutants [196].
- 3D models in unsaturated soils starting from the hypothesis that, during the application of an electric gradient, electric current lines in the soil are similar to those of a current of water produced by a hydraulic gradient [197, 198].

A rather common approach in the proposed models is the use of simplified geochemical systems that, taking into account the reagents and products of the chemical and electrochemical reactions that may occur, contain less than ten species. This limits the chemical speciation due to the electrokinetic processes, but also the ability to reproduce faithfully the behavior observed in real applications [199].

More recently, López-Vizcaíno et al. [200] have developed a model called M4EKR (Multiphysiscs for ElectroKinetic Remediation) for the removal of ionic species present in the pore water of unsaturated natural soils, which allow taking into account up to 34 species (involved in 24 chemical equilibria) and including their transport due to electroosmosis, electromigration, diffusion, and advection.

For more details on the mathematical models available and their features, please refer to the dedicated chapter.

## 5 Conclusions

The smart coupling of all electrokinetic phenomena with the heating of the soil (due to ohmic effects) and the various chemical (e.g., ion exchange, formation and dissolution of precipitates) and electrochemical processes (oxidation and reduction of water, deposition of metallic species) makes electrochemically assisted soil remediation a very versatile and efficient approach. However, the variety of the phenomena involved is also the reason behind its complexity, which hinders the development and large-scale application of the technology.

In disciplines such as chemical and environmental engineering, any scaling-up requires a deep understanding of processes, which can initially be characterized on a small scale using small devices (for which operating conditions can be more easily controlled and process details more easily clarified) [201]. The key factor for scalability is therefore the definition and understanding of "control mechanisms" on a real scale, rather than the study of the fundamentals of a process that, on a smaller scale and with more controlled conditions, can certainly be achieved with greater precision.

For this reason, there are numerous aspects that should be taken into consideration when performing laboratory tests aimed at studying treatment processes for which an effective field application of the technology is desirable [202]. For instance, in the first EK prototypes in the field, the electric heating of the soil was the main control mechanism, which produced a great volatilization of the organic contaminants contained in the soil. Unfortunately, the relevance of such a phenomenon in a laboratory-scale plant is very low because the low current intensity does not produce any significant change in temperature profile.

As widely discussed in this chapter, numerous studies have focused on the treatment of soils polluted by a wide variety of compounds, including heavy metals [17], PAHs [203], PCBs [204], pesticides and herbicides [124], among many others. From a scientific point of view, these works have generated a remarkable increase in knowledge in the sector. In addition, some researchers have taken a further step forward and started studying the scaling-up of the processes [205–208]. Among the scientific conclusions that have been obtained, it has been confirmed that large-scale studies cannot be avoided because otherwise the conclusions drawn could be significantly incorrect [88, 126, 128].

In any case, and despite research efforts, the level of technology readiness (TRL) for many of these technologies remains very low; although most are considered promising, many are far from being marketed as efficient processes. Important barriers need to be overcome to achieve high TRLs and, as a result, very few technologies are currently being applied on a large scale.

A key requirement for the applicability of electrochemical technology is the optimization of all processes to maximize treatment results [126, 128, 209]. This includes the implementation of suitable anodic materials but also the synergistic use of the cathodic reaction and the promotion of mediated oxidation processes during the treatment.

The high number of processes and the strong interactions of the parameters involved make each application a unique case, from which it is not possible to extend the application directly to other cases; however, it is possible to learn important lessons applicable to many other situations. As a result, many electrochemical technologies, including electrochemical oxidation of wastewater, electro-disinfection, electro-coagulation, and electrokinetic soil remediation, are currently at a high TRL [171], with many companies already offering some commercial solutions.

## References

1. R.A. Wuana, F.E. Okieimen, Heavy metals in contaminated soils: a review of sources, chemistry, risks and best available strategies for remediation. ISRN Ecol. **2011**, 1–11 (2011)
2. J. Ye, X. Chen, C. Chen, B. Bate, Emerging sustainable technologies for remediation of soils and groundwater in a municipal solid waste landfill site – a review. Chemosphere **227**, 681–702 (2019)

3. European Environmental Agency (EEA), 2007. Progress in Management of Contaminated Sites (CSI 015/LSI 003). http://www.eea.europa.eu/data-andmaps/indicators
4. P. Panagos, M. Van Liedekerke, Y. Yigini, L. Montanarella, Contaminated sites in Europe: review of the current situation based on data collected through a European network. J. Environ. Public Health **2013** (2013)
5. EPA, U.S.: Protecting and Restoring Land: Making a Visible Difference in Communities: OSWER FY13 End of Year Accomplishments Report (2013)
6. R. Naidu, V. Birke (eds.), *Permeable Reactive Barrier: Sustainable Groundwater Remediation*, vol 1 (CRC Press, Boca Raton, 2014)
7. Z. He, J. Shentu, X. Yang, V.C. Baligar, T. Zhang, P.J. Stoffella, Heavy metal contamination of soils: sources, indicators and assessment. J. Environ. Indic. **9**, 17–18 (2015)
8. E. Smith, P. Thavamani, K. Ramadass, R. Naidu, P. Srivastava, M. Megharaj, Remediation trials for hydrocarbon-contaminated soils in arid environments: evaluation of bioslurry and biopiling techniques. Int. Biodeterior. Biodegradation **101**, 56–65 (2015)
9. R. Naidu, Recent advances in contaminated site remediation. Water Air Soil Pollut. **224**(12), 1705 (2013)
10. M. Gavrilescu, L.V. Pavel, I. Cretescu, Characterization and remediation of soils contaminated with uranium. J. Hazard. Mater. **163**(2–3), 475–510 (2009)
11. C.N. Mulligan, R.N. Yong, B.F. Gibbs, Remediation technologies for metal-contaminated soils and groundwater: an evaluation. Eng. Geol. **60**(1–4), 193–207 (2001)
12. M. Farhadian, C. Vachelard, D. Duchez, C. Larroche, In situ bioremediation of monoaromatic pollutants in groundwater: a review. Bioresour. Technol. **99**(13), 5296–5308 (2008)
13. L.V. Pavel, M. Gavrilescu, Overview of ex situ decontamination techniques for soil cleanup. Environ. Eng. Manag. J. **7**(6), 815–834 (2008)
14. S.J. Varjani, Microbial degradation of petroleum hydrocarbons. Bioresour. Technol. **223**, 277–286 (2017)
15. B.M. da Silva, L.T. Maranho, Petroleum-contaminated sites: decision framework for selecting remediation technologies. J. Hazard. Mater. **378**, 120722 (2019)
16. F.I. Khan, T. Husain, R. Hejazi, An overview and analysis of site remediation technologies. J. Environ. Manag. **71**(2), 95–122 (2004)
17. A. Ferrucci, M. Vocciante, R. Bagatin, S. Ferro, Electrokinetic remediation of soils contaminated by potentially toxic metals: dedicated analytical tools for assessing the contamination baseline in a complex scenario. J. Environ. Manag. **203**, 1163–1168 (2017)
18. L. Liu, W. Li, W. Song, M. Guo, Remediation techniques for heavy metal-contaminated soils: principles and applicability. Sci. Total Environ. **633**, 206–219 (2018)
19. M.A. Rodrigo, N. Oturan, M.A. Oturan, Electrochemically assisted remediation of pesticides in soils and water: a review. Chem. Rev. **114**(17), 8720–8745 (2014)
20. S. Rodrigo, C. Saez, P. Cañizares, M.A. Rodrigo, Reversible electrokinetic adsorption barriers for the removal of organochlorine herbicide from spiked soils. Sci. Total Environ. **640**, 629–636 (2018)
21. D. Rosestolato, R. Bagatin, S. Ferro, Electrokinetic remediation of soils polluted by heavy metals (mercury in particular). Chem. Eng. J. **264**, 16–23 (2015)
22. M. Vocciante, A. Caretta, L. Bua, R. Bagatin, S. Ferro, Enhancements in ELECTROKINETIC remediation technology: environmental assessment in comparison with other configurations and consolidated solutions. Chem. Eng. J. **289**, 123–134 (2016)
23. Y.B. Acar, A.N. Alshawabkeh, Principles of electrokinetic remediation. Environ. Sci. Technol. **27**(13), 2638–2647 (1993)
24. M. Pazos, E. Rosales, T. Alcántara, J. Gómez, M.A. Sanromán, Decontamination of soils containing PAHs by electroremediation: a review. J. Hazard. Mater. **177**(1–3), 1–11 (2010)
25. A.B. Ribeiro, E.P. Mateus, J.M. Rodríguez-Maroto, Removal of organic contaminants from soils by an electrokinetic process: the case of molinate and bentazone. Experimental and modeling. Sep. Purif. Technol. **79**(2), 193–203 (2011)

26. C. Risco, S. Rodrigo, R. López-Vizcaíno, A. Yustres, C. Sáez, P. Cañizares, V. Navarro, M.A. Rodrigo, Electrochemically assisted fences for the electroremediation of soils polluted with 2,4-D: a case study in a pilot plant. Sep. Purif. Technol. **156**, 234–241 (2015)
27. R.E. Saichek, K.R. Reddy, Electrokinetically enhanced remediation of hydrophobic organic compounds in soils: a review. Crit. Rev. Environ. Sci. Technol. **35**(2), 115–192 (2005)
28. F.F. Reuss, Notice sur un nouvel effet de l'électricité galvanique. Memoires de la Société Impériale des Naturalistes de Moscou **2**, 327–337 (1809)
29. K.R. Reddy, R.E. Saichek, Effect of soil type on electrokinetic removal of phenanthrene using surfactants and cosolvents. J. Environ. Eng. **129**(4), 336–346 (2003)
30. M. Schnarr, C. Truax, G. Farquhar, E. Hood, T. Gonullu, B. Stickney, Laboratory and controlled field experiments using potassium permanganate to remediate trichloroethylene and perchloroethylene DNAPLs in porous media. J. Contam. Hydrol. **29**(3), 205–224 (1998)
31. K.R. Reddy, S. Danda, R.E. Saichek, Complicating factors of using ethylenediamine tetraacetic acid to enhance electrokinetic remediation of multiple heavy metals in clayey soils. J. Environ. Eng. **130**(11), 1357–1366 (2004)
32. A. McBratney, D.J. Field, A. Koch, The dimensions of soil security. Geoderma **213**, 203–213 (2014)
33. M. Vocciante, R. Bagatin, S. Ferro, Enhancements in electrokinetic remediation technology: focus on water management and wastewater recovery. Chem. Eng. J. **309**, 708–716 (2017)
34. Y.B. Acar, R.J. Gale, A.N. Alshawabkeh, R.E. Marks, S. Puppala, M. Bricka, R. Parker, Electrokinetic remediation: basics and technology status. J. Hazard. Mater. **40**(2), 117–137 (1995)
35. K.R. Reddy, C. Cameselle, *Electrochemical Remediation Technologies for Polluted Soils, Sediments and Groundwater* (John Wiley & Sons, Hoboken, 2009)
36. I. Llorente, S. Fajardo, J.M. Bastidas, Applications of electrokinetic phenomena in materials science. J. Solid State Electrochem. **18**(2), 293–307 (2014)
37. I.M.V. Rocha, K.N.O. Silva, D.R. Silva, C.A. Martínez-Huitle, E.V. dos Santos, Coupling electrokinetic remediation with phytoremediation for depolluting soil with petroleum and the use of electrochemical technologies for treating the effluent generated. Sep. Purif. Technol. **208**, 194–200 (2019)
38. R.L. López-Vizcaíno, A. Yustres, L. Asensio, C. Saez, P. Cañizares, M.A. Rodrigo, V. Navarro, Enhanced electrokinetic remediation of polluted soils by anolyte pH conditioning. Chemosphere **199**, 477–485 (2018)
39. A.N. Alshawabkeh, A.T. Yeung, M.R. Bricka, Practical aspects of in-situ electrokinetic extraction. J. Environ. Eng. **125**(1), 27–35 (1999a)
40. A.N. Alshawabkeh, R.J. Gale, E. Ozsu-Acar, M.R. Bricka, Optimization of 2-D electrode configuration for electrokinetic remediation. J. Soil Contam. **8**(6), 617–635 (1999b)
41. K.R. Reddy, Technical challenges to in-situ remediation of polluted sites. Geotech. Geol. Eng. **28**(3), 211–221 (2010)
42. H.R. Kruyt, G.H. Jonker, J.T.G. Overbeek, Colloid Science, in *Irreversible Systems*, vol. 1, (Elsevier, New York, 1952)
43. M.M. Page, C.L. Page, Electroremediation of contaminated soils. J. Environ. Eng. **128**(3), 208–219 (2002)
44. J. Yuan, M.A. Hicks, Numerical evaluation of optimal approaches for electro-osmosis dewatering. Dry. Technol. **36**(8), 973–989 (2018)
45. E.V. dos Santos, F. Souza, C. Saez, P. Cañizares, M.R.V. Lanza, C.A. Martinez-Huitle, M.A. Rodrigo, Application of electrokinetic soil flushing to four herbicides: a comparison. Chemosphere **153**, 205–211 (2016)
46. E. Mena, C. Ruiz, J. Villaseñor, M.A. Rodrigo, P. Cañizares, Biological permeable reactive barriers coupled with electrokinetic soil flushing for the treatment of diesel-polluted clay soil. J. Hazard. Mater. **283**, 131–139 (2015)
47. R. Porrett Jr., Curious galvanic experiments. Ann. Philos. **8**, 74–76 (1816)
48. L. Casagrande, Electro-osmosis in soils. Géotechnique **1**, 159–177 (1949)

49. Hamnett, R.: A Study of the Processes Involved in the Electro-Reclamation of Contaminated Soils. M.Sc. Thesis, University of Manchester (1980)
50. B.A. Segall, J.A. Matthias, C.E. O'Bannon, Electro-osmosis chemistry and water quality. J. Geotech. Eng. Div. **106**(10), 1148–1152 (1980)
51. S.K. Puppala, A.N. Alshawabkeh, Y.B. Acar, R.J. Gale, M. Bricka, Enhanced electrokinetic remediation of high sorption capacity soil. J. Hazard. Mater. **55**(1–3), 203–220 (1997)
52. Z. Li, J.W. Yu, I. Neretnieks, Electroremediation: removal of heavy metals from soils by using cation selective membrane. Environ. Sci. Technol. **32**(3), 394–397 (1998)
53. L.M. Ottosen, H.K. Hansen, S. Laursen, A. Villumsen, Electrodialytic remediation of soil polluted with copper from wood preservation industry. Environ. Sci. Technol. **31**(6), 1711–1715 (1997)
54. J.S. Wong, R.E. Hicks, R.F. Probstein, EDTA-enhanced electroremediation of metal-contaminated soils. J. Hazard. Mater. **55**(1–3), 61–79 (1997)
55. S.-O. Ko, M.A. Schlautman, and E.R. Carraway, Cyclodextrin-enhanced electrokinetic removal of phenanthrene from a model clay soil. Environ. Sci. Technol. **34**(8), 1535–1541 (2000). https://doi.org/10.1021/es990223t
56. K.-J. Kim, D.-H. Kim, J.-C. Yoo, K. Baek, Electrokinetic extraction of heavy metals from dredged marine sediment. Sep. Purif. Technol. **79**, 164–169 (2011)
57. F. Rozas, M. Castellote, Electrokinetic remediation of dredged sediments polluted with heavy metals with different enhancing electrolytes. Electrochim. Acta **86**, 102–109 (2012)
58. R. Iannelli, M. Masi, A. Ceccarini, M.B. Ostuni, R. Lageman, A. Muntoni, ... R. Pomi, Electrokinetic remediation of metal-polluted marine sediments: experimental investigation for plant design. Electrochim. Acta **181**, 146–159 (2015)
59. A. Giannis, D. Pentari, J.Y. Wang, E. Gidarakos, Application of sequential extraction analysis to electrokinetic remediation of cadmium, nickel and zinc from contaminated soils. J. Hazard. Mater. **184**(1e3), 547e554 (2010). https://doi.org/10.1016/j.jhazmat.2010.08.070
60. Y. Song, M.T. Ammami, A. Benamar, S. Mezazigh, H. Wang, Effect of EDTA, EDDS, NTA and citric acid on electrokinetic remediation of As, Cd, Cr, Cu, Ni, Pb and Zn from dredged marine sediment. Environ. Sci. Pollut. Res. **23**(11), 10577e10589 (2016). https://doi.org/10.1007/s11356-015-5966-5
61. Q.Y. Wang, D.M. Zhou, L. Cang, T.R. Sun, Application of bioassays to evaluate a copper contaminated soil before and after a pilot-scale electrokinetic remediation. Environ. Pollut. **157**(2), 410e416 (2009). https://doi.org/10.1016/j.envpol.2008.09.036
62. D.T. Moussa, M.H. El-Naas, M. Nasser, M.J. Al-Marri, A comprehensive review of electrocoagulation for water treatment: potentials and challenges. J. Environ. Manag. **186**, 24–41 (2017)
63. A. Mukimin, N. Zen, A. Purwanto, K.A. Wicaksono, H. Vistanty, A.S. Alfauzi, Application of a full-scale electrocatalytic reactor as real batik printing wastewater treatment by indirect oxidation process. J. Environ. Chem. Eng. **5**(5), 5222–5232 (2017)
64. M. Munoz-Morales, C. Sáez, P. Cañizares, M.A. Rodrigo, A new strategy for the electrolytic removal of organics based on adsorption onto granular activated carbon. Electrochem. Commun. **90**, 47–50 (2018)
65. R. Araújo, A.C.M. Castro, J.S. Baptista, A. Fiúza, Nanosized iron based permeable reactive barriers for nitrate removal–systematic review. Phys. Chem. Earth A/B/C **94**, 29–34 (2016)
66. A.S. Naje, S. Chelliapan, Z. Zakaria, M.A. Ajeel, P.A. Alaba, A review of electrocoagulation technology for the treatment of textile wastewater. Rev. Chem. Eng. **33**(3), 263–292 (2017)
67. S.D. Warner, S. Ferro, A. De Battisti, Barriere permeabili reattive – Considerazioni generali e stato dell'arte. Chim. Ind. **88**(4), 44–46 (2006)
68. P.V. Nidheesh, M. Zhou, M.A. Oturan, An overview on the removal of synthetic dyes from water by electrochemical advanced oxidation processes. Chemosphere **197**, 210–227 (2018)
69. A. Colacicco, G. De Gioannis, A. Muntoni, E. Pettinao, A. Polettini, R. Pomi, Enhanced electrokinetic treatment of marine sediments contaminated by heavy metals and PAHs. Chemosphere **81**(1), 46–56 (2010)

70. J.N. Hahladakis, W. Calmano, E. Gidarakos, Use and comparison of the non-ionic surfactants Poloxamer 407 and Nonidet P40 with HP-β-CD cyclodextrin, for the enhanced electroremediation of real contaminated sediments from PAHs. Sep. Purif. Technol. **113**, 104–113 (2013)
71. R. López-Vizcaíno, C. Sáez, P. Cañizares, V. Navarro, M.A. Rodrigo, Influence of the type of surfactant on the mobility of flushing fluids for electro-remediation processes. Sep. Sci. Technol. **46**(13), 2148–2156 (2011)
72. M.T. Alcántara, J. Gómez, M. Pazos, M.A. Sanromán, Combined treatment of PAHs contaminated soils using the sequence extraction with surfactant–electrochemical degradation. Chemosphere **70**(8), 1438–1444 (2008)
73. G. Fan, L. Cang, G. Fang, D. Zhou, Surfactant and oxidant enhanced electrokinetic remediation of a PCBs polluted soil. Sep. Purif. Technol. **123**, 106–113 (2014)
74. K. Maturi, K.R. Reddy, Cosolvent-enhanced desorption and transport of heavy metals and organic contaminants in soils during electrokinetic remediation. Water Air Soil Pollut. **189**(1–4), 199–211 (2008)
75. S. Cappello, C. Cruz Viggi, M. Yakimov, S. Rossetti, B. Matturro, L. Molina, A. Segura, S. Marqués, L. Yuste, E. Sevilla, F. Rojo, A. Sherry, O.K. Mejeha, I.M. Head, L. Malmquist, J.H. Christensen, N. Kalogerakis, F. Aulenta, Combining electrokinetic transport and bioremediation for enhanced removal of crude oil from contaminated marine sediments: results of a long-term, mesocosm-scale experiment. Water Res. **157**, 381–395 (2019)
76. F. Aulenta, A. Canosa, P. Reale, S. Rossetti, S. Panero, M. Majone, Microbial reductive dechlorination of trichloroethene to ethene with electrodes serving as electron donors without the external addition of redox mediators. Biotechnol. Bioeng. **103**, 85–91 (2009)
77. C. Cruz Viggi, E. Presta, M. Bellagamba, S. Kaciulis, S.K. Balijepalli, G. Zanaroli, M. Petrangeli Papini, S. Rossetti, F. Aulenta, The "Oil-Spill Snorkel": an innovative bioelectrochemical approach to accelerate hydrocarbons biodegradation in marine sediments. Front. Microbiol. **6**, 881 (2015)
78. M. Daghio, E. Vaiopoulou, S.A. Patil, A. Suárez-Suárez, I.M. Head, A. Franzetti, K. Rabaey, Anodes stimulate anaerobic toluene degradation via sulphur cycling in marine sediments. Appl. Environ. Microbiol. **82**, 297–307 (2016)
79. M. Daghio, A. Espinoza Tofalos, B. Leoni, P. Cristiani, M. Papacchini, E. Jalilnejad, G. Bestetti, A. Franzetti, Bioelectrochemical BTEX removal at different voltages: assessment of the degradation and characterization of the microbial communities. J. Hazard. Mater. **341**, 120–127 (2018)
80. A. Domínguez-Garay, J.R. Quejigo, U. Dörfler, R. Schroll, A. Esteve-Núñez, Bioelectroventing: an electrochemical-assisted bioremediation strategy for cleaning-up atrazine-polluted soils. Microb. Biotechnol. **11**, 50–62 (2018)
81. X. Li, X. Wang, Z.J. Ren, Y. Zhang, N. Li, Q. Zhou, Sand amendment enhances bioelectrochemical remediation of petroleum hydrocarbon contaminated soil. Chemosphere **141**, 62–70 (2015)
82. D. Mao, L. Lu, A. Revil, Y. Zuo, J. Hinton, Z.J. Ren, Geophysical monitoring of hydrocarbon-contaminated soils remediated with a bioelectrochemical system. Environ. Sci. Technol. **50**, 8205–8213 (2016)
83. O. Modin, F. Aulenta, Three promising applications of microbial electrochemistry for the water sector. Environ. Sci. Water Res. Technol. **3**, 391–402 (2017)
84. J. Rakoczy, S. Feisthauer, K. Wasmund, P. Bombach, T.R. Neu, C. Vogt, H.H. Richnow, Benzene and sulfide removal from groundwater treated in a microbial fuel cell. Biotechnol. Bioeng. **110**, 3104–3113 (2013)
85. J. Rodrigo Quejigo, U. Dörfler, R. Schroll, A. Esteve-Núñez, Stimulating soil microorganisms for mineralizing the herbicide isoproturon by means of microbial electroremediating cells. Microb. Biotechnol. **9**, 369–380 (2016)
86. H. Wang, H. Luo, P.H. Fallgren, S. Jin, Z.J. Ren, Bioelectrochemical system platform for sustainable environmental remediation and energy generation. Biotechnol. Adv. **33**, 317–334 (2015)

87. T. Zhang, S.M. Gannon, K.P. Nevin, A.E. Franks, D.R. Lovley, Stimulating the anaerobic degradation of aromatic hydrocarbons in contaminated sediments by providing an electrode as the electron acceptor. Environ. Microbiol. **12**, 1011–1020 (2010)
88. S. Barba, R. López-Vizcaíno, C. Saez, J. Villaseñor, P. Cañizares, V. Navarro, M.A. Rodrigo, Electro-bioremediation at the prototype scale: what it should be learned for the scale-up. Chem. Eng. J. **334**, 2030–2038 (2018)
89. F. Yan, D.D. Reible, PAH degradation and redox control in an electrode enhanced sediment cap. J. Chem. Technol. Biotechnol. **87**, 1222–1228 (2012)
90. F. Yan, D. Reible, Electro-bioremediation of contaminated sediment by electrode enhanced capping. J. Environ. Manag. **155**, 154–161 (2015)
91. F. Yan, W. Chen, D. Reible, Electrochemical stimulation of PAH biodegradation in sediment. Soil Sediment Contam. **24**, 143–156 (2015)
92. M. Bellagamba, C. Cruz Viggi, N. Ademollo, S. Rossetti, F. Aulenta, Electrolysis-driven bioremediation of crude oil-contaminated marine sediments. Nat. Biotechnol. **38**, 84–90 (2017)
93. C. Cameselle, S. Gouveia, Electrokinetic remediation for the removal of organic contaminants in soils. Curr. Opin. Electrochem. **11**, 41–47 (2018)
94. R.T. Gill, M.J. Harbottle, J.W.N. Smith, S.F. Thornton, Electrokinetic-enhanced bioremediation of organic contaminants: a review of processes and environmental applications. Chemosphere **107**, 31–42 (2014)
95. A.T. Yeung, Y.Y. Gu, A review on techniques to enhance electrochemical remediation of contaminated soils. J. Hazard. Mater. **195**, 11–29 (2011)
96. S.H. Kim, H.Y. Han, Y.J. Lee, C.W. Kim, J.W. Yang, Effect of electrokinetic remediation on indigenous microbial activity and community within diesel contaminated soil. Sci. Total Environ. **408**(16), 3162–3168 (2010)
97. N. Velasco-Alvarez, I. González, P. Damian-Matsumura, M. Gutiérrez-Rojas, Enhanced hexadecane degradation and low biomass production by Aspergillus niger exposed to an electric current in a model system. Bioresour. Technol. **102**(2), 1509–1515 (2011)
98. D.D. Aćimović, S.D. Karić, Ž.M. Nikolić, T.P. Brdarić, G.S. Tasić, M.P.M. Kaninski, V.M. Nikolić, Electrochemical oxidation of the polycyclic aromatic hydrocarbons in polluted concrete of the residential buildings. Environ. Pollut. **220**, 393–399 (2017)
99. H. Huang, J. Tang, Z. Niu, J.P. Giesy, Interactions between electrokinetics and rhizoremediation on the remediation of crude oil-contaminated soil. Chemosphere **229**, 418–425 (2019)
100. S. Lemström, *Electricity in Agriculture and Horticulture* ("The Electrician" Printing & Publishing Company, Limited, London, 1904)
101. G. Acosta-Santoyo, C. Cameselle, E. Bustos, Electrokinetic–enhanced ryegrass cultures in soils polluted with organic and inorganic compounds. Environ. Res. **158**, 118–125 (2017)
102. C.S. O'connor, N.W. Lepp, R. Edwards, G. Sunderland, The combined use of electrokinetic remediation and phytoremediation to decontaminate metal-polluted soils: a laboratory-scale feasibility study. Enviro. Monit. Assess. **84**(1–2), 141–158 (2003)
103. A. Denvir, D. Hodko, J. Van Hyfte, J.W. Magnuson, *U.S. Patent No. 6,145,244* (U.S. Patent and Trademark Office, Washington, DC, 2000)
104. H. Aboughalma, R. Bi, M. Schlaak, Electrokinetic enhancement on phytoremediation in Zn, Pb, Cu and Cd contaminated soil using potato plants. J. Environ. Sci. Health A **43**(8), 926–933 (2008)
105. C. Cameselle, S. Gouveia, S. Urréjola, Benefits of phytoremediation amended with DC electric field. Application to soils contaminated with heavy metals. Chemosphere **229**, 481–488 (2019)
106. R.A. Chirakkara, K.R. Reddy, C. Cameselle, Electrokinetic amendment in phytoremediation of mixed contaminated soil. Electrochim. Acta **181**, 179–191 (2015)
107. R. Bi, M. Schlaak, E. Siefert, R. Lord, H. Connolly, Alternating current electrical field effects on lettuce (*Lactuca sativa*) growing in hydroponic culture with and without cadmium contamination. J. Appl. Electrochem. **40**(6), 1217–1223 (2010)
108. R. Bi, M. Schlaak, E. Siefert, R. Lord, H. Connolly, Influence of electrical fields (AC and DC) on phytoremediation of metal polluted soils with rapeseed (*Brassica napus*) and tobacco (*Nicotiana tabacum*). Chemosphere **83**(3), 318–326 (2011)

109. M.R. Cho, H.S. Thatte, M.T. Silvia, D.E. Golan, Transmembrane calcium influx induced by ac electric fields. FASEB J. **13**(6), 677–683 (1999)
110. Y. Cong, Q. Ye, Z. Wu, Electrokinetic behaviour of chlorinated phenols in soil and their electrochemical degradation. Process Saf. Environ. Prot. **83**(2), 178–183 (2005)
111. B. Murillo-Rivera, I. Labastida, J. Barrón, M.T. Oropeza-Guzman, I. González, M.M.M. Teutli-Leon, Influence of anolyte and catholyte composition on TPHs removal from low permeability soil by electrokinetic reclamation. Electrochim. Acta **54**(7), 2119–2124 (2009)
112. B.C. Martin, S.J. George, C.A. Price, M.H. Ryan, M. Tibbett, The role of root exuded low molecular weight organic anions in facilitating petroleum hydrocarbon degradation: current knowledge and future directions. Sci. Total Environ. **472**, 642–653 (2014)
113. F. Rozas, M. Castellote, Selecting enhancing solutions for electrokinetic remediation of dredged sediments polluted with fuel. J. Environ. Manag. **151**, 153–159 (2015)
114. C. Cameselle, R.A. Chirakkara, K.R. Reddy, Electrokinetic-enhanced phytoremediation of soils: status and opportunities. Chemosphere **93**(4), 626–636 (2013)
115. Jamari, S., Embong, Z., & Bakar, I. (2014). Elemental Composition Study of Heavy Metal (Ni, Cu, Zn) in Riverbank Soil by Electrokinetic-Assisted Phytoremediation Using XRF and SEM/EDX. In AIP Conference Proceedings (1584, 1, pp. 221-227). American Institute of Physics, College Park
116. R.S. Putra, Y. Ohkawa, S. Tanaka, Application of EAPR system on the removal of lead from sandy soil and uptake by Kentucky bluegrass (*Poa pratensis* L.). Sep. Purif. Technol. **102**, 34–42 (2013)
117. J. Vidal, M. Carvela, C. Saez, P. Cañizares, V. Navarro, R. Salazar, M.A. Rodrigo, Testing different strategies for the remediation of soils polluted with Lindane. Chem. Eng. J., 122674 (2019)
118. I. Pikaar, E.M. Likosova, S. Freguia, J. Keller, K. Rabaey, Z. Yuan, Electrochemical abatement of hydrogen sulfide from waste streams. Crit. Rev. Environ. Sci. Technol. **45**(14), 1555–1578 (2015)
119. M. Resch, J. Bühler, B. Schachler, R. Kunert, A. Meier, A. Sumper, Technical and economic comparison of grid supportive vanadium redox flow batteries for primary control reserve and community electricity storage in Germany. Int. J. Energy Res. **43**(1), 337–357 (2019)
120. C. Trellu, E. Mousset, Y. Pechaud, D. Huguenot, E.D. Van Hullebusch, G. Esposito, M.A. Oturan, Removal of hydrophobic organic pollutants from soil washing/flushing solutions: a critical review. J. Hazard. Mater. **306**, 149–174 (2016)
121. Y. Song, L. Cang, G. Fang, S.T. Ata-Ul-Karim, H. Xu, D. Zhou, Electrokinetic delivery of anodic in situ generated active chlorine to remediate diesel-contaminated sand. Chem. Eng. J. **337**, 499–505 (2018)
122. S.J. Seo, J.H. Kim, J.W. Shin, J.Y. Park, Treatment of artificial and real co-contaminated soil by an enhanced electrokinetic-Fenton process with a soil flushing method. Water Air Soil Pollut. **226**(4), 86 (2015)
123. I.C. Paixão, R. López-Vizcaíno, A.M.S. Solano, C.A. Martínez-Huitle, V. Navarro, M.A. Rodrigo, E.V. dos Santos, Electrokinetic-Fenton for the remediation low hydraulic conductivity soil contaminated with petroleum. Chemosphere (2020). https://doi.org/10.1016/j.chemosphere.2020.126029
124. E.V. dos Santos, C. Sáez, P. Cañizares, C.A. Martínez-Huitle, M.A. Rodrigo, Reversible electrokinetic adsorption barriers for the removal of atrazine and oxyfluorfen from spiked soils. J. Hazard. Mater. **322**, 413–420 (2017)
125. A. Karagunduz, A. Gezer, G. Karasuloglu, Surfactant enhanced electrokinetic remediation of DDT from soils. Sci. Total Environ. **385**(1–3), 1–11 (2007)
126. R. López-Vizcaíno, A. Yustres, C. Sáez, P. Cañizares, M.A. Rodrigo, V. Navarro, Effect of polarity reversal on the enhanced electrokinetic remediation of 2, 4-D-polluted soils: a numerical study. Electrochim. Acta **258**, 414–422 (2017a)
127. Z. Li, S. Yuan, J. Wan, H. Long, M. Tong, A combination of electrokinetics and Pd/Fe PRB for the remediation of pentachlorophenol-contaminated soil. J. Contam. Hydrol. **124**(1–4), 99–107 (2011)

128. R. López-Vizcaíno, C. Risco, J. Isidro, S. Rodrigo, C. Saez, P. Cañizares, V. Navarro, M.A. Rodrigo, Scale-up of the electrokinetic fence technology for the removal of pesticides. Part II: Does size matter for removal of herbicides? Chemosphere **166**, 549–555 (2017b)
129. A.N. Alshawabkeh, Electrokinetic soil remediation: challenges and opportunities. Sep. Sci. Technol. **44**(10), 2171–2187 (2009)
130. R.D. Fan, S.Y. Liu, Y.J. Du, K.R. Reddy, Y.L. Yang, Impacts of presence of lead contamination on settling behavior and microstructure of clayey soil-calcium bentonite blends. Appl. Clay Sci. **142**, 109–119 (2017)
131. C. Cameselle, Enhancement of electro-osmotic flow during the electrokinetic treatment of a contaminated soil. Electrochim. Acta **181**, 31–38 (2014)
132. A.T. Yeung, C.N. Hsu, R.M. Menon, Physicochemical soil-contaminant interactions during electrokinetic extraction. J. Hazard. Mater. **55**, 221–237 (1997)
133. A.A. Hussein, M.J.A. Alatabe, Remediation of lead-contaminated soil, using clean energy in combination with electro-kinetic methods. Pollution **5**(4), 859–869 (2019)
134. M. Villen-Guzman, J. Paz-Garcia, G. Amaya-Santos, J.M. Rodriguez-Maroto, C. Vereda-Alonso, C. Gomez-Lahoz, Effects of the buffering capacity of the soil on the mobilization of heavy metals. Equilibrium and kinetics. Chemosphere **131**, 78–84 (2015a)
135. M. Villen-Guzman, J. Paz-Garcia, G. Amaya-Santos, J.M. Rodriguez-Maroto, C. Vereda-Alonso, C. Gomez-Lahoz, Scaling-up the acid-enhanced electrokinetic remediation of a real contaminated soil. Electrochim. Acta **181**, 139–145 (2015b)
136. Klouche, F., Bendani, K., Benamar, A., Missoum, H., Maliki, M., & Laredj, N., in *Electrokinetic Restoration of Local Saline Soil*. Materials Today: Proceedings (2019) https://doi.org/10.1016/j.matpr.2019.08.082
137. A. Giannis, A. Nikolaou, D. Pentari, E. Gidarakos, Chelating agent-assisted electrokinetic removal of cadmium, lead and copper from contaminated soils. Environ. Pollut. **157**(12), 3379–3386 (2009)
138. A. Giannis, E. Gidarakos, A. Skouta, Application of sodium dodecyl sulfate and humic acid as surfactants on electrokinetic remediation of cadmium-contaminated soil. Desalination **211**(1–3), 249–260 (2007)
139. Y.S. Ng, B.S. Gupta, M.A. Hashim, Remediation of Pb/Cr co-contaminated soil using electrokinetic process and approaching electrode technique. Environ. Sci. Pollut. Res. **23**(1), 546–555 (2016)
140. R. Fu, D. Wen, X. Xia, W. Zhang, Y. Gu, Electrokinetic remediation of chromium (Cr)-contaminated soil with citric acid (CA) and polyaspartic acid (PASP) as electrolytes. Chem. Eng. J. **316**, 601–608 (2017)
141. Y. Gu, A.T. Yeung, H. Li, Enhanced electrokinetic remediation of cadmium-contaminated natural clay using organophosphonates in comparison with EDTA. Chin. J. Chem. Eng. **26**(5), 1152–1159 (2018a)
142. M. Masi, R. Iannelli, G. Losito, Ligand-enhanced electrokinetic remediation of metal-contaminated marine sediments with high acid buffering capacity. Environ. Sci. Pollut. Res. **23**(11), 10566–10576 (2016)
143. F. Meng, H. Xue, Y. Wang, B. Zheng, J. Wang, Citric-acid preacidification enhanced electrokinetic remediation for removal of chromium from chromium-residue-contaminated soil. Environ. Technol. **39**(3), 356–362 (2018)
144. A.T. Yeung, Y.Y. Gu, *Use of Chelating Agents in Electrochemical Remediation of Contaminated Soil. Chelating Agents for Land Decontamination Technologies* (American Society of Civil Engineers, Reston, 2012), pp. 212–280
145. R. Bassi, S.O. Prasher, B.K. Simpson, Extraction of metals from a contaminated sandy soil using citric acid. Environ. Prog. **19**(4), 275–282 (2000)
146. K. Fischer, H.P. Bipp, P. Riemschneider, P. Leidmann, D. Bieniek, A. Kettrup, Utilization of biomass residues for the remediation of metal-polluted soils. Environ. Sci. Technol. **32**(14), 2154–2161 (1998)
147. J.W. Neilson, J.F. Artiola, R.M. Maier, Characterization of lead removal from contaminated soils by nontoxic soil-washing agents. J. Environ. Qual. **32**(3), 899–908 (2003)

148. F. Liu, Non-uniform electrokinetic removal of heavy metals from contaminated soil with permeable reactive composite electrodes. Appl. Mech. Mater. **260**, 1145–1150 (2013)
149. H. Hu, An enhanced remediation technology for contaminated groundwater. Appl. Mech. Mater. **253–255**, 1089–1092 (2013)
150. R.B. Fu, X.X. Liu, F. Liu, J. Ma, Y.M. Ma, C.B. Zhang, J. Zhu, Electrokinetic remediation of heavy metal contaminated soil using permeable reactive composite electrodes. Adv. Mater. Res. **518**, 361–368 (2012)
151. N. Shariatmadari, C.H. Weng, H. Daryaee, Enhancement of hexavalent chromium [Cr (VI)] remediation from clayey soils by electrokinetics coupled with a nano-sized zero-valent iron barrier. Environ. Eng. Sci. **26**(6), 1071–1079 (2009)
152. Y. Yan, F. Xue, F. Muhammad, L. Yu, F. Xu, B. Jiao, Y. Shiau, D. Li, Application of iron-loaded activated carbon electrodes for electrokinetic remediation of chromium-contaminated soil in a three-dimensional electrode system. Sci. Rep. **8**(1), 1–11 (2018)
153. Y. García, C. Ruiz, E. Mena, J. Villaseñor, P. Cañizares, M.A. Rodrigo, Removal of nitrates from spiked clay soils by coupling electrokinetic and permeable reactive barrier technologies. J. Chem. Technol. Biotechnol. **90**(9), 1719–1726 (2015)
154. S. Adikesavan, A. Rajasekar, A statistical approach of zinc remediation using acidophilic bacterium via an integrated approach of bioleaching enhanced electrokinetic remediation (BEER) technology. Chemosphere **207**, 753–763 (2018)
155. K.N. Silva, S.S. Paiva, F.L. Souza, D.R. Silva, C.A. Martínez-Huitle, E.V. dos Santos, Applicability of electrochemical technologies for removing and monitoring $Pb^{2+}$ from soil and water. J. Electroanal. Chem. **816**, 171–178 (2018)
156. Y.Y. Gu, C. Zhao, H. Li, H. An, The enhancement of synthesized wastewater on non-uniform electrokinetic remediation of a Cd-spiked natural clayey soil. Environ. Sci. Pollut. Res. **25**(2), 1103–1114 (2018b)
157. C. Dias-Ferreira, G.M. Kirkelund, L.M. Ottosen, Ammonium citrate as enhancement for electrodialytic soil remediation and investigation of soil solution during the process. Chemosphere **119**, 889–895 (2015)
158. L. Ding, W. Lv, K. Yao, L. Li, M. Wang, G. Liu, Remediation of Cd (II)-contaminated soil via humin-enhanced electrokinetic technology. Environ. Sci. Pollut. Res. **24**(4), 3430–3436 (2017)
159. A. Giannis, E. Gidarakos, Washing enhanced electrokinetic remediation for removal cadmium from real contaminated soil. J. Hazard. Mater. **123**(1–3), 165–175 (2005)
160. I.F. Poulsen, H.C.B. Hansen, Soil sorption of nickel in presence of citrate or arginine. Water Air Soil Pollut. **120**(3–4), 249–259 (2000)
161. O. Karaca, C. Cameselle, M. Bozcu, Opportunities of electrokinetics for the remediation of mining sites in Biga peninsula, Turkey. Chemosphere **227**, 606–613 (2019)
162. R. Ortiz-Soto, D. Leal, C. Gutierrez, A. Aracena, A. Rojo, H.K. Hansen, Electrokinetic remediation of manganese and zinc in copper mine tailings. J. Hazard. Mater. **365**, 905–911 (2019)
163. L. Wang, L. Huang, H. Xia, H. Li, X. Li, X. Liu, Application of a multi-electrode system with polyaniline auxiliary electrodes for electrokinetic remediation of chromium-contaminated soil. Sep. Purif. Technol. **224**, 106–112 (2019a)
164. X.J. Chen, Z.M. Shen, T. Yuan, S.S. Zheng, B.X. Ju, W.H. Wang, Enhancing electrokinetic remediation of cadmium-contaminated soils with stepwise moving anode method. J. Environ. Sci. Health A **41**, 2517–2530 (2006)
165. I. Hassan, E. Mohamedelhassan, E.K. Yanful, Solar powered electrokinetic remediation of Cu polluted soil using a novel anode configuration. Electrochim. Acta **181**, 58–67 (2015)
166. P. Lu, Q. Feng, Q. Meng, T. Yuan, Electrokinetic remediation of chromium-and cadmium-contaminated soil from abandoned industrial site. Sep. Purif. Technol. **98**, 216–220 (2012)
167. Z. Shi, W. Cong, X. Song, S. Jia, L. Wang, Stepwise simulation on the motion of a single cathode spot of vacuum arc in external transverse magnetic field. IEEE Trans. Plasma Sci. **43**, 472–479 (2015)

168. Y. Wang, L. Huang, Z. Wang, L. Wang, Y. Han, X. Liu, T. Ma, Application of polypyrrole flexible electrode for electrokinetic remediation of Cr(VI)-contaminated soil in a main-auxiliary electrode system. Chem. Eng. J. **373**, 131–139 (2019b)
169. J. Virkutyte, M. Sillanpää, P. Latostenmaa, Electrokinetic soil remediation—Critical overview. Sci. Total Environ. **289**(1–3), 97–121 (2002)
170. F.C. Walsh, L.F. Arenas, C. Ponce de León, Developments in electrode design: structure, decoration and applications of electrodes for electrochemical technology. J. Chem. Technol. Biotechnol. **93**(11), 3073–3090 (2018)
171. E. Lacasa, S. Cotillas, C. Saez, J. Lobato, P. Cañizares, M.A. Rodrigo, Environmental applications of electrochemical technology. What is needed to enable full-scale applications? Curr. Opin. Electrochem. **16**, 149–156 (2019)
172. K.-Y. Lee, K. Woongkim, Heavy metal removal from shooting range soil by hybrid electrokinetics with bacteria and enhancing agents. Environ. Sci. Technol. **44**, 9482–9487 (2010)
173. K.R. Reddy, C. Cameselle, P. Ala, Integrated electrokinetic-soil flushing to remove mixed organic and metal contaminants. J. Appl. Electrochem. **40**(6), 1269–1279 (2010)
174. B.-G. Ryu, G.-Y. Park, J.-W. Yang, K. Baek, Electrolyte conditioning for electrokinetic remediation of As, Cu, and Pb-contaminated soil. Sep. Purif. Technol. **79**, 170–176 (2011)
175. S.Y. Shin, S.M. Park, K. Baek, Soil moisture could enhance electrokinetic remediation of arsenic-contaminated soil. Environ. Sci. Pollut. Res. **24**(10), 9820–9825 (2017)
176. A.N. Alshawabkeh, R.M. Bricka, D.B. Gent, Pilot-scale electrokinetic cleanup of lead-contaminated soils. J. Geotech. Geoenviron. Eng. **131**(3), 283–291 (2005)
177. M. Castellote, S. Botija, C. Andrade, Assessment of electrophoresis and electroosmosis in construction materials: effect of enhancing electrolytes and heavy metals contamination. J. Appl. Electrochem. **40**(6), 1195–1208 (2010)
178. G. Peng, G. Tian, Using electrode electrolytes to enhance electrokinetic removal of heavy metals from electroplating sludge. Chem. Eng. J. **165**(2), 388–394 (2010)
179. F.R. Segura, E.A. Nunes, F.P. Paniz, A.C.C. Paulelli, G.B. Rodrigues, G.Ú.L. Braga, W. dos Reis Pedreira Filho, F. Bardosa Jr., F. Cerchiaro, F.F. Silva, B.L. Batista, Potential risks of the residue from Samarco's mine dam burst (Bento Rodrigues, Brazil). Environ. Pollut. **218**, 813–825 (2016)
180. V. da Penha Rhodes, J.C. de Lena, C.V.A. Santolin, T. da Silva Pinto, L.A. Mendes, C.C. Windmöller, Speciation and quantification of Hg in sediments contaminated by artisanal gold mining in the Gualaxo do Norte River, Minas Gerais, SE, Brazil. Environ. Monit. Assess. **190**(1), 49 (2018)
181. M. Marta-Almeida, R. Mendes, F.N. Amorim, M. Cirano, J.M. Dias, Fundão Dam collapse: oceanic dispersion of River Doce after the greatest Brazilian environmental accident. Mar. Pollut. Bull. **112**(1–2), 359–364 (2016)
182. M. Dubovina, D. Krčmar, N. Grba, M.A. Watson, D. Rađenović, D. Tomašević-Pilipović, B. Dalmacija, Distribution and ecological risk assessment of organic and inorganic pollutants in the sediments of the transnational Begej canal (Serbia-Romania). Environ. Pollut. **236**, 773–784 (2018)
183. J.Y. Wang, X.J. Huang, J.C. Kao, O. Stabnikova, Simultaneous removal of organic contaminants and heavy metals from kaolin using an upward electrokinetic soil remediation process. J. Hazard. Mater. **144**(1–2), 292–299 (2007)
184. M.T. Ammami, A. Benamar, H. Wang, C. Bailleul, M. Legras, F. Le Derf, F. Portet-Koltalo, Simultaneous electrokinetic removal of polycyclic aromatic hydrocarbons and metals from a sediment using mixed enhancing agents. Int. J. Environ. Sci. Technol. **11**(7), 1801–1816 (2014)
185. M. Pazos, O. Iglesias, J. Gómez, E. Rosales, M.A. Sanromán, Remediation of contaminated marine sediment using electrokinetic–Fenton technology. J. Ind. Eng. Chem. **19**(3), 932–937 (2013)
186. M.T. Alcántara, J. Gómez, M. Pazos, M.A. Sanromán, Electrokinetic remediation of lead and phenanthrene polluted soils. Geoderma **173**, 128–133 (2012)

187. Z.Y. Dong, W.H. Huang, D.F. Xing, H.F. Zhang, Remediation of soil co-contaminated with petroleum and heavy metals by the integration of electrokinetics and biostimulation. J. Hazard. Mater. **260**, 399–408 (2013)
188. T. Suzuki, K. Kawai, M. Moribe, M. Niinae, Recovery of Cr as Cr(III) from Cr (VI)-contaminated kaolinite clay by electrokinetics coupled with a permeable reactive barrier. J. Hazard. Mater. **278**, 297–303 (2014b)
189. A.P. Shapiro, Preliminary studies on the removal of chemical species from saturated porous media by electroosmosis. Phys. Chem. Hydrodyn. **11**(5), 785–802 (1989)
190. A.P. Shapiro, R.F. Probstein, Removal of contaminants from saturated clay by electroosmosis. Environ. Sci. Technol. **27**(2), 283–291 (1993)
191. R.A. Jacobs, M.Z. Sengun, R.E. Hicks, R.F. Probstein, Model and experiments on soil remediation by electric fields. J. Environ. Sci. Health A **29**(9), 1933–1955 (1994)
192. J.M. Paz-Garcia, K. Baek, I.D. Alshawabkeh, A.N. Alshawabkeh, A generalized model for transport of contaminants in soil by electric fields. J. Environ. Sci. Health A **47**(2), 308–318 (2012)
193. R.A. Jacobs, R.F. Probstein, Two-dimensional modeling of electroremediation. AICHE J. **42**(6), 1685–1696 (1996)
194. C. Vereda-Alonso, J.M. Rodrıguez-Maroto, R.A. Garcıa-Delgado, C. Gómez-Lahoz, F. Garcıa-Herruzo, Two-dimensional model for soil electrokinetic remediation of heavy metals: application to a copper spiked kaolin. Chemosphere **54**(7), 895–903 (2004)
195. C. Vereda-Alonso, C. Heras-Lois, C. Gomez-Lahoz, F. Garcia-Herruzo, J.M. Rodriguez-Maroto, Ammonia enhanced two-dimensional electrokinetic remediation of copper spiked kaolin. Electrochim. Acta **52**, 3366–3371 (2007)
196. A.B. Ribeiro, J.M. Rodriguez-Maroto, E.P. Mateus, H. Gomes, Removal of organic contaminants from soils by an electrokinetic process: the case of atrazine. Experimental and modeling. Chemosphere **59**, 1229–1239 (2005)
197. E.D. Mattson, R.S. Bowman, E.R. Lindgren, Electrokinetic ion transport through unsaturated soil: 1. Theory, model development, and testing. J. Contam. Hydrol. **54**(1–2), 99–120 (2002a)
198. E.D. Mattson, R.S. Bowman, E.R. Lindgren, Electrokinetic ion transport through unsaturated soil: 2. Application to a heterogeneous field site. J. Contam. Hydrol. **54**(1–2), 121–140 (2002b)
199. A.Z. Al-Hamdan, K.R. Reddy, Electrokinetic remediation modeling incorporating geochemical effects. J. Geotech. Geoenviron. **134**(1), 91–105 (2008)
200. R. López-Vizcaíno, A. Yustres, M.J. León, C. Saez, P. Cañizares, M.A. Rodrigo, V. Navarro, Multiphysics implementation of electrokinetic remediation models for natural soils and porewaters. Electrochim. Acta **225**, 93–104 (2017c)
201. R. López-Vizcaíno, C. Risco, J. Isidro, S. Rodrigo, C. Saez, P. Cañizares, V. Navarro, M.A. Rodrigo, Scale-up of the electrokinetic fence technology for the removal of pesticides. Part I: some notes about the transport of inorganic species. Chemosphere **166**, 540–548 (2017d)
202. R. López-Vizcaíno, V. Navarro, M.J. León, C. Risco, M.A. Rodrigo, C. Sáez, P. Cañizares, Scale-up on electrokinetic remediation: Engineering and technological parameters. J. Hazard. Mater. **315**, 135–143 (2016)
203. E. Bocos, C. Fernández-Costas, M. Pazos, M.Á. Sanromán, Removal of PAHs and pesticides from polluted soils by enhanced electrokinetic-Fenton treatment. Chemosphere **125**, 168–174 (2015)
204. K.B. Pedersen, T. Lejon, P.E. Jensen, L.M. Ottosen, Simultaneous electrodialytic removal of PAH, PCB, TBT and heavy metals from sediments. J. Environ. Manag. **198**, 192–202 (2017)
205. M. Gupta, N.K. Garg, H. Joshi, M.P. Sharma, Persistence and mobility of 2,4-D in unsaturated soil zone under winter wheat crop in sub-tropical region of India. Agric. Ecosyst. Environ. **146**, 60–72 (2012)
206. E.-K. Jeon, J.-M. Jung, W.-S. Kim, S.-H. Ko, K. Baek, In situ electrokinetic remediation of As-, Cu-, and Pb-contaminated paddy soil using hexagonal electrode configuration: a full scale study. Environ. Sci. Pollut. Res. **22**, 711–720 (2015)

207. D.-H. Kim, S.-U. Jo, J.-H. Choi, J.-S. Yang, K. Baek, Hexagonal two-dimensional electrokinetic systems for restoration of saline agricultural lands: a pilot study. Chem. Eng. J. **198**, 110–121 (2012)
208. Y.J. Lee, J.H. Choi, H.G. Lee, T.H. Ha, J.H. Bae, Pilot-scale study on in situ electrokinetic removal of nitrate from greenhouse soil. Sep. Purif. Technol. **79**, 254–263 (2011)
209. L. Lu, H. Yazdi, S. Jin, Y. Zuo, P.H. Fallgren, Z.J. Ren, Enhanced bioremediation of hydrocarbon-contaminated soil using pilot-scale bioelectrochemical systems. J. Hazard. Mater. **274**, 8–15 (2014)

# Fenton Processes for Remediation of Polluted Soils

**Aida María Díez, María Ángeles Sanromán, and Marta Pazos**

## 1 Introduction

Fenton reaction was first reported by Henry John Horstman Fenton in 1894 when oxidizing tartaric acid by hydrogen peroxide in the presence of iron. He observed the fast removal of the carboxylic acid when iron was present in the media. Haber and Weiss established in 1934 [1] the chain reactions that are the base of the Fenton process. The technology is based on the generation of hydroxyl radicals ($HO^{\bullet}$) by the catalytic decomposition of hydrogen peroxide ($H_2O_2$) [2, 3]. This radical, after $F^-$, is considered one of the most powerful oxidants (2.8 *vs* SHE) [4].

The Fenton reaction takes place under acid conditions where $H_2O_2$ suffers the homolytic O–O bond cleavage in presence of iron ($Fe^{2+}$) following Eq. (1). The maximum catalytic activity was established at pH 2.8–3.0, which drastically diminishes with an increase or a reduction of this pH value [5]. At pH higher than 3, the $H_2O_2$ breaks down into $O_2$ and $H_2O$ and $Fe^{3+}$ precipitates as $Fe(OH)_3$ [6]. Moreover, the formation of Fe(II) complexes at high pH values leads to a drop in the $Fe^{2+}$ concentration [7]. The generated $HO^{\bullet}$ starts a series of chain reactions [1] with the organic matter (RH) present in the aqueous media (Eq. 2), which would lead ultimately to complete mineralization, that is, the conversion of all the organic matter into $CO_2$, $H_2O$ and other gases ($N_2$, $Cl_2$, etc.) or ions ($SO_4^{2-}$, $PO_4^{3-}$, etc.). The $Fe^{3+}$ generated by the Eq. (1) is then reduced back to $Fe^{2+}$ by another molecule of $H_2O_2$, forming a hydroperoxyl radical ($HO_2^{\bullet}$) (Eq. 3) [8]. After the discovery of Fenton reaction, Norman and West [9] demonstrated that not only $Fe^{2+}$ was useful for decomposing $H_2O_2$ to $HO^{\bullet}$, but also other transition metals (M, Eq. 4) such as $Ti^{3+}$ or $Cu^{2+}$ can generate the radicals on the so-called Fenton-like process [3]. In addition, the Fenton process has also associated some scavenger reactions, which

A. M. Díez · M. Á. Sanromán · M. Pazos (✉)
BIOSUV Group, Department of Chemical Engineering, University of Vigo, Vigo, Spain
e-mail: mcurras@uvigo.es



M. A. Rodrigo, E. V. Dos Santos (eds.), *Electrochemically Assisted Remediation of Contaminated Soils*, Environmental Pollution 30,
https://doi.org/10.1007/978-3-030-68140-1_8

diminish the availability of $HO^{\bullet}$ or the $H_2O_2$ towards the generation of other less powerful oxidants or by-products with less oxidative capability [8] such as $HO_2\bullet$, $O_2^{-}\bullet$ or $HO_2^-$ (Eqs. 5–10) [10].

$$H_2O_2 + Fe^{2+} \rightarrow HO^{\bullet} + HO^- + Fe^{3+} \quad (1)$$

$$RH + HO^{\bullet} \rightarrow R^{\bullet} + H_2O + HO_2^{\bullet} \quad (2)$$

$$H_2O_2 + Fe^{3+} \rightarrow Fe^{2+} + HO_2^{\bullet} + H^+ \quad (3)$$

$$H_2O_2 + M^+ \rightarrow HO^{\bullet} + HO^- + M^{+1} \quad (4)$$

$$HO^{\bullet} + H_2O_2 \rightarrow HO_2^{\bullet} + H_2O \quad (5)$$

$$HO_2^{\bullet} \rightarrow O_2^{\bullet-} + H^+ \quad (6)$$

$$HO_2^{\bullet} + Fe^{2+} \rightarrow HO_2^- + Fe^{3+} \quad (7)$$

$$HO_2^{\bullet} + O_2^{\bullet-} \rightarrow HO_2^- + O_2 \quad (8)$$

$$HO^{\bullet} + Fe^{2+} \rightarrow HO^- + Fe^{3+} \quad (9)$$

$$HO^{\bullet} + HO^{\bullet} \rightarrow H_2O_2 \quad (10)$$

The use of this technology for the removal of organic matter in aqueous media has been deeply studied, and it has several advantages to be considered as a feasible treatment alternative such as (1) quick degradation, because the $HO^{\bullet}$ radicals attack the organic matter at diffusion controlled rate of $10^{10}$ $M^{-1}s^{-1}$ (in the case of aromatic compounds) [11], (2) nonselective degradation, being a process suitable for the treatment of whichever organic pollutant, and (3) the reaction products generated are fairly environmentally nonthreatening [12], converting hazardous pollutants into nonhazardous or less toxic compounds which are more stable, less mobile or inert [13]. For instance, in their review, Bautista et al. [7] reported the Fenton process as a solution for the toxicity reduction and biodegradability enhancement of diverse industrial wastewaters (chemical, pharmaceutical, textile, cosmetic, etc.).

In the 1980s, the outstanding results of Fenton process for the remediation of effluents opened the possibility for its application in the treatment of polluted soils. Hence, Bove et al. [14] reported, in 1983, the use of the Fenton process for ex situ treatment of soils polluted with organic compounds, surpassing the performance of biological processes, as they are usually extremely slow and ineffective in the presence of toxic initial products or generated by-products. Actually, Oturan and Aaron presented years later the usage of AOPs in soil remediation as a suitable alternative [6].

However, the expansion on the use of $H_2O_2$ for in situ soil remediation was performed during the 1990s [15]. Thus, based on the principles of in situ chemical oxidation process (ISCO) the Fenton reagent can be injected into the soil producing the $HO^{\bullet}$ inside the porous matrix and degrading the organic pollutant. [16]. In addition, the natural iron present in the soils evades the necessity of adding the catalyst, reducing the cost associated to the dose of reagents.

The process was presented as an interesting approach for soil decontamination when compared to ex situ treatments [17] because it avoids the difficult remediation and economic costs of other soil remediation alternatives. These alternatives encompass soil washing, incineration, vacuum extraction, biodigestion, or dig and dump [18], which may generate additional wastes that would need secondary treatment or even disposal. However, in order to achieve an efficient treatment, the controlled delivery and homogenous distribution of oxidants in the soil are required [19]. Thus, while several researches reported very encouraging results, the feasibility of the process was questioned for its application to some types of soils such those with low permeability. For instance, Watts et al. [20] reported total degradation of pentachlorophenol from a natural sandy soil collected from the Moreno Valley University of California in 5 h, by adding $H_2O_2$ and iron. However, in fine-grained soils with low permeability, the movement of $H_2O_2$ through the soil is limited and thus the efficiency of the process is reduced [12, 21]. Accordingly, Oonnittan et al. [12] reported how $H_2O_2$ may have a scavenger effect, as it reacts undesirably with $HO^{\bullet}$ instead of following the Fenton reaction. Actually, $H_2O_2$ is usually decomposed before reaching the polluted site because of the slow rate in the soil [8]. Nevertheless, Bryand and Wilson [22] applied for the first time the Fenton ISCO, treating chlorinated and petroleum compounds in real soils by in situ application of the Fenton process. They treated different soils, which varied from 222 to 28,889 tons although not in all cases the technology could be labelled as cost-effective.

Hence, the reported drawbacks in the use of Fenton reaction for soil remediation, appeared as new milestones in which the scientific community focused its attention (Fig. 1). As a result of these research works, Clarke et al. [18] patented in January 1999 the novel idea of combining ElectroKinetic (EK) Process and Fenton reaction (F), the so-called ElectroKinetic-Fenton (EK-F) Treatment. Regarding the EK process, it had been developed in 1960 for the remediation of metal polluted soils [23] whereas this new EK-F process was proposed as an efficient soil remediation alternative, which would be useful for sites polluted with both organic and inorganic species [16].

Few months after the publication of the Clarke et al. patent [18], the first article of the proposed technology, entitled "Removal and degradation of phenol in a saturated flow by **in situ** ElectroKinetic remediation and Fenton-Like process," was presented by Yang and Long [24]. In this study, a phenol polluted sandy soil obtained

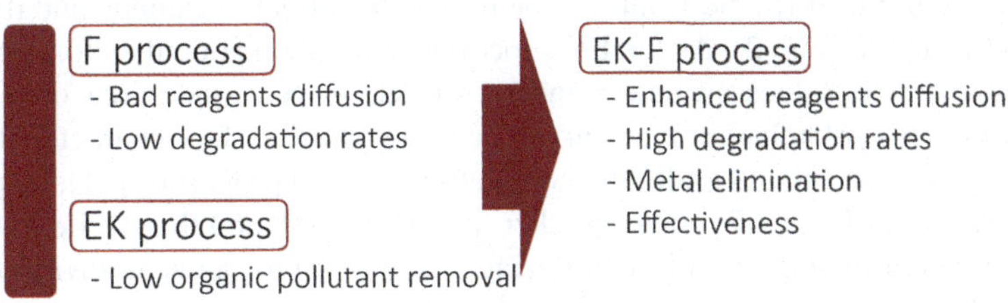


**Fig. 1** Disadvantages of the independent processes and advantages of the EK-F process combination

from a farmland was 99% remediated within 10 days by this new EK-F treatment. The process is within the so-called EK-ISCO processes, where the penetration capability of oxidants is accelerated by electric field action [19]. The synergy effect of the EK-F process is given not only because of the easier delivery of $H_2O_2$ through the soil which leads to the generation of $HO^{\bullet}$ but also by the promotion of oxidation–reduction reactions into the soil [25]. Thus, this combination enhances the EK process performance (Fig. 1), which has showed low efficiency for the treatment of organic polluted soils. This is why the Fenton process is the most common chemical oxidation method to be added to EK [26]. Thus, this new alternative overcame some of the limitations of EK technology for the remediation of organically polluted soils. For instance, fine grained soils have a large specific area and thus they have abundant sites for soil–pollutant interactions, making the mobility of the latter more difficult with the typical EK process [27]. Besides, the EK process could only could eliminate soluble pollutants present on the pore fluid or sorbed substances on colloidal particulates suspended in the soil pore fluid [28].

All of this made the EK treatment an inadequate remediation alternative, which could be synergistically coupled to other techniques [27]. On this context, during the last years EK-F technology has been successfully applied for in situ remediation of polluted soils. As mentioned earlier, it is an interesting methodology based on the combination of EK process and the Fenton reaction, enhancing the treatment of both organic and inorganic pollutants [16, 29, 30].

The principle of EK remediation relies upon the application an applied electric potential gradient through the soil between a couple of inert electrodes (Fig. 2). The main transport mechanisms of this technology include electrophoresis (movement of charged particles), electro-osmosis (movement of interstitial fluid of a porous matrix), and electromigration (movement of ionic species). Among them, the latter two are the dominant phenomena in EK remediation [31]. In addition, water electrolysis reactions take place over the electrodes. In the anode, an acid front is produced and $O_2$ is liberated, while in the cathode the reduction of water takes place and $H_2$ and $OH^-$ are produced [32, 33] as depicted in Eqs. (11) and (12) [34].

$$2H_2O \rightarrow O_2 + 4H^+ + 4e^- \tag{11}$$

$$2H_2O + 2e^- \rightarrow H_2 + 2HO^- \tag{12}$$

Both fronts, acid and basic, are transported by the electromigration to their opposite electrode; thus, the acid front will be moved towards the cathode and the basic front to the anode (Fig. 2). As the EK process occurs, the pH on the close-to-anode area decreases to values around 3 and on the close-to-cathode area increases to environ 11. This pH behavior may vary depending on the surface characteristics and on the interactions such as hydraulic conductivity, or ion properties [31]. Hopefully, the movement of $H^+$ is 1.75 times quicker than $HO^-$ [34], and the profile of the soil pH is not equal in soils with low buffering capacity. Thus, an acid environment is generated in most of the soils to be remediated.

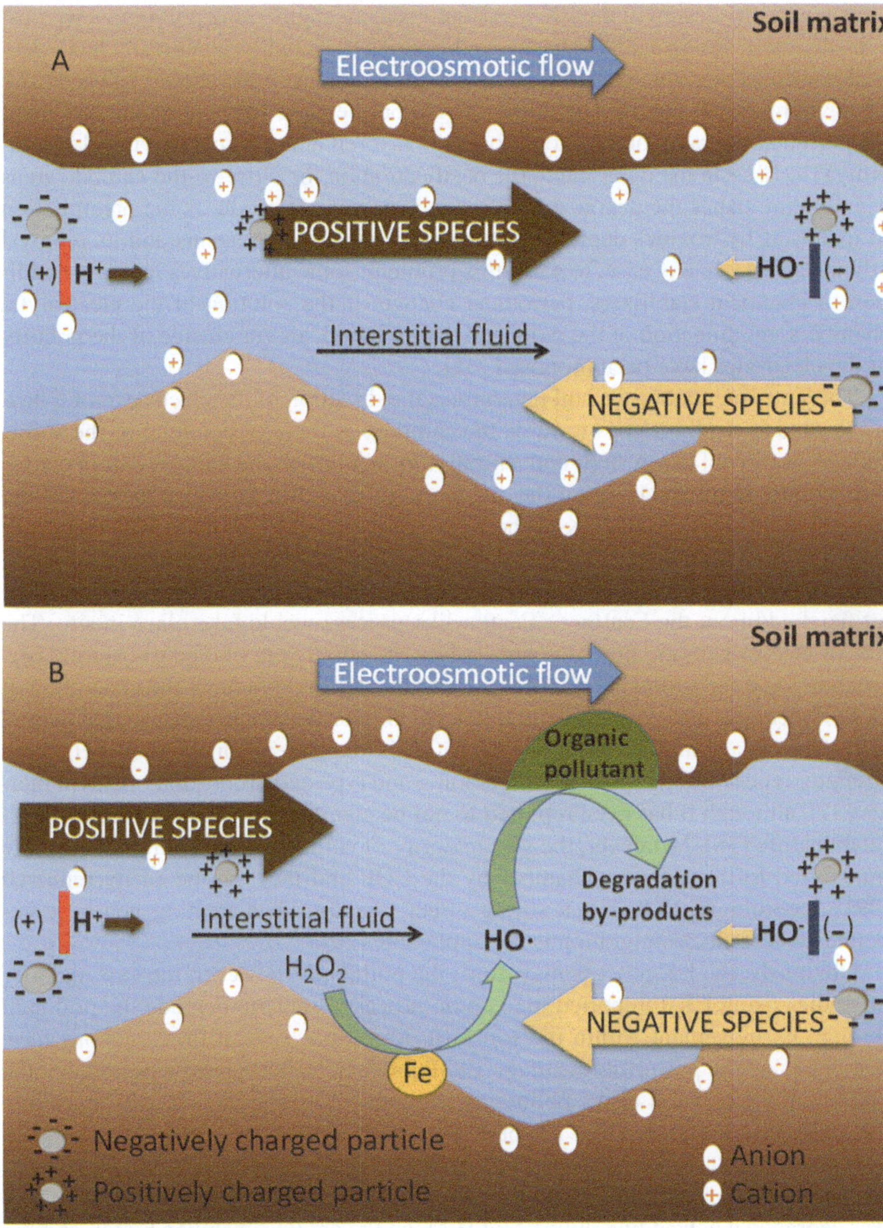


**Fig. 2** Scheme of the phenomena involved on the pollutant removal during EK (**a**) and EK-F (**b**) treatment

Anyway, the design of the EK-F system, from an engineering point of view, must consider these strong acid and basic conditions [24]. Actually, chemically inert electrodes such as graphite, coated titanium, or platinum electrode should be used [31]. Thus, a proper pH control should be carried out because the acid conditions favor the pollutant degradation (Fenton process is favored at acidic pH and the $H_2O_2$ stability) [8, 10]. On the other hand, the basification on the close-to-the-cathode areas usually diminishes the degradation efficiency because it produces the precipitation of metals as hydroxides depleting the mobility of Fenton reagents and its removal efficiency [34]. In any case, to solve this problem, some alternatives including addition of chemical stabilizers, periodical change of the solution in the electrolytic chambers, modification of the oxidant delivery mode, and/or change of the polarity of the electrodes have been proposed [35].

The zeta potential of the soil determines the direction of the electroosmotic flow (EOF) and typically it goes towards the cathode, as the zeta potential of the soil is usually negative [12]. A high concentration of cations in the soil pore fluid causes the decrease in the thickness of the diffuse double layer and thus, an increase in the ionic strength [27]. In the EK-F, the EOF is the responsible of the distribution of $H_2O_2$ [29] homogeneously through the soil, independently of its hydraulic conductivity. Thus, if the soil has iron, or other transition metals, there by nature or by an artificial addition, the Fenton or Fenton-like process would take place in situ. The rate of electro-osmotic flow is controlled by the coefficient of electro-osmotic permeability and the soil pH [36]. All these englobed processes favor the generation of an EOF because of the soil pH variations which increase the negative zeta potential [10]. The zeta potential and thus the EOF is dependent on solution pH, ionic strength, types of ionic species, temperature, and type and proportion of clay minerals [37], although it has been reported to not be significantly dependent on the pollutant nature [38]. Moreover, the vast majority of soluble no-charged pollutants are transferred to the cathodic chamber by the EOF and they can be easily removed [39]. Accordingly, EOF has a strong effect, mainly when treating neutral compounds where electromigration is not applicable [26].

Obviously, the EK process may move the pollutants; however, the vast majority of studies using EK-F reported the Fenton process is able to degrade the pollutant through its movement within the soil [40]. Moreover, several investigations have demonstrated the powerfulness of the EK-F process, as it has been reported that the sequential degradation of the pollutants while the Fenton's reagent is diffused into the soil is more efficient that the pure extraction movement on EK systems [12], thus avoiding a waste effluent. For instance, Reddy et al. [16] reported how phenanthrene was not eliminated from the soil in the classical EK process whereas 76% was degraded when adding only 5% of $H_2O_2$ on the anolyte chamber. So far, organic pollutants have been treated under EK process when being partially dissociated, such as phenol [41], whereas the EK-F technology has opened a path for the treatment of more strongly attached compounds such as hydrophobic pollutants [32].

In Fig. 3, the timeline and the main milestones by the EK and Fenton process and their combination are provided. The major events on the development of the combined process are also represented. From the first patent of the EK-F presented

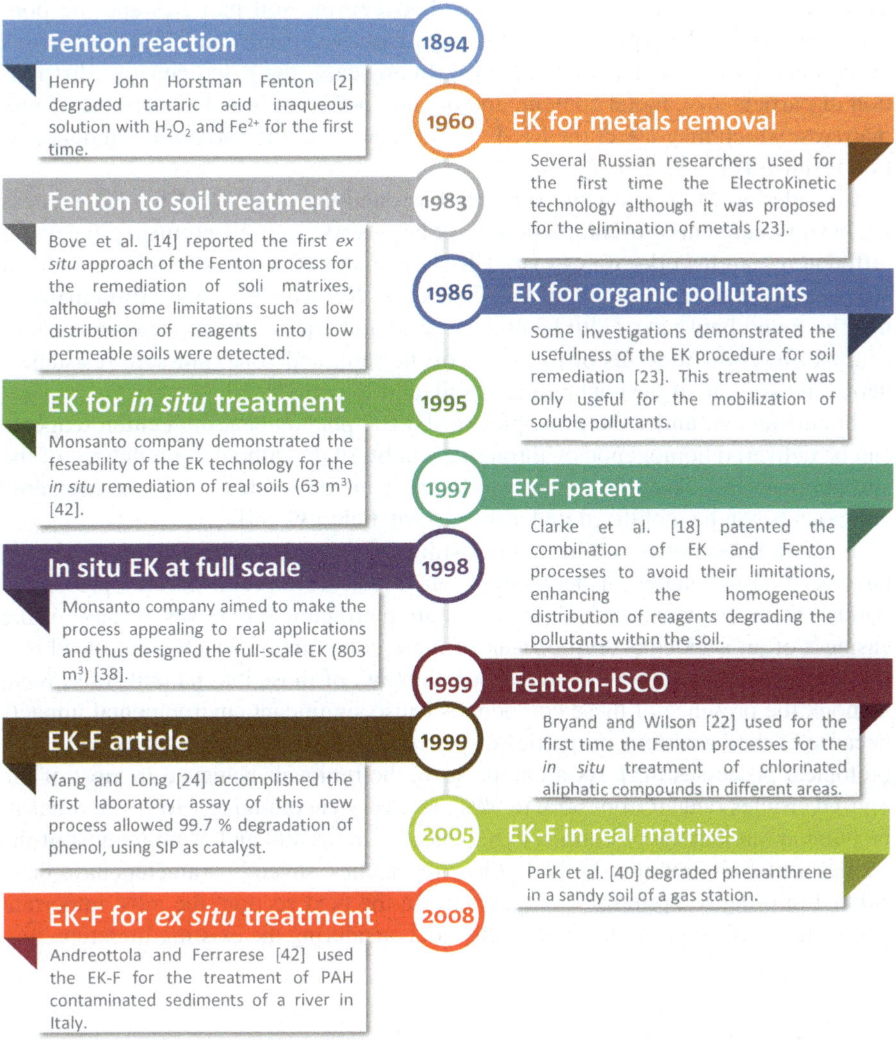


**Fig. 3** Timeline and milestones of the EK and F investigations towards their combination

by Clark et al. [18] in 1997, several findings were reached by the study of polluted model matrixes or artificially polluted real matrixes. However, it was not until 2008 that Andreottola and Ferrarese [42] used the technology for the treatment of a real sediments coming from a river in Italy. These authors highlighted the necessity of high oxidant dosages in order to cope with the natural organic load present in those sediments, attaining, under the optimal conditions, 90% of polycyclic aromatic hydrocarbon (PAH) removal. On the other hand, the treatment of real matrixes may favor the redox reactions, not only in the close-to-the-electrode areas like in simulated soils but also in all interfaces between the soil particles and the pore water because of the presence of substances with electronic conducting properties (microconductors) such as minerals [43]. In any case, the heterogeneous distribution of particle size, metal content, and organic pollutant is a characteristic of real matrixes, which may affect the EK-F performance and the variability on degradation results [42].

Considering Monsanto Company demonstrated the viability of the EK process for in situ treatment of trichloroethylene at large scale, and the promising results of different research works after ex situ EK-F treatment of real matrixes [42], the next step would be to test the in situ EK-F process. Nevertheless, some discrepancies may be found between ex situ laboratory scale and real in situ application because of the differences in terms of thickness of the saturated zone, pressure resistance, depth and volume of the contaminated soil, and so on [31].

In conclusion, under the principles of the EK phenomena, the Fenton reagents can be delivered homogenously through a soil by EOF, without the influence of the soil permeability. Then the organic pollutant is in situ oxidized, and the inorganic compounds can be mobilized under the electric field [39, 44].

In Table 1, a summary of the main existing articles, where different polluted soils have been treated under this technology, and the achieved removals, are presented. Among them, the most studied pollutants are petroleum derivatives because of the easiness of an accidental spill or leak of these compounds, which would lead to a soil contamination due to the high absorption rate of those into natural soils. What is more, the presence of these compounds cause significant environmental impacts because of the hazardous properties of petroleum, which are not usually solved by biological processes [45]. As it can be seen, the removals achieved in most of the revised articles ranged from 60% to 99%, fact very encouraging. However, it has to be pointed out that the performed experiences were ex situ and using in most of the studies artificially polluted soils. In these researches, several parameters have been studied in order to optimize the treatment. In the next section, the most important parameters will be presented and discussed, considering the existing literature.

## 2 Influencing Parameters

In the combined system EK-F, several parameters affect the treatment efficiency and the synergistic effect of different operational conditions should be evaluated in order to achieve the complete soil remediation (Fig. 4). Thus, it is expected that EK-F process effectiveness strongly varies depending on the nature of both, the contamination and the properties of soil matrix. In fact, Ng et al. [8] reported the organic pollutant removal from soils using the EK-F process varies within the range 26–99.7% for different pollutants which may be found in soils, such as petroleum derivatives. In addition, the presence and mode of addition of Fenton reagents, as well as environmental conditions are key factors that should be optimized. Furthermore, the electric field applied, the electrolyte, the type of electrodes, and

**Table 1** Main studies in the application of the EK-F

| Ref. | Operational conditions | | | | | | Soil | | Removal (%/days) |
|---|---|---|---|---|---|---|---|---|---|
| | Electric field | A | C | Anolyte | Catholyte | Catalyst (mg/g) | Type | Pollutant (mg/Kg) | |
| [24] | 1 V/cm | G | G | 0.3% $H_2O_2$ | DW | Inherent Fe 2.6 + 1.05 g SIP | Sandy loam | Phenol, 90–115 | 99.7/10 |
| [46] | 1 V/cm | G | G | $H_2O_2$ 4000 mg/L | DW | Inherent Fe 2.6 + SIP | Loamy sand | Trichloroethylene, 233 | 88.9/10 |
| [32] | 1.5 V/cm | G | G | 0.3% $H_2O_2$, 0.5 M $H_2SO_4$ | DW | Inherent Fe, 4.2 | Kaolin | Phenanthrene, 200 | 74/13 |
| [40] | 140 V | CP | 2 SS | 5% $H_2O_2$, 0.02 M NaCl | 0.02 M NaCl | Fe, 4.4 | Sandy soil | Phenanthrene, 200 | 81.6/21 |
| [10] | 30 V | G | G | 7% $H_2O_2$, 0.5 M $H_2SO_4$ | DW | Fe, 4.2 | Kaolin | Phenanthrene, 200 | 75/13 |
| [17] | 30 V | G | G | 7% $H_2O_2$, 0.03 N SDS | DW | Inherents Fe, 3.1; ca, 11.6 | Clay | Phenanthrene, 200 | 30/14 |
| [16] | 1 V/cm | G | G | 20% $H_2O_2$ | DW | Inherent Fe, 805 | Kaolin | Phenanthrene, 200 | 85/28 |
| [12] | 30 V | G | G | 30% $H_2O_2$ | DW | Inherent Fe | Kaolin | Hexachlorobenzene, 100 | 76/15 |
| [34] | 30 V | G | G | 15% $H_2O_2$ | DW | $FeSO_4$ (10%) | Kaolin | Hexachlorobenzene, 100 | 62/12 |
| [34] | 30 V | G | G | 15% $H_2O_2$ $FeSO_4$ (10%) | DW | – | Kaolin | Hexachlorobenzene, 100 | 64/14 |
| [47] | 30 V | G | G | 7% $H_2O_2$, 30 mM HCl | DW | Inherent Fe, 3.15 | Clay | Phenanthrene, 200 | 23/10 |
| [21] | 40 V | 3 Fe | 6 Fe | 8%$H_2O_2$ 0.1 M NaCl | 8%$H_2O_2$, 0.1 M NaCl | Inherent Fe, 45.8 | Clay | Diesel, 10,000 | 95/60 |
| [33] | 3 V/cm | G | G | 10% $H_2O_2$, 0.1 M $Na_2SO_4$ | 10% $H_2O_2$, 0.1 M $Na_2SO_4$ | Inherent Fe, 6 | Kaolin | Phenanthrene, 500 | 99/14 |
| [48] | 2 V/cm | PG | PG | 8% $H_2O_2$, $FeSO_4$10% | DW | – | Kaolin | Hexachlorobenzene, 100 | 57 |
| [49] | 1 V/cm | G | G | 1 g/L $H_2O_2$ $Na_2S_2O8$ | $Na_2S_2O_8$ | Inherent Fe | Sandy soil | Trichloroethylene, 136.45 | 100/14 |
| [29] | 3 V/cm | G | G | 10% $H_2O_2$, 0.1 M $Na_2SO_4$ | 0.5 M $HNO_3$ | Inherent Fe, 18.4 | Loamy sand | Diesel, 5000 | 87/30 |
| [50] | 30 V | G | G | 7% $H_2O_2$, SDS 30 mM | DW | Inherent Fe, 11.6 | Clay | Phenanthrene, 200 | 70/18 |

(continued)

**Table 1** (continued)

| | Operational conditions | | | | | | Soil | | |
|---|---|---|---|---|---|---|---|---|---|
| Ref. | Electric field | A | C | Anolyte | Catholyte | Catalyst (mg/g) | Type | Pollutant (mg/Kg) | Removal (%/days) |
| [30] | 3 V/cm | G | G | 10% $H_2O_2$, EDTA 0.1 M | 10% $H_2O_2$, EDTA 0.1 M | Inherent Fe, 20.9 | Marine sediment | Petroleum hydrocarbon, 11.68 | 90/15 |
| [51] | 2 V/cm | G | G | 20% $H_2O_2$, 20 mM SDS | 20 mM SDS, 20 mM NaOH | Inherents Fe, 20,914;Ca19,000 | Sandy soil | Petroleum hydrocarbons 3440,<br>PAHs 1.99, zinc 120 | 19.9/7.9<br>18.0/7.9<br>9.14/7.9 |
| [52] | 3 V/cm | G | G | 10% $H_2O_2$, 0.1 M $Na_2SO_4$, 0.1 M citric acid, 1% triton X100 | DW | Inherents Fe, 18,220; Ca, 2849<br>Mg, 3218 | Sandy soil | Petroleum hydrocarbons, 45,557 | 10–60/15 |
| [26] | 1 V/cm | Ir | Ir | 5% $H_2O_2$, 0.1 M NaCl | 0.1 M NaCl | Inherent Fe, 23.6% | Sandy soil | Lubricant, 6441 | 45.2/20 |

*A:* anode, *C:* cathode, *G:* graphite, *CP:* carbon plate, *SS:* stainless steel, *PG:* platinum-coated graphite, *Ir:* Ir-coated, *EDTA:* ethylenediaminetetraacetic acid, *SDS:* sodium dodecyl sulfate, *DW:* distilled water

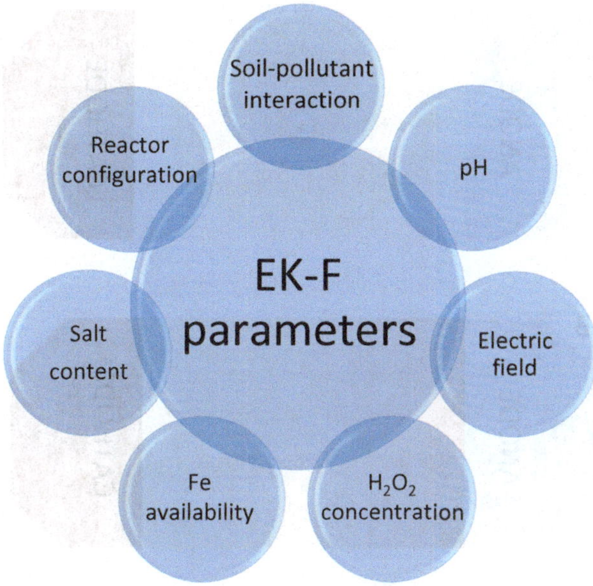


**Fig. 4** Key factors affecting EK-F process performance

their disposition are also important factors that the scientific community have studied in depth for a successful implementation at field scale.

## *2.1 Fenton Reagents Dosage*

In order to produce the Fenton reaction (Eq. 1), the presence of $H_2O_2$ and $Fe^{2+}$ is mandatory; as well as having them at adequate concentrations to avoid scavenger reactions (Eqs. 4–9). $H_2O_2$ is consumed due to (1) (and mainly) its reaction with $Fe^{2+}$, (2) its reaction with $Fe^{3+}$ or other oxidants and (3) reactions with organic pollutants [53]. Thus, the efficacy of the Fenton reaction is principally delimited by the $H_2O_2$ concentration and by the ratio $Fe^{2+}/H_2O_2$ [3].

### 2.1.1 $H_2O_2$ Concentration

During EK-F treatment, a solution containing $H_2O_2$ at acid pH is flushed into the soil from anode to cathode by the application of an electric field. Typically, the concentration of $H_2O_2$ used varies from 5% to 20% (Table 1). Although other studies with even lower $H_2O_2$ concentrations had been done (3%), their poor results in terms of pollutant degradation made increasing the reagent concentration necessary [26]. Thus, Reddy and Chandhuri [54] enhanced phenanthrene degradation from 49.8%

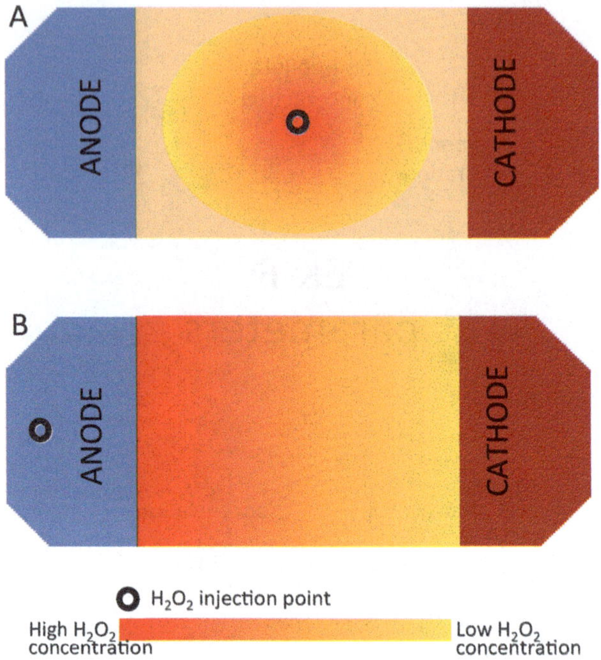


**Fig. 5** Oxidant distribution depending on the $H_2O_2$ injection point (black hole): (**a**) central injection, (**b**) close-to-the-anode injection

to 82.3% by increasing $H_2O_2$ concentration from 5% to 10%. They defended these strong oxidizing conditions may also be effective for mineralization of the generated by-products.

The flushing solution is diffused throughout the soil where mainly the iron available in the soil or other transitional metals promote $H_2O_2$ decomposition to $HO^•$, which in situ oxidizes the organic pollutants [55]. Actually, Kim et al. [10] treated two different soils under the EK-F process, and when measuring the residual $H_2O_2$ concentration, it was lower in the soil with higher content of iron. As it might be expected, the pollutant degradation is higher in the close-to-the-anode areas, because of the higher $H_2O_2$ concentration due to the following reasons: (1) $H_2O_2$ is usually added in the anolyte, (2) the pH increase in the close-to-the-cathode areas makes the $H_2O_2$ instable, and (3) it requires a longer transportation time [8].

The reactants content not only affects the Fenton reaction but also the EK process. For instance, it is important to highlight the fact that $HO^•$ are generated in aqueous solution; thus, Fenton treatment of adsorbed pollutants seem to be defined by the desorption equilibria. However, the direct oxidation of adsorbed pollutants can be favored when using high $H_2O_2$ concentrations (>2%) [56] because of the powerfulness of the generated reagents through the Fenton process [12] or for the possible generation of non-hydroxyl radicals [12]. In this sense, not only the $H_2O_2$ presence causes the generation of more $HO^•$ by the Fenton reaction, but also its stability is enhanced [54].

Moreover, Kim et al. [32] reported that the electrical current increases with the $H_2O_2$ concentration due to the generation of ionic compounds as a result of the oxidation reactions and the liberation of salts from the soil. However, there is some oxidant concentration values above from further increase is not justified by the modest increase on the pollutant degradation [16]. What is more, an extremely high concentration may lead to scavenger effect (Eqs. 4 and 8) [3, 12]. For instance, Chang et al. [57] reduced approximately 40% their diesel degradation when augmenting the $H_2O_2$ concentration from 15% to 30%, because of the increase on the $H_2O_2$ self-decomposition. Actually, Park and Kim [58] only needed 0.35% $H_2O_2$ to attain the best results for EK-F degradation of phenanthrene, enhancing the performance by other means such as adding an anionic surfactant and $FeCl_2$. Increasing the $H_2O_2$ concentration too much also may reduce EOF generation. For instance, Kang et al. [26] augmented $H_2O_2$ concentration from 3.3% until 10% causing a reduction of EOF from 196 to 84 mL. Despite the smaller EOF flow, the degradation performance increased from 25% to 39% with this $H_2O_2$ concentration augmentation, demonstrating the degradation performance was more affected by the oxidation capacity than by the EOF produced.

### 2.1.2 $H_2O_2$ Injection Point

Some researchers have denoted the degradation of $H_2O_2$ in the electrode chamber, so the injection of this reagent into the soil have been postulated as a solution. Therefore, Issosaari et al. [59] introduced $H_2O_2$ in the middle of the soil (Fig. 5a) (between anode and cathode), which allowed them to obtain approximately 13% of PAH removal in 8 weeks because of low EOF generated. In fact, Kang et al. [26] highlighted that the EOF and its oxidizing capacity determined the EK-F process removal efficiency. However, this configuration was further studied and Oonnittan et al. [48] highlighted the importance of the selection of the $H_2O_2$ injection point, being the central addition the optimal configuration which allowed a hexachlorobenzene degradation of 60%. This configuration reduces the $H_2O_2$ decomposition, enhances the EOF generation and favors a more equal distribution of oxidants (Fig. 5a). Nevertheless, the in situ application is likely to be more difficult so more studies should be carried out. Actually, Ho et al. [60] developed an in situ $H_2O_2$ injection system at pilot scale which allowed them to vary the injection pressure which augmentation favored the decomposition of nitrobenzene, reaching 50% in 15 days at 20% of $H_2O_2$.

### 2.1.3 Metal Content

The metal content is also an important parameter as it delimits Eq. (1) efficacy. Iron is one of the most common elements found in soils and at many sites, and there is an abundance of naturally occurring heterogeneous forms of iron which serve as the main source of catalyst for the Fenton process [61]. For instance, Reddy et al. [54]

reported how the inherent iron in kaolin was useful for the degradation of phenanthrene, as the oxidation of this compound reached 49.8% in presence of 5% $H_2O_2$ whereas the oxidation was practically negligible without it.

Hopefully, quite high iron concentrations have been detected in soil, ranging from 3 to 20,914 mg/g (Table 1). Under low pH and/or reduced conditions, part of the total iron may be Fe(II), which could act a as catalyst yielding $HO^{•}$. Although the naturally available metals in soils can cause, to some extent, the Fenton process, it has been reported that the addition of an external catalyst enhances the pollutant degradation [46]. In this context and in order to facilitate the EK-F application, other iron sources may be added such as magnetite ($Fe_2O_3$), goethite (FeOOH) or wustite (FeO), which would lead to the addition of lower iron concentrations, the possible reutilization of this catalyst and the self-regulation of iron concentration within the soil [25]. Moreover, as mentioned earlier, other transition metals can be also considered, which, as well as $Fe^{2+}$, may be also naturally present in the soil ($Mn^{2+}$, $Ca^{2+}$, $Cu^{2+}$, etc.).

### 2.1.4 Metal Addition

The way in which the Fenton reagent is present into the soil also affects the performance of the EK-F process. Thus, when needing an extra supply of catalyst, it is better to add it directly to the soil, as it increases the catalyst content since the beginning of the process instead of adding it to the anolyte, in which case it is necessary to wait for the electroosmotic movement, the ionic strength and with that, the EOF generation [34]. In addition, the possible movement of the metals through the soil should be considered. For example, Cameselle and Reddy [44] reported an iron elimination of 17% where the remaining iron was enough for act as Fenton catalyst.

Accordingly, ferrous sulfate ($FeSO_4$) and other salts of Fe(II) have been co-injected with $H_2O_2$ to facilitate the Fenton reaction, as they would react in that moment. In the EK-F process, the catalyst can be artificially added by different means.

1. By electromigration using an iron solution added on the anolyte for the first days of the treatment so that iron–substrate mass ratio is 1:10 [48] or the pumped solution is between 0.01 and 0.02 M [46]. Then the anolyte is changed for a $H_2O_2$ solution. With this procedure, Oonnittan et al. [48] ensured the presence of iron throughout the soil before adding the oxidant.
2. By addition of an iron metal solution into the soil. Hence, Alcantara et al. [33] adsorbed iron into kaolinite and afterwards they treated adsorbed phenanthrene, being able to apply the EK-F process thanks to the 10% $H_2O_2$ added, degrading 99% of the polycyclic aromatic hydrocarbon. Another example was brought by Iglesias et al. [62] who added iron to kaolinite in order to EK-F remediate reactive black 5.
3. Through the addition of a scrap iron powder (SIP), an iron piece which is placed within the soil (Fig. 6c). The use of SIP favors the pollutant degradation, because

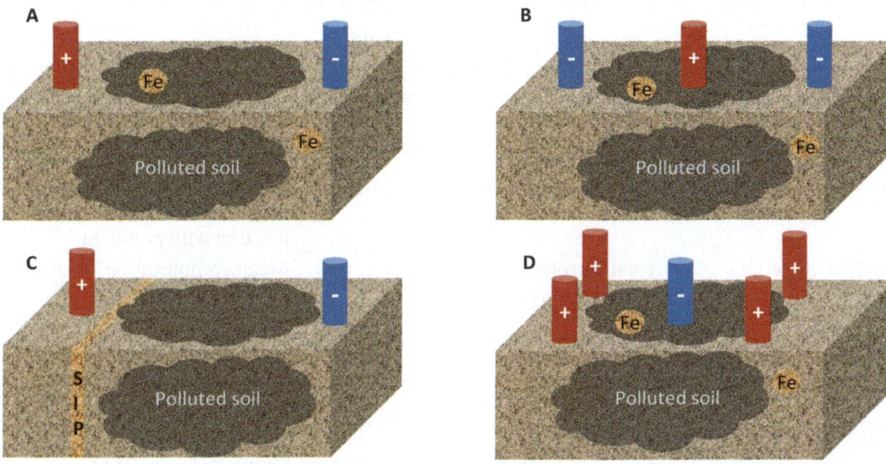


**Fig. 6** Electrodes location for EK-F application: (**a**) face to face (**b**); two-dimensional; (**c**) face to face with SIP as iron source and (**d**) cyclic configurations

using soluble iron makes the HO• to be generated at the very beginning of the soil section, losing their oxidizing potential before reaching all the pollutants [46]. Yang and Liu [46] stated utilization of SIP makes the pollutant degradation to be the dominant mechanism whereas the addition of $FeSO_4$ would favor the removal towards the cathode.

4. By the use of iron as sacrificial electrode (anode). Using this alternative, Tsai et al. [21] enhanced EK-F oxidation to remediate diesel contaminated soils. Thus, they reported a degradation of 27% and 55% of diesel hydrocarbons with, respectively, Fenton and EK processes whereas the combination of both processes caused 97% of removal.

The simultaneous addition of $Fe^{2+}$ and $H_2O_2$ is not recommended, as the Fenton reaction takes place and both reagents are consumed before electroosmosis [48]. Actually, if using the native iron of the soil or adsorbing it previously, avoids the unnecessary $H_2O_2$ consumption, reducing the costs and enhancing the process performance [8]. On the other hand, the presence of the Fenton catalyst causes the ionic strength of the pore water to slowly increase, increasing the current, then this value is stabilized [9, 12, 34]. This is why the initial addition of the metal into the soil is recommended rather than putting it on the anode, to avoid the initial delay that is caused because of the time that the ions need to be moved through the soil.

On the other hand, an extremely fast $H_2O_2$ decomposition is favored when the catalyst is exceedingly available [17], thus natural iron oxides would help to slow down the $H_2O_2$ decomposition. As an alternative, some stabilizers can be added, to complex the metal so $H_2O_2$ can be moved thought the soil as the reaction happens [12, 57]. Some examples of this procedure are depicted Table 2, which is further explained in Sect. 2.6. In any case, the best alternative would be to compare both approaches for a given process, as well as the $H_2O_2$ concentration [57].

**Table 2** Examples of the most widely used enhancing agents in EK-F processes

| Ref. | Enhancing agent | Compound | Degradation (%) | | Effect caused |
|---|---|---|---|---|---|
| | | | Without the agent | With the agent | |
| [69] | Hydroxypropyl-b-cyclodextrin | Pentachlorophenol | 49 | 67 | Ternary pollutant-cyclodextrin-iron → direct HO˙ towards reaction with pollutant |
| [57] | Citrate | Total petroleum hydrocarbons | ≈65 | ≈80 | Metal complexation → slower Fenton reaction ($H_2O_2$ stability) |
| [52] | Tween 80 | Total petroleum hydrocarbons | 26 | 81.8 | Solubilize hydrocarbons, avoids the precipitation of metals |
| [17] | Phosphate | Phenanthrene | ≈5 | ≈22 | Metal complexation → slower Fenton reaction ($H_2O_2$ stability) |
| [17] | SDS | Phenanthrene | ≈5 | ≈28 | Metal complexation → slower Fenton reaction ($H_2O_2$ stability). Dissolve iron oxides at high pH |
| [12] | β-Cyclodextrin | Hexachlorobenzene | 64 | 33 | Enhanced pollutant solubility in pore fluid. Trap the pollutant |
| [54] | Fe-EDTA | Phenanthrene | 50 | 70 | Metal complexation → enhances EOF |
| [50] | SDS | Phenanthrene | ≈17 | ≈50 | $H_2O_2$ stabilizer → overcome the short life time of HO• |

Obviously, the external addition of iron can produce the EOF to diminish due to metal precipitation when compared with traditional EK processes, because of the pH basification on the close-to-the-cathode areas. However, the EK-F process favors the pollutant chemical oxidation more than the pure removal [24], overcoming this drawback. Actually, Yang and Yeh [49] reported a smaller EOF generation after having added $Fe_3O_4$ as a catalyst. This is why the metal content should be carefully selected. Actually, the phenol elimination in Yang and Long [24] investigation increases from 53 to 99.7% when diminishing the SIP content from 32.69 to 1.05 g, as not only they favored generation of more EOF but also avoided some scavenger reactions (Eqs. 7 and 9).

## 2.2 Types of Electrode Materials and Placement

The selection of material and placement of the electrodes are very important parameters as it may affect strongly the efficiency of the process. The typical ex situ or laboratory scale reactor are designed with an electrode configuration face to face.

This configuration is useful for the application of EK-F process used to study and optimize the parameters before its real EK-ISCO application [16, 17, 24]. Thus, anode electrode wells are located opposite to cathode electrode wells (Fig. 6a). However, other configurations have been studied with the aim of increasing the process performance. Accordingly, Park et al. [40] suggested for the first time a two-dimensional EK-F reactor configuration, where two cathodes were placed between the whole amount of the polluted soil and the anode was placed in the middle (Fig. 6b). This configuration allowed them to produce a high EOF which was, in any case, favored by the high electric field applied (140 V). Tsai et al. [21] proposed the configuration of a circular EK-F process where four anodes were inserted on the surroundings and the EOF was extracted from the very center of the soil, where the cathode is (Fig. 6d). Similarly, Risco et al. [63] set up a circular six-anode system where the cathode was placed in the middle and they also tried the inverse situation with six cathodes and the anode put in the middle for the removal of oxyfluorfen. The former promote the generation of high quantities of EOF although both of them kept a good moisture in the soil.

In all these cases, the electrodes chamber may have a removable lid for putting the electrodes which would be also useful as gas vents [21] for $CO_2$, $N_2$, or other decomposition gases which may be formed during the pollutant degradation.

Other configurations have been proposed; for instance, the introduction of passive electrodes into the soil to monitor the resistance and intensity fluctuations [29]. However, Ng et al. [8] underlined the lack of references to electrodes cost and life span, pointing what should be considered for real scale applications.

The aim of this process is to degrade all the pollutant adsorbed into the soil, thus, the utilization of carbonaceous electrodes is recommended as metallic electrodes may favor the $H_2O_2$ degradation [40] only on the close-to-the-electrodes area. Nevertheless, the selection of iron electrodes, which would be dissolved though the process, may also favor the EK-F [31]. Thus, Tsai et al. reported that the switch from graphite to iron electrodes caused an amelioration of 10% on the degradation of diesel [21].

## 2.3 Electric Field

The applied voltage should be carefully selected because increasing it favors the EOF generation due to the higher number of ions in the pore solution according to Helmholtz–Smoluchowski theory [44]; and with that, the oxidation process. Oppositely, a too high voltage diminishes the EOF [8] and the electrophoretic action can cause the displacement away of the soil from the cathode which causes the EOF to diminish and thus, the pollutant removal [12].

In spite of the importance of this parameter, at the present time there is a lack of investigations regarding the influence of the voltage applied during EF-K and the typical values used for the application of EF-K vary between 1–3 V/cm (Table 1). In this issue, the scientific community has focused its attention in the way to apply the electric field. Therefore, Park and Kim [50] reported the switching of electrode polarity as a good option for avoiding extreme pH fluctuations in order to favor the EOF and the Fenton degradation. With this procedure, the typical low degradation of the pollutant on the cathode section is avoided [8]. What is more, the resistance of the media tend to increase as the time passes by, after an initial reduction because of the dissolution of the ionic species and the mobilization of the ions of the soil matrix [44], those are moved and thus gradually eliminated on the catholyte. Switching the electrodes favors that the ionic species do not reach the catholyte and thus, from the close-to-the-cathode areas are moved towards the new cathode [50].

Additionally, other researches evaluated the effect of switching on and off the electric current so the pollutant is transferred from the soil towards the interstitial fluid and the pollutant can diffuse through the soil pores, having the time necessary for mass transfer, diffusion and pollutant degradation, all of that reducing the energy consumption [44].

## 2.4 pH Control and Electrolytes

Several researches have reported the significant influence of the soil pH on the EK Treatment, affecting the pollutant retention in the soil and the EOF [64]. As general rule, pH control in the anode reservoir, to maintain neutral conditions, is necessary to keep high EOF [65]. Regarding the Fenton reaction, the soil pH also plays an important role [3], as it has been reported $H_2O_2$ stability is enhanced at acidic pH whereas at alkaline pH it is decomposed to water and oxygen [21]. Thus, Kim et al. [32] reported the $H_2O_2$ residual concentration increased over time due to the acid front (because of the quicker movement of $H^+$ regarding $HO^-$ [34]) on the soil, and with an increased life time for $H_2O_2$, the pollutant degradation is likely to increase [66]. In fact, Kim et al. [10] detected lower $H_2O_2$ concentrations in soils with high acid buffering capacity.

However, very low pH can diminish the degradation performance. For instance, Kim et al. [32] increased the phenanthrene degradation from kaolinite when adding 0.05 M of $H_2SO_4$ in the anode whereas Alcantara et al. [33] used a 0.1 M $H_2SO_4$ solution to keep the pH on the catholyte at values around 4. This high concentration caused all the soil sections to have a pH lower than 2.5, eliminating the negative zeta potential and with that, the EOF generation. Moreover, extreme acid conditions lead to $HO^{\bullet}$ scavenge (Eq. 11) [12]. In addition, the $H_2O_2$ stability under acid conditions should be considered. Hence, $H_2SO_4$ produces reductive species which degrade $H_2O_2$ whereas HCl did not report any significant $H_2O_2$ decay [47].

Different approaches have been carried out to control pH in the process. The addition of a clay such as bentonite, between the electrolyte solution and the polluted soil acts as buffer, reducing the impact of the acid and basic fronts [40]. Another option is to recirculate the catholyte towards the anolyte so the negative effect of an excess of $H^+$ can be neutralized [44], which would augment the EOF. pH may also be controlled by addition of buffers such as tris-acetate [30], citric acid [53], and acetic acid [27], these two latter are, respectively, also useful as complexing agent and chelant. In addition, the presence of high concentrations of $H_2O_2$ in the electrode solutions decrease the electrochemical decomposition of water and avoided formation of extremely high/low pH environments in the electrode chambers [33]. All these strategies of control should be done carefully, as if the soil pH is reduced at values lower than its point of zero charge, the direction of the EOF would be inversed and in that case, the $H_2O_2$ addition should be done consequently [67].

$$HO^{\bullet} + H^{+} + e^{-} \rightarrow H_2O \tag{11}$$

The presence of salts on the soil and/or on the electrolytes is mandatory in order to diminish the media resistance and then, the energy consumption for a given degradation rate. Usually, the resistance is lower at the beginning of the EK-F treatments because the high salt content and the generation of dissolved ionic species [17]; then, it increases because of the electromigration of the ions [10, 40]. In order to cope with that, the supplement rate of electrolyte from the anode tank should be quicker [40]. Hence, when the intensity is practically constant, the electromigration processes and the supplement of electrolyte rates are essentially equal.

The addition of salts such as NaCl or $Na_2SO_4$ on the electrolytes usually favors the EK-F performance [8], mainly due to the reduction of the media resistance, a high EOF and the enhancement on $H_2O_2$ transportation, although there is the risk of forming chlorinated by-products when using salts with $Cl^-$ ions. In this sense, the type of electrolyte added may affect the performance of the EK-F process and the effect of the electrolyte composition in these systems has been elsewhere reported. Actually, Kang et al. [26] compared the effect of the addition of NaCl, $MgSO_4$ and $HNO_3$ and the best results were reported with NaCl, which produced an EOF of almost 300 mL whereas $MgSO_4$ and $HNO_3$ caused, more or less and respectively, 100 and 200 mL. This can be explained by the $Cl_2$ generation in the anode which is converted to HClO which has a high oxidizing capacity, favoring the pollutants degradation along with the Fenton oxidation [26].

Electrolyte concentration has also importance in removal efficiency. Tsai et al. [21] degraded 56% diesel when using as electrolyte 0.1 M NaCl in their EK-F process, whereas 35 and 23% was, respectively, achieved after the usage of 0.01 M NaCl or tap water.

Accordingly, in order to accomplish EK-F processes, the presence of electrolyte and neutral or slightly acid conditions in the electrode chambers are required. Under these conditions, the EOF is favored and the decomposition of $H_2O_2$ to HO• is preferential.

## 2.5 Solubilizing and Stabilizer Agents

Several substances can be added to the EK-F process in order to enhance, in the vast majority of cases, its performance (Table 2). For example, during the Fenton treatment through the soil matrix, the solubility of the pollutant on the pore water is a key factor, because a faster degradation is produced if the pollutant is in aqueous phase [68]. Thus, when having hydrophobic organic compounds (low $K_{OW}$) some agents should be added in order to increase the pollutant solubility and transport trough the soil [12, 38]. On the other hand, in order to cope with the EK-F difficulties such as $H_2O_2$ decomposition and metal movement through the cathode, some $H_2O_2$ stabilizers such as phosphate, surfactants (for instance sodium dodecyl sulphate (SDS)) or metal complexing agents can be added, which ameliorate the EK-F performance [17] (Fig. 7).

Solubilizing agents, such as cyclodextrins, chelates or surfactants (Fig. 7), can be used in the flushing solution to favor the pollutant desorption, to enhance catalyst/pollutant availability or to reduce the surface tension [64]. Huang et al. [39] reported that these substances can affect the organic pollutant behavior within the soil, as it may desorb, chelate, dissolve or complex the pollutant.

Cyclodextrins can form inclusion complexes of hydrophobic compounds, which would be trapped into their cyclic chain structure (Fig. 7), enhancing their solubility

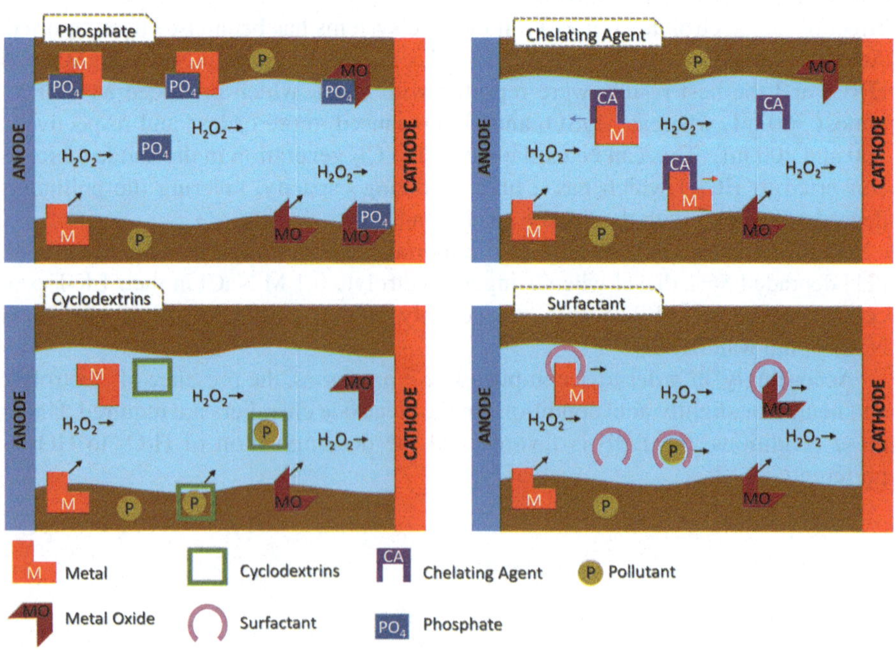


**Fig. 7** Schema of the stabilizers effect. The arrows symbolize the movement of the species. (Based on Ng et al. [8])

[12]. Hopefully, these compounds are non-toxic, biodegradable and have a low affinity to be adsorbed into soils in a wide pH range [27]. Several authors have reported the improvement caused by the addition of substances which can favor the mobility of the pollutant through the soil. Oonnittan et al. [12] added β-cyclodextrin to enhance permeability and hence the movement of hexachlorobenzene and ionic species, which enhanced the higher developed current in comparison with the results in the absence of it, thus reaching 33% of degradation when using 15% of $H_2O_2$. However, the oxidation of hexachlorobenzene was slower when this complexant agent was added, maybe because of the stability of the formed complexes and the fact that $HO^{\bullet}$ may attack this new added compound (Table 2). Thus, the results were favored in the absence of β-cyclodextrin, achieving 64% of hexachlorobenzene degradation. Nevertheless, the utilization of this enhancing agent avoid the accumulation of undegraded hexachlorobenzene in the close-to-the-anode area. Similarly, Reddy and Chandhuri [54] tried to enhance phenanthrene degradation under the EK-F process (49.8% of degradation in 7 days) by adding chelants such as DTPA or EDTA, reaching then, and respectively, 44% and 27.6%. This fact demonstrates that inherent iron into some soils may be enough for promoting the Fenton process. However, if adding iron with either DTPA or EDTA, the degradation attained was 40.8% and 70%, demonstrating the suitability of EDTA to work as iron chelant and EOF promoter. These authors [54] proposed using ethanol as a flushing agent prior to the EK-F application, in order to favor phenanthrene solubilization by using this agent. After that, and by using Fe-EDTA, they achieved 90.5% of phenanthrene degradation.

Surfactants such as Triton X-100 [52], SDS [50] or Tween 80 [52] can also increase the water solubility of the pollutant (Fig. 7), by forming micelles, and in order to enhance the EOF by the variation on the zeta potential, a suitable control pH should be done [29]. Surfactant addition may enhance not only the pollutant solubility, surface tension reduction and wetting capability [27] but also the $H_2O_2$ stability, which enhance the EK-F performance [51]. This can be explained because of the interaction of the pollutant with the surfactant, reducing the interfacial tension at the soil surface. The addition of not only surfactant but also NaOH in the catholyte have been reported as an option to increase even more the pollutant solubility. Seo et al. [51] observed that a combination of SDS and NaOH in the cathode chamber leads to an increase in solubility of the total petroleum hydrocarbons and then to an increase of their removal efficiency by EK-F process. In fact, surfactants have been widely used for the remediation of polluted sites with low solubility compounds such as polycyclic aromatic hydrocarbons. Park and Kim [58] demonstrated how the addition of an anionic surfactant ameliorated the $H_2O_2$ stability and also demonstrated how the surfactant concentration was lineally related with the dissolved phenanthrene. Increasing the surfactant concentration may favor the results but economic considerations should be taken as well as environmental aspects. However, they also noticed that anionic surfactants, when combined with $H_2O_2$, dissolve and transport iron ions in basic pH regions [50], which could diminish even more the Fenton performance in basic sites.

Besides, Chang et al. [57], have investigated the effect of different chemical stabilizers such as citrate, $H_3PO_4$, EDTA or Ethylenediamine-*N*,*N′*-disuccinic acid (EDDS) on the removal of total petroleum hydrocarbons by Fenton-like oxidation. These authors determined that total petroleum hydrocarbons removal was enhanced by the use of these complexing agents following the order: citrate > $H_3PO_4$ > EDDS > EDTA, as with the former the detected residual $H_2O_2$ concentration was maximum. Nevertheless, the authors found difficult to explain the high citrate performance taking into consideration it presents a low chelating efficiency. Thus, other causes, such as the reduction of the media resistance may explain the good performance of citrate. In any case, the addition of chelating agents not only aids on the movement of pollutants but also affects the zeta potential, usually, decreasing it which would enhance the generation of EOF and therefore the EK-F performance [27]. In the case of adding a chelating agent, its concentration should be carefully selected, as increasing it too much may cause a scavenger effect [50]. What is more, the mobility of iron in soil should be studied as it is likely to change [17]. In fact, Pazos et al. [30] highlighted the importance of using chelating agents to maintain soluble iron. Accordingly, Sandu et al. [52] added citric acid to complex natural iron present in a polluted soil and to generate an acid environment inside the soil. These authors, attained between 70% and 80% (for Romanian and Spanish soils, respectively) thanks to the citric and surfactant (Triton X100 or Tween 80, respectively) addition. In this case, the different performance between those surfactants was caused as result of the unalike interactions of surfactants with the metals present in soils. Kim et al. [17] noted how differently the stabilizers can work; thus, the addition of phosphate blocked iron on soil, retarding its transportation, however causes a stabilization of the $H_2O_2$ (it is decomposed more slowly). Thus, its addition is worth it when EK-F processes of long time are required. On the other hand, the use of an anionic surfactant increases the iron content in closer-to-the cathode areas, due to the pH variation and the dissolution of metal oxides. Following Pazos et al. [30] experiences, some considerations should be taken before adding complexing agents. For instance, EDTA forms a negative complex with iron which it is moved from the anode, which can lead to diminish its availability. Ng et al. [8] highlighted the importance of the system pH, as it varies the characteristics of the complexes formed and of the stabilizer concentration, as an excess of stabilizer may cause scavenger reactions.

The addition of persulfate can cause the EK-F if having iron in the soil, because persulfate oxidizes water to $H_2O_2$ in acid conditions. What is more, the persulphate addition diminishes the media resistance and thus, higher current intensities are applied to the soil [49].

## 2.6 Soil Properties

The soil physicochemical properties also modified significantly the performance of EK-F systems [51]. The interaction soil-pollutant delimits the efficiency of the process. For instance, the sorption on the soil surface retards the transport of the species and the transfer to liquid phase, diminishing the Fenton degradation [32]. Nevertheless, the Fenton process has been proved to be a potent oxidation process for the degradation of strongly adsorbed pollutants [70].

On this context, Kim et al. [10] demonstrated how influent are the soil properties on the degradation mechanism. Thus, when they treated phenanthrene polluted soil when being in Hadong clay the residual concentration of phenanthrene in the close-to-the-cathode areas increased, not only because of the farther distance from the $H_2O_2$ supply, but because of the desorption of phenanthrene in the close-to-the-anode areas which favored this compound was transported along with the EOF. Thus, the transportation rate towards the cathode was quicker than the degradation rate, being the desorption-transportation, the dominant removal mechanism, that is the EK process. However, in kaolin soil, the aforementioned gradient concentration was not detected within the soil, indicating the phenanthrene degradation rate was as fast as the desorption rate, having the synergy EK-F process.

Soils with low acid buffering capacity and organic content are easily treated under EK-F processes because of, respectively, the easy acidification of the polluted soil and the reduction of the inefficient $H_2O_2$ consumption [8]. Actually, this latter is the reason why the remediation of real polluted soils are more difficult treated [51]. In fact, Sandu et al. [52] treated different historically hydrocarbon contaminated soils because the pollutants may be more strongly attached to the soil and more heterogeneity and physicochemical properties changes can be found.

The increase on the $H_2O_2$ concentration diminishes the pH and with that, the metal movement through the EK cell is favored [16]. The soil pH is lowered as time passes because of the acid front and thus the surface of the soil is more positive, decreasing the zeta potential which reduces the EOF [44]. On the close-to-the-cathode areas, the pH augments and that may cause the precipitation of the present metals in form of hydroxides, which can block the EOF movement [44].

On the other hand, the presence of carbonates on the soil diminishes the EK-F performance due to several factors: (1) carbonates act as acid buffer, thus reducing the pH fluctuation within the soil and with that, the negative zeta potential, which causes the EOF to diminish; (2) the Fenton degradation may be lessened as the carbonate content overcomes the pH acidification and thus the pH is not acid to favor the Fenton process and (3) the presence of carbonates reduce the stability of $H_2O_2$ [10].

Other soil properties, such as soil permeability and conductivity are less important because, respectively, EK-F processes are suitable for low permeability soils and the addition of some electrolytes can augment the soil conductivity, although in both cases the treatment times are likely to be high [8].

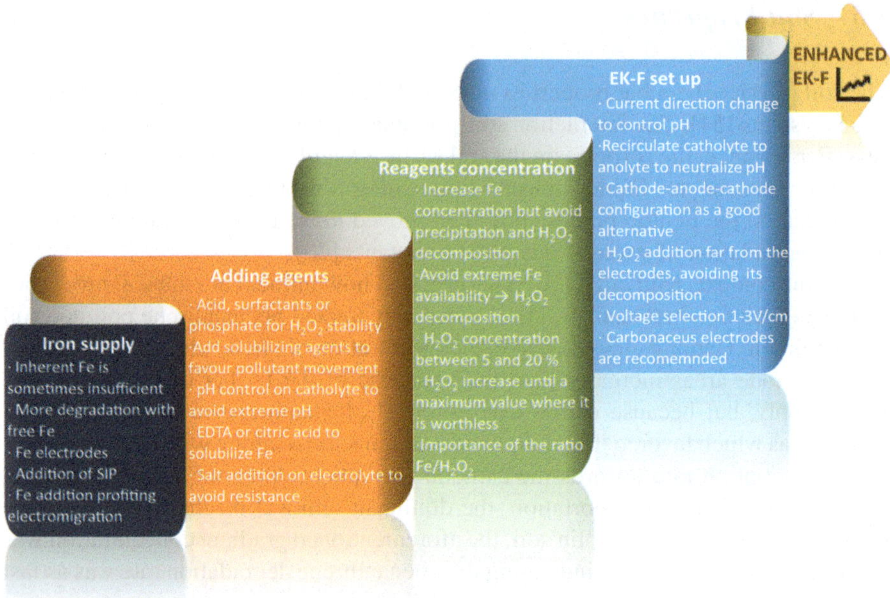


**Fig. 8** Progresses on the EK-F application with regard to the typical configuration

According to the information reported in the previous sections the reported parameters should be studied in detail previous to a field implementation. In summary, the most relevant suggestions that were previously discussed for enhancing the EK-F are depicted in Fig. 8.

## 3 Future Challenges

Some discrepancies may be found between ex situ laboratory scale and real in situ application because of the differences in terms of thickness of the saturated zone, pressure resistance, depth and volume of contaminated soil, etc. [31]. The in situ EK-F application depends on different parameters which should be assessed. For instance, the concentration of ions in soil and their distribution, the soil type, moisture and its buffering capacity as aforementioned. The vast majority of them are deeply related; for instance, the percentage and type of clay in the soil affects the sorption and buffering capacity [31]. The pH of the soil varies the valence, solubility and absorption of the present ions in the soil and therefore the conductivity which modifies the voltage gradient [71].

The optimization of the EK-F operational conditions for in situ application is mandatory in order to avoid the Fenton process to not cause any amelioration to the EK treatment [59]. Additionally, the energy consumption should also be taken into account

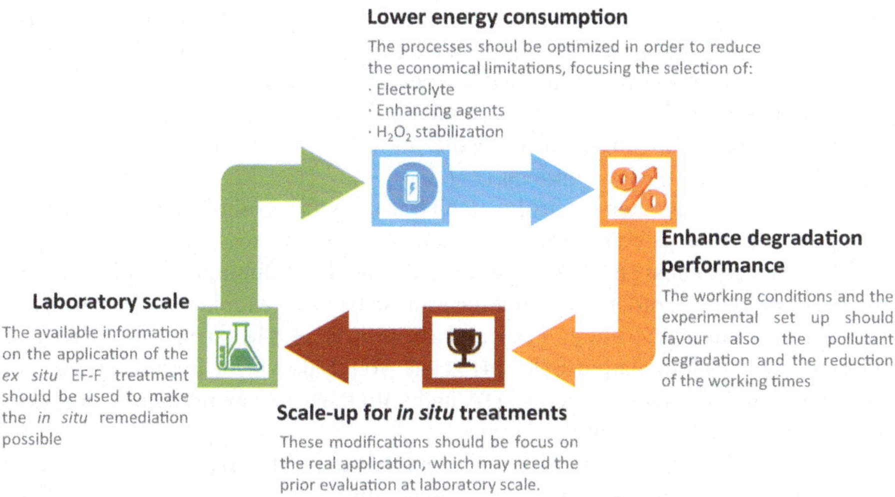


**Fig. 9** Cyclic process to reach the successful application of EK-F processes

because it is considered as one of the main factors to select one EK-F procedure or another [34]. As in all oxidation processes, treatment time affects the efficiency of the process [3]. For example, Pazos et al. [30] noticed how applying 15 or 30 days the EK-F process provided virtually the same petroleum hydrocarbon removal.

The process of switching from laboratory scale to in situ application should then follow a well-defined cyclic procedure, as several tests at laboratory scale may be necessary, where both parameters, energy consumption and overall degradation performance are a priority (Fig. 9). Ng et al. [8] summarized the EK-F optimization alternatives into four big families: (1) delay of $H_2O_2$ decomposition mediated by chemical stabilizers, (2) increase of oxidant availability by modifying the Fenton reagents injection, (3) type of catalyst and electrodes and (4) operating conditions, mainly voltage, electrolytes or reagents concentration. In all cases, the EK-F process optimization should be done in terms of efficiency, costs and environmental impact [8].

The reagents cost should also be considered as Kang et al. [26] did, they attained 7% improvement on the lubricant removal by doubling the $H_2O_2$ concentration until 10%, however, they selected 5% as the optimal concentration to work with considering the cost of the EK-F process application.

The finding of an enhancing agent, which can solubilize the pollutants and stabilize the Fenton reagents and thus improve the EK-F process, without using high $H_2O_2$ concentrations would be a good alternative. However, they are usually organic molecules so the oxidant species can firstly attack the complex agent. What is more, the selection of the stabilizer should be done considering its degree of biodegradability, cost, and efficiency [57]. For instance, if adding surfactants to enhance the EK-F process, it is an interesting solution to select Tween 80 or Triton X-100 because of their low cost and high biodegradability [52]. Moreover, the fate of these agents and its by-products should be assessed in order to avoid a counter-productive process. Thus, after the EK-F treatment, the generation of bioavailable oxidation by-products may affect the local biota and thus cause a toxic effect on the receiving environment [42]. Furthermore, for an in situ application, vent tubes [44], with an external system for gas treatment, will be necessary because the decomposition of pollutants could produce toxic gases. In addition, although an acid pH is beneficial for the EK-F process, extreme acid conditions should be avoided as the low pH in the soil may impact the environment [27].

In the case of adding an electrolyte which usually favors the EK-F performance, the optimization of its concentration is mandatory as an excess of it may not only cause an excess of cost and wastes generation, but also a diminution on the process performance [26]. In order to avoid industrial costs, the recirculation of electrolytes can be a good alternative although special attention should be taken to avoid electric shortcuts in the recirculation line [44]. That would also favor the reutilization of some used chelating agents such EDTA and with that, the environmental problems they would cause because of their toxicity [27].

When EK-F process produces residual wastewater with organic content (due to some pollutant desorption or by-products generation), this effluent can be treated by Fenton based processes as proposed by Ochoa et al. [72]. On this context, the generation of by-products and inorganic wastes which can be eliminated through the EOF should be considered, mainly in the case of toxic by-products or iron sludge which can be generated when an excess of iron was added [8].

Considering the real application of these processes, not only extreme pH conditions should be considered, but also special attention should be focused on the selection of non-reactive material such as polyethylene or Teflon [60] to avoid its reaction

**Fig. 10** SWOT schema of the EK-F process

with $H_2O_2$ on the electrolyte tanks and junctions. Moreover, the junctions between wire and electrode should be covered with water-resistant glue in order to avoid corrosion in the contact area [21].

To sum up, the SWOT schema of the EK-F process is presented in Fig. 10, where the main strengths and opportunities of the application of this technology are highlighted. Moreover, some of the difficulties this process application can yield are also pointed, although the proper application of the aforementioned considerations may diminish their weight on the overall process, making its in situ application at large scale a feasible option for the remediation of polluted soils.

## 4 Conclusions

The combination of EK and Fenton processes has been proposed as an alternative for enhancing the removal of organic pollutants from soils. In order to open a path for in situ soil remediation, some specific considerations should be made. The concentration of the Fenton reagents ($H_2O_2$ and $Fe^{2+}$), the pH of the soil and the ways of controlling its evolution through time, the electrode selection and disposition within the reactor, and so on are parameters that should be assessed. The addition of chelating agents for mobilizing the iron, or surfactants to favor the pollutant mobility, is also encouraged. Moreover, working parameters should be selected in order to not only optimize the degradation capability of the process but also consider the final conditions of the soil, avoiding extreme acid conditions or toxicity. Finally, energy consumption should be considered in order to evaluate the feasibility of these processes for real applications.

**Acknowledgements** The authors are grateful to Spanish Government (Project CTQ2017-90659-REDT (MEIC/AEI) and Aida M. Díez grant BES-2015.071317) and Xunta de Galicia (ED431C 2017/47).

## References

1. F. Haber, J. Weiss, The catalytic decomposition of hydrogen peroxide by iron salts. Proc. R. Soc. Lond. A **147**, 332–351 (1934)
2. H.J.H. Fenton, LXXIII.-Oxidation of tartaric acid in presence of iron. J. Chem. Soc. **65**, 899–910 (1894)
3. K. Barbusinski, Fenton reaction – controversy concerning the chemistry. Ecol. Chem. Eng. S-Chemia i Inzynieria Ekologiczna S **16**, 347–358 (2009)
4. M. Bobu, S. Wilson, T. Greibrokk, E. Lundanes, I. Siminiceanu, Comparison of advanced oxidation processes and identification of monuron photodegradation products in aqueous solution. Chemosphere **63**, 1718–1727 (2006)
5. M. Kremer, The Fenton reaction. Dependence of the rate on pH. Phys. Chem. A **107**, 1734–1741 (2003)
6. M.A. Oturan, J. Aaron, Advanced oxidation processes in water/wastewater treatment: principles and applications. A review. Crit. Rev. Environ. Sci. Technol. **44**, 2577–2641 (2014)

7. P. Bautista, A.F. Mohedano, J.A. Casas, J.A. Zazo, J.J. Rodriguez, An overview of the application of Fenton oxidation to industrial wastewaters treatment. J. Chem. Technol. Biotechnol. **83**, 1323–1338 (2008)
8. Y.S. Ng, B. Sen Gupta, M.A. Hashim, Stability and performance enhancements of electrokinetic-Fenton soil remediation. Rev. Environ. Sci. Biotechnol. **13**, 251–263 (2014)
9. R.O.C. Norman, P.R. West, Electron spin resonance studies. Part XIX. Oxidation of organic radicals, and the occurrence of chain processes, during the reactions of some organic compounds with the hydroxyl radical derived from hydrogen peroxide and metal ions. J. Chem. Soc. B.: Physic. Org. **0**, 389–399 (1969)
10. J. Kim, S. Han, S. Kim, J. Yang, Effect of soil chemical properties on the remediation of phenanthrene-contaminated soil by electrokinetic-Fenton process. Chemosphere **63**, 1667–1676 (2006)
11. W. Haag, C. Yao, Rate constants for reaction of hydroxyl radicals with several drinking-water contaminants. Environ. Sci. Technol. **26**, 1005–1013 (1992)
12. A. Oonnittan, R.A. Shrestha, M. Sillanpää, Effect of cyclodextrin on the remediation of hexachlorobenzene in soil by electrokinetic Fenton process. Sep. Purif. Technol. **64**, 314–320 (2009)
13. H.I. Gomes, C. Dias-Ferreira, A.B. Ribeiro, Electrokinetic remediation of organochlorines in soil: enhancement techniques and integration with other remediation technologies. Chemosphere **87**, 1077–1090 (2012)
14. L.J. Bove, C.L. Cundall, W.P. Lambert, P.J. Marks, J.F. Martino, Removal of contaminants from soil phase I: identification and evaluation of technologies. Roy F. Weston Inc. DRXTH-TE-CR-83249, 1–235 (1983)
15. B.G. Petri, R.J. Watts, A.L. Teel, S.G. Huling, R.A. Brown, Fundamentals of ISCO using hydrogen peroxide: in situ chemical oxidation for groundwater remediation. SERDP ESTCP Environ. Remediat. Technol. (2015)
16. K.R. Reddy, M.R. Karri, Effect of oxidant dosage on integrated electrochemical remediation of contaminant mixtures in soils. J. Environ. Sci. Health Part A-Toxic/Hazard. Subs. Environ. Eng. **43**, 881–893 (2008)
17. J.H. Kim, S.S. Kim, J.W. Yang, Role of stabilizers for treatment of clayey soil contaminated with phenanthrene through electrokinetic-Fenton process—some experimental evidences. Electrochim. Acta **53**, 1663–1670 (2007)
18. R.L. Clarke, S.L. Smedley, S. Kimmel, In situ electrochemical remediation of organically contaminated soil, sediments and ground water using electrochemically generated and delivered Fenton's reagent. Electrochem. Des. Associat. **5,861,090**, 1–7 (1997)
19. M.Z. Wu, D.A. Reynolds, A. Fourie, H. Prommer, D.G. Thomas, Electrokinetic in situ oxidation remediation: assessment of parameter sensitivities and the influence of aquifer heterogeneity on remediation efficiency. J. Contam. Hydrol. **136–137**, 72–85 (2012)
20. R. Watts, M. Udell, P. Rauch, S. Leung, Treatment of pentachlorophenol-contaminated soils using Fentons reagent. Hazard. Waste Hazard. Mater. **7**, 335–345 (1990)
21. T. Tsai, J. Sah, C. Kao, Application of iron electrode corrosion enhanced electrokinetic-Fenton oxidation to remediate diesel contaminated soils: a laboratory feasibility study. J. Hydrol. **380**, 4–13 (2010)
22. J.D. Bryant, J.T. Wilson, Fenton's in-situ reagent chemical oxidation of hydrocarbon contamination in soil and groundwater. Remediation, 13–25 (1999)
23. N. Mosavat, E. Oh, G. Chai, A review of electrokinetic treatment technique for improving the engineering characteristics of low permeable problematic soils. Int. J. GEOMATE **2**, 266–272 (2012)
24. G. Yang, Y. Long, Removal and degradation of phenol in a saturated flow by in-situ electrokinetic remediation and Fenton-like process. J. Hazard. Mater. **69**, 259–271 (1999)
25. L. Ren, H. Lu, L. He, Y. Zhang, Enhanced electrokinetic technologies with oxidization–reduction for organically-contaminated soil remediation. Chem. Eng. J. **247**, 111–124 (2014)
26. H. Kang, J. Lee, B. Kim, T. Kwon, Application of the electrokinetic-Fenton process for the treatment of lubricant-contaminated railroad soil. Adv. Mater. Res. **955–959**, 2248–2253 (2014)
27. A.T. Yeung, Y. Gu, A review on techniques to enhance electrochemical remediation of contaminated soils. J. Hazard. Mater. **195**, 11–29 (2011)

28. A. Yeung, Contaminant extractability by electrokinetics. Environ. Eng. Sci. **23**, 202–224 (2006)
29. M. Pazos, M.T. Alcantara, E. Rosales, M. Angeles Sanroman, Hybrid technologies for the remediation of diesel fuel polluted soil. Chem. Eng. Technol. **34**, 2077–2082 (2011)
30. M. Pazos, O. Iglesias, J. Gómez, E. Rosales, M.A. Sanromán, Remediation of contaminated marine sediment using electrokinetic–Fenton technology. J. Indust. Eng. Chem. **19**, 932–937 (2013)
31. S. Kim, S. Han, Y. Cho, Electrokinetic remediation strategy considering ground strata: a review. Geosci. J. **6**, 57–75 (2002)
32. S. Kim, J. Kim, S. Han, Application of the electrokinetic-Fenton process for the remediation of kaolinite contaminated with phenanthrene. J. Hazard. Mater. **118**, 121–131 (2005)
33. T. Alcantara, M. Pazos, S. Gouveia, C. Cameselle, M.A. Sanroman, Remediation of phenanthrene from contaminated kaolinite by electroremediation-Fenton technology. J. Environ. Sci. Health Part A-Toxic/Hazard. Subst. Environ. Eng. **43**, 901–906 (2008)
34. A. Oonnittan, R.A. Shrestha, M. Sillanpää, Removal of hexachlorobenzene from soil by electrokinetically enhanced chemical oxidation. J. Hazard. Mater. **162**, 989–993 (2009)
35. C. Trellu, E. Mousset, Y. Pechaud, D. Huguenot, E.D. van Hullebusch, G. Esposito, M.A. Oturan, Removal of hydrophobic organic pollutants from soil washing/flushing solutions: a critical review. J. Hazard. Mater. **306**, 149–174 (2016)
36. A. Asadi, B.B.K. Huat, H. Nahazanan, H.A. Keykhah, Theory of electroosmosis in soil. Int. J. Electrochem. Sci. **8**, 1016–1025 (2013)
37. L.M. Vane, G.M. Zang, Effect of aqueous phase properties on clay particle zeta potential and electro-osmotic permeability: implications for electro-kinetic soil remediation processes. J. Hazard. Mater. **55**, 1–22 (1997)
38. E. Vieira dos Santos, F. Souza, C. Saez, P. Cañizares, M.R.V. Lanza, C.A. Martinez-Huitle, M.A. Rodrigo, Application of electrokinetic soil flushing to four herbicides: a comparison. Chemosphere **153**, 205–211 (2016)
39. D. Huang, Q. Xu, J. Cheng, X. Lu, H. Zhang, Electrokinetic remediation and its combined technologies for removal of organic pollutants from contaminated soils. Int. J. Electrochem. Sci. **7**, 4528–4544 (2012)
40. J. Park, S. Kim, Y. Lee, K. Baek, J. Yang, EK-Fenton process for removal of phenanthrene in a two-dimensional soil system. Eng. Geol. **77**, 217–224 (2005)
41. Y. Acar, H. Li, R. Gale, Phenol removal from kaolinite by electrokinetics. J. Geotech. Eng.-ASCE **118**, 1837–1852 (1992)
42. G. Andreottola, E. Ferrarese, Application of advanced oxidation processes and electrooxidation for the remediation of river sediments contaminated by PAHs. J. Environ. Sci. Health Part A-Toxic/Hazard. Subst. Environ. Eng. **43**, 1361–1372 (2008)
43. D. Rahner, G. Ludwig, J. Röhrs, Electrochemically induced reactions in soils—a new approach to the in-situ remediation of contaminated soils? Part 1: The microconductor principle. Electrochim. Acta **47**, 1395–1403 (2002)
44. C. Cameselle, K.R. Reddy, Effects of periodic electric potential and electrolyte recirculation on electrochemical remediation of contaminant mixtures in clayey soils. Water Air Soil Pollut. **224**, 1636 (2013)
45. M. Lu, Z. Zhang, W. Qiao, X. Wei, Y. Guan, Q. Ma, Y. Guan, Remediation of petroleum-contaminated soil after composting by sequential treatment with Fenton-like oxidation and biodegradation. Bioresour. Technol. **101**, 2106–2113 (2010)
46. G.C.C. Yang, C. Liu, Remediation of TCE contaminated soils by in situ EK-Fenton process. J. Hazard. Mater. **85**, 317–331 (2001)
47. J. Kim, J.Y. Kim, S. Kim, Effect of $H_2SO_4$ and HCl in the anode purging solution for the electrokinetic-Fenton remediation of soil contaminated with phenanthrene. J. Environ. Sci. Health Part A-Toxic/Hazard. Subst. Environ. Eng. **44**, 1111–1119 (2009)
48. A. Oonnittan, P. Isosaari, M. Sillanpaa, Oxidant availability in soil and its effect on HCB removal during electrokinetic Fenton process. Sep. Purif. Technol. **76**, 146–150 (2010)
49. G.C.C. Yang, C. Yeh, Enhanced nano-$Fe_3O_4/S_2O_{82}$- oxidation of trichloroethylene in a clayey soil by electrokinetics. Sep. Purif. Technol. **79**, 264–271 (2011)

50. J.Y. Park, J.H. Kim, Switching effects of electrode polarity and introduction direction of reagents in electrokinetic-Fenton process with anionic surfactant for remediating iron-rich soil contaminated with phenanthrene. Electrochim. Acta **56**, 8094–8100 (2011)
51. S. Seo, J. Kim, J. Shin, J. Park, Treatment of artificial and real co-contaminated soil by an enhanced electrokinetic-Fenton process with a soil flushing method. Water Air Soil Pollut. **226**, 86 (2015)
52. C. Sandu, M. Popescu, E. Rosales, E. Bocos, M. Pazos, G. Lazar, M.A. Sanromán, Electrokinetic-Fenton technology for the remediation of hydrocarbons historically polluted sites. Chemosphere **156**, 347–356 (2016)
53. M. Popescu, E. Rosales, C. Sandu, J. Meijide, M. Pazos, G. Lazar, M.A. Sanromán, Soil flushing and simultaneous degradation of organic pollutants in soils by electrokinetic-Fenton treatment. Process Saf. Environ. Protect. **108**, 99–107 (2017)
54. K.R. Reddy, K.S. Chandhuri, Fenton-like oxidation of polycyclic aromatic hydrocarbons in soils using electrokinetics. J. Geotech. Geoenviron. Eng. **135**, 1429–1439 (2009)
55. M.A. Rodrigo, N. Oturan, M.A. Oturan, Electrochemically assisted remediation of pesticides in soils and water: a review. Chem. Rev. **114**, 8720–8745 (2014)
56. R.J. Watts, P.C. Stanton, Mineralization of sorbed and NAPL-phase hexadecane by catalyzed hydrogen peroxide. Water Res. **33**, 1405–1414 (1999)
57. Y. Chang, H. Roh, J. Yang, Improving the clean-up efficiency of field soil contaminated with diesel oil by the application of stabilizers. Environ. Technol. **34**, 1481–1487 (2013)
58. J. Park, J. Kim, Role of sol with iron oxyhydroxide/sodium dodecyl sulfate composites on Fenton oxidation of sorbed phenanthrene in sand. J. Environ. Manag. **126**, 72–78 (2013)
59. P. Isosaari, R. Piskonen, P. Ojala, S. Voipio, K. Eilola, E. Lehmus, M. Itävaara, Integration of electrokinetics and chemical oxidation for the remediation of creosote-contaminated clay. J. Hazard. Mater. **144**, 538–548 (2007)
60. C.L. Ho, M.A. Shebl, R.J. Watts, Development of an injection system for in situ catalyzed peroxide remediation of contaminated soil. Harzard. Waste Harzard. Mat. **12**, 15–25 (1995)
61. Huling, S.G., Pivetz, B.: In situ chemical oxidation – engineering issue. (EPA/600/R-06/072), 1–60 (2006).
62. O. Iglesias, M.A. Fernandez de Dios, M. Pazos, M.A. Sanroman, Using iron-loaded sepiolite obtained by adsorption as a catalyst in the electro-Fenton oxidation of Reactive Black 5. Environ. Sci. Pollut. Res. **20**, 5983–5993 (2013)
63. C. Risco, H. Rubí-Juárez, S. Rodrigo, R. López-Vizcaíno, C. Saez, P. Cañizares, C. Barrera-Díaz, V. Navarro, M.A. Rodrigo, Removal of oxyfluorfen from spiked soils using electrokinetic soil flushing with the surrounding arrangements of electrodes. Sci. Total Environ. **559**, 94–102 (2016)
64. M. Pazos, E. Rosales, T. Alcantara, J. Gomez, M.A. Sanroman, Decontamination of soils containing PAHs by electroremediation: a review. J. Hazard. Mater. **177**, 1–11 (2010)
65. S. Ko, M. Schlautman, E. Carraway, Cyclodextrin-enhanced electrokinetic removal of phenanthrene from a model clay soil. Environ. Sci. Technol. **34**, 1535–1541 (2000)
66. R. Baciocchi, M.R. Boni, L. D'Aprile, Hydrogen peroxide lifetime as an indicator of the efficiency of 3-chlorophenol Fenton's and Fenton-like oxidation in soils. J. Hazard. Mater. **96**, 305–329 (2003)
67. G. Sposito, On points of zero charge. Environ. Sci. Technol. **32**, 2815–2819 (1998)
68. F. Pardo, J.M. Rosas, A. Santos, A. Romero, Remediation of soil contaminated by NAPLs using modified Fenton reagent: application to gasoline type compounds. J. Chem. Technol. Biotechnol. **90**, 754–764 (2015)
69. K. Hanna, S. Chiron, M.A. Oturan, Coupling enhanced water solubilization with cyclodextrin to indirect electrochemical treatment for pentachlorophenol contaminated soil remediation. Water Res. **39**, 2763–2773 (2005)
70. P. Kakarla, R. Watts, Depth of Fenton-like oxidation in remediation of surface soil. J. Environ. Eng.-ASCE **123**, 11–17 (1997)

71. US EPA, Recents develpments for in situ treatment of metal contaminated soils. EPA/542/R/97/004, 1–2 (1997).
72. B. Ochoa, L. Ramos, A. Garibay, M. Perez-Corona, M.C. Cuevas, J. Cardenas, M. Teutli, E. Bustos, Electrokinetic treatment of polluted soil at pilot level coupled to an advanced oxidation process of its wastewater. Phys. Chem. Earth **91**, 68–76 (2016)

# Coupling of Anodic Oxidation and Soil Remediation Processes

**Soliu O. Ganiyu and Carlos A. Martínez-Huitle**

## 1 Introduction

Industrial revolution and increased agricultural mechanization in the last decades have had several consequences, one of which is the environmental issues and sustainable development. Decision and policy makers in industries, economics and politics take careful consideration of these topics and several stringent rules and regulations on environmental pollution issues are being annually promulgated, especially on hazardous waste, atmospheric pollution and wastewater [1, 2]. Soil, an important component of ecosystem and a key resource for the survival of humans and animals, has been under constant contamination due to different human activities such as excessive use of pesticides and fertilizer on farmlands, infiltration from livestock impoundments, accidental discharge of harmful pollutants, industrial wastewater, landfill leachates, petroleum spillage and oil rigging, improper waste disposal and stockpiles, all of which contribute to serious soil pollution and deterioration of soil quality [3, 4]. As such, soil contamination is a global issue and it is considered as one of the barriers for sustainable development. The major soil contaminants include heavy metals and toxic anions, toxic organic compounds and radionuclide [5]. Heavy metal contaminants such as Cr, Hg, Cd and Pb and organic pollutants like volatile chlorinate solvent, polycyclic aromatic hydrocarbons

S. O. Ganiyu (✉)
Department of Civil and Environmental Engineering, University of Alberta, Edmonton, AB, Canada

Institute of Chemistry, Federal University of Rio Grande do Norte, Campus Universitário, Natal, RN, Brazil
e-mail: ganiyu@ualberta.ca

C. A. Martínez-Huitle
Institute of Chemistry, Federal University of Rio Grande do Norte, Campus Universitário, Natal, RN, Brazil

 
M. A. Rodrigo, E. V. Dos Santos (eds.), *Electrochemically Assisted Remediation of Contaminated Soils*, Environmental Pollution 30,
https://doi.org/10.1007/978-3-030-68140-1_9

(PAHs), polychlorinated biphenyls (PCBs) and total petroleum hydrocarbons (TPHs) have been reportedly found at different contamination levels in soil across the globe [6–8]. These contaminants pose a great threat to the safety of ecosystems and human health via food chain and direct exposure to contaminated soils. Moreover, organic contaminants are hardly degraded in soil by abiotic/biotic agents, leading to their persistence and accumulation [2, 3].

Due to their hazardous and potential risk on human health and safety of ecosystem as well as land reuse, contaminated soils require effective remediation for their reclamation and reuse. In-depth works have been devoted for the development of soil remediation techniques, and several new and innovative solutions for effective abatement of contaminants from soils have been investigated either to completely eliminate the pollutants or reduce their concentrations to tolerable and safe level [3, 9, 10]. Most of the soil remediation processes, in existence, have at least one major "bottleneck" like high cost (thermal treatment), low efficiency (pump and treat), long treatment time (biological treatment), low treatment capacity and high chemical requirement (in situ chemical oxidation) or necessity for post-treatment decontamination of the effluents (soil washing/flushing system) [2, 11]. Sequential or combined remediation techniques in which two or more treatment techniques are applied either simultaneously or sequentially has the advantage of each technique complementing the merit and overcoming the challenges of each other [1]. Among the existing combined techniques, soil washing/flushing/electrokinetic soil remediation coupled with advanced oxidation processes (AOPs) is a versatile and time-efficient treatment for the remediation organic pollutants contaminated soil and it has been globally used over the years [2, 12]. Soil washing is a mechanical process, which involves the use of liquids (extracting agents) to remove chemical pollutants from the soils [2, 6, 13, 14]. However, the contaminants are only separated into solution, which implies that necessary treatment is still required for washing effluents before discharge into environment. AOPs based on hydroxyl radical productions ($^{\bullet}OH$) such as ozonation, photocatalysis, Fenton oxidation and electrochemical AOPs have recently been investigated for the treatment of soil washing effluents with best removal efficiency observed with electrochemical AOPs [2, 3, 12, 15–17]. Among the electrochemical AOPs, electrooxidation, otherwise known as anodic oxidation, has considerable advantage in that it requires very limited or no chemicals for generation of strong oxidants (reactive species) needed for oxidation of the organic pollutants [18–21]. In anodic oxidation (AO), reactive oxygen species (ROS) and/or reactive chloride species (RCS) are electrogenerated at the anode region via water/chlorine oxidation [22–25]. The produced reactive oxygen species especially hydroxyl radical ($^{\bullet}OH$) is a very strong oxidant ($E = -2.8$ eV vs. SHE) and can react non-selectively with any class of organic pollutants until their total conversion to $CO_2$ or at least achieve high mineralization of the organic [26–29]. The fundamental principles and application of this process are available in the literature [30–34].

Coupling of AO and soil remediation for the treatment of organic contaminated soil has been studied via two configurations (Fig. 1): (1) AO as post-treatment stage for the treatment of soil washing/flushing/electrokinetic soil remediation effluents and (2) in-situ/simultaneous AO and soil washing/flushing process. The former is

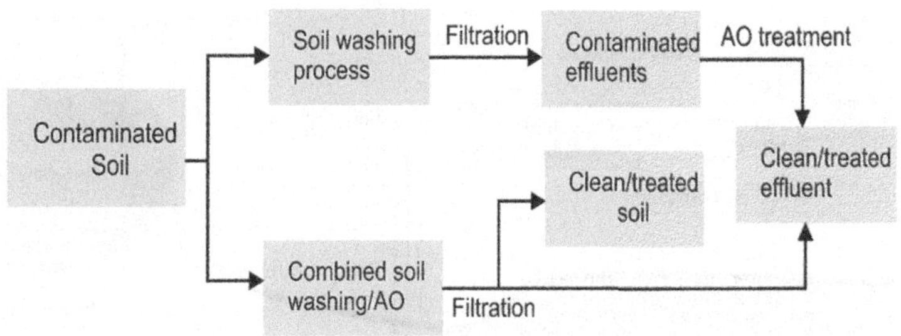


**Fig. 1** Schematic of different configurations of coupling soil washing with AO

the most widely investigated configuration of the combined soil remediation-AO processes, whereas the latter is less common owing to the higher diffusion dependent of electrode process like AO. In this chapter, the concept of soil washing/flushing/electrokinetic soil remediation is briefly explained before different configurations of combined anodic oxidation and soil remediation were discussed. In the final section a brief future perspective and concluding remarks are provided.

## 2 Soil Remediation Processes

Remediation of contaminated soils has become a global issue not only due to the hazardous effect of the contaminants on the ecosystem but also due to land shortage that necessitates reclamation of contaminated sites for reuse. Among the existing soil remediation techniques, soil washing/flushing is a versatile, cost- and time-efficient method and has attracted increasing attention in recent years across the globe [2, 6, 13]. Another rising and recently wide studied technique is electrokinetic soil remediation process in which an electric field produced by low DC current is applied across the soil surface, which separates the contaminants from soil with the aid of added electrolytes [12, 35]. Unfortunately, both processes generate effluents which contain contaminants extracted from the soil and require extensive and proper treatment before disposal.

Soil washing/flushing processes: In soil washing/flushing, liquid extracting agents usually in aqueous solutions are used to mechanically remove chemical pollutants from contaminated soils [6, 12]. Contaminants usually have low solubility and adhere strongly to the surface of the soil in real contaminated sites, as in practical soil washing processes, in which additives such as acids, surfactants, chelating agents are often added to washing solution to solubilize the contaminants from the soil [6]. Soil flushing is an in situ process where extracting agents are added to the contaminated soils to improve the mobility of pollutants by reducing the interfacial tension between them and the groundwater [2, 12, 36]. The mobilized pollutants can then be evacuated in the extraction wells (Fig. 2). However, soil flushing is more adapted to light-weight polycyclic aromatic hydrocarbons like naphthalene

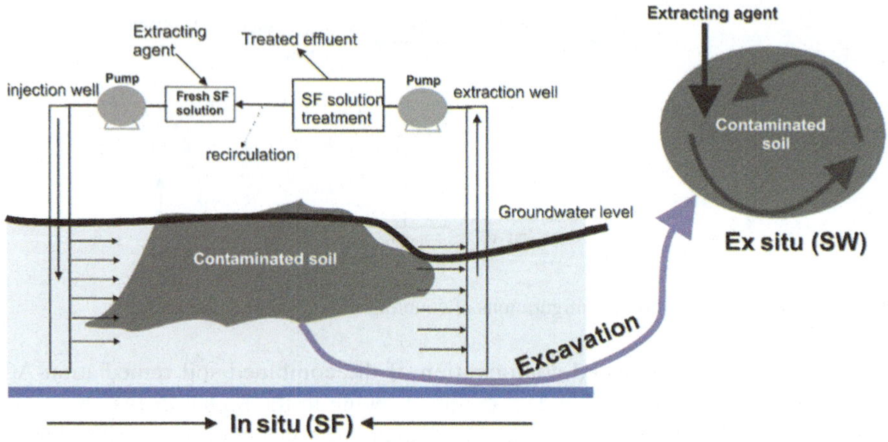


**Fig. 2** Scheme of a typical soil flushing and soil washing processes. (Adapted from ref. [2])

remediation because the pumping can be easily operated from the surface of the groundwater table [2]. The contaminated site/field characteristics such as soil heterogeneity, contaminant nature, naphthalene saturation and others strongly influence the efficiency of the soil flushing process [37]. On the other hand, soil washing is an ex situ process (Fig. 2) where the contaminated soil/site is excavated, transported and treated with a certain soil–liquid ratio, usually between 5 and 45% [14, 38]. The contaminants sorbed to the soil are removed by adding extracting agents to the washing solution and there is always enhanced contact between the extracting agents and the soil contaminants in soil washing, thereby allowing for better treatment efficiency assessment compared to soil flushing process [14].

Extracting agents: The aqueous solutions with or without additives used to mobilize the contaminants from the soil to the soil washing solution is termed extracting agents. Aside having good extracting and solubilizing ability, the extracting agents should possess excellent biodegradability and low eco-toxicity to the soil organisms as well as environmental compartments where it is disposed after usage [36]. Extracting agents can also mobilize non-targeted contaminants especially heavy metals and toxic anions such as lead, cadmium, chromium, copper and arsenic; thus, the soil washing effluents require proper treatments and toxicity assessment before disposal [2, 6]. Some of the commonly used extracting agents are briefly described below.

1. Water and organic solvents: Organic solvents are the earliest extracting agents used for the removal of polycyclic aromatic hydrocarbons (PAHs) from contaminated soils, both on bench and field scales [13, 39]. Hydrophobic soil contaminants such as PAHs with lower octanol-water partition coefficient (log $K_{OW}$) have been removed from contaminated soils using non-polar organic solvents including 1-pentanol, *n*-hexane, benzene, toluene and dichloromethane [39]. To date, water and several organic solvents including alcohols, esters, ketones, alkylamines and aromatics have been studied to extract PAHs from soils, however,

organic solvents such as ethanol, 2-propanol, 1-pentanol and ethyl acetate have been reported to be only effective for the removal of lower-molecular weight (LMW) PAHs but less efficiency for higher molecular weight (HMW) PAHs, which continue to persist in soil after extraction [40, 41].

2. Synthetic surfactants: Surfactants are a class of amphiphilic chemicals composed of a hydrophilic water-soluble head and a hydrophobic or water-insoluble tail. Their unique molecular structure gives them the ability to solubilize relatively insoluble xenobiotics including hydrophobic soil contaminants [6]. They are characterized by their chemical structures, hydrophobic–lipophobic balance and critical micellar concentration (CMC), which is defined as the surfactant concentration above which micelles are formed and all additional surfactants added to the solution go to the micelles (Fig. 3) [36, 42, 43]. The solubility of hydrophobic organic compounds (HOCs) is strongly enhanced at surfactant concentrations above CMC along with decreased surface tension [36].

   Three different mechanisms are reported to be involved in surfactant enhanced removal of hydrophobic organics sorbed to soils, which include decrease of interfacial tension, phase transfer of HOCs from soil–liquid interface to micellar pseudo-aqueous phase and solubilization of the HOCs inside the hydrophobic enclosure formed by micelles [44]. Surfactants can be classified as anionic (e.g. sodium dodecyl sulfate and linear alkylbenzene sulfonate), cationic (quaternary ammonium derivatives), non-ionic (Brij 35, Tween 80, Triton X-100) and amphoteric (cocoamidopropyl hydroxylsultaine), but non-ionic surfactants are preferred due to their lower soil sorption ability, cost-effectiveness and higher solubilization capacity [36]. Several studies have reported the use of different synthetic surfactants for remediation of PAHs from contaminated soils both on laboratory and pilot scales and some review papers have summarized these studies in the literature [2, 6, 12–14].

3. Biosurfactants: These are amphiphilic chemicals similar to synthetic surfactant but having microbial origin. They are capable of forming micelles and are manu-

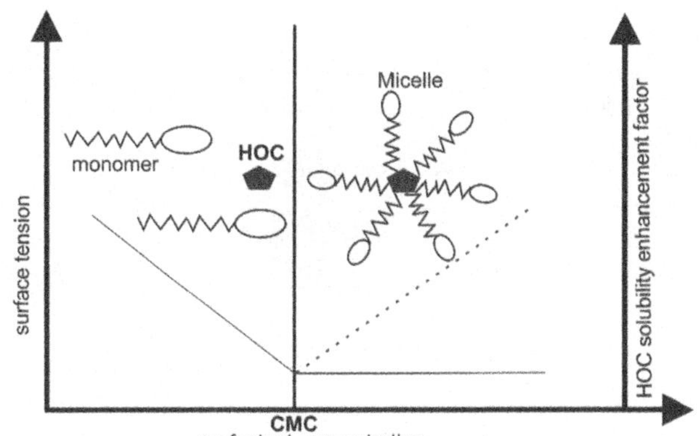


**Fig. 3** Surface tension and hydrophobic organic compound (HOC) solubility enhancement factor as a function of surfactant concentration. (Printed with the permission of ref. [2])

factured from renewable resources such as water-soluble carbon sources, water-immiscible substrates and nitrogen sources [45]. Biosurfactants have some distinguish advantage such as high extraction efficiency, extremely biodegradable, ecological safety, lower toxicity and possibility of produced in situ during soil washing process [13]. Indeed, Lai et al. [46] reported higher petroleum hydrocarbon removal efficiency from contaminated soil with biosurfactants rhamnolipid and saponin compared to synthetic surfactants Tween 80 and TX100 during soil washing process. However, the ability to produce sufficient quantities during the soil washing/flushing process at economical rate is a major challenge of using biosurfactants.

4. Microemulsion: These are optically transparent and thermodynamically stable single phase, usually prepared from a ternary mixture of water, water-immiscible oil and a cosurfactant [13, 47]. Water based on the microemulsions behave like a separate bulk phase which is capable of desorbing and concentrating pollutants from soil [48]. Unlike synthetic surfactants where extent of solubility enhancement sharply increased at CMC, the extent of solubility enhancement is linearly proportional to the concentration of microemulsions [13]. Sodium castor oil sulfate, fatty ester–water–non-ionic surfactants (methyl ester from babassus oil and unsaturated fraction of palm oil), 1-butanol–oil and other microemulsions based on vegetable oil have been demonstrated to showed higher extraction efficiency for several organic pollutants especially PAHs and TPHs [47–49].
5. Cyclodextrins (CDs): CDs have been proposed as non-toxic and highly biodegradable alternative to organic solvents and surfactants due to the environmental concern, for removal of PAHs from contaminated soils [50, 51]. They consist of hydrophilic groups on the external side of their ring, which can dissolve in water and a low-polarity cavity providing hydrophobic matrix which can entrap many organic compounds into the rings. This characteristic provides CDs with a larger capacity in solubilizing hydrophobic contaminants like PAHs [50, 52, 53]. CDs such as β-cyclodextrins, hydroxylpropyl-β-cyclodextrins, methyl-β-cyclodextrins

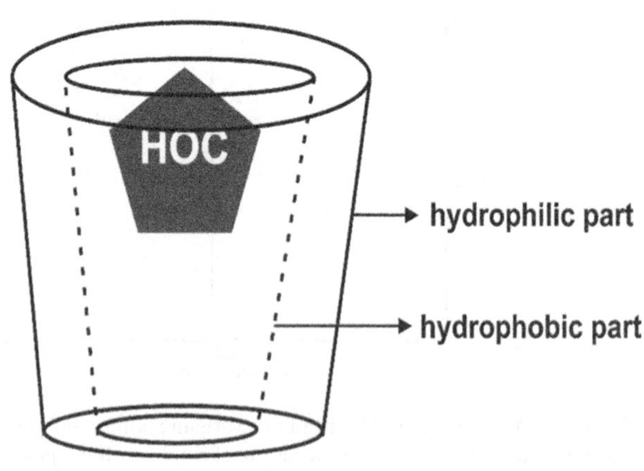


**Fig. 4** Inclusion complex formed by CDs with HOCs during soil washing. (Reprinted with the permission of ref. [2])

and recently chemically modified CDs have been investigated on batch experiments for desorption of PAHs from soils [50, 54–57] (Fig. 4).
6. Humic acids (HAs): HAs are the fraction of humic substances composes in soil that is insoluble in water under acidic conditions [13]. Conte et al. [58] were the first to hypothesize that HAs is capable of reducing sorption of organic contaminants onto the soil. Subsequently, the same author [59] reported the removal of PAHs from soil using HAs as a natural surfactant. The HA as an extracting agent showed similar extraction efficiencies to those of synthetic surfactants, with more than 80% removal of PAHs from the contaminated soil. Some studies have utilized low molecular weight organic acids mainly released by plants [60] and soil nanoparticles which composed mainly organic contents (along some inorganic clays) [61] to absorb organic compound from the soil and enhanced its water solubility during the subsequent soil washing process.
7. Vegetable oils and organic cosolvents: Due to several drawbacks such as high cost, risk of handling and storing, toxicity and soil permeability disturbance, organic co-solvents are no longer considered promising extracting agents for soil remediation process. Vegetable oils which composed high triglycerides and have high affinity for PAHs are favourable alternative/replacement to costly, toxic and non-biodegradable solvents and surfactants for removing PAHs from soil [13, 62]. The planner aromatic rings of PAH molecules bind to the triglycerides structure of the vegetable oils during the soil washing. This is possible because vegetable oils are characterized by their hydrophobicity and long aliphatic carbon chain structure which form hydrophobic interactions with non-polar molecules [13]. The use of vegetable oils for removal of PAHs from contaminated soils has been reported in the literature [63–66].

**Electrokinetic soil remediation (EKSR):** Like soil washing/flushing processes, EKSR desorbed pollutants from contaminated soils and produced contaminants loaded effluents that require proper treatment before disposal. In EKSR, contaminants are desorbed from the soil by the electric field created within the contaminated soil by the application of direct current via electrodes located at the subsurface [12, 67, 68]. Besides electrokinetic processes such as electro-osmosis, electromigration and electrophoresis, the applied current simultaneously initiates many physical processes (heating, change in viscosity etc.), electrochemical processes (water oxidation and reduction, $H_2$ evolution), and chemical processes (ion exchange, dissolution of precipitates, etc.) which tremendously change the soil [12]. These processes can be systematically combined by setting optimum configurations and operation conditions in a soil treatment process which can promote certain process such as electrochemical soil flushing to ensure the removal of many inorganic and organic contaminants from the soil and minimize other non-beneficial processes (heating, $H_2$ evolution, etc.) [69–73]. This technology has been investigated for the remediation of different type of contaminated soils and is particularly more effective for fine-grained soils with low hydraulic conductivities and large specific surface areas [68]. The main phenomenon during the treatment of such fine-grained soil is electro-osmosis, which involves accumulation of net electric charge at the surface of the solids in contact with the electrolyte solution and the accumulation of a thin counterion layer (electrical double layer/Debye layer) of the liquid surrounding the solid surface. Since the Debye layer is charged, this portion of the fluid is mobile within the electric field between the electrodes due to attraction and repulsion from opposite and the same charge respectively [12].

Electrokinetic soil flushing involves the driving of the ground water/added aqueous solutions (chemicals) in the soil to mobilize the pollutants in the soil. The pollutants are washed out of the soil with the aid of the water/solution via dissolution of precipitates, ionic exchange, desorption or by simple mechanic dragging during washing, thus the pollutants are transferred from the soil into the water/solution and solving the soil contamination issues. The process is only economical for treatment of low permeability soils with small hydraulic flux where conventional soil flushing is ineffective, however for highly permeable soils, conventional soil flushing driving by pressure gradients is sufficient in order to avoid cost/expenses associated with application of electrochemical technologies [12]. In a typical EKSR, the electro-osmotic flux mobilizes groundwater from anode to cathode and the water at the cathode can be recycled by pumping it to the anode to begin new flushing process [74]. Electrodes materials such as platinum, graphite, platinized titanium, carbon-felt, stainless steel and platinum-coated graphite have been applied as either anode or cathode in the electrokinetic flushing process; however, the configuration of the electrodes depends on the reactor designs and the nature of the contamination in soil [12].

Several flushing fluids have been used with or without additives to mobilize pollutants in the contaminated soils. Among the flushing fluids that have utilized, fluids that are capable of soil pH regulation are more beneficial because they can compensate for the influence of the acidic and/or basic fronts created in the soil during electrokinetic remediation [12, 75, 76]. These fluids can use alone or along with surfactants such Tween 80, SDS, β-CD and others for soil flushing. Buffer solution of $Na_2CO_3$/$NaHCO_3$ which neutralizes the acid fronts and acetic acid which neutralizes the basic fronts and lower the treated soil pH are the important reagents used in soil flushing [77, 78]. Other fluids such as citric acid, $NaNO_3$ and $NaHPO_4$ have been studied either alone or with surfactants for the remediation of soil contaminated with organic pollutants [12, 79–81].

As stated earlier, the soil washing/flushing and ESKR processes effluents are loaded with both the extracting agents and the extracted organic contaminants. Therefore, necessary treatments are required to remediate the organic pollutants to harmless or biodegradable substances prior to disposal to environment. Note that, some studies have reported in-situ remediation using peroxidation with $H_2O_2$ and Fenton oxidation ($FeSO_4$ and $H_2O_2$) simultaneously along with electrokinetic soil flushing for the treatment of contaminated soil [82, 83]. However, effluents of such combined processes still contain significant quantities of extracted contaminants, which explains the inadequacy of such oxidation process for complete decontamination of electrokinetic soil flushing effluents.

# 3 Coupling of SW/EKSR and Anodic Oxidation

The effluents of the SW/SF and EKSR processes are loaded with contaminants extracted from the soil. As such, post-treatment decontamination and detoxification of the effluents are needed prior to disposal into the environment. Besides, selective removal of the contaminants from the effluents, the combined treatment ensures the reusability and recyclability of the extracting solution. Indeed, some studies have employed different remediation processes such as selective electrochemical adsorption [84], photocatalysis [2, 3, 85, 86], Fenton's reaction–based oxidation [2, 3, 87, 88] and electrochemical AOPs [2, 12, 54] for treatment of effluents of SW/SF and EKSR processes and possible recycling of the extracting agents. Among the studied treatment techniques, electrooxidation using "non-active" anodes especially boron-diamond electrode is an efficient and effective method for both remediation of the contaminants loaded SW/SF and EKSR effluents as well as recycling of the extracting agents for possible reuse. Two configurations of coupling AO and soil remediation process can be proposed: (1) AO as post-treatment stage—ex situ electrooxidation treatment of SW/SF and EKSR effluents and (2) in situ or simultaneous SW/SF/EKSR and AO. As stated earlier, the former is the most widely investigated configuration and has been reported by many researchers, whereas the latter is less common in the literature. Both configurations are vividly discussed in the following sections.

Ex situ treatment of SW/SF/EKSR effluent by AO: The treatment of SW/SF/EKSR effluents by AO is a well-investigated combined process for complete removal of contaminants and detoxification of effluents. AO using different electrode materials and cell configurations has been demonstrated to achieve excellent degradation and mineralization of the organic and organo-metallic pollutants contained in the SW/SF/EKSR effluents especially when "non-active" electrode such as BDD, doped $PbO_2$ and $SnO_2$ are utilized. In most cases, both the contaminants and the extracting agents contained in the effluents are degraded and mineralized, since hydroxyl radicals is a non-selective oxidizing agent that react with any class of organic pollutants. However, few studies have reported selective degradation of the pollutants encapsulated in the micelles formed by the extracting agents and possible reuse of the extracting agents. Most of the research works on application of AO for the treatment of SW/SF/EKSR effluents were reported by Rodrigo's group [89–92] (see Table 1).

For instance, dos Santos [91, 99] studied the removal of atrazine from soils using combined soil washing and conductive diamond electrode electrooxidation. Atrazine was removed from spiked soils by surfactant fluids (SDS) assisted soil washing and the resulting effluents were treated by electrooxidation using BDD electrode. The authors [90] showed that the combined technologies were efficient for the removal and total mineralization of atrazine from soils and soil washing effluents. Surfactant–soil ratio (Fig. 5) was identified as the key parameter for the removal of the atrazine from the soil and it affects significantly, the characteristics of the effluents, mostly importantly the total organic loading and the size of the micelles. Beside applied current density, the size of the particles in the soil washing effluents (reaction media)

**Table 1** Some examples of SW effluents treated by ex situ electrochemical processes

| Compound | Concentration | Experimental conditions | *J* | Removal | Refs |
|---|---|---|---|---|---|
| Organochlorines | 160 mg $L^{-1}$ of chloride | Platinum and BDD anode; graphite cathode; *T*: 25 °C; *V*: 150 $cm^3$ | 300 A $m^{-2}$ | 0.22 mmol $Ah^{-1}$ | [93] |
| Clopyralid | 100 mg $kg^{-1}$ of soil | BDD anode and stainless-steel cathode; *T*: 25 °C; *V*: 1 L | 25 mA $cm^{-2}$ | Complete | [94] |
| Clopyralid | 30 mg $dm^{-3}$ | BDD anode and stainless-steel cathode; *T*: 40 °C; *V*: 2 L | 25 mA $cm^{-2}$ | Complete | [95] |
| Lindane | 100 mg $kg^{-1}$ of soil | BDD anode and stainless-steel cathode; *T*: 25 °C; *V*: 1 L | 50 mA $cm^{-2}$ | Complete | [96] |
| Oxyfluorfen | 100 mg $kg^{-1}$ of soil | BDD anode and stainless-steel cathode; *T*: 25 °C; *V*: 1 L | 930 W $cm^{-2}$ | Complete | [97] |
| Oxyfluorfen | 100 mg $kg^{-1}$ of soil | BDD anode and stainless-steel cathode; *T*: 25 °C; *V*: 1 L | 30 mA $cm^{-2}$ | Complete | [98] |
| Phenanthrene | 500 mg $kg^{-1}$ of soil | Aluminium and iron as anodes and cathodes; *T*: 25 °C; *V*: 5000 $dm^3$ | 7.5 mA $cm^{-2}$ | Values close to zero for COD | [75] |
| Atrazine | 100 mg $kg^{-1}$ of soil | BDD anode and stainless-steel cathode; *T*: 25 °C; *V*: 1 L | 30 mA $cm^{-2}$ | Complete | [99] |
| Clopyralid | 0.02 mg $g^{-1}$ of soil | BDD anode and stainless-steel cathode; *T*: 25 °C; | 12.8 mA $cm^{-2}$ | Complete | [92] |
| Oxyfluorfen | 100 mg $kg^{-1}$ of soil | BDD anode and stainless-steel cathode; *T*: 25 °C; *V*: 700 $cm^3$ | 300 A $m^{-2}$ | Complete | [89] |

was the key parameter that influences the efficiency of the electrooxidation process, which continuously decreased during the electrolysis.

The same authors, have reported the removal of herbicide oxyfluorfen from soil washing fluids using either electrolysis [89], sono-electrolysis [90, 98] or UV-assisted electrolysis [97] with BDD electrode. Although the electrolysis using BDD electrode was quite efficient for the total degradation and mineralization of the

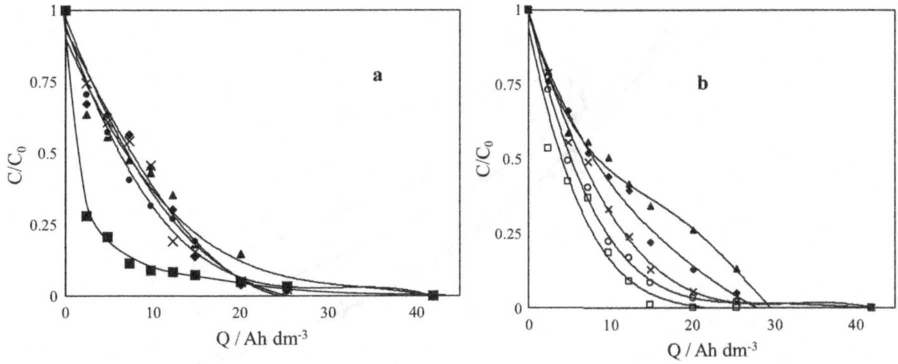


**Fig. 5** Effect of [SDS]–soil ratio: (▲) 0.5, (◆) 2.5, (●) 5, (×) 12.5 and (■) 25 on the degradation profiles of (**a**) SDS and (**b**) atrazine during the CDEO treatment of the soil washing effluent at current density of 30 mA $cm^{-2}$. (Reprinted with the permission of ref. [99])

herbicide, sono and UV-assisted electrolysis with BDD electrode achieve faster and better decontamination of the treated effluent. It was demonstrated that prolong sonolysis and UV photolysis treatment without electrolysis could also achieve degradation (with very poor mineralization) of oxyfluorfen but at very slow rate. The same group [96] has investigated the treatment of soils polluted with lindane by surfactants aided soil washing and AO using BDD. The processes were efficient for removing the hazardous substance from the soil and mineralization from the effluents with over 70% recovery of the surfactant solution after electrolysis for reuse in soil washing. Effluents of soil washing containing other contaminants such as pendimethalin [100], clopyralid [92], PAHs and petroleum [101] have also be treated by electrooxidation, sono or irradiation assisted electrooxidation with BDD electrode. In all cases, the electrolysis with BDD with or without sonolysis/photolysis was observed to achieve excellent degradation and mineralization of both the herbicides and the surfactants in the washing effluents and the total decontamination of the eluents could be achieved in 8 h of electrolysis.

Other relevant studies on treatment of SW/SF effluents by AO were reported by Oturan's group. For example, the group [102] has investigated the treatment of synthetic soil washing effluent containing phenanthrene and CD by AO with BDD electrode. Complete toxicity removal and 100% biodegradability enhancement was achieved during the electrolysis at 1 A for 8 h. The same group [16] has utilized a combination of anodic oxidation and biological treatment for the removal of phenanthrene and Tween 80 from synthetic soil washing solution. The AO with BDD was able to achieve 95% of phenanthrene and Tween 80 as well as 71% COD removal at 1 A for 5 h (Fig. 6). Biological treatment was reported to achieve complete phenanthrene and Tween 80 degradation but could only remove 44% COD. Due to higher energy consumption during AO at this condition, synergistic effect was achieved by performing AO at low current and treatment time (3 h) as either pre- or post-treatment to biological process. In this way, a cost-effective combined process where AO degraded the organic in the washing solution to biodegradable organic

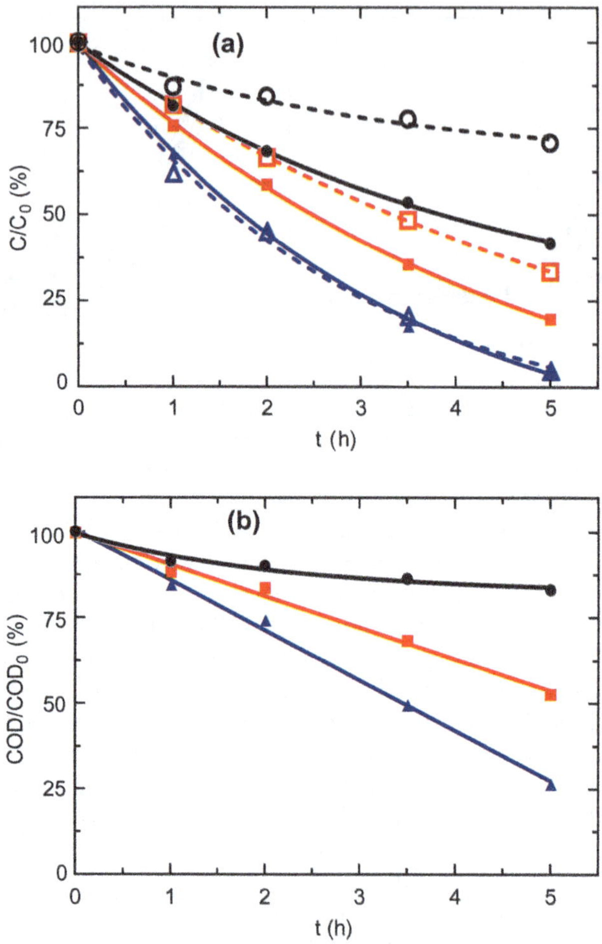


**Fig. 6** Decay of normalized (**a**) concentration: (○, □, Δ) TW80 and (●, ■, ▲) phenanthrene and (**b**) COD during the electrooxidation of 330 mL soil washing solution at applied current of (○, ●) 200 mA, (□, ■) 500 mA and (Δ, ▲) 1000 mA. (Reprinted with the permission of ref. [16])

compounds (short-chain carboxylic acids) followed by biological treatment or initial biological treatment of washing solution to remove the biodegradables followed by AO to degrade recalcitrant organic pollutants remaining in the solution.

Treatment of effluents of EKSR by AO has not receive much attention like SW/SF effluents possible because EKSR has widely been utilized for the remediation of heavy metals/toxic anions contaminated soils compared to organic pollutants contaminated soils. However, a recently study by da Silva et al. [103] showed the possibility of applying AO using BDD electrode for the treatment of produced water from EKSR treatment of synthetic petroleum contaminated soil. The EKSR using graphite electrodes was able to achieve excellent removal of the petroleum products from the soil as revealed by the gradual accumulation of TOC (2250 and 250 mg $L^{-1}$)

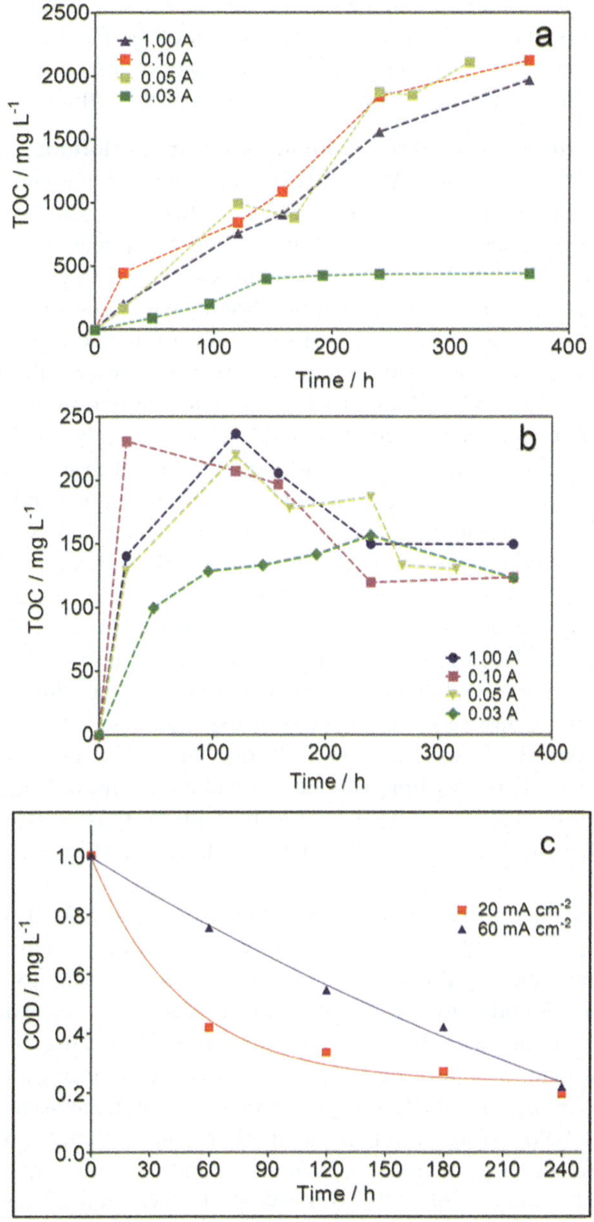


**Fig. 7** TOC evolution as a function of time at (**a**) anodic and (**b**) cathodic compartments during the ESR applying different current values at 25 °C with 0.1 M $Na_2SO_4$ as supporting electrolyte by using graphite electrodes; and (**c**) decay of normalized COD with time obtained during the electrochemical treatment of effluents produced after ESR of soil polluted with petroleum at applied current density of (■) 20 mA $cm^{-2}$ and (▲) 60 mA $cm^{-2}$. (Reprinted with the permission of ref. [103])

(Fig. 7a, b) in both solutions at the anodic and cathodic chambers respectively. Electrolytic treatment of the organic loaded effluent of the EKSR using BDD electrode achieved over 80% COD removal at either 20 or 60 mA $cm^{-2}$ in 4 h, demonstrating the efficacy of the process for the treatment of the effluent.

**In-situ treatment of soil washing effluent by electrooxidation:** This configuration involves performing both SW/SF/EKSR and AO simultaneously for the remediation of organic pollutants contaminated soils. This approach is very rare in the literature, but the concept is very exciting for field application. The electrolyte chambers of the EKSR or flushing fluid channels may serve as an electrolytic cell where the electrolysis process is carried out. Some challenges are envisage including agitation problem, possibility of wearing of electrode due to friction impacted by the soil particles, sampling and longer treatment time especially in SW, due to time requires for efficient washing process. As stated earlier, in-situ AO with SW/SF/EKSR has not been extensively explored, but recent study by Rodrigo's group [95] reported the combination of soil washing, zero valent iron (ZVI) dehalogenation and anodic oxidation in single assemble reactor for removal and degradation of clopyralid spiked contaminated soil. The assemble which consist of the electrochemical cell, the soil washing and dehalogenation tank made of a rigid silicon tube of 1 m × 69 mm, and continuously circulated with the 2 L washing solution at 40 $dm^3$ $h^{-1}$. The electrochemical cell was equipped with a BDD anode and a stainless-steel cathode, both of geometric surface area of 78 $cm^2$. The combined treatment was efficient, achieving complete removal of the chlorinated organics from the soil and complete mineralization of the organics in the generated liquid waste effluent during electrooxidation at 25 mA $cm^{-2}$. The authors observed that combined process of soil washing and electrooxidation achieved similar efficiency when compared to system operated with iron addition; thus, ZVI dehalogenation may not be necessary for the remediation of the clopyralid (Fig. 8).

**Future perspective and conclusion remarks:** Combined SW/SF/EKSR and AO using BDD electrode is an exciting technology that have been studied for remediation soils contaminated by different organic pollutants, with very promising results obtained for bench and some pilot-scale experiments. However, extensive studies are still required to advance the technologies for possible license and commercialization. Future studies should be tailored towards the optimization of operation parameters, modelling and automation, toxicity assessment, environmental and economic analysis, field testing as well as potential of reuse of both the remediated soil and effluents for other purposes (i.e. agriculture and construction). The operation parameters that influence both the soil SW/SF/EKSR such as physicochemical properties of the soils, nature of the contaminants and contamination levels, washing solution composition and concentration as well as washing solution–soil ratio and electrooxidation such as current density, electrode materials and inorganic ions in the treated solution requires proper optimization to ensure optimum efficiency at lowest economic cost possible. Problems or challenges associated with each parameter like high organic loading and cost associated with using higher washing

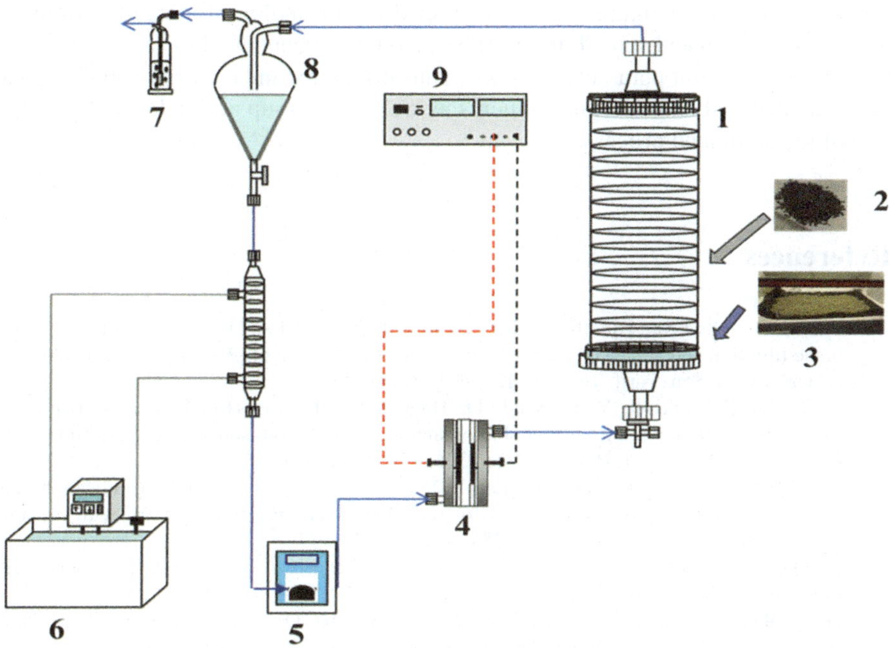


**Fig. 8** Schematic of concurrent soil washing–ZVI dehalogenation and electrochemical oxidation processes for remediation of soil contaminated by herbicide clopyralid: (1) soil washing reactor, (2) ZVI, (3) contaminated soil, (4) electrochemical reactor, (5) peristatic pump (6) thermostatic cooling system (7) secondary tank, (8) gas exchange and (9) DC power generator. (Adapted from ref. [95]

solution/soil ratio during SW/SF process, heating and hydrogen evolution at higher current density as well as non-uniform current distribution in large surface area electrode during AO should be carefully considered during optimization and scale-up of the combined SW/SF/EKSR and AO technologies.

Toxicity assessment and environmental impact of combined technologies on the ecosystem should be thoroughly studied as the literature currently available is scarce. Since both the treated soil and washing effluents are disposed into the environments, thorough assessment of the toxicity of both soils and effluents are necessary to avoid secondary pollution and hazardous effect on ecosystem. Additionally the economic assessment of the technologies is necessary in order to compare the benefit and disadvantage of the combined technologies with other existing soil remediation processes. Extensive studies on scale-up/pilot and field experiments are still needed before the full-scale implementation of the combined technologies. Most of the studies reported in the literature are either bench or pre-pilot scales, and pilot and field studies are essential before certification of the process.

Conclusively, SW/SF/EK/SR coupled with AO is an efficient technology for the remediation of organic polluted contaminated soils. The AO process is applied as post-treatment stage to treat the organic loads in the effluents of soil SW/SF/EK/SR process. In-situ SW/SF/EK/SR coupled with electrooxidation is very rare in the literature, but the configuration is a very exciting technology for treating organic polluted contaminated soils. Future studies should be channelled towards optimization of the operation parameters, assessment of toxicity and environmental impact of the treated soil and washing effluents as well as scale-up and field implementation of the combined process.

## References

1. S.O. Ganiyu, E.D. van Hullebusch, M. Cretin, G. Esposito, M.A. Oturan, Coupling of membrane filtration and advanced oxidation processes for removal of pharmaceutical residues: a critical review. Sep. Purif. Technol. **156**, 891–914 (2015)
2. C. Trellu, E. Mousset, Y. Pechaud, D. Huguenot, E.D. van Hullebusch, G. Esposito, M.A. Oturan, Removal of hydrophobic organic pollutants from soil washing/flushing solutions: a critical review. J. Hazard. Mater. **306**, 149–174 (2016)
3. M. Cheng, G. Zeng, D. Huang, C. Lai, P. Xu, C. Zhang, Y. Liu, Hydroxyl radicals based advanced oxidation processes (AOPs) for remediation of soils contaminated with organic compounds: a review. Chem. Eng. J. **284**, 582–598 (2016)
4. M. Manz, K.-D. Wenzel, U. Dietze, G. Schüürmann, Persistent organic pollutants in agricultural soils of Central Germany. Sci. Total Environ. **277**, 187–198 (2001)
5. H.H. Tabak, P. Lens, E.D. van Hullebusch, W. Dejonghe, Developments in bioremediation of soils and sediments polluted with metals and radionuclides – 1. Microbial processes and mechanisms affecting bioremediation of metal contamination and influencing metal toxicity and transport. Rev. Environ. Sci. Biotechnol. **4**, 115–156 (2005)
6. X. Mao, R. Jiang, W. Xiao, J. Yu, Use of surfactants for the remediation of contaminated soils: a review. J. Hazard. Mater. **285**, 419–435 (2015)
7. R.A. Wuana, F.E. Okieimen, Heavy metals in contaminated soils: a review of sources, chemistry, risks and best available strategies for remediation. Int. Sch. Res. Notices. Article ID 402647, 20 pages (2011). https://doi.org/10.5402/2011/402647
8. M. Arias-Estévez, E. López-Periago, E. Martínez-Carballo, J. Simal-Gándara, J.-C. Mejuto, L. García-Río, The mobility and degradation of pesticides in soils and the pollution of groundwater resources. Agric. Ecosyst. Environ. **123**, 247–260 (2008)
9. R.D. Villa, A.G. Trovó, R.F.P. Nogueira, Environmental implications of soil remediation using the Fenton process. Chemosphere **71**, 43–50 (2008)
10. X. He, A.A. de la Cruz, K.E. O'Shea, D.D. Dionysiou, Kinetics and mechanisms of cylindrospermopsin destruction by sulfate radical-based advanced oxidation processes. Water Res. **63**, 168–178 (2014)
11. P. Roudier, Techniques de réhabilitation des sites et sols pollués - Fiches de synthèse, Ref TIP254WEB - Trav. Publics Infrastruct (2005)
12. M.A. Rodrigo, N. Oturan, M.A. Oturan, Electrochemically assisted remediation of pesticides in soils and water: a review. Chem. Rev. **114**, 8720–8745 (2014)
13. E.V. Lau, S. Gan, H.K. Ng, P.E. Poh, Extraction agents for the removal of polycyclic aromatic hydrocarbons (PAHs) from soil in soil washing technologies. Environ. Pollut. **184**, 640–649 (2014)
14. E. Mousset, M.A. Oturan, E.D.V. Hullebusch, G. Guibaud, G. Esposito, Soil washing/flushing treatments of organic pollutants enhanced by cyclodextrins and integrated treatments: state of the art. Crit. Rev. Environ. Sci. Technol. **44**, 705–795 (2014)
15. C. Trellu, N. Oturan, Y. Pechaud, E.D. van Hullebusch, G. Esposito, M.A. Oturan, Anodic oxidation of surfactants and organic compounds entrapped in micelles – selective degradation mechanisms and soil washing solution reuse. Water Res. **118**, 1–11 (2017)

16. C. Trellu, O. Ganzenko, S. Papirio, Y. Pechaud, N. Oturan, D. Huguenot, E.D. van Hullebusch, G. Esposito, M.A. Oturan, Combination of anodic oxidation and biological treatment for the removal of phenanthrene and Tween 80 from soil washing solution. Chem. Eng. J. **306**, 588–596 (2016)
17. D. Huang, C. Hu, G. Zeng, M. Cheng, P. Xu, X. Gong, R. Wang, W. Xue, Combination of Fenton processes and biotreatment for wastewater treatment and soil remediation. Sci. Total Environ. **574**, 1599–1610 (2017)
18. S.O. Ganiyu, N. Oturan, S. Raffy, M. Cretin, R. Esmilaire, E. van Hullebusch, G. Esposito, M.A. Oturan, Sub-stoichiometric titanium oxide (Ti4O7) as a suitable ceramic anode for electrooxidation of organic pollutants: a case study of kinetics, mineralization and toxicity assessment of amoxicillin. Water Res. **106**, 171–182 (2016)
19. C.N. Brito, M.B. Ferreira, Suzana M. L. de O. Marcionilio, E.C.M. de Moura Santos, J.J.L. Léon, S.O. Ganiyu, C.A. Martínez-Huitle, Electrochemical oxidation of acid Violet 7 dye by using Si/BDD and Nb/BDD electrodes, J. Electrochem. Soc. **165**, E250–E255 (2018)
20. M. Panizza, G. Cerisola, Direct and mediated anodic oxidation of organic pollutants. Chem. Rev. **109**, 6541–6569 (2009)
21. C.A. Martínez-Huitle, M.A. Rodrigo, I. Sirés, O. Scialdone, Single and coupled electrochemical processes and reactors for the abatement of organic water pollutants: a critical review. Chem. Rev. **115**, 13362–13407 (2015)
22. C.A. Martinez-Huitle, S. Ferro, Electrochemical oxidation of organic pollutants for the wastewater treatment: direct and indirect processes. Chem. Soc. Rev. **35**, 1324 (2006)
23. S.O. Ganiyu, T.X. Le Huong, M. Bechelany, N. Oturan, S. Papirio, G. Esposito, E. van Hullebusch, M. Cretin, M.A. Oturan, Electrochemical mineralization of sulfamethoxazole over wide pH range using FeIIFeIII LDH modified carbon felt cathode: degradation pathway, toxicity and reusability of the modified cathode. Chem. Eng. J. **350**, 844–855 (2018)
24. S.O. Ganiyu, N. Oturan, S. Raffy, M. Cretin, C. Causserand, M.A. Oturan, Efficiency of plasma elaborated sub-stoichiometric titanium oxide (Ti4O7) ceramic electrode for advanced electrochemical degradation of paracetamol in different electrolyte media. Sep. Purif. Technol. **208**, 142–152 (2018)
25. I. Sirés, E. Brillas, M.A. Oturan, M.A. Rodrigo, M. Panizza, Electrochemical advanced oxidation processes: today and tomorrow. A review. Environ. Sci. Pollut. Res. **21**, 8336–8367 (2014)
26. C.A. Martínez-Huitle, M. Panizza, Electrochemical oxidation of organic pollutants for wastewater treatment. Curr. Opin. Electrochem. **11**, 62 (2018)
27. F.E. Durán, D.M. de Araújo, C. do Nascimento Brito, E.V. Santos, S.O. Ganiyu, C.A. Martínez-Huitle, Electrochemical technology for the treatment of real washing machine effluent at pre-pilot plant scale by using active and non-active anodes. J. Electroanal. Chem. **818**, 216–222 (2018)
28. S.O. Ganiyu, N. Oturan, S. Raffy, G. Esposito, E.D. van Hullebusch, M. Cretin, M.A. Oturan, Use of sub-stoichiometric titanium oxide as a ceramic electrode in anodic oxidation and electro-Fenton degradation of the Beta-blocker propranolol: degradation kinetics and mineralization pathway. Electrochim. Acta **242**, 344–354 (2017)
29. I. Sirés, E. Brillas, Remediation of water pollution caused by pharmaceutical residues based on electrochemical separation and degradation technologies: a review. Environ. Int. **40**, 212–229 (2012)
30. H. Särkkä, A. Bhatnagar, M. Sillanpää, Recent developments of electro-oxidation in water treatment — a review. J. Electroanal. Chem. **754**, 46–56 (2015)
31. M.A. Rodrigo, P. Cañizares, C. Buitrón, C. Sáez, Electrochemical technologies for the regeneration of urban wastewaters. Electrochim. Acta **55**, 8160–8164 (2010)
32. F.C. Moreira, R.A.R. Boaventura, E. Brillas, V.J.P. Vilar, Electrochemical advanced oxidation processes: a review on their application to synthetic and real wastewaters. Appl. Catal. B Environ. **202**, 217–261 (2017)
33. M.A. Oturan, Electrochemical advanced oxidation technologies for removal of organic pollutants from water. Environ. Sci. Pollut. Res. **21**, 8333–8335 (2014)

34. M.A. Oturan, J.-J. Aaron, Advanced oxidation processes in water/wastewater treatment: principles and applications. A review. Crit. Rev. Environ. Sci. Technol. **44**, 2577–2641 (2014)
35. E. Vieira dos Santos, F. Souza, C. Saez, P. Cañizares, M.R.V. Lanza, C.A. Martinez-Huitle, M.A. Rodrigo, Application of electrokinetic soil flushing to four herbicides: a comparison. Chemosphere **153**, 205–211 (2016)
36. C.N. Mulligan, R.N. Yong, B.F. Gibbs, Surfactant-enhanced remediation of contaminated soil: a review. Eng. Geol. **60**, 371–380 (2001)
37. O. Atteia, E. Del Campo Estrada, H. Bertin, Soil flushing: a review of the origin of efficiency variability. Rev. Environ. Sci. Biotechnol. **12**, 379–389 (2013)
38. K. Yang, L. Zhu, B. Xing, Enhanced soil washing of phenanthrene by mixed solutions of TX100 and SDBS. Environ. Sci. Technol. **40**, 4274–4280 (2006)
39. S.-J. Park, J.-M. Back, Prediction of partition coefficients for some organic compounds using UNIFAC. J. Ind. Eng. Chem. **6**, 100–106 (2000)
40. A.P. Khodadoust, R. Bagchi, M.T. Suidan, R.C. Brenner, N.G. Sellers, Removal of PAHs from highly contaminated soils found at prior manufactured gas operations. J. Hazard. Mater. **80**, 159–174 (2000)
41. P.-H. Lee, S.K. Ong, J. Golchin, G.L. Nelson, Use of solvents to enhance PAH biodegradation of coal tar-contaminated soils. Water Res **35**, 3941–3949 (2001)
42. A. Vishnyakov, M.-T. Lee, A.V. Neimark, Prediction of the critical micelle concentration of nonionic surfactants by dissipative particle dynamics simulations. J. Phys. Chem. Lett. **4**, 797–802 (2013)
43. M.-T. Lee, A. Vishnyakov, A.V. Neimark, Calculations of critical micelle concentration by dissipative particle dynamics simulations: the role of chain rigidity. J. Phys. Chem. B **117**, 10304–10310 (2013)
44. B.-K. Kim, K. Baek, S.-H. Ko, J.-W. Yang, Research and field experiences on electrokinetic remediation in South Korea. Sep. Purif. Technol. **79**, 116–123 (2011)
45. T.T. Nguyen, N.H. Youssef, M.J. McInerney, D.A. Sabatini, Rhamnolipid biosurfactant mixtures for environmental remediation. Water Res. **42**, 1735–1743 (2008)
46. C.-C. Lai, Y.-C. Huang, Y.-H. Wei, J.-S. Chang, Biosurfactant-enhanced removal of total petroleum hydrocarbons from contaminated soil. J. Hazard. Mater. **167**, 609–614 (2009)
47. B. Zhao, L. Zhu, Y. Gao, A novel solubilization of phenanthrene using Winsor I microemulsion-based sodium castor oil sulfate. J. Hazard. Mater. **119**, 205–211 (2005)
48. M. Bragato, O.A. El Seoud, Formation, properties, and "ex situ" soil decontamination by vegetable oil-based microemulsions. J. Surfactant Deterg. **6**, 143–150 (2003)
49. G. Song, C. Lu, J.-M. Lin, Application of surfactants and microemulsions to the extraction of pyrene and phenanthrene from soil with three different extraction methods. Anal. Chim. Acta **596**, 312–318 (2007)
50. T. Badr, K. Hanna, C. de Brauer, Enhanced solubilization and removal of naphthalene and phenanthrene by cyclodextrins from two contaminated soils. J. Hazard. Mater. **112**, 215–223 (2004)
51. C. Viglianti, K. Hanna, C. de Brauer, P. Germain, Removal of polycyclic aromatic hydrocarbons from aged-contaminated soil using cyclodextrins: experimental study. Environ. Pollut. **140**, 427–435 (2006)
52. A. Petitgirard, M. Djehiche, J. Persello, P. Fievet, N. Fatin-Rouge, PAH contaminated soil remediation by reusing an aqueous solution of cyclodextrins. Chemosphere **75**, 714–718 (2009)
53. J. Szejtli, Downstream processing using cyclodextrins. Trends Biotechnol. **7**, 170–174 (1989)
54. J. Gómez, M.T. Alcántara, M. Pazos, M.Á. Sanromán, Soil washing using cyclodextrins and their recovery by application of electrochemical technology. Chem. Eng. J. **159**, 53–57 (2010)
55. Z. Li, J. Chen, J. Yang, Y. Su, X. Fan, Y. Wu, C. Yu, Z.L. Wang, β-cyclodextrin enhanced triboelectrification for self-powered phenol detection and electrochemical degradation. Energy Environ. Sci. **8**, 887–896 (2015)

56. C. Yang, Q. Zeng, Y. Wang, B. Liao, J. Sun, H. Shi, X. Chen, Simultaneous elution of polycyclic aromatic hydrocarbons and heavy metals from contaminated soil by two amino acids derived from β-cyclodextrins. J. Environ. Sci. **22**, 1910–1915 (2010)
57. M.A. Sánchez-Trujillo, E. Morillo, J. Villaverde, S. Lacorte, Comparative effects of several cyclodextrins on the extraction of PAHs from an aged contaminated soil. Environ. Pollut. **178**, 52–58 (2013)
58. P. Conte, A. Zena, G. Pilidis, A. Piccolo, Increased retention of polycyclic aromatic hydrocarbons in soils induced by soil treatment with humic substances. Environ. Pollut. **112**, 27–31 (2001)
59. P. Conte, A. Agretto, R. Spaccini, A. Piccolo, Soil remediation: humic acids as natural surfactants in the washings of highly contaminated soils. Environ. Pollut. **135**, 515–522 (2005)
60. W. Li, X. Zhu, Y. He, B. Xing, J. Xu, P.C. Brookes, Enhancement of water solubility and mobility of phenanthrene by natural soil nanoparticles. Environ. Pollut. **176**, 228–233 (2013)
61. W. Ling, L. Ren, Y. Gao, X. Zhu, B. Sun, Impact of low-molecular-weight organic acids on the availability of phenanthrene and pyrene in soil. Soil Biol. Biochem. **41**, 2187–2195 (2009)
62. C.C. Parrish, Dissolved and particulate marine lipid classes: a review. Mar. Chem. **23**, 17–40 (1988)
63. Z. Gong, B.-M. Wilke, K. Alef, P. Li, Influence of soil moisture on sunflower oil extraction of polycyclic aromatic hydrocarbons from a manufactured gas plant soil. Sci. Total Environ. **343**, 51–59 (2005)
64. Z. Gong, B.-M. Wilke, K. Alef, P. Li, Q. Zhou, Removal of polycyclic aromatic hydrocarbons from manufactured gas plant-contaminated soils using sunflower oil: laboratory column experiments. Chemosphere **62**, 780–787 (2006)
65. Z. Gong, K. Alef, B.-M. Wilke, P. Li, Activated carbon adsorption of PAHs from vegetable oil used in soil remediation. J. Hazard. Mater. **143**, 372–378 (2007)
66. E.V. Lau, S. Gan, H.K. Ng, Extraction of phenanthrene and fluoranthene from contaminated sand using palm kernel and soybean oils. J. Environ. Manag. **107**, 124–130 (2012)
67. A.T. Yeung, Contaminant extractability by electrokinetics. Environ. Eng. Sci. **23**, 202–224 (2005)
68. A.T. Yeung, Y.-Y. Gu, A review on techniques to enhance electrochemical remediation of contaminated soils. J. Hazard. Mater. **195**, 11–29 (2011)
69. M.M. Page, C.L. Page, Electroremediation of contaminated soils. J. Environ. Eng. **128**, 208–219 (2002)
70. H.I. Gomes, C. Dias-Ferreira, A.B. Ribeiro, Electrokinetic remediation of organochlorines in soil: enhancement techniques and integration with other remediation technologies. Chemosphere **87**, 1077–1090 (2012). https://doi.org/10.1016/j.chemosphere.2012.02.037
71. K.R. Reddy, R.E. Saichek, Effect of soil type on electrokinetic removal of phenanthrene using surfactants and cosolvents. J. Environ. Eng. **129**, 336–346 (2003). https://doi.org/10.1061/(ASCE)0733-9372(2003)129:4(336)
72. K.R. Reddy, C. Supraja, Enhanced electrokinetic remediation of heavy metals in glacial till soils using different electrolyte solutions. J. Environ. Eng. **130**, 442–455 (2004). https://doi.org/10.1061/(ASCE)0733-9372(2004)130:4(442)
73. K.R. Reddy, M. Kranti, C. Claudio, Sequential electrokinetic remediation of mixed contaminants in low permeability soils. J. Environ. Eng. **135**, 989–998 (2009)
74. R. López-Vizcaíno, C. Sáez, E. Mena, J. Villaseñor, P. Cañizares, M.A. Rodrigo, Electroosmotic fluxes in multi-well electro-remediation processes. J. Environ. Sci. Health Part A. **46**, 1549–1557 (2011)
75. R. Lopez-Vizcaíno, C. Sáez, P. Cañizares, M.A. Rodrigo, Electrocoagulation of the effluents from surfactant-aided soil-remediation processes. Sep. Purif. Technol. **98**, 88–93 (2012)
76. R. López-Vizcaíno, C. Sáez, P. Cañizares, M.A. Rodrigo, The use of a combined process of surfactant-aided soil washing and coagulation for PAH-contaminated soils treatment. Sep. Purif. Technol. **88**, 46–51 (2012)

77. S. Yuan, M. Tian, X. Lu, Electrokinetic movement of hexachlorobenzene in clayed soils enhanced by Tween 80 and β-cyclodextrin. J. Hazard. Mater. **137**, 1218–1225 (2006)
78. S. Yuan, J. Wan, X. Lu, Electrokinetic movement of multiple chlorobenzenes in contaminated soils in the presence of β-cyclodextrin. J. Environ. Sci. **19**, 968–976 (2007)
79. S.A. Jackman, G. Maini, A.K. Sharman, G. Sunderland, C.J. Knowles, Electrokinetic movement and biodegradation of 2,4-dichlorophenoxyacetic acid in silt soil. Biotechnol. Bioeng. **74**, 40–48 (2001)
80. J.W. Ma, F.Y. Wang, Z.H. Huang, H. Wang, Simultaneous removal of 2,4-dichlorophenol and Cd from soils by electrokinetic remediation combined with activated bamboo charcoal. J. Hazard. Mater. **176**, 715–720 (2010)
81. A.B. Ribeiro, E.P. Mateus, J.-M. Rodríguez-Maroto, Removal of organic contaminants from soils by an electrokinetic process: the case of molinate and bentazone. Experimental and modeling. Sep. Purif. Technol. **79**, 193–203 (2011)
82. A. Oonnittan, R.A. Shrestha, M. Sillanpää, Effect of cyclodextrin on the remediation of hexachlorobenzene in soil by electrokinetic Fenton process. Sep. Purif. Technol. **64**, 314–320 (2009)
83. A. Oonnittan, P. Isosaari, M. Sillanpää, Oxidant availability in soil and its effect on HCB removal during electrokinetic Fenton process. Sep. Purif. Technol. **76**, 146–150 (2010)
84. S. Cotillas, C. Sáez, P. Cañizares, I. Cretescu, M.A. Rodrigo, Removal of 2,4-D herbicide in soils using a combined process based on washing and adsorption electrochemically assisted. Sep. Purif. Technol. **194**, 19–25 (2018)
85. Z. Zhou, Y. Zhang, H. Wang, T. Chen, W. Lu, The comparative photodegradation activities of pentachlorophenol (PCP) and polychlorinated biphenyls (PCBs) using UV alone and TiO2-derived photocatalysts in methanol soil washing solution. PLoS One **9**, e108765 (2014)
86. X. Zhu, D. Zhou, Y. Wang, L. Cang, G. Fang, J. Fan, Remediation of polychlorinated biphenyl-contaminated soil by soil washing and subsequent TiO2 photocatalytic degradation. J. Soils Sediments **12**, 1371–1379 (2012)
87. M.E. Lindsey, G. Xu, J. Lu, M.A. Tarr, Enhanced Fenton degradation of hydrophobic organics by simultaneous iron and pollutant complexation with cyclodextrins. Sci. Total Environ. **307**, 215–229 (2003)
88. M.A. Manzano, J.A. Perales, D. Sales, J.M. Quiroga, Catalyzed hydrogen peroxide treatment of polychlorinated biphenyl contaminated sandy soils. Water Air Soil Pollut. **154**, 57–69 (2004)
89. E.V. dos Santos, C. Sáez, C.A. Martínez-Huitle, P. Cañizares, M.A. Rodrigo, Removal of oxyfluorfen from ex-situ soil washing fluids using electrolysis with diamond anodes. J. Environ. Manag. **171**, 260–266 (2016)
90. E.V. dos Santos, C. Sáez, P. Cañizares, M.A. Rodrigo, C.A. Martínez-Huitle, Coupling photo and Sono technologies with BDD anodic oxidation for treating soil-washing effluent polluted with atrazine. J. Electrochem. Soc. **165**, E262–E267 (2018)
91. E.V. dos Santos, C. Sáez, C.A. Martínez-Huitle, P. Cañizares, M.A. Rodrigo, The role of particle size on the conductive diamond electrochemical oxidation of soil-washing effluent polluted with atrazine. Electrochem. Commun. **55**, 26–29 (2015)
92. M.J. Martín de Vidales, M.P. Castro, C. Sáez, P. Cañizares, M.A. Rodrigo, Radiation-assisted electrochemical processes in semi-pilot scale for the removal of clopyralid from soil washing wastes. Sep. Purif. Technol. **208**, 100–109 (2019)
93. A. Karaçali, M. Muñoz-Morales, S. Kalkan, B.K. Körbahti, C. Saez, P. Cañizares, M.A. Rodrigo, A comparison of the electrolysis of soil washing wastes with active and non-active electrodes. Chemosphere **225**, 9–26 (2019)
94. C.C. de Almeida, M. Muñoz-Morales, C. Saez, P. Cañizares, C.A. Martínez-Huitle, M.A. Rodrigo, Electrolysis with diamond anodes of the effluents of a combined soil washing – ZVI dechlorination process. J. Hazard. Mater. **369**, 577–583 (2019)

95. C. Carvalho de Almeida, M. Muñoz-Morales, C. Sáez, P. Cañizares, C.A. Martínez-Huitle, M.A. Rodrigo, Integrating ZVI-dehalogenation into an electrolytic soil-washing cell. Sep. Purif. Technol. **211**, 28–34 (2019)
96. M. Muñoz-Morales, M. Braojos, C. Sáez, P. Cañizares, M.A. Rodrigo, Remediation of soils polluted with lindane using surfactant-aided soil washing and electrochemical oxidation. J. Hazard. Mater. **339**, 232–238 (2017)
97. E.V. dos Santos, C. Sáez, P. Cañizares, C.A. Martínez-Huitle, M.A. Rodrigo, UV assisted electrochemical technologies for the removal of oxyfluorfen from soil washing wastes. Chem. Eng. J. **318**, 2–9 (2017)
98. E. Vieira dos Santos, C. Sáez, P. Cañizares, C.A. Martínez-Huitle, M.A. Rodrigo, Treating soil-washing fluids polluted with oxyfluorfen by sono-electrolysis with diamond anodes. Ultrason. Sonochem. **34**, 115–122 (2017)
99. E.V. dos Santos, C. Sáez, C.A. Martínez-Huitle, P. Cañizares, M.A. Rodrigo, Combined soil washing and CDEO for the removal of atrazine from soils. J. Hazard. Mater. **300**, 129–134 (2015)
100. P.T. Almazán-Sánchez, S. Cotillas, C. Sáez, M.J. Solache-Ríos, V. Martínez-Miranda, P. Cañizares, I. Linares-Hernández, M.A. Rodrigo, Removal of pendimethalin from soil washing effluents using electrolytic and electro-irradiated technologies based on diamond anodes. Appl. Catal. B Environ. **213**, 190–197 (2017)
101. E.V. dos Santos, C. Sáez, P. Cañizares, D.R. da Silva, C.A. Martínez-Huitle, M.A. Rodrigo, Treatment of ex-situ soil-washing fluids polluted with petroleum by anodic oxidation, photolysis, sonolysis and combined approaches. Chem. Eng. J. **310**, 581–588 (2017)
102. E. Mousset, N. Oturan, E.D. van Hullebusch, G. Guibaud, G. Esposito, M.A. Oturan, Treatment of synthetic soil washing solutions containing phenanthrene and cyclodextrin by electro-oxidation. Influence of anode materials on toxicity removal and biodegradability enhancement. Appl. Catal. B Environ. **160–161**, 666–675 (2014)
103. E.B.S. da Silva, Electrokinetic treatment of polluted soil with petroleum coupled to an advanced oxidation process for remediation of its effluent. Int. J. Electrochem. Sci. **12**, 1247–1262 (2017)

# Persulfate in Remediation of Soil and Groundwater Contaminated by Organic Compounds

Aurora Santos, David Lorenzo, and Carmen M. Dominguez

## 1 Introduction

Contamination of soils by organic compounds caused by local sources is a widespread problem that still represents a challenge for many industrialized countries. This kind of contamination takes place, among other reasons, due to inadequate waste disposal (unsecured landfills), point-source oil and chemical releases, and spills during industrial activities (accidental or not), with many of these spills being nonaqueous phase liquids, NAPLs (BTEX, chlorinated solvents, mineral oils, etc.).

The local contamination of soils can also have a great impact on the groundwater due to the transport of the pollutants through the subsurface, increasing the risk for the receptors and causing human health problems or negative ecological impacts, as schematized in Fig. 1.

Remediation of contaminated sites must be carried out by applying the best available techniques, taking into account the specific characteristics of each case. Often, the contamination is not located in the superficial soil layer but in the saturated zone, at depths of up to several meters. In this scenario, the "in situ" technologies are preferred by the environmental policies since they avoid soil excavation and transport of the contaminants. Environmental legislation for soil and groundwater protection prioritizes in situ treatment techniques that avoid the generation, transfer, and elimination of wastes. Among these technologies, in situ chemical oxidation (ISCO) has been proved to be a good option for the cleanup of contaminated sites [1]. In situ remediation techniques avoid the need of removing the contaminated matrix (groundwater or soil) and the further on site or even off site treatment. The chemicals injected get directly in contact with the contamination source, as shown

A. Santos (✉) · D. Lorenzo · C. M. Dominguez
Chemical Engineering and Materials Department, Universidad Complutense de Madrid, Madrid, Spain
e-mail: aursan@ucm.es



M. A. Rodrigo, E. V. Dos Santos (eds.), *Electrochemically Assisted Remediation of Contaminated Soils*, Environmental Pollution 30,
https://doi.org/10.1007/978-3-030-68140-1_10

**Fig. 1** Risk associated with spills of toxic organics in the environment

in Fig. 2, and abate the contaminants. It has been proved that ISCO reduces the high cost and treatment times of traditional techniques, such as Pump and Treat.

Contaminants amenable to treatment by ISCO include the following [2].

- Benzene, toluene, ethylbenzene, and xylenes (BTEX).
- Methyl *tert*-butyl ether (MTBE).
- Total petroleum hydrocarbons (TPH).
- Chlorinated solvents (ethenes and ethanes).
- Polyaromatic hydrocarbons (PAHs).
- Polychlorinated biphenyls (PCBs).
- Chlorinated benzenes (CBs).
- Phenols.
- Organic pesticides (insecticides and herbicides).
- Munition constituents (RDX, TNT, HMX, etc.).

The oxidant injected reacts with the organic contaminants, producing harmless substances, ideally carbon dioxide and water, with the oxidant by-products also not being dangerous for the environment. Several oxidants have been tested in ISCO applications, like hydrogen peroxide, potassium and sodium permanganate, sodium persulfate, and ozone. The effectiveness of some of these oxidants can be enhanced through their activation, as occurs in catalyzed hydrogen peroxide, CHP (Fenton's reagent) and activated persulfate (PS), in order to produce radical species with high redox potential [1–7].

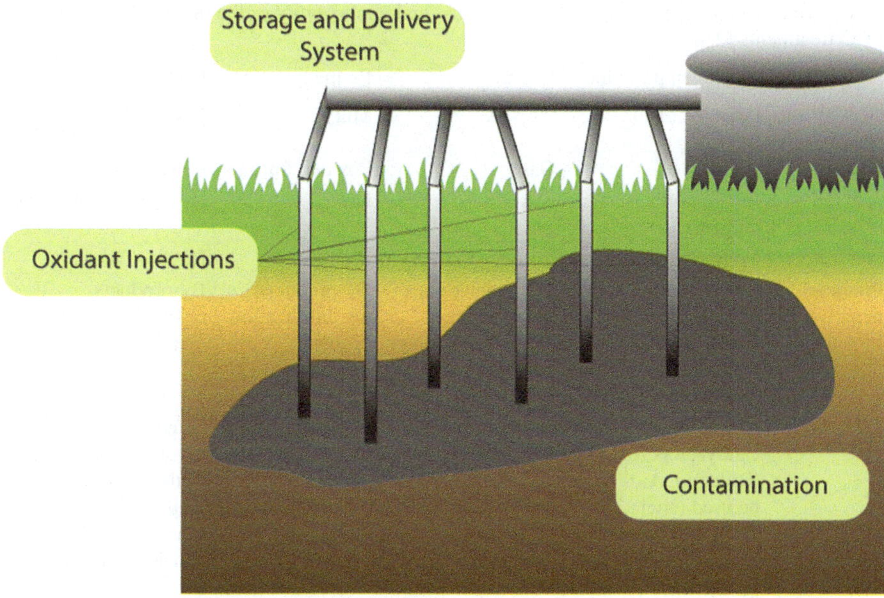


**Fig. 2** Delivery of the oxidant in the ISCO treatment

The selection of the oxidant for a specific site application will depend on the nature of the pollutants and the lithological and mineralogical characteristics of the site. The oxidant must be able to degrade the pollutant to less harmful and more biodegradable compounds (the complete mineralization, that is, the transformation of the pollutant into carbon dioxide, water, and inorganic salts, is not necessary) with high enough chemical oxidation rates. A comparison of the main oxidants and activators used for ISCO applications is summarized in Table 1, including some considerations about the applications of the oxidant system to the contaminated site. The effectivity of the oxidant against some groups of organic pollutants is shown in Table 2 [1, 8].

As it was indicated before, not only chemical aspects but transport issues should also be considered when selecting the oxidant. The oxidant–activator system injected has to be transported through the subsurface to react with the pollutants sorbed onto the soil or solubilized in the groundwater plumes. Therefore, the transport and stability of the oxidant–activator system in the subsurface is a key point in the implementation of ISCO technology. The stability of the oxidant and the activators in the subsurface can be highly affected by the chemical composition of the soil (organic and inorganic), pH, and grain size of the soil.

A high unproductive consumption of the oxidant due to the production of non-desired reactions between the oxidant and the soil components will limit the availability of the oxidant for pollutant degradation. A natural pH of the soil different than that required to keep the oxidant–activator in solution will also produce a loss of efficiency of the oxidation system. In both cases, a quick depletion of the oxidant

**Table 1** Reactive species, oxidant form, stability, and applicability of the main oxidants used for ISCO treatments [2, 4]

| | Hydrogen peroxide | Ozone | Permanganate | Persulfate |
|---|---|---|---|---|
| Reactive species | $OH^{-\bullet}$, $O_2^{-}$, $HO_2^{-}$ | $O_3$, $OH^{-\bullet}$ | $MnO_4^{-}$ | $SO_4^{-}$, $OH^{-\bullet}$, $O_2^{-}$ |
| Form | Liquid | Gas | Powder, liquid | Powder, liquid |
| Activation | Fe (Fenton's reagent) | Not | Not | Heat, $Fe^{2+}$, $H_2O_2$, base |
| Persistence | Minutes to hours | Minutes to hours | Months | Days to weeks |
| Potential negative impacts | Gas formation, heat generation, by-products, resolubilization of metals | Gas formation, by-products, resolubilization of metals | By-products, resolubilization of metals | By-products, resolubilization of metals |
| pH range | Acid pH for catalyzed peroxide oxidation. Aids must be used (chelating agents) to keep Fe in solution at neutral pH. Not adequate for carbonated soils | Effective over a wide pH range, carbonate in the media decreases notably the efficiency | Effective over a wide range | Effective over a wide pH range, depending of the activation method. Carbonate must be taken into consideration for thermal or Iron activated Persulfate. Aids are required for Fe-PS at neutral pH |

**Table 2** Effectiveness of alkaline activation of PS in the oxidation of hydrocarbon contaminants [8] (PS + Base from Siegrist et al. [1])

| COC | Benzene | TEX | Phenols | PAHs | Munition | PCBs | Pesticides |
|---|---|---|---|---|---|---|---|
| CHP | G | G | G | M | M | L | L |
| $KMnO_4$ | NR | G | G | G | G | L | M |
| $NaMnO_4$ | NR | G | G | G | G | L | M |
| PS + Iron | G | G | G | M | M | L | M |
| PS + heat | G | G | G | G | G | G | G |
| PS + base | G | G | G | G | G | M | M-G |
| Ozone | M | M | G | G | G | G | G |

*NR* non-reactive, *G* good, *M* medium, *L* low

or activator from the injection point will be produced and the radius of influence (ROI) of the oxidant or activator would dramatically decrease the efficiency of the ISCO technology in groundwater remediation.

The oxidant concentration at a generic radial distance ($x$) from the injection point can be calculated using the following expression [4, 9].

$$C(x) = C_0 \exp\left(-\frac{k\pi x^2 Zn}{Q}\right) \tag{1}$$

where $C_0$ is the oxidant concentration at the injection point, $k$ the oxidant decomposition rate constant, $Z$ the vertical thickness of the injection well, $n$ the soil porosity, and $Q$ the well flow rate.

Equation (1) can be reorganized to calculate the radius of influence (ROI) of the oxidant injected.

$$\mathrm{ROI} = \sqrt{\frac{Q}{k\pi Zn}\ln\left(\frac{C(x)}{C_0}\right)} \tag{2}$$

As the decomposition kinetic constant of the oxidant increases, the ROI decreases, and the oxidant injected is unproductively decomposed (without reacting with the plume of contaminants). In addition, the higher the unproductive consumption, the lower the ROI and the smaller the required distance between the injection points, as shown in Fig. 3. This fact has also been associated with an increase in installation and operation costs.

Among all ISCO techniques, activated persulfate is becoming an effective alternative to Fenton's reagent, permanganate oxidation, or ozonation. Persulfate anion is a powerful oxidant and one of the strongest oxidants used in soil remediation. The standard oxidation–reduction potential for the reaction,

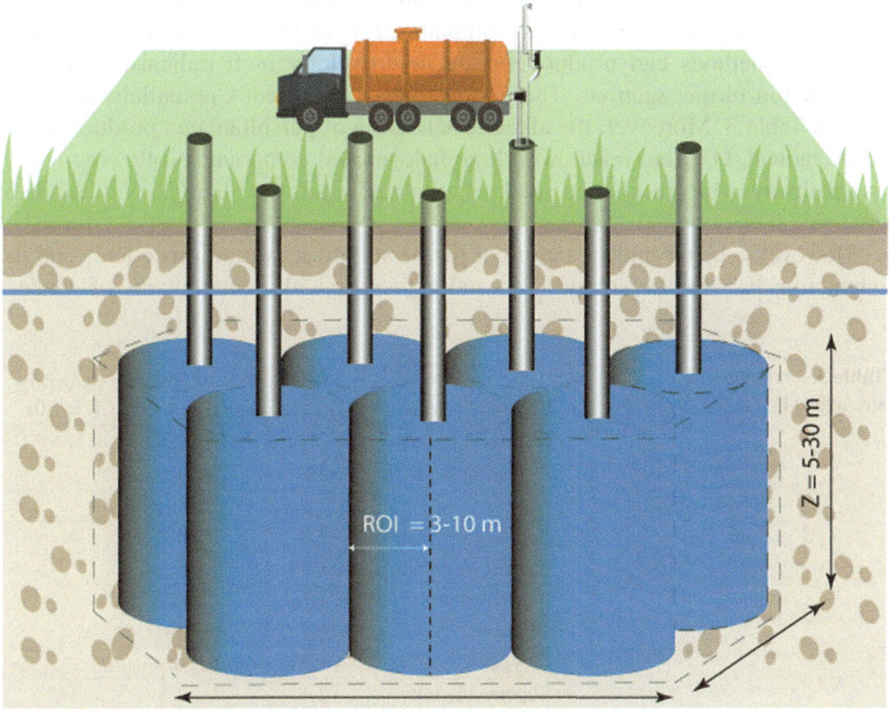


**Fig. 3** Design of the injection points taken into account the ROI of the oxidant

$$S_2O_8^{2-} + 2e^- \rightarrow 2SO_4^{2-} \quad (3)$$

is estimated to be 2.01 V [10]. Persulfate is usually supplied as a solid sodium salt ($Na_2S_2O_8$) for ISCO applications (solubility 420 g $L^{-1}$ at 25 °C). Ammonium persulfate and potassium persulfate solutions are also available but less suitable for ISCO treatments. Ammonium persulfate could produce ammonium/nitrate contamination in the subsurface, and potassium persulfate is less soluble (52 g $L^{-1}$ at 25 °C) than sodium persulfate and more expensive [1, 11].

Persulfate presents a high stability in water and/or soil, is relatively inexpensive, and produces benign end-products. It can be applied to a wide range of pH in soil and shows less affinity for natural soil organics than the permanganate ion. Consequently, persulfate has a lower natural oxidant demand (NOD) than permanganate, with a lower unproductive consumption of the oxidant in soils with not negligible soil organic matter (SOM) content. Moreover, it is more stable than hydrogen peroxide and can be transported at greater distances in the subsurface, achieving a higher radius of influence that CHP. This fact is particularly relevant when the soil has significant carbonate content. All these advantages have resulted in an increase in the use of persulfate in ISCO applications at field scale in the last 10 years, as well as in a remarkable increase in the research about this topic [11–14].

As previously cited, persulfate can be activated to generate free radicals, with a remarkable increase of the oxidation rates of pollutants. Activation of PS can be accomplished by heat, transition metals, hydrogen peroxide and bases (this last also known as "alkaline activation"), as summarized in the literature above cited. These activation methods can produce sulfate, hydroxyl, or both radicals, as will be explained in further sections. These radicals have high redox potentials, as can be seen in Table 3. Moreover, the alkaline activation of persulfate can produce superoxide radical, $O_2^{-\bullet}$ that reacts with high halogenated compounds following reductive pathways. As a result, activated persulfate allows for oxidizing a wide range of contaminants.

However, a better understanding of the persulfate activation procedures and the interaction between persulfate (PS) and activators with contaminants of concern

**Table 3** Oxidant strengths [1]

| Chemical species | Oxidation potential (V) | Relative oxidizing power ($Cl_2$ = 1.0) |
|---|---|---|
| Hydroxyl radical | 2.8 | 2.0 |
| Sulfate radical | 2.5 | 1.8 |
| Ozone | 2.1 | 1.5 |
| Sodium persulfate | 2.0 | 1.5 |
| Hydrogen peroxide | 1.8 | 1.3 |
| Permanganate (Na/K) | 1.7 | 1.2 |
| Chlorine | 1.4 | 1.0 |
| Oxygen | 1.2 | 0.9 |
| Superoxide ion | 2.4 | 1.8 |

(COC) and aquifer materials is required. Depletion of PS and/or aids used as activating agents can greatly limit the persistence of the activated persulfate system. The efficiency of ISCO treatments using persulfate in the degradation of organic contaminants depends on the competition kinetics between contaminants, activation aids, and reactive species present in the soil and groundwater. The oxidant demand for persulfate in soil is variable (from low values as 0.1–0.3 g $kg^{-1}$ [15] to high values about 3 g $kg^{-1}$ [8]).

In this chapter, the main activation strategies for PS and their application for soil and groundwater remediation are summarized, analyzing the main limitations and advantages of each of them.

## 2 Thermal Activation of Persulfate (TAP)

Persulfate can be activated by heat to generate sulfate radicals ($SO_4^{-}\cdot$) with a stronger oxidation potential than persulfate anion, as shown in Table 3.

$$S_2O_8^{2-} \xrightarrow{\text{heat or } h\nu} 2SO_4^{-} \tag{4}$$

The thermal activation of PS (TAP) has been proved to be a highly effective method for the degradation of organic contaminants in aqueous phase. Huang at al. [16] studied the effectiveness of thermally activated persulfate for 59 volatile organic compounds (VOCs) listed in the EPA SW-846 Method 8260B, frequently detected in contaminated soils and groundwater. The study was carried out in aqueous phase at moderate temperature (from 20 to 40 °C). They found that aromatic and unsaturated VOCs were generally degraded at higher rates than saturated hydrocarbon and halogenated alkanes. This can be explained considering that $SO_4^{-}\cdot$ has a greater ability for removing electrons than to abstract hydrogen atoms from an organic molecule [17, 18].

Ma et al. [19] studied the BTEX degradation with TAP at 50 °C. They found that the most recalcitrant compound was benzene ($X_{benzene}$ = 70% at 6 h) while at the same reaction time ($t$ = 6 h), a conversion higher than 98% was achieved for the others BTEX tested.

Heat-activated PS has been also applied successfully for the treatment in aqueous phase of other less volatile pollutants that can be usually found in the environment. Tan et al. [20] achieved the total abatement of diuron with TAP completely at temperatures from 40 to 70 °C. At 70 °C, diuron conversion was higher than 95% in 185 min using an initial molar ratio of PS and diuron = 10). Santos et al. [21] found that TAP degraded perfluorooctanoic acid (PFOA) at temperatures from 50 to 70 °C in less than 6 h.

Qian et al. [22] found that cefalexin was degraded with TAP at temperatures from 50 to 65 °C with pollutant conversions around 90% at 2 h ($T^a$ = 65 °C). Norzaee et al. [23] have studied the penicillin G degradation at temperatures between 40 and

80 °C, at several pH, the optimal pH being around 5. They found that penicillin conversion was above 99%, while COD (chemical oxygen demand) removal was around 90% at 80 °C and 75 min. Wang et al. [24] studied and reviewed the degradation of emerging pollutants in aqueous phase with TAP at temperatures between 50 and 80 °C varying the initial pH between 4 and 8. At 80 °C, the conversion of most of the pollutants studied was usually above 90% in less than 3 h. Feng et al. [25] studied the ketoprofen degradation by TAP in the temperature range 40–70 °C. A total conversion of the pollutant was obtained at 70 °C in only 30 min.

Ma et al. [26] studied the phenol abatement at several initial pH values (from 1.3 to 14) and temperatures (30–70 °C). Higher activation energies were obtained at pH < 7. At stronger alkaline conditions, lower reaction times were required working at the same activation temperature. This fact was explained because of the radicals involved in pollutants oxidations were hydroxyl radicals. Moreover, as noticed by Ahmad et al. [27], phenolate anion can act also as a PS activator. Manz et al. [28] studied the oxidation of furfural by TAP in the range 30–60 °C achieving a pollutant conversion higher than 95% at 5 h and 60 °C.

Wang et al. [29] studied the degradation of 18 PAHs by TAP in aqueous phase. Due to the low solubility of these compounds in water, several surfactants were added (nonionic and anionic type). The selected temperature range was between 20 and 60 °C. It was found that the surfactants competed with the pollutants for the oxidant, resulting in a higher consumption of PS. Sodium dodecyl diphenyl ether disulfonate (C12-MADS) and sodium dodecyl benzene sulfonate (SDBS) were the surfactant-producing the lowest unproductive consumption of PS. A conversion around 60% of PAH was obtained at 3 days at 60 °C.

In the works cited above, an excess of the oxidant species has been usually employed and first-order kinetic models have been achieved in aqueous phase; the main radical generated by this activation method at pH < 7 *is* the persulfate radical [24].

The effect of water matrix has been also studied by analyzing the effect of the most abundant ions in groundwater (mainly chloride ions and carbonate species) on the pollutant degradation rates. However, ambiguous results have been obtained. Tan et al. [20] found that diuron degradation rate was remarkably decreased by the presence of bicarbonate ($HCO_3^-$) from 0 to 37.5 mM, using a molar ratio of anion/PS between 0 and 100, in the pH range from 5 to 8.6. The effect of chloride anion at the same pH interval and anion/PS molar ratio was almost negligible. The carbonate addition yielded a lower effect on the pollutant kinetic constant than bicarbonate addition, but it has to be pointed out that the pH after $CO_3^{2-}$ addition was around 10–11. On the other hand, Quian et al. [22] found that at pH = 7 and 60 °C, cephalexin degradation was affected by natural organic matter (NOM), as well as the presence of bicarbonate and chloride anions in groundwater. They found a positive effect of chloride anion (between 0 and 300 mM) on the kinetic constant, while the influence of bicarbonate, between 0 and 300 mM, was contradictory. The presence of NOM between 0 and 20 mg $L^{-1}$ had a slight negative effect on the kinetic constant. Other common anions as sulfate, nitrate, and phosphate had a negligible effect on pollutant degradation. On the contrary, Norzaee [23] found that both chloride and carbonate (10 mg $L^{-1}$) had a remarkable negative impact on Penicillin G degradation at pH = 5.

Ma et al. [26] found that chlorides inhibited benzene degradation while slightly promoted the degradation of *m*-xylene and *p*-xylene. High concentrations of $HPO_4^{2-}$ slightly inhibited the oxidation of BTEX by PS, but concentrations lower than 100 mM had a negligible influence on the kinetic constant. Bicarbonate concentration higher than 100 mM inhibited all the BETX degradation. NOM over 100 mg $L^{-1}$ had a remarkable negative effect on benzene degradation while other BTEX were less affected. The effect of sulfate and nitrate anions was negligible. In this study, the pH was in the neutral–acidic range. In general, it is assumed that chloride, bicarbonate, carbonate, and NOM have a scavenging effect of sulfate radicals by means of the following reactions.

$$SO_4^{\cdot-} + Cl^- \rightarrow Cl^{\cdot} + SO_4^{2-} \tag{5}$$

$$Cl^{\cdot} + Cl^- \rightarrow Cl_2^{\cdot-} \tag{6}$$

$$Cl_2^{\cdot-} + Cl_2^{\cdot-} \rightarrow 2Cl^- + Cl_2 \tag{7}$$

$$SO_4^{\cdot-} + HCO_3^- \rightarrow HCO_3^{\cdot} + SO_4^{2-} \tag{8}$$

$$SO_4^{\cdot-} + CO_3^{2-} \rightarrow CO_3^{\cdot-} + SO_4^{2-} \tag{9}$$

$$HCO_3^{\cdot} \rightarrow CO_3^{\cdot-} \rightarrow +H^+ \tag{10}$$

$$NOM + SO_4^{\cdot-} \rightarrow Products \tag{11}$$

A summary of the kinetic constant of sulfate radicals with water matrices components was summarized by Waclawek et al. [12], as shown in Table 4. The reaction rate constants of these radical scavengers with hydroxyl radicals, taken from literature, have been also included. Most of the values collected in Table 4 have been obtained at room temperature. However, the activation energies of these reactions have been scarcely studied [30].

**Table 4** Second-order reaction rate constants of hydroxyl and sulfate radicals with common water matrix components [12]

| Radical | Compound | Rate constants ($M^{-1}\cdot s^{-1}$) |
|---|---|---|
| $^{\cdot}OH$ | $Cl^-$ | $3.0–4.3 \times 10^9$ |
| $SO_4^{\cdot-}$ | $Cl^-$ | $1.3–6.6 \times 10^8$ |
| $^{\cdot}OH$ | $Br^-$ | $1.9 \times 10^9$ |
| $SO_4^{\cdot-}$ | $Br^-$ | $3.5 \times 10^9$ |
| $^{\cdot}OH$ | $HCO_3^-$ | $n \times 10^7$ |
| $SO_4^{\cdot-}$ | $HCO_3^-$ | $2.6–9.1 \times 10^6$ |
| $^{\cdot}OH$ | $CO_3^{2-}$ | $4 \times 10^8$ |
| $SO_4^{\cdot-}$ | $CO_3^{2-}$ | $4.1 \times 10^6$ |
| $^{\cdot}OH$ | Humic acid | $1.4 \times 10^4$ (mg of $C\cdot L^{-1})^{-1}$ $s^{-1}$ |
| $SO_4^{\cdot-}$ | Humic acid | $6.8 \times 10^3$ (mg of $C\cdot L^{-1})^{-1}$ $s^{-1}$ |

The effect of pH on the degradation of different pollutants by using thermally activated persulfate has been studied, among others, by Tan et al. [20] in the pH range 5–8, and Norzaee et al. [23] in the pH range from 3 to 11. In both works, it has been found that the optimal pH for pollutant abatement was around 5.

The lower reaction rate obtained at more acidic conditions can be explained considering that the following reactions take place Johnson et al. [32].

$$S_2O_8^{2-} + H^+ \rightarrow HSO_4^- + SO_4\,(\text{slow}) \tag{12}$$

$$SO_4 + H_2O \rightarrow H_2SO_5\,(\text{fast}) \tag{13}$$

Peroxymonosulfuric acid ($H_2SO_5$) is less reactive than sulfate radical and therefore the oxidation rate of the pollutant decreases. On the other hand, the bicarbonate and carbonate concentration in the media increases at pH values in the range from 8 to 11, with the associated negative scavenging effect both for sulfate and hydroxyl radicals. It should be noted that the use of pH higher than 12 can result in the alkaline activation of PS, that will be further explained in the subsequent sections, with the hydroxyl radical being the predominant species in this case.

Few works deal with the stability of PS with temperature in aqueous phase. Results from Wang [29] at 20, 40 and 60 °C and Domínguez et al. [31] at 30, 40, and 50 °C have been plotted in Fig. 4. A first-order reaction rate was obtained for the decomposition of PS in distilled water at temperatures from 20 to 60 °C (Eq. 14).

$$\frac{dX_{Ps}}{dt} = k_d\left(1 - X_{Ps}\right) \tag{14}$$

where $X_{PS}$ is the conversion of PS, $t$ is the time, and $k_d$ is the kinetic constant of PS decomposition.

Values of the kinetic constant for PS decomposition in aqueous phase, $k_d$, have been calculated from the authors' data at each temperature and the estimated values

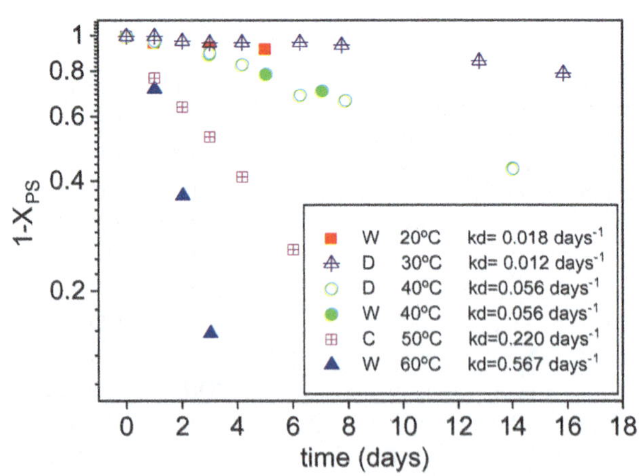


**Fig. 4** Persulfate stability in aqueous phase with temperature *W*: data obtained from Wang et al. [29] and *D*: Data obtained by Dominguez et al. [31]

are given in the legend. These values are similar to those obtained by Johnson et al. [32] ($k_{d,30\,°C} = 0.0048\ days^{-1}$, $k_{d,50\,°C} = 0.17\ days^{-1}$, $k_{d,70\,°C} = 1.92\ days^{-1}$), with the main differences obtained at the lowest temperatures tested.

To the best of our knowledge, only a few works have dealt with the application of TAP to soil–water systems. Moreover, most of them have been carried out under lab conditions and in batches. Johnson et al. [32] have compared the thermal decomposition of PS in soil and deionized water (temperature from 30 to 70 °C). The soil employed was a medium-grained Columbia River sand of basaltic origin with an organic carbon content of about 0.3% by weight. They found that PS decomposition was higher in soil slurries (1 g soil/1 g water) than in deionized water. This was attributed to the NOD of the soil (NOD = 0.019 reducing equivalents per soil gram, assuming three equivalents per permanganate ion).

Other authors [33] have studied the persulfate consumption at 30 and 50 °C in sand soil (Total Organic Carbon, TOC = 0.29%), clay soil (TOC = 0.47%) and paddy soil (TOC = 0.94%), at batch conditions, with an initial PS concentration of 30 g $L^{-1}$. Soils with and without chlorinated volatile organic compounds (CVOCs) addition were used (7800 mg NAPL $kg^{-1}$ soil, 1 kg soil/2 kg aqueous phase). It was obtained that the higher the TOC content of the soil, the higher the unproductive consumption of PS. In addition, the presence of CVOCs decreases the unproductive consumption of PS. Results obtained by these authors have been summarized in Fig. 5. They studied the removal of *cis*-1,2-dichloroethene (cis-DCE), 1,2-dichloroethane (1,2-DCA), trichloroethylene (TCE), and 1,1,1-trichloroethane (1,1,1-TCA) [33] with persulfate (30 g $L^{-1}$) at $T = 30$ and 50 °C in slurry systems, using the three soils previously cited. The most recalcitrant compound was 1,1,1-TCA, followed by 1,2-DCA. TCE and cis-DCE were more quickly oxidized. At 14 days, the maximum conversion was obtained at 50 °C in clay soil (40% for 1,1,1-TCA, 90% for 1,2 DCA), while at this reaction time the conversion of TCE and cis-DCE was almost complete.

Peng et al. [34] have studied the abatement of decabromodiphenyl ether (BDE209) spiked in soil (10–50 mg $kg^{-1}$) by TAP. At 70 °C and PS = 125 g $L^{-1}$ they found a pollutant conversion around 50% in 6 h reaction time. The optimal pH found was around 5. The same authors [35] have studied the abatement of lindane ($\gamma$-HCH) spiked in soil by TAP at 40 °C. Total conversion of the $\gamma$-HCH isomer was obtained at 15 days of treatment. Peng et al. [36] have studied the remediation of a soil polluted by phenanthrene by microwave activated persulfate. They have obtained a good removal of this pollutant at 80 °C, but the pH was close to 2 at the end of the treatment.

Zhao et al. [37] have studied the abatement of PAHs in soil samples collected from the former coking plant (Beijing, China) at 30, 50, and 60 °C in batch way. They found that, among the activation methods tested for PS, the thermal activation was the most effective in PAHs removal, achieving a PAH conversion around 95% at 50 °C and 72 h.

To select the optimal temperature range in TAP it should be considered that the rate of PS decomposition increases more rapidly than the rate of organic substrates oxidation with the increase of temperature, especially if refractory pollutants are the

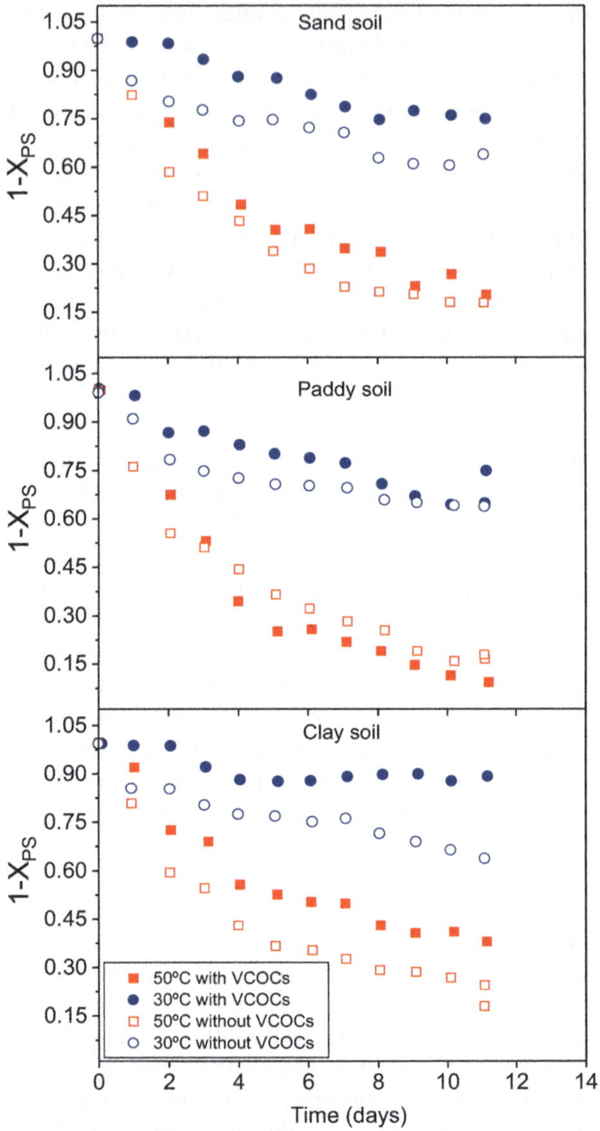


**Fig. 5** Persulfate conversion in different soils unpolluted and polluted with chlorinated volatile organic compounds (CVOCs). $C_{PSo} = 30$ g/L, 7800 mg NAPL/kg soil, $W_{soil}/W_{aq\ phase} = 0.5$. Data obtained from Liu et al. [33]

objective [38]. Therefore, for each case, there will be an optimal temperature interval, that should consider the effect of the temperature in both kinetic constants, pollutants oxidation and unproductive reactions consuming PS. In order to take this aspect into account, the following model has been proposed.

$$\frac{-\mathrm{d}C_{PS}}{\mathrm{d}t} = k_u C_{PS} + k_{ox} C_{COCs} C_{PS} \quad (15)$$

$$\frac{-\mathrm{d}C_{\mathrm{COCs}}}{\mathrm{d}t} = \alpha k_{\mathrm{ox}} C_{\mathrm{COCs}} C_{\mathrm{Ps}} \qquad (16)$$

where $k_u$ is a lumped kinetic constant of the unproductive reactions consuming PS and $k_{OX}$ is the lumped kinetic constant for COCs oxidation, with α being a stoichiometric coefficient. The lumped kinetic constant $k_u$ should also include the radical scavenging reactions. As a result, the maximum temperature that can be employed is usually 70–80 °C in aqueous phase, and 50–60 °C in soil or slurry systems, as was summarized in the works above reviewed.

In spite of TAP has proved to be applicable to a wide range of contaminants at lab scale, achieving high mineralization degrees of the pollutants, this technology presents important limitations for field application. Heat activation requires the installation of a parallel heating system to heat the aquifer matrix up to the desired temperature. This fact entails both capital expenditures as well as additional operating costs. In situ heating can be achieved by injecting steam or hot air, by using electrical resistances, radio frequency heating, etc., as shown in Fig. 6. There have been only a few numbers of documented field applications of heat-activated persulfate to date. Thompson et al. [40] successfully used steam injections to thermally activate persulfate for in situ applications. However, generally, heating is best applied for source treatment where the target area is limited. In situ heating, with an external heating source, is impractical for treating large groundwater plumes [41].

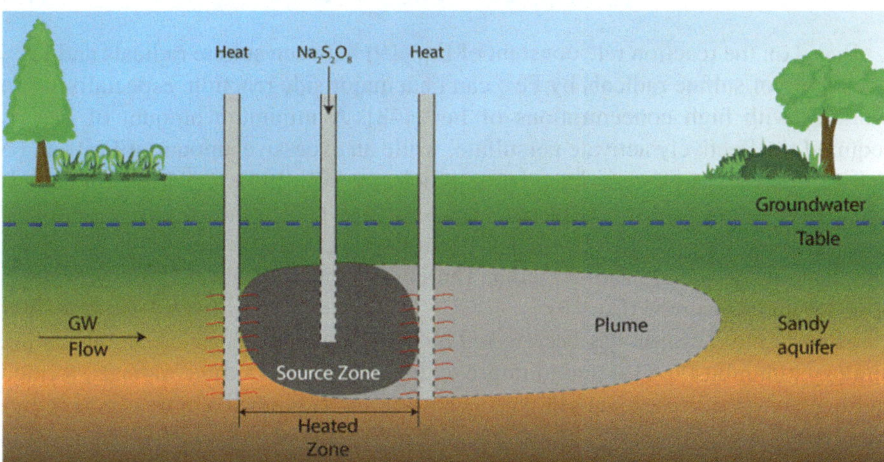


**Fig. 6** Example of TAP application in the field. Heat is used for activation (adapted from [39])

## 3 Persulfate Activated by Transition Metals

Persulfate can be activated to obtain sulfate radicals through electron transfer, which can be achieved using several metals (M) as iron, copper, manganese, iron, zinc, cobalt, cerium, vanadium, and silver [24, 42–45], following the reaction in Eq. (17).

$$S_2O_8^{2-} + M^{n^+} \rightarrow SO_4^{2-} + SO_4^{\cdot -} + M^{(n+1)^+} \qquad (17)$$

Among the metal ions above cited, ferrous iron, $Fe^{2+}$, is the most commonly used [2, 4, 6, 7, 11]. This is explained due to iron is cheaper than other transition metals, readily available and not dangerous for the environment. Ferrous iron reacts with persulfate to form the sulfate radical, as shown in Eq. (18):

$$S_2O_8^{2-} + Fe^{2^+} \rightarrow SO_4^{2-} + SO_4^{\cdot -} + Fe^{3^+} \qquad (18)$$

The homogeneous activation of PS using dissolved $Fe^{2+}$ requires a relatively lower activation energy (i.e., 14.8 kcal $mol^{-1}$) than thermal activation (33.5 kcal $mol^{-1}$) [43]. Moreover, Fe-PS has higher applicability at field scale than TAP [46].

According to Eq. (18), sulfate radicals are generated by the reaction between PS and ferrous iron. However, if $Fe^{2+}$ is present in large amounts, this species will significantly scavenge the formed radical $SO_4^{\bullet -}$, as shown in Eq. (19), inhibiting therefore the oxidation of the contaminants by this radical [47].

$$SO_4^{\cdot -} + Fe^{2^+} \rightarrow SO_4^{2-} + Fe^{3^+} \qquad (19)$$

Based on the reaction rate constant of Eq. (19) between sulfate radicals and $Fe^{2+}$, quenching of sulfate radicals by $Fe^{2+}$ can be a major side reaction, especially when working with high concentrations of $Fe^{2+}$ [48]. A minimum amount of $Fe^{2+}$ is required to effectively activate persulfate, while an excessive amount of $Fe^{2+}$ can be detrimental for the process degradation efficiency [47]. In fact, plateaus in persulfate consumption values, pollutant conversion and TOC removal have been often found in literature [47, 49–51] when persulfate and its activator, $Fe^{2+}$, are added simultaneously to the reaction media. Thereby, sulfate radicals are quickly formed (Eq. 18) and consumed (Eq. 19).

Slow addition of iron has been tested by some authors [48, 51] as an interesting strategy to inhibit Eq. (19) and to improve the efficiency of sulfate radicals, formed in Eq. (18) towards organic pollutants. These authors obtained a more efficient diuron oxidation when $Fe^{2+}$ was gradually adding. For this purpose, an iron source was continuously fed into the reactor from a concentrate solution of iron. Higher pollutant conversion, as well as higher TOC removal, were obtained by this procedure compared to a single addition (employing the same amount of $Fe^{2+}$).

Other strategy tested to minimize the extension of reaction (19) has been the use of zero valent iron instead of $Fe^{2+}$ [52, 53]. Rodriguez et al. [53] studied the oxidation of emerging and priority pollutants with $Fe^{2+}$ and ZVI as PS–activators in aqueous phase. They found that when $Fe^{2+}$ was employed, the undesired reaction (19) had a significant role, while in the case of ZVI, a slow release of $Fe^{2+}$ to the media occurred through the acid corrosion of iron at the ZVI particles surface due to its reaction with PS, as shown in Eq. (20):

$$Fe^0 + S_2O_8^{2-} \rightarrow Fe^{2+} + 2SO_4^{2-} \quad (20)$$

The controlled production rate of $SO_4^{2-}$ through the slow release of $Fe^{2+}$ from the ZVI surface inhibits the extension of the reaction (19), increasing the efficiency of the oxidant. Better results were obtained when ZVI was used instead of an initial addition of $Fe^{2+}$, especially when the regeneration of $Fe^{3+}$ to $Fe^{2+}$, through oxidation intermediates, did not occur.

Both strategies, feeding $Fe^{2+}$ in semi-batch way, or using ZVI as the iron source, lead to a slow release of $Fe^{2+}$ to the media, improving the use of sulfate radicals in the desired reaction with the organic pollutant.

However, the strategies before cited cannot be easily applied in soil and groundwater remediation as occurs in aqueous phase. For a site application, ZVI can be injected as nanoparticles, nZVI [54–57], although the stability and transport of nZVI trough the subsurface remains yet as a problem to solve [55].

The activation of persulfate by iron has been widely employed in both aqueous and soil phase. In this chapter, the study will be focused on the application to soil and groundwater. As described in Fenton's reagent, iron must be kept in solution to produce the sulfate radicals. This can be accomplished by working at acidic pH values. However, reaching a low pH value is often difficult or impracticable for an in situ remediation project, since many soils and groundwater behave as a buffer system, showing neutral–alkaline conditions. Moreover, the production of an acid pH could yield non-desirable side effects as metal mobilization, loss of soil properties, etc. [1]. In order to add $Fe^{2+}$ at neutral or light-alkaline pH, chelating agents (CA) have been applied [58]. Chelating agents are substances able to keep the iron in solution by complexing the transition metals. Examples of CA include ethylenediaminetetraacetic acid (EDTA) and (*S*,*S*)-ethylenediamine-*N*,*N*0-disuccinic acid (EDDS), organic acids or their salts (citric acid, oxalic acid, tartaric acid), and polyphosphates [37, 50, 58–72].

Low-molecular weight organic acids (LMWOA) have the advantage of being more biodegradable in the media. In Fig. 7 a scheme of the iron chelated by EDTA and citrate is shown.

Results obtained with chelating agents found in the literature are not conclusive. Moreover, the chelating agent is usually an organic compound and competes with the organic pollutants for the oxidant or the radicals generated. If the organic targets are recalcitrant compounds to oxidation (as is the case of diesel compounds), the chelating agent can increase notably the persulfate consumption, as noticed by Pardo et al. [54] using citrate as CA. On the contrary, others authors found a positive effect of this CA when more readily oxidizable compounds were studied, such as ethylbenzene [68] or aniline [58, 62]. Killian et al. [60] found that the degradation of BTEX and PAH compounds in MGP contaminated soil was improved by the

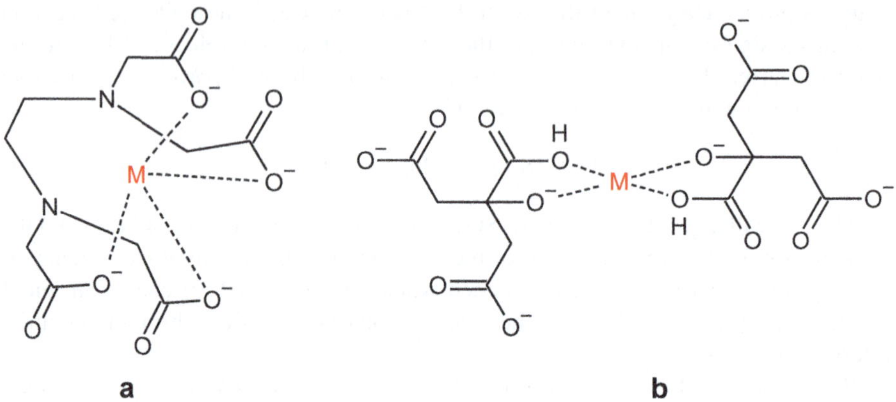


**Fig. 7** (**a**) Scheme of EDTA and (**b**) citrate as chelating agents of transition metals

addition of citric acid to the $Fe^{2+}$ solution. Liang et al. [59] found that the use of citric acid, as a chelating agent, resulted in the almost complete TCE destruction in comparison with observations under the same experimental conditions but without the presence of citric acid.

The availability of ferrous ion appears to be controlled by adjusting the molar ratio of chelate/$Fe^{2+}$. The amount of chelate provided should be sufficient to ensure the chelation of ferrous ion present. In general, higher chelated ferrous ion concentrations resulted in faster target compound degradation and higher persulfate decomposition. However, if the organic pollutant is more recalcitrant to oxidation than the CA, the addition and selection of the CA should be carefully considered.

The molar ratio between $S_2O_8^{2-}$/chelate/$Fe^{2+}$ is usually in the range 20:2:10 to 20:50:10. The best results were obtained by Han et al. [58] with the ratio 20:10:10. However, the optimal dosage must be found experimentally for each case.

On the other hand, chelating agents such as citric acid (or citrate) have been proven to be effective in mobilizing and removing metals from soils and sediments. In this case, the citric acid (or citrate) served to extract native metals from soils making them available to activate the production of sulfate free radicals and promote the subsequent target compound destruction [59]. This fact was also noticed in the studies carried out by Vicente et al. [64], in which the addition of citrate (3000 mg $kg^{-1}$) as a chelating agent mobilized the iron from the soil, with the iron concentration in the aqueous phase being approximately 40 mg $kg^{-1}$.

The use of chelated ferrous ion for in situ applications is far superior to the use of unchelated ferrous ion as PS–activator, as can be seen in the registered patents [4, 63]. Therefore, it appears that in situ chemical oxidation of organic pollutants using chelated ferrous ion to activate persulfate can be a viable method for aquifer remediation. However, the interaction of the CA with the iron and the oxidant in this system is a critical factor that must be evaluated in each ISCO project.

On the other hand, one of the main differences of activated PS by iron with respect to the use of iron in Fenton reagent is that the regeneration of $Fe^{3+}$ by PS is a very low process. PS reacts with ionic iron by oxidizing it (Eq. (18)), but the reduction of $Fe^{3+}$ to $Fe^{2+}$, proposed by Wu et al. [73, 74], and shown in Eq. (21), is unfavorable thermodynamically [14].

$$Fe^{3+} + S_2O_8^{2-} \rightarrow Fe^{2+} + S_2O_8^{-} \tag{21}$$

Another mechanism considered for the $Fe^{3+}$ reduction is the formation of a more oxidized species ($FeO^{2+}$) from $Fe^{2+}$ with the corresponding sulfate radical formation [14] (Eq. 22).

$$Fe^{3+} + S_2O_8^{2-} + \frac{1}{2}O_2 \rightarrow SO_4^{\cdot-} + SO_4^{2-} + FeO^{2+} \tag{22}$$

where the ferrate anion behaves also as an oxidant in the pollutant abatement [75, 76].

However, the fastest reaction is still that between $Fe^{2+}$ and PS. This may result in the loss of oxidation capacity of the $Fe^{2+}$/PS system once the $Fe^{2+}$ is exhausted. Therefore, the iron does not behave usually as a catalyst in the activation of PS. Because of this, high molar ratios of $Fe^{2+}$/PS are used in the iron activated PS (usually higher than 1/2 Fe/PS), in comparison to the ratios $Fe^{2+}$/hydrogen peroxide used in the Fenton reagent.

The reaction of $Fe^{2+}$ with PS (Eq. 18) is quite fast. This means that sulfate radicals are generated quickly with a quick rise in the pollutant conversion at short reaction times. However, a plateau for pollutant and PS conversion with reaction time is soon noticed once $Fe^{2+}$ has been oxidized to $Fe^{3+}$ [49, 51]. As an example of these plateaus and the results obtained by Liang et al. [49] in the abatement of TCE are shown in Fig. 8.

Liang et al. [49] promoted the regeneration of $Fe^{2+}$ by adding thiosulfate to the reaction media. In this case, the following reaction was proposed.

$$2Fe^{3+} + 2S_2O_3^{2-} \rightarrow 2Fe^{2+} + S_4O_6^{2-} \tag{23}$$

However, the thiosulfate could also consume the remaining persulfate, and this fact could increase the cost of the process since a higher amount of oxidant would be required.

On the other hand, smaller differences were found using $Fe^{2+}$ or $Fe^{3+}$ to activate PS in the abatement of the dye orange G [77]. This was explained by the oxidation route proposed for orange G degradation, showed in Fig. 9. As can be seen, phenol and benzoquinone were obtained as oxidation intermediates. Quinone intermediates generated during pollutant oxidation may act as electron shuttles, allowing for the reduction of $Fe^{3+}$ into $Fe^{2+}$ in the redox cycling of iron [77, 78], as shown in Fig. 10. Therefore, the activation of PS by $Fe^{3+}$ allowed the complete orange G removal.

In this line, the regeneration of $Fe^{3+}$ to $Fe^{2+}$ by quinone-type compounds has been also described by Peluffo et al. [56] when the oxidation of four PAHs in soil was carried out by using three types of iron sources (nZVI, $Fe^{2+}$, $Fe^{3+}$) to activate PS. No differences were found in PAHs abatement with $Fe^{2+}$ and $Fe^{3+}$, with this fact being

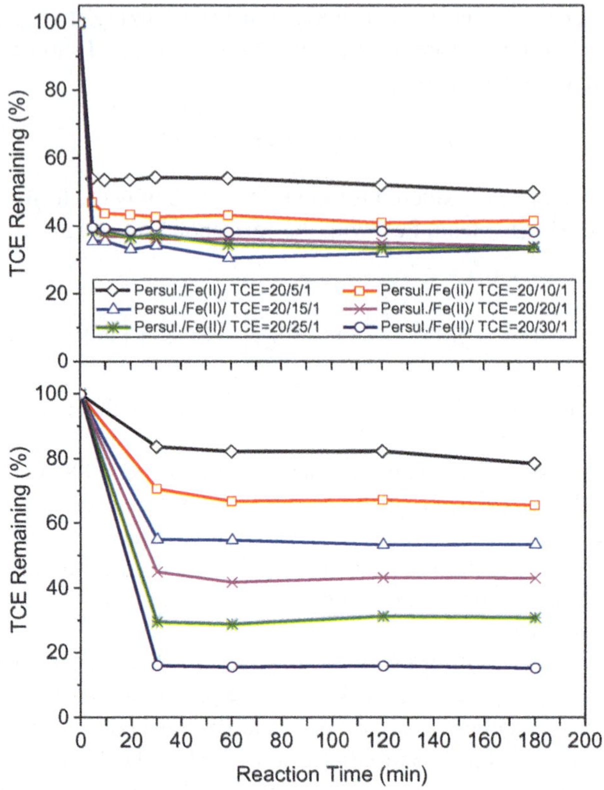


**Fig. 8** Results obtained by Liang et al. in TCE abatement with iron-activated PS [49]. $C_{TCE} = 0.45$ mM, $C_{PSo} = 9.10$ mM, 20 °C

explained by taking into consideration the appearance of quinone-type organic compounds as oxidation degradation by-products from their parent compounds. This is the case of 9,10-anthraquinone, coming from the oxidation of anthracene, capable of reducing $Fe^{3+}$ to $Fe^{2+}$, the activator of PS to produce sulfate radicals by means of Eq. (18).

The regeneration of $Fe^{3+}$ to $Fe^{2+}$ has been also found when hydroxylamine (HA) was added to the reaction media. Jin et al. [79] found that ibuprofen was quickly oxidized by the system $Fe^{2+}$-nitrilotriacetic acid-PS (NTA was used as a chelating compound) when hydroxylamine was added, with this fact being explained by the quick regeneration of $Fe^{2+}$-NTA complex by HA.

Wu et al. [80] found also a positive effect when using HA in the perchloroethylene (PCE) removal by PS activated by iron. At the operating conditions tested (PS–Fe–PCE molar ratio 30:4:1) they found that the optimal Fe–HA molar ratio was 4:4, as can be seen in Fig. 11. The $Fe^{2+}$-$Fe^{3+}$ cycle in the presence of HA was schematized by Wu et al. [80] (Fig. 12).

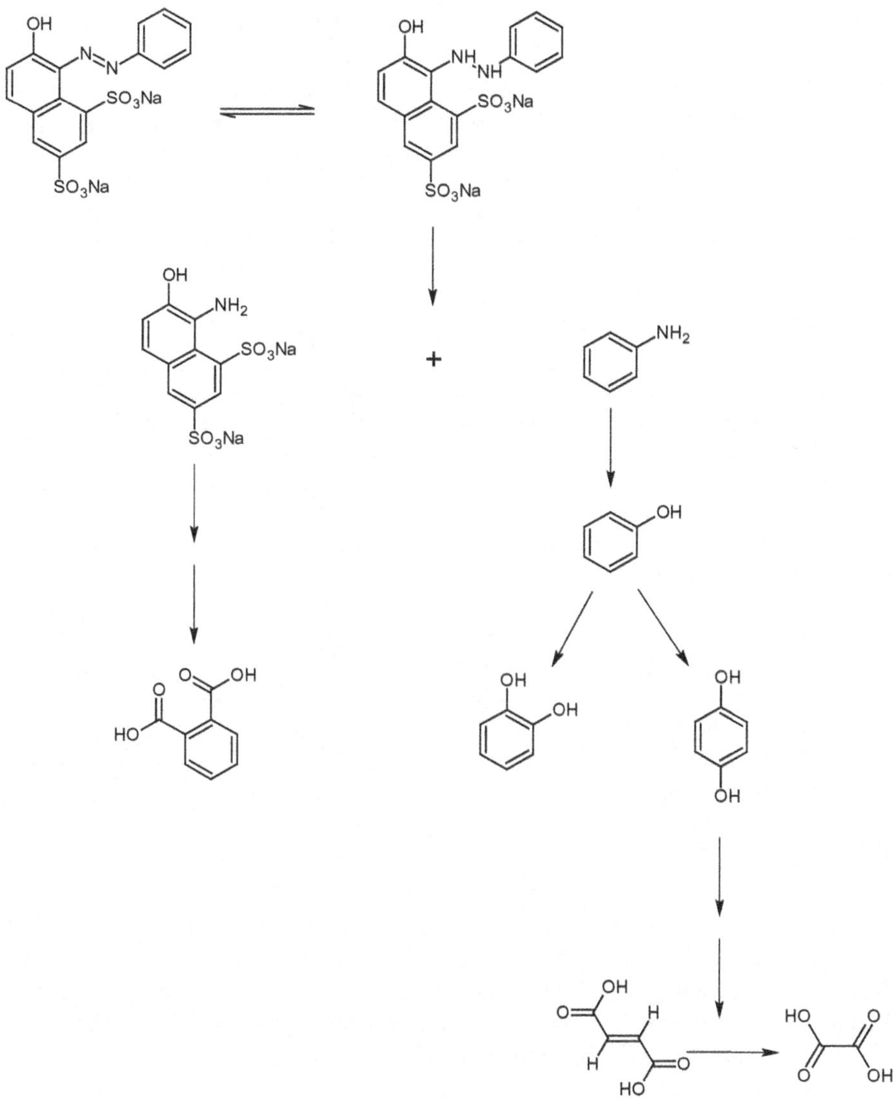


**Fig. 9** Oxidation route of orange G by PS activated by iron [77]

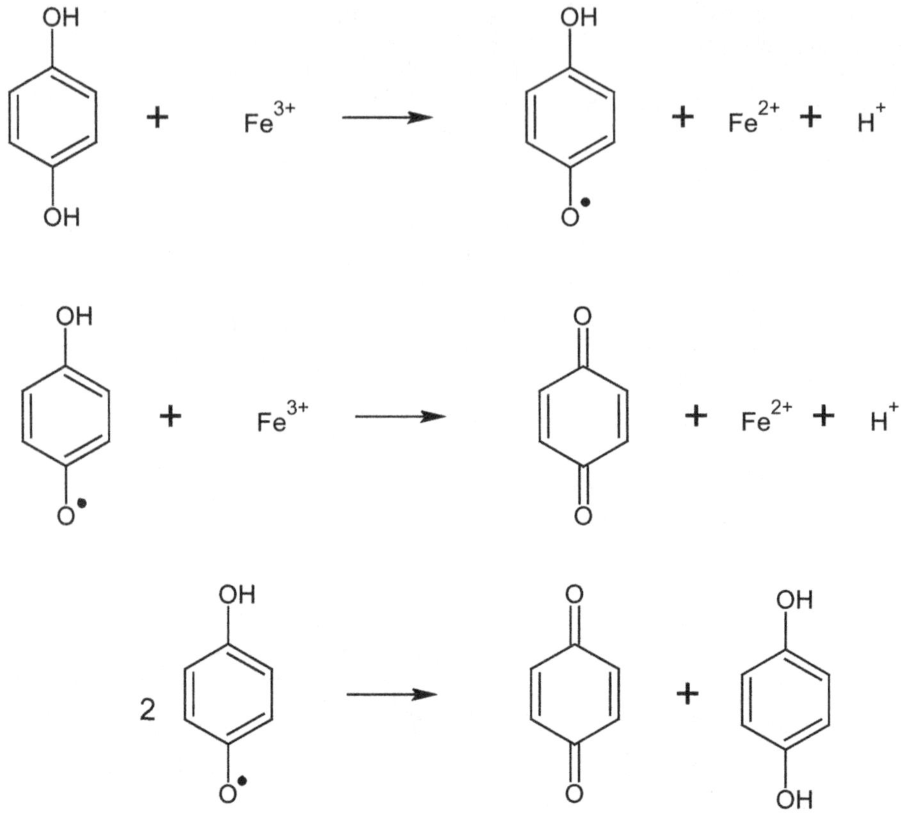


**Fig. 10** Reduction of $Fe^{3+}$ by dihydroxyphenols [77, 78]

Han et al. [81] also found a positive effect of adding HA to the reaction media when the oxidation of orange G was investigated by using $Fe^{2+}$/(*S*,*S*)-ethylenediamine-*N*,*N*0-disuccinic acid (EDDS)–persulfate (PS) system. In this case, the optimal HA–Fe molar ratio was 8:5. The role of HA in iron regeneration has been explained by Han et al. [81] by the following reactions.

$$Fe^{3+} + NH_2OH \rightarrow NH_2O^{\cdot} + Fe^{2+} + H^+ \quad (24)$$

$$Fe^{3+} + NH_2O^{\cdot} \rightarrow NHO^{\cdot} + Fe^{2+} + H^+ \quad (25)$$

$$5Fe^{3+} + 2H_2O + NH_2O^{\cdot} \rightarrow NO_3^- + 5Fe^{2+} + 6H^+ \quad (26)$$

However, in spite of the good results obtained when HA was added to regenerate $Fe^{3+}$, the toxicity of this reducing agent in the media, as well as the production of other unproductive reactions between this compound and the oxidant, should be better explored before its application in site remediation.

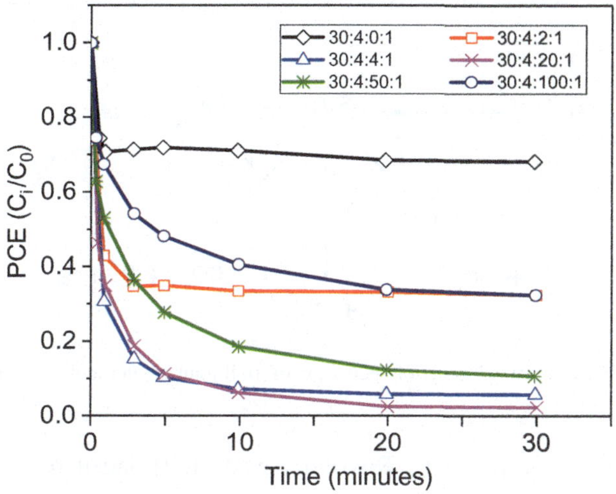


**Fig. 11** Effect of HA dosage on PCE degradation, in Fe(II)/HA activated PS process ($[PCE]_0 = 0.15$ mmol $L^{-1}$, $[PS]_0 = 4.5$ mmol $L^{-1}$, $[Fe(II)]_0 = 0.6$ mmol $L^{-1}$, $T = 20 \pm 0.5$ ∘C). PS–$Fe^{2+}$–HA–PCE molar ratios. Adapted from [80]

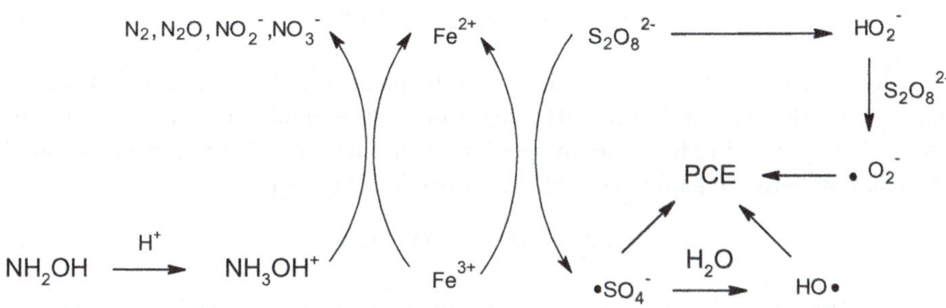


**Fig. 12** Schematic diagram of PCE degradation mechanism in the PS/$Fe^{2+}$/HA system. Adapted from [80]

The reduction of iron $Fe^{3+}$ to $Fe^{2+}$, due to the presence of some organic compounds in the reaction media, has been generalized by Rodriguez et al. [53, 82] by the following lumped reaction:

$$A + v_{Fe} Fe^{3+} \rightarrow (1-x) CO_2 + xB + v_{Fe} Fe^{2+} \tag{27}$$

representing $A$ the oxidizable organic matter in the TOC, able to reduce $Fe^{3+}$ to $Fe^{2+}$ and $B$, the refractory organic compounds. Therefore, $x$ refers to the overall fractional yield of the organic carbon of $A$ that is not mineralized but reacts with refractory compounds, $B$.

A simplified reaction pathway, that summarizes the overall reaction pathway of the pollutant abatement with PS activated by ZVI, which takes also into account the

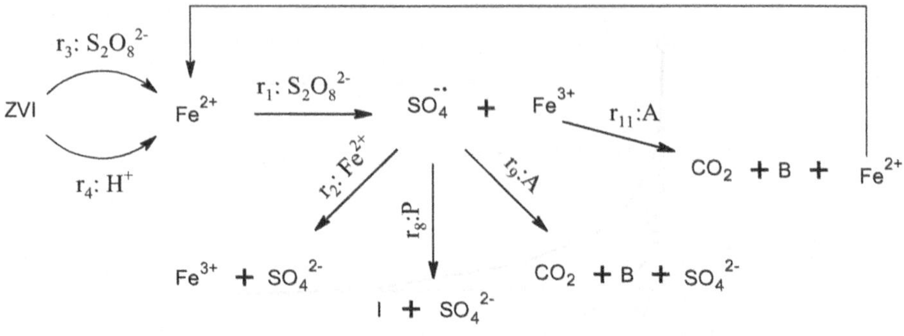


**Fig. 13** Overall summarized reaction pathway of pollutant abatement with PS activated by ZVI [53]

reaction (27), was proposed by Rodriguez et al. [53] and it has been shown in Fig. 13.

Another important point to be considered in the iron activated PS is that the sulfate radical generated can produce other active species, in particular, hydroxyl radicals, by the following reaction.

$$SO_4^{\cdot-} + H_2O \rightarrow OH^{\cdot} + H^+ + SO_4^{2-} \tag{28}$$

It has been proved that both radicals can be present in the iron activated persulfate system (Fe-AP) [58] at the pH range from 3 to 9, while at alkaline conditions ($pH > 11$) hydroxyl radicals are the predominant ones [66], due to the extension of the reaction between sulfate radical and hydroxide (Eq. 29):

$$SO_4^{\cdot-} + OH^- \rightarrow OH^{\cdot} + SO_4^{2-} \tag{29}$$

The efficiency of iron-activated PS in the abatement of organic pollutants has been tested not only at lab scale but also at field conditions [1, 46, 63]. However, most of the studies found in the literature, as those previously cited, correspond to experiments carried out at lab scale using spiked soils, and have been accomplished in batch operation mode. Only a few number of investigations have been carried out for column operation ([83–86]), although column conditions represent the field scale much better.

Among the works carried out at column operation, Pardo et al. [55] have studied the effectiveness of activated PS in the remediation of a contaminated soil with phenanthrene (PHE), anthracene (ANT), pyrene (PYR), and benzo(a)pyrene (BaP). The soil was artificially spiked and aged for 3 months. Five runs were carried out: blank (without PS), control (only PS was fed to the column), CPS1 (two aqueous solutions were simultaneously fed to the column for 25 days, one containing PS and the other containing a $Fe^{3+}$ salt), CPS2 (the same amount of iron that was introduced during 25 days in CPS1 was fed into the column as an aqueous suspension of nZVI. Afterwards, PS was fed for 25 days), and CPS3 (a more concentrated suspension of nZVI was fed to the column. Subsequently, PS solution was added for 25 days).

After 25 days of treatment, samples of soil were taken at different column lengths and profiles of labile iron and PAHs were measured. The removal of PAHs was negligible in blank and control runs. Higher efficiencies were reached using nZVI in comparison to those obtained at the same concentration of labile $Fe^{3+}$ in the column soil. Moreover, a remarkable iron adsorption into the soil was noticed in spite of the acidic pH of the aqueous phase fed to the column soil. The highest conversions of PAHs were found in the nearest sections to the injection source, with this fact being related to the iron concentration profile found along the column. Results obtained for phenanthrene, the most refractory PAH among those tested, are shown in Fig. 14. The profiles of labile iron along the soil column are shown in Fig. 15. By comparison of the results found in CPS1 and CPS2, shown in Figs. 14 and 15, respectively, it can be seen that the labile iron in the column was higher in CPS1 (feeding $Fe^{3+}$) than in CPS2 (feeding nZVI). However, the conversion of PAHs was slightly higher in CPS2. In CPS3, a higher conversion of phenanthrene was obtained, due to the higher labile iron present in the column (ten times higher than in CPS2). It was also noticed that when the nZVI solution was fed to the column, the agglomeration of the iron nanoparticles was noticed at the entrance of the column, with the transport of nZVI being hindered through the column. Therefore, the stability of the nZVI emulsion in the subsurface is a critical issue for field applications.

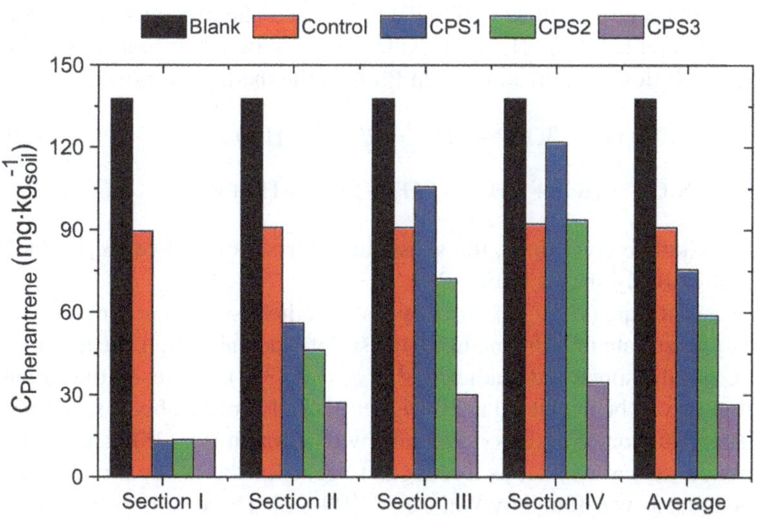


**Fig. 14** Remaining concentrations of phenanthrene obtained at sections I: entrance, II: medium, and III: top of the column after 25 days of treatment for phenanthrene, soil length = 17.5 cm, interstitial velocity = 13.9 cm day$^{-1}$. Data from Pardo et al. [55]. The concentration of labile iron by EDTA in the different sections is indicated in the legend

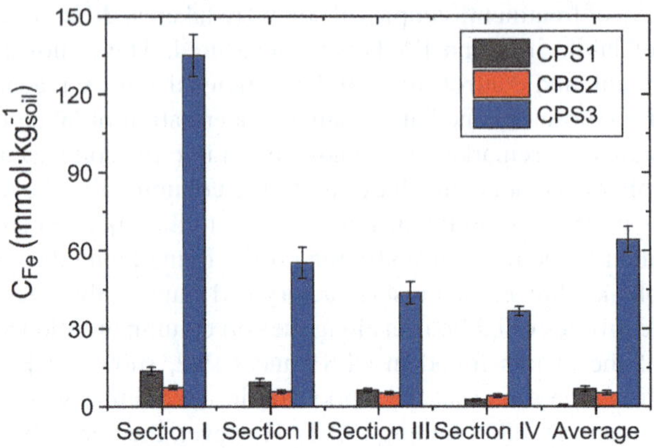


**Fig. 15** Distribution of labile iron present in each of the different sections in the soil column after the oxidation treatment. Section I: entrance, section II: medium, and section III: top of the column, after 25 days of treatment for phenanthrene, soil length = 17.5 cm, interstitial velocity = 13.9 cm·day$^{-1}$. Data from Pardo et al. [55]

# 4 Alkaline (Base) Activation of PS (BAP)

The alkaline activation of persulfate (pH > 10) has gained attention recently both in literature and field applications. The hydroxide ($OH^-$) is used to produce alkaline pH, promoting the following reactions when PS is in the media [41, 66, 87–89]:

$$S_2O_8^{2-} + 3OH^- \rightarrow HO_2^- + 2SO_4^{2-} + H_2O \tag{30}$$

$$S_2O_8^{2-} + HO_2^- + OH^- \rightarrow SO_4^{\cdot-} + SO_4^{2-} + H^+ + O_2^{\cdot-} \tag{31}$$

Moreover, at alkaline conditions, the subsequent formation of hydroxyl radical takes place, as previously shown in Eq. (29).

As can be seen in Eqs. (29)–(31), three types of radical species are involved in the activation of persulfate in basic medium: the sulfate radical $\left(SO_4^{\cdot-}\right)$, the hydroxyl radical ($OH^{\cdot-}$), and the superoxide radical ($O_2^{\cdot-}$, $E^0 = 0.33$ V). However, the sulfate radical contribution can be neglected at pH higher than 11 [66, 87], because at these conditions, the production of hydroxyl radicals by reaction in Eq. (28) is favored.

The superoxide radical can act as a reductant agent for the halide atoms in the organic compounds, as was found by Watts et al. [90–92]. These authors have studied the abatement of carbon tetrachloride in groundwater by using Fenton reagent, oxidation system in which the superoxide radical is also produced. The high nucleophilicity of $O_2^{\cdot-}$ can explain the nucleophilic substitution mechanism in reaction with halides, as shown in Eq. (32) [93].

$$O_2^{\cdot-} + AX \rightarrow AO_2^{\cdot} + X^- \tag{32}$$

Due to the fact that both types of radicals (hydroxyl and superoxide radicals) are produced in the alkaline activation of PS, this method is effective for the abatement of a broad type of organic pollutants, in which oxidation and reduction mechanisms are involved in the attack of these radicals to the organic pollutants. Halogenated contaminants, as the halomethanes, have weak reactivity with strong oxidants, including hydroxyl radicals or sulfate radicals, but are more reactive against radicals with high nucleophilicity, as $O_2^{\cdot-}$.

The base usually employed as PS–activator is sodium hydroxide [66, 88, 94–98], due to the lower solubility of other salts like potassium persulfate, although other reagents like lime (calcium oxide—CaO) [99, 100] or potassium hydroxide [41] have been also applied.

The dissolution of PS in aqueous medium generates an acidic pH due to the reaction PS with water, shown in Eq. (33):

$$S_2O_8^{2-} + H_2O \rightarrow 2HSO_4^- + \frac{1}{2}O_2 \quad (33)$$

Therefore, a sufficient amount of activator should be added to adjust the initial pH of the PS solution to the required alkalinity values. Moreover, taking into account that hydroxide is not a catalyst but a reagent, and it will be consumed during the PS decomposition into radical species (as previously indicated in Eqs. (29)–(31)), extra activator addition is needed to maintain the alkaline pH [41]. The base to PS molar ratio (NaOH–PS) usually used in literature is within the range 1:1 to 8:1 being the ratios 2:1 and 4:1 the most frequently applied [94–96, 98, 99].

One of the advantages of alkaline activation of PS at pH > 11 is that the carbonate scavenging is not a concern. Moreover, it seems than PS activation at highly alkaline conditions had a comparatively lower impact than the other activation strategies in soil–water systems [66]. However, as occurs when using other PS activation methods, the depletion of persulfate and the activating agents, due to their interaction among them and/or with the aquifer materials, should be taken into account while they can greatly limit the persistence of the oxidant–activator system.

Moreover, the cost required for the implementation at field scale of alkaline activation of persulfate in a ISCO remediation project should consider the costs associated to the reagents and the specific equipment needed to inject large volumes of hydroxide solutions, which is usually in carried out in several steps, in order to maintain the alkaline conditions in the subsurface during the degradation process. However, when dealing with soils and groundwater with neutral–slightly alkaline pH and/or with high carbonate/bicarbonate content, the alkaline activation of PS is very interesting since the use of high amounts of chelating agents (as required in Fe-PS activation) is avoided and minimizes the negative effect of radical scavenging by carbonates and bicarbonates on the process efficiency.

Another relevant issue that it has to be taken into account in the alkaline activation of PS is that when working at pH values above 11, a simple alkaline hydrolysis can take place for some organic molecules, resulting in the breakdown of these compounds without the participation of radical species. This fact has been noticed for non-aromatic halogenated compounds such as hexachlorocyclohexane [98], as shown in Fig. 16. Santos et al. [98] found than lindane (γ-hexachlorocyclohexane) and other HCHs in groundwater from a landfill were completely hydrolyzed to 1,2,4 and 1,2,3 trichlorobenzenes in less than 1 day at pH 12. Moreover, at this pH, calcium and magnesium carbonates precipitated, inhibiting the scavenging of radicals by carbonate/bicarbonate in the original groundwater.

Trichlorobenzenes formed from HCHs at alkaline conditions, as well as other chlorobenzenes present in the groundwater, were gradually eliminated by the radicals generated in the alkaline activation of PS, as shown in Fig. 17 [98]. A kinetic equation for each chlorinated compound *i* (mono-, di-, tri-, and tetrachlorobenzene) was proposed by using a first-order kinetic rate expression for the pollutants *(i)* and the oxidant (PS) concentrations, as follows (Eq. 34).

$$r_i = k_i \cdot C_i \cdot C_{PS} \tag{34}$$

In Fig. 17, symbols correspond to experimental data and lines correspond to predicted values using the kinetic model proposed in Eq. (34). As can be seen, the same results were obtained with groundwater (initial conductivity = 6640 μS cm$^{-1}$ at pH = 7) and milliQ water spiked with lindane, confirming the absence of scavenging reactions after the hydroxide addition, which is explained attending to the carbonate precipitation at pH = 12–13. Moreover, the molar ratio NaOH/PS tested (in the range 2/1 to 4/1) did not influence the pollutant conversion profiles obtained with time. No other aromatic or chlorinated by-products were found in the oxidation of chlorobenzenes by PS activated by alkali. Weisner et al. [100] found also an alkaline hydrolysis of PCBs, TNT, and DDT when lime was added and a pH value higher than 10–11 was achieved. Crimi et al. [101] have noticed also a total depletion of lindane when PS activated by alkali was used. However, this cannot be

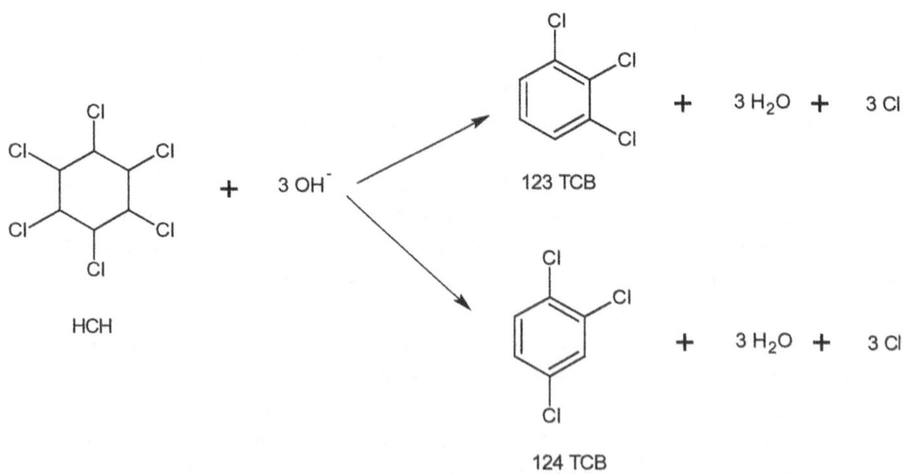


**Fig. 16** Alkaline hydrolysis of hexachlorocyclohexane at pH 12 [98]

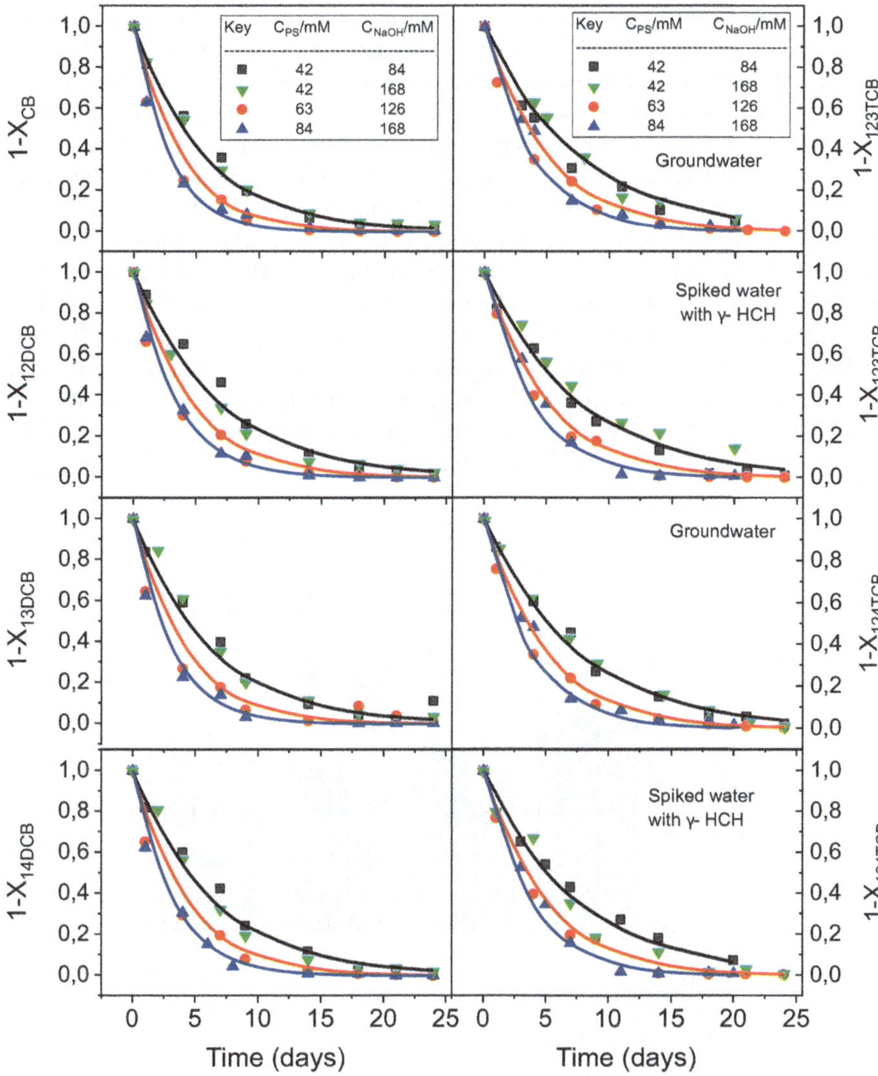


**Fig. 17** Conversion of chlorobenzenes: mono (CB), di (12CB, 13CB, 14CB) tri (124TCB, 123 TCB) in groundwater with reaction time at different PS and NaOH doses. Experimental data (symbols) and predicted values (lines) [98]

attributed to lindane oxidation reaction but to the alkaline hydrolysis of lindane to trichlorobenzene.

Lominchar et al. [94] have studied the effectiveness of alkaline activated persulfate for the remediation of a soil polluted with aged diesel spill located in a train maintenance facility in Madrid (Spain). The fuel had leaked over time from an underground storage tank. Soil samples were collected and analyzed and the alkaline activation of persulfate was chosen as the activation method due to the natural pH of the natural soil, close to 8, with a remarkable content of carbonates. Different initial concentrations of oxidant (PS) and activator (NaOH) were tested. The total petroleum hydrocarbons (TPH) removal efficiency and the elimination of the most representative compounds (aliphatic and aromatic) present in the contaminated soil was studied along the treatment time (56 days). To do this, 30 compounds present in the aged diesel were identified and quantified in the soil samples at different reaction conditions and reaction times. The main results obtained from this study are shown in Fig. 18. As can be seen, the highest TPH elimination achieved was 98%, obtained when working with the highest concentrations of PS (100 g $L^{-1}$) and with a NaOH/PS molar ratio equal to 4. Aromatic compounds mainly alkylated naphthalenes and phenanthrenes were found to be easily oxidized during the first 4 days of treatment without generating other detectable aromatic compounds, as oxidation intermediates. On the other hand, aliphatic compounds with carbon numbers above 16 were more refractory to oxidation.

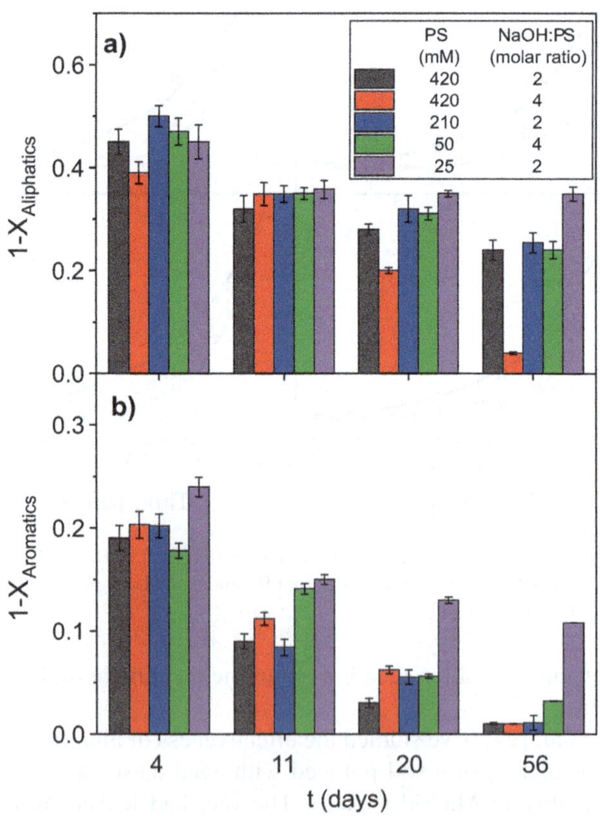


**Fig. 18** Conversion of aliphatic (**a**) and aromatic (**b**) compounds in runs P1–P5 with increasing reaction time. $T = 20$ °C, $V_L/W_{soil} = 2$ mL $g^{-1}$, $TPH_0 = 4930 \pm 50$ mg $kg^{-1}$ [96]

Lominchar et al. [94] have also studied the effect of adding commercial nonionic and biodegradable surfactant (Verusol-3, from Verutek) to the oxidant system PS + NaOH to remediate a soil polluted with a commercial home heating oil and 4-ring PAHs. The objective of the study was to test the possible enhancement of the oxidation rate of the pollutants sorbed into the soil by improving their solubilization in the aqueous phase, where the oxidation reaction takes place, by adding a surfactant. The study was carried out at column scale. Four columns were built and operated: Blank (C-B, water flushing), C-ISCO (persulfate activated by alkali but without surfactant), C-S (a surfactant aqueous solution was fed to the column), and C-SISCO (an oxidant–surfactant mixture was fed to the column). The aqueous effluent collected and the column soil were analyzed during and after column runs, respectively. A fingerprint of the fuel was obtained before and after different treatments. The compounds present in the fuel, including PAHs, were analyzed and quantified. Persulfate and pH were monitored for 25 days in the aqueous effluent. The results are shown in Fig. 19. As can be seen, the column flushing with water (blank) did not produce any change in the pollutant concentration in the soil during the time interval evaluated. In the run C-ISCO, it was found that, in the absence of surfactant, 30% of the TPHs remained in the soil after 25 days, which can be explained by the strong adsorption of the pollutants to the soil, probably located in the soil micropores, producing a residual contamination difficult to eliminate without surfactant addition. In run C-S the complete removal of TPH and PAHs from the soil was achieved at 25 days, but the pollutants were only transferred to the aqueous phase and not degraded. In run C-SISCO, the complete removal of the organic

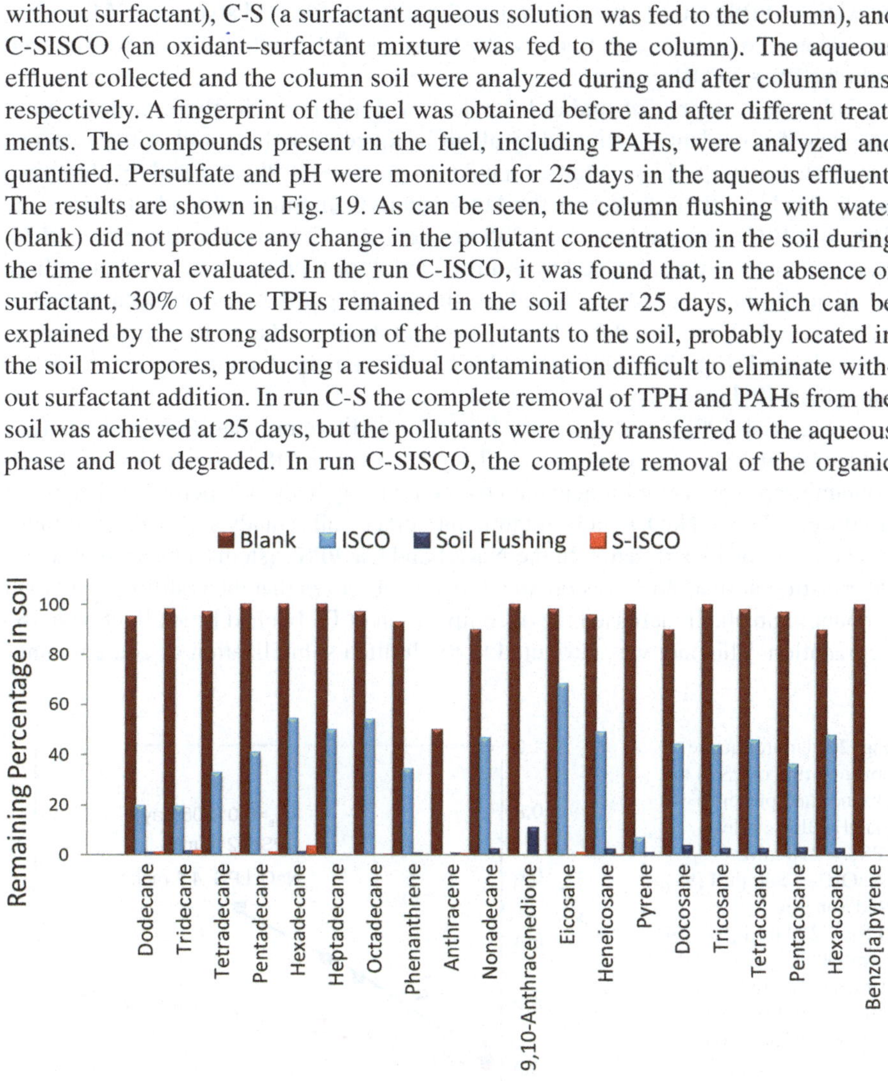


**Fig. 19** Remaining aliphatic compounds and PAHs in column soil after 25 days of treatment (as a fraction of the initial concentration in soil): C-B (blank), C-ISCO (PS + NaOH), C-S (Surfactant), and C-SISCO (Surfactant + PS + NaOH) [94]

pollutants in soil was obtained at 25 days. MoreoverMoreover, in C-SISCO no organic pollutants were transferred to the effluent, since they were oxidized by the action of PS-NaOH system in the aqueous phase. The main concern that has to be considered was the unproductive consumption of the oxidant by the surfactant. The process S-ISCO should be optimized by adjusting the surfactant dosage fed into the column, in order to decrease the unproductive oxidant consumption, due to the side reaction of the surfactant oxidation. Moreover, comparing the results obtained in this study with those obtained by using iron activated persulfate [56], the alkaline activation of PS seems to have a similar effectiveness for TPH abatement. Moreover, an enhanced oxidation of refractory PAHs such as 9,10-anthraquinone was noticed with the PS + NaOH system.

Only a few number of studies have dealt with the unproductive consumption of PS when this oxidant is activated by alkali in aqueous or slurry soil–water systems. Lominchar et al. [94] have studied the decomposition of PS activated by alkali in a sandy clay loam Bt horizon from the Autonomous Community of Madrid (pH = 7.22, TOC = 0.196% and labile iron = 7235 mg kg$^{-1}$, d$p$ < 2 mm). The conversion of PS was compared to that obtained by Peluffo et al. [56] with the same soil but without alkali addition. The results obtained are shown in Fig. 20. As can be seen, the addition of an alkali increased the decomposition rate of PS due to the reactions previously showed (Eqs. 29–31). Despite this fact, PS remained in the reaction media for more than 40 days at the conditions tested.

Liang et al. [99] is another work found in the literature that deals with alkaline activation of PS in soil remediation. They have studied the remediation of diesel-contaminated soil using PS activated by a base or by hydrogen peroxide. The bases used were CaO or NaOH, and contaminant-spiked soils (sandy soil) were employed to carry out the experiments. In the NaOH and CaO/PS systems a maximum diesel degradation around 30% was obtained. It was observed that the addition of a larger amount of an alkali increased the decomposition of PS but had little effect on diesel degradation. This fact was attributed to the limited solubilization of contaminants,

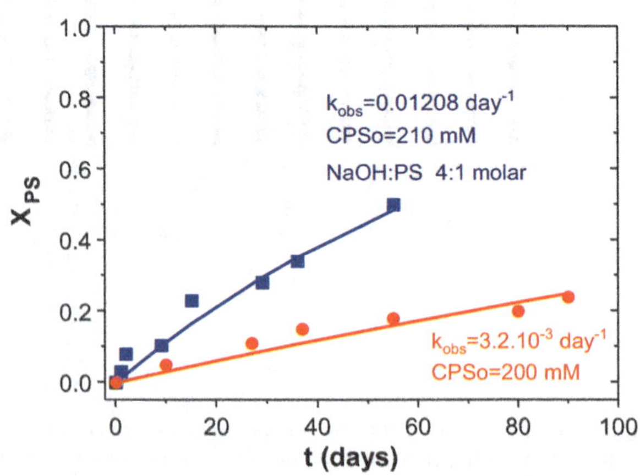


**Fig. 20** Unproductive consumption of PS in the absence and presence of NaOH. Blue circles $[PS]_0 = 210$ mM, $[NaOH]_0 = 840$ mM [94]. Red Squares: $[PS]_0 = 210$ mM [54, 56]. Symbols: Experimental data. Lines: predicted values with the kinetic constant in the legend

decreasing the effectiveness of the oxidation in the aqueous phase. Other treatments that improve the desorption of hydrophobic pollutants from soils (as hydrogen peroxide addition) resulted in an increase in the degree of pollutant removal achieved.

Weisner [100] has studied the abatement of PCBs, TNT, and DDT in two soils collected in a former explosive factory, using lime and temperature as PS–activators. They found a significant removal of pollutants by alkaline hydrolysis. Moreover, the conversions of TNT, DNT, and PCBs achieved with the lime-persulfate treatment were smaller than those achieved with the alkali addition alone. Looking at these results a possible interpretation of the authors of this chapter is that the PS used consumed the alkali and, therefore, the alkaline hydrolysis took place in a lower extension. Therefore, the results found in literature dealing with persulfate activated by alkali should be carefully examined when the target organic compounds are susceptible to alkaline hydrolysis degradation mechanism may be discriminated. On the other hand, the by-products generated as a result of hydrolysis and their removal by further oxidation have been scarcely studied in the literature.

## 5 Persulfate Activation by Hydrogen Peroxide (HAP)

The combination of persulfate with hydrogen peroxide is also often referred to as peroxide activation of PS. Actually, the knowledge about the interaction between hydrogen peroxide and persulfate is limited [1, 11, 43]. Liu et al. [102] proposed that hydroxyl radicals can initiate persulfate radicals, as well as sulfate radicals can increase the formation of hydroxyl radicals, as summarized in Eqs. (33)–(36).

$$HO^{\cdot} + S_2O_8^{2-} \rightarrow SO_4^{\cdot-} + HSO_4^{-} + \frac{1}{2}O_2 \quad (35)$$

$$H_2O_2 + S_2O_8^{2-} \rightarrow SO_4^{\cdot-} + HOO^{\cdot} + HSO_4^{-} \quad (36)$$

$$SO_4^{\cdot-} + H_2O \rightarrow SO_4^{2-} + HO^{\cdot} + H^{+} \quad (37)$$

$$SO_4^{\cdot-} + OH^{-} \rightarrow SO_4^{2-} + HO^{\cdot} \quad (38)$$

Therefore, the combination of hydrogen peroxide and persulfate could result in a multiradical pool in the reaction media. However, the mechanism behind this activation process is still uncertain. Several soil minerals containing Fe and Mn oxides (goethite, pyrolusite, ferrihydrite) can activate the hydrogen peroxide [103], starting the production of hydroxyl radicals in a Fenton-like reaction. These hydroxyl radicals can react with persulfate to produce sulfate radical as indicated in Eq. (35). Moreover, the heat released from the exothermic hydrogen peroxide reactions could improve the radical generation form persulfate. Therefore, the behavior of hydrogen peroxide as activator will be greatly dependent on the soil characteristics. The behavior of HAP in aqueous phase, slurries, or subsurface cannot be easily compared or predicted.

There are few works found in the literature dealing with hydrogen peroxide activation of PS (HAP), and most of them, make a comparison with other activation methods. The comparison of hydrogen peroxide as activator vs. other PS–activator in soil–water systems has been carried out in several works but contradictory results have been published, which can be probably attributed, at least partially, to the different interactions of the oxidants with the soil.

Lominchar et al. [95] have studied the abatement of phenol by PS activated by alkali, nZVI, and hydrogen peroxide in batch runs. Phenol and PS conversions and aromatic by-product concentration profiles during 168 h of reaction time were measured and compared, as well as the degree of mineralization achieved and the ecotoxicity of the samples. It was found that both phenol and aromatic by-products (catechol and hydroquinone) completely disappeared using PS activated by alkali before 24 h, while a significant amount of remaining aromatic intermediates was obtained when the activating agent was nZVI and hydrogen peroxide. Moreover, the highest unproductive consumption of PS was obtained in HAP, as can be seen in Fig. 21.

Liu et al. [102] have studied the effectiveness of hydrogen peroxide, PS and hydrogen peroxide + PS in the abatement of *ortho*-nitrochlorobenzene (o-NCB) in soil. The soil used was an aquic cinnamon soil and the runs were carried out in batches (slurry systems). They found an improvement in o-NCB degradation when both oxidants were applied simultaneously. On the other hand, it was pointed out that natural iron species, present in soil, could effectively facilitate the degradation of organic pollutants in the presence of hydrogen peroxide + PS.

Block et al. [41] found that the combined hydrogen peroxide–persulfate reaction was effective for chloroethanes and chloroethenes removal in aqueous phase after 7 days of treatment. These authors also tested the efficiency of the dual system against BTEX and other VOCs present in a soil from an MGP site. A remarkable abatement of these pollutants was found after 14 days of treatment. The unproductive consumption of the oxidants was not studied in this work. Moreover, also good results were found with PS activated with alkali at molar ratios KOH/PS about

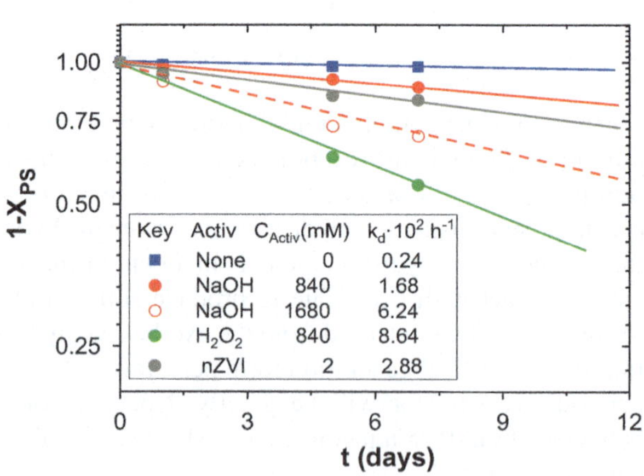


**Fig. 21** Results obtained in the absence of phenol: Persulfate consumption (experimental results as symbols, predicted values using Eq. (13) as lines). $[PS]_0 = 420$ mM [95]

1/1 in the abatement of BTEX and some chlorinated VOCs in aqueous phase after 7 days of treatment.

Ferrarese et al. [104] studied the effectiveness of using several chemical systems to oxidize sorbed PAHs from a sediment with aged contamination. Hydrogen peroxide, modified Fenton's reagent (chelating agent was added to prevent iron precipitation), activated sodium persulfate, potassium permanganate, and the dual systems potassium permanganate + hydrogen peroxide and sodium persulfate-Fe + hydrogen peroxide, were tested. Reaction samples of each oxidation treatment were analyzed (the reaction time was not indicated). Several dosages of oxidant were tested (from 50 to 200 mM) but the consumption of the oxidant was not reported. The pollutants removal at 50 mM of oxidant dose was above 80% for all the system tested except hydrogen peroxide alone. The highest TOC conversions were obtained with the systems potassium permanganate and modified Fenton's reagent.

Zhao et al. [37] have studied the influence of persulfate activation methods on PAHs degradation. Activation by heat, citrate-chelated-iron, base, and hydrogen peroxide were tested. Runs were carried out in batch way (slurry system) with soils contaminated with PAHs (340 mg $kg^{-1}$) collected from a coking plant in Beijing, China. The properties of the soil were not provided. By analysis of the soils at 72 h, it was found that thermal activation of PS was the most effective way for PAHs removal, followed by citrate-chelated-ferrous iron activation of PS. Lower PAH removals were obtained with hydrogen peroxide–persulfate binary mixture and with alkaline activation. The lower conversion achieved by these last methods was explained by the low concentration of hydroxyl radicals generated in the hydrogen peroxide-persulfate binary mixture system and the superoxide radicals predominated in the alkaline activated system, respectively, in agreement with the low oxidative potential (Eh), measured in the aqueous solution.

Ko et al. [105] have studied the oxidation of chlorinated ethanes and ethenes, forming a DNAPL, in soil samples from a fire training area. The oxidation systems tested were catalyzed hydrogen peroxide (CHP), activated persulfate by iron (AP), and the dual system hydrogen peroxide–persulfate (HP). In all cases, citrate was added or not as a chelating agent (CA). Runs were carried out in slurry systems and at batch operation, and reaction samples were analyzed at 48 h. The highest pollutant removal was obtained with CHP in the absence of CA at acid pH—around 2.5. In this case, the complete consumption of the oxidant was measured at the end of the experiment. When using the AP system, the oxidant conversion was lower than the obtained with CHP, but also a lower pollutant conversion was noticed in this case. Therefore, the addition of CA did not improve the removal efficiency in CHP system but increased the conversion of pollutant in the AP system. The dual system was more effective in pollutant removal at moderate concentrations of both oxidants. The addition of CA to the dual system decreased the pollutant removal and the oxidant consumption, related probably to the pH increase noticed when citrate (CA) was added. As a conclusion, the authors indicated that testing different oxidative conditions can be useful in order to determine the most optimal conditions for possible large-scale deployments.

Sra et al. [66] have investigated the persistence of activated persulfate in the subsurface, using citric acid (CA), chelated ferrous (Fe(II)), hydrogen peroxide (hydrogen peroxide), or alkaline ($OH^-$) activation in four well-characterized aquifer solids with low TOC content but a significant concentration of available Fe. They found that the chelated-Fe(II) and the hydrogen peroxide-activation showed an increase on the overall reduction in persulfate. At the conditions tested by these authors, chelation of Fe(II) by citric acid was not effective controlling the reaction between Fe(II) and persulfate. Consequently, a fast initial loss of persulfate took place. This can be explained taking into account that the addition of citric acid increased the unproductive consumption of PS. It should be noticed that citric acid produces an acidic pH and therefore, enhances the radical production rate and the persulfate consumption. If citrate was used, instead of citric acid, the pH could be kept in a neutral range, modulating the unproductive consumption of the oxidant by the CA [64].

Hydrogen peroxide was quickly decomposed due to the highly available iron content of the aquifer materials. Therefore, the effect of hydrogen peroxide on PS stability was not relevant. However, when hydrogen peroxide concentration increased, a high impact on persulfate stability was noticed. Authors conclude that, in general, a higher concentration of the Fe(II) or hydrogen peroxide, led to a higher depletion of persulfate.

However, the effect of the activators on the stability of PS depends on the persistence of these activators on the subsurface. A quick depletion of the activating agent, by interaction with the aquifer, will imply that the further decomposition of PS will correspond to unactivated persulfate degradation kinetics.

# 6 Conclusions

The application of in situ chemical oxidation (ISCO) technologies has proved to be a good alternative for the remediation of contaminated sites with organic compounds. Among them, persulfate (PS) technologies constitute an important development in the field of soil and groundwater remediation and the number of works in literature dealing with this topic, as well as its application to field scale, were continuously increasing in the last decade.

The main advantages of PS are the easy handling, the high aqueous solubility, high stability, relatively low cost, longer lifetime in the subsurface than hydrogen peroxide, and production of benign-end products, which makes it very competitive against other oxidants. PS can be applied "in situ" for groundwater remediation or "on site" for soil remediation after excavation or for the treatment of the groundwater pumped. It is effective for the treatment of a broad type of organic pollutants, including halogenated compounds (pesticides, PCBs, chloroalkanes, etc.), BTEXs, perfluorinated chemicals, phenols, PAHs, dyes, and pharmaceuticals. Moreover, PS can be applied to wastewater decontamination, sludge conditioning, and so on, although these last purposes are out of the scope of this chapter.

However, in spite of persulfate alone is a very powerful oxidant, its reaction with common organic contaminants is relatively slow and the activation of PS (AP) is recommended in order to produce reactive radicals such as sulfate, hydroxyl and superoxide, that considerably increase the oxidation rate of the pollutants. Because of the different reactivity of the formed radical in AP, a wide range of pollutants can be removed. Moreover, since a wide pH range can be used with AP, a high number of water matrices can be treated, which represents an important advantage over other treatments.

The main activation PS technologies include heating (thermal activation or TAP), the use of transition metals as $Fe^{2+}$ (Fe-AP), base addition (BAP) to get strong alkaline conditions, and hydrogen peroxide addition (HAP). Each method can produce different types of radical species and shows particular interactions with the soil and/or groundwater matrixes. In fact, the stability of the activators in the subsurface is a key point to determine the feasibility for their application and must be taken into account when choosing the most adequate method for each particular case.

The thermal activation of persulfate (TAP) has been carried out at temperatures usually above 30 °C. In aqueous phase, temperatures as high as 80 °C can be applied and high pollutant removal was noticed in short reaction times for a wide variety of organic contaminants. However, the maximum temperature allowed in the subsurface is much lower (40 °C) and the high cost associated to the energy requirements and the equipment needed to heat the aquifer can make unaffordable the application of this technology to the treatment of large groundwater plumes. When thermal activation is used, the type of radicals involved depends on the pH of the reaction media. At pH lower than 5 the sulfate radical is the predominant species, while at pH from 5 to 9 both hydroxyl and sulfate radicals contribute to the pollutant abatement. This method produces a significant unproductive consumption of PS, which increases significantly with increasing temperature, even in pure milliQ water. Moreover, in real water matrices, the scavenging effect of radicals by carbonates, bicarbonates, chlorides and NOM should be taken into account since these compounds compete with the pollutants for the oxidant. The scavenging of hydroxyl and sulfate radicals by bicarbonate is a relevant issue for the viability of groundwater remediation by this technology.

Activation by $Fe^{2+}$, Fe-AP, is an efficient and relatively cheap method to generate sulfate radicals. However, some important limitations should be taken into account for this method. The first concern is the pH of the media: to keep the iron in solution, the pH should be kept in the acid region. However, for an in situ treatment, this requirement is often difficult or impracticable since many soils and groundwater behave as a buffer system showing neutral–alkaline conditions. Moreover, an acid pH could yield non-desirable side effects as metal mobilization and loss of soil properties. Moreover, the adsorption of the external iron added into the soil has been noticed, even al low pH values. To avoid the acidification of the media and keep the iron in solution at neutral pH, chelating agents (CA) can be added. Contradictory effects on the use of CA on both pollutant removal and oxidant stability have been described in the literature. This fact can be probably explained considering that the CA is an organic molecule (EDTA, citric acid, and citrate are the most usually CA applied) that can compete with the pollutants for the oxidant. If the pollutants are more refractory towards oxidation than the CA, the oxidant will be consumed in unproductive reactions with the CA rather than in pollutant abatement. Moreover, at pH lower than 5—that is, when citric acid instead of citrate is used—the lower the pH, the higher the unproductive consumption of the oxidant. On the contrary, if pollutants are more easily oxidized than the CA, it will be noticed an enhancement in the pollutants oxidation rate due to a higher concentration of soluble iron in the media. Moreover, the use of CA allows extracting the labile iron from the soil avoiding the need of an external activator addition. Another relevant issue to take into account with iron activation of PS is that $Fe^{2+}$ acts as a reagent and not as a catalyst. Therefore, a higher amount of iron is needed in comparison with that required in Fenton's reagent, or different strategies to increase the regeneration of $Fe^{3+}$ to $Fe^{2+}$ must be implemented. Some pollutants (as quinone type compounds) have shown a particular activity in the reduction of $Fe^{3+}$.

Alkaline activation of PS (BAP), usually by sodium hydroxide, promotes the formation of both hydroxyl and superoxide radicals at pH > 10–11. This pool of radicals is able to react with a wide range of pollutants of different electronegativity, as TPHs, PAHs, and chlorinated compounds.

Moreover, as carbonates and bicarbonates precipitate at high alkaline pH, their scavenging effect is minimized; therefore, this method can be successfully applied to soils with a high carbonate content. A disadvantage of the BAP is the high amount of base that is usually required and the high concentration of PS (much greater than the stoichiometric one) usually needed for the pollutant abatement. In this sense, 1 mole of PS generates 2 moles of sulfate radicals in TAP, 1 mole of sulfate radicals in Fe-AP and 0.5 mole of hydroxyl radical in BAP. However, when the interaction with the subsurface dramatically affects the stability of the oxidant system, in TAP and Fe-AP systems, the activation of PS by alkali highlights as the optimal strategy. On the other hand, the reactivity of superoxide radical towards some pollutants with high halide content, can sustain the application of this activation method. Another relevant issue is that at pH above 11–12, a simple alkaline hydrolysis can take place for some organic molecules, resulting in the breakdown of these organic compounds without the participation of radical species. Moreover, it seems that activation by highly alkaline conditions had a lower impact than the other activation strategies in soil–water systems.

The use of hydrogen peroxide as activator of PS in a dual oxidant system has been scarcely studied and the mechanism for the interaction between hydrogen peroxide and persulfate is still unclear. It seems that the behavior of hydrogen peroxide as activator will greatly depend on the soil characteristics. The behavior of HAP in aqueous phase, slurries or subsurface cannot be easily compared or predicted.

Many of the studies with AP have been carried out at lab scale with batch-operation mode and spiked soils. The activation efficiency in this laboratory experiments is usually higher than that found in the application to real aged polluted soils. Moreover, at field conditions, there are scavengers that have not been considered in the lab works. This can lead to an underestimation of the actual oxidant demand in the field. Therefore, it is greatly encouraged the study of real polluted soils, at column or pilot scale in order to evaluate the viability of the AP application at field scale.

Depending on the oxidation rate of the organic contaminants by PS, a specific contact time between the oxidant system and the pollutants will be required for a given concentration of the oxidant and the activator. Moreover, the stability of the oxidant system in the water matrix has to be taken into account in order to determine the PS decomposition rate and the oxidant concentration depletion with the distance from the injection point. In the case of in situ applications, both kinetic and transport issues, rule the injection strategy. Direct push or recirculation of the oxidant should be considered in order to complete the remediation goals.

**Acknowledgments** The authors acknowledge financial support from the Comunidad Autonoma of Madrid (Project S2013-MAE-2739 CARESOIL-CM) and from the Spanish MINECO (Project CTM2016-77151-C2-1-R). Carmen M. Dominguez acknowledges the Spanish MINECO for the "Juan de la Cierva" postdoctoral contract (FJCI-2016-28462).

## References

1. R.L. Siegrist, M. Crimi, T.J. Simpkin, in *In Situ Chemical Oxidation for Groundwater Remediation*, ed. by R. U. C. Herb Ward, (Springer, New York, 2011)
2. ITRC, *Technical and regulatory guidance for in situ chemical oxidation of contaminated soil and groundwater* (Interstate Technology and Regulatory Council, Washington, DC, 2005)
3. R. Baciocchi, L. D'Aprile, I. Innocenti, F. Massetti, I. Verginelli, Development of technical guidelines for the application of in-situ chemical oxidation to groundwater remediation. J. Clean. Prod. **77**, 47–55 (2014)
4. S.G. Huling, B.E. Pivetz, *In-Situ Chemical Oxidation* (Environmental Protection Agency, Washington DC Office of Water, 2006)
5. F.J. Krembs, R.L. Siegrist, M.L. Crimi, R.F. Furrer, B.G. Petri, ISCO for groundwater remediation: analysis of field applications and performance. Ground Water Monit. Remidiat. **30**, 42–53 (2010)
6. R. Baciocchi, Principles, developments and design criteria of in situ chemical oxidation. Water Air Soil Pollut. **224**, 1717 (2013)
7. J.T.V.S. de Albergaria, H.P.A. Nouws, *Soil Remediation: Applications and New Technologies* (CRC Press, Boca Raton, 2016)
8. Brown, R., In situ chemical oxidation: performance, practice, and pitfalls, in AFCEE Technology Transfer Workshop, San Antonio, 2003
9. D. Zingaretti, I. Verginelli, R. Baciocchi, Catalyzed hydrogen peroxide combined with CO2 sparging for the treatment of contaminated groundwater. Chem. Eng. J. **300**, 119–126 (2016)
10. D.A. House, Kinetics and mechanism of oxidations by peroxydisulfate. Chem. Rev. **62**, 185–203 (1962)
11. A. Tsitonaki, B. Petri, M. Crimi, H. Mosbæk, R.L. Siegrist, P.L. Bjerg, In situ chemical oxidation of contaminated soil and groundwater using Persulfate: a review. Crit. Rev. Env. Sci. Technol. **40**, 55–91 (2010)
12. S. Wacławek, H.V. Lutze, K. Grübel, V.V. Padil, M. Černík, D.D. Dionysiou, Chemistry of persulfates in water and wastewater treatment: a review. Chem. Eng. J. **330**, 44–62 (2017)
13. L.W. Matzek, K.E. Carter, Activated persulfate for organic chemical degradation: a review. Chemosphere **151**, 178–188 (2016)

14. I.A. Ike, K.G. Linden, J.D. Orbell, M. Duke, Critical review of the science and sustainability of persulphate advanced oxidation processes. Chem. Eng. J. **338**, 651–669 (2018)
15. M.A. Dahmani, K. Huang, G.E. Hoag, Sodium persulfate oxidation for the remediation of chlorinated solvents (USEPA Superfund Innovative Technology Evaluation Program). Water Air Soil Pollut. Focus. **6**, 127–141 (2006)
16. K.C. Huang, Z. Zhao, G.E. Hoag, A. Dahmani, P.A. Block, Degradation of volatile organic compounds with thermally activated persulfate oxidation. Chemosphere **61**, 551–560 (2005)
17. P. Neta, R.E. Huie, A.B. Ross, Rate constants for reactions of inorganic radicals in aqueous solution. J. Phys. Chem. Ref. Data **17**, 1027–1284 (1988)
18. R.E. Huie, C.L. Clifton, P. Neta, Electron transfer reaction rates and equilibria of the carbonate and sulfate radical anions. Int. J. Radiat. Appl. Instrum. Part C. Radiat. Phys. **38**, 477–481 (1991)
19. J. Ma, Y. Yang, X. Jiang, Z. Xie, X. Li, C. Chen, H. Chen, Impacts of inorganic anions and natural organic matter on thermally activated persulfate oxidation of BTEX in water. Chemosphere **190**, 296–306 (2018)
20. C. Tan, N. Gao, Y. Deng, N. An, J. Deng, Heat-activated persulfate oxidation of diuron in water. Chem. Eng. J. **203**, 294–300 (2012)
21. A. Santos, S. Rodríguez, F. Pardo, A. Romero, Use of Fenton reagent combined with humic acids for the removal of PFOA from contaminated water. Sci. Tot. Environ. **563**, 657–663 (2016)
22. Y. Qian, G. Xue, J. Chen, J. Luo, X. Zhou, P. Gao, Q. Wang, Oxidation of cefalexin by thermally activated persulfate: kinetics, products, and antibacterial activity change. J. Hazard. Mat. **354**, 153–160 (2018)
23. S. Norzaee, M. Taghavi, B. Djahed, F.K. Mostafapour, Degradation of penicillin G by heat activated persulfate in aqueous solution. J. Eviron. Manage. **215**, 316–323 (2018)
24. J.L. Wang, S.Z. Wang, Activation of persulfate (PS) and peroxymonosulfate (PMS) and application for the degradation of emerging contaminants. Chem. Eng. J. **334**, 1502–1517 (2018)
25. Y. Feng, Q. Song, W. Lv, G. Liu, Degradation of ketoprofen by sulfate radical-based advanced oxidation processes: kinetics, mechanisms, and effects of natural water matrices. Chemosphere **189**, 643–651 (2017)
26. J. Ma, H. Li, L. Chi, H. Chen, C. Chen, Changes in activation energy and kinetics of heat-activated persulfate oxidation of phenol in response to changes in pH and temperature. Chemosphere **189**, 86–93 (2017)
27. M. Ahmad, A.L. Teel, R.J. Watts, Mechanism of persulfate activation by phenols. Environ. Sci. Technol. **47**, 5864–5871 (2013)
28. K.E. Manz, K.E. Carter, Investigating the effects of heat activated persulfate on the degradation of furfural, a component of hydraulic fracturing fluid chemical additives. Chem. Eng. J. **327**, 1021–1032 (2017)
29. L. Wang, L. Peng, L. Xie, P. Deng, D. Deng, Compatibility of surfactants. and thermally activated persulfate for enhanced subsurface remediation. Env. Sci. Technol. **51**, 7055–7064 (2017)
30. S. Padmaja, P. Neta, R.E. Huie, Rate constants for some reactions of inorganic radicals with inorganic-ions - temperature and solvent dependence. Int. J. Chem. Kinet. **25**, 447–455 (1993)
31. C. M. Dominguez, M. A. Lominchar, F. Bertel, A. Romero, A. Santos, Eliminaci6n de HCHs y clorobencenos mediante oxidaci6n química in situ: persulfato activado t6rmicamente, in XIII Congreso Español de Tratamiento de Aguas (META), Leon, 2018
32. R.L. Johnson, P.G. Tratnyek, R.O.B. Johnson, Persulfate persistence under thermal activation conditions. Environ. Sci. Technol. **42**, 9350–9356 (2008)
33. J. Liu, Z. Liu, F. Zhang, X. Su, C. Lyu, Thermally activated persulfate oxidation of NAPL chlorinated organic compounds: effect of soil composition on oxidant demand in different soil-persulfate systems. Water Sci. Technol. **75**, 1794–1803 (2017)

34. H. Peng, W. Zhang, L. Xu, R. Fu, K. Lin, Oxidation and mechanism of decabromodiphenyl ether (BDE209) by thermally activated persulfate (TAP) in a soil system. Chem. Eng. J. **306**, 226–232 (2016)
35. L. Peng, D. Deng, M. Guan, X. Fang, Q. Zhu, Remediation HCHs POPs-contaminated soil by activated persulfate technologies: feasibility, impact of activation methods and mechanistic implications. Sep. Purif. Technol. **150**, 215–222 (2015)
36. L.B. Peng, D.Y. Deng, F.T. Ye, Efficient oxidation of high levels of soil-sorbed phenanthrene by microwave-activated persulfate: implication for in situ subsurface remediation engineering. J. Soils Sediments **16**, 28–37 (2016)
37. D. Zhao, X. Liao, X. Yan, S.G. Huling, T. Chai, H. Tao, Effect and mechanism of persulfate activated by different methods for PAHs removal in soil. J. Hazard. Mater. **254**, 228–235 (2013)
38. P.D. Goulden, D.H.J. Anthony, Kinetics of uncatalyzed peroxydisulfate oxidation of organic material in fresh-water. Anal. Chem. **50**, 953–958 (1978)
39. A. Tsitonaki, *Treatment Trains for the Remediation of Aquifers Contaminated with MTBE and Other Xenobiotic Compounds* (Department of Environmental Engineering. Technical University of Denmark, Denmark, 2008)
40. S. Thompson, J. Riggenbach, R.A. Brown, J. Hines, J. Haselow, Catalyzed persulfate remediation of chlorinated and recalcitrant compounds in soil, in Fifth International Conference on Remediation of Chlorinated and Recalcitrant Compounds, Monterey, www.battelle.org/bookstore. (Battelle Press, Columbus, 2006)
41. P.A. Block, R.A. Brown, D. Robinson Novel Activation Technologies for Sodium Persulfate In Situ Chemical Oxidation, in Proceedings of the Fourth International Conference on the Remediation of Chlorinated and Recalcitrant Compounds (Battelle Press, Columbus, 2004)
42. G.V. Buxton, T.N. Malone, G. Arthur Salmon, Reaction of SO4- with Fe2+, Mn2+ and Cu2+ in aqueous solution. J. Chem. Soc. Faraday Trans. **93**, 2893–2897 (1997)
43. P. Devi, U. Das, A.K. Dalai, In-situ chemical oxidation: principle and applications of peroxide and persulfate treatments in wastewater systems. Sci. Total Environ. **571**, 643–657 (2016)
44. R.M. Félix-Navarro, M. Heredia-Alarcón, S. Pérez-Sicairos, M.I. Salazar-Gastélum, A.F. Diaz, S.W. Lin, Kinetic parameter determination for MTBE degradation with persulfate and Ag+ ions. Rev. Mex. Ing. Quim. **16**, 873–882 (2017)
45. Z. Wei, T. Gao, J. Wang, H. Liu, C. Liu, J. Zhu, M. Chen, Mn(II)-activated persulfate for oxidative degradation of DDT. Fresenius Environ. Bull. **27**, 4598–4605 (2018)
46. N. Boulos, D. Carvel, J. Muessig, Ex Situ and in Situ Remediation with Activated Persulfate, Google Patents, 2008
47. A. Romero, A. Santos, F. Vicente, C. González, Diuron abatement using activated persulphate: effect of pH, Fe(II) and oxidant dosage. Chem. Eng. J. **162**, 257–265 (2010)
48. F. Vicente, A. Santos, A. Romero, S. Rodriguez, Kinetic study of diuron oxidation and mineralization by persulphate: effects of temperature, oxidant concentration and iron dosage method. Chem. Eng. J. **170**, 127–135 (2011)
49. C. Liang, C.J. Bruell, M.C. Marley, K.L. Sperry, Persulfate oxidation for in situ remediation of TCE. I. Activated by ferrous ion with and without a persulfate–thiosulfate redox couple. Chemosphere **55**, 1213–1223 (2004)
50. C. Liang, C.F. Huang, Y.J. Chen, Potential for activated persulfate degradation of BTEX contamination. Water Res. **42**, 4091–4100 (2008)
51. S. Rodriguez, A. Santos, A. Romero, F. Vicente, Kinetic of oxidation and mineralization of priority and emerging pollutants by activated persulfate. Chem. Eng. J. **213**, 225–234 (2012)
52. C.J. Liang, M.C. Lai, Trichloroethylene degradation by zero valent iron activated persulfate oxidation. Environ. Eng. Sci. **25**, 1071–1077 (2008)
53. S. Rodriguez, A. Santos, A. Romero, Oxidation of priority and emerging pollutants with persulfate activated by iron: effect of iron valence and particle size. Chem. Eng. J. **318**, 197–205 (2017)

54. F. Pardo, J.M. Rosas, A. Santos, A. Romero, Remediation of a biodiesel blend-contaminated soil with activated persulfate by different sources of iron. Water Air Soil Pollut. **226**, 1–12 (2015)
55. F. Pardo, A. Santos, A. Romero, Fate of iron and polycyclic aromatic hydrocarbons during the remediation of a contaminated soil using iron-activated persulfate: a column study. Sci. Tot. Environ. **566**, 480–488 (2016)
56. M. Peluffo, F. Pardo, A. Santos, A. Romero, Use of different kinds of persulfate activation with iron for the remediation of a PAH-contaminated soil. Sci. Total Environ. **563**, 649–656 (2016)
57. M.A. Al-Shamsi, N.R. Thomson, Treatment of organic compounds by activated persulfate using nanoscale zerovalent iron. Ind. Eng. Chem. Res. **52**, 13564–13571 (2013)
58. D. Han, J. Wan, Y. Ma, Y. Wang, Y. Li, D. Li, Z. Guan, New insights into the role of organic chelating agents in Fe(II) activated persulfate processes. Chem. Eng. J. **269**, 425–433 (2015)
59. C. Liang, C.J. Bruell, M.C. Marley, K.L. Sperry, Persulfate oxidation for in situ remediation of TCE. II. Activated by chelated ferrous ion. Chemosphere **55**, 1225–1233 (2004)
60. P.F. Killian, C.J. Bruell, C. Liang, M.C. Marley, Iron (II) activated persulfate oxidation of MGP contaminated soil. Soil Sediment Contam. **16**, 523–537 (2007)
61. A. Rastogi, S.R. Al-Abed, D.D. Dionysiou, Effect of inorganic, synthetic and naturally occurring chelating agents on Fe(II) mediated advanced oxidation of chlorophenols. Water Res. **43**, 684–694 (2009)
62. J. Anotai, W.S. Bunmahotama, M.-C.J.E.E. Lu, Oxidation of aniline with sulfate radicals in the presence of citric acid. Environ. Eng. Sci. **28**, 207–215 (2011)
63. G.E. Hoag, J. Collins, Soil remediation method and composition, Google Patents (2011)
64. F. Vicente, A. Santos, E.G. Sagüillo, Á.M. Martínez-Villacorta, J.M. Rosas, A. Romero, Diuron abatement in contaminated soil using Fenton-like process. Chem. Eng. J. **183**, 357–364 (2012)
65. D.Y.S. Yan, I.M.C. Lo, Removal effectiveness and mechanisms of naphthalene and heavy metals from artificially contaminated soil by iron chelate-activated persulfate. Environ. Pollut. **178**, 15–22 (2013)
66. K.S. Sra, N.R. Thomson, J.F. Barker, Stability of activated Persulfate in the presence of aquifer solids. Soil Sediment Contam. **23**, 820–837 (2014)
67. Y. Lei, H. Zhang, J. Wang, J. Ai, Rapid and continuous oxidation of organic contaminants with ascorbic acid and a modified ferric/persulfate system. Chem. Eng. J. **270**, 73–79 (2015)
68. F. Pardo, J.M. Rosas, A. Santos, A. Romero, Remediation of soil contaminated by NAPLs using modified Fenton reagent: application to gasoline type compounds. J. Chem. Technol. Biotechnol. **90**, 754–764 (2015)
69. M. Danish, X. Gu, S. Lu, X. Zhang, X. Fu, Y. Xue, A.S. Qureshi, The effect of chelating agents on enhancement of 1,1,1-trichloroethane and trichloroethylene degradation by Z-nZVI-catalyzed percarbonate process. Water Air Soil Pollut **227**, 301 (2016)
70. X. Fu, M.L. Brusseau, X. Zang, S. Lu, X. Zhang, U. Farooq, Q. Sui, Enhanced effect of HAH on citric acid-chelated Fe(II)-catalyzed percarbonate for trichloroethene degradation. Environ. Sci. Pollut. Res. **24**, 24318–24326 (2017)
71. H. Peng, L. Xu, W. Zhang, L. Liu, F. Liu, K. Lin, Q. Lu, Enhanced degradation of BDE209 in spiked soil by ferrous-activated persulfate process with chelating agents. Environ. Sci. Pollut. Res. **24**, 2442–2448 (2017)
72. S. Yu, X. Gu, S. Lu, Y. Xue, X. Zhang, M. Xu, Z. Qui, Q. Sui, Degradation of phenanthrene in aqueous solution by a persulfate/percarbonate system activated with CA chelated-Fe(II). Chem. Eng. J. **333**, 122–131 (2018)
73. Y. Wu, R. Prulho, M. Brigante, W. Dong, K. Hanna, G. Mailhot, Activation of persulfate by Fe (III) species: Implications for 4-tert-butylphenol degradation. J. Hazard. Mater **322**, 380–386 (2017)
74. H. Liu, T.A. Bruton, W. Li, J.V. Buren, C. Prasse, F.M. Doyle, D. Sedlak, L oxidation of benzene by persulfate in the presence of Fe (III)-and Mn (IV)-containing oxides: stoichiometric efficiency and transformation products. J. Environ. Sci. **50**, 890–898 (2016)

75. K. Manoli, G. Nakhla, A.K. Ray, V.K. Sharma, Enhanced oxidative transformation of organic contaminants by activation of ferrate (VI): possible involvement of FeV/FeIV species. Chem. Eng. J. **307**, 513–517 (2017)
76. V.K. Sharma, R. Zboril, R.S. Varma, Ferrates: greener oxidants with multimodal action in water treatment technologies. Acc. Chem. Res. **42**, 182–191 (2015)
77. S. Rodriguez, L. Vasquez, D. Costa, A. Romero, A. Santos, Oxidation of Orange G by persulfate activated by Fe(II), Fe(III) and zero valent iron (ZVI). Chemosphere **101**, 86–92 (2014)
78. R. Chen, J. Pignatello, Role of quinone intermediates as electron shuttles in Fenton and photoassisted Fenton oxidations of aromatic compounds. J. Environ. Sci. Technol. **31**, 2399–2406 (1997)
79. Y. Jin, S.P. Sun, X. Yang, X.D. Chen, Degradation of ibuprofen in water by Fe-II-NTA complex-activated persulfate with hydroxylamine at neutral pH. Chem. Eng. J. **337**, 152–160 (2018)
80. X. Wu, X. Gu, S. Lu, Z. Qiu, Q. Sui, X. Zang, M. Danish, Accelerated degradation of tetrachloroethylene by Fe (II) activated persulfate process with hydroxylamine for enhancing Fe (II) regeneration. J. Chem. Technol. Biotechnol. **91**, 1280–1289 (2016)
81. D. Han, J. Wan, Y. Ma, Y. Wang, M. Huang, Y. Chen, D. Li, Z. Guan, Y. Li, Enhanced decolorization of Orange G in a Fe(II)-EDDS activated persulfate process by accelerating the regeneration of ferrous iron with hydroxylamine. Chem. Eng. J. **256**, 316–323 (2014)
82. S. Rodriguez, L. Vasquez, A. Romero, A. Santos, Dye oxidation in aqueous phase by using zero-valent iron as persulfate activator: kinetic model and effect of particle size. Ind. Eng. Chem. Res. **53**, 12288–12294 (2014)
83. J. Bolobajev, N.B. Öncü, M. Viisimaa, M. Trapido, I. Balcıoğlu, A. Goi, Column experiment on activation aids and biosurfactant application to the persulphate treatment of chlorophene-contaminated soil. Environ. Technol. **36**, 348–357 (2015)
84. F. Jousse, O. Atteia, P. Höhener, G. Cohen, Removal of NAPL from columns by oxidation, sparging, surfactant and thermal treatment. Chemosphere **188**, 182–189 (2017)
85. V. Rybnikova, N. Singhal, K. Hanna, Remediation of an aged PCP-contaminated soil by chemical oxidation under flow-through conditions. Chem. Eng. J. **314**, 202–211 (2017)
86. M. Usman, O. Tascone, V. Rybnikova, P. Faure, K. Hanna, Application of chemical oxidation to remediate HCH-contaminated soil under batch and flow through conditions. Environ. Sci. Pollut. Res. **24**, 14748–14757 (2017)
87. O.S. Furman, A.L. Teel, M. Ahmad, M.C. Merker, R.J. Watts, Effect of basicity on persulfate reactivity. J. Environ. Eng. **137**, 241–247 (2011)
88. O.S. Furman, A.L. Teel, R.J. Watts, Mechanism of base activation of persulfate. Environ. Sci. Technol. **44**, 6423–6428 (2010)
89. C.J. Liang, J.H. Lei, Identification of active radical species in alkaline persulfate oxidation. Water Environ. Res. **87**, 656–659 (2015)
90. B.A. Smith, A.L. Teel, R.J. Watts, Identification of the reactive oxygen species responsible for carbon tetrachloride degradation in modified Fenton's systems. Environ. Sci. Technol. **38**, 5465–5469 (2004)
91. A.L. Teel, R.J. Watts, Degradation of carbon tetrachloride by modified Fenton's reagent. J. Hazard. Mater **94**, 179–189 (2002)
92. R.J. Watts, J. Howsawkeng, A.L. Teel, Destruction of a carbon tetrachloride dense nonaqueous phase liquid by modified Fenton's reagent. J. Environ. Eng. **131**, 1114–1119 (2005)
93. M. Hayyan, M.A. Hashim, I.M. AlNashef, Superoxide ion: generation and chemical implications. Chem. Rev. **116**, 3029–3085 (2016)
94. M.A. Lominchar, D. Lorenzo, A. Romero, A. Santos, Remediation of soil contaminated by PAHs and TPH using alkaline activated persulfate enhanced by surfactant addition at flow conditions. J. Chem. Technol. Biotechnol. **93**, 1270–1278 (2018)
95. M.A. Lominchar, S. Rodríguez, D. Lorenzo, N. Santos, A. Romero, A. Santos, Phenol abatement using persulfate activated by nZVI, H2O2 and NaOH and development of a kinetic model for alkaline activation. Environ. Technol. **39**, 35–43 (2018)

96. M.A. Lominchar, A. Santos, E. de Miguel, A. Romero, Remediation of aged diesel contaminated soil by alkaline activated persulfate. Sci. Total Environ. **662**, 41–48 (2018)
97. S.Y. Oh, D.S. Shin, Remediation of explosive-contaminated soils: alkaline hydrolysis and subcritical water degradation. Soil Sediment Contam. **24**, 157–171 (2015)
98. A. Santos, J. Fernandez, S. Rodriguez, C.M. Dominguez, M.A. Lominchar, D. Lorenzo, A. Romero, Abatement of chlorinated compounds in groundwater contaminated by HCH wastes using ISCO with alkali activated persulfate. Sci. Tot. Environ. **615**, 1070–1077 (2018)
99. C.J. Liang, Y.Y. Guo, Remediation of diesel-contaminated soils using persulfate under alkaline condition. Water Air Soil Pollut. **223**, 4605–4614 (2012)
100. S. Waisner, V.F. Medina, A.B. Morrow, C.C. Nestler, Evaluation of chemical treatments for a mixed contaminant soil. J. Environ. Eng. ASCE **134**, 743–749 (2008)
101. M. Crimi, A comparison of methods to activate sodium persulfate for lindane destruction, Final project report prepared for FMC Corporation, Colorado School of Mines (2005)
102. Z. Liu, W. Guo, X. Han, X. Li, K. Zhang, Z. Qiao, In situ remediation of ortho-nitrochlorobenzene in soil by dual oxidants (hydrogen peroxide/persulfate). Environ. Sci. Pollut. Res. **23**, 19707–19712 (2016)
103. A.L. Teel, D.D. Finn, J.T. Schmidt, L.M. Cutler, R.J. Watts, Rates of trace mineral-catalysed decomposition of hydrogen peroxide. J. Environ. Eng. ASCE **133**, 853–858 (2007)
104. E. Ferrarese, G. Andreottola, I.A. Oprea, Remediation of PAH-contaminated sediments by chemical oxidation. J. Hazard. Mater. **152**, 128–139 (2008)
105. S. Ko, M. Crimi, B.K. Marvin, V. Holmes, S.G. Huling, Comparative study on oxidative treatments of NAPL containing chlorinated ethanes and ethenes using hydrogen peroxide and persulfate in soils. J. Environ. Manag. **108**, 42–48 (2012)

# Electro-Phytoremediation of Cropland and Mine Tailings Polluted by Mercury, Using $IrO_2$-$Ta_2O_5$/Ti Electrodes, *Lavandula vera*, and *Solanum tuberosum*

**D. I. Trejo, V. E. Herrera, S. Solís, M. V. Paz, L. Chávez-Guerrero, S. Sepúlveda-Guzmán, J. Manríquez, and E. Bustos**

## 1 Introduction

Soil pollution is currently a serious environmental issue, mostly because soil is considered a "universal sink" and bears the greatest burden of environmental pollution. Soil contamination is a problem at multiple levels: contaminants can end up in plants that are growing in the soil; groundwater that interacts with the soil becomes contaminated as a result of the soil contamination; and animals (including humans) that eat the vegetation growing in the soil can absorb its contaminants [1, 2]. Human activities such as agriculture, mining, and industrial activities typically lead to contamination of soils in many ways. Among the most common is deposition of heavy metals (e.g., Cd, Hg, Ni, Cu, and Cr), inorganic compounds (e.g., $F^-$, $CN^-$, and arsenic compounds), and a wide variety of organic compounds (e.g., hydrocarbons; benzene, toluene, ethylbenzene, and xylene isomers (BTEX); polycyclic aromatic hydrocarbons (PAHs); polychlorinated biphenyls (PCBs); pesticides; and energetic compounds) [3]. In the case of heavy metals, contamination of soil is a major problem due to potential biohazards associated with sites contaminated by humans, mainly because of the potential health impacts of consumption of contaminated produce [4, 5]. Heavy metals occur naturally in soil as a result of weathering of parent materials at levels that are regarded as trace levels (less than 1000 mg $kg^{-1}$) [6].

D. I. Trejo · V. E. Herrera · M. V. Paz · J. Manríquez · E. Bustos (✉)
Centro de Investigación y Desarrollo Tecnológico en Electroquímica, S.C. Parque Tecnológico Querétaro, Querétaro, Mexico
e-mail: ebustos@cideteq.mx

S. Solís
Centro de Geociencias, Universidad Nacional Autónoma de México, Querétaro, Mexico

L. Chávez-Guerrero · S. Sepúlveda-Guzmán
Facultad de Ingeniería Mecánica y Eléctrica (FIME), Universidad Autónoma de Nuevo León, San Nicolás de los Garza, Mexico

 
M. A. Rodrigo, E. V. Dos Santos (eds.), *Electrochemically Assisted Remediation of Contaminated Soils*, Environmental Pollution 30,
https://doi.org/10.1007/978-3-030-68140-1_11

As a consequence, different thermal, biological, and chemical techniques to remove pollutants from soil have been explored. Of particular interest is phytoremediation (PhyR), a process in which plants and their associated microorganisms can be used to remove, degrade, or isolate toxic substances in the environment. This type of process has been employed to remediate both soil and water pollution [7, 8]. Some plants can accumulate various organic or inorganic chemicals from environmental media and degrade (or otherwise process) them by assimilation and use in their physiological processes. The different PhyR processes listed in Table 1 are natural ways in which plants and microbes in the rhizosphere degrade or stabilize pollutants [9, 10].

In the specific case of heavy metals in soil, they can be lixiviated in surface water and groundwater, absorbed by plants, and incorporated into the trophic chain. When they are present in huge concentrations, they may be bioavailable as essential elements (e.g., Cu, Zn, Mn, Fe, Ni, and Mo) or nonessential elements (e.g., Cd, Pb, Cr, and Hg), causing problems because of their toxicity in the environment. In this situation, depollution of heavy metals in soil can be accomplished using phytoextraction, phytostabilization, or phytoimmobilization processes (Table 1) [9–19].

These PhyR processes have been studied in relation to bioabsorption, bioaccumulation, and bioconcentration coefficients [20]. PhyR can be an in situ cost-effective technology for treatment of mine tailings by introduction of pollutant-tolerant plant species [11], particularly in comparison with more

**Table 1** Different phytoremediation mechanisms for transformation or elimination of pollutants from soil [9–16]

| Process | Mechanism |
|---|---|
| Phytoextraction | Plant roots absorb and remove pollutants from the soil by transportation and accumulation in their roots, stems, and leaves. Once the plants are fully grown, they are harvested and either composted or incinerated (followed by removal of the ashes to a disposal site). This process is used in phytoaccumulation, phytoabsorption, phytosequestration, and phytodesalination |
| Phytovolatilization | Pollutants can be transported via the radicular system to the plant surface, where they are degraded and transformed or merely volatilized and released into the atmosphere |
| Phytodegradation | Plants take up pollutants and metabolize them into materials without environmental risk by breaking down and destroying the pollutants. Associated microorganisms and plant enzymes degrade the pollutants within the plant tissues. This process is used for phytotransformation of highly toxic contaminants into less toxic chemical compounds |
| Rhizodegradation | In the plant root zone (rhizosphere), pollutants are removed from flowing water and transformed or destroyed in the radicular zone through the action of rhizospheric microorganisms, which help to degrade organic xenobiotic compounds |
| Phytostabilization | Plant roots sequester and stabilize pollutants. In general, this process uses pollutant-tolerant plants to stabilize the pollutants by reducing their mobility and bioavailability in the environment. This process is used in phytoimmobilization of pollutants |

conventional methods of physical treatment (such as excavation and landfilling), thermal treatment (such as incineration and desorption), and physicochemical treatments (such as soil washing, leaching, vitrification, and electrokinetic treatment (EKT)) [14, 21, 22].

Different metals and metalloids need to be removed from soil, such as mercury (Hg). The main problem of mercury contamination is the different oxidation states in which it can exist in the environment [23]. Environmental mercury contamination is a result of poor mining practices. Mercury has a high market value, but if its waste is treated poorly, it may cause pollution and damage in the environment, as well as posing risks to human health. Mercury occurs naturally in concentrations ranging between 0.003 and 4.6 mg $kg^{-1}$; however, at contaminated sites, the observed concentrations can be between 11,500 and 14,000 mg $kg^{-1}$ [24, 25] and it can be found in different chemical forms: elemental (metallic ($Hg^{\circ}$)), inorganic ($Hg^{2+}$), and organic forms (HgS, HgCl, and $HgCH_3$) [26].

Mercury contamination is not simply a local problem in the vicinity of mine tailings: according to modeling results published by Tørseth [27], mercury compounds can remain in the atmosphere for up to 2 years and can be transported over large distances, extending their harmful impacts to the global scale. Mercury is a heavy metal, is highly toxic to living beings, and can be bioaccumulated in organisms without degradation of its overall system [28, 29]. In soils, mercury can persist for long periods, and it can be transported into the atmosphere, hydrosphere, and biosphere many years after its initial deposition as waste [30]. Hg can be removed by PhyR (Table 2)—specifically by phytovolatilization ($[Hg]_{soil}$ = 0.55–1605 mg $kg^{-1}$; $\eta$ = 77–81%), usually after methylation. During this process, plants convert methylmercury or dimethylmercury (through the action of anaerobic bacteria in the soil) into ionic $Hg^{+2}$ via organomercurial lyase, and then into nonreactive elemental $Hg^{\circ}$ via mercuric reductase, whereupon it can be volatilized [14, 15, 21, 32, 68–74]. Phenylmercury can also be microbially converted into diphenylmercury [15, 75, 76]. There have been reports describing engineering of transgenic plants with the ability to phytoaccumulate and/or phytostabilize Hg [14, 21, 22, 77, 78].

In the case of PhyR of soils polluted by mercury, *Lavandula vera* and *Solanum tuberosum* have proved to be very useful as hyperaccumulative plants with adequate removal efficiency (>80%). Considering bioelectricity, living things are analogous to an electrolyte container filled with millions of small chemical batteries [17]. Many renewable and sustainable energy sources have been developed in recent years in attempts to offset the use of fossil fuels, and use of plants to generate electricity is now being investigated [18], which is a potential side benefit of PhyR.

Additionally, EKT has been applied successfully in a variety of polluted soils. This technology refers to application of an electrical field or direct current through a pair of electrodes (an anode and a cathode). These electrodes are inserted into the soil, where an electrolyte improves the conductive properties of the electrical field [79, 80]. EKT involves different mass transport mechanisms, as shown in Table 3. During EKT, electrolysis of water occurs at the electrodes, producing hydrogen ($H_2$) and protons ($^{+}H$) close to the anode, and producing oxygen ($O_2$) and hydroxyl ions ($^{-}OH$) close to the cathode, and thereby generating acid and base frontiers, respectively [16, 36, 81–84].

**Table 2** Plant species used in phytoremediation of mercury-contaminated soil, as reported in the literature

| Plant species | References |
|---|---|
| *Alfalfa* and black locust | [31] |
| *Alyssum* flowers | [31] |
| *Arabidopsis thaliana* | [12, 22, 32–35] |
| *Artemisia douglasiana* | [14, 36, 37] |
| *Armoracia lapathifolia* | [36, 38] |
| *Brassica juncea* | [36, 39–45] |
| *Caulanthus* sp. | [14, 36, 37] |
| *Cestrum buxifolium* | [12, 46, 47] |
| *Chenopodium glaucum* L. | [36, 37, 45] |
| *Enterobacter cloacae* | [37] |
| *Eucalyptus globulus* | [14, 36, 37, 48] |
| *Euglena gracilis* | [49] |
| *Festuca rubra* | [36, 38] |
| *Fragaria vesca* | [14, 36, 37] |
| *Halimione portulacoides* | [50] |
| *Helianthus tuberosus* | [36, 38] |
| Indian mustard | [31] |
| *Juncus maritimus* | [36, 51, 52] |
| *Lactuca sativa* ssp. *capitata* | [39, 43] |
| *Lepidium latifolium* | [14, 36, 37] |
| *Ludwigia peploides* | [12, 53–55] |
| *Nicotiana tabacum* | [12, 32, 34, 46, 56] |
| Orange pulp | [39, 44] |
| *Oryza sativa* | [39, 41, 42] |
| Pennycress | [31] |
| *Phaseolus vulgaris* ssp. *nanus* | [36, 39, 43, 45] |
| *Phragmites australis* | [57, 58] |
| *Physalis peruviana* | [12, 59, 60] |
| *Poa pratensis* | [36, 38] |
| Poplar | [31] |
| *Portulaca oleracea* | [12, 61, 62] |
| *Pseudomonas fluorescens* | [63, 64] |
| *Salix viminalis* | [14, 64, 65] |
| *Salix schwerinii* | [14, 64, 65] |
| *Salvinia auriculata* | [36, 66] |
| *Sarcocornia fructicosa* | [50] |
| *Silene vulgaris* | [36, 67] |
| *Spartina maritima* | [50] |
| *Taraxacum officinale* | [12, 60] |
| *Triticum aestivum* | [15] |

**Table 3** Different mass transport mechanisms that transform or eliminate pollutants in soil during electrokinetic treatment [16, 36, 81–84]

| Process | Mechanism |
|---|---|
| Electromigration | Movement of ions in a solution. In this process, a current is carried by ions, with anions being transported toward the anode and cations being transported toward the cathode |
| Electrophoresis | Movement of charged, dissolved, or suspended particles in a pore fluid. The negative charge on the surface of most soil particles causes accumulation of positively charged cations near particle surfaces in the diffuse electrical double layer. Under the action of the electrical field, these cations provide a net flow of ions in the direction of the cathode and, through this action, water also moves toward the cathode |
| Electro-osmosis | Bulk movement of fluid though pores. This process generates movement of chemical compounds without a charge (such as organic matter), microorganisms, or surfactants in the soil during application of an electrical field |

Enhancing solutions can be added to the soil to improve the efficiency of the treatments [85, 86]. The most commonly used enhancing solutions are ethylenediaminetetraacetic acid (EDTA), KI, and NaCl [87, 88]. EDTA is a compound with four carboxylic and two associated amino groups, which can act as electron pair donors or Lewis bases. In an aqueous solution, EDTA forms an octahedral complex with mostly divalent metal cations ($M^{2+}$). A very strong mercury–EDTA ($[Hg\text{-}EDTA]^{2-}$) complex is formed, with a log *b* value of 21.8 [89]. Considering the negative charge of this complex, most mercury moves toward the anode side, where it can be removed or recovered [90]. The key element is electrokinetic mobilization of metals in the soil, which increases the availability of the metals to plant roots and therefore increases the absorption of the metals by plants and their subsequent extraction using PhyR [91].

The problem of mercury contamination is exacerbated when polluted croplands and mining areas close to people are repurposed, introducing mercury into the food chain, where it can cause serious damage to human health [23]. EKT and PhyR are different remediation techniques used to remove mercury from cropland and mine areas. EKT has been considered the most efficient alternative to treat soil polluted by mercury in some mining areas, because laboratory test results suggest that it can remove more than 75% of mercury from polluted soil within 72 h of treatment [92]. Table 4 lists different EKT processes that have been studied for removal of mercury from soil. It is necessary to add chloride, oxidizing agents, and chelates to the soil in order to mobilize the mercury and increase the rate of its removal [36, 97].

A technique combining EKT and PhyR (EKT + PhyR) has been developed for electro-phytoremediation of contaminated soil. In this technique, a low-intensity electrical field is applied to contaminated soil in the vicinity of growing plants, which increases the bioavailability of the pollutants close to the plant roots, generating hyperaccumulation of the pollutants in the plants through the different EKT and PhyR mechanisms listed in Tables 1 and 3 [3].

**Table 4** Experimental conditions used in electrokinetic treatment (EKT) to reduce soil mercury concentrations ($[Hg]_{soil}$), and efficiency of mercury removal from polluted soil ($\eta$), as reported in the literature

| Experimental conditions | $[Hg]_{soil}$ or $\eta$ | References |
|---|---|---|
| Complexing agent: different ligands such as $HO^-$, $Cl^-$, and $I^-$, as well as the chelating agent EDTA<br>Electrical field: 1 V $cm^{-1}$<br>Soil type: spiked kaolin and glacial till soils | $\eta = 97\%$ from kaolin<br>$\eta = 56\%$ from glacial till | [93] |
| Extracting agent: $I_2/I^-$ | $\eta = 99\%$ | [94] |
| Electrodes: approaching cathodes<br>Extracting agent: $I_2/I^-$<br>Soil type: Wanshan mercury mine<br>EKT duration: 5 days | $\eta = 89–92\%$ | [95] |
| Cathode: Ti<br>Anode: Ti<br>$E_{dc}$: 0.4, 0.7, 1.4, 1.6, 1.9, 2.0 and 5.0 V<br>Complexing agent: 0.1 M EDTA<br>EKT duration: 1–72 h | $[Hg]_{soil} = 30$ mg $kg^{-1}$<br>$\eta = 75\%$ | [37, 80, 96] |

$E_{dc}$ potential with direct current
*EDTA* ethylenediaminetetraacetic acid

In this research, EKT + PhyR, using *Lavandula vera* and *Solanum tuberosum*, was evaluated for its ability to increase the efficiency of $Hg^{2+}$ removal from cropland, mine tailings, and control soils. Additionally, this research evaluated the potential of potato tubers to be used as a candidate plant to remove $Hg^{2+}$ from either liquid or solid matrices, and as a "biopile" for generation of small amounts of electrical energy. Because it is necessary to look toward "greener" and less expensive forms of energy to meet the power demands of the growing world population, scientists around the globe are investigating different energy sources such as air, water, and vegetal sources. Bioelectricity is fundamental to all life processes. Biopotentials result from complex biochemical processes, which are intimately associated with transfer of electrical charges. From the standpoint of bioelectricity, living things are analogous to an electrolyte container filled with millions of small chemical batteries [17]. Many renewable and sustainable energy sources have been developed in recent years in attempts to combat pollution from fossil fuels, and generation of electricity from living plants is now under investigation [18, 98].

We proposed a model in which potato tubers could be used in the EKT + PhyR process, involving use of plants coupled with electrokinetic methods to potentiate the extent of $Hg^{2+}$ removal from contaminated soils by different parts of *Lavandula vera,* mainly its leaves. These techniques were subsequently coupled with the possibility of generating bioelectricity with *Solanum tuberosum,* which is an excellent electrical conductor because of its solid matrix, its high water level (80%), its elevated content of starch and potassium, and the large size of its tubers in comparison with other vegetables [99].

## 2 Methodology

Three sites were chosen for soil sampling: a cropland in the vicinity of an abandoned mining site, a deposition site for mining waste, and a reference site with no expected contamination. The samples from the cropland and mining waste deposition sites were obtained near the Otatal mercury mine, located at the coordinates 20°56′51.23″ N and 99°32′54″ W, in the municipality of San Joaquín in the state of Querétaro, Mexico. The three different types of soil sample were collected from the following areas: (1) cropland at 20°56′38.46″ N and 99°32′42.63″ W, at an altitude of 1941 meters above sea level (masl), on a farming site (located at the periphery of the Otatal mine) that had been used for growing apples, pumpkins, corn, and beans with temporary irrigation, and for producing compost from cattle manure; (2) a mine-tailing site at 20°56′31.70″ N and 99°32′59.36″ W, at an altitude of 2045 masl, which contained residue from cinnabar (HgS) incineration, deposited at the periphery of the mine without any further treatment, some of it dating back more than 50 years; and (3) a distant (control) site at 20°56′57.97″ N and 99°33′00.60″ W, at an altitude of 2336 masl, which was not expected to contain high levels of mercury.

All samples were stored in tightly sealed plastic containers during their transfer. Pedological characterization of each sample in this study was carried out in accordance with the methods described in Official Mexican Standard NOM-021-SEMARNAT-2000 [100] and the International Soil Reference and Information Center (ISRIC) *Procedures Manual* [101]. The physicochemical characteristics of the samples were determined using the same methods as those used for the soil analyses.

After transport to the laboratory, one part of each sample was refrigerated for a period of 15 days prior to its use in the EKT, PhyR, and EKT + PhyR processes. Another part was dried at room temperature in a designated area of the laboratory for subsequent physicochemical characterization. To monitor the concentrations of mercury ($Hg^{2+}$) in solution, the technique of anodic stripping voltammetry (ASV) [92] was used. Two calibration curves were constructed: one with EDTA and another with HCl.

ASV was performed using an Epsilon potentiostat (BASi, West Lafayette, IN, USA). The resulting data were used to construct a calibration curve, using 1 M HCl as a support electrolyte and 1 mM $HgCl_2$ as the precursor of $Hg^{2+}$. The electrochemical cell had working, reference, and counter electrodes of vitreous carbon (VC), Ag/AgCl, and Pt wire, respectively, which contained 10 mL of 0.1 mM EDTA as a support electrolyte. Between ASV tests, the VC electrode was polished with alumina (1.0, 0.3, and 0.05 μm). The polished electrode was then rinsed with deionized water before being introduced into the electrochemical cell. The deposition time for ASV was 300 s, with a standing time of 40 s. The potential of the deposit was −600 mV versus Ag/AgCl, using a potential window from −500 to 600 mV versus Ag/AgCl. Linear voltammetry was obtained from each addition of an aliquot of 1 mM $HgCl_2$ to 0.1 M HCl and 0.1 mM EDTA.

Calibration curves were constructed using the current values found after each $HgCl_2$ addition (Fig. 1). These calibration curves were then used for quantification of $Hg^{2+}$ concentrations in the various analytical samples. Mercury concentrations in different parts of *Lavandula vera* and *Solanum tuberosum* were determined by mercury atomic absorption spectroscopy (HgAAS), using methods 3051 and 3120 B from Official Mexican Standard NOM-004-SEMARNAT-2002 [101–104].

EKT of the $Hg^{2+}$-contaminated soil, employing a two-dimensional circular electrode array, was implemented (Fig. 2). The anode (Ti) was placed at the center of the cell and six cathodes (Ti) were set up around it at a separation distance of 6 cm. They were 6 cm in length and 0.5 cm in diameter (Torres Estructuras y Postes de Puebla SA de CV (TEPSA), Mexico City, Mexico). In the center of the electrochemical cell, two pieces of polyvinyl acetate (PVC) were placed as spacers in the anodic area. The resulting solution was subsequently collected from the area between them [90, 105].

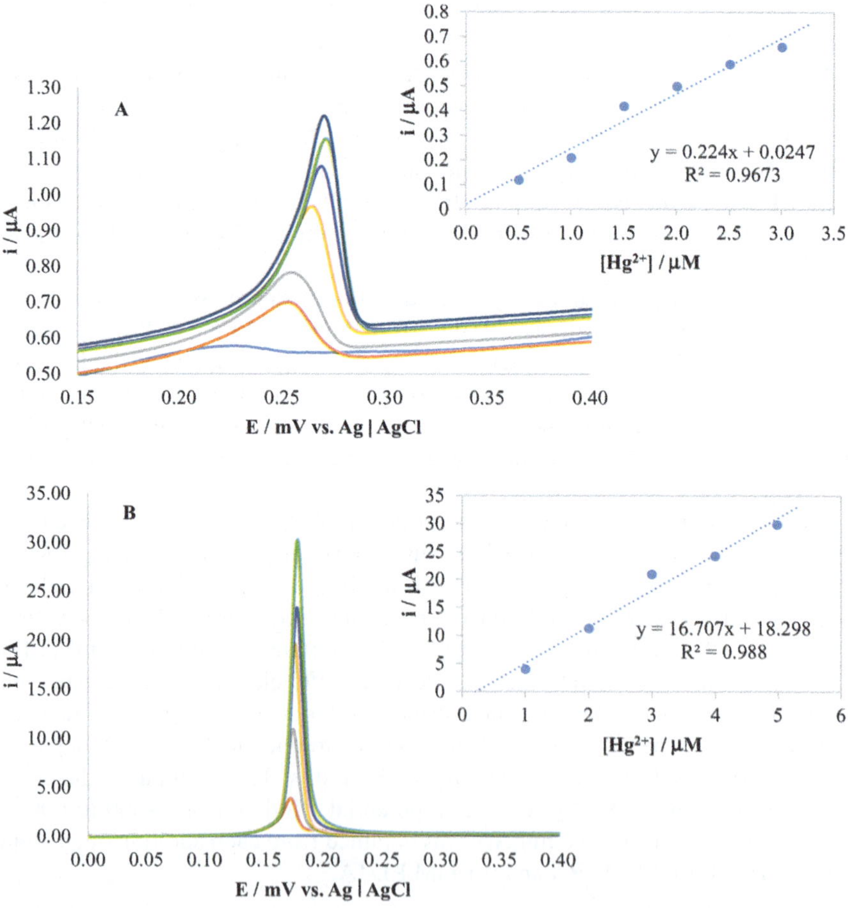


**Fig. 1** Anodic stripping voltammetry (ASV) of $HgCl_2$ in 0.1 M ethylenediaminetetraacetic acid (EDTA) (**A**) and 0.1 M HCl (**B**), with their corresponding calibration curves

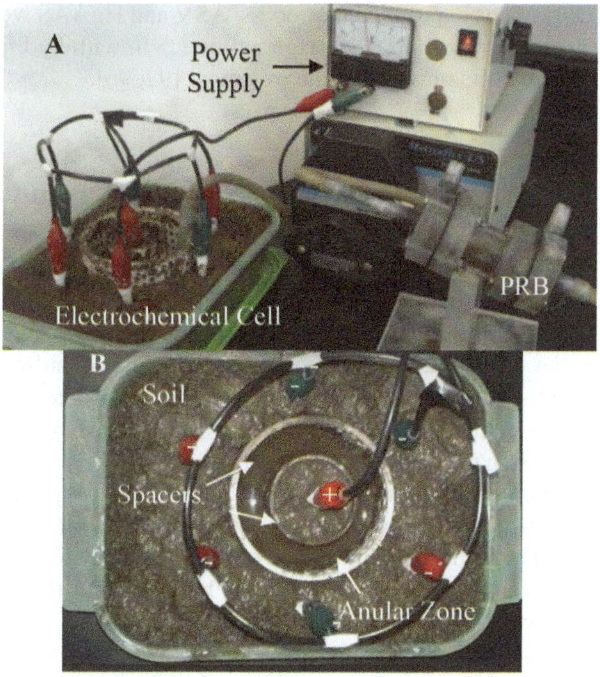


**Fig. 2** Electrokinetic system of soil (**A**) with a circular arrangement of electrodes in the soil. At the center is the anode (+), surrounded by six cathodes (−) in the ring zone, delimited by circular polyvinyl chloride spacers in the anodic area (**B**). Wastewater containing ethylenediaminetetraacetic acid ions is pumped out to the permeable reactive barrier (PRB) for cleaning

EKT was performed on all three sample types: cropland soil, mine-tailing soil, and off-site control soil. The EKT process used samples weighing 1.2 kg. They were saturated with 350 mL of 0.1 M EDTA for 24 h, and then the circular PVC spacers were inserted into the center of each system [105]. Later, electrodes were inserted into each sample and connected to a GP-4303DU direct current (DC) power supply (EZ Digital Company, Busan City, Korea), with application of 20 V for 90 min. Meanwhile, the electrical current was measured using a multimeter (Steren, San Diego, CA, USA), at 0, 30, 60, and 90 min during the cycle. After 90 min, the experiment was stopped and the aqueous solution was collected and removed using a Masterflex peristaltic pump (Cole-Parmer, Vernon Hills, IL, USA).

The polluted solution was pumped through a reactor containing permeable reactive barriers (PRBs) made of different materials. These materials contained iron ($Fe^{\circ}$) and carbon (C): $Fe^{\circ}$ (Aldrich, St. Louis, MO, USA), chemically reduced graphene (QRGO), *Agave* biomass sponge (CGOO), C tissue (0.5 mm) (Alfa Aesar, Ward Hill, MA, USA), C fiber (1.12 cm) (Alfa Aesar), chemically reduced $Fe^{\circ}$ + C (FeQRGO), and thermically reduced $Fe^{\circ}$ + C (FeTRGO) to treat 250 mL of $Hg^{2+}$-contaminated solution (Fig. 2a). The final solution that was obtained was collected by pumping (at 120 mL $min^{-1}$ with a peristaltic pump) for analysis by ASV [90].

The different PRB materials were analyzed by scanning electron microscopy (SEM) (Bruker, Billerica, MA, USA). ASV was used to evaluate the efficiency of mercury removal from samples of wastewater that had passed through the PRBs.

During this experimental process, ASV and HgAAS were used to analyze the aqueous samples of $[EDTA—Hg]^{2-}$ after 1.5 h, with hydride generation, and then this solution was returned to the treatment [92, 96].

## 2.1 *Phytoremediation of Soil with* Lavandula vera

Propagation of seedlings of *Lavandula vera* was accomplished through a process that started with cuttings obtained from a primogenital plant. The cuttings were subsequently placed singly in a germination box, using peat moss as a growth substrate. This type of substrate is used for seeds or cuttings to achieve optimal penetration for germination and to help them properly take root, since it provides moisture, temperature, and porosity suitable for plant development.

During the tests, 500 g of dry and sieved soil samples were placed in a 10 × 12 cm plastic container wetted with tap water. The samples were placed inside a growth chamber illuminated with artificial light for 8 h followed by 16 h of darkness per day. The cuttings were watered every second day with running water. The *Lavandula vera* samples were matured in the box from the time of transplantation until their roots grew, which took between 21 and 25 days.

Once the cuttings developed roots, the PhyR process began with transplantation into the different soil types and continued with assessment of their development and growth for 5 weeks. At the end of the treatment, the mercury concentrations in the different parts of the plants were analyzed; additionally, the final mercury concentration in each soil sample was tested. The rooted plants were distributed into one of the containers with the test soil samples. A total of 12 *Lavandula vera* plants were placed in each type of soil sample: mine-tailing soil, adjacent cropland soil, and remotely sourced soil. The plants were watered every second day with running water and were kept in the growth chamber again, receiving 8 h of light and 16 h of darkness daily during the 5-week experiment.

During the testing period, the growth of *Lavandula vera* was assessed weekly by use of a vernier caliper to measure the stem diameter. Every seventh day, a plant was taken from each soil sample container and then dried at room temperature and dry weighed. With these data, the rates of growth and the tolerance of the pollutant were determined for *Lavandula vera*.

After completion of the EKT processing of the soil samples, *Lavandula vera* cuttings were transplanted into the soils. To evaluate the effect of EDTA as a support electrolyte in the EKT, a series of experiments was developed in which $Hg^{2+}$ was introduced in different concentrations (0.001, 0.01, 0.1, and 1.0 mM) into 100 mL of EDTA. These mixtures were introduced into 30 g of substrate (peat moss) in five different individual pots and held for 24 h before rooted cuttings of *Lavandula vera* were transplanted, one into each individual pot (Fig. 3). All plants were watered with tap water every third day, were kept in the growth chamber, and received 8 h of daylight and 16 h of darkness each day.

**Fig. 3** *Lavandula vera* growth without electro-phytoremediation (*A1* and *A2*) and with electro-phytoremediation (*B1* and *B2*) in cropland soil (*A1* and *B1*) and mine-tailing soil (*A2* and *B2*)

## *2.2 Phytoremediation of Soil with* Solanum tuberosum

Tubers of the Alpha potato (*Solanum tuberosum*) variety were pretreated before being planted in electroremediated soil polluted by mercury to induce phytoremediation of the soil by the potatoes. During this study, three different potato groups were compared: (1) electroporated potatoes, (2) boiled potatoes, and (3) raw potatoes.

Electroporation was achieved by application of electrical pulses, which allowed observation of intracellular and intercellular communication within plant tissues [106]. Electroporation creates openings in cell membranes, resulting in a layer of micropores, which increase the presence of ions and the electrical conductivity of the tissue [107]. The objective of performing electrophoresis on the potatoes was to apply an alternating (changing) electrical field in order to create pores in their cell membranes and allow greater flow of electricity and phosphoric acid [108].

The opening of these pores in the potato cell membranes allowed evaluation of the effect that would occur during immersion in $HgCl_2$ solution, so that we could determine if this condition was useful for accumulation and extraction of mercury from soil in mining areas. In this process, the first reaction that happened was decomposition of organic material, since the potato tubers contained large amounts of nutrients and phosphoric acid [109, 110], allowing transport between Cu/Zn electrodes, which occurred as shown in the following reaction under conditions unaffected by any external activity other than exposure to the Cu/Zn electrodes:

$$Zn^{\circ} \rightarrow Zn^{+2} + 2e^{-}$$

$$Cu^{+2} + 2e^{-} \rightarrow Cu^{\circ}$$

$$2H_3PO_4 \rightarrow 3H_2 + 2PO_4^{-2}$$

$$Zn^{\circ} + 2H_3PO_4 + Cu^{2+} \rightarrow Zn^{2+} + Cu^{\circ} + 3H_2 + 2PO_4^{-2}$$

The condition of boiled potatoes was used in order to determine what occurred before any physicochemical change caused by external factors and the way in which the potato behaved in contact with the solution. This potato condition ensured that with the passage of time, the solution would not be affected in terms of smell and color, as occurred with the raw potatoes. This was because boiled foods go through a process of solubilization of components as a result of the Maillard reaction, which occurs during the cooking process. In this process, the permeability of the vegetable tissue changes and pores are opened; moreover, some components such as starch are transformed into another component called acrylamide, resulting in a characteristic smell [111, 112].

To cause irreversible electroporation, an HP 3310A function generator (Hewlett Packard, Palo Alto, CA, USA) was used to treat some potato tubers with a 2 V, 500 Hz alternating current (AC) applied for 1 h with two electrodes (Cu and Zn) inserted 0.5 cm deep and 1 cm apart [17]. These potatoes constituted the "electroporated" treatment. Other potato tubers were left untreated (or fresh) and constituted the "raw" treatment [108, 113] (Fig. 4).

ASV was performed to detect the amount of $Hg^{2+}$ using 10 mL of each solution. Additional measurements of pH and electrical conductivity (EC) were performed on the solutions, using an HI 2550 pH meter (Hanna Instruments, Woonsocket, RI, USA) and a YSI 3200 conductivity instrument (YSI, Yellow Springs, OH, USA). Additionally, information on the color and smell of the potato tubers was recorded to determine how the tuber–solution interaction was progressing.

At the end of the 10 days of analysis, the potato tubers were removed from the $HgCl_2$ solution and placed in an open controlled environment to dry at room temperature, protected from sunlight. The dried tubers were then weighed and submitted to acid digestion for measurement of their final $Hg^{2+}$ concentrations, in accordance with Official Mexican Standard NOM-147-SEMARNAT-SSA1-2004, in each of the studied conditions [114]. These tests were performed as follows: 0.5 g of dried potato tuber was cut, weighed, and placed in a Teflon vessel with deionized water (2 mL), concentrated $HNO_3$ (4 mL), and concentrated HCl (1 mL). The Teflon vessels were then placed in a MARS microwave digestion system (CEM, Charlotte, NC, USA) with use of the following experimental conditions: heating to 150–200 °C at 800 W for 20 min, with application of the biosolids method. At the end of the acid

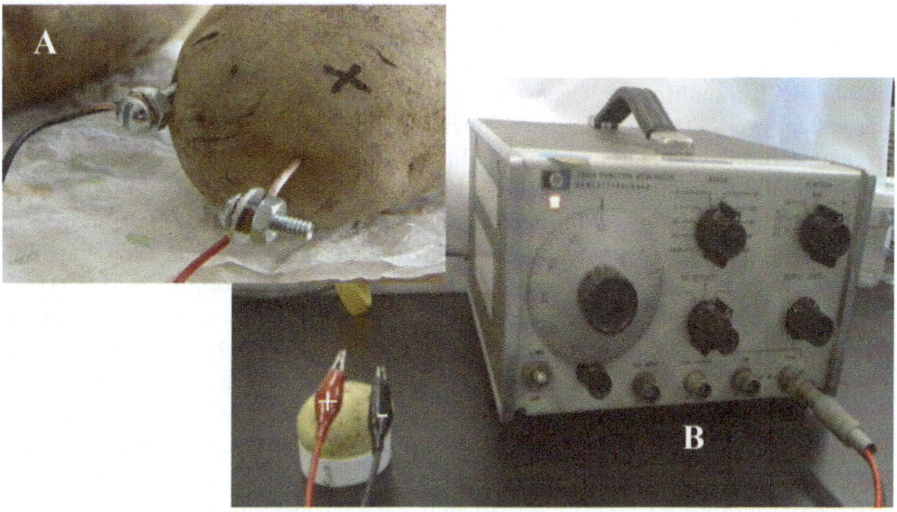


**Fig. 4** Electroporation of a potato (**A**), using Cu (+) and Zn (−) electrodes with a function generator (**B**)

digestion of the samples, the final solution was placed in 100 mL flasks and analyzed in an Epsilon potentiostat, with the same conditions being applied as those used for the calibration method to determine the Hg content [115].

Finally, to validate the results, the Hg content was determined with use of the HgAAS technique with hydride generation, in accordance with Official Mexican Standard NMX-SCT-051-2001 [116], using an AANALYST 400 spectrophotometer (Perkin Elmer, Waltham, MA, USA). Additionally, the potato tubers were characterized using Fourier transform infrared spectroscopy (FTIR) using a NEXUS Nicolet spectrophotometer (Thermo Scientific, Waltham, MA, USA) to assess the changes produced in the previously dried potato tubers [117–121].

Potato tubers taken from the $HgCl_2$ solutions and an untreated potato tuber were used to determine if the current and voltage were sufficient to produce usable energy. Potato tubers were sown in triplicate in solid soil collected from the Otatal mine. The potato tubers remained in the soil for 12 weeks so we could determine which soil condition best favored plant growth. At the end of this period, the best soil condition was selected on the basis of the pH and EC data [122].

After EKT, the potatoes were grown in electroremediated soil. After 15 days, we analyzed each potato by obtaining impedance spectra and a discharge curve, using a potentiostat (Zahner-Elektrik, Kronach, Germany) and Cu/Zn electrodes placed in each potato to study their behavior over time. For these experiments, we used resistance of 52 kΩ to assess the performance of each potato biocell.

## 3 Discussion

During EKT, a multimeter was used to quantify the electrical current generated in the system every 30 min during a 72-h test period. The electrical current decreased over time in the cropland and mine-tailing samples. A small rise was observed at about the 30-h treatment mark, but there was a tendency for the current to decrease after 50 h. This variation in the electrical current was associated with the motion of the ions migrating to the anode and cathode in the electrochemical cell, depending on their charge, as a result of the applied electrical field (electromigration) [123]. To verify this movement of ions (including $Hg^{2+}$) in solution, samples were extracted from the electrochemical cell after 1 day of EKT and analyzed by ASV. In the evaluation of the current peak that was generated (between −0.1 and 0.1 V versus Ag/AgCl), its changes were attributed to the presence of small amounts of organic matter, as has been reported previously [80]. Similarly, it was observed that the current signal was higher in the EKT-treated mine-tailing sample obtained from the extracted solution after 1 day (approximately 8 μA). This result indicated an increase in the removal efficiency of EKT, which was verified with HgAAS analysis. The HgAAS analysis results showing the efficiency of mercury removal from the cropland and mine-tailing samples treated with EKT are presented in Table 5 and Fig. 5. The highest efficiency of mercury removal was reached in the mine-tailing sample treated with EKT. The $Hg^{2+}$ ($\eta_{[Hg2+]}$) results showed that the order of removal efficiency was as follows: close to the cathodes (87.2%) > middle cell (84.0%) > close to the anode (79.0%).

High concentrations of ions (complexes of EDTA + $Hg^{2+}$ and EDTA + cations) were observed in the mine-tailing sample, with an initial [$Hg^{2+}$] concentration of 2061.771 parts per million (ppm), and there were higher concentration of ions with a negative charge that migrated close to the anode [80] in comparison with the cropland sample. Additionally, the organic matter content in the cropland sample was reduced by the movement of the ions (with an [$Hg^{2+}$] concentration of

**Table 5** Mercury concentrations ([Hg]) in cropland and mine-tailing samples after electrokinetic treatment, and efficiency of mercury removal ($\eta$) in comparison with control samples

| Sample | $[Hg]_{initial}$ (ppm) | $[Hg]_{final}$ (ppm) | $\eta$ (%) |
|---|---|---|---|
| Cropland | | | |
| Control | 440.46 | – | – |
| Close to anode | – | 360.17 | 18.23 |
| Middle cell | – | 212.53 | 51.75 |
| Close to cathodes | – | 354.87 | 19.43 |
| Mine tailing | | | |
| Control | 2061.77 | – | – |
| Close to anode | – | 431.69 | 79.06 |
| Middle cell | – | 324.01 | 84.29 |
| Close to cathodes | – | 272.60 | 86.78 |

*ppm* parts per million

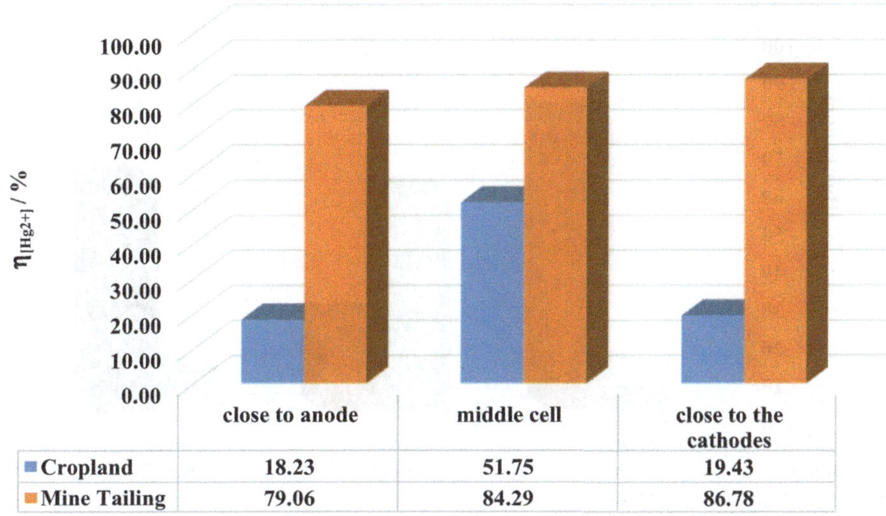


**Fig. 5** Efficiency of $Hg^{2+}$ removal ($\eta_{[Hg2+]}$) from cropland and mine-tailing samples, using a two-dimensional circular arrangement of electrochemical cells for electrokinetic treatment close to the anode, in the middle cell, and close to the cathodes

440.462 ppm—just over one fifth of that observed in the mine-tailing sample) and the complexation arising between mercury and organic matter [115]. In contrast, in the cropland sample, the major $\eta_{[Hg2+]}$ was observed in the middle cell (51.7%), followed by the zones close to the cathodes (24.1%) and close to the anode (18.2%). In both samples, the lowest $\eta_{[Hg2+]}$ was close to the anode because the polluted solution was pumped from the central zone where there was a separate compartment between the circular PVC spacers surrounding the anodic area (Fig. 2B). In contrast, the $\eta_{[Hg2+]}$ was greatest in the middle cell, with both samples influenced by different transport phenomena during EKT: electromigration, electrophoresis, and electro-osmosis [80, 90, 105, 124]. As expected, more $Hg^{2+}$ was found in *Lavandula vera* plants grown in mine-tailing soil than in those grown in cropland soil (Fig. 6), with mercury retained in the following order: stems (77.303 ppm) > roots (60.798 ppm) > leaves (24.659 ppm), whereas in plants grown in the cropland soil, the mercury retention was lower and was distributed as follows: leaves (57.914 ppm) > roots (10.527 ppm) > stems (2.802 ppm).

The physicochemical characteristics of the samples after EKT are reported in Table 6. In these results, a small increase in particle density was observed. This could have indicated improvements in both compaction and retention of water, which could be used by the plants [125], or it could have indicated disaggregation of particles by the organic compound liberation after the treatment.

The soil texture (considering the existing percentages of silt, clay, and sand), was unchanged in the mine-tailing soil, but minor changes were observed in the cropland soil. These results showed that EKT caused no significant changes in the physical properties of the soils. The pH values of the cropland soil increased slightly as a

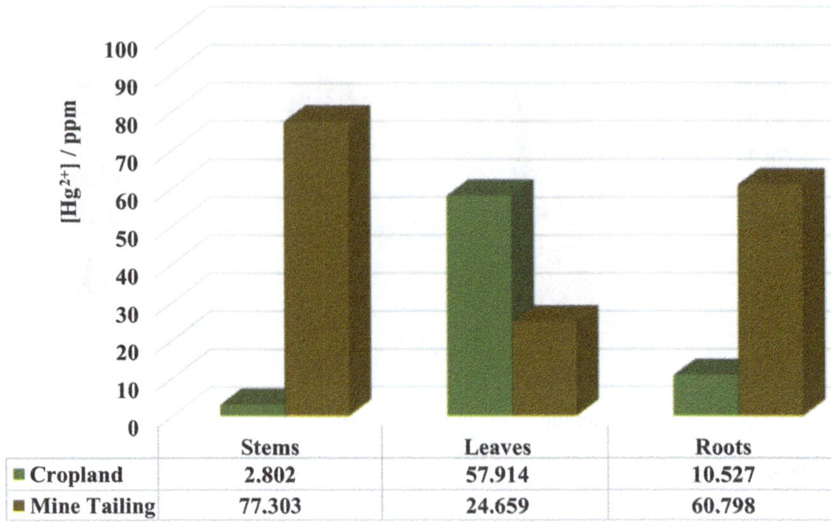


**Fig. 6** Mercury concentrations in stems, leaves, and roots after phytoremediation using *Lavandula vera* in cropland and mine-tailing soil samples. *ppm* parts per million

result of migration of $^{-}OH$ ions promoting formation of an alkaline border [3] and migration to the area with a positive charge (the anode), showing the greatest increase in pH. This was consistent with results reported previously for this type of treatment. In contrast, the mine-tailing soil showed a decrease in pH values after EKT, indicating greater presence and/or greater mobility of $H^+$ ions in the electrokinetic system.

It should also be recalled that the composition of the mine tailings drove this behavior because this type of sample was very dissimilar to natural soil. The EC observed in both samples increased sharply after application of EKT because of the addition of EDTA as a complexing agent, and the salt concentration of the soil was increased. The addition of EDTA to the system and the change in salinity were the two reasons why the growth of the plants was so greatly affected. Since the plants transplanted into the soils after EKT did not grow, the team decided to assess the effect of the addition of EDTA separately.

The results showed that with EDTA concentrations higher than 10 mM, the plants suffered from its toxic effects, whereas with EDTA concentrations of the order of 1 mM, EDTA growth was promoted, and when an EDTA concentration of 0.1 mM was used, growth was improved in comparison with the control soil tests. It is known that plants exhibit poor development when there is an increase in EC [126]. This parameter changes with the effects of salinity. Growth efficiency is restricted in very salt-sensitive crops at an EC value of 2 dS $cm^{-1}$, in the majority of crops at an EC value of 4 dS $cm^{-1}$, in salt-tolerant crops at an EC value of 8 dS $cm^{-1}$, and in highly salt-tolerant crops at an EC value of 16 dS $cm^{-1}$. The EC results from the test samples are presented in Table 7. The highest EC values were observed in

**Table 6** Physicochemical characteristics of cropland and mine-tailing samples close to the anode, in the middle cell, and close to the cathodes after electrokinetic treatment in comparison with control samples

| Sample | pH | EC (dS $cm^{-1}$) | CEC (mEq/100 g) | AD (g $mL^{-1}$) | RD (g $mL^{-1}$) | Porosity (%) | OM (%) | $Ca^{2+}$ (mg $L^{-1}$) | $Mg^{2+}$ (mg $L^{-1}$) | $Na^{+}$ (mg $L^{-1}$) | $K^{+}$ (mg $L^{-1}$) | Texture |
|---|---|---|---|---|---|---|---|---|---|---|---|---|
| Cropland | | | | | | | | | | | | |
| Control | 8.07 | 0.134 | 113.4 | 1.042 | 2.457 | 57.58 | 5.29 | 85.60 | 42.27 | 77.4 | 1.38 | Loam |
| Close to anode | 8.45 | 2.47 | 118.2 | 1.097 | 2.457 | 56.69 | 6.99 | 51.59 | 58.14 | 71.9 | 1.25 | Sandy loam |
| Middle cell | 8.25 | 2.32 | 116.6 | 1.135 | 2.621 | 56.71 | 6.80 | 73.65 | 37.86 | 62.7 | 0.91 | Loam |
| Close to cathodes | 8.35 | 2.36 | 125.5 | 1.132 | 2.615 | 55.35 | 7.43 | 88.43 | 51.71 | 72.3 | 1.13 | Loam |
| Mine tailing | | | | | | | | | | | | |
| Control | 9.03 | 1.55 | 299.4 | 1.195 | 2.420 | 50.66 | 1.80 | 48.35 | 57.26 | 19.0 | 1.63 | Sandy loam |
| Close to anode | 8.21 | 4.32 | 128.7 | 1.211 | 2.520 | 51.95 | 3.21 | 111.88 | 51.64 | 30.8 | 1.41 | Sandy loam |
| Middle cell | 8.74 | 3.97 | 194.3 | 1.248 | 2.610 | 52.20 | 3.63 | 89.57 | 62.57 | 28.5 | 1.86 | Sandy loam |
| Close to cathodes | 8.83 | 3.68 | 246.3 | 1.248 | 2.550 | 51.06 | 2.70 | 60.07 | 77.60 | 22.8 | 1.57 | Sandy loam |

*AD* apparent density, *CEC* cation exchange capacity, *EC* electrical conductivity, *OM* organic matter, *RD* real density

**Table 7** Electrical conductivity (EC) of cropland and mine-tailing samples after electrokinetic treatment in comparison with control samples

| Sample | EC (mS $cm^{-1}$) |
|---|---|
| Cropland | |
| Control | 0.13 |
| Close to anode | 2.47 |
| Middle cell | 2.32 |
| Close to cathodes | 2.36 |
| Mine tailing | |
| Control | 1.55 |
| Close to anode | 4.32 |
| Middle cell | 3.96 |
| Close to cathodes | 3.68 |

the mine-tailing sample (close to the anode (4.32 mS $cm^{-1}$) > middle cell (3.96 mS $cm^{-1}$) > close to the cathodes (3.68 mS $cm^{-1}$)). In the cropland sample, the corresponding values were as follows: close to the anode (2.47 mS $cm^{-1}$) > close to the cathodes (2.36 mS $cm^{-1}$) > middle cell (2.32 mS $cm^{-1}$).

On the other hand, the percentages of organic matter were increased, which indicated that application of an electrical field and consequential transport processes increased the possibility of utilizing this matter as a nutrient. This was considered a positive result because organic matter is essential for plant growth and development, and it improves the physical properties of the soil, increases the cation exchange capacity (CEC), and improves moisture retention [125]. In the mine-tailing soil, the tests indicated an increase in the availability of organic matter, which would greatly improve remediation performance if the soil was subsequently treated with a biological process such as PhyR. The CEC results showed that the organic matter content increased only a little in the cropland soil. However, a measured increase in the level of negative charges present on the surfaces of minerals and organic components of the soil indicates an increase in the capacity of the soil to give these sites to cations, such as those found in macronutrients, and this could indicate that rehabilitation of land would be possible.

The concentrations of $Ca^{2+}$, $Mg^{2+}$, $Na^{+}$, and $K^{+}$ (the so-called exchangeable bases) present in the soil were determined. These ions can be interchangeable between plants and the medium in which they are found. In the cropland soil, a general decrease in their concentrations occurred and a slight increase was observed in the area near the anode, which showed that movement of ions was occurring. This was associated with the ability of EDTA to complex $Ca^{2+}$ ($\log K_f = 10.70$) and $Mg^{2+}$ ($\log K_f = 13.79$). This capability has been proved and depends directly on the constants of complexation, as shown in Table 8. These cations can be mobilized within the matrix, similarly to $Hg^{2+}$. In the case of the mine tailings, the behavior was different: the concentrations of $Ca^{2+}$ and $Na^{+}$ increased considerably, but those of $Mg^{2+}$ and $K^{+}$ did not; they remained almost constant during the process, with a slight decline in the zone close to the anode. After EKT, the $Hg^{2+}$-contaminated wastewater was treated using a PRB with different electrodes of $Fe^{\circ}$ and carbon, with EKT-PRB being performed over 10 min (Table 9). This testing showed that the efficiency of

**Table 8** Stability constants with ethylene-diaminetetraacetic acid at 20 °C and ionic strength of 0.1 M

| Cation | log $K_f$ |
|---|---|
| $Ag^+$ | 7.32 |
| $Mg^{2+}$ | 8.69 |
| $Ca^{2+}$ | 10.70 |
| $Sr^{2+}$ | 8.63 |
| $Ba^{2+}$ | 7.76 |
| $Mn^{2+}$ | 13.79 |
| $Fe^{2+}$ | 14.33 |
| $Co^{2+}$ | 16.31 |
| $Or^{2+}$ | 18.62 |
| $Cu^{2+}$ | 18.80 |
| $Zn^{2+}$ | 16.50 |
| $Cd^{2+}$ | 16.46 |
| $Hg^{2+}$ | 21.80 |
| $Pb^{2+}$ | 18.04 |
| $Mn^{3+}$ | 16.13 |
| $Fe^{3+}$ | 25.40 |
| $V^{3+}$ | 25.90 |
| $Th^{4+}$ | 2320 |

**Table 9** Efficiency of $Hg^{2+}$ removal ($\eta$) using different permeable reactive barrier materials and electrokinetic treatment for 10 min

| Material | $\eta$ (%) |
|---|---|
| C fiber | 97.1 |
| C tissue | 91.2 |
| CGOO | 90.6 |
| FeTRGO | 89.9 |
| Fe° | 80.0 |

*CGOO Agave* biomass sponge, Fe*QRGO* chemically reduced Fe° + C, *FeTRGO* thermically reduced Fe° + C

mercury removal was greatest using C fibers ($\eta = 97.1\%$), followed by CGOO, with intermediate removal ($\eta = 90.6\%$), and the lowest rate was seen with Fe° ($\eta = 80.0\%$), with the rates of removal using additional materials containing mixtures of Fe° and C falling between these values.

When the timed behavior was studied, Fe° showed the highest removal efficiency (>90%) over the first 10 min of EKT-PRB (Fig. 7), but this then decreased to 45% after 15 min and to <10% after 40 min. Although the removal efficiency of C fiber was initially high, the $\eta_{[Hg2+]}$ value decreased from 87% to less than 10% after 10 min, before increasing again to 30–40% after 40 min. In contrast, CGOO showed an $\eta_{[Hg2+]}$ value close to 50% throughout the entire 120-min duration of the

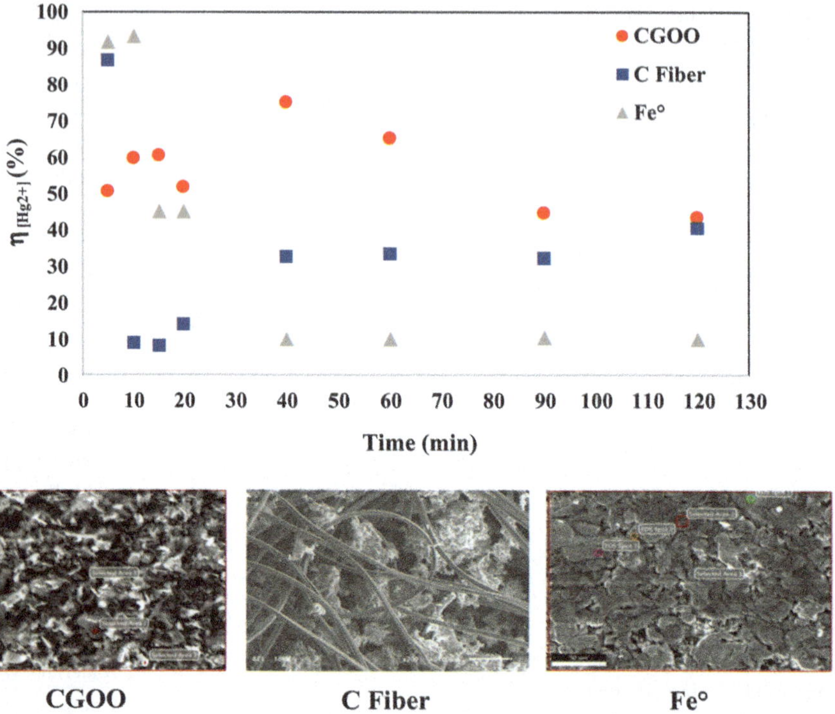


**Fig. 7** Efficiency of $Hg^{2+}$ removal ($\eta_{[Hg2+]}$) using different permeable reactive barrier materials consisting of *Agave* biomass sponge (CGOO), carbon fiber (C fiber), and iron (Fe°) from 10 to 120 min, and the corresponding scanning electron microscopy images (×20,000)

experiment, with the highest efficiency being observed at 45 min ($\eta$ = 75.47%). Because of this behavior, this material was selected for EKT-PRB (but with a treatment period reduced to 45 min) in remediation, followed by PhyR with *Lavandula vera* and *Solanum tuberosum* in subsequent experiments.

## *3.1 Electro-Phytoremediation of Soil with* Lavandula vera

The results of the process of PhyR with *Lavandula vera* were evaluated in terms of plant growth in mercury-polluted soils for 5 weeks (Fig. 8A, B). The height and dry weight of the plants that were transplanted into cropland, mine-tailing, and control soils (from off-site) were measured. During the test period, the dry weight (Fig. 8A) and normalized height (Fig. 8B) of *Lavandula vera* were greatest in the cropland soil. In specific cases when cropland and off-site samples were compared, the changes in both properties showed logarithmic increases with respect to time; in contrast, logarithmic decreases were observed in the plants grown in mine-tailing soil.

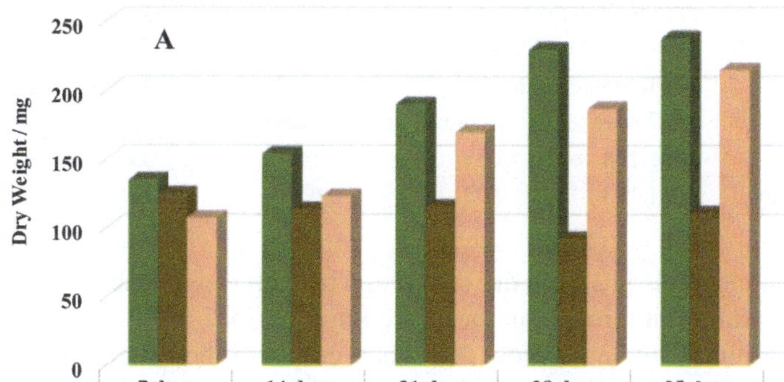


| | 7 days | 14 days | 21 days | 28 days | 35 days |
|---|---|---|---|---|---|
| Cropland | 134.2 | 153.15 | 188.85 | 228.45 | 236.7 |
| Mine Tailing | 124.05 | 112.75 | 114.65 | 91.2 | 110 |
| Control | 106.7 | 122.45 | 168.75 | 185.7 | 214.05 |

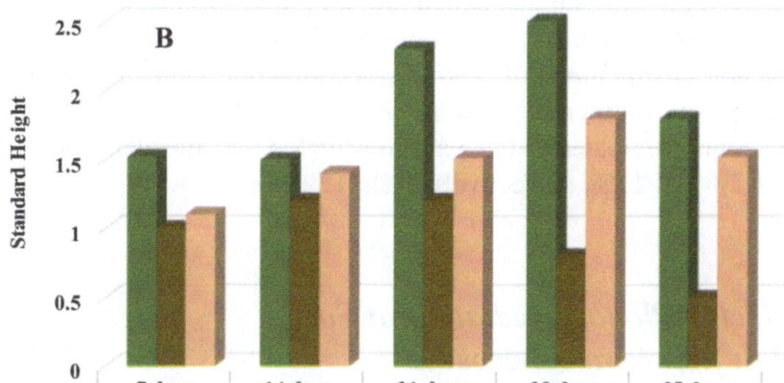


| | 7 days | 14 days | 21 days | 28 days | 35 days |
|---|---|---|---|---|---|
| Cropland | 1.52 | 1.5 | 2.3 | 2.5 | 1.8 |
| Mine Tailing | 1 | 1.2 | 1.2 | 0.8 | 0.5 |
| Control | 1.1 | 1.4 | 1.51 | 1.8 | 1.52 |

**Fig. 8** Dry weight (**A**) and standard height (**B**) of *Lavandula vera* after 7, 14, 21, 28, and 35 days of phytoremediation of cropland, mine-tailing, and control soil samples

The tolerance index (TI) is defined as the relation between the dry weight of biomass produced by plants grown in polluted soil and the dry weight of biomass produced by plants grown in uncontaminated soil [127]. Here, then, the TI was equal to the dry weight of biomass produced by *Lavandula vera* plants grown in cropland and mine-tailing soils, both polluted by Hg, with respect to the dry weight of biomass of *Lavandula vera* grown in unpolluted soil (off-site control soil) (Fig. 8A). According to Audet and Charest [127], a TI value $<1$ signifies stress on the plant from metal pollution, as was shown by the TI of plants grown in mine-tailing soil for 14 days ($TI < 1$). If the TI value is equal to 1, the soil is considered to exhibit no difference in comparison with the control soil; none of our experiments showed this. On the other hand, a TI value $>1$ means that the plants are not affected by the contaminant and that the plant material may be a hyperaccumulator of the

**Table 10** Changes in the tolerance index of *Lavandula vera* plants used for phytoremediation of mine-tailing and cropland soil samples

| | Tolerance index | | | | |
|---|---|---|---|---|---|
| Sample | 7 days | 14 days | 21 days | 28 days | 35 days |
| Cropland | 1.258 | 1.251 | 1.119 | 1.230 | 1.106 |
| Mine tailing | 1.163 | 0.921 | 0.679 | 0.491 | 0.514 |

metal contaminants [127]. In our study, this behavior was observed in plants growing in cropland soil throughout the entire PhyR period.

The results in Table 10 show that when grown in the extremely polluted mine-tailing soil, *Lavandula vera* grew rapidly in the first week after transplantation (7 days), but the plant TI value decreased to <1 thereafter. These plants started growing but then suffered stress, suggesting that their further development was stunted, as they were not able to include the pollutant in their metabolism when they were exposed to the mine-tailing soil.

Conversely, the plants transplanted into cropland soil had an excellent response; their TI values remained >1 throughout the 5 weeks of PhyR. This result further suggested that *Lavandula vera* is a hyperaccumulator and can be used for phytoremediation of soil. Furthermore, the mine-tailing and cropland soil samples used in this study contained material that was bioavailable to some plant species, in concentrations of 7% and 15%, respectively [128, 129].

## 3.2 *Electro-Phytoremediation of Soil with* **Solanum tuberosum**

When potatoes were exposed to the different types of soil sample (cropland, mine-tailing, and control soils), their growth was greater when they were exposed to the soils containing mercury than when they were exposed to the control soil. The samples had a pH close to neutral (6.7) and EC of 149.3 μS $cm^{-1}$; these results were obtained in triplicate. Later, 21 days after the potatoes were planted, they were extracted to obtain the corresponding discharge curves in order to assess whether they were able to generate energy after the EKT + PhyR process. It has previously been reported that during EKT, electrical charges generated by electrical pulses into the soil can influence the physiology of microorganisms, the behavior of ionic components, and the potential oxide reduction, which would benefit the growth of plant species [130].

In this way, the coupling of EKT + PhyR could enhance plant growth through movement of nutrients by application of a direct electrical current to remove contaminants from the soil. After EKT, PhyR allows accumulation of contaminants that may still be in the soil, with accumulation of metals in plant roots, stems, or leaves. The key element is electrokinetic mobilization of metals in the soil, which increases the availability of the metals to plant roots and therefore increases metal absorption by plants and corresponding extraction of the metal from the soil by PhyR [131].

Potency generation was evaluated 30 days after the potato tubers were extracted from solutions containing 10 μM $HgCl_2$ in 0.1 M HCl, with constant stirring, in order to verify the adsorption capacity of raw, boiled, and electroporated potatoes. After the exposure of the potato to the mercury-containing solution, a Cu/Zn electrode pair was inserted the tuber, and the voltage and current generated by this arrangement were measured immediately with help of a multimeter, using the group of raw potato tubers. The results shown in Table 11 suggest that all of the potato samples generated potency containing $Hg^{2+}$, as follows: raw potato (17.15 μW) ≫ electroporated potato (8 μW) > boiled potato (5 μW). This result could have been due to the greater amounts of starch in the tissue of the raw potato tubers, which remained mostly available to take up free mercury [132, 133], than in the electroporated and boiled potato tissue.

Figure 9 shows the discharge curves (Fig. 9A) and cell power curves (Fig. 9B) curves. The raw potato showed higher values than the boiled and electroporated potato; according to the literature on this type of experiment, untreated (raw) potato tubers generate more energy than pretreated (boiled or electroporated) potato [99, 134, 135]. After the different pretreatments were performed on the potatoes, it was concluded that raw potatoes generated the greatest amount of energy, followed by those pretreated with electroporation [99, 134–137].

After the capacity of the potato tubers to generate an electrical potential was evaluated, the raw potatoes were placed in a solid matrix of mercury-polluted soil for EKT + PhyR to be performed. This experiment involved application of an electrical potential of 20 V, using seven electrodes immersed in the soil in a two-dimensional configuration in which titanium cathodes were placed in a circular arrangement and connected in series to a power source. In the middle of the electrochemical cell, a single Ti anode was placed and connected to a power source. All electrodes were placed equidistant from one another (with a 6 cm distance between electrodes), as shown in Fig. 2B. EKT was applied for 72 h, and a coadjutant solution of 0.1 M EDTA was used to facilitate removal of mercury through formation of complexes. The results presented in Table 12 and Fig. 10 show that the $Hg^{2+}$ concentrations in the mine-tailing soil after EKT + PhyR were 14.93 mg $kg^{-1}$ in the position designed as the middle cell, 7.15 mg $kg^{-1}$ in the zone close to the cathodes, and 2.85 mg $kg^{-1}$ in the zone close to the anode. Meanwhile, the $Hg^{2+}$ concentrations in the cropland soil were 8.57 mg $kg^{-1}$ in the zone close to the anode, 28.3 mg $kg^{-1}$ in the zone designed as the middle cell, and 2.575 mg $kg^{-1}$ in the zone close to the cathode. All cropland samples showed a $\eta$ value of >93%, and the mine-tailing samples showed an $\eta$ value of >99%, with the final concentrations in soil

**Table 11** Energy generation in raw, boiled, and electroporated potato samples

| Sample | *E* (V) | *i* (μA) | $IV_{max}$ | ff | *P* (μW) |
|---|---|---|---|---|---|
| Raw | 0.910 | 58.00 | 0.0000159500 | 0.30 | 17.15 |
| Boiled | 0.732 | 34.00 | 0.0000048752 | 0.20 | 5.00 |
| Electroporated | 0.852 | 39.00 | 0.0000071765 | 0.22 | 8.00 |

*E* potential, *ff* fill factor, *i* current, *I* ampere, *P* cell power, *max* maximum, *V* potential

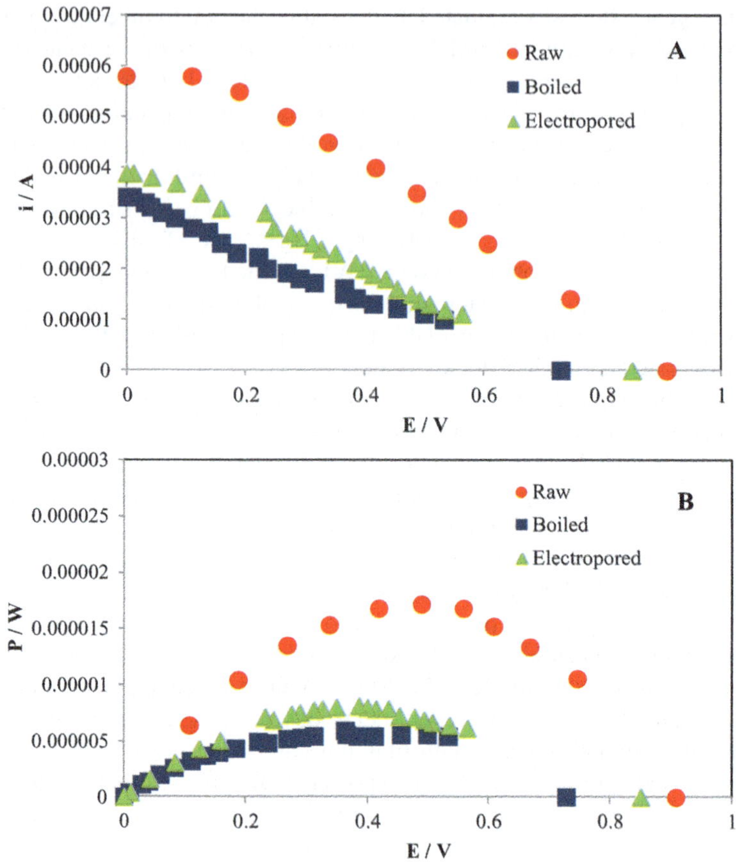


**Fig. 9** Discharge curves (**A**) and cell power curves (**B**) constructed for raw, boiled, and electroporated potatoes

remediated by EKT + PhyR being in accordance with Official Mexican Standard NOM 147-SEMARNAT/SSA1-2004 [107], where the maximum permissible limit in the sample is 23 ppm.

As a consequence, application of EKT + PhyR with raw potatoes was demonstrated to be a good alternative for extraction of $Hg^{2+}$ from contaminated cropland and mine-tailing soils. This may have been due to physiological and anatomical characteristics of the potato tubers that led to accumulation of mercury in the plant tissue. After EKT + PhyR was carried out, at 21 days, the potatoes were extracted, and the corresponding discharge curves were constructed to evaluate whether they were able to generate energy despite having been in contact with soil treated with EKT + PhyR. These results showed that the potatoes grown in mine-tailing soil could generate current after EKT + PhyR, and they were compared with the potatoes exposed to cropland soil (Table 12).

**Table 12** Mercury concentrations ([Hg]) in cropland and mine-tailing samples after electrokinetic treatment (EKT) and electro-phytoremediation (EKT + PhyR), and efficiency of mercury removal ($\eta$) in comparison with control samples

| Sample | $[Hg]_{EKT}$ (ppm) | $[Hg]_{EKT+PhyR}$ (ppm) | $\eta$ (%) |
|---|---|---|---|
| Cropland | | | |
| Control | 440.46 | – | – |
| Close to anode | 360.17 | 8.57 | 98.05 |
| Middle cell | 212.53 | 28.30 | 93.57 |
| Close to cathodes | 354.87 | 2.58 | 99.42 |
| Mine tailing | | | |
| Control | 2061.77 | – | – |
| Close to anode | 431.69 | 2.85 | 99.86 |
| Middle cell | 324.01 | 14.93 | 99.28 |
| Close to cathodes | 272.60 | 7.15 | 99.65 |

*ppm* parts per million

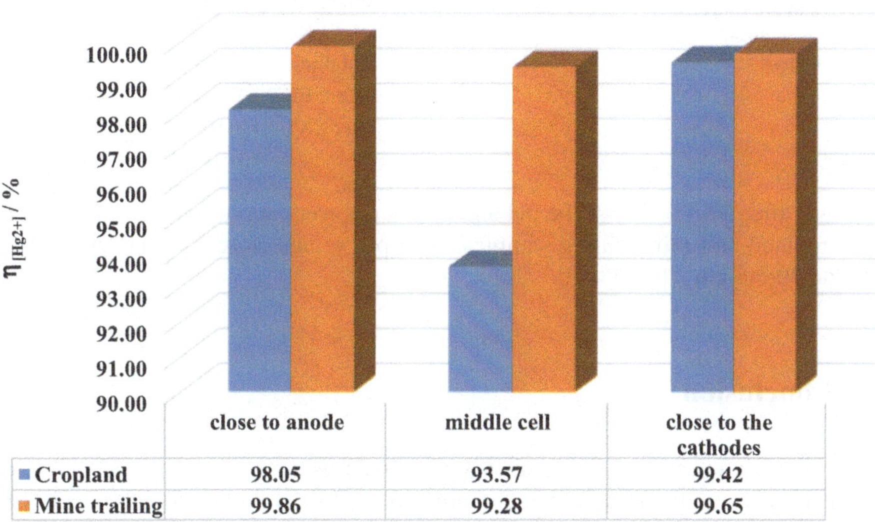


**Fig. 10** Efficiency of $Hg^{2+}$ removal ($\eta_{[Hg2+]}$) from cropland and mine-tailing samples, using a two-dimensional circular arrangement of electrochemical cells for electro-phytoremediation close to the anode, in the middle cell, and close to the cathodes

In the potato exposed to mine-tailing soil, the zone close to the cathodes during EKT + PhyR generated the most power (8.53 μW), with values of 0.7993 V and 51 μA at close to cathodes (Table 13), consistent with the high mercury concentration in the soil (2061.77 ppm). In contrast, in the cropland sample, the highest maximum power generation (11.18 μW), with values of 0.4709 V and 30 μA at close to anode, occurred close to the anode, consistent with the lower mercury concentration and the differences in $Ca^{2+}$, $Mg^{2+}$, $K^{+}$, and $Na^{+}$ concentrations relative to the mine-tailing sample (Table 6). This behavior was related to changes in the pH and EC of

**Table 13** Energy generation in raw potatoes grown in cropland and mine-tailing soil after electro-phytoremediation

| Raw potato exposed to | *E* (V) | *i* (μA) | $IV_{max}$ (μW) |
|---|---|---|---|
| Cropland soil | | | |
| Close to cathodes | 0.4012 | 1.00 | 0.05 |
| Close to anode | 0.4709 | 30.00 | 11.18 |
| Middle cell | 0.7302 | 49.00 | 6.75 |
| Mine-tailing soil | | | |
| Close to cathodes | 0.7993 | 51.00 | 8.53 |
| Close to anode | 0.444 | 17.00 | 1.57 |
| Middle cell | 0.4675 | 20.00 | 1.98 |

*E* potential, *i* current, *I* ampere, *max* maximum, *V* potential

**Table 14** pH and electrical conductivity (EC) in cropland and mine-tailing samples before and after electro-phytoremediation (EKT + PhyR)

| Sample | pH | EC (S cm$^{-1}$) |
|---|---|---|
| Cropland | | |
| Before EKT + PhyR | 7.4 | 0.0002 |
| After EKT + PhyR | 7.8 | 0.0026 |
| Mine-tailing | | |
| Before EKT + PhyR | 8.5 | 2.2970 |
| After EKT + PhyR | 7.9 | 0.0030 |

the soil; after EKT + PhyR, the pH and EC values were approximately the same in the cropland and mine-tailing samples, with pH values close to 8 and EC values around 30 mS cm$^{-1}$ (Table 14).

## 4 Conclusion

Electrokinetic treatment (EKT) was implemented using a circular arrangement of six titanium cathodes around one titanium anode. Tests were developed in which 20 V was applied for 72 h in the presence of ethylenediaminetetraacetic acid as a support electrolyte to remove mercury from cropland and mine-tailing soil samples. A permeable reactive barrier of *Agave* biomass sponge was then used to reduce the mercury content in the wastewater generated during EKT. After electro-phytoremediation (EKT + PhyR), the physical and chemical characteristics of the cropland and mine-tailing samples showed significant differences. After EKT + PhyR using *Lavandula vera*, quantification of the mercury content in the plant structure showed that in the mine-tailing samples, mercury had accumulated in the stems and roots, whereas in the cropland samples, it had accumulated in the leaves. When EKT + PhyR was performed using raw, boiled, and electroporated *Solanum tuberosum*, raw potato performed best for generating bioelectricity in terms of the discharge curves that were constructed; thus, it could be used to bioaccumulate mercury during the EKT + PhyR process.

**Acknowledgements** The authors wish to thank Consejo Nacional de Ciencia y Tecnología (CONACyT) (Mexico City, Mexico) for financial support; Richard Lindeke, PhD (Professor Emeritus, University of Minnesota Duluth (Duluth, MN, USA) and US Peace Corps Volunteer, Centro de Innovación Aplicada en Tecnologías Competitivas (CIATEC) (Leon, Mexico)), for his English-language revision of this manuscript; and R. Gutiérrez, G. Acosta-Santoyo, G. Olvera, and G. Hernández for their experimental contributions to this research.

## References

1. M.A. Ashraf, M.J. Maah, Soil contamination, risk assessment and remediation, in *Environmental Risk Assessment of Soil Contamination*, ed. by M. C. Hernandez Soriano, (InTech, Rijeka, 2014), pp. 3–56
2. A. Heinegg, P. Maragos, M. Edmund, J. Rabinowicz, G. Straccini, H. Waish, *Soil Contamination and Urban Agriculture: A Practical Guide to Soil Contamination Issues for Individuals and Groups* (McGill University, Montreal, 2002)
3. C. Cameselle, R.A. Chirakkara, K.R. Reddy, Electrokinetic-enhanced phytoremediation of soils: status and opportunities. Chemosphere **93**, 626–636 (2013)
4. R.A. Wuana, F.E. Okieimen, Heavy metals in contaminated soils: a review of sources, chemistry, risks and best available strategies for remediation. Int. Sch. Res. Notices Ecol. **2011**, 402647 (2011)
5. H.D. Sharma, K.R. Reddy, *Geoenvironmental Engineering: Site Remediation, Waste Containment, and Emergency Waste Management Technologies* (Wiley, Hoboken, 2004)
6. Q. Zhao, J.J. Kaluarachchi, Risk assessment at hazardous waste–contaminated sites with variability of population characteristics. Environ. Int. **28**, 41–53 (2002)
7. M. Patra, A. Sharma, Mercury toxicity in plants. Bot. Rev. **66**, 379–422 (2000)
8. P.J.C. Favas, J. Pratas, M. Varun, R.D. Souza, M.S. Paul, Phytoremediation of soils contaminated with metals and metalloids at mining areas: potential of native flora. Environ. Risk Assess. Soil Contam. **17**, 485–516 (2014)
9. O.P. Dhankher, E.A.H. Pilon-Smits, R.B. Meagher, S. Doty, Biotechnological approaches for phytoremediation, in *Plant Biotechnology and Agriculture: Prospects for the 21st Century*, ed. by A. Altman, P. M. Hasegawa, (Academic, Oxford, 2012), pp. 309–328
10. S. López-Martínez, M.E. Gallegos-Martínez, L.J. Pérez Flores, M.G. Rojas, Mecanismos de fitorremediación de suelos contaminados con moléculas orgánicas xenobióticas. Rev. Intern. Contam. Amb. **21**, 91–100 (2005)
11. L. Wang, B. Ji, Y. Hu, R. Liu, W. Sun, A review on in situ phytoremediation of mine tailings. Chemosphere **184**, 594–600 (2017)
12. A. Mayerly, L. Oyuela, D. Wilmar, G. Fernández, M.C. Gutiérres Sarmiento, Native herbaceous plant species with potential use in phytoremediation of heavy metals, spotlight on wetlands—a review. Chemosphere **168**, 1230–1247 (2017)
13. A. Wiszniewska, E. Hanus-Fajerska, E. Muszynska, K. Ciarkowska, Natural organic amendments for improved phytoremediation of polluted soils: a review of recent progress. Pedosphere **26**, 1–12 (2016)
14. A. Cristaldi, G. Oliveri Conti, E. Hea Jho, P. Zuccarello, A. Grasso, C. Copat, M. Ferrante, Phytoremediation of contaminated soils by heavy metals and PAHs: a brief review. Environ. Technol. Innov. **8**, 309–326 (2017)
15. A. Ullah, S. Heng, M.F.H. Munis, S. Fahad, X. Yang, Phytoremediation of heavy metals assisted by plant growth promoting (PGP) bacteria: a review. Environ. Exp. Bot. **117**, 28–40 (2015)
16. R. Gill, T. Harbottle, M.J. Smith, J.W.N. Smith, S.F. Thornton, Electrokinetic-enhanced bioremediation of organic contaminants: a review of processes and environmental applications. Chemosphere **107**, 31–42 (2014)

17. B.C. Towe, Bioelectricity and its measurement, in *Standard Handbook of Biomedical Engineering and Design*, ed. by M. Kutz, (McGraw-Hill, New York, 2003), pp. 17.1–17.50
18. M. Helder, W.-S. Chen, E.J.M. Van der Harst, D.P.B.T. Strik, H.B.V.M. Hamelers, C.J.N. Buisman, Electricity production with living plants on a green roof: environmental performance of the plant–microbial fuel cell. Biofuel. Bioprod. Bioref. **7**, 52–64 (2013)
19. P.J. Weinberg, *Introducción a la contaminación por mercurio para las ONG* (Gothenburg, International Pollutants Elimination Network, 2002)
20. A. Buscaroli, An overview of indexes to evaluated terrestrial plants for phytoremediation purposes (review). Ecol. Indic. **82**, 367–380 (2017)
21. K.E. Gerhardt, D.G. Perry, B.M. Greenberg, Opinion: taking phytoremediation from proven technology to accepted practice. Plant Sci. **256**, 170–185 (2017)
22. J. Wang, X. Feng, C.W.N. Anderson, Y. Xing, L. Shang, Remediation of mercury contaminated sites—a review. J. Hazard. Mater. **221–222**, 1–18 (2012)
23. S. Martinez-Trinidad, G. Hernández, M.E. Ramirez, J. Martínez, G. Solorio, S. Solís, R. García, Total mercury in terrestrial systems (air–soil–plant–water) at the mine region of San Joaquín, Querétaro. Mexico. Geofis. Int. **52**, 43–58 (2013)
24. J.E. Gray, J.G. Crock, B.K. Lasorsa, Mercury methylation at mercury mines in the Humboldt River basin, Nevada, USA. Geochem. Explor. Environ. Anal. **2**, 143–149 (2002)
25. C.-M. Neculita, G.J. Zagury, L. Deschênes, Mercury speciation in highly contaminated soils from chlor-alkali plants using chemical extractions. J. Environ. Qual. **34**, 255–262 (2005)
26. A. Cain, *Binational Toxics Strategy Mercury Progress Report* (Environmental Protection Agency, Chicago, 2006)
27. K. Tørseth, Introduction to the European Monitoring and Evaluation Programme (EMEP). Atmos. Chem. Phys. **12**, 5447–5481 (2012)
28. J. Weinberg, *An NGO Introduction to Mercury Pollution* (Gothenburg, International POPs Elimination Network, 2010)
29. M.A. Hawk, J. Castro, The contamination by mercury in Mexico. Gac. Ecol. **72**, 21–34 (2004)
30. A.T. Reis, C.B. Lopes, C.M. Davidson, A.C. Duarte, E. Pereira, Extraction of available and labile fractions of mercury from contaminated soils: the role of operational parameters. Geoderma. **259–260**, 213–223 (2015)
31. L. Ho-Man, W. Zhen-Wen, Y. Zhi-Hong, Y. Kin-Lam, P. Xiao-Ling, C. Kwai-Chung, Interactions between arbuscular mycorrhizae and plants in phytoremediaton of metal-contaminanted soils: a review. Pedosphere **23**, 549–563 (2013)
32. S. Jeevanantham, A. Saravanan, R.V. Hemavathy, P. Senthil Kumar, P.R. Yaashikaa, D. Yuvaraj, Removal of toxic pollutants from water environment by phytoremediation: a survey on application and future propects. Environ. Technol. Innov. **13**, 264–276 (2019)
33. F. Battke, D. Ernst, F. Fleischmann, S. Halbach, Phytoreduction and volatilization of mercury by ascorbate in *Arabidopsis thaliana,* European beech and Norway spruce. Appl. Geochem. **23**, 494–502 (2008)
34. S. Khalid, M. Shahid, N. Kahn Niazi, B. Murtaza, I. Bibi, C. Dumat, A comparison of technologies for remediation of heavy metal contaminated soils. J. Geochem. Explor. **182**, 247–268 (2017)
35. H. Yang, J. Naim, P. Ozias-Akins, Transformation of peanut using a modified bacterial mercuric ion reductase gene driven by an actin promoter from *Arabidopsis thaliana*. J. Plant Physiol. **160**, 945–952 (2003)
36. G. Wang, Z. Guo, W. Liu, Interfacial effects of superhydrophobic plant surfaces: a review. J. Bionic Eng. **11**, 325–345 (2014)
37. J.X. Wang, X.B. Feng, C.W.N. Anderson, G.L.L. Qiu, Z.D. Ping, A. Bao, Ammonium thiosulphate enhanced phytoextraction from mercury contaminated soil—results from a greenhouse study. J. Hazard. Mater. **186**, 119–127 (2011)
38. A. Sas-Nowosielska, R. Galimska-Stypa, R. Kucharski, U. Zielonka, E. Małkowski, L. Gray, Remediation aspect of microbial changes of plant rhizosphere in mercury contaminated soil. Environ. Monit. Assess. **137**, 101–109 (2008)

39. V. Antoniadis, E. Levizou, S.M. Shaheen, Y. Sik Ok, A. Sebastian, C. Baum, M.N.V. Prasad, W.W. Wenzel, J. Rinklebe, Trace elements in the soil–plant interface: phytoavailability, translocation, and phytoremediation: a review. Earth Sci. Rev. **171**, 621–645 (2017)
40. S. Mounicou, M. Shah, J. Meija, J.A. Caruso, A.P. Vonderheide, J. Shan, Localization and speciation of selenium and mercury in *Brassica juncea*—implications for Se-Hg antagonism. J. Anal. At. Spectrom. **21**, 404–412 (2006)
41. R.-H. Wang, X.-F. Zhu, W. Qian, Y.-C. Yu, R.-K. Xu, Effect of pectin in adsorption of Cu(II) of two variable-charge soils from southern China. Environ. Sci. Pollut. Res. **22**, 19687–19694 (2015)
42. X. Wang, N.F. Tam, H. He, Z. Ye, The role of root anatomy, organic acids and iron plaque on mercury accumulation in rice. Plant Soil. **394**, 301–313 (2015)
43. V. Antoniadis, S.M. Shaheen, J. Boersch, T. Frohne, G. Du Laing, J. Rinklebe, Bioavailability and risk assessment of potentially toxic elements in garden edible vegetables and soils around a highly contaminated former mining area in Germany. J. Environ. Manage. **186**, 192–200 (2017)
44. J. Cheng, C. Ding, X. Li, T. Zhang, X. Wang, Heavy metals in navel orange orchards of Xinfeng county and their transfer from soils to navel oranges. Ecotoxicol. Environ. Saf. **122**, 153–158 (2015)
45. F.N. Moreno, C.W.N. Anderson, R.B. Stewart, B.H. Robinson, R. Nomura, M. Ghomshei, J.A. Meech, Effect of thioligands on plant-Hg accumulation and volatilisation from mercury-contaminated mine tailings. Plant Soil. **275**, 233–246 (2005)
46. R. Bennicelli, Z. Stepniewska, A. Banach, K. Szajnocha, J. Ostrowski, The ability of *Azolla caroliniana* to remove heavy metals (Hg(II), Cr(III), Cr(VI)) from municipal wastewater. Chemosphere **55**, 141–146 (2004)
47. M. Del Río, R. Font, C. Almela, D. Velez, R. Montoro, A. De Haro Bailon, Heavy metals and arsenic uptake by wild vegetation in the Guadiamar River area after the toxic spill of the Aznalcollar mine. J. Biotechnol. **98**, 125–137 (2002)
48. J. Luo, L. Cai, S. Qi, J. Wu, X.W.S. Gu, The interactive effects between chelator and electric fields on the leaching risk of metals and the phytoremediation efficiency of *Eucalyptus globulus*. J. Clean. Prod. **202**, 830–837 (2018)
49. J.D. García-García, R. Sánchez-Thomas, R. Moreno-Sánchez, Bio-recovery of non-essential heavy metals by intra- and extracellular mechanisms in free-living microorganisms. Biotechnol. Adv. **34**, 859–873 (2016)
50. J. Canário, C. Vale, L. Poissant, M. Nogueira, M. Pilote, V. Branco, Mercury in sediments and vegetation in a moderately contaminated salt marsh (Tagus Estuary, Portugal). J. Environ. Sci. **22**, 1151–1157 (2010)
51. N.A. Anjum, I. Ahmad, M. Valega, M. Pacheco, E. Figueira, A.C. Duarte, E. Pereira, Impact of seasonal fluctuations on the sediment-mercury, its accumulation and partitioning in *Halimione portulacoides* and *Juncus maritimus* collected from Ria de Aveiro coastal lagoon (Portugal). Water Air Soil Pollut. **222**, 1–15 (2011)
52. A. Marques, E. Lillebu, A. Pereira, A. Duarte, Mercury cycling and sequestration in salt marshes sediments: an ecosystem service provided by *Juncus maritimus* and *Scirpus maritimus*. Environ. Pollut. **159**, 1869–1876 (2011)
53. M. Das, S. Maiti, Comparison between availability of heavy metals in dry and wetland tailing of an abandoned copper tailing pond. Environ. Monit. Assess. **137**, 343–350 (2008)
54. M. Kamal, A. Ghaly, N. Mahmoud, E.R. Coteot, Phytoaccumulation of heavy metals by aquatic plants. Environ. Int. **29**, 1029–1039 (2004)
55. N. Mganga, M. Manoko, Z. Rulangaranga, Classification of plants according to their heavy metal content around North Mara gold mine, Tanzania: implication for phytoremediation. Tanz. J. Sci. **37**, 109–119 (2011)
56. T. Nedelkoska, P. Doran, Characteristics of heavy metal uptake by plant species with potential for phytoremediation and phytomining. Miner. Eng. **13**, 549–561 (2000)

57. R.A. Chirakkara, C. Cameselle, K.R. Reddy, Assessing the applicability of phytoremediation of soils with mixed organic and heavy metal contaminants. Rev. Environ. Sci. Biotechnol. **15**, 299–326 (2016)
58. J.S. Weis, P. Weis, Metal uptake, transport and release by wetland plants: implications for phytoremediation and restoration. Environ. Int. **30**, 685–700 (2004)
59. M. Maobe, E. Gatebe, L. Gitu, H. Rotich, Profile of heavy metals in selected medicinal plants used for the treatment of diabetes, malaria and pneumonia in Kisii region, southwest Kenya. Glob. J. Pharmacol. **6**, 245–251 (2012)
60. M. Malawska, B. Wiołkomirski, An analysis of soil and plant (*Taraxacum officinale*) contamination with heavy metals and polycyclic aromatic hydrocarbons (PAHs) in the area of the railway junction Iława Głowna, Poland. Water Air Soil Pollut. **127**, 339–349 (2001)
61. P. Thangavel, V. Subburam, Effect of trace metals on the restoration potential of leaves of the medicinal plant *Portulaca oleracea* Linn. Biol. Trace Elem. Res. **61**, 313–321 (1998)
62. P. Thangavel, A. Shahira, V. Subburam, Interactive effects of selenium and mercury on the restoration potential of leaves of the medicinal plant *Portulaca oleracea* Linn. Sci. Total Environ. **243–244**, 1–8 (1999)
63. A. Gupta, V. Rai, N. Bagdwal, R. Goel, In situ characterization of mercury resistant growth promoting fluorescent pseudomonads. Microbiol. Res. **160**, 385–388 (2005)
64. A. Munees, Remediation of metalliferous soils through the heavy metal resistant plant growth promoting bacteria: paradigms and prospects. Arab. J. Chem. **12**, 1365–1377 (2014)
65. G. Dermont, M. Bergeron, G. Mercier, M. Richer-Lafleche, Metal-contaminated soils: remediation practices and treatment technologies. Pract. Period. Hazard. Tox. Radioact. Waste Manage. **12**, 188–210 (2008)
66. M.M. Molisani, R. Rocha, W. Machado, R.C. Barreto, L.D. Lacerda, Mercury contents in aquatic macrophytes from two reservoirs in the Paraíba do Sul: Guandú River system, SE Brazil. Braz. J. Biol. **66**, 101–107 (2006)
67. A. Perez-Sanz, R. Millan, M.J. Sierra, R. Alarcon, P. Garcia, M. Gil-Diaz, S. Vazquez, M.C. Lobo, Mercury uptake by *Silene vulgaris* grown on contaminated spiked soils. J. Environ. Manage. **95**, 233–237 (2012)
68. S.P. Bizily, C.L. Rugh, R.B. Meagher, Phytodetoxification of hazardous organomercurials by genetically engineered plants. Nat. Biotechnol. **18**, 213–217 (2000)
69. N. Sarwar, M. Imran, M. Rashid Shaheen, W. Ishaque, M. Asif Kamran, A. Matloob, A. Rehim, S. Husain, Phytoremediation strategies for soils contaminated with heavy metals: modifications and future perspectives. Chemosphere **171**, 710–721 (2017)
70. D. Osman, J. Cavet, Copper homeostasis in bacteria. Adv. Appl. Microbiol. **65**, 217–247 (2008)
71. V. Pavel, D. Sobariu, I. Fertu, F. Statescu, Symbiosis in the environment biomanagement of soils contaminated with heavy metals. Eur. J. Sci. Theol. **9**, 211–224 (2013)
72. M. Arslan Ashraf, I. Hussain, R. Rasheed, M. Iqbal, M. Riaz, M. Saleem Arif, Advances in microbe-assisted reclamation of heavy metal contaminated soils over the last decade: a review. J. Environ. Manage. **198**, 132–143 (2017)
73. C. Adriano, W.W. Wenzel, J. Vangronsveld, N.S. Bolan, Role of assisted natural remediation in environmental cleanup. Geoderma. **122**, 121–144 (2004)
74. J. Xu, A. Garcia Bravo, A. Lagerkvist, S. Bertilsson, R. Jurate Kumpiene, Sources and remediation techniques for mercury contaminated soil. Environ. Int. **74**, 42–53 (2015)
75. T. Barkay, I. Wagner-Dobler, Microbial transformations of mercury: potentials challenges, and achievements in controlling mercury toxicity in the environment. Adv. Appl. Microbiol. **57**, 1–52 (2005)
76. G.M. Gadd, Metals: minerals and microbes: geomicrobiology and bioremediation. Microbiology. **156**, 609–643 (2010)
77. T. Vamerali, M. Bandiera, G. Mosca, Field crops for phytoremediation of metal-contaminated land: a review. Environ. Chem. Lett. **8**, 1–17 (2010)
78. S.L. Doty, Enhancing phytoremediation through the use of transgenics and endophytes. New Phytol. **179**, 318–333 (2008)
79. R. Stegmann, G. Brunner, W. Calmano, G. Matz, *Treatment of Contaminated Soil: Fundamentals, Analysis, Applications* (Springer, Berlin, 2001)

80. I. Robles, T. Serrano, J.J. Pérez, G. Hernández, S. Solis, R. García, T. Pi, E. Bustos, Influence of EDTA on the electrochemical removal of mercury(II) in soil from San Joaquín, Queretaro, Mexico. J. Mex. Chem. Soc. **58**, 333 (2014)
81. G. Acosta-Santoyo, S. Solís, G. Hernández Silva, J. Cárdenas, Z. Plank, E. Bustos, Analysis of the biological recovery of soils contaminated with hydrocarbons using an electrokinetic treatment. J. Hazard. Mater. **371**, 625–633 (2019)
82. K.R. Reddy, C. Cameselle, *Electrochemical Remediation Technologies for Polluted Soils, Sediments and Groundwater* (Wiley, Hoboken, 2009)
83. R. Chirakkara, K.R. Reddy, C. Cameselle, Electrokinetic amendment in phytorremediation of mixed contaminated soil. Electrochim. Acta. **181**, 179–191 (2014)
84. Y.B. Acar, A.N. Alshawabkeh, Principles of electrokinetic remediation. Environ. Sci. Technol. **27**, 2638–2647 (1993)
85. J. Virkutytea, M. Sillanpää, P. Latostenmaa, Electrokinetic soil remediation critical overview. Sci. Total Environ. **289**, 97 (2002)
86. H. Niroumand, R. Nazir, K.A. Kassim, The performance of electrochemical remediation technologies in soil mechanics. Int. J. Electrochem. Sci. **7**, 5708 (2012)
87. H.E. Allen, P.H. Chen, Remediation of metal contaminated soil by EDTA incorporating electrochemical recovery of metal and EDTA. Environ. Prog. **12**, 284 (1993)
88. D. Lestan, C.X. Luo, The use of chelating agents in the remediation of metal contaminated soils: a review. Environ. Pollut. **153**, 3 (2008)
89. M. Haitzer, G.R. Aiken, J.N. Ryan, Binding of mercury(II) to dissolved organic matter: the role of the mercury-to-DOM concentration ratio. Environ. Sci. Technol. **36**, 3564 (2002)
90. I. Robles, M.J. Lozano, S. Solís, G. Hernández, M.V. Paz, M.G. Olvera, E. Bustos, Electrokinetic treatment of mercury-polluted soil facilitated by ethylenediaminetetraacetic acid coupled with a reactor with a permeable reactive barrier of iron to recover mercury(II) from water. Electrochim. Acta. **181**, 68–72 (2015)
91. C.S. Oapos, O. Connor, N.W. Lepp, R. Edwards, G. Sunderland, Tñhe combined use of electrokinetic remediation and phytoremediation to decontaminate metal-polluted soils: a laboratory-scale feasibility study. Environ. Monit. Assess. **84**, 141–158 (2003)
92. I. Robles, J. Lakatos, P. Scharek, S. Solís, Z. Planck, G. Hernandez, E. Bustos, Characterization and remediation of soils and sediments polluted with mercury: occurrence, transformations, environmental considerations and San Joaquin's Sierra Gorda case, in *Environmental Risk Assessment of Soil Contamination*, ed. by M. C. Hernandez Soriano, (InTech, Rijeka, 2014), pp. 827–850
93. K.R. Reddy, C. Chaparro, R.E. Saichek, Removal of mercury from clayey soils using electrokinetics. J. Environ. Sci. Health A Tox. Hazard. Subst. Environ. Eng. **38**, 307–338 (2003)
94. C.D. Cox, M.A. Shoesmith, M.M. Ghosh, Electrokinetic remediation of mercury contaminated soils using iodine/iodide lixiviant. Environ. Sci. Technol. **30**, 1933–1938 (1996)
95. Z. Shen, J. Zhang, L. Qu, Z. Dong, S. Zheng, W. Wang, A modified EK method with an $I^-$/$I_2$ lixiviant assisted and approaching cathodes to remedy mercury contaminated field soils. Environ. Geol. **57**, 1399–1407 (2009)
96. I. Robles, M.G. García, S. Solís, G. Hernández, Y. Bandala, E. Juaristi, E. Bustos, Electroremediation of mercury polluted soil facilitated by complexing agents. Int. J. Electrochem. Sci. **7**, 2276–2287 (2012)
97. H.K. Hansen, L.M. Ottosen, B.K. Kliem, A. Villumsen, Electrodialytic remediation of soils polluted with Cu, Cr, Hg, Pb and Zn. J. Chem. Technol. Biotechnol. **70**, 67–73 (1997)
98. V.C. Pandey, O. Bajpai, N. Singh, Energy crops in sustainable phytoremediation. Renew. Sustain. Energy Rev. **54**, 58–73 (2016)
99. A. Golberg, H.D. Rabinowitch, B. Rubinsky, Zn/Cu–vegetative batteries, bioelectrical characterizations, and primary cost analyses. J. Renew. Sustain. Energy. **2**, 1–11 (2010)
100. Secretaría de Medio Ambiente y Recursos Naturales, *Norma Oficial Mexicana NOM-021-SEMARNAT-2000: especificaciones de fertilidad, salinidad y clasificación de suelos: estu-*

*dios, muestreo y análisis* (Secretaría de Medio Ambiente y Recursos Naturales, Mexico City, 2000)

101. V. W. P. van Engelen, J. A. Dijkshoorn (eds.), *Global and National Soils and Terrain Databases (SOTER). Procedures Manual, Version 2.0* (ISRIC—World Soil Information, Wageningen, 2013)
102. Secretaría de Medio Ambiente y Recursos Naturales, *Norma Oficial Mexicana NOM-004-SEMARNAT-2002: protección ambiental—lodos y biosólidos—especificaciones y límites máximos permisibles de contaminantes para su aprovechamiento y disposición final* (Secretaría de Medio Ambiente y Recursos Naturales, Mexico City, 2002)
103. Environmental Protection Agency, Method 3051: microwave assisted acid digestion of sediments, sludges, soils and oils. EPA test methods for evaluating solid waste, in *Physical/Chemical Methods*, (Environmental Protection Agency, Washington, DC, 2007)
104. American Public Health Association, *Method 3120 B: Metals by Plasma Emission Spectroscopy. Standard Methods for the Examination of Water and Wastewater*, 20th edn. (American Public Health Association, Washington, DC, 1998)
105. M. Perez-Corona, B. Ochoa, J. Cardenas, G. Hernández, S. Solís, R. Fernández, M. Teutli, E. Bustos, Comparison of different arrangements of electrodes during the electrokinetic treatment of polluted soil with hydrocarbons and its final application in situ. Transw. Res. Netw. **661**, 2 (2013)
106. A. Pavlovič, The effect of electrical signals on photosynthesis and respiration, in *Plant electrophysiology: signaling and responses*, ed. by A. Volkov, (Springer, Berlin, 2012), pp. 33–62
107. S.M. Talai, Z.O. Siagi, S.K. Kimutai, S.S. Simiyu, W.T. Ngigi, A.B. Makokha, C. Engineering, Comparative energy generation of Irish-potato, tomato and pineapple Zn/Cu vegetative batteries. Res. J. Appl. Sci. Eng. Technol. **8**(1), 9–19 (2014)
108. A. Golberg, H.D. Rabinowitch, B. Rubinsky, Zn/Cu–vegetative batteries, bioelectrical characterizations, and primary cost analyses. J. Renew. Sustain. Energy **2**, 3 (2010)
109. E.N. La, P.D.E. Curicó, R.D.E.L. Maule, P.O.F. Curicó, M. Región, Evaluación del contenido de almidón en papas. Evaluation **2002**, 41–52 (2010)
110. N. Loyola, E. Oyarce, C. Acuña, Evaluación del contenido de almidon en papas (*Solanum tuberosum* cv. *desirée*), producidas en forma orgánica y convencional, en la provincia de Curicó, región del Maule. Idesia (Arica) **28**, 41–52 (2010)
111. H. Moon, M. Haruki, K. Takahashi, Soil respiration in different forest ecosystems established after volcanic eruptions on Mt. Showa-Shinzan. Res. Bull. Hokkaido Univ. For. **55**, 87–96 (1998)
112. E.R. Saldaña Reacción de Maillard. Cocina Cientifica (2001), https://sites.google.com/site/cocina4ingenieros/ciencia-y-tecnologia/conceptos-basicos/Alimentacion/reaccion-de-maillard
113. E. Jovanov, A. G. Kolvov, Plant electrostimulation and data acquisition, Chapter 2, book of Plant Electrophysiology, A. G. Volkov (ed.), Springer – Verlag Berlin Heidelberg, pp. 45–67 (2012)
114. Secretaria de Medio Ambiente y Recursos Naturales, *Norma Oficial Mexicana NOM-147-SEMARNAT-SSA1-2004: Criterios para determinar las concentraciones de remediación de suelos contaminados por arsénico, bario, berilio, cadmio, cromo hexavalente, mercurio, níquel, plata, plomo, selenio, talio y/o vanadio* (Secretaria de Medio Ambiente y Recursos Naturales, Mexico City, 2004)
115. T.R. Copeland, R.K. Skogerboe, Anodic stripping voltammetry. Anal. Chem. **46**, 1257–1268 (1974)
116. S. de Economía, *NMX-AA-051-SCFI-2001: Análisis de agua—determinación de metales por absorción atómica en aguas naturales, potables, residuales y residuales tratadas—método de prueba (cancela a la NMX-AA-051-1981)* (Mexico City, Secretaría de Economía, 2001)
117. C. Douny, J. Widart, G. Maghuin-Rogister, E. De Pauw, M.-L. Scippo, Quantification of acrylamide in various Belgian potato products using solid phase extraction and liquid chromatography tandem mass spectrometry detection. Food Public Health **2**, 137–141 (2012)

118. A. Khoddami, M.A. Wilkes, T.H. Roberts, Techniques for analysis of plant phenolic compounds. Molecules. **18**, 2328–2375 (2013)
119. A. Panighel, R. Flamini, Applications of solid-phase microextraction and gas chromatography/mass spectrometry (SPME-GC/MS) in the study of grape and wine volatile compounds. Molecules. **19**, 21291–21309 (2014)
120. International Chemical Safety Cards (ICSCs) 0056 – Mercurio, CAS: 7439-97-6 (2019)
121. M. Clugston, R. Flemming, *Advanced chemistry* (Oxford University Press, Oxford, 2000)
122. W.H.O. Ernst, Bioavailability of heavy metals and decontamination of soils by plants. Appl. Geochem. **11**(1–2), 163–167 (1996)
123. FAO/UNESCO, *World Reference Base for Soil Resources* (Food and Agriculture Organization of the United Nations, Rome, 1998)
124. W.T. McGeorge, Diagnosis and improvement of saline and alkaline soils. Soil Sci. Soc. Am. J. **18**, 348 (1954)
125. P. Audet, C. Charest, Heavy metal phytoremediation from a goal-analytical perspective. Environ. Pollut. **147**, 231–237 (2007)
126. S. Solís-Valdez, O. Pérez-Arvizu, M.E. Anzar-Garces, *Content of Total Mercury in Agricultural Crops of the Mine Area of San Joaquín, in the South of the Sierra Gorda, Qro. Querétaro, Qro.* (Mexico City, UNAM, 2009)
127. L. Band, *They Reactivate 12 Mines of Mercury in San Joaquin* (Querétaro, El Diario de Querétaro, 2013)
128. H.–Sh. Moon, M. Haruki, K. Takahashi, Soil respiration in different forest ecosystems established after volcanic eruptions on Mt. Showa – Shinzan. Res. Bull. Hokkaido Univ. Forests. **55**(1), 87–96 (1998)
129. E.R. Saldaña Brisuela, Reacción de Maillard - cocina científica. Cocina Cientifica (2001)
130. J.Y. Yi, Effects of a low-voltage electric pulse charged to culture soil on plant growth and variations of the bacterial community. Agric. Sci. **03**, 339–346 (2012)
131. C.S. Oapos, N.W. Connor, R.E. Lepp, G. Sunderland, The combined use of electrokinetic remediation and phytoremediation to decontaminate metal-polluted soils: a laboratory-scale feasibility study. Environ. Monit. Assess. **84**(1–2), 141–158 (2003)
132. A. Pavlovic, The effect of electrical signals on photosynthesis, Chapter 2, book of Plant Electrophysiology, A. G. Volkov (ed.), Springer – Verlag, Berlin Heidelberg, pp. 33–62 (2012)
133. S.M. Talai, Z.O. Siagi, S.K. Kimutai, S.S. Simiyu, W.T. Ngigi, A.B. Makokha, Comparative energy generation of Irish-potato, tomato and pineapple Zn/Cu vegetative batteries. Res. J. Appl. Sci. Eng. Technol. **8**, 9–19 (2014)
134. S. Abdalla, A.A. Al-Ghamdi, Electrical energy from foods. J. Renew. Sustain. Energy. **3**, 17 (2011)
135. J.Y. Yi, Effects of a low-voltage electric pulse charged to culture soil on plant growth and variations of the bacterial community. Agric. Sci. **3**, 339–346 (2012)
136. J. Dacey, Potato battery could help meet rural energy needs, 20 years Sci. Dev. Net. (2010). https://www.scidev.net/global/news/potato-battery-could-help-meet-rural-energy-needs/
137. A. Golberg, H. D. Rabinowitch, B. Rubinsky, Zn/Cu – vegetative batteries, bioelectrical characterization, and primary cost analyses. J. Renew. Sustain. Energ. **2**(3) 1–11 (2010)

# Electrobioremediation of Polluted Soils

**José Villaseñor Camacho**

## 1 Bioremediation of Polluted Soil

Bioremediation is usually known as the technology used for restoration of polluted sites through the biodegradation of the soil organic pollutants. Its fundamental concept does not differ from the fundamentals or principles of conventional water-suspended biodegradation technology using selected microbes or mixed microbial cultures [1]. Because of the usual occurrence of soil pollution problems, bioremediation is mainly applied to eliminate chemicals such as petroleum hydrocarbons, polyaromatic hydrocarbons, organochlorinated compounds (solvents, pesticides, or herbicides), polychlorinated biphenyls (PCBs), and heavy metals [2]. The pollutant biodegradability, by means of adapted microorganisms, is the key factor affecting the performance of such technology, which in turn depends mainly on the pollutant chemical structure. However, the fact that the pollutant is included in a heterogeneous matrix that could involve up to four different phases (solid, water, organic, and gas phases) supposedly causes the pollutant to be transported and distributed between such phases, making bioremediation a more complex process than a conventional biodegradation process in water [3].

Bioremediation can be achieved by in situ or ex situ options. Ex situ bioremediation is the most commonly used technology: the polluted soil is excavated and removed from its original contaminated site and subsequently soil is transported to be treated using an external bioreactor [4]. In situ bioremediation is not so common and it consists on remediating the soil in its original site, and thus soil excavation and transport is not necessary. The in situ treatments have several advantages among which it can be highlighted the minimal disruption to activities on site or on

J. Villaseñor Camacho (✉)
Chemical Engineering Department, Institute for Chemical and Environmental Technology (ITQUIMA), University of Castilla-La Mancha, Ciudad Real, Spain
e-mail: jose.villasenor@uclm.es



M. A. Rodrigo, E. V. Dos Santos (eds.), *Electrochemically Assisted Remediation of Contaminated Soils*, Environmental Pollution 30,
https://doi.org/10.1007/978-3-030-68140-1_12

adjacent land. It usually involves the movement of air or water through the polluted soil, which is favored by more permeable media and by lower heterogeneity of physical conditions and pollution distribution. However, one of the main disadvantages of in situ treatments is the reduced mobility of microorganisms, pollutants and nutrients into the soil, especially in low permeable soil, causing reduced bioavailability for biodegradation, while ex situ approaches generally offer greater scope for managing conditions to optimize treatment efficiency and for controlling potential spread of pollutants [5].

Bioavailability of nonsoluble pollutants to microorganisms is also one of the main factors that influences soil bioremediation performance. The low water solubility and the adsorption to particulate matter in soil and sediments are important factors that can reduce in situ biodegradation of organic pollutants such as hydrocarbons or nonpolar organochlorines. The rates of desorption and dissolution of such pollutants in interstitial water in soil can be improved by adding surfactants (either biosurfactants or synthetic detergents) to the contaminated zone [6].

There are additional environmental factors such as pH, temperature, salinity, inorganic nutrients and electron acceptors availability that influence bioremediation [7]. Most natural environments have pH values between 5.0 and 9.0 because of the natural buffering capacity of soil that contains carbonates and other minerals. pH values out of this interval can inhibit microbial growth. Regarding temperature, bioremediation is generally carried out under mesophilic conditions (20–40 °C). Low temperatures can kill or inactivate microbes. The biodegradation rate increases with temperature up to a maximum above which the rate declines as enzyme denaturation occurs. Requirements of inorganic nutrients (N and P) depend of the nutrient availability in soil (N and P are usually present in agricultural soils), the nature of pollutants and the type of metabolism (aerobic, anoxic, anaerobic). Thus, electron acceptors availability is also important. The amount of available oxygen will determine whether bioremediation is carried out under aerobic or anaerobic conditions. Hydrocarbons are readily degraded under aerobic conditions while organochlorines can be degraded under anaerobic ones [7]. The depth of pollution in soil is an additional factor that conditions the oxygen availability. Other electron acceptors such as nitrate could be used. Megharaj and Naidu [2] deepen the effect of different conditions and physicochemical characteristics in soil bioremediation technology.

There are different bioremediation approaches that can be selected and applied either in situ or ex situ depending on the characteristics of pollutants and site conditions [1]. Additionally, three possible general strategies can be considered in order to enhance the biological process: natural attenuation, biostimulation, and bioaugmentation [1, 4]. *Natural attenuation* involves slow microbial degradation only if adapted microorganisms are present into the soil, together with processes such as volatilization, sorption and immobilization, but no actions to stimulate the biological process are considered. *Biostimulation* consists of providing favourable conditions for the enhancement of the biological process (through homogenization, addition of nutrients, electron acceptors, or pH buffering) as the polluted soil already contains a native population of microorganisms adapted to biodegradation of the soil pollutants (usually because the pollution episode happened long time ago). Finally, if soil microbial populations to efficiently degrade the pollutants do not exist (what happens in recent spills), inoculation of enriched/acclimated consortia or strains can be provided, and this operation is called *bioaugmentation*. Mixed cultures with a large variety of microorganisms are utilized in bioaugmentation practice [8]. Additional updated information about species and experimental conditions used in bioremediation is available in a recent reported review [9].

Taking into account the influence of the abovementioned factors, in situ bioremediation can be considered as an adequate and cost-effective treatment to eliminate organic pollutants in soil although the transport and contact between microorganisms, water, pollutants, nutrients and electron acceptors in order to stimulate biodegradation is still a challenge, especially in low permeable soil.

## 2 Electrobioremediation of Polluted Soil. Concepts and Objectives

Electrobioremediation (EBR) is a generic name that can be used for different technological approaches focused on biodegradation of pollutants combined with electrochemical methods. The electrobioremediation concept in the present chapter refers to the electrokinetic-enhanced in situ biodegradation of soil pollutants. Its objective is to improve in situ bioremediation by the enhancement of different transport processes by means of the application of low-voltage direct electric currents through the soil. This method would previously assume that the limiting step in the bioremediation process would be the transport processes rather than the pollutants biodegradability.

The basic method to achieve EBR is schemed in Fig. 1, and it consists of the insertion of appropriately distributed electrodes into the soil (or inside electrolyte wells) so that the polluted zone is located between them. The most widely studied arrangement of electrodes is facing two linear rows of them with different polarities because this configuration has been related to better electric current distribution lines, which would produce a well-distributed pattern of electrokinetic flows. However, this is not the only possibility of electrodes arrangement in soil, and configurations in which various electrodes surround a central electrode of the opposite polarity have also been checked [10, 11]. The results reported in these studies suggest that the most effective electrode configuration (from a pollutant transport viewpoint) is a hexagonal arrangement, that is, a ring of six cathodes with a central anode, or vice versa. When DC electric current is applied, many transport processes previously described in previous chapters may occur (such as electroosmosis, electromigration, and electrophoresis). These transport phenomena could help (or disturb, if not correctly applied) biological processes by allowing contact between pollution, nutrients and microorganisms. Electric fields (between 1 and 5 V $cm^{-1}$ approximately) induce microscale dispersion rather than macroscale movement,

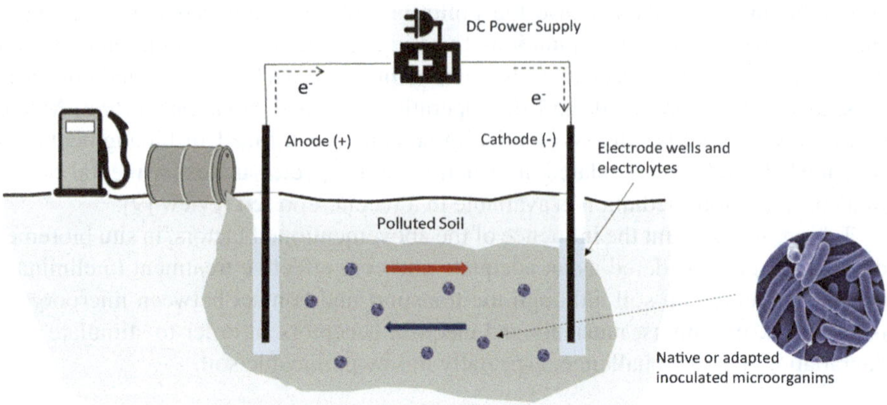


**Fig. 1** Electrokinetic enhanced in situ bioremediation or "electrobioremediation" concept

which was reported to be favourable to the stimulation of the microbial activity [9]. In general, the potential benefits of such technology may include (a) the enhancement of contaminant bioavailability; (b) increase of bacteria mobility; (c) electrokinetic-induced transport of nutrients and electron acceptors and (d) in situ electrochemical generation of electron donors and acceptors [12]. EBR is specially recommended for low permeable soil, as clay soil, where the hydraulic conductivity is very low and hydrodynamic transport would not be suitable. Regarding pollutants, EBR is used mainly for hydrocarbons and organochlorines removal, but also metal remediation has been reported [13]. The first EBR fundamental studies were reported in the 1990s by Marks et al. [14], Ho et al. [15] and Chillingar et al. [16], and subsequently numerous works have expanded and deepened the study of this technology. Wick [17] reported an interesting review about the first decade of EBR works and offered a conceptualization of the "electrobioremediation tetrahedron" which described the critical factors influencing the biotransformation of hydrophobic organic compounds during subsurface electrokinetic treatment. The following sections in the present chapter focus on the main factors that influence EBR, including some relevant findings by many authors.

## 3 Microbial Transport and Activity Under Electrokinetics

One important aspect that must be considered in EBR is the feasibility of microbial mobility and transport into the soil under electrokinetic treatment. This point is important as it opens the possibility of microbial inoculation in bioaugmentation processes (Sect. 5), or just because it enables the possibility of moving microorganisms into the soil to contact pollutants and nutrients in biostimulation processes

(Sect. 4). Microorganisms in the soil under electrokinetic treatment can move by electrophoresis (EF) and electroosmosis (EO), and they have a strong tendency to adhere onto the surface of soil particles [18]. Research works about microbial electrokinetic mobility in soil are usually carried out at bench-scale setups (laboratory model aquifers or soil columns) and batch mode, using electric fields between 0.5 and 4.0 V $cm^{-1}$ approximately. Figure 2 shows an example of a bench-scale installation used to study microbial transport through a soil sample under electrokinetics using different voltages and different soil textures [19].

It has been observed in most of the published works that the negative electric charge on the microbial surface causes the movement of microorganisms towards the anode by EF whereas EO simultaneously moves them towards the cathode. Different works tried to study which is the predominant mechanism. Lee and Lee [20] supplied *Pseudomonas* to a diesel-contaminated soil bed of 15 cm under 40 mA. They observed the transport of diesel-degrading microorganisms towards the anode mainly by EF. The cells acted as negatively charged particles at neutral pH, and they concluded that pH and ionic concentration played an important role. Da Rocha et al. [21] performed electrokinesis on a low hydraulic reconstituted clayey soil column subjected to a 5 mA electrical current for 24 h. They studied the

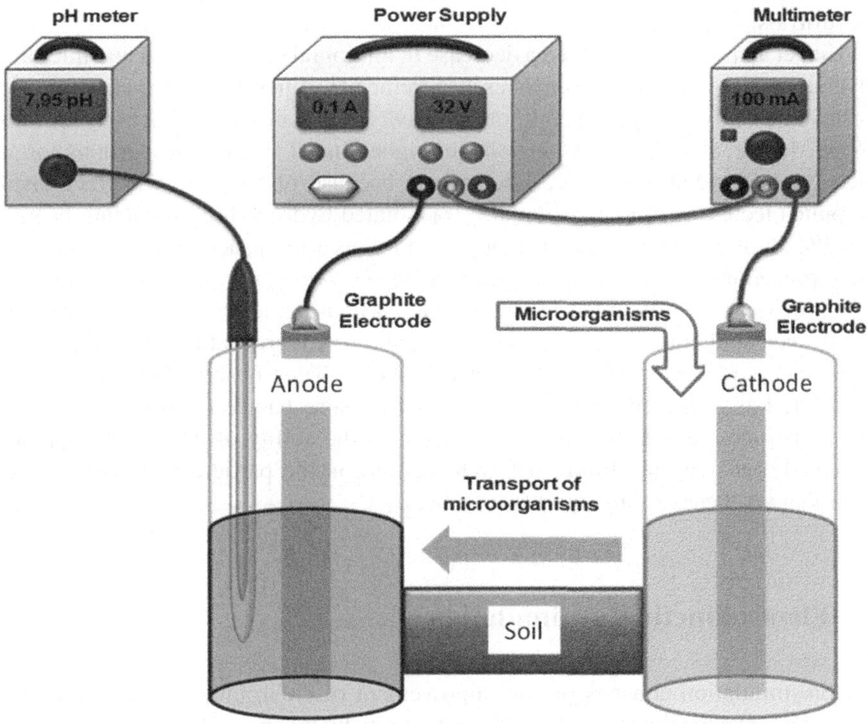


**Fig. 2** Bench-scale installation used to study microbial transport through a soil sample under electrokinetics. (Adapted from Mena et al. [19])

efficacy of EF against the electroosmotic flow to transport endospores of *Bacillus subtilis* LBBMA 155 and nitrogen-starved cells of *Pseudomonas* sp. LBBMA 81. They observed EF to be the predominant mechanism, and they observed that the negative charge on the cells surface played an important influence.

On the contrary, Suni and Romantschuk [22] reported that the microbial mobility of phenol-degrading bacteria in three types of soil (garden soil, fine sand, and clay) was mainly produced by EO, and the transport velocities ranged between 0.1 cm $h^{-1}$ (clay soil) and 1 cm $h^{-1}$ (fine sand). Wick et al. [23] reported the mobility of different strains of hydrocarbon-degrading bacteria that presented different adhesion potential to soil particles, and they found that EO transport mechanism was more important than EF if bacteria were not strongly adhered to soil. Shi et al. [24] observed the important role of EO. They reported microbial velocity of 0.6 cm $min^{-1}$ using hydrocarbon-degrading bacteria under 1 V $cm^{-1}$, and they proposed that the different electrokinetic behaviour of individual cells could be solely attributed to intra-population heterogeneity of the cell surface charge.

There is also an important aspect to be studied such as the possible influence of the electric current on the soil microbial activity, physiology and the microbial diversity. Wick et al. [25] did not observe harmful effects on the soil microbial population when using 1.4 V $cm^{-1}$ and 1 mA $cm^{-2}$ except in areas near the electrodes because of the extreme pH values (<1.5 pH units) caused by water electrolysis.

Kim et al. [26] also observed a decrease in microbial concentration and microbial diversity under 0.6 mA $cm^{-2}$ in zones at extreme pH values, but on the contrary, they observed positive effects on microbial activity and soil enzyme activity in other areas. Mena et al. [27] found that voltages higher than 2 V $cm^{-1}$ caused an important increase in the endogenous cell decay rate because of harmful pH effects using graphite electrodes. Finally, Li et al. [28] isolated hydrocarbon-degrading bacteria capable of growing under electrokinetic conditions after an acclimation and enrichment procedure. They found that strains PB4 (*Pseudomonas fluorescens*) and FB6 (*Kocuria* ***sp.***) were the most efficient hydrocarbon degraders under electrokinetic conditions, and that their degradation capabilities were enhanced compared to experiments without application of electric fields. They observed that the electric field acted as a selective pressure for isolating those bacteria capable of growing under such conditions. However, to understand the ability of these particular species, authors suggested future research focused on the particular biological functions that set these species apart from others.

## 4 Electrokinetic Biostimulation

EK-biostimulation consists on the improvement of environmental conditions into the soil to enhance the in situ bioremediation rate by means of the positive influence of electrokinetic phenomena. However, electrokinetics should be used under controlled conditions as they also could produce disadvantages. It has been previously indicated (Sect. 3) that soil microorganisms can move by EF and EO and that low-voltage DC does not cause negative effects to microorganisms (i.e., no considerable increase in the endogenous cell decay rate is observed), but it is very important to avoid extreme temperatures or pH values, or lack of nutrients and electron acceptors. It is currently assumed that EK will improve the transport and contact between the different species involved in the biological mechanism into the soil although it is necessary to maintain environmental conditions in values adequate for microbial life. Lohner et al. [12] reported data about transport rates between 0.4 and 5 cm $h^{-1}$ approximately for microbial nutrients and electron acceptors (nitrate, sulfate, phosphate, and ammonium) but also limitations because of undesirable side reactions, electroporation, irreversible permeabilization of cell membranes, oxidative stress, and cell death due to electrochemically generated oxidants and electrochemical oxidation of vital cellular constituents.

Most of the scientific works that study EK-biostimulation use bench-scale installations and, because of the experimental conditions in microcosms should be perfectly controlled (approximately: temperatures between 10 and 30 °C, pH close to neutrality, electrical conductivity between 500 and 3000 μS $cm^{-1}$, constant values of moisture and porosity), many of these works use artificially polluted soil in order to simulate real pollution, and they also inoculate acclimated microorganisms into the soil before the experiments in order to simulate the presence of a native microbial population. Then, the objective of batch experimental studies is to know the influence of variables such as pH, temperature, and nutrient concentrations. A scheme of a typical bench-scale setup for EK-biostimulation studies is shown in Fig. 3.

Some works just study the feasibility of the combination of bioremediation and electrokinetics. Yuan et al. [29] studied EBR in soil microcosms that were spiked with *n*-hexadecane at 1.0% (v/w) and inoculated with a mixture of petroleum-degrading bacteria ($10^7$–$10^8$ CFU $g^{-1}$) before being subjected to a constant voltage gradient of 1.3 V $cm^{-1}$ for 42 days. They observed that the degradation rate of *n*-hexadecane by electrobioremediation was up to 53.7%, representing an increase of 20.3% compared to conventional in situ bioremediation without an electric field.

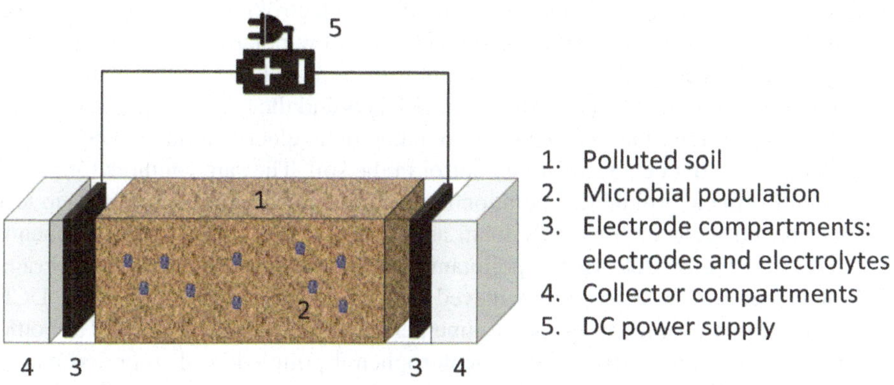


**Fig. 3** Typical bench-scale set-up for batch EK-biostimulation studies

Guo et al. [30] observed the positive effect of the biological–electrochemical combination in EBR experiments (100 days duration, 1 V $cm^{-1}$ and using polarity changes every 5 min) for petroleum hydrocarbons removal in soil. Most of the experimental studies carried out also use simultaneous reference tests (only biological test or only electrokinetic tests) in order to evaluate the feasibility of the combination of both technologies. The following subsections describe specific reported works regarding the influence of some factors such as pH, nutrients, and pollutant bioavailability.

## 4.1 pH Control into the Soil

Previous chapters in the present issue have described the formation of extreme pH fronts into the soil (low pH near the anode and high pH near the cathode) because of water electrolysis. These extreme pH values cause inhibition of the biological mechanisms and the microbial activity is drastically reduced in a few days [31]. Different strategies have been proposed to maintain a suitable pH in the soil during EBR processes [32]. One of the most interesting methods is the periodic change of the polarity of the electric field (the so-called periodic polarity reversal).

Different works have been reported in the last decade regarding the application of polarity reversal in order to control pH during EBR treatment, and some of them also showed beneficial effects in temperature and moisture control. Kim and Han [33] used EK (12.5 V $cm^{-1}$) for clay soil remediation and proposed circulation of anolyte and catholyte between electrode wells. pH was maintained continuously only by circulation of electrolytes ($H_2SO_4$ and $NH_4OH$) in each chamber without any buffering solutions, and pH in soil showed a difference not greater than 0.2 of initial pH. Alternatively, they used periodic polarity reversal and they found that electrode polarity reversal prevented the development of pH gradient, and it was inferred that electrode polarity reversal enabled an effective ion injection, and ions were distributed more uniformly in soil. Luo et al. [34] studied EBR of sandy loam spiked with phenol. They found that nonuniform electrokinetics could accelerate the movement and in situ phenol biodegradation. Low polarity-reversing intervals (every 3 h) induced a higher and more uniform removal of phenol (a maximum removal efficiency of 58% was achieved in 10 days and the bioremediation rate was increased about five times as compared to that with no electric field applied), and it was also observed a better moisture control in the soil. The same authors [35] used 2D nonuniform electrokinetic operation modes (bidirectional and rotational) to test EBR at bench-scale with a sandy loam as the model soil and 2,4-dichlorophenol (2,4-DCP) as the model organic pollutant. At the bidirectional mode, an average 2,4-DCP removal of 73.4% was achieved in 15 days, whereas 34.8% of 2,4-DCP was removed on average in the same time period at the rotational mode. Harbottle et al. [36] studied EBR of pentachlorophenol-polluted soil (approximately 100 mg $kg^{-1}$) and inoculated with a specific pentachlorophenol-degrading bacterium (*Sphingobium* sp. UG30) and subjected to constant and regularly reversed electric currents (10 mA) during different experimental periods (between 36 and 95 days). When both pH and moisture content were controlled using a regularly reversed electric field, instead of unidirectional field, it was found a positive effect on biodegradation of PCP.

Li et al. [37] studied the influence of polarity reversal and electrical intensity on oil removal from soil by EBR. Soil pH remained at around 6.6 obtaining nearly 30% removal rate after 6 weeks when using polarity reversal (1 $h^{-1}$) and 1 V $cm^{-1}$. These authors also studied the biodegradation of Pyrene in contaminated soil under electrokinetic treatment. Three strategies were conducted: In situ conventional bioremediation (Bio), electrobioremediation without polarity-reversal (EK-Bio), and electrobioremediation with polarity-reversal every 2 h (EK-Bio-PR). Pyrene degradation efficiency was 55.9% after 6 weeks under EK-Bio-PR at the end of experiment.

Despite polarity reversal has been proved to be an efficient method for pH control during EBR, no studies were reported regarding the optimization of the reversal frequency. Barba et al. [38] studied EBR of pesticide (oxyfluorfen) polluted clay. Two-weeks duration batch experiments were carried out and used different reversal frequencies between 1 and 6 $day^{-1}$, and they found 2 $day^{-1}$ as optimal value that would produce the optimum pH control and mixture effect in soil.

## *4.2 Availability and Supply of Inorganic Nutrients and Electron Acceptors*

The availability of electron acceptors and nutrients is often a key factor influencing the success of microbiological remediation at contaminated sites. Despite nutrients requirements may be very low (because of the low organic pollutant concentration or because of the anaerobic mechanism) a lack of nutrients can occur in EBR. Moreover, adsorption in soil, chemical precipitation or ion exchange can reduce nutrients bioavailability, and thus nutrient replacement is needed.

Nutrient injection into low permeable soil could be done using electrokinetics as the most important inorganic nutrients (ammonium, nitrate, and phosphate) are ionic substances that could be transported by electromigration, and also by water electroosmosis. However, electrokinetics also could produce nutrients depletion into the soil in long-time EBR processes because of their transport to the electrode wells. Barba et al. [39] found that the biological process could be inactivated in a clay soil EBR study after 11 weeks because of nutrients depletion, despite they were partially supplemented at intermediate operation times.

Some works have been reported regarding electrokinetic injection of nutrients in order to avoid possible inactivation of the bioremediation mechanisms. Schmidt et al. [40] studied the feasibility of injecting inorganic nutrients (ammonium, nitrate, and phosphate) to low permeable clay soil by the addition of prepared solutions into the soil in an electrokinetic cell. Ammonium moved but decreased in the system during the tests because of chemical reactions. Nitrate showed greater mobility and a behaviour less reactive, remaining in the system and only moving towards the anode. Phosphate was not successfully transported, probably because of reactions with calcium carbonate and precipitation; thus, the injection of phosphorous did not prove to be successful. Suni et al. [41] studied enhanced bioremediation of creosote-contaminated soil with a combination of electric heating and infiltration-electrokinetic introduction of oxygenated, nutrient-rich liquid. Nutrient and oxygen levels in the soils were elevated by hydraulic and electrokinetic pumping of urea and phosphate amended, aerated water into the soil. Total hydrocarbon concentrations decreased by 50–80% during 12 weeks of treatment. Xu et al. [42] investigated an EK injection method, which combined electrolyte circulation between electrodes and electrode polarity reversal in bioremediation of phenanthrene-polluted low permeable soil. As expected, soil pH was successfully controlled, but additionally it was also possible supply and distribution of nutrients and electron acceptors (ammonium and nitrate) uniformly in soil. Over 80% of phenanthrene was removed in 20 days.

Regarding the availability of electron acceptors, it is important to notice that oxygen (the most usual acceptor necessary for the most efficient aerobic biodegradation mechanisms) will not be easily available in the low permeable soil. Mena Ramírez et al. [43] studied the feasibility of supply oxygen to a soil by electrokinetics. Oxygen was generated by water electrolysis in the anode and transported to the soil in a bench-scale electrokinetic cell. It was found that oxygen transport was only available in silty and sandy soils by oxygen diffusion, while transport was not possible in clay soil. Moreover, electroosmotic flow in clay soil did not contribute to the transport of oxygen, and only a minimum fraction of the electrolytically generated oxygen was efficiently used.

Nitrate could be alternatively used instead of oxygen if denitrifying microorganisms were able to be used in EBR of organics-polluted soil, and it would be an important advantage as nitrate is easily transported by EK. Additionally, nitrate-polluted soil remediation could be also achieved. Choi et al. [44] studied nitrate removal in soil by electrobioremediation using reducing bacteria (*Bacillus* spp.) and humic substances as electron donor. Thiem et al. [45] studied electrokinetic nitrate transport and toluene biodegradation under denitrifying conditions and different voltage gradients. A denitrifying microbial mixed culture was inoculated. Application of the electric field allowed nitrate migration into toluene-polluted areas and resulted in toluene biodegradation.

Finally, anaerobic treatment has also been reported. Wu et al. [46] studied in situ EBR of tetrachloroethylene low permeable soil by electrokinetic injection of lactate, a common electron donor for anaerobic biodegradation. The soil was inoculated with KB-1® dechlorinators. They concluded that ionic migration delivered organic additives and induced biological activity and complete tetrachloroethylene transformation in soil.

### 4.3 *Bioavailability of Nonpolar Pollutants. Surfactant Applications*

Previous chapters in the present issue discussed the role of surfactants in electrokinetic soil washing to improve the nonpolar pollutant transport and it is known that surfactant addition is a common strategy in electroremediation. However, there are few studies regarding the use of surfactants in electrobioremediation. From the biological perspective, surfactants are also necessary to allow accessibility of microorganisms to nonsoluble organic substrates. Mena et al. [47] used sodium dodecyl sulfate (SDS) to improve diesel hydrocarbons removal by EBR. The same authors [48] used again this surfactant to improve oxyfluorfen removal: SDS was introduced in the electrode compartments and transported into the soil. An optimum amount (2.5 g $L^{-1}$) was found and it increased the oxyfluorfen removal efficiency by 52%.

## 5 Electrokinetic Bioaugmentation

Biodegradation of soil pollutants is usually carried out by soil indigenous microorganisms as they developed the ability to consume the organic pollutants after a long period of time since the spill or pollution episode happened. However, if the spill was recent, and thus the soil was recently polluted, it is possible that no indigenous acclimated microbial population able to degrade the pollutants was present. In these occasions, it is necessary to develop a microbial culture adapted to the biodegradation of such pollutants in external bioreactors, and then introduce it into the soil.

Another situation could be that, although the soil indigenous population is adapted to the soil pollutants biodegradation, the microbial concentration in soil is very low. Bioremediation could be accelerated under high microbial concentration. Thus, it is possible to obtain and isolate indigenous microbial seeds from the soil, in order to develop again external growth processes in bioreactors, and then introduce high amounts of such cultures into the soil.

Both situations correspond to the bioaugmentation option [8]. Electrokinetic bioaugmentation consists on the delivery of microorganisms to the soil by using the EK transport mechanisms, and it is again recommended in low permeable soils where hydraulic advection is not feasible. On one hand, it is necessary to know the performance of the microbial culture under low DC electric fields (which has been previously discussed in Sect. 3). On the other hand, it is necessary to study what are the options to deliver external microorganisms into the soil. EK injection of inorganic nutrients and electron acceptors has been successfully studied (Sect. 4.2) but EK injection of microorganisms is not so common.

Only some previous works has attempted to use EK as the sole delivery mechanism to inoculate nonnative bacteria into soils for bioremediation purposes, and they tried to do this through the injection in the electrode wells or in soil positions far from the electrodes, and they always tried to avoid extreme pH values in the delivery point.

Mao et al. [49] studied EK-enhanced bioaugmentation for remediation of clays contaminated with chlorinated solvents in laboratory experiments under 5 A $m^{-2}$. *Dehalococcoides* (*Dhc*) bacterial strain and lactate ions were uniformly injected in contaminated clay. To bioaugment the soil, the power was turned off and the microbial culture solution was added to the electrode compartments and to a central injection well. After 2 days of acclimation following bioaugmentation, the power supply was turned on to resume the EK operation. The distribution of *Dhc* within the clay suggested that electrokinetic microbial transport was primarily driven by electroosmosis. The injected bacteria were able to survive and grow, and complete effective dechlorination of chlorinated ethene was observed after 94 days.

Secord et al. [50] evaluated the possibility of using EK as a delivery mechanism to introduce organic pollutant degrading bacteria, *Sphingomonas paucimobilis* EPA505 and *Mycobacterium vanbaalenii* PYR-1, into hydrocarbons-polluted low permeability soils. Bacterial cultures were previously grown to exponential phase using hydrocarbons as carbon source, and then transferred to EK reservoirs (one or both bacterial strains were inoculated into either the anolyte or catholyte). It was demonstrated that in situ inoculation of nonnative bacterial species using EK is possible. Although hydrocarbon degradation was not monitored in this study, it was hypothesized that the presence of carbon sources enabled bacteria to thrive in the polluted soil.

One alternative option for EK-bioaugmentation is the application of biological PRBs (permeable reactive barriers), also called "biobarriers." Generally, a PRB is a reactive material that is placed into the soil in the direction of the groundwater flow to help intercept a pollution plume that is carried within an aquifer by degrading or retaining the pollutants [51]. The subsurface pollution plume can be moved through the PRB using the natural hydraulic gradient, or forced by a pump-and-treat method. However, for low permeability soils, the EK mobilization of water is recommended (in that case, EK-PRB is used).

PRBs can be built using different materials that are based on different mechanisms (adsorption with porous high-surface materials, ion exchange with resin-based materials, reduction using elemental metals, biological degradation, etc.). In particular, a barrier based on biological degradation (*Bio*PRB, or biobarrier) is a fixed culture bioreactor that includes a porous supporting material and a microbial biofilm attached on its surface. The working principle of a biobarrier is the same as that of a conventional biofilm reactor. In fact, the only difference is that it is inserted in the soil during EK treatment. Figure 4 shows a conceptual scheme of an EK-*Bio*PRB process and a typical experimental setup used in bench-scale studies.

Although numerous works have been previously reported regarding the use of EK-PRB technology, that is, different types of PRBs coupled to electrokinetics [52], only a limited number of publications regard biological barriers. Fonseca et al. [53] developed EK-*Bio*PRB to treat soils contaminated with hexavalent chromium. The electric field promoted the electromigration of chromium oxyanions towards the anode while the biobarriers, placed before the anode electrode, promoted the reduction and retention of the chromium migrating in its direction. The reactive biobarriers were composed by *Arthrobacter viscosus* bacteria, supported either in activated

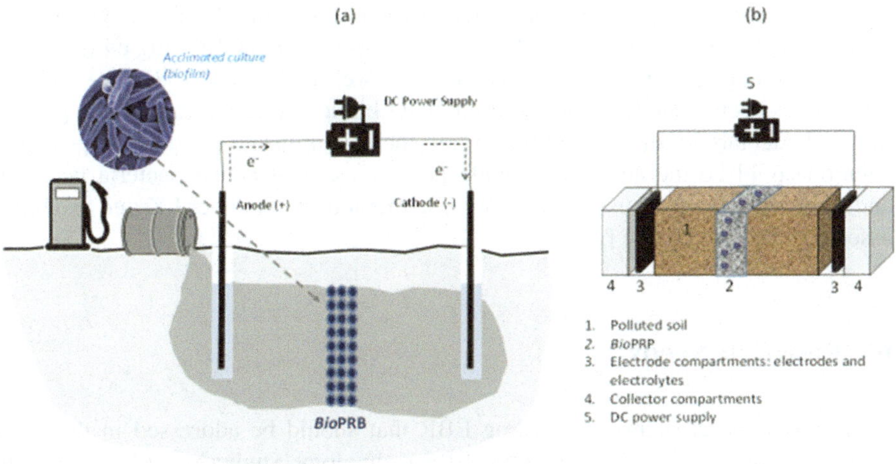


**Fig. 4** (**a**) A conceptual scheme of an EK-*Bio*PRB process and (**b**) a typical experimental set-up used in bench-scale studies

carbon or zeolite. They reported removal values of 60% and 79% under 10 V and after 18 days when electrokinetic treatment was coupled with zeolite and activated carbon biobarriers, respectively.

Mena et al. [54] studied electrobioremediation of diesel hydrocarbons–polluted clay soil by means of coupled electrokinetic soil flushing and biobarriers, using bench scale setups under 0.5 and 1.0 V $cm^{-1}$ and 14 day-long tests. Biobarriers were introduced in a central position of the soil to be treated in order to prevent extreme pH values near the electrodes. They evaluated two types of biobarriers: one of them (BB1) was a fixed-bed biofilm reactor previously developed in the laboratory, with a culture of diesel-degrading microorganisms supported on gravel particles; the other one (BB2) was obtained by mixing clean clay soil with activated sludge obtained from a wastewater treatment plant. Results showed diesel removal rates of 19.36% and 27.36% (using BB1 under 0.5 and 1.0 V $cm^{-1}$, respectively) and of 23.33% and 29.10% (when using BB2) indicating that despite the nonspecific barrier BB2 did not contain an acclimated culture, it reached similar results than BB1. The same authors [55] studied the application of EK-*Bio*PRB for the elimination of organochlorinated pesticides in clay soil. Two compounds were used as model pollutants: oxyfluorfen (a nonpolar pollutant) and 2,4-dichlorophenoxyacetic acid (2,4-D, polar pollutant). The two types of biobarriers previously mentioned were tested again (BB1 and BB2). Compared to the EK-Biostimulation option, it was found that the inclusion of a central biobarrier slightly decreased the pollutant removal rates probably because a decrease in EO flow. However, the EK-*Bio*PRB technology was checked to be a feasible option when no native microbial population is available and there is no option to apply EK-biostimulation.

However, in general, bioaugmentation studies (not only EK-bioaugmentation) have not been successful. It has been repeatedly reported that it is often difficult to maintain the survival of exogenous microbes introduced to a foreign environment [56]. Often the inoculum shows a dramatic decrease in colony forming units (CFU) upon soil inoculation but this behaviour is not well-understood [57]. The lack of success has been related to the formation of antibiotics by native bacteria, or predation and adaptability of external bacteria to the contaminated soil [58]. A possible solution could be the addition of endospores instead of active bacteria because endospores are more robust and migrate faster than bacteria under EK due to a high associated surface charge [21].

## 6 Research Needs

There are some challenges regarding EBR that should be addressed in the near future. One of them is scaling up of the EBR technology. Studies regarding scale-up are scarce. Mena et al. [59] studied EBR of diesel hydrocarbons polluted soil in a prototype (650 kg, 0.25 $m^3$) using *Bio*PRBs, and the bioremediation performance was strongly affected by the high temperatures reached because of ohmic heating. Similar conclusions were obtained by Barba et al. [60]. They used a large prototype (32 $m^3$) and they studied EBR of organochlorinated pollutants. Most of the pollutant removal was caused by volatilization because of the high temperatures reached. It is clear that maintenance of adequate conditions for microbial life should be one of the future research efforts.

Regarding the lack of success of bioaugmentation, successive periodic inoculations after several days of treatment could be applied to compensate microbial mortality in the soil. The use of enzymes instead of microorganisms has also been proposed [58]. The use of enzymes in bioaugmentation can result in avoiding the competition between indigenous bacteria and the new strains. Using enzymes in bioaugmentation could offer additional advantages such as the simplification of the process (they do not generate by-products), it is easier to work with enzymes than with the whole microorganism, and enzyme capabilities can be improved at the production stage. However, the cost of enzyme production is high. Enzyme delivery via electrokinetics transport mechanisms has not been investigated to date.

In conclusion, it is important to note that once the viability of the soil microbial metabolism was assured, electrobioremediation does not increase costs and energy consumption compared to conventional electrokinetic remediation; moreover, the pollutant is not transferred to another phase or matrix but eliminated in situ.

**Acknowledgments** Financial support from the Spanish Government and European Union through projects CTM2016-76197-R (AEI/FEDER, UE) from Ministry of Economy, Industry and Competitiveness, and EQC2018-004240-P from Ministry of Science, Innovation and Universities is gratefully acknowledged.

## References

1. A.A. Juwarkar, S.K. Singh, A. Mudhoo, A comprehensive overview of elements in bioremediation. Rev. Environ. Sci. Biotechnol. **9**, 215–288 (2010)
2. M. Megharaj, R. Naidu, Soil and brownfield bioremediation. Microbial Biotechnol. **10**, 1244–1249 (2017)
3. E.L. Fernández, E.M. Merlo, L.R. Mayor, J.V. Camacho, Kinetic modelling of a diesel-polluted clayey soil bioremediation process. Sci. Total Environ. **557–558**, 276–284 (2016)
4. M.C. Tomei, A.J. Daugulis, Ex situ bioremediation of contaminated soils: an overview of conventional and innovative technologies. Crit. Rev. Environ. Sci. Technol. **43**, 2107–2139 (2013)
5. J. Scullion, Remediating polluted soils. Naturwissenschaften **93**, 51–65 (2006)
6. C.N. Mulligan, Recent advances in the environmental applications of biosurfactants. Curr. Opin. Colloid Interface Sci. **14**, 372–378 (2009)
7. M. Vidali, Bioremediation. An overview. Pure Appl. Chem. **73**, 1163–1172 (2001)
8. S. Di Toro, G. Zanaroli, F. Fava, Intensification of the aerobic bioremediation of an actual site soil historically contaminated by polychlorinated biphenyls (PCBs) through bioaugmentation with a nonacclimated, complex source of microorganisms. Microbial. Cell Fact. **5**, 11–20 (2006)
9. C.M. Quintella, A.M.T. Mata, L.C.P. Lima, Overview of bioremediation with technology assessment and emphasis on fungal bioremediation of oil contaminated soils. J. Environ. Manage. **241**, 156–166 (2019)
10. C. Risco, H. Rubí-Juárez, S. Rodrigo, R. López-Vizcaíno, C. Saez, P. Cañizares, C. Barrera-Díaz, V. Navarro, M.A. Rodrigo, Removal of oxyfluorfen from spiked soils using electrokinetic soil flushing with the surrounding arrangements of electrodes. Sci. Total Environ. **559**, 94–102 (2016)
11. G.L. Grundmann, Spatial scales of soil bacterial diversity—the size of a clone. FEMS Microbiol. Ecol. **48**, 119–127 (2004)
12. S.T. Lohner, A. Tiehm, S.A. Jackman, P. Carter, Coupled Electrokinetic-bioremediation: applied aspects, in *Electrochemical Remediation Technologies for Polluted Soils, Sediments and Groundwater*, ed. by K. R. Reddy, C. Cameselle, (Wiley, Hoboken, 2009), pp. 389–416
13. K.Y. Lee, H.A. Kim, B.T. Lee, S.O. Kim, Y.H. Kwon, K.W. Kim, A feasibility study on bioelectrokinetics for the removal of heavy metals from tailing soil. Environ. Geochem. Health **33**, 3–11 (2011)
14. R.E. Marks, Y.B. Acar, R.J. Gale, In situ remediation of contaminated soils containing hazardous mixed wastes by bioelectric remediation and other competitive technologies, in *Remediation of Hazardous Waste Contaminated Soils*, ed. by D. L. Wise, D. J. Trantolo, (Marcel Dekker, New York, 1994), pp. 405–436
15. S.V. Ho, P.W. Sheridan, C.J. Athmer, M.A. Heitkamp, J.M. Brackin, D. Weber, P.H. Brodsky, Integrated in situ soil remediation technology: the Lasagna process. Environ. Sci. Technol. **29**, 2528–2534 (1995)
16. G.V. Chillingar, W.W. Loo, L.F. Khilyuk, S.A. Katz, Electrobioremediation of soils contaminated with hydrocarbons and metals: progress report. Energy Source. **19**, 129–146 (1997)
17. L.Y. Wick, L. Shi, H. Harms, Electro-bioremediation of hydrophobic organic soil-contaminants: a review of fundamental interactions. Electrochim. Acta **52**, 3441–3448 (2007)
18. M.F. Deflaun, C.W. Condee, Electrokinetic transport of bacteria. J. Hazard. Mater. **55**, 263–277 (1997)
19. E. Mena, P. Rubio, P. Cañizares, J. Villaseñor, M.A. Rodrigo, Electrokinetic transport of diesel-degrading microorganisms through soils of different textures using electric fields. J. Environ. Sci. Health A **47**, 274–279 (2012)
20. H.S. Lee, K. Lee, Bioremediation of diesel-contaminated soil by bacterial cells transported by electrokinetics. J. Microbiol. Biotechnol. **11**, 1038–1045 (2001)

21. U.N. Da Rocha, M.R. Tótola, D.M.M. Pessoa, J.T.A. Júnior, J.C.L. Neves, A.C. Borges, Mobilisation of bacteria in a fine-grained residual soil by electrophoresis. J. Hazard. Mater. **161**, 485–491 (2009)
22. S. Suni, M. Romantschuk, Mobilisation of bacteria in soils by electro-osmosis. FEMS Microbiol. Ecol. **49**, 51–57 (2004)
23. L.Y. Wick, P.A. Mattle, P. Wattiau, H. Harms, Electrokinetic transport of PAH-degrading bacteria in model aquifers and soil. Environ. Sci. Technol. **38**, 4596–4602 (2004)
24. L. Shi, S. Müller, H. Harms, L.Y. Wick, Factors influencing the electrokinetic dispersion of PAH-degrading bacteria in a laboratory model aquifer. Appl. Microbiol. Biotechnol. **80**, 507–515 (2008)
25. L.Y. Wick, F. Buchholz, I. Fetzer, S. Kleinsteuber, C. Härtig, L. Shi, A. Miltner, H. Harms, G.N. Pucci, Responses of soil microbial communities to weak electric fields. Sci. Total Environ. **408**, 4886–4893 (2010)
26. S.H. Kim, H.Y. Han, Y.J. Lee, C.W. Kim, J.W. Yang, Effect of electrokinetic remediation on indigenous microbial activity and community within diesel contaminated soil. Sci. Total Environ. **408**, 3162–3168 (2010)
27. E. Mena, J. Villaseñor, P. Cañizares, M.A. Rodrigo, Effect of a direct electric current on the activity of a hydrocarbon-degrading microorganism culture used as the flushing liquid in soil remediation processes. Sep. Purif. Technol. **124**, 217–223 (2014)
28. F. Li, S. Guo, N. Hartog, Y. Yuan, X. Yang, Isolation and characterization of heavy polycyclic aromatic hydrocarbon-degrading bacteria adapted to electrokinetic conditions. Biodegradation **27**, 1–13 (2016)
29. Y. Yuan, S.H. Guo, F.M. Li, T.T. Li, Effect of an electric field on n-hexadecane microbial degradation in contaminated soil. Int. Biodeter. Biodegr. **77**, 78–84 (2013)
30. S. Guo, R. Fan, T. Li, N. Hartog, F. Li, X. Yang, Synergistic effects of bioremediation and electrokinetics in the remediation of petroleum-contaminated soil. Chemosphere **109**, 226–233 (2014)
31. E. Mena Ramírez, J. Villaseñor Camacho, M.A. Rodrigo Rodrigo, P. Cañizares Cañizares, Combination of bioremediation and electrokinetics for the in-situ treatment of diesel polluted soil: a comparison of strategies. Sci. Total Environ. **533**, 307–316 (2015)
32. A.T. Yeung, Y.Y. Gu, A review on techniques to enhance electrochemical remediation of contaminated soils. J. Hazard. Mater. **195**, 11–29 (2011)
33. S.S. Kim, S.J. Han, Application of an enhanced electrokinetic ion injection system to bioremediation. Water Air Soil Pollut. **146**, 365–377 (2003)
34. Q. Luo, H. Wang, X. Zhang, X. Fan, Y. Qian, In situ bioelectrokinetic remediation of phenol-contaminated soil by use of an electrode matrix and a rotational operation mode. Chemosphere **64**, 415–422 (2006)
35. X. Fan, H. Wang, Q. Luo, J. Ma, X. Zhang, The use of 2D non-uniform electric field to enhance in situ bioremediation of 2,4-dichlorophenol-contaminated soil. J. Hazard. Mater. **148**, 29–37 (2007)
36. M.J. Harbottle, G. Lear, G.C. Sills, I.P. Thompson, Enhanced biodegradation of pentachlorophenol in unsaturated soil using reversed field electrokinetics. J. Environ. Manage. **90**, 1893–1900 (2009)
37. T. Li, S. Guo, B. Wu, L. Zhang, Y. Gao, Effect of polarity-reversal and electrical intensity on the oil removal from soil. J. Chem. Technol. Biotechnol. **90**, 441–448 (2015)
38. S. Barba, J. Villaseñor, M.A. Rodrigo, P. Cañizares, Effect of the polarity reversal frequency in the electrokinetic-biological remediation of oxyfluorfen polluted soil. Chemosphere **177**, 120–127 (2017)
39. S. Barba, J. Villaseñor, M.A. Rodrigo, P. Cañizares, Can electro-bioremediation of polluted soils perform as a self-sustainable process? J. Appl. Electrochem. **48**, 579–588 (2018)
40. C.A.B. Schmidt, M.C. Barbosa, M.D.S.S. de Almeida, A laboratory feasibility study on electrokinetic injection of nutrients on an organic, tropical, clayey soil. J. Hazard. Mater. **143**, 655–661 (2007)

41. S. Suni, E. Malinen, J. Kosonen, H. Silvennoinen, M. Romantschuk, Electrokinetically enhanced bioremediation of creosote-contaminated soil: laboratory and field studies. J. Environ. Sci. Health A **42**, 277–287 (2007)
42. W. Xu, C. Wang, H. Liu, Z. Zhang, H. Sun, A laboratory feasibility study on a new electrokinetic nutrient injection pattern and bioremediation of phenanthrene in a clayey soil. J. Hazard. Mater. **184**, 798–804 (2010)
43. E. Mena Ramírez, J. Villaseñor Camacho, M.A. Rodrigo Rodrigo, P. Cañizares Cañizares, Feasibility of electrokinetic oxygen supply for soil bioremediation purposes. Chemosphere **117**, 382–387 (2014)
44. J.H. Choi, S. Maruthamuthu, H.G. Lee, T.H. Ha, J.H. Bae, Nitrate removal by electro-bioremediation technology in Korean soil. J. Hazard. Mater. **168**, 1208–1216 (2009)
45. A. Tiehm, T. Augenstein, D. Ilieva, H. Schell, C. Weidlich, K.M. Mangold, Bio-electro-remediation: electrokinetic transport of nitrate in a flow-through system for enhanced toluene biodegradation. J. Appl. Electrochem. **40**, 1263–1268 (2010)
46. X. Wu, D.B. Gent, J.L. Davis, A.N. Alshawabkeh, Lactate injection by electric currents for bioremediation of tetrachloroethylene in clay. Electrochim. Acta **86**, 157–163 (2012)
47. E. Mena, C. Ruiz, J. Villaseñor, M.A. Rodrigo, P. Cañizares, Biological permeable reactive barriers coupled with electrokinetic soil flushing for the treatment of diesel polluted clay soil. J. Hazard. Mater. **283**, 131–139 (2015)
48. S. Barba, M. Carvela, J. Villaseñor, M.A. Rodrigo, P. Cañizares, Improvement of the electro-bioremediation process of a non-polar herbicide polluted soil by means of surfactant addition. Sci. Total Environ. **650**, 1961–1968 (2019)
49. X. Mao, J. Wang, A. Ciblak, E.E. Cox, C. Riis, M. Terkelsen, D.B. Gent, A.N. Alshawabkeh, Electrokinetic-enhanced bioaugmentation for remediation of chlorinated solvents contaminated clay. J. Hazard. Mater. **213–214**, 311–317 (2012)
50. E.L. Secord, A. Kottara, P. Van Cappellen, A.T. Lima, Inoculating bacteria into polycyclic aromatic hydrocarbon-contaminated oil sands soil by means of electrokinetics. Water Air Soil Pollut. **227**, 288–301 (2016)
51. E.K. Nyer, *In situ treatment technology* (Lewis Publishers, Boca Raton, 2001)
52. H.I. Chung, M. Lee, A new method for remedial treatment of contaminated clayey soils by electrokinetics coupled with permeable reactive barriers. Electrochim. Acta **52**, 3427–3431 (2007)
53. B. Fonseca, M. Pazos, T. Tavares, M.A. Sanromán, Removal of hexavalent chromium of contaminated soil by coupling electrokinetic remediation and permeable reactive biobarriers. Environ. Sci. Pollut. Res. **19**, 1800–1808 (2012)
54. E. Mena, J. Villaseñor, P. Cañizares, M.A. Rodrigo, Influence of electric field on the remediation of polluted soil using a biobarrier assisted electro-bioremediation process. Electrochim. Acta **190**, 294–304 (2016)
55. S. Barba, M. Carvela, J. Villaseñor, M.A. Rodrigo, P. Cañizares, In-situ electro-bioremediation of soil polluted with organochlorinated compounds, in: *7th European Bioremediation Conference & 11th ISEB Conference*, Chania, 25–28 June 2018
56. M. Megharaj, B. Ramakrishnan, K. Venkateswarlu, N. Sethunathan, R. Naidu, Bioremediation approaches for organic pollutants: a critical perspective. Environ. Int. **37**, 1362–1375 (2011)
57. T.T. Fida, S.K. Moreno-Forero, P. Breugelmans, H.J. Heipieper, W.F.M. Röling, D. Springael, Physiological and transcriptome response of the polycyclic aromatic hydrocarbon degrading *Novosphingobium sp.* LH128 after inoculation in soil. Environ. Sci. Technol. **51**, 1570–1579 (2017)
58. I. Hassan, E. Mohamedelhassan, E.K. Yanful, Z.C. Yuan, A review article: electrokinetic bioremediation current knowledge and new prospects. Adv. Microbiol. **6**, 57–72 (2016)
59. E. Mena, S. Barba, C. Sáez, V. Navarro, J. Villaseñor, M.A. Rodrigo, P. Cañizares, Pre scale-up of electrobioremediation processes, in *Geo-Chicago 2016. Sustainable Waste Management and Remediation, Geotechnical Special Publication, 273*, (American Society of Civil Engineers, Reston, 2016), pp. 264–273
60. S. Barba, R. López-Vizcaíno, C. Saez, J. Villaseñor, P. Cañizares, V. Navarro, M. Rodrigo, Electro-bioremediation at the prototype scale: what it should be learned for the scale-up. Chem. Eng. J. **334**, 2030–2038 (2018)

# The Soil and Groundwater Remediation with Zero-Valent Iron Nanoparticles

**Jorge Rodrigues Gonçalves and Margarida Delgado Alves**

## 1 Introduction

The present study was organized and planned in order to understand the behavior of the nanoparticles of zero-valent iron (nZVI) when applied in the natural environment where the soil was contaminated by heavy metals. The main knowledge about the nZVI behavior is mostly dedicated to the organic solvents; however, the intention of this project was to test the effective reaction of the nanoparticles in contaminated environments, taking into account the geological context and the target contaminating elements, but also all the logistical and technical aspects that a decontamination intervention of this type involves.

It was not intended, with this project, to study in detail all the chemical processes associated with the chemical reactions that may be involved. The main objective was focused on the effectiveness of the applied methodology, and the possible replication in other geological and chemical contexts.

The possibility of mobilizing less logistical means, no soil mobilization needed due to its in situ treatment and the option of changing its future use in a different industrial context, were the objectives of this project. The simplification of the processes and environmental protection by the minimization of side effects were also considered.

The Barreiro Industrial area, currently managed by the Baia do Tejo Society, is of vital importance in the context of maritime and river transport and consequently on the financial aspects that an infrastructure of this nature brings to the region. Despite the successive improvements that this old industrial area has been suffering,

J. R. Gonçalves (✉)
JG.Geologia-Design & Consulting, Lisbon, Portugal

M. D. Alves
LYNX – Engineering, Lisbon, Portugal



M. A. Rodrigo, E. V. Dos Santos (eds.), *Electrochemically Assisted Remediation of Contaminated Soils*, Environmental Pollution 30,
https://doi.org/10.1007/978-3-030-68140-1_13

there are deep scars in terms of environmental liabilities and any intervention to make this industrial space will involve, not only improving the access conditions and port infrastructure, but also the need of a careful soils management.

Despite the superficial removal of old deposits of ash and pyrite slag that occurred between the years 2008 and 2010, contaminated subsurface soils, with laboring residues, remained on this extensive industrial area.

The groundwater level reveals the presence of several of those contaminants, resulting from the continuous leaching that these contaminants have been suffering over the years. The groundwater level varies between 4.5 and 5.0 m below the ground surface, due to the proximity of the Tagus River.

The migration of the contaminants dissolved in the groundwater creates a significant environmental problem in this old industrial site, moving continuously to the Tagus river, affecting also the surrounding aquifer areas. The absence of effective decontamination measures allowed the maturation phenomenon of those contaminants in the environment, delaying the soil's treatment and consequently of their existing aquifers. The environmental impact, whether by these soils removal for hazardous or non-waste landfill, or treatment centers, or by the incapability, without prior treatment, to reuse these soils in construction processes, has led to this concern being addressed in a different way, using alternative methodologies for the remediation of those contaminated soils. The use of specific nZVI technology, which promotes the soil and groundwater decontamination, avoiding the soils removal, enhancing a minimum environmental impact with a maximization of the remediation process, is thus a clear advantage.

The creation/installation of reactive barriers that would limit the pollutants mobilization and could keep low their respective concentrations in case of continuous mobilization, were also the objectives of this project.

Several countries in Europe, but not limited to, have a positive track record, where several patents have been implemented (e.g., Nanoiron®), and where the soils remediation using nZVI is an emerging technology.[1,2]

In Portugal this new green technology made the first steps, under an investigation and experimental process performed in Barreiro's industrial area. In the present report, the methodology, the obtained data, and further conclusions are presented.

## 2 The Zero-Valent Iron Nanoparticles

The soil's remediation with the nanoparticles is based on the application of an aqueous suspensions to change the structure and/or degrade contaminants in soil and/or groundwater. Several definitions for the term "nanoparticle" can be found as a

[1] https://nanoiron.cz/en/application.

[2] https://nanoiron.cz/en/news.

particle having one or more dimensions of 100 nm or less, which can be written as $10^{-9}$ m [1].

Nanoparticles of $Fe^{(0)}$ have recently become a strategic material with great application potential in the broad range of modern nanotechnologies. Due to their extraordinary reduction capabilities (Fig. 1), small size in the range of several tens of nanometers, and high reactivity with a broad spectrum of toxic substances, these ultrafine particles are highly applicable in the reduction technologies of groundwater remediation and wastewater treatment.

In comparison with other frequently used procedures for water treatment, the treatment exploiting of $Fe^{(0)}$ nanoparticles represents environmentally friendly technology since non-toxic and nature-abundant iron oxides (mainly magnetite—$Fe_3O_4$) are the products of transformation of $Fe^{(0)}$.

This Fig. 1 also shows the porosity of the nanoparticle, which is also an important factor in physicochemical reactivity. Zero-valent nano-iron can also be coupled with trace metals (e.g., Pt, Pd, Ag), showing significantly enhanced reaction. Nano-iron is potentially benign to the environment and, ultimately, is mainly transformed into $Fe_3O_4$ and $Fe_2O_3$, which are abundant on earth (Copyright © 2007 GeoNano Environ. Tech., Inc., in [1]). Elemental iron slowly oxidizes to ferrous iron and releases two electrons. These electrons begin to function in a variety of reactions that lead to the transformation of target contaminants. In the environment and in accordance with the half reaction (Eq. 1), elemental iron is oxidized by several substances under the following oxidations half reaction.

$$Fe^0(s) \rightarrow Fe^{2+}(aq) + 2^{e-}(aq) \tag{1}$$

NANOFER 25S® was applied for the Barreiro project. An inorganic and a biodegradable organic modifier were added as stabilizers for the dispersion of the $Fe^{(0)}$ in an aqueous form. The zero-valent iron nanoparticles have, in average, a size of

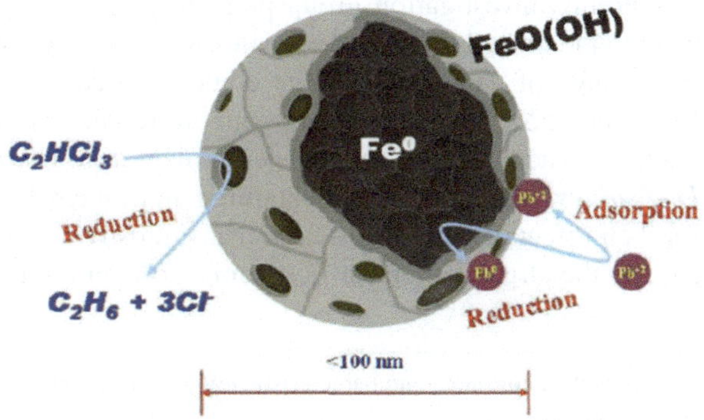


**Fig. 1** Schematic three-dimensional presentation of the nZVI regarding the double effect of reduction process and adsorption of the heavy metal on its surface. (Copyright © 2007 GeoNano Environ. Tech., Inc., in [1])

**Table 1** Nano zero-valent iron specifications

| | |
|---|---|
| NANOFER25S® chemical composition | Core (Fe)/Capsule (FeO) |
| Shape of the particle | Spherical |
| Solid fraction (FeO mass) | 80% |
| Solution density | 1210 kg/m$^3$ |
| Solution (%/mass) | 20% |
| Substances (liquid fraction) | Organic stabilizer |
| Substances (solid fraction) | FeO, $Fe_3O_4$, C |
| Particle size (FeO) | d50 nm < 50 |
| Color | Black |
| Density (FeO) | 7870 kg/m$^3$ |
| Density ($Fe_3O_4$) | 5700 kg/m$^3$ |
| Specific surface | >25 m$^2$/g |

50 nm, the surface area varies between 20 and 25 m$^2$/g, a high content of iron between 80 and 90 wt%,[3] and a particle size distribution of 20–100 nm.

The NANOFER25S®, giving the producer[4] data, has the following composition summarized in Table 1.

## 2.1 *The Environmental Impact of the Use of Nanoparticles*

The European NanoRem project[5] addressed this issue and as part of the project, increasing knowledge, confidence and providing regulators and other stakeholders evidences whether the technology is environmentally harmful to the natural ecosystem functions of soil and groundwater. Uncertainties about the environmental impact of reactive nanoparticles' use to ecosystems is identified as a key factor to the restricted use of them in soil and groundwater remediation. According to recent NanoRem's investigation group press release, no significant toxicological effects were found on soil or water organisms when ecotoxicological tests were undertaken for a range of nanoparticles that could be used for remediation projects, namely ***NanoFer 25S***, made from nanoscale zero-valent iron; ***Carbo-Iron***, a composite made from activated carbon and zero-valent iron; ***Fe-Oxide***, nanoscale goethite and ***Fe-Zeolites***, aluminosilicate containing an iron catalyst.[6] Nanoparticles were tested for their effects on a range of organisms, mostly using standard methods published, e.g., by the Organization for Economic Co-operation and Development (OECD).[7]

[3] http://www.nanoiron.cz/en/characteristics-of-iron-nanoparticles.

[4] www.nanoiron.cz.

[5] http://www.nanorem.eu/Displaynews.aspx?ID=824.

[6] http://www.nanorem.eu/displayworkpackage.aspx?id=3.

[7] http://www.nanorem.eu/displayworkpackage.aspx?id=5.

Ecotoxicity testing will continue for any new nanoparticles or formulations developed as the NanoRem project progresses. The project has also been looking at how nanoparticles reactivity and toxicity change with time. It is believed that as nanoparticles interact within the soil matrices they become less reactive, and therefore less toxic with time. NanoRem's findings confirm this anticipated trend which is similar to how chemicals react in soil. As chemical contaminants age in the soil, their reactivity is reduced along with their bioavailability and toxicity.

## 3 The Tested Area Framework

The investigation program was conducted inside a brownfield site, located in southern Lisbon, in the left bank of the Tagus River (Fig. 2), in the town of Barreiro, between December 2010 and November 2011. Given the estuarine conditions, the tidal effect influences the water table intersected in the test area, which is positioned approximately 4.5–5.0 m deep, from the ground surface.

### *3.1 Geologic and Hydrogeological Local Context*

The ground of the testing area belongs to the Pliocene detrital formations (PSM-Formação de Santa Marta; [2]) (Fig. 2).

From the base to the top occurs a discontinuous conglomerate layer, followed by fine to coarse arkosic sand of fluvial genesis. The colors range from white to red or yellow. Blocks of Cretaceous sandstones and chert nodules occur within the sands. Kaolinite and illite predominate in the clay fraction.

Boreholes performed in this area intersected coarse to fine yellow to orange sand, with a silty-clay matrix, and some interbedded centimetric clay levels with reddish color [3].

In terms of effective porosity, sustained in the fact that the local soils are mainly coarse and medium sands with silt and clays, a value of 15% [4] was considered.

This area is part of the left margin aquifer system of the Tagus-Sado Basin, a multi aquifer, free, confined or semi-confined, where the lateral and vertical facies variations are responsible for significant changes in hydrogeological conditions.

Given the proximity of the mouth of the Tagus River, occurs in the testing area the influence of tidal effect on the groundwater level, which is positioned between 4.0 and 5.0 m deep. The observations of the tidal effect on the piezometric levels in the tested area allowed estimating the hydraulic diffusivity of the aquifer.

The determination of the hydraulic diffusivity in confined aquifers is possible thanks to the work of Jacob [5], Ferris [6], van der Kamp [7], and de Cazenove [8]. The latter also presents solutions for semi-confined aquifers. Regarding the free aquifers, the development of appropriate solutions to describe the tides propagation phenomenon has been the subject of recent work: Ataie-Ashtiani et al. [9, 10], Chen

Extract from the Portuguese Geological Map – Part 34-D from Lisbon, scale 1:50 000

**Fig. 2** Location of the Barreiro site area in the geological map. (Extract from the Portuguese Geological Map – Part 34-D from Lisbon, scale 1:50,000)

et al. [11], Li and Jeng [12], Li and Jiao [13], Pandit et al. [14], Smith and Hick [15], and Wang and Tsay [16]. A summary of different studies and methods that analyze this phenomenon can be found in Li and Jiao [17].

Almeida and Silva [18] applied the equations previously developed in Algarve aquifers. The diffusivity can be obtained from the aquifer oscillations amplitude or from the discrepancy between the maximum (or minimum) of the tide and the maximum (or minimum) of the aquifer oscillation.

From the ratio between amplitudes, the diffusivity can be obtained by the expression:

$$D = \frac{x^2 \pi}{t_0 \cdot \ln^2 \left( \Delta h_0 / \Delta H_0 \right)} \tag{2}$$

From the discrepancy, the following expression can be used:

$$D = \frac{x^2 t_0}{4\pi t_L^2} \tag{3}$$

where $D$ is the diffusivity (m$^2$/h), $x$ the distance between the piezometer and the coast, $t_0$ the tide period, $t_L$ the discrepancy (h), $\Delta h_0$ and $\Delta H_0$, the oscillation semi-amplitudes in the piezometer and tide, respectively.

Although the tide results from the combination of several harmonics, short period of observation is sufficient to consider the most important one, with a period of 745′.

The expression given by de Cazenove [8]:

$$D = \frac{x^2}{2t_L \cdot \left| \ln\left(\Delta h_0 / \Delta H_0\right) \right|} \tag{4}$$

allows to obtain the diffusivity by simultaneously using the amplitude and discrepancies.

In this case it is possible to obtain an additional parameter, $\rho$, related with the drenancy factor, $\lambda = \sqrt{KBB' / K'}$, where $K$ is the aquifer hydraulic conductivity, $K'$ is the aquitard hydraulic conductivity, $B$ is the thickness of the aquifer, and $B'$ is the thickness of the aquitard $x_0 = \sqrt{Tt_0 / \pi S}$),

$$\rho^2 - 1/\rho^2 = x_0^2 / \lambda^2 \tag{5}$$

In order to obtain a better accuracy in the determination of the oscillation amplitude in the river and piezometer, and the discrepancies between the maximum (or minimum) in both locations, equivalent sinusoids were adjusted.

From the ratio between amplitudes and the discrepancies value, diffusivity calculations were performed assuming a 580 m distance between the testing area and the river. This distance is only an approximation given the irregular contour of the river front in the nearest area of the site test.

The equivalent sinusoids were obtained by a nonlinear optimization method, minimizing the squares of the differences between the observed values and the corresponding values of the equivalent sinusoid (MARSINUS program, Almeida and Silva [18]).

According to Caldeira et al. [3], the diffusivity of this aquifer system, considering the drenancy, is 18.344 m$^2$/h, the drenancy factor is 432 m$^2$, and $\rho$ is 2.1416. The obtained result is compatible with a hydraulic conductivity of $5 \times 10^{-5}$ m/s (equivalent to 0.18 m/h), a 10 m thickness, and a storage coefficient of $10^{-4}$.

The obtained diffusivity value is clearly compatible with a semi-confined aquifer system.

Groundwater studies defined the local hydrological system as an aggressive environment due to its low pH varying between 2.9 and 4.1, which is mainly related with the ground and groundwater high concentration of sulfates.

# 4 The Site Investigation

In the area where the tests took place, an industrial complex worked for more than a half-century, having as its core business the manufacture of fertilizers and sulfuric acid (from massive polymetallic sulfides). The in situ methodology of soil and groundwater remediation used zero-valent iron nanoparticles (nZVI) NANOFER25S [3] (Fig. 3).

According to the available data, the area where the pilot tests were conducted has an industrial background with a strong presence of heavy metals, namely zinc, copper, lead, arsenic, sulfates, and nitrates. The implementation of the in situ testing was preceded by the physical and chemical characterization of soils, groundwater, and lixiviates produced by contaminated soils.

Some of the recorded values significantly exceed the concentrations considered to limit the use of land, requiring that it will be transported to landfill of hazardous waste, dangerous or not depending on the effective concentration of the different compounds.

The soil's contaminants presence investigation, was also performed in water samples taken in the piezometers installed in the various testing areas, working as blank values for further comparison after the application of nZVI.

With the perspective of being able to reduce or minimize the presence of these compounds in that soil matrix, or even to stabilize their chemical behavior in order to reduce the leaching process and the presence in the aqueous phase, it was developed the remediation project with the use of nZVI. The implementation plan and respective control will be described in the following chapters.

**Fig. 3** General view of the pilot area, inside de Barreiro Industrial Park

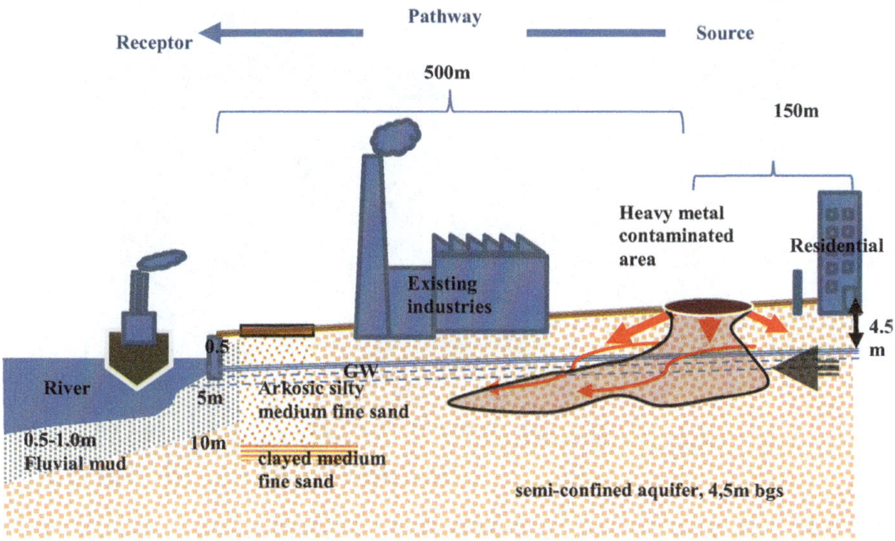


**Fig. 4** Conceptual site model of the contaminated area

According to the available data, the conceptual model of the site can be summarized as presented in Fig. 4, regarding the water flow direction, the position of the contaminated area, and the progressive contaminants dispersion.

From the conceptual model for the contaminated zone, an injection plan/monitoring was implemented not only to allow the assessment of the downstream effect and through the groundwater flow direction, but also in the opposite direction, upstream, evaluating the mobilization capacity of the nanoparticles in different directions from the injection point.

## *4.1 Baseline Situation: Soil Contaminants*

In order to make a preliminary assessment of the main contaminants in the soil (characterization of existent contaminated situation), three drillings boreholes (S1, S2, and S3) were conducted at the pilot test site (Fig. 5), with full and continuous soil sampling by "Direct Push" method (ASTM D6282 05).

The soil samples were collected in the unsaturated zone, until 4.0 m deep. The samples were collected inside transparent liner tubes of 100 cm long (Fig. 6). Further investigations were made and samples from selected liner sections were submitted to chemical evaluation.

Taking into account the history of the local contamination, and that the processes of migration of the contaminating compounds were made in the first few meters below the surface, it was decided to investigate the more superficial horizons. The

**Fig. 5** Location of the preliminary drilling investigation campaign

**Fig. 6** Different phases during the sampling process using the direct push soil sampling procedure

horizons to investigate were selected considering the depth and proximity to the water table, which varies from 4.5 to 5.0 m deep. Two layers of analysis were considered:

1. a shallower horizon between 0.5 and 1.0 m
2. a more profound horizon within the 3.0–3.5 m

In the following Table 2, it can be seen the main heavy metal pollutant compounds distribution, present in the soils of the investigated area.

**Table 2** Main pollutant compounds at different depths for the drilling investigation points

| Pollutant compounds | Boreholes | | | | | |
|---|---|---|---|---|---|---|
| | S1 (0.5–1.0 m) | S1 (3.0–3.5 m) | S2 (0.5–1.0 m) | S2 (3.0–3.5 m) | S3 (0.5–1.0 m) | S3 (3.0–3.5 m) |
| | mg/kg | mg/kg | mg/kg | mg/kg | mg/kg | mg/kg |
| Ar | 72 | 11 | 8.1 | 220 | 4 | 3700 |
| Ba | 300 | 14 | 67 | 17 | 16 | 19 |
| Cr | 5.3 | 9.7 | 5.8 | 7.2 | 9.4 | 15 |
| Pb | 190 | 18 | 220 | 44 | 6.3 | 320 |
| Zn | 270 | 46 | 83 | 49 | 110 | 6600 |
| V | 6.2 | 16 | 7.8 | 12 | 10 | 16 |
| Cu | 920 | 210 | 440 | 270 | 160 | 4300 |
| Co | 23 | N.I | N.I | N.I | N.I | 15 |
| Sb | 9.2 | N.I | N.I | N.I | N.I | N.I |
| Sn | N.I | N.I | 11 | N.I | N.I | N.I |
| Ni | N.I | 3.9 | N.I | N.I | 3.4 | N.I |

*N.I* not identified

## 4.2 Laboratory Evaluation on the nZVI Application Effectiveness

Prior to the field work, laboratory tests were developed that could anticipate the behavior of nZVI when applied in the field, while at the same time assessing the most appropriate concentrations to be applied, taking into account the type of contaminants and the effect of reducing these contaminants downstream of the circuit.

The aim of the laboratory work was to perform long-term (2 months) kinetic and concentration tests on the nZVI effectiveness in degrading a contaminant cocktail (inorganic contamination by heavy metals) present in water and soil samples from the study area.

The laboratory tests have been performed over four soil samples collected at two depth intervals (1.0–1.5 m and 3.0–3.5 m) and over groundwater samples taken from two wells—PZ2 and PZ4 (Fig. 12).

The soil was stored at a temperature of 8 °C until it was used in analytical preparations for chemicals tests. Approximately 12 L of water was collected from the site, transported to the laboratory and subsequently stored in cold storage at a temperature of 8 °C.

The laboratory tests were performed on soil samples collected in two boreholes made during the instrumentation of the field test area, taking up two depth ranges (1.0–1.5 m and 3.0–3.5 m) to obtain four soil samples and on groundwater samples collected from two installed piezometers (PZ2 and PZ4, see Fig. 12).

The contaminated soil, without previous drying, was dosed with water in the mass ratio of 1:2, i.e., approximately 350 g of soil for 700 mL of water. The mixture was homogenized in a mechanical shaker [19] (Fig. 7).

**Fig. 7** General aspect of the mechanical shaker device

Kinetic batch tests with water and soil were performed for three different concentrations of nano-iron (0.3, 1.4, and 7.1 g/L). A sufficient number of sampling containers were prepared so that enough samples could be collected to perform analysis in a total of four time steps from the nZVI application—i.e., after 24 h, and then after 6, 26, and 58 days. Liquid phase samples were taken from the blank samples[8] and from the sample container with the nano-iron, in each time step. During the sample collection, pH and Oxidation Reduction Potential (ORP) measurements were taken in the liquid fraction.

The nZVI particles used in the laboratory tests were supplied by NANOIRON, s.r.o, having a trade name of NANOFER 25S. These nanoparticles were also used in the site tests performed in Barreiro. The technical specifications of the NANOFER25S are summarized in Table 1.

The laboratory tests showed that an important decrease in heavy metals concentrations on the samples was achieved when 1.41 and 7.1 g/L nZVI concentrations were used. As an example of the heavy metal concentrations changes, the following graphs reveal the effective evolution along 58 days (Figs. 8 and 9).

All metals (except Pb) are initially reduced by the initial dose of iron in the amount of 1.4 g/L. Some of metals are even reduced below the detection limit. However, the observed concentration decrease was not permanent, and the concentration rebound has been monitored up to their initial values for a major part of metals.

[8] Blank sample, as considered in the tests performed by Aquatest. a.s (www.aquatest.cz), refers to the water samples collected on site, before nZVI injections.

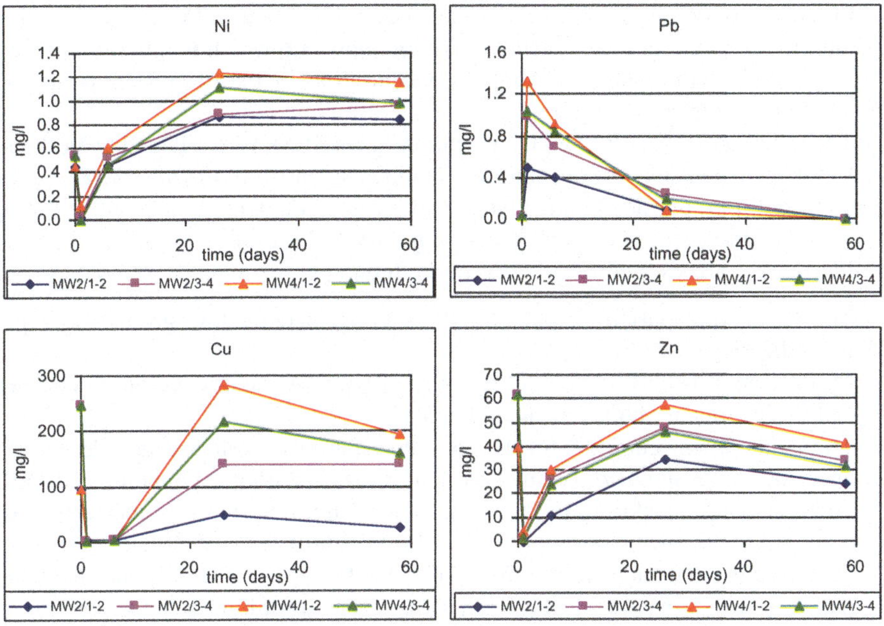


Kinetics of heavy metal concentrations changes in samples with 1,4 g Fe/l.

**Fig. 8** Results for a 58 days period, for 1.4 g/L nZVI addition in the four lab tested samples. Kinetics of heavy metal concentrations changes in samples with 1.4 g Fe/L

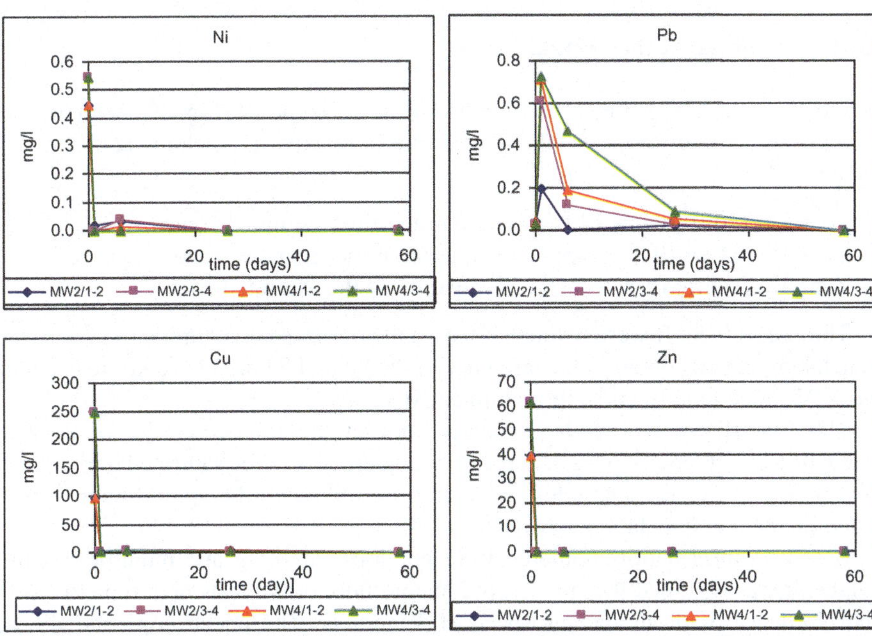


Kinetics of heavy metal concentrations changes in samples with 7,1 g Fe/l.

**Fig. 9** Results for a 58 days period, for 7.1 g/L nZVI addition in the four lab tested samples. Kinetics of heavy metal concentrations changes in samples with 7.1 g Fe/L

Figure 9 shows the result of the efficient and permanent nZVI dose (7.1 g/L). Again the exception in heavy metals behavior shows Pb which is initially released from the soil to the water (the initial concentration increase) and afterwards is slowly removed from the water.

The initial increasing of the Pb concentration could not be explained by polluted nZVI addition because its concentration is independent from the dosed nZVI concentration. Based on the obtained laboratory results, a few recommendations were issued considering the next step by applying the nZVI on site.

From the evaluation of the analysis of the liquid phase, it can be stated that:

- The substantial part of applied nZVI is consumed on low groundwater pH neutralization.
- In this case it would be more efficient to provide pH neutralization by cheaper and more efficient, buffering amendment.
- The efficient concentration for the present system was evaluated in order to balance between 3 and 4 g/L (with equivalent values 6–8 g/kg of contaminated soil), however and considering the renewal of the natural system, by the reentry of more contaminants, resulting from the dynamism of the system itself, it is prudent to admit that the concentration of nZVI to be applied can be higher. And this has been demonstrated in the effectiveness revealed for higher concentrations of nZVI in the system.

## *4.3 Site Injection Method and Monitoring*

### 4.3.1 Definition of the Testing Areas

The test site was divided into three zones—Zone I, II, and III (Fig. 10), with distinct dimensions and purposes, resulting from the interactive procedures and obtained results:

- Zone I—72 $m^2$ (6 m × 12 m)—Evaluation of the saturated zone
- Zone II—54 $m^2$ (9 m × 6 m)—Evaluation of the saturated zone
- Zone III—36 $m^2$ (6 m × 6 m)—Evaluation of the saturated zone

The Zones I and II were defined based on the observed contaminant's concentration, taking in consideration the influence of the S2 and S3 drilling areas, to evaluate the nZVI injection effects in the saturated zone.

The Zone III was defined later, located between the first two zones, for an nZVI evaluation effect, considering also the injection process in the saturated horizon, but now with a higher nanoparticles concentration, following the main directives provided by the previous laboratory phase.

Based on information, gathered from previous sampling and monitoring campaigns, it was assumed that the hydrodynamic flow in the tests area runs to North and Northeast, with an average speed of 30 m/year. The injection points and the piezometers position in Zones I, II, and III were defined considering these data.

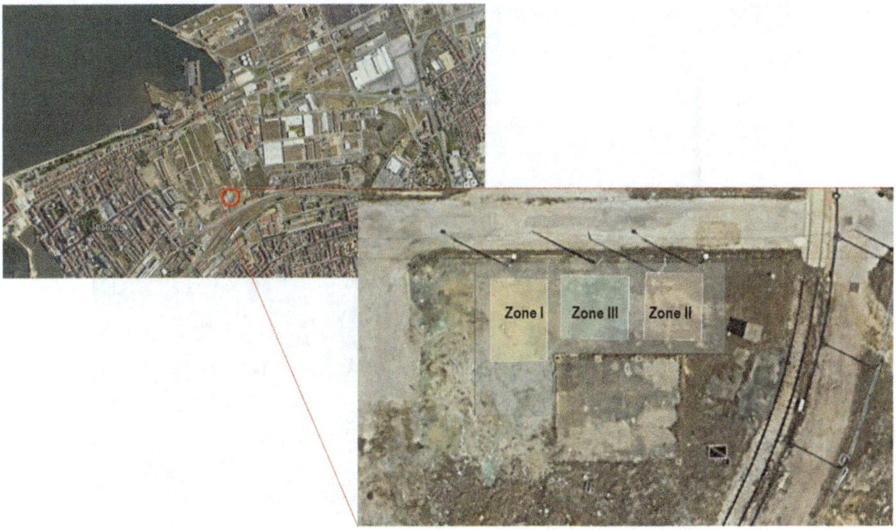


**Fig. 10** Three different zones have been prepared for the nZVI application

For each area two distinct sections were defined: (1) one for ***monitoring points***, which consisted of installed piezometers, from which water samples were collected, which would monitor the evolution of contaminants over time, after injection of nZVI, and (2) another relative to ***injection points*** that would be different in relation to the previous ones, depending on the type of evaluation that was intended for each zone.

### 4.3.2 Instrumentation

Piezometers were installed in the test areas, with a specific distribution in order to assess:

1. The nanoparticles mobilization capacity in saturated environments
2. The local hydrodynamic flow influence in the dispersion of nanoparticles
3. The relationship between nZVI concentrations in the injected solution and the decreasing of contaminants concentrations

The piezometers were installed in 6″ diameter boreholes, opened with a hollow auger (Fig. 11). The groundwater level was found about 5 m deep. Piezometers were installed to a depth of 10.0 m, leaving the screen openings positioned between 4.0 and 9.0 m deep. HDPE 2″ diameter pipes were used in the construction of the piezometers, with factory-set threaded caps and screen openings.

The space around the tubes was filled with calibrated gravel (fraction between 2 and 4 mm) from the base of piezometers to about 2.5 m from the surface. The sealing of the piezometers was carried out with bentonite, from 2.5 m depth to the

**Fig. 11** Monitoring point's installation and general aspect of the Zone I field testing area

surface. Tight plastic lids were placed in the tops of the piezometers with the corresponding reference. All piezometers were developed before the nZVI injection, to ensure that the aquifer system balance was achieved soon after the installation process.

Knowing the limitations of the nZVI mobility and considering the effect of the low groundwater flow, it was decided to install the monitoring points at a maximum distance from the injection point of 3 m. The following Fig. 12 indicates the general distribution of the injection and monitoring points, for the three zones.

For the Zone I, nine piezometers (PZ) were installed downstream of the injection points (IP), and three piezometers were installed upstream. The piezometers were arranged in a square mesh of 3.0 m × 3.0 m.

In Zone II, considering the possibility of occurring radial dispersion from the injection points, the monitoring points were arranged on a semi-circle, leaving a single monitoring point upstream.

In Zone III, six piezometers were installed, three of them downstream of the injection point and the other three upstream. The piezometers were spaced 3.0 m in between, and arranged in two rows, 6.0 m distanced. The assembly of only six piezometers in the immediate vicinity of the injection points took into account the results of injections made previously in Zones I and II, which proved the reduced nZVI mobility slightly beyond 3 m distances.

In order to validate the local geological and hydrogeological model and to determine the factors that control the transport mechanisms and dispersion parameters, the use of lithium/bromine tracers took place in Zone I.

It should be noted that before the injection of nZVI, all the monitoring points were properly developed so that the aquifer system could reach the equilibrium and effectively translated the underground water circulation, after the piezometers

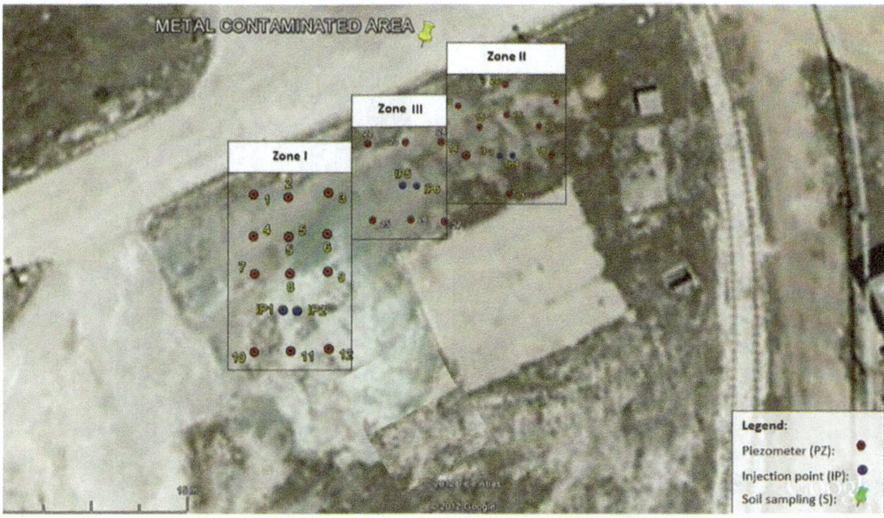


**Fig. 12** General aspect of the field testing area, and monitoring point's distribution for the three zones

installation. In this way, any cross-disturbance effects imposed by the drilling process were reduced.

## *4.4 The nZVI Application*

### 4.4.1 Equipment

The nZVI injection was made using a rig mounted on a tracked chassis, equipped with percussion devices for crimping a probe. The injection probe used was pressure activated (provided by GeoProbe) with four nozzles and one non-return valve. It was coupled to hollow rods of 1.5 m long with threaded sections. The extent necessary to achieve the programmed depth for each injection was obtained by coupling rods in sequence (Fig. 13).

The nZVI solution was injected under pressure, with the support of a GeoProbe pump GP300, keeping an injection pressure of about 5 bar (Fig. 13a).

Using a mechanical drilling rig adapted for the injection process, the nZVI were injected at different depths, in the saturated horizon using different slurry concentrations for the different zones—1, 3, and 7 g nZVI/L of water (Fig. 13b) (see Sect. 4.4.3).

Considering the direction of the groundwater flow, it was decided to inject in a middle point which would be at the same distance between the upstream and downstream monitoring points (Fig. 14).

**Fig. 13** (**a**, **b**) General aspect of the nZVI injection device

**Fig. 14** General view of the Zones I and III, respectively

Therefore the distribution of the monitoring points to evaluate the injection method effect was 3 m distance between monitoring wells (Fig. 15).

It was intended, with this methodology, to create a reactive barrier that could somehow contain the downstream migration of the heavy metals: (1) by their immobilization when aggregating into multidimensional structures, (2) or by their decomposition or transformation into less reactive and therefore less polluting elements.

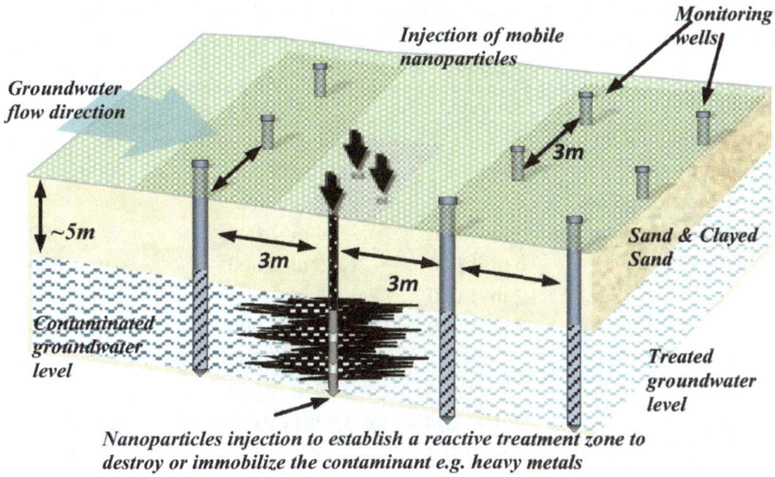


**Fig. 15** Scheme of nZVI application procedure [20]

### 4.4.2 Dosage

As previously referred, different nZVI concentrations were injected in the three described areas, to evaluate mobility and contaminants concentration decrease on soil and groundwater.

As mentioned before and taking into account the laboratory tests results, the solutions injected below the saturated zone had 1 g nZVI/L water, 3 g nZVI/L water, and 7 g nZVI/L water.

The process of preparing the solution to inject took place at the test site, using bottom agitator pumps to keep the solution in the tanks thoroughly mixed before injection. The water used in the solution preparation was water from the public supply system, in order to not introduce into the system another variable that would make the entire application process more complex and expensive.

### 4.4.3 Injection

The injection process ran continuously at each point until the pre-set amount was injected. For this procedure, three technician teams were deployed, working continuously in 8-h shifts.

A synthesis of the injection quantities procedure in each test area is presented in the following Table 3. It should be noted that the nanoparticles transport efficiency depends on the characteristics of the groundwater flow. That is why the nZVI injection occurred in the saturated zone.

Table 3 summarizes the characteristics of the solutions injected into each of the test zones.

**Table 3** Description of the quantities, dosages, and date of the nZVI injection process, on each zone

| | Zone I | Zone II | Zone III |
|---|---|---|---|
| Target | Saturated zone | Saturated zone | Saturated zone |
| Dosage (g nZVI/L water) | 3 | 1 | 7 |
| Injected quantity (L) | 8000 | 12,000 | 8000 |
| Number of injection points | 2 | 2 | 2 |
| Depth of injection and amount solution injected (L) | 5 m–1.000 | 4 m–1.000 | 5 m–1.000 |
| | 6 m–1.000 | 5 m–1.000 | 6 m–1.000 |
| | 7 m–1.000 | 6 m–1.000 | 7 m–1.000 |
| | 8 m–1.000 | 7 m–1.000 | 8 m–1.000 |
| | | 8 m–1.000 | |
| | | 9 m–1.000 | |
| Injection date | 2010.12.11 up to 2010.12.28 | 2011.01.04 up to 2011.01.05 | 2011.07.23 up to 2011.07.27 |

### 4.4.4 Monitoring

The groundwater quality monitoring plan for I, II, and III Zones was based on the following sampling scheme:

- First sampling: reference situation/baseline—before the nZVI injection start, on the day of injection
- Second sample: 14 days after nZVI injection
- Third sample: 28 days after nZVI injection
- Fourth sampling: 56 days after nZVI injection
- Fifth sampling: 112 days after nZVI injection

Water sampling was carried out at 7.0 m deep within each of the installed piezometers and done by *Low Flow Sampling method* using a peristaltic pump. The following physical parameters were monitored by the *Flow Through Cell method*: temperature, pH, dissolved oxygen, and conductivity (Fig. 16).

In order to validate the local hydrogeological model and determine the factors that control the transport mechanisms and dispersion parameters, lithium/bromine tracers were applied in Zone I.

Given the disposition of the different monitoring points (PZ) and considering that the groundwater flows from south to north, parallel to the alignment set by the injection point and the central PZs, the tracer was added inside PZ8. A sampling program was carried out in PZs 4, 5, and 6, on April 21st 2011, 37 days after the tracer injection, not having been detected any change in lithium/bromine concentrations before and after injection in those PZs. The observed fact supports the local hydrogeological data that refers to a water flow average speed of ≈30 m/year.

**Fig. 16** Site water sampling procedure and water parameters monitoring

**Table 4** Main activities calendar

| Actions | Zone I (GW samples) | Zone II (GW samples) | Zone IV (GW samples) |
|---|---|---|---|
| Sampling for the reference situation | 2010.12.21 | 2010.12.27 | 2011.07.22 |
| nZVI injection date (not continuous) | 2010.12.22 up to 2010.12.28 | 2011.01.04 up to 2011.01.05 | 2011.07.23 up to 2011.07.27 |
| First sampling after injection (14 days) | 2011.01.10 | 2011.01.25 | 2011.08.10 |
| Second sampling after injection (28 days) | 2011.01.24 | 2011.02.08 | 2011.08.24 |
| Third sampling after injection (56 days) | 2011.02.21 | 2011.03.09 | 2011.09.22 |
| Fourth sampling after injection (112 days) | 2011.04.21 | 2011.05.09 | 2011.11.16 |

*GW* groundwater

### 4.4.5 Main Actions Calendar

Table 4 evidences the main activities performed, and the subsequent sampling program.

## 4.5 *Results*

The results of water samples tests collected at Zones I, II, and III show significant differences in metals concentrations (before and after nZVI injections), particularly in the monitoring wells located nearby the injection points (3 m). By contrast, in the piezometers located more than 6 m from the injection points, the nanoparticles injection effect did not show a definite trend, which allows to conclude that the reactive effect of nZVI particles is only significant for distances which do not exceed 3 m from the injection point.

The average differences of the metal concentrations in water samples collected in Zones I, II, and III are slightly significant downstream the injection point, confirming the influence of water flow in the underground transport of nanoparticles. These slight differences are probably related with the low transmissivity in the testing area.

Table 5 shows the proportions between the elements average value before and after the nZVI injection on three piezometers located downstream of the injection points, in water samples collected in the I, II, and III zones.

It appears that Zone I is the most suitable to establish the chemical mechanisms sequence that occurs with the nZVI injection (this area is also the better instrumented and where, therefore, can be better established the connection with the underground water flow). In Zone I there is, even slightly, an initial decrease of sulfates, although the values will recover at the end of 56 and 112 days. Correspondingly, the values of Co and, secondarily, Cd, decrease, recovering thereafter. In Zn case, this effect is less pronounced, but the pattern is similar. Cu tends to initially increase and then decreased. In this area, there is the notable exception of As, which undergoes a sharp increase after the nZVI injection.

It appears that there is a temporal sequence relatively constant on each sampled point in Zone II, tending to decrease regarding the metal concentration over time, also with the Arsenic exception.

In Zone III it is observed the arsenic reduction but, instead, other elements rise dramatically in terms of concentrations (e.g., Cu), presented some of them an erratic behavior over time. A justification could probably focus on the fact that this metal could have been released from soil to water, but its complete reduction by nZVI has not occurred. In this context, a new injection could have facilitated its continuous reduction, effectively creating a reactive barrier to its mobilization.

In this pilot study it was found that there is a decrease in tendency of a part of the sulfate concentration into the aquifer system after the injection of nZVI, which usually correlates with a decrease of several metals in solution. This correlation may suggest that the effect of nZVI induces the ion sulfate reduction and subsequent precipitation of metals as metastable sulfides. However, it is noticed that the pH tends, however, to decrease or maintain a low level so that the primary controlling mechanism will be, with high probability, the progressive oxidation of the Fe particles and consequent precipitation of the Fe(III) in hydroxide form. More than hypothetical sulfate reduction, the Fe hydroxide arrangement ensures the existence of solid phase with a high surface area and adsorption capacity, with which the soluble metals have high affinity.

The Fe concentrations consistently show an initial peak (injection) followed by a more or less pronounced break, to consistently low levels. This pattern indicates the progressive Fe oxidation and dispersion, which, depending on the redox potential, tends to precipitate from the solution. These effects tend to spread in different areas in response to the groundwater flow direction.

One of the observed characteristics is a pronounced reduction of the chemical oxygen demand in the water samples, which suggests that organic compounds are

**Table 5** Ratios between average elements concentrations before (index 100) and after nZVI injection, in water samples from three downstream monitoring wells (values in %, except pH)

| Area | Days | Ammonia as N | Ammonia and ammonium ions | Chemical Oxygen Demand (COD) | Nitrates | Nitrites | Sulfate as ($SO_4^{2-}$) | Aluminum (Al) | Arsenic (As) | Cobalt (Co) | Nickel (Ni) | Lead (Pb) | Manganese (Mn) | Copper (Cu) | Zinc (Zn) | Cadmium (Cd) | Selenium (Se) | Iron (Fe) | pH (absolute value) |
|---|---|---|---|---|---|---|---|---|---|---|---|---|---|---|---|---|---|---|---|
| Zone I | 0 | 100 | 100 | 100 | 100 | 100 | 100 | 100 | 100 | 100 | 100 | 100 | 100 | 100 | 100 | 100 | 100 | 100 | 4.63 |
| | 14 | 0 | 0 | 115 | 5 | 461 | 79 | 54 | 631 | 0 | 0 | 24 | 52 | 114 | 61 | 78 | 35 | 40 | 5.03 |
| | 28 | 0 | 0 | 77 | 18 | 41 | 78 | 46 | 600 | 0 | 0 | 21 | 48 | 82 | 51 | 62 | 27 | 37 | 5.04 |
| | 56 | 0 | 0 | 54 | 9 | 12 | 90 | 36 | 631 | 0 | 0 | 15 | 39 | 75 | 45 | 47 | 29 | 27 | 4.86 |
| | 112 | 45 | 45 | 62 | 48 | 73 | 105 | 74 | 131 | 65 | 78 | 45 | 85 | 68 | 72 | 97 | 101 | 26 | 4.64 |
| Zone II | 0 | 100 | 100 | 100 | 100 | 100 | 100 | 100 | 100 | 100 | 100 | 100 | 100 | 100 | 100 | 100 | 100 | 100 | 3.65 |
| | 14 | 142 | 142 | 85 | 812 | 158 | 81 | 54 | 66 | 60 | 70 | 46 | 68 | 44 | 70 | 63 | 85 | 0 | 3.59 |
| | 28 | 142 | 142 | 46 | 655 | 180 | 96 | 93 | 133 | 93 | 106 | 54 | 101 | 76 | 110 | 102 | 117 | 0 | 3.61 |
| | 56 | 117 | 117 | 38 | 334 | 86 | 66 | 85 | 215 | 90 | 103 | 38 | 97 | 68 | 103 | 97 | 115 | 0 | 3.59 |
| | 112 | 125 | 125 | 85 | 707 | 32 | 78 | 87 | 233 | 80 | 94 | 62 | 90 | 63 | 95 | 82 | 88 | 319 | 3.65 |
| Zone III | 0 | 100 | 100 | 100 | 100 | 100 | 100 | 100 | 100 | 100 | 100 | 100 | 100 | 100 | 100 | 100 | 100 | 100 | 4.79 |
| | 14 | 32 | 32 | 111 | 113 | 126 | 138 | 202 | 110 | 201 | 96 | 88 | 68 | 438 | 127 | 105 | 135 | 14 | 4.07 |
| | 28 | 34 | 34 | 133 | 68 | 102 | 139 | 209 | 50 | 211 | 100 | 119 | 70 | 432 | 130 | 101 | 100 | 22 | 4.01 |
| | 56 | 37 | 37 | 100 | 107 | 38 | 135 | 225 | 50 | 209 | 102 | 123 | 66 | 459 | 136 | 102 | 110 | 15 | 3.84 |
| | 112 | 47 | 47 | 89 | 253 | 42 | 120 | 171 | 50 | 156 | 74 | 73 | 49 | 356 | 100 | 60 | 100 | 0 | 3.74 |

being degraded by reduction. The degradation of these compounds may explain the increase of the soluble Cu, since this metal has a high affinity for combining with organic compounds, unlike other metals such as Zn.

Other observed peculiarities, which still requires explanation, are that the injection tests in Zone I (3 g/L nZVI) have given better average results than observed in Zone III (7 g/L nZVI), contradicting the prior nZVI dosing laboratory study, where it was found that excessive Fe in the system would be essential to keep the redox balance, which stabilizes the solid phases where part of the soluble metals is fixed. Furthermore, the mechanisms for subsequent Fe oxidation and consequent Fe hydroxide form precipitation, decreases the environment's pH (particularly well noticed in Zone III results).

However, the pH decrease tends to disfavor the metals (cationic) adsorption on the hydroxides surfaces, with the exception of the As, which forms anionic complexes in solution. The metals adsorption is significantly more favorable at alkaline pH and, in this sense, the pH values recorded for the Zone I (3 g/L) are slightly higher than those measured in Zone III (7 g/L). It follows that, on average, the best results came, from other motives, by the environment pH value.

The evolution of some major pollutants was favorable with a sharp decrease in the first days after the application of nZVI, assuming that this effect was mainly due to the significant reactivity which occurred immediately after injection (see tendency lines, Fig. 17), clearly noted on the piezometers near the injection point [3].

Despite the significant amount values for all the monitoring points available for consultation, in this work it was intended to highlight the most relevant aspects, considering the initial value as the baseline obtained from collected water samples in the three zones, between the average concentrations of elements before and after the nZVI injection in the three piezometers downstream of the injection point.

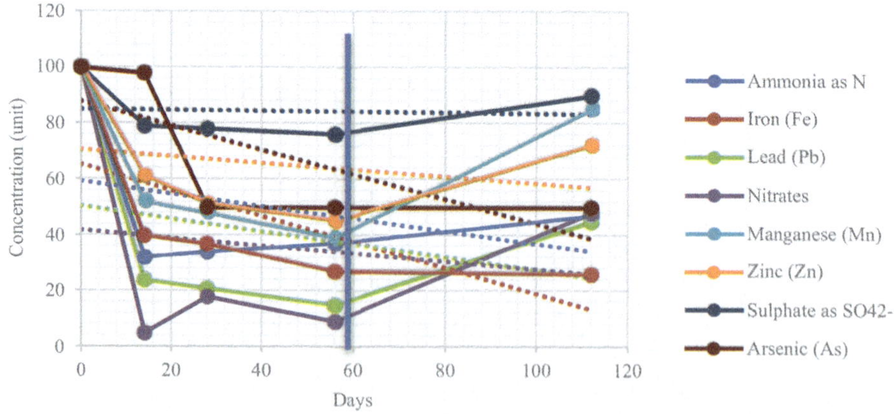


**Fig. 17** Time evolution of some of the contaminants' concentration, and half period milestone

It is observed also after a significant decrease in the concentration of most heavy metals, there was a stabilization of the values which reflects also a stabilization of the reducing effect caused by nZVI.

Considering the half time of the investigation process and for almost all the referred elements, the reducing concentration of each element passes over 60% below the initial concentration. However it is observed in a comprehensive way, and after several weeks of monitoring, an elevation of contaminants to levels close to the initial, but still below baseline values, contributed to the renewal effect that this type of hydrogeological system provides (Fig. 7). In this pilot study, it was found that there is a tendency to decrease the concentration of a part of the sulfate in the aquifer after the injection of nZVI, which usually correlates with a decrease of various metals in solution. The effect of the nZVI was reflected in the sulfate ion reduction and subsequent precipitation of the metals as sulfides.

As mentioned before, the progressive oxidation of the Fe particles and consequent precipitation of Fe(III) in the hydroxide form, rather than the hypothetical sulfate reduction, iron hydroxides formation ensures the existence of a solid phase with high surface area and adsorption capacity with which the metals in solution have high affinity.

The metal's adsorption is more favorable in alkaline pH. The recorded values revealed that the pH increase has influenced the nanoparticles' degrading action for some of the present metals.

Moreover, the pH increase is due to the reductive action of nZVI, acting not as the metals reducing agent but as a hydrogen source.

In these more aggressive environments, the corrosion effect of nanoparticles occurs, limiting their functionality.

## 5 Conclusions

Despite the minor effect of the nanoparticles in some contaminants elements, it was observed that this kind of methodology for the soils and groundwater remediation purposes act in real time reducing the available concentration of these heavy metals. Some recent studies have described the surrounding environment as the main responsible for the less expected effect of the nZVI in the remediation process, whether related with the renovation of the general environmental system or with the remobilization of the chemical elements due to the injection or simple delivery process of the nZVI solution.

The increasing of the contaminant elements concentration, in this particular case, after 4 months of monitoring, is due to the renovation of the hydraulic system which have contributed to the renovation of the leachate process and so to the heavy metals transport into the aquifer. The corrosion effect of the low pH of the Barreiro groundwater aquifer, degraded rapidly the nZVI which have contributed also for the less expected effect on the nanoparticles heavy metal removal. It is possible to increase the local pH by adding a hydroxide as the $Ca(OH)_2$ or Na(OH), creating a better environment for the effective reaction of the nanoparticles and consequently a more effective remediation process of the natural ground and aquifer system, which will also increase the costs of all remediation process.

Despite all described actions and satisfactory preliminary results, it is important to continue these investigation processes by using different nanoparticles in eventual different contaminant contexts and the Industrial Barreiro site could be a large-scale laboratory for these kind of investigation purposes.

An optimization of the nanoparticles effect on the heavy metal soil and groundwater contamination may pass by using a cocktail of different products based on nZVI or other, but with more resistance and maintenance in the environment, increasing the possibility of a more durable effect on the degradation of contaminants.

Recent laboratory studies gave quite interesting answers and guidelines for this important and useful method for soils and groundwater remediation.

The urgent need for low-cost and effective solutions that enable the in situ recovery of contaminated sites, implies the necessity of using new technologies with efficiency and low impact on the surrounding environment. The use of nZVI is one of the possible new approach options.

## References

1. S. Cook, *Assessing the Use and Application of Zero-Valent Iron Nanoparticle Technology for Remediation at Contaminated Sites* (US EPA, Office of Solid Waste and Emergency Response Office of Superfund Remediation and Technology Innovation, Washington, DC, 2009)
2. J. Pais, C. Moniz, J. Cabral, J. Cardoso, P. Legoinha, S. Machado, M.A. Morais, C. Lourenço, M.L. Ribeiro, P. Henriques, P. Falé, *Carta Geológica de Portugal na escala de 1:50 000. Notícia explicativa da folha 34-D* (Depart. Geologia, INETI, Lisboa, 2006)
3. L. Caldeira et al., In situ remediation of ground and groundwater using zero-valent iron nano particles: preliminary evaluation, in *XIII Geotechnics National Congress, Lisbon*, (2012)
4. C.W. Fetter, *Contaminant Hydrogeology* (Macmillan Pub. Co, New York, NY, 1993)
5. C.E. Jacob, Flow of groundwater, in *Engineering Hydraulics*, ed. by H. Rouse, (John Wiley & Sons, New York, NY, 1950)
6. J. Ferris, Cyclic fluctuations of water level as a basis for determining aquifer transmissibility. Ass. Int. Hydrol. Sci. Assemb. Génerale Bruxelles **11** (1951). 148 p
7. G.S.J.P. Van der Kamp, *Periodic Flow of Groundwater* (Editions Rodopi N. V, Amsterdam, 1973). 121 p
8. E. de Cazenove, Ondes phréatiques sinusoïdales. La Houille Blanche **26**, 610–616 (1971)
9. B. Ataie-Ashtiani, R.E. Volker, D.A. Lockington, Tidal effects on sea water intrusion in unconfined aquifers. J. Hydrol. **216**, 17–31 (1999)
10. B. Ataie-Ashtiani, R.E. Volker, D.A. Lockington, Tidal effects on groundwater dynamics in unconfined aquifers. Hydrol. Process. **15**, 655–669 (2001)
11. Y.-J. Chen, G.-Y. Chen, H.-D. Yeh, D.-S. Jeng, Estimations of tidal characteristics and aquifer parameters via tide- induced head changes in costal observation wells. Hydrol. Earth Syst. Sci. Discuss. **7**, 9155–9171 (2010)
12. L. Li, D.-S. Jeng, Tidal fluctuations in a leaky confined aquifer; Dynamic effects of an overlying phreatic aquifer. Water Resour. Res. **37**(4), 1095–1098 (2001)

13. H. Li, J.J. Jiao, Analytical solutions of tidal groundwater flow in coastal two-aquifer system. Adv. Water Resour. **25**, 417–426 (2002)
14. A. Pandit, C.C. El-Khazen, S. Sivaramapillai, Estimation of hydraulic conductivity values in a coastal aquifer. Ground Water **29**(2), 175–180 (1991)
15. A.J. Smith, W.P. Hick, *Hydrogeology and Aquifer Tidal Propagation in Cockburn Sound, Western Australia, CSIRO Land and Water Technical Report 6/01* (CSIRO, Collingwood, ON, 2001)
16. J. Wang, T.-K. Tsay, Tidal effects on groundwater motions. Transp. Porous Media **43**, 159–178 (2001)
17. H. Li, J.J. Jiao, Review of analytical studies of tidal groundwater flow in coastal aquifer systems, in *Water Resources and the Urban Environment*, (2008), pp. 86–91
18. C. Almeida, M.L. Silva, Novas observações sobre o efeito de maré em aquíferos costeiros do Algarve. Bol. da Soc. Geol. de Portugal **24**, 289–293 (1987)
19. P. Kvapil, L. Lacinová, *Technical Report Laboratory Tests NZVI, Geoplano Consultores* (Aquatest a.s, Prague, 2011). 15 p
20. J. Gonçalves, Soil and groundwater remediation with zero-valent iron nanoparticles: Barreiro pilot test, in *International Workshop "Nanoparticles for Soil Remediation - from Green Synthesis to Ecotoxicological Effects", Escola Superior Agrária de Coimbra*, (2014)

# Adsorption and Ion Exchange Permeable Reactive Barriers

**Deborah C. de Andrade, João M. M. Henrique, E. V. Dos Santos, and Fernanda L. de Souza**

## 1 Introduction

The soil is the base for human development and survival due to its capability to generate food and air through plants [1]. In this context, has been increased the number of industry that contributes to the contamination of the soil. In order to improve the results of this sector, in the early twentieth century, the commercialization of chemicals as fertilizers and pesticides was introduced [2]. Besides this type of industry, there are many other like pharmaceutical, textile, oil etc. Because of the diversity of human services such as industrialization and expansion of social activities, a huge amount of hazardous materials are wasted in the environment without appropriated treatment [3, 4]. Among those materials are heavy metals, organic and inorganic compounds which causes a great negative impact directly in the soil, groundwater and associated ecosystems [5].

Heavy metals are found in large quantities in the soil, for example, chromium (VI) is a highly toxic material which is soluble in water in practically all pH ranges but can be reduced to its less harmful form as Cr(III) in acid conditions in the presence of ferrous iron, sulfide, soil organic matter and electrical potential [6]. Nitrates are added in overdose to the soil as fertilizers, and high nitrogen levels induce

D. C. de Andrade · J. M. M. Henrique
Laboratório de Eletroquímica Ambiental e Aplicada, Instituto de Química, Universidade Federal do Rio Grande do Norte, Lagoa Nova, RN, Brazil

E. V. Dos Santos
School of Science and Technology, Federal University of Rio Grande Do Norte, Natal, RN, Brazil

F. L. de Souza (✉)
Faculty of Chemical Sciences and Technologies, Department of Chemical Engineering, University of Castilla-La Mancha, Ciudad Real, Spain



M. A. Rodrigo, E. V. Dos Santos (eds.), *Electrochemically Assisted Remediation of Contaminated Soils*, Environmental Pollution 30,
https://doi.org/10.1007/978-3-030-68140-1_14

serious problems to humans, animals, and plants due to its transformation into nitrites and nitrosamines, which are related to some tumors [2]. Different technologies have been developed over the years based on physicochemical methods [7–9], phytoremediation [10, 11], adsorption [12, 13], electrokinetic remediation [13], and bioremediation [2, 14–19]. These methods can be classified as in situ and ex situ [20].

The purpose of this chapter book is to present the state of the art on adsorption and ion exchange permeable reactive barriers (PRB), discussing its fundamentals, PRBs mechanisms in the soil, applicability for removing organic, inorganic and mixtures contaminants, and some cases of integrate electrokinetic remediation (EK) to describe the applicability.

## 2 Fundamentals of Permeable Reactive Barriers (PRBs)

The PRB is one of the most effective in situ approaches for the removal of pollutants present in soil, is considered an innovative and green engineering approach used to remediate contaminated groundwater [2]. PRBs have been used with varying degrees of success to help meet groundwater standards. It is a passive, in situ technology that has demonstrated high potential to treat aquifers with advantages over conventional pump-and-treat methods, such as (1) degrade or immobilize pollutants in situ without any need to bring them up to the surface, avoiding potential cross media contamination; (2) no need of transportation, storage or other disposal; (3) no require continuous input of energy; (4) degradation of the contaminants is achieved not only the change of phase; (5) no require discharge of effluent and consequently technical and regulatory problems; (6) relative lower cost but due to a lack of long-term data, its cost-effectiveness has not been proven [21, 22]. According to Environmental Protection Agency (EPA), PRB is defined as an *emplacement of reactive media in the subsurface designed to intercept a contaminant plume, provide a flow path through the reactive media, and transform the contaminants into environmentally acceptable forms to attain remediation concentration,* as can be observed in Fig. 1 [24]. The concept of the PRB was first developed by the University of Waterloo in the early 1990s. The PRBs are typically constructed in two ways: digging a trench ranging over the whole width of the contaminated plume or a funnel and gate system that uses impermeable gates forcing water through a reactive barrier [25]. In this case, the funnel and gate system increases flow rate through a gate system 2–5 times. The first pilot-scale PRB was installed in 1991 at Borden, Ontario, to treat a plume of chlorinated solvents.

PRBs can be achieved as replaceable, semi-permanent, or permanent units. Two main configuration types have been used, the funnel-and-gate and the continuous gate designs. Continuous wall or curtain is the basic configuration of barriers that stands up and transversely faces the direction of the contaminant front [26]. The main advantages of this configuration are it relies on conventional methods of installation, is easy to conceptualize and construct, is less expensive and creates fewer disturbances to the groundwater flow [12, 13]. However, this design is only suitable for plumes with narrow widths [27].

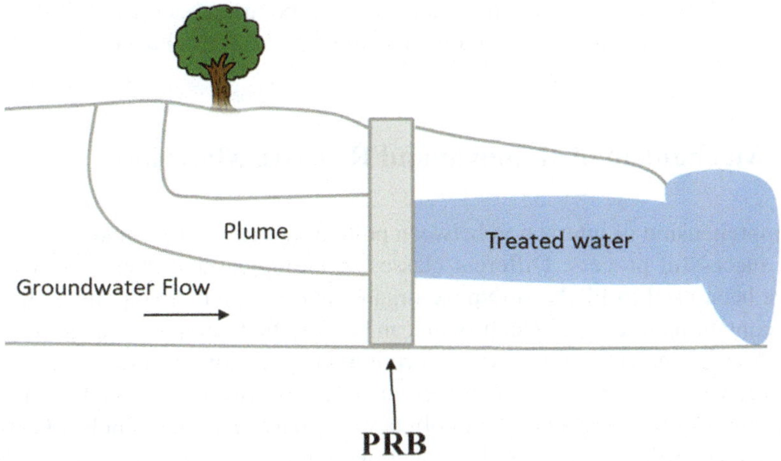


**Fig. 1** Permeable reactive barrier [23]

McMurtry and Cherry [25], introduced the concept "funnel and gate" which was used interchangeably with PRBs; nevertheless, funnel and gate configuration consisted of impermeable walls that directed groundwater to the reactive middle gate panel. However, the choice between the two configurations depends on both the reactive cost and hydrogeological characteristics of the site [23]. When a high cost reactive material is used, funnel and gate configuration is preferred since the reactive zone requires less material. Nevertheless, construction cost of continuous type barrier is much cheaper than funnel and gate system. Furthermore, the adoption of funnel and gate configuration promotes the use of double or multi-reactive barriers for multi-action, improving the efficiency of treatment for more than one type of contaminants. Usually, a balance must be struck between the cost of reactive material and the construction cost of the barrier, in accordance with the target pollutant and level of removal to be achieved [12, 13].

The configurations more used in this field are Continuous and Funnel-and –gate PRB. The Continuous installations consists of a single reactive zone installed across the contaminant plume, while the funnel-and-gate system consist of a permeable gate placed between two impermeable walls that direct the contaminated plume towards the reactive zone. The choice between the two configurations depends on both the hydrogeological characteristics of the site and the reactive material cost even as to be in accordance with the target pollutant and level of removal to be achieved.

The most PRB used is single barrier type. This kind of PRB design is fill with a single reactive material applied to decontamination of plumes containing one contaminant or contaminants of single nature. However, this kind of barrier are ineffective to remediate multi-contaminant plumes with different physical, chemical, and thermodynamic properties. Therefore, multi-barrier concept was introduced in the last years and have been received considerable attention. A multi-barrier system is defined as PRBs consisting of two or more barriers filled with the same or different reactive materials in which contaminants are removed sequentially. A single barrier filled with different reactive materials also can be interpreted as multi-barrier system. However, in this case, contaminants are removed simultaneously [28].

## 3 Mechanism of Removal and Reactive Materials

Comprehension of interaction between pollutants and reactive material is crucial to the successful process. Different classes of biological and chemical mechanisms have been used in PRBs to remove organic and inorganic compounds. In general, the contaminant removal mechanism can be classified into three categories: (1) degradation, (2) precipitation, and (3) sorption. Degradation mechanism also called as destructive processes occurs through chemical or biological reaction that lead to decomposition or degradation of pollutant into harmless compounds, like as abiotic reduction–oxidation and biotic reduction–oxidation, [29–31]. Abiotic reduction consists of chemical decomposition reaction of organic compounds resulting in the formation of non-toxic products that are either immobilized in the barrier or permeated through the barrier in a reduced form [32, 33].

On the other hand, biotic reduction in PRBs is initiated by supplying electron donor (leaf mulch, saw dust, wheat straw, and alfalfa hay) and nutrient materials (municipal waste and compost) that are used by microorganisms [34]. The presence of dissolved sulfate can performance as an electron acceptor that consumes $H^+$ contributing to the oxidation of organic material acidic, coupled to reduction and precipitation of metal as outlined by Eqs. (1) and (2)[35].

$$2CH_2O_{[\text{solid organic}]} + SO_4^{2-} + H^+ \rightarrow 2CO_2 + 2H_2O + HS^- \quad (1)$$

$$M^{2+} + HS^- \rightarrow MS_{[\text{solid}]} + H^+, \quad \text{where } M = \text{metal} \quad (2)$$

However, in groundwater lives bacteria which are organisms that require suitable physical and chemical conditions to exist and survive, also the temperature will affect the population of microorganisms and the rate sulfate reduction [36–38].

*Adsorption and ion exchange* also are important mechanisms that contribute for removing contaminants by adsorption a chemical species attaches to a solid surface. Moreover, the ion exchange process involves reversible reactions in which a contaminant ion in solution replaces a similar ion on the surface of an immobile solid. Regarding to adsorption reactions are reversible and occur at relativity rapid rates, and particular sites [39, 40]. Reactions chemistry based by adsorption process can occur due to surface complexation models. Most material are naturally occurring inorganic zeolites, but there are also synthetically produced organic resins that can be altered for specific remediation as can be observed in Table 1. The particularity associated of zeolites is due to internal surface are and can treat inorganics and/or organic by both adsorption and cation exchange [49]. *Chemical precipitation* consists of the removal of contaminants by mineral precipitation associated with an increase in the pH value. Limestone ($CaCO_3$) and apatite $Ca_5(PO_4)(OH)$ are the

**Table 1** Operational conditions for organics removal by applying EK coupled with different PRBs

| Pollutant | PRB | Experimental conditions | Observations—best results | Ref. |
|---|---|---|---|---|
| Clopyralid 30 mg $kg^{-1}$ | Granulated activated carbon | Silty loam soil, 11% initial moisture, 1 V $cm^{-1}$ (38 cm long), polarity reversal, 30 days duration | 45% EKAB and 57% REKAB | [41] |
| Pentachlorophenol (PCP) 100 mg $kg^{-1}$ | Pd-Fe particles (reduction) | Turf soil (0.355 mm sieve), soil pH = 7.8, 30% initial moisture, 0.025 M $Na_2SO_4$ as anolyte, 20 V, 5–15 days duration | (5 days): 80.7–90.6% (10 days): 51–89.7% (15 days): 53.7–75.9% | [42] |
| Hexachlorobenzene (HCB) 55 mg $kg^{-1}$ | Pd-Fe particles (reduction) | Spiked turf soil (0.355 mm sieve), 10 mM TX-100 as anode flushing solution, 30 V, 0.025 M $Na_2CO_3$ as anolyte and 0.025 M $Na_2SO_4$ as catholyte, polarity reversal, 5 and 10 days duration | 60.1% | [43] |
| Atrazine and Oxyfluorfen 30 mg $kg^{-1}$ | Granulated activated carbon (adsorption) | Kaolinite soil, SDS 1000 mg $dm^{-3}$ as flushing fluid, 1 V $cm^{-1}$ (20 cm long), 30% initial moisture, polarity reversal, 15 days duration | Atrazine: EKSF: 80%. REKAB: 85% Oxyfluorfen: EKSF: 78% REKAB: 90% | [13] |
| Diesel 10,000 mg $kg^{-1}$ | Biological (degradation) | Kaolinite soil (pH = 4.9), 1 V $cm^{-1}$ (20 cm long), SDS as anionic surfactant in cathode, diesel chain: 10–25 carbons (882 g $L^{-1}$ density), 40% initial moisture, 30.36 mg $L^{-1}$ $NaNO_3$ + 70 mg $L^{-1}$ $NaHCO_3$ + 88.75 mg $L^{-1}$ $Na_2SO_4$ as anolyte, 14 days duration | 27% | [38] |
| Diesel 10,000 mg $kg^{-1}$ | EK/biological (degradation) | Kaolinite soil (pH = 4.9), 1 V $cm^{-1}$ (20 cm long), SDS as anionic surfactant in cathode, diesel chain: 10–25 carbons (882 g $L^{-1}$ density), 40% initial moisture, 30.36 mg $L^{-1}$ $NaNO_3$ + 70 mg $L^{-1}$ $NaHCO_3$ + 88.75 mg $L^{-1}$ $Na_2SO_4$ as anolyte, 14 days duration | 30% | [44] |

(continued)

**Table 1** (continued)

| Pollutant | PRB | Experimental conditions | Observations—best results | Ref. |
|---|---|---|---|---|
| Diesel 10,000 mg $kg^{-1}$ | EK/biological (degradation) | Kaolinite soil (pH = 4.9), 1 V $cm^{-1}$ (20 cm long), SDS as anionic surfactant in cathode, diesel chain: 10–25 carbons (882 g $L^{-1}$ density), 40% initial moisture, 30.36 mg $L^{-1}$ $NaNO_3$ + 70 mg $L^{-1}$ $NaHCO_3$ + 88.75 mg $L^{-1}$ $Na_2SO_4$ as anolyte, 14 days duration | 36% | [37] |
| 2,4,6-Trichlorophenol 100 mg $kg^{-1}$ | Granulated activated carbon (adsorption) | Kaolinite soil (pH = 4.9), 27.3% initial moisture, 5, 15, 25, 40, and 80 V, 7 days duration | 80% | [45] |
| Phenol 90–115 mg $kg^{-1}$ | EK-Fenton/ scrap iron powder (reduction) | Sandy loam natural soil (10-mesh), soil pH = 4.67, 0.3% $H_2O_2$ as anolyte and ASTM type II de-ionized water as catholyte, 1 V $cm^{-1}$ (20 cm long), 10 days duration | 53–99% | [46] |
| Trichloroethylene (TCE) 233–266 mg $kg^{-1}$ | Fenton/scrap iron powder (reduction) | Loamy sand (pH = 4.4) and sandy loam (pH = 7.64) soils (10 mesh sieve), 1 V $cm^{-1}$ (20 cm long), 0.098 M and 0.0196 M $FeSO_4$ as Fenton catalyst, $H_2O_2$ as anolyte and de-ionized water as catholyte, 10 d duration. | 49.71–88.91% | [47] |
| 1,2-dichlorobenzene 245–261 mg $kg^{-1}$ | Carbon nanotube (adsorption) | Farmland spiked clay soil (2 mm sieve), soil pH = 8.5, 2 V $cm^{-1}$, SDS and PANNOX 110 as processing fluids, 5 days duration | 45.9–75.5% | [48] |
| Chlorosulfuron (CLSF)/2,4-Dichlorophenoxyacetic acid (2,4-D) 30 mg $kg^{-1}$ | Activated carbon (adsorption) | Kaolinite soil, 1 V $cm^{-1}$ (20 cm long), polarity reversal every 24 h, 30% initial moisture, 15 days duration | CSLF: EKSF: 61%. REKAB: 88% 2,4-D: EKSF: 95% REKAB: 72% | [12] |

most common materials used as PRBs for the treatment of soil [24], and are often reduced to a less-species. In this context, contaminants are removed by mineral precipitation as hydroxides (Eq. 3) and carbonates (Eq. 4).

$$M^{2+} + 2(OH)^{-} \rightarrow M(OH)_2 \tag{3}$$

$$M^{2+} + HCO_3^- \rightarrow MCO_3 + H^+ \tag{4}$$

PRBs can be used to remediation organic and inorganic compounds from soil, and the choice of a reactive medium is dependent on site characteristics and contaminants properties. Reactive media is determined based on the characteristics for removing volatile organic compounds (VOCs), semi-volatile organic compounds (SVOCs), and inorganics. The types of reactive can be characterized by reactivity and stability. The reactivity is associated with reaction rate and equilibrium constant of the contaminant with the reactive material used to determine the required residence time, and therefore, the size of the PRB. Moreover, the barriers must show high stability to persist in the subsurface environment for an extended period of time as secondary precipitates [40, 49–55]. Furthermore, the cost and the environmental sustainability is important to promote the applicability of PRBs [56].

The reactive media is main responsible to destroy or immobilize contaminants through physical contact or altering the biogeochemical process. Many materials have been studied in order to find more suitable and cost-effective materials. The choice of the reactive materials depend on (1) if the contaminant to be removed is organic (volatile organic compounds (VOCs), semi-volatile organic compounds (SVOCs)) and inorganic, ii) their concentrations, and the mechanisms needed for their removal (2) the hydrogeological and biogeochemical conditions of the aquifer; (3) the environmental/health impacts; (4) mechanical stability (in order to persist in the subsurface environment for an extended period of time as secondary precipitates [40, 49–55], and (5) the availability and environmental sustainability and (6) the cost of the material [57]. Suitable materials currents employed for use in a PRB are presented in Table 2. These materials have been reported in the literature to increase the removal efficiency of organic and inorganic compounds [58–61]. The most common of them is ZVI. Besides, activated carbon, zeolites, peat, saw dust, oxygen releasing compounds, and so on [56, 62, 63].

The first of them, Zero-valent iron (ZVI), was Gillahm and O'Hannesin who determined that zero-valent iron could be used in PRBs to remediate groundwater contaminated by halogenated organic solvents [24]. The first ZVI-based PRB was installed in 1991 at the Canadian Forces Base Borden site at Ontario, Canada that obtained success to the removal chlorinated organic solvents between 1993 and

**Table 2** Summary of reactive materials for organic and inorganic contaminants in soil

| Reactive materials | Process |
|---|---|
| Activated carbon | Adsorption |
| Transformed red mud (TRM) | Adsorption or precipitation |
| Bagasse fly ash | Adsorption |
| Zeolites | Adsorption |
| Atomizing slag | Adsorption |
| Carbon nanotubes (CNT) | Adsorption |
| Bacterial | Microbial degradation |
| Zero-valent Iron | Reduction and precipitation |

2001 [64]. ZVI tend to be oxidized, passing its electron to contaminants (organic, halogenated hydrocarbons, inorganic some metal, etc.,) which undergo reductive mechanism resulting in precipitation or degradation [65]. The efficiency of ZVI is attributed by three phenomena: dissolution of the iron, mineral precipitation leading to permeability reduction, and passivation of the ZVI from alteration of ZVI grain surfaces. Moreover, ZVI dissolves in the presence of oxygenated groundwater (Eq. 5). Additionally, if $O_2$ is available promotes the ZVI oxidizes further and forms a mixture of ferric oxyhydroxides or rust (Eq. 6) [66, 67].

$$2Fe^{0}{}_{[ZVI]} + 4H^{+} + O_2 \rightarrow 2H_2O + 2Fe^{2+} \quad (5)$$

$$4Fe^{0}{}_{[ZVI]} + 6H_2O + 3O_2 \rightarrow 4Fe(OH)_{3[solid]} \quad (6)$$

In the case absence of $O_2$, ZVI will dissolve [68], where water is an oxidant and gaseous $H_2$ is produced by the reduction of aqueous protons. Moreover, hydrogen gas and an increase in pH values are often observed in ZVI based PRBs, confirming the anaerobic dissolution of ZVI [66].

$$Fe^{0}{}_{[ZVI]} + 2H^{+} \rightarrow Fe^{2+} + H_2 \quad (7)$$

## 4 Activated Carbon

Activated carbons are carbonaceous materials possessing chemically heterogeneous surfaces with high adsorption capacity. They are chemically stable materials and are widely considered as suitable adsorbent for on-site or off-site treatment of polluted groundwater [21, 69]. The treatment potential of active carbon is presented by adsorption of contaminant particles in a physically manner on its high lattice surface area (about 1000 $m^2/g$) and presence of different types of surface functional groups (hydroxyl, carbonyl, lactone, carboxylic acid, etc.). Therefore, they have been applied widely for the removal of pollutants such as phenols, BTEX, PCE, and TCE [54, 70]. Active carbon is considered a conventional material; however, in passive groundwater remediation technology its use as a reactive material is somewhat limited because it is a fabricated material and expensive.

## 5 Zeolite

Zeolites are aluminosilicate minerals with three-dimensional structure containing water molecules, alkali and alkaline earth metals in their structural framework, and have high cation-exchange capacity (200–400 meq/100 g) and large surface area (up to 145 $m^2\ g^{-1}$) [56]. According to Shoumkova and coworkers [71], zeolite can be classified according to its source into natural zeolites such as clinoptilolite, chabazite, analcime, eronite, faujasite, laumontite, phillipsite, ferrierite, mordenite, and heulandite; synthetic zeolites such as those from natural materials, waste, coal fly ash, municipal solid waste incineration ash, oil shale ash, rice husk, or modified natural and synthetic zeolites [50]. About capacities of ion-exchange are attributed to their permanent negative charges, which develop from isomorphic substitution. In this case, these charges are not pH dependent and are usually balanced by alkali and alkaline earth metals such as $Na^{+}$, $Ca^{2+}$, and $Mg^{2+}$ [72].

## 6 Mixed Materials

Different blends has been used to remove organic and inorganic compounds using PRBs to control the flow of contaminated water that the reactive material was placed in a shallow trench so that groundwater through the burred PRBs [24]. The use of various naturally occurring oxides such as basic oxygen furnace oxide, alumina ($Al_2O_3$), amorphous ferric oxide ($Fe(OH)_3$), goethite (a–FeOOH), magnetite (a–$Fe_2O_3$), and hematite ($Fe_3O_4$), and hydrous titanium oxide ($Ti(OH)_4$) for groundwater remediation has also received much attention as PRBs [73]. Besides that, other complex organic materials have been used as materials for permeable reactive barriers, such as organic mulch [74] and nutrient release [75].

## 7 Coupling PRBs and Electrokinetics for Soil Remediation

In recent years, the PRBs and EK approach have shown to improve soil remediation results through the electrically induced mobilization of contaminants present in the soil (Fig. 2). Some articles have been published coupling PRBs and electrokinetic remediation of soil contaminated by complex organic compounds such as pesticides [13, 45], petroleum hydrocarbon [44], organochlorinated species [41], and polycyclic aromatic hydrocarbons (PAH) [52, 76]. Table 2 collects data related to the coupling PRBs and EK to remediation of soils polluted by organic compounds, under selected conditions. It is important to remark that the treatment of organic and inorganic contaminants can be unique for each other. Organic compounds can be broken down into innocuous elements and compounds, such as carbon dioxide and water because they are molecules consisting of carbon, hydrogen, halogens, oxygen, sulfur, phosphorus, and nitrogen [23]. On the other hand, inorganic pollutants usually are themselves elements; this way, they cannot be destroyed but can only change speciation. Consequently, remediation strategies must focus on integrate technologies for transforming inorganics and organic compounds into forms that are non-toxic, not bioavailable, immobile, or capable of being removed from the subsurface [77].

**Advantages**

- The degradation or mobilization of pollutants in situ without necessity to bring them up to the surface, avoiding potential cross media contamination
- No need of transportation, storage or other disposal;
- No require continuous input of energy;
- Degradation of the contaminants is achieved not only the change of phase;
- No require discharge of effluent and consequently technical and regulatory problems;
- Relative lower cost but due to a lack of long-term data, its cost-effectiveness has not been proven

**Disadvantage**

- Management and monitoring of risks resulted from persistent contaminant source required long period of time;
- The capital investment of PRB is very high then the treatment of its counterpart technologies;
- It is depend exclusively relies on site qualities (hydrology, geochemistry and geography);
- Underground structures, geological conditions, and site characterization are common constrains for construction of this technology;
- Frequently, reactive media are removed at the end of operation or changed during operation.

**Fig. 2** Schematic diagram of coupled EK-PRB for soil remediation

## 8 Pesticides

In the last years, the production and consumption of pesticides has increased considerably. Consequently, theses pollutants have received great attention thanks to their wide range of chemical properties as well as to the high risk associated with their hazardousness. For this reason, different research groups are investigating the PRB-EK process [4, 45, 48, 78–80].

Wan and coworkers [43] investigated the couple of surfactant-enhanced electrokinetics (SEEK) and PRB composed of Pd/Fe particles in micro-scale to remove hexachlorobenzene (HCB) with aid of Triton X-100 surfactant. TX-100 was chosen due to its capacity to improve HCB's solubility, promoting a reductive dechlorination. The authors observed that this combination increased the HCB removal by a factor of 4 compared with EK alone. They also proposed that using EK-PRB system, the HCB was removed from soil through several sequential processes, that is, the movement driven by EOF in the anode column, the complete adsorption/degradation by the Pd/Fe PRB, and the consequent movement by EOF.

Pentachlorophenol (PCP) is an important organochlorine compound used as a pesticide, general wood preservative, and broad-spectrum biocide, for this reason, PCB has been found from soil. The removal of PCP is difficult due to dissociates at different pH values along soil. Li and coworkers [42] installed reactive Pd/Fe particles as PRB to remove PCP into the soil. They coupling Pd/Fe-EK and investigated the positioning PRB in a distance from anode of 3–5 cm and amount of Pd/Fe (0–50 g), the time of treatment (5–15 days) and electrolyte injected during Pd/Fe-EK. Elimination of the PCP trough EK alone is difficult due to molecular PCP in the acidic region near anode was moved towards cathode, dissociated to anionic PCP in the basic region neat cathode. The investigation of PRB located at the position of 0.3 cm (normalized distance from anode), while PCB at the anode side of PRB could transport through PRB, and PCP at the cathode side of PRB could not transport through PRB and was accumulated near PRB, that could be hardly dechlorinated by Pd/Fe in PRB because of high pH attained in the compartment (pH around 10–12). The role of electrolyte addition was tested by considering acetic acid and $Na_2SO_4$, in order to decrease of pH and enhance PCP extraction. When acetic acid and $Na_2SO_4$ was injected from anode reservoir, the removal of PCP attained 49% and 22.9% respectively.

Souza and coworkers [12] examined the performance of a PRB filled with active carbon, installed between electrodes, with the aim of remove chlorosulfuron (CLSF) and 2,4-dichlorophenoxyacetic acid (2,4-D) from a kaolinite soil by applying combined with reversible electrokinetic adsorption barriers (REKAB). CLSF and 2,4-D are classified as herbicides and have very different volatility values (0.02 MPa and $3.0 \cdot 10^{-6}$ MPa at 25 °C for 2,4-D and CLSF, respectively). So, evaporation rate is important parameter during EK remediation that can affect the contribution of soil treatment. They observed that REKAB promote for the removal of soluble pesticides from soils and after 15 day more than 70% of the pollution can be removed. Moreover, the adsorption contributes to decrease of evaporation fluxes during soil treatment. On the other hand, removal of 2,4D and CLSF in electrolyte wells is less important in REKAB due to the reversion of polarity that effect on the pH regulation and on the prevention of the washing up of salts contained from soil.

Ruiz and coworkers [45] investigated the applicability of granular activated carbon (used as PRB) coupling with EK to remediate a soil contaminated with 2,4,6-trichlorophenol (2,4,6-TCP). As expected, phenolates present negative charge, so EM direction was from cathode to anode and EOF have the opposite guidance. Temperature was monitored is this work because of pollutant's volatility. This variant increased with electrical current increase which promotes heating. Observing its results, the authors noticed a cooling effect near electrodes wells due to add of flushing fluids in it and by greater electrolyte resistance (lower ionic conductivity). Eighty percent of TCP was removed from soil in less than 1 week consuming below 200 $kWh/m^3$.

## 9 Petroleum

The displacement and subsequent removal of petroleum can be associated to those of PAH and TPH compounds that, under an electric field, are rather limited due to the fact that these pollutants are poorly soluble in water, and therefore also in the soil solution [44]. Diesel are represented by TPHs compounds present in the petroleum. TPH are hydrophobic compounds; for this reason, is difficult to remove in the soil, consequently in same cases is necessary to use surfactants [37, 38, 44].

Mena and coworkers [37, 44] proposed the use of biobarrier consists of mixing of non-polluted kaolinite with the corresponding volume of raw biological activated sludge to obtain a similar moisture level as in the polluted soil. Remediation of clay soil polluted diesel using a combined biological permeable reactive barriers (BPRB) with EK technology. In addition to increase of solubility of diesel (low-solubility organics) was used sodium dodecyl sulfate as solubilizing agent, this compound modifies the properties of the pore fluid, influencing its dielectric constant, pH and viscosity. They observed that the increase of field applied (0.5, 1.0, and 1.5 V $cm^{-1}$) enhance of removal efficiency (36%) after 2 weeks. The polarity reversal was observed contributed avoiding acid or basic fronts on the viability of microorganisms.

Yang and Long [46], investigated the removal of phenol in sandy loam soil treated by in-situ PRB-Electrokinetic-Fenton process and as PRB was used of scrap iron powder (SIP). The authors noticed that SIP amount is determine relating to remediation. For example, EOF decreased when SIPs amount increased; the cumulative, $H_2O_2$ consumed mass present in anode increased as SIPs concentration decreased. In opposite, the cumulative, phenol increased mass in cathode had a reversal effect, for example, increased with a decreasing amount of SIP. Relating to the residual phenol concentration in soil, this concentration decreased with a decreasing amount of SIP. Removal of pollutants is associated with EO permeability (Kc), so, to increase this rate, it was noticed that is favorable by decreasing SIP. Comparing the Kc results in EK alone process with Fenton coupled with EK, the first one showed greater values, but, EK alone just promotes phenol removal and by applying Fenton both removal and chemical oxidation contributions occur.

The removal of trichloroethylene from spiked two types of soil were studied: sandy loam and loamy sand was investigated by Yang and Liu [47] by coupling an electrokinetic-Fenton process and (SIP) as PRB constituted only of magnetite (type I) and a mix of magnetite, hematite and maghemite (type II). When $FeSO_4$ (0.098 and 0.0196 M) and then replaced $H_2O_2$ in the anodic reservoir to promote the formation of hydroxyl radicals. In particular, destruction potential of TCE is related with SIP size, once a smaller particle improves this reaction, but, on the other hand reduce the overall treatment efficiency. In fact, smaller granular size provides a greater surface area, thus more $Fe(OH)_3$ precipitates forming on the iron wall and lowering Kc. Moreover, EOF also is affected due to size of SPI. Results showed a better removal of TCE when one wall of SIP type I (UPF-030) was placed (88.91%) in soil type I (loamy sand) comparing to SIP type II (86.83%) and $FeSO_4$ (77.41%); all were performed with graphite electrode.

The electrokinetic remediation have been investigated coupling PRB constituted of active carbon by Santos and coworkers [13]. A comparison was realized by considering clopyralid removal through adsorption. Two types of electrokinetically assisted soil flushing (EKSF) were applied in this study: one using permeable reactive barrier (EKAB) and a second one performing a reversal EKAB (REKAB).

Results presented clopyralid removal through adsorption, EK transport and evaporation. REKAB is characterized by a polarity reversal, mechanism also applied by [3, 13]. The advantage of reverse electrodes polarity relays in controlling its pH—neutralization of $H^+$ and $OH^-$ ions produced—without adding buffer solutions and provide a greater retain capacity of the organic in the GAC. Besides that, exchanging electrodes promotes a change in electroosmosis and electromigration flux of clopyralid, favoring its mobilization. During EKAB tests, a graphite consumption of the electrodes due to electric current was noticed, showing the necessity of periodical replacement. In REKAB tests, this effect did not occurred but current intensity decreased over time, probably due to changes in medium resistance. Comparing removal rates of a reference process (without applying electrokinetics), EKAB and REKAB, only 3% of clopyralid was evaporated for process number 1, remaining 97% in soil. For EKAB and REKAB, a greater extraction was identified (45% vs. 57%).

## 10 Inorganic Contaminants

Soil contamination associate with heavy metals represents a worldwide problem for human health and the environment. Heavy metals are unintentionally introduced to the soil as by-products of various anthropogenic activities, which include metals extraction through smelting, waste disposal, and agrochemical applications. Currently, estimates expose that 20 million hectares of land are contaminated heavy metals (Pb, Ni, Cr, Co, Cd, As, Fe, and others), at soil concentrations higher than the normal geochemical baseline or regulatory levels [81].

In recent years, the PRB approach has been shown to improve the contaminated soil with heavy metals [77]. However, several research groups have observed that PRB-EK can be help eliminating the pollutants from the soil by promoting ion exchange, the redissolution of precipitates or simply by dragging pollutants through complexation reactions. In order to decrease the problem associate with decontamination of the wastewater generated during the soil treatment.

Cappai and coworkers [22] studied the removal of Cr(VI) and As(V) from soils by an enhance PRB (it was used transformed red mud (TRM)) integrate with EK. When the TRM was used as PRB near anode influenced on the electromigration of Cr(VI) and AS(V), electroosmotic flow (EOF) and removal percentage these species from soil. In chromium runs, was observed that the current profile was higher than (25–28 mA compared with 11–17 mA without the reactive barrier) when TRM was present because of Cr movement towards anode and its accumulation in the anodic chamber. Besides that, there was a decrease in the ion strength due to possible TRM dissolution in Mg, Ca, aluminates, sodium silicates, and aluminium hydroxides. The EOF when the transformed red mud was present was less than in tests performed with RB owing to Cr migration in direction of the anode which is opposite to electroosmotic flow. The removal rate was higher at 12 days of duration and with increasing TRM mass from 12 to 50% in the RB, leading to 60.6% of total

Cr trapped in. This percentage removal was due to an acidic front which decreased soil adsorption capacity for chromium oxyanions, increasing EM towards anode. In arsenic runs, the current profile increased rapidly, then decreased, stabilizing in a constant value. Unlike chromium, arsenic EOF with RB in the runs was higher than without. As(V) removal was very low during EK (~29%) in 12 days. This result can be explained by acidification in the anode which promotes As(V) protonation, thus decreasing its movement by electromigration. Beyond that, this environment can lead to $H_3AsO_4$ formation, an uncharged compound, which is transported only by EO.

Gioannis and coworkers [39] also worked with TRM as the reactive barrier integrate to EK. They also noticed that electromigration to anode and to TRM PRB was enhanced by the contaminated soil acidification suppress. This was due to acid neutralizing, resulting in an oxyanions desorption under alkalinity present in the cathode. Hydrogen ions production on anode decreased pH, allowing oxyanion sorption capacity of the TRM PRB. Process duration and pH conditions were determining to total chromium removal efficiency (~93%). In relation to alkaline non-extractable Cr, the absence of RB resulted in higher percentage due to chromium reduction under acidic conditions, once every Cr(VI) is alkaline extractable. As $Fe^{2+}$ is present in the soil, redox reactions occurs, decreasing the contaminant removal by Cr(III) accumulation near anode, in the acidified soil. Another effect also observed in this study compared with Cappai and coworkers [22] was the increasing percentage removal when TRM mass increased (from 15 to 30 g). This can be explained through a strong binding between chromium on the RB.

Ferritization method also was used by Kimura and coworkers [20] as the treatment zone (FTZ), placed near cathode, for copper remove from soil due to its metal ions insolubilization capacity and by changing this components to the magnetic material (copper-ferrite). Acidic conditions, once more, are highly needed for metal removal. The authors observed that this kind of environment enables metal and metal oxide transformation into metal ions, which are strongly adsorbed by ion-exchange mechanism on a clay mineral soil. This method can remove several hazardous metal ions at the same time and treat Cr(VI) and Mn(VII) without especial pretreatment [82]. Those metals are separated easily by sedimentation or magnetic separation. Also, ferritization does not need a treatment for water produced during EK [20].

Zhang and coworkers [6] incorporated a calcined hydrotalcite–based PRB for Cr(VI) removal due to its high anion exchange and regeneration capacities. The authors understood the synergistic effect of this coupled process through increasing in anionic chromate concentration in the anode region by EK with PRB absorption and immobilization capacities. Another characteristic which enables hydrotalcite as a PRB is its antacid type of material, capable of suppress acidification and enhance EK. It was observed that no DC voltage applied resulted in almost 100% removal of chromium in only 3 h. Applying 20–30 V, this percentage was achieved after 250 h. When placing RB near anode, chromate anions as $CrO_4^{2-}$ are quickly absorbed, reducing Cr(VI) concentration gradient and increasing its removal.

Suzuki and coworkers [83] compared the efficiency of two PRBs —magnetite and zero-valent iron—for Cr(VI) removal. According to Zhang and coworkers [6], zero-valent iron as a reactive barrier has the disadvantage of consequent precipitation because of iron corrosion, thus decreasing its permeability and prohibiting its filler activity. As a result, Cr(VI) was reduced but Cr(III) recovery in the PRB was low. This study showed that magnetite PRB reduces Cr(VI) to Cr(III), favoring acidic conditions, without releasing $Fe^{2+}$ in the aqueous phase. By controlling anode pH at 6.8, the authors could prevent Cr(VI) sorption onto the kaolinite and consequently its migration towards magnetite PRB and the oxidation state that block it conversion to Cr(III). The negative point of this was that 18% of Cr(VI) passed through magnetite PRB. The solution was to increase reaction time by applying a cation exchange membrane between this PRB and the anode. The removal rate for magnetite was almost 70% and for zero-valent iron was near 5%.

A polyacrylonitrile nanofiber (PANN) membrane as PRB was developed by Peng and coworkers [78]to remove $Zn^{2+}$, $Fe^{3+}$ and $Ca^{2+}$ because of its excellent physicochemical properties, for example, heat and corrosion resistance. This nanofiber membrane can be improved by narrowing its nanopore diameter and by applying electrospinning technology to get high molecular weight polymers [84]. This membrane helped to improve acidic-basic zone, thus enhancing the remediation process. It was noticed that removal rates with PRB were higher when a lower initial ions concentration and pH were set. The main reason to this effect is due to limited adsorption and complexation capacities of PANN, which had influenced in the ionic movement as well. In relation to runs without PRB, increasing voltage resulted in a higher removal efficiency. Increasing voltage from 25 to 50 V, ions penetration rate into PANN increased. It can be explained by a stronger EM, accelerating $ZnO_2^{2-}$ and $FeO_3^{3-}$ to anode and $Ca^{2+}$ to cathode. Besides that, an increase in migration function makes metal ion binding to the nanofiber not so strong, leading to the formation of $CN^-$ complexes in PANN. Electroosmotic flow towards electrodes enhanced metal ion removal because of a higher current and, consequently, ion movement. At 50 V, the efficiency was increased but DC power and temperature operation also increased. Considering all statements, the best conditions to remove these metals were at 25 V and pH = 1.2 and the results were: $Zn^{2+}$ 99.15% $Fe^{3+}$ 98.03% and $Ca^{2+}$ 99.73%.

Yuan and coworkers [85] removed arsenic using two types of carbon nanotube: one purified with 0.3 M $HNO_3$ (CNT) and another coated with cobalt (CNT-Co) as PRB. The removal of this contaminant is related to sorption on a surface area and so CNT-Co has a great surface area. EDTA was applied as processing fluid due to its low molecular weight and formation capacity in water-soluble complex with several metal ions. Groundwater was also used as a processing fluid, but its removal results were lower (66% compared with 74% applying EDTA). As arsenic mainly exists in anionic form, more of it was collected at anode compartment. Comparing results without PRB and groundwater as processing fluid, there was an improvement from 35 to 74%, showing how EK was enhanced. When CNT adopted, the removal rate was similar to the one achieved without a PRB. An As(V) uptake was identified in the acidic pH due to a lower potential release of Co from CNT. Although this event occurred, less of this metal was removed by electroosmotic flow near anode. EOF was dominant when groundwater was the processing fluid and EM dominated when EDTA was applied.

Saeedi and coworkers [4] performed EK remediation by employing an activated carbon (ACB) as permeable reactive barrier, placed beside cathode well, to remove nickel from a kaolinite soil. This PRB presented a successful result of reversed EOF prevention, which has a negative effect on Ni removal. The reverse EO occurrence may be explained by a high Ni (500 mg/kg) content, which increases ionic strength, in the tests without PRB, thus influences soil zeta potential. Once clays surface charge depends on pH and ionic strength, a change in net surface charge results in an opposite EO direction. Furthermore, nickel in its high concentration may interacted with clay's surface. As consequence, the reverse EO flow occurred the end of the second process day. When ACB was placed, some $Ni^{2+}$ was absorbed. Hence, ionic strength diminished and this reversal effect did not happen. KNO (0.05 M) was used as conducting fluid because of its higher conductivity compared to distilled water, once facilitates ions migration. The best rate removal was 50.6% applying 1.25 V/cm during 3 days.

From an economic, social and environmental points of view, recycling unusable materials are very important due to its large production and its negative impact when are disposal at wrong places. Many researchers are studying ways to reuse these resources and some of them from Korea applied carbonized foods waste (CFW—85% oxygen, calcium and carbon) as an option for PRB filling material. Han and coworkers [3] compared this kind of material with zeolites in order to remove Cu from a low permeable soil—kaolinite, achieving 84.6% of efficiency for CFW, which was 4–8 times greater than zeolites potential, through adsorption isotherms examining. CFW was placed in the increasing point of pH, near cathode, to remove Cu by adsorption and those heavy metals in soil which could not be removed with PRB were separated through electrodes exchange. Five solvents were tested to be the EK enhancer in the anode: HCl, acetic acid, citric acid, ethylene diamine tetra acetic acid (EDTA), and sodium dodecyl sulfate (SDS). The first two are considered hydration sedimentation preventers which are caused by electrical current. Citric acid and EDTA has the same capacity of HCl and acetic acid for metal ions for the sake of the surfactant and the chelating of chelate-zero metal ions. SDS showed more constant pH values but acetic acid was the chosen one owing to its electrical conductivity stability and biodegradable capacity. Two effects were noticed during EK-PRB experiments: first, EOF increased with increasing process time and electric current density; second, the electrical current density compared between the following polarity reversal and before that was lower. The electrode exchange was performed to have a better removal of precipitated Cu near cathode. This response can be elucidated due to a great number of Cu being remediate by PRB-CFW through adsorption or sedimentation. An economic evaluation was made and by adding acetic acid and energy cost, to remediate 84.6% of Cu was spent 32.2 US$/$m^3$, once CFW cost is close to zero.

Huang and coworkers [52] removed several heavy metals, for example, Zn, Pb, Cu, and Cd from Municipal Solid Waste Incineration (MSWI) fly ash with activated charcoal as the PRB material. Once more, pH control has a significant role for metal ions extracting, since it affect adsorbent electrical properties, the ionization degree and adsorbate speciation. The authors observed that a basic pH promotes a better percentage withdraw of the heavy metals because there are reductions in competing ions ($Al^{3+}$ and $Ca^{2+}$), monovalent cations or non-soluble hydroxides formation and heavy metals combining with functional groups on adsorbent surface. Other effect was the concentration of activated charcoal in remediation which increased Zn, Pb, Cu, and Cd removal by increasing its quantity. Higher percentage removal was obtained using a adsorvent concentration at 10 g/100 mL. It happened once more surface area was available, thus more active sites were accessible to bind these metal ions. To enhance EK-PRB tests, oxalic acid (0.05 M) was added every 2 days, promoting a called ENEKR-PRB, to decrease pH gradient between sections S2 and S3. Removal rates were greater at 15 days, 2 V/cm, and 0.1 mol/L; Zn: 78.34%, Pb: 62.45%, Cu: 80.14%, and Cd: 46.25%.

Fonseca and coworkers [51] used a different approach by applying a bio barrier constituted of *Arthrobacter viscosus* supported in reactive materials, e.g., zeolites or an activated carbon to reduce Cr(VI) to Cr(III) and so entrap in the physical support through adsorption or ion exchange. This bacterium is a great enhancer due to its capacity of exopolysaccharides producing which makes this microorganism be able to attach itself to a support. Cr(VI) reduction is a positive point relating to chromium removal according to the "adsorption coupled reduction" theory, once trivalent cations entrapped more effectively on zeolites and activated carbon supports [40, 49, 50]. The electric field applied in the experiments promoted a decrease in Cr(III) concentration from the bio barriers to the cathode, elucidating EM influence. It was detected an amount of Cr(III) in the anode when the activated carbon was the PRBB higher than with zeolites (1.6 vs. 0.3 mg). The explanation relays in the bio barrier saturation with chromium trivalent, thus Cr(VI) was transported through it and because activated carbon has a higher porosity and specific area. Zeolites present a higher compact structure which interrupts Cr(VI) oxyanion ionic migration and its dispersion on bio barriers, making its conversion difficult.

Zanjani and coworkers [54] implemented activated carbon near anode, in the middle section and near cathode to remove nickel from a kaolin soil. Results showed that Ni migration was bigger when PRB was placed beside the cathode. The same reason presented by Saeedi and coworkers [4] to use activated carbon as PRB was considerable, which is to prevent reverse EOF. Weng and coworkers [53], although worked with zero-valent iron (ZVI) as PRB, also detected a reverse EOF during chromium remediation when ZVI was not used. The authors attributed this effect to anions existing in the soil due to high Cr(VI) concentration and so anions migration was faster than cations. In Zanjani and coworkers [54] study, this occurrence when activated carbon was not applied is owing to high nitrates ions content. However, when PRB experiments were carried out, due to an increase in sorption capacity [5], reverse EOF was extinct. Because initial soil pH was 8.2, a relatively low removal rate was observed (47%), but applying a constant voltage of 1 or 1.25 V/cm resulted in a powerful acid front causing Ni removal increase.

Xu and coworkers [86] described Cr(VI)-contaminated farmland soil remediation using an PRB constituted by hydrocalumite (CaAl-LDH)—a low-cost and long term feasible material. This kind of reactive material promoted a decrease in Cr(VI) content when results showed more oxidizable and residual fractions of it in soil.

Relating to the migration rate observed, when applied PRB, it was a little faster than without PRB because the reactive barrier capture the metal ion in-time, which reduced chromium accumulation influence on its migration. Removal rates from Cr(VI) for EK and EK-PRB test were, respectively: 85.5% and 96.49%; and for total Cr were: 40.97% and 69.34%. The low removal rate achieved of total Cr is due to Cr(III) production. Table 3 presents a summary of operational conditions to remove heavy metals from soil by applying EK coupled with different PRB found in literature.

## 11 Simultaneous Inorganic and Organic Remediation

Atomizing slag—a composite of CaO, FeO, and $Fe_2O_3$—was employed by Chung and Lee [88] as PRB to remove TCE and Cd from a clay soil, since it dechlorinates TCE, enhancing its removal. Atomizing slag is produced when conventional slag is cooled in a slow manner. It was chosen due to its good strength and its characteristic of not releasing toxic materials. Besides that, it is a low-cost material. The removal of Cd was slightly higher than TCE because of electro-osmosis and electromigration (positive charge). The recovery rate for Cd was lower than TCE's because of the difference between adsorption for Cd and the depletion for TCE.

A new type of bamboo charcoal was used by Ma and coworkers [89] as PRB for 2,4-dichlorophenol (2,4-DCP) and Cd removals from soil because of its excellent adsorption characteristic for nitrate-nitrogen, heavy metals, harmful gases and its capacity for water or air purification. Two factors influence positively EK when soil type is taken into account: first, only 5.33% of organic matter and 11.21% of soil particles were lower than 2 μm were present in soil and secondly a coarse grain structure soil improves metal movement. The authors observed that Cd removal was greater than the organic pollutant owing to coupled EO and EM. It is known that electromigration is the main process during electrokinetics remediation; thus, when pH is neutral, positively charged ions of Cd ($Cd^{2+}$ and $Cd(OH)^+$) migrate to cathode. Sandy soil was chosen to be contaminated with these two pollutants regardless shows low total porosity and weak EO due to its high permeability and low water-holding capacity. Although neutral pH is favorable to remove Cd, it is a disadvantage for 2,4-DCP removal, once it exists as non-dissociated form, turning its movement difficult. In addition, 2,4-DCP electroosmosis decreases with a decrease in soil water content. Another result found was related to the electric consumption. A highest electric consumption was noticed when there was 12 h interval, once polarity reversal changed migration direction of charged contaminants repeatedly. When 24 h intervals were applied, it was perceived a greater efficiency in removal rates and less energy consumption compared to 12 h, probably due to Cd and 2,4-DCP did not reach the treatment zone before polarity changed in the second case. As a result, with 12 h of interval, the efficiency was 40.13 and 24.98% and with 24 h of interval it was 75.97 and 54.92% of Cd and 2,4-DCP, respectively. Energy consumption values were 12.24 and 11.61 kWh $m^{-3}$ for 12 and 24 h of interval, respectively [90, 91].

**Table 3** Operational conditions of heavy metals removal from soils using PRB-EK

| Pollutant | PRB | Experimental conditions | Observations—best results | Ref. |
|---|---|---|---|---|
| Cr(VI)<br>1000 mg $kg^{-1}$<br>As (V)<br>100 mg $kg^{-1}$ | Transformed red mud (adsorption) | Illitic-kaolinitic soil (2 mm sieve), 1 V $cm^{-1}$ (30 cm long), 20% initial moisture, soil pH = 5.49, 6 and 12 days duration | Total Cr:<br>(6 days): 19.4% in the anolyte and 20.6% trapped in PRB<br>(12 days): 60.8% in the anolyte and 25.5% trapped in PRB<br>As:<br>(6 days): 16%<br>(12 days): 29% | [22] |
| Cr (VI)<br>1000 mg $kg^{-1}$ | Transformed red mud (adsorption) | Kaolinite, illite, and quartz soil, 1 V $cm^{-1}$ (30 cm long), 15% initial moisture, soil pH = 5.5, 6 and 12 days duration, 15 and 30 g of TRM | Total Cr, non-extractable Cr and trapped in PRB: (6 days) and 15 g of TRM: 33.6, 42.4, and 54.4%<br>(12 days) and 30 g of TRM: 40.3, 59.5, and 93.2 | [39] |
| Cu(II)<br>50 mg $kg^{-1}$ | Ferrite treatment zone (ferritization) | Kaolin soil +0.1 M HCl + 0.1 M NaOH, 2 V $cm^{-1}$, 30% initial moisture, soil pH = 7, 1000 mg $kg^{-1}$ of ferrite in FTZ, 4 days duration | 92% was converted to copper-ferrite | [20] |
| Cr (VI)<br>0.16–1.65 mg $g^{-1}$ | Calcined-hydrotalcite (adsorption) | Loam and kaolin soil (100 mesh sieve), 0.7–2 V $cm^{-1}$, soil pH = 6, flowing solution: 0.1 M HCl, flow rates: 0, 0.1, 0.2, and 0.3 ml $min^{-1}$ | 80% under 2 V $cm^{-1}$ in 4 h<br>>99% under 30 V in 3 h | [6] |
| Cr (VI)<br>8.3 mg $kg^{-1}$ | Magnetite (reduction) | Kaolinite +0.001 M HCl in 1 M $NaNO_3$, NaClO and 0.01 M $NaNO_3$, 1 V $cm^{-1}$ (20 cm long), 50% initial moisture, 5 days duration | 70% | [83] |
| Zinc 5, 25 and 100 mg $kg^{-1}$<br>Iron 10, 25 and 150 mg $kg^{-1}$<br>Ca 100, 150 and 300 mg $kg^{-1}$ | Polyacrylonitrile nanofiber membrane (adsorption) | Coal gangue heap soil, 10, 25 and 50 V, pH = 1.2, 4, 7, and 10 with HCl or NaOH, 13 days duration | 10 V: 86.81, 69.16, 88.1%<br>25 V and pH = 1.2: 99.15, 98.03, 99.73%<br>From 25 to 50 V: 6 to 44%, 31 to 34% and 2 to 16% | [78] |

(continued)

**Table 3** (continued)

| Pollutant | PRB | Experimental conditions | Observations—best results | Ref. |
|---|---|---|---|---|
| Arsenic (V) 855–972 mg $kg^{-1}$ | Carbon nanotube coated with cobalt (adsorption) | Clay soil (2 mm sieve), processing fluid: Local groundwater and 0.25 M EDTA, soil pH = 8.5, 2 V $cm^{-1}$, 5 days duration | With groundwater: 66% With EDTA: 71–77% | [48] |
| Nitrate 100 mg $kg^{-1}$ | Anion exchange resin (Purolite A-520-E) (ion exchange) | Spiked kaolinite soil, 1 to 2 V $cm^{-1}$, soil pH = 4.9, tap water as processing fluid, 37.5% initial moisture, 15, 25, and 40 V, 7 days duration | 90% | [2] |
| Ni 500 mg $kg^{-1}$ | Activated carbon (adsorption) | Kaolinite soil, 1 and 1.25 V $cm^{-1}$ (15 cm long), soil pH = 8.2, conductive solution: 0.05 M KNO, 3 and 7 days duration | 37.06–50.6% | [4] |
| Cu 500 mg $kg^{-1}$ | Carbonized foods waste (adsorption) | Kaolinite soil, acetic acid as enhancer, soil pH = 4.5–5.5, 60% initial moisture, polarity reversal after 8 and 10 days, 20 days duration | 53.4–84.6% | [3] |
| Zinc 5256.25 mg $kg^{-1}$ Pb 2047.26 mg $kg^{-1}$ Cu 593.33 mg $kg^{-1}$ Cadmium 149.02 mg $kg^{-1}$ | Activated charcoal (adsorption) | MSWI fly ash soil (200 mesh sieve), 1, 1.5, and 2 V $cm^{-1}$, oxalic acid as enhancer (0.05, 0.1 and 0.2 M), 5, 10, and 15 days duration | Zn: 54.32–78.34% Pb: 42.12–62.45% Cu: 47.16–80.14% Cd: 34.91–49.23% | [52] |
| Cr (VI) 50 mg $kg^{-1}$ | Activated carbon (reduction) *Arthrobacter viscosus* (adsorption) | Spiked kaolinite soil, 30% initial moisture, 10 V, 0.1 M NaOH to adjust anode pH and 2 M $HNO_3$ to adjust cathode pH at 5, 9 and 18 days duration | (9 days): 22% and 17% conversion (18 days): 79% and 44% conversion | [51] |
| Cr(VI) 50 mg $kg^{-1}$ | Zeolite 13X (reduction) *Arthrobacter viscosus* (adsorption) | Spiked kaolinite soil, 30% initial moisture, 10 V, 0.1 M NaOH to adjust anode pH and 2 M $HNO_3$ to adjust cathode pH at 5, 9 and 18 days duration | (9 days): 47% and 44% conversion (18 days): 60% and 45% conversion | [51] |

(continued)

**Table 3** (continued)

| Pollutant | PRB | Experimental conditions | Observations—best results | Ref. |
|---|---|---|---|---|
| Cu 1000–2000 mg kg$^{-1}$ | Activated carbon (adsorption) | Kaolinite soil, 40% initial moisture, 0.1 M $NaNO_3$ + citric acid-sodium citrate buffer solution (pH = 5), 1 V cm$^{-1}$, 4 days duration | 96.6% for 2000 mg kg$^{-1}$ initial concentration | [87] |
| Ni 500 mg kg$^{-1}$ | Activated carbon (adsorption) | Kaolinite soil, 1 and 1.25 V cm$^{-1}$, soil pH = 8.2, 0.05 M KNO3 + distilled water as conductive solution, stainless steel electrodes (A316), 3 days duration | 47% with 1.25 V cm$^{-1}$ | [54] |
| Cr 1017 mg kg$^{-1}$ Cr(VI) 965.1 mg kg$^{-1}$ | Hydrocalumite (reduction) | Farmland clay soil, 1, 1.5 and 2 V cm$^{-1}$, 30% initial moisture, 5 days duration | Total Cr with 1 V cm$^{-1}$: 69.34% Cr(VI) with 1 V cm$^{-1}$: 96.49% | [86] |

# References


1. H.C.J. Godfray, J.R. Beddington, I.R. Crute, L. Haddad, D. Lawrence, J.F. Muir, J. Pretty, S. Robinson, S.M. Thomas, C. Toulmin, Food security: The challenge of feeding 9 billion people. Science **327**(5967), 812–818 (2010)
2. Y. García, C. Ruiz, E. Mena, J. Villaseñor, P. Cañizares, M.A. Rodrigo, Removal of nitrates from spiked clay soils by coupling electrokinetic and permeable reactive barrier technologies. J. Chem. Technol. Biotechnol. **90**, 1719–1726 (2015)
3. J.G. Han, K.K. Hong, Y.W. Kim, J.Y. Lee, Enhanced electrokinetic (E/K) remediation on copper contaminated soil by CFW (carbonized foods waste). J. Hazard. Mater. **177**, 530–538 (2010)
4. M. Saeedi, A. Jamshidi, N. Shariatmadri, A. Falamaki, An investigation on the efficiency of electro kinetic coupled with carbon active barrier to remediate nickel contaminated clay. Int. J. Environ. Res. **3**, 629–636 (2009)
5. J. Virkutyte, M. Sillanpää, P. Latostenmaa, Electrokinetic soil remediation - critical overview. Sci. Total Environ. **289**, 97–121 (2002)
6. J. Zhang, Y. Xu, W. Li, J. Zhou, J. Zhao, G. Qian, Z.P. Xu, Enhanced remediation of Cr(VI)-contaminated soil by incorporating a calcined-hydrotalcite-based permeable reactive barrier with electrokinetics. J. Hazard. Mater. **239–240**, 128–134 (2012)
7. P. Bhattacharya, A.B. Mukherjee, G. Jacks, S. Nordqvist, Metal contamination at a wood preservation site: Characterisation and experimental studies on remediation. Sci. Total Environ. **290**, 165–180 (2002)
8. Z.S. Chen, G.J. Lee, J.C. Liu, The effects of chemical remediation treatments on the extractability and speciation of cadmium and lead in contaminated soils. Chemosphere **41**, 235–242 (2000)
9. H.N. Lim, H. Choi, T.M. Hwang, J.W. Kang, Characterization of ozone decomposition in a soil slurry: kinetics and mechanism. Water Res. **36**, 219–229 (2002)
10. M.S. Hong, W.F. Farmayan, I.J. Dortch, C.Y. Chiang, S.K. McMillan, J.L. Schnoor, Phytoremediation of MTBE from a groundwater plume. Environ. Sci. Technol. **35**, 1231–1239 (2001)

11. S. Susarla, V.F. Medina, S.C. McCutcheon, Phytoremediation: an ecological solution to organic chemical contamination. Ecol. Eng. **18**, 647–658 (2002)
12. F.L. Souza, C. Sáez, M.R.V. Lanza, P. Cañizares, M.A. Rodrigo, Removal of chlorsulfuron and 2,4-D from spiked soil using reversible electrokinetic adsorption barriers. Sep. Purif. Technol. **178**, 147–153 (2017)
13. E. Vieira dos Santos, C. Sáez, P. Cañizares, C.A. Martínez-Huitle, M.A. Rodrigo, Reversible electrokinetic adsorption barriers for the removal of atrazine and oxyfluorfen from spiked soils. J. Hazard. Mater. **322**, 413–420 (2017)
14. M.D.L. Bellinaso, C.W. Greer, M. do Carmo Peralba, J.A.P. Henriques, C.C. Gaylarde, Biodegradation of the herbicide trifluralin by bacteria isolated from soil. FEMS Microbiol. Ecol. **43**, 191–194 (2003)
15. S.R. Kanel, H. Choi, J. Kim, S. Vigneswaran, W.G. Shim, Removal of arsenic (III) from groundwater using low-cost industrial by-products — blast furnace slag. Water Qual Res J Canada **41**, 130–139 (2006)
16. I.S. Kim, J.-S. Park, K.-W. Kim, Enhanced biodegradation of polycyclic aromatic hydrocarbons using nonionic surfactants in soil slurry. Appl. Geochem. **16**, 1419–1428 (2001)
17. G. Nano, A. Borroni, R. Rota, Combined slurry and solid-phase bioremediation of diesel contaminated soils. J. Hazard. Mater. **100**, 79–94 (2003)
18. K.S.M. Rahman, I.M. Banat, J. Thahira, T. Thayumanavan, P. Lakshmanaperumalsamy, Bioremediation of gasoline contaminated soil by a bacterial consortium amended with poultry litter, coir pith and rhamnolipid biosurfactant. Bioresour. Technol. **81**, 25–32 (2002)
19. A.G. Rike, K.B. Haugen, B. Marion, B. Engene, P. Kolstad, In situ biodegradation of petroleum hydrocarbons in frozen arctic soils. Cold Reg. Sci. Technol. **37**, 97–120 (2003)
20. T. Kimura, K.I. Takase, N. Terui, S. Tanaka, Ferritization treatment of copper in soil by electrokinetic remediation. J. Hazard. Mater. **143**, 662–667 (2007)
21. M. Arora, I. Snape, G.W. Stevens, Toluene sorption by granular activated carbon and its use in cold regions permeable reactive barrier: Fixed bed studies. Cold Reg. Sci. Technol. **69**, 59–63 (2011)
22. G. Cappai, G. De Gioannis, A. Muntoni, D. Spiga, J.J.P. Zijlstra, Combined use of a transformed red mud reactive barrier and electrokinetics for remediation of Cr/As contaminated soil. Chemosphere **86**, 400–408 (2012)
23. R. Thiruvenkatachari, S. Vigneswaran, R. Naidu, Permeable reactive barrier for groundwater remediation. J. Ind. Eng. Chem. **14**, 145–156 (2008)
24. K. Bronstein, *Permeable Reactive Barriers for Inorganic and Radionuclide Contamination* (U.S. Environmental Protection Agency, Washington, DC, 2005)
25. D.C. McMurtry, R.O. Elton, New approach to in-situ treatment of contaminated groundwaters. Environ. Prog. **4**, 168–170 (1985)
26. S.J. Morrison, D.L. Naftz, J.A. Davis, C. Fuller, Introduction to groundwater remediation of metals, radionuclides, and nutrients with permeable reactive barriers, in *Handbook of Groundwater Remediation using Permeable Reactive Barriers Applications to Radionuclides, Trace Metals, and Nutrients* (2003), pp. 1–15
27. A. Gavaskar, N. Gupta, B. Sass, R. Janosy, J. Hicks, Design Guidance for Application of Permeable Reactive Barriers for Groundwater Remediation. Columbus, 2000, p. 157
28. R. Köber, D. Schäfer, M. Ebert, A. Dahmke, Coupled in situ reactors using Fe0 and activated carbon for the remediation of complex contaminant mixtures in groundwater, in *Proceedings of the Groundwater Quality 2001 Conference*, ed. by S. F. Thornton, S. E. Oswald, (IAHS, Sheffield, 2002), pp. 18–21
29. S. Fiorenza, C.L. Oubre, C. Ward, *Sequenced Reactive Barriers for Groundwater Remediation* (CRC Press, Boca Raton, 2000)
30. Interstate Technology & Regulatory Council, ITRC, *Permeable Reactive Barriers: Lessons Learned/New Directions* (PRB: Technology Update Team, Washington, DC, 2005)
31. Interstate Technology & Regulatory Council, ITRC, *Permeable Reactive Barrier: Technology Update* (PRB: Technology Update Team, Washington, DC, 2011)

32. S.G. Benner, D.W. Blowes, W.D. Gould, R.B. Herbert, C.J. Ptacek, Geochemistry of a permeable reactive barrier for metals and acid mine drainage. Environ. Sci. Technol. **33**, 2793–2799 (1999)
33. R.W. Gillham, S.F. O'Hannesin, W.S. Orth, Metal enhanced abiotic degradation of halogenated aliphatics: Laboratory tests and field trials. Proc., 1993 HazMat Central Conference, Chicago, Illinois, (1993)
34. M.A. Carey, B.A. Fretwell, N.G. Mosley, J.W.N. Smith, *Guidance on the use of permeable reactive barriers for remediating contaminated groundwater* (Environment Agency, National Groundwater and Contaminated Land Centre, Olton, 2002)
35. S.E. Tappert, M. Ishihara, Permeable reactive barrier installation, Fairfield New Jersey, Presented at the Remediation Technology Development Forum Permeable Reactive Barriers Action Team Meeting, Oak Ridge, 2008
36. B. Gu, L. Liang, J. Zhou, D. Phillips. In situ microbial investigation of a Fe(0) reactive barrier, in: Proceedings American Chemical Society National Meeting, San Diego, vol. 41, no. 1, 2011, pp. 1173–1180
37. E. Mena, J. Villaseñor, M.A. Rodrigo, P. Cañizares, Electrokinetic remediation of soil polluted with insoluble organics using biological permeable reactive barriers: Effect of periodic polarity reversal and voltage gradient. Chem. Eng. J. **299**, 30–36 (2016)
38. E.M. Ramírez, C.S. Jiménez, J.V. Camacho, M.A.R. Rodrigo, P. Cañizares, Feasibility of coupling permeable bio-barriers and electrokinetics for the treatment of diesel hydrocarbons polluted soils. Electrochim. Acta **181**, 192–199 (2015)
39. G. De Gioannis, A. Muntoni, R. Ruggeri, J.J.P. Zijlstra, Chromate adsorption in a transformed red mud permeable reactive barrier using electrokinesis. J. Environ. Sci. Heal. Part A Toxic Hazard. Subst. Environ. Eng. **43**, 969–974 (2008)
40. B. Silva, H. Figueiredo, I.C. Neves, T. Tavares, The role of pH on Cr (VI) reduction and removal by arthrobacter viscosus. Int. J. Chem. Biol. Eng. **2**, 100–103 (2009)
41. S. Rodrigo, C. Saez, P. Cañizares, M.A. Rodrigo, Reversible electrokinetic adsorption barriers for the removal of organochlorine herbicide from spiked soils. Sci. Total Environ. **640–641**, 629–636 (2018)
42. Z. Li, S. Yuan, J. Wan, H. Long, M. Tong, A combination of electrokinetics and Pd/Fe PRB for the remediation of pentachlorophenol-contaminated soil. J. Contam. Hydrol. **124**, 99–107 (2011)
43. J. Wan, Z. Li, X. Lu, S. Yuan, Remediation of a hexachlorobenzene-contaminated soil by surfactant-enhanced electrokinetics coupled with microscale Pd/Fe PRB. J. Hazard. Mater. **184**, 184–190 (2010)
44. E. Mena, C. Ruiz, J. Villaseñor, M.A. Rodrigo, P. Cañizares, Biological permeable reactive barriers coupled with electrokinetic soil flushing for the treatment of diesel-polluted clay soil. J. Hazard. Mater. **283**, 131–139 (2015)
45. C. Ruiz, E. Mena, P. Cañizares, J. Villaseñor, M.A. Rodrigo, Removal of 2,4,6-trichlorophenol from spiked clay soils by electrokinetic soil flushing assisted with granular activated carbon permeable reactive barrier. Ind. Eng. Chem. Res. **53**, 840–846 (2014)
46. G.C.C. Yang, Y.W. Long, Removal and degradation of phenol in a saturated flow by in-situ electrokinetic remediation and Fenton-like process. J. Hazard. Mater. **69**, 259–271 (1999)
47. G.C.C. Yang, C.Y. Liu, Remediation of TCE contaminated soils by in situ EK-Fenton process. J. Hazard. Mater. **85**, 317–331 (2001)
48. C. Yuan, C.H. Hung, W.L. Huang, Enhancement with carbon nanotube barrier on 1,2-dichlorobenzene removal from soil by surfactant-assisted electrokinetic (SAEK) process - the effect of processing fluid. Sep. Sci. Technol. **44**, 2284–2303 (2009)
49. B. Silva, H. Figueiredo, C. Quintelas, I.C. Neves, T. Tavares, Zeolites as supports for the biorecovery of hexavalent and trivalent chromium. Microporous Mesoporous Mater. **116**, 555–560 (2008)

50. H. Figueiredo, B. Silva, C. Quintelas, I.C. Neves, T. Tavares, Effect of the supporting zeolite structure on Cr biosorption: Performance of a single-step reactor and of a sequential batch reactor — A comparison study. Chem. Eng. J. **163**, 22–27 (2010)
51. B. Fonseca, M. Pazos, T. Tavares, M.A. Sanromán, Removal of hexavalent chromium of contaminated soil by coupling electrokinetic remediation and permeable reactive biobarriers. Environ. Sci. Pollut. Res. **19**(5), 1800–1808 (2012)
52. T. Huang, D. Li, L. Kexiang, Y. Zhang, Heavy metal removal from MSWI fly ash by electrokinetic remediation coupled with a permeable activated charcoal reactive barrier. Sci. Rep. **5**, 15412 (2015)
53. C. Weng, Y. Lin, T.Y. Lin, C.M. Kao, Enhancement of electrokinetic remediation of hyper-Cr (VI) contaminated clay by zero-valent iron. J. Hazard. Mater. **149**, 292–302 (2007)
54. A.J. Zanjani, M. Saeedi, C.-H. Weng, An electrokinetic process coupled activated carbon barrier for nickel removal from kaolinite. Environ. Asia **5**, 28–35 (2012)
55. M. Zhou, S. Zhu, Y. Yi, T. Zhang, An electrokinetic/activated alumina permeable reactive barrier-system for the treatment of fluorine-contaminated soil. Clean Technol. Environ. Policy **18**, 2691–2699 (2016)
56. F. Obiri-Nyarko, S.J. Grajales-Mesa, G. Malina, An overview of permeable reactive barriers for in situ sustainable groundwater remediation. Chemosphere **111**, 243–259 (2014)
57. K.E. Roehl, K. Czurda, T. Meggyes, F.G. Simon, D.I. Stewart, Chapter 1 Permeable reactive barriers. Trace Metals Other Contam. Environ. **7**, 1–25 (2005)
58. S.M. Testa, D.L. Winegardner, *Restoration of Contaminated aquifers: Petroleum hydrocarbons and organic compounds*, 2nd edn. (CRC Press, Boca Raton, 2000)
59. U.S. EPA, *Field Applications of In Situ Remediation Technologies: Permeable Reactive Barriers* (US Environmental Protection Agency, Office of Solid Waste and Emergency Response, Washington, DC, 2002)
60. R.D. Vidic, F.G. Pohland, *Treatment Walls. Technology Evaluation Report TE-96-01* (Ground-Water Remediation Technologies Analysis Center, Pittsburgh, 1996)
61. R.T. Wilkin, R.W. Puls, Capstone Report on the Application Monitoring and Performance of Prbs for Ground Water Remediation. Report EPA/600/R-03/045a, Cincinnati, 2003
62. T. McGovern, T.F. Guerin, S. Horner, B. Davey, Design, construction and operation of a funnel and gate in situ permeable reactive barrier for remediation of petroleum hydrocarbons in groundwater. Water Air Soil Pollut. **136**, 11–31 (2002)
63. J. Muegge, An Assessment of Zero Valence Iron Permeable-Reactive Barrier Projects In California. Office of Pollution Prevention and Green Technology, Document No. 1219, California, 2008, p. 147
64. R.W. Gillham, S.F. O'Hannesin, Metal-catalysed abiotic degradation of halogenated organic compounds, in IAH Conference, Modern Trends in Hydrogeology, Ontario, 1992
65. T. Meggyes, F. Simon, Removal of organic and inorganic pollutants from groundwater using permeable reactive barriers: Part 2. Engineering of permeable reactive barriers. Land Contam. Reclam. **8**(3), 103–116 (2000)
66. L. Cang, D.-M. Zhou, D.-Y. Wu, A.N. Alshawabkeh, Coupling Electrokinetics with permeable reactive barriers of zero-valent iron for treating a chromium contaminated soil. Sep. Sci. Technol. **44**, 2188–2202 (2009)
67. C. Yuan, T.-S. Chiang, The mechanisms of arsenic removal from soil by electrokinetic process coupled with iron permeable reaction barrier. Chemosphere **67**, 1533–1542 (2007)
68. S.R. Kanel, J.-M. Grenèche, H. Choi, Arsenic (V) removal from groundwater using nano scale zero-valent iron as a colloidal reactive barrier material. Environ. Sci. Technol. **40**, 2045–2050 (2006)
69. G. Hornig, K. Northcott, G. Stevens, Assessment of sorbent materials for treatment of hydrocarbon contaminated ground water in cold regions. Cold Reg. Sci. Technol. **53**, 83–91 (2008)
70. A. Erto, R. Andreozzi, A. Lancia, D. Musmarra, Factors affecting the adsorption of trichloroethylene onto activated carbons. Appl. Surf. Sci. **256**, 5237–5242 (2010)
71. A. Shoumkova, Zeolites for water and wastewater treatment: an overview. Australian Institute High Energetic Materials (2011) http://www.ausihem.org

72. J. Fronczyk, K. Pawluk, M. Michniak, Application of permeable reactive barriers near roads for chloride ions removal. Ann. Warsaw Univ. Life Sci. SGGW L Reclam **42**(2), 249–259 (2010)
73. S.J. Morrison, R.R. Spangler, Extraction of uranium and molybdenum from aqueous solutions: a survey of industrial materials for use in chemical barriers for uranium mill tailings remediation. Environ. Sci. Technol. **26**, 1922–1931 (1992)
74. F. Ahmad, P.S. Stephen, C.J. Newell, Remediation of RDX- and HMX-contaminated groundwater using organic Mulch permeable reactive barriers. J. Contam. Hydrol. **90**, 1–20 (2007)
75. B.L. Freidman, G. Sally, I. Snape, G. Stevens, K. Mumford, Application of controlled nutrient release to permeable reactive barriers. J. Environ. Manage. **169**, 145–154 (2016)
76. R. López-Vizcaíno, J. Alonso, P. Cañizares, M.J. León, V. Navarro, M.A. Rodrigo, C. Sáez, Removal of phenanthrene from synthetic kaolin soils by eletrokinetic soil flushing. Sep. Purif. Technol. **132**, 33–40 (2014)
77. N. Ott, Permeable Reactive Barriers for Inorganics (USEPA, Washington, DC, 2000), http://www.clu-in.org
78. L. Peng, X. Chen, Y. Zhang, Y. Du, M. Huang, J. Wang, Remediation of metal contamination by electrokinetics coupled with electrospun polyacrylonitrile nanofiber membrane. Process Saf. Environ. Prot. **98**, 1–10 (2015)
79. Z. Zheng, S. Yuan, Y. Liu, X. Lu, J. Wan, X. Wu, J. Chen, Reductive dechlorination of hexachlorobenzene by Cu/Fe bimetal in the presence of nonionic surfactant. J. Hazard. Mater. **170**, 895–901 (2009)
80. B.W. Zhu, T.T. Lim, J. Feng, Influences of amphiphiles on dechlorination of a trichlorobenzene by nanoscale Pd/Fe: adsorption, reaction kinetics, and interfacial interactions. Environ. Sci. Technol. **42**, 4513–4519 (2008)
81. L. Liu, W. Li, W. Song, M. Guo, Remediation techniques for heavy metal-contaminated soils: principles and applicability. Sci. Total Environ. **633**, 206–219 (2018)
82. M. Erdem, F. Tumen, Chromium removal from aqueous solution by the ferrite process. J. Hazard. Mater. **109**, 71–77 (2004)
83. T. Suzuki, K. Kawai, M. Moribe, M. Niinae, Recovery of Cr as Cr(III) from Cr(VI)-contaminated kaolinite clay by electrokinetics coupled with a permeable reactive barrier. J. Hazard. Mater. **278**, 297–303 (2014)
84. J. Chen, Z. Li, D. Chao, W. Zhang, C. Wang, Synthesis of size-tunable metal nanoparticles based on polyacrylonitrile nanofibers enabled by electrospinning and microwave irradiation. Mater. Lett. **62**, 692–694 (2008)
85. C. Yuan, C.H. Hung, K.C. Chen, Electrokinetic remediation of arsenate spiked soil assisted by CNT-Co barrier-the effect of barrier position and processing fluid. J. Hazard. Mater. **171**, 563–570 (2009)
86. Y. Xu, W. Xia, H. Hou, J. Zhang, G. Qian, Remediation of chromium-contaminated soil by electrokinetics and electrokinetics coupled with CaAl-LDH permeable reaction barrier. Environ. Sci. Pollut. Res. **24**(20), 479–20486 (2017)
87. S. Zhao, L. Fan, M. Zhou, X. Zhu, X. Li, Remediation of copper contaminated kaolin by electrokinetics coupled with permeable reactive barrier. Proc. Environ. Sci. **31**, 274–279 (2016)
88. H.I. Chung, M.H. Lee, A new method for remedial treatment of contaminated clayey soils by electrokinetics coupled with permeable reactive barriers. Electrochim. Acta **52**, 3427–3431 (2007)
89. J.W. Ma, F.Y. Wang, Z.H. Huang, H. Wang, Simultaneous removal of 2,4-dichlorophenol and Cd from soils by electrokinetic remediation combined with activated bamboo charcoal. J. Hazard. Mater. **176**, 715–720 (2010)
90. P.S. Calabrò, N. Moraci, P. Suraci, Estimate of the optimum weight ratio in zero-valent iron/pumice granular mixtures used in permeable reactive barriers for the remediation of nickel contaminated groundwater. J. Hazard. Mater. **207–208**, 111–116 (2012)
91. S.V.R. Thiruvenkatachari, R. Naidu, Permeable reactive barrier for groundwater remediation. J. Ind. Eng. Chem. **14**, 145–156 (2008)

[illegible]

# Electrochemically Assisted Thermal-Based Technologies for Soil Remediation

**Sonia Lanzalaco and Ignasi Sirés**

## 1 Introduction

Hazardous waste discharge into the environment is unceasing in modern societies, which is caused not only by the ever growing waste generation from a myriad of human activities, but it also derives from improper waste disposal, deficient maintenance of infrastructures and accidental leakage and spillage. As a result, all the environmental compartments of biota (air, water, soil, and aquatic sediments) are nowadays seriously jeopardized. Soil contamination is particularly worrisome, not only because it directly affects the viability of plantations and the quality of crops, but also because percolation of chemicals can indirectly poison the aquifers. This is a quite serious problem, since groundwater constitutes the main source of freshwater for human and animal consumption in many countries, being also employed by small producers in the agri-food sector. Based on a recent report entitled 'Progress in the management of contaminated sites in Europe' by the European Soil Data Centre (ESDAC) of the European Environmental Agency (EEA), as of December 2019, there are an estimated 340,000 contaminated land areas in Europe [1]. Heavy metals [2] and organic molecules like pesticides [3] and polycyclic aromatic

S. Lanzalaco (✉)
Departament d'Enginyeria Química, EEBE, Universitat Politècnica de Catalunya, Barcelona, Spain

Barcelona Research Center in Multiscale Science and Engineering, Universitat Politècnica de Catalunya, Barcelona, Spain
e-mail: sonia.lanzalaco@upc.edu

I. Sirés
Laboratori d'Electroquímica dels Materials i del Medi Ambient, Departament de Química Física, Facultat de Química, Universitat de Barcelona, Barcelona, Spain
e-mail: i.sires@ub.edu

 
M. A. Rodrigo, E. V. Dos Santos (eds.), *Electrochemically Assisted Remediation of Contaminated Soils*, Environmental Pollution 30,
https://doi.org/10.1007/978-3-030-68140-1_15

hydrocarbons (PAHs) [4] are the main target pollutants. In Europe, the management of polluted land currently involves a total estimated cost of €6.5 billion per year.

Currently, a cornucopia of technologies is available to cleanup contaminated sites. Remediation technologies include a vast range of operations to reduce the pollution of the site and its associated toxic effects. The primary strategies to remediate soil, sediment or sludge, which can be used in the form of single or combined technologies, can entail one or more of the following effects for the contaminants: (a) immobilization; (b) extraction or separation; and (c) destruction or alteration. Among them, the biological, physical/chemical, and thermal treatments can be performed in situ (i.e., treatment of contamination performed in the original place, without significant alteration of the soil matrix) or ex situ (assuming excavation of affected soils and its treatment on site or off-site; soil flushing/washing techniques can be alternatively suggested) [5]. According to the United States Environmental Protection Agency (USEPA), other methods include containment, retrieval, and off-site disposal.

Whenever possible, in situ technologies are the recommended choice, since they avoid the dangers and costs related to transportation, thus becoming more ecological and cost-efficient than their ex situ counterparts. Among them, in situ thermal methods give rise to the so-called In Situ Thermal Remediation (ISTR) or In Situ Thermal Treatment (ISTT). Several methods have been developed over time, since it does not exist a universally adequate thermal treatment technology that can be considered as optimum against any type of pollutant and soil. The main effect of ISTR is the increase of the subsurface temperature, which may favor different processes experienced by the pollutant and/or the water (vaporization, volatilization, dissolution, desorption, and transport/mobilization), depending on the variation of physical properties like solubility, density, viscosity, interfacial tension, etc. Final liquid pumping is an option, although vapor extraction is more typical. Therefore, these technologies are especially well suited to remove volatile organic carbons (VOCs) and semi-volatile organic carbons (sVOCs) such as petroleum hydrocarbons and halogenated solvents. For example, it has been reported that a suitable ISTR attains over 99% hydrocarbon removal in a relatively quick manner (from hours to months depending on the magnitude of the problem) [6]. Worth mentioning, cleanup of dense non-aqueous phase liquid (DNAPL) source zones (i.e., subsurface portion) at heterogeneous sites via ISTR technologies is gaining importance [7]. DNAPL may consist of chlorinated solvents, creosote, polychlorinated biphenyls (PCBs) and coal tar. It frequently sinks in groundwater, further forming a pool above a low permeability area or becoming trapped as residual. This is a particularly challenging case to manage because these zones can undergo flow bypassing of injected amendments, causing back diffusion from zones with lower permeability [8]. Thermal methods may overcome this limitation thanks to the smaller spatial variations in thermal and electrical properties as compared to 3D changes in aqueous phase (i.e., hydraulic) permeability. As will be presented, non-volatile pollutants can also be treated by these methods in more advanced setups. Potential drawbacks of thermal treatments are the energy consumption and possible damage of soil because minerals and organic matter may decompose, although more research is needed to quantify the impact on soil quality and hence, on agricultural characteristics [6]. As a rule of thumb, the lowest effective temperature should be employed in ISTR to avoid such collateral effects.

From an electrochemistry point of view, ISTR technologies encompass two different categories, namely non-electrochemical and electrochemical thermal methods. Some brief considerations of the former ones are presented in the first subsection below, whereas the main insights types of the latter methods are summarized in the second. Literature sources classify some of these technologies under the global name of dynamic underground stripping (DUS) when their purpose is to remediate soil with organic pollutants.

The readers of this chapter will realize about the range of possibilities that are currently available to induce the removal of contaminants from soil upon temperature changes experienced by the medium, focusing the attention on applications in which an electrical current is applied. This is an updated survey, descriptive rather than critical, to offer a comprehensive view of this niche group of technologies. They have received more attention from industrial partners than from researchers, which is justified by the complexity and high cost of experimental setups needed to emulate real polluted sites. Also worth highlighting, electrochemically-assisted soil remediation gathers a group of techniques for which the extrapolation from laboratory to real scale is less consistent, as will be discussed below. Significant differences have been observed in pollutant removal when comparing small synthetic soil samples, pilot plants, and actual sites. Therefore, sound data and robust models must be directly obtained from field trials.

## *1.1 Non-electrochemical ISTR*

At present, a large series of non-electrochemical ISTR technologies is available for soil remediation. Triplett Kingston et al. identified the geological settings in which they can be applied, and results were presented assuming five idealized geologic scenarios [7]. This subsection summarizes the key details and applications of the following methods: thermal desorption (TD), hot air injection, steam-based heating, smoldering, incineration, and vitrification.

TD consists in heating soils with the intention of volatilizing/desorbing the contaminants, thereby transporting the off-gases by a sweep gas or vacuum, followed by their destruction or adsorption [6]. Although volatilization/desorption are the most frequent effects of soil heating, TD systems often achieve oxidation/incineration and pyrolytic reactions like thermal cracking. The predominance of one or another mechanism depends on the temperature and oxygen distribution. TD applications are frequently carried out at more than 110 °C, being divided into low-temperature thermal desorption (LTTD, at 100–300 °C) and high temperature thermal desorption (HTTD, at 300–550 °C). The removal efficiency of LTTD and HTTD can be greater than 99%, although treatment time varies significantly with configuration and contaminant. Current research involves the optimization through temperature refinement, effective vapor treatment and vapor-scrubbing issues to remove by-products generated [9].

TD can be performed ex situ or in situ (ISTD, also called conductive heating or thermal conduction heating (TCH) [7]). Reported costs per metric ton range from \$46 to \$99, and between \$70 and \$460 (considering 2016 USD), respectively [6]. Due to lengthy heating times, ISTD can take weeks to years, while ex situ TD can be completed within several minutes.

In ex situ TD, soil is excavated, transported, and heated in a TD facility. Alternatively, the excavated soil can be placed into covered piles that are heated via gas/diesel burners or heater rods to desorb contaminants (so-called thermopiles) [9]. A closely related technology is pyrolysis, which is based on the same setup but maintaining an anoxic atmosphere. On the other hand, TD can be operated in situ employing dual heater/vacuum wells to desorb and remove contaminants via vapor extraction (Fig. 1). Although thermal conductivity is not much affected by soil heterogeneity and contaminant dispersal, thermal conduction heaters allow a uniform heating of the whole target area. ISTD is typically performed at low temperature (250–300 °C) [10]. However, to ensure sufficient heating in a quick manner throughout the contaminant zone, soil near the heaters may attain high temperatures (800–900 °C). Initial soil heating to promote desorption in the contaminated zone may be rather slow because of the relatively low soil heat capacity. Radiative heat transfer dominates in the zone close the heaters, whereas thermal conduction (i.e., heat transfer via direct contact of soil particles) prevails far from the heating sources [6].

In practice, heating and removal mechanisms in ISTD vary spatially depending on the distance to the heat/vacuum wells. In general, anoxic conditions are preferred

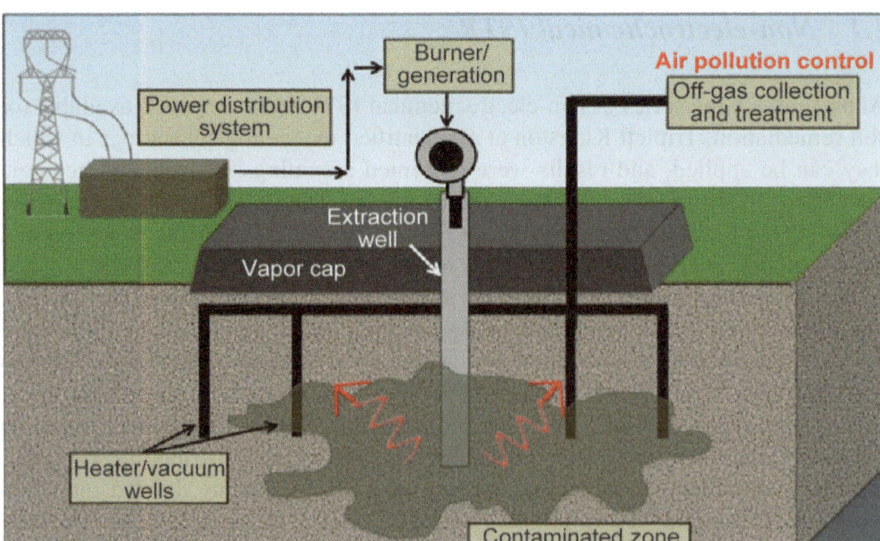


**Fig. 1** In situ TD utilizes dual heater/vacuum wells to heat soils and remove contaminants. Off-gases are collected for reuse or disposal. (Reproduced with permission from [6]. Copyright 2016 Elsevier)

to avoid combustion, but this is not always ensured. Heavy hydrocarbons in overheated soil preferentially undergo thermal cracking (usual at 300–500 °C) over desorption if oxygen content is low, and it is rather incinerated/combusted if oxygen content is high. This is not easy to control because oxygen concentration changes due to gas flow and smoldering.

Hot air injection is an in situ method used to enhance the contaminant mobility and its subsequent extraction. It is an energy-intensive process because, due to the low heat capacity of air, large volumes of air at high temperatures are required to heat soils up to pollutant desorption temperature. Steam is often combined with hot air to increase the desorption efficiency. Hot air injection is typically used with bioremediation or other processes, and its most common application is hydrocarbon removal, from light fuels to crude oils and creosotes. Treatment costs range from $54 to $82 per metric ton [6].

Steam-based heating, so-called steam injection or steam-enhanced extraction (SEE) [7], was first developed by the energy industry for enhanced oil recovery. In general, injected steam causes a decrease in contaminant viscosity alongside soil heating. It is a more efficient alternative than hot air injection because steam has a higher heat capacity. As in TD, steam also causes the desorption and evaporation of volatile hydrocarbons. There exist three primary steam-delivery methods for large-scale application: (a) Steam/vacuum wells, (b) steam injection through drill bits, and (c) steam injected beneath the contaminant zone, which condenses and flows upward as hot water [6]. The setups are analogous to that shown for ISTD in Fig. 1, also requiring gas/vapor post-treatment. SEE is particularly effective against organic contaminants with boiling points under 250 °C, but the efficiencies vary depending on soil type, contaminant polarity and vapor pressure. Treatment costs range from $37 to $380 per metric ton [6].

Note that some author described the combination of in situ soil mixing with steam and hot air injection as a distinguishable technology [7].

Smoldering (or smoldering combustion) is a slow, flameless combustion process that can be self-sustained if fuel and oxygen demand is met. Combustion transforms the contaminants into heat, carbon dioxide and water. Heat is thus transferred through the soil matrix. The typical temperatures in smoldering are considered low, ranging from 600 to 1000 °C, although they are discontinuous in space and time. The ignition of contaminants to initiate smoldering remediation may take several hours. Once started, heat supply can be stopped and the process can be controlled via modulation of the air injection rate [6]. The dominant mechanism for decontamination is the exothermic combustion, which co-exists with desorption and pyrolysis (endothermic). Heater/vacuum wells, as those employed in TD, can be also utilized for smoldering remediation. The application of this technique is limited at field-scale, being the treatment of DNAPL the most addressed case. Costs are estimated to be $260–$330 per metric ton [6].

In-site incineration, which does not require excavation, known as on-land burning or open burning, can be difficult, costly, and unpredictable [11]. Therefore, incineration is preferably applied as an ex situ technology. Finally, in situ vitrification involves the melting (1600–2000 °C) and fusion of contaminants and soil into a glass-like solid, providing heat via molybdenum electrodes. It has been especially tested for treating radioactive waste in addition to petroleum. As mentioned for the majority of non-electrochemical ISTR technologies, volatiles desorb and are treated afterwards.

The three latter technologies summarized can effectively treat nearly all hydrocarbons, but the high temperatures needed can cause extensive soil damage such as decomposition of clays, carbonates and organic content, eventually decreasing the soil fertility [6]. Conversely, the other ISTR methods minimize such drawback.

## 1.2 Electrochemically-Assisted Thermal-Based Technologies

The core of this chapter is specifically devoted to this approach, aiming at describing the existing in situ electrochemical thermal (i.e., electro-thermal) methods. The primary aim of the alternative energy sources described here is that explained for heat in the previous subsection: volatilize and desorb low molecular weight hydrocarbons and decrease viscosity.

Radio-frequency heating (RFH) or radio-wave heating (RWH) is a first technology that can be placed in this subsection, since it is based on the application of electric current to soil through the use of electrodes (which simultaneously act as extraction wells). It is classified as a dielectric heating method, being more efficient than microwave heating. RFH was first developed to enhance oil recovery in shale and tar sands in the 1970s [6]. In soils, radio frequencies are transformed into heat, with an efficiency of more than 90% [12, 13], by acting on electric dipoles (unbalanced charges in soil, contaminants and water). Heat is thus transferred on a molecular level, being very effective in the case of water molecules, which suggests that soil moisture contributes positively to decontamination. The different dielectric properties of soil, water and organic matter have to be considered in the calculations prior to remediation. Amendments such as carbon fiber and nanoparticles have been proven to modify the dielectric heating properties of soil [6]. Since both, dry and wet soils, can be treated by this technique, final soil temperatures can be as high as 400 °C, which adds flexibility to this technology. For the treatment of large contaminated soil areas, radio waves with wavelengths and penetration depths in the meter range and frequencies of 1–50 MHz are used [14]. In situ RFH (ISRFH) is especially well suited for the treatment of highly contaminated source areas, or remediation near or under buildings. The energy is transferred to selected depths in order to remediate the soil in a defined manner.

RFH is often applied in situ (Fig. 2). Heat is supplied by electrodes and antennae powered by a radio-frequency generator. Several days are required to treat a polluted soil, but this varies according to the specifications of each case study. RFH can be applied as a stand-alone technology to remediate low molecular weight hydrocarbons. Cost ranges from $400 to $7500 per metric ton [6]. This cost is similar or higher to that of vitrification and incineration. In situ oxidation is feasible in RFH when catalysts are directly placed within the electrode.

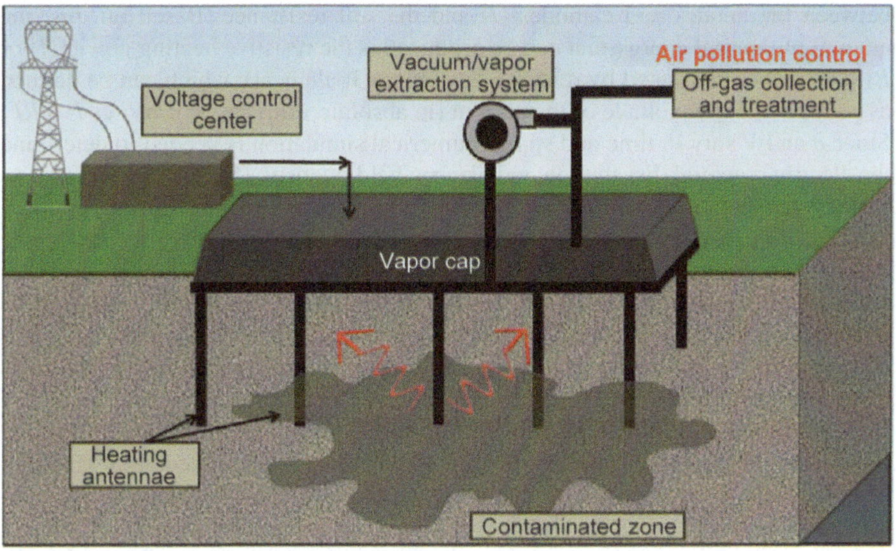


**Fig. 2** In situ RFH uses heating antennae to heat the subsurface. Off-gases are collected for reuse or disposal. (Reproduced with permission from [6]. Copyright 2016 Elsevier)

Mimicking the more common and better established electrochemical water treatment, the main idea when electrochemistry is employed to treat a solid or semisolid is to convert such medium in an electrochemical/electrolytic cell by inserting at least two electrodes (i.e., an anode and a cathode). This is in contrast to RFH, an electrical/thermal rather than electrochemical process because individual electrodes are employed instead of anode-cathode pairs. Two types of pure electrochemically-assisted ISTR technologies can be distinguished: direct current (DC) and alternating current (AC). By comparison with RFH, they are classified as resistive heating methods because they rely on the electrical conductivity of soil. A certain water content in the soil is required to enable heating at sufficiently high voltages. However, this condition limits resistive heating to 100 °C as maximum in practice, unlike dielectric methods.

The fundamental idea of DC and AC methods is that the application of an electric field (in the form of current or voltage) causes an increase of the temperature of subsurface soil and fluids (i.e., Joule effect), especially in the zone near the electrodes [15]. Such heating arises from soil resistance to current flow anode to cathode (thus so-called ohmic heating). In other words, electric charge in soil, in the form of hydrated positive and negative ions (cations and anions, respectively), cannot be transported as easily as through water as occurs in conventional electrochemistry. This difficulty is much higher as the soil to be treated is or becomes drier. Therefore, it is evident that electrical conductivity ($\sigma$, also known as subsurface electrical conductivity) or its inverse, the electrical resistance ($R$) of subsurface soil is a key parameter that determines the performance of these methods, since the increase in temperature is proportional to the intensity of the current that flows externally between the anodes and cathodes ($I$) and the soil resistance. Based on this, the amount of thermal energy that can be produced is the resistive heating power ($U$ or $P$) (i.e., power dissipated by resistive heating, or Joule heat), which can be defined as a function of the voltage or the current (in absolute value) as: $P = \sigma V^2$ or $P = RI^2$. Since $\sigma$ and $V$ vary in time and space, numerical simulation is needed to determine the electric potential distribution and electric field intensity (Sect. 2).

Since temperature is the critical parameter that defines the performance in these treatments, it must be accurately monitored, which is easy through thermocouple bundles. The temperature rise affects the desorption and transport properties of the volatile compounds [3]. As mentioned above, the maximum heat-up that can be attained is limited by the boiling point of water. If water boils off, liquid water is converted into steam, $\sigma$ tends to zero and heating ceases, which means that it is important sometimes to add water at each electrode well to prevent total drying [16].

Soil remediation employing mild DC power is known as electrokinetic remediation (EKR) or treatment (EKT), or even electrokinetic soil remediation (EKSR). Usually, they cause a smaller temperature variation than AC techniques. In EKR, current or voltage is thus applied between at least one pair of electrodes, namely cathode and anode, which are in the subsurface or partially inserted into the soil [17]. Several charge transport (i.e., electrokinetic) phenomena are stimulated, acting the soil humidity as hydraulic and ion conductor. In addition, chemical (precipitation, dissolution, etc.), electrochemical (electrolytic, i.e., redox processes of water or chemicals at the electrode surface) and physical (viscosity change, temperature rise, etc.) phenomena are concomitantly induced. EKR is convenient to treat heterogeneous soils with low permeability [4], such as clayey soil, for which conventional techniques are not effective to move the underground water retained [18].

EKR allows the removal of heavy metals and organic pollutants like hydrocarbons, as demonstrated in studies at laboratory, pilot and field scale described in other chapters of this book. The former ones are mobilized throughout the soil thanks to the generation of an acidic front appearing in the anodes that dissolves and transports the metal ions toward the cathode wells under the effect of the applied electric field, further being extracted. The organic pollutants are mobilized via electroosmosis, electromigration and electrophoresis [2], and their removal percentages (from 20% to 90%) are strongly dependent on several factors, such as the type of electrodes and their configuration, the inter-electrode distance and the magnitude of the applied electric field. The major advantages of EKR are [19]: (a) unique applicability to low permeability soils, which tend to adsorb pollutants, thus becoming resistant to standard non-electrochemical ISTR technologies; (b) high degree of control and direction; (c) wide range of pollutants that can be treated; and (d) relatively low electric power consumption.

An interesting alternative to well-known EKR, aimed at enhancing the pollutant removal degree and even at shortening the treatment, consists in its integration with chemical oxidation. The implementation of chemical oxidation gives rise to an in situ chemical oxidation (ISCO) process [20]. These are beneficial for in situ soil remediation because permit that a certain percentage of transformation of the targeted pollutants is reached, avoiding their simple transfer from subsurface to surface. Among the most powerful oxidants known (hydrogen peroxide, Fenton's reagent, permanganate, ozone, etc.), the persulfate anion ($S_2O_8^{2-}$) is the most widely employed in EKR due to its chemical stability and fast reaction, high aqueous solubility, relatively low cost and eco-friendliness because innocuous products are released. When activated, $S_2O_8^{2-}$ acts as a much stronger oxidant due to its decomposition into sulfate radicals ($SO_4^{\bullet}$), which have a high redox potential and show high kinetic rate constants that confer them the possibility to break-up the organic contaminants into non-toxic aliphatic organic fragments [21]. The activation of $S_2O_8^{2-}$ can be made via different routes, being thermal heating one of them, thereby originating the so-called electrokinetic thermal activated persulfate (EK-TAP) process [22]. From this, it is evident that among the several factors that have to be considered to ensure the large effectiveness of EKR/$S_2O_8^{2-}$, the temperature value attained upon application of DC power is the most relevant one (~40 °C is considered a high temperature) [23]. Worth highlighting, the electric field has an additional role as transporter of $S_2O_8^{2-}$ into the contaminated region. EK-TAP is currently undergoing field testing at a number of locations in Europe and the USA.

Amongst the AC techniques, electrical resistance heating (ERH) is the most widely applied [24]. ERH was developed by Battelle Northwest Laboratories in the early 1990s for the Office of Science and Technology of the US Department of Energy (DOE). It became commercially available for use as stand-alone technology in 1997. The technology was first field-tested at the DOE Handford in Washington. Initially, ERH was devised by the petroleum industry to improve subsurface oil recovery and, nowadays, it is recognized as a reliable and cost-effective remediation tool.

As in the case of EKR, the electrodes are installed in recovery wells (to collect steam and contaminant vapors that are further treated in specialized units) throughout the contaminated soil and groundwater volume. The electrode array, typically in a triangular or hexagonal pattern, is connected to a power supply (i.e., power control unit) that uses (unlike EKR) standard three-phase power from the grid [25] or a six-phase AC voltage source [24]. Three-phase heating is generally more applicable for full-scale treatment and six-phase heating is generally more applicable to the pilot scale (note that six-phase heating was a more widespread term than ERH in the beginning, before field tests became popular). In both cases, electromagnetic effects can be neglected because ERH typically operates at low frequency (60 Hz), yielding quasi-static electric fields. Current then flows between the electrodes through moisture, causing the increase in soil temperature (i.e., subsurface heating). This mainly favors the generation of a vapor phase containing volatile organic compounds (other effects of heating are discussed in Sect. 2) and, in some cases, degradation products. This vapor can be extracted through the electrode wells and through soil (with produced steam acting as carrier gas), cooled down and treated. In some setups, ERH is combined with air stripping to enhance the recovery of the volatile contaminants; for example, Buettner and Daily (1995) presented the first engineering-scale (89 $m^3$) system to decrease the concentration of trichloroethylene in soil from 140 to 5 ppm using 6 electrodes for 25 days, thanks to a temperature rise from 16 to 38 °C [26].

Regarding the operation temperatures, ERH can be carried out either at boiling or sub-boiling temperature, usually between 80 and 110 °C [7].

Since the ERH heat source is stationary, boiling temperatures may be reached near the electrodes, although significant infrastructure is needed. If current flow through soil warms the soil moisture to the maximum, steam is produced. This in situ steam generation occurs in all soil types as well as in fractured or porous rock. The ability of ERH to produce steam in situ represents its most significant advantage over other subsurface heating techniques [25]. Also typical ERH for DNAPL remediation raises the subsurface temperature to the boiling point of water. In this case, the primary removal mechanisms involve volatilization and/or steam stripping, being the contaminant vapors and fluids collected and treated. As subsurface temperatures begin to rise, contaminant vapor pressure and the corresponding rate of contaminant extraction, typically increases by a factor of about 30 [7].

Nonetheless, the high temperatures can result in unfavorable conditions for other important remediation processes, such as bioremediation. Operating ERH at sub-boiling temperatures may address these issues, since lower temperatures require less power (i.e., lower cost) and infrastructure and may enhance microbial and/or abiotic degradation.

Worth highlighting, even subsurface heating is not easy to reach in ERH, regardless of the operation temperature. Soils around the electrodes receive more heat, because the spatial electric power density decreases from the electrode as the inverse of the distance raised to the second power.

So far, ERH has successfully permitted the cleanup of highly contaminated sites that were impossible to treat adequately by using other technologies like ISTR dependent on advective flow. Examples include DNAPL sources, soils with heterogeneous lithologies, and low permeability soil (silts, clay). In fact, vendors claim that ERH is particularly well suited for low permeability soils, as electricity is conducted primarily through water in porous clays. However, slower heating and additional design and operational changes needed in those sites result in higher time, energy and cost. Luckily, ERH operates indistinctly regardless of the soil permeability and heterogeneity. Few remediation technologies can offer equal levels of DNAPL cleanup in the same timespan at an equivalent price as that of ERH.

Some advantages of ERH are [16]:

(a) Excellent heat transfer when purpose-made electrodes are employed, since their output is higher and the resulting heating is easier and more effective via real-time computerized power control;
(b) Convective energy delivery is feasible in permeable soils, being possible to inject hot air or steam (see Sect. 1) for additional heating;
(c) The technology can be employed around and under operating buildings, roads, and public premises with limited impact on daily activities and usage.

As in the case of EKR, ISCO is also feasible in ERH. Again, the most suitable oxidant to be thermally activated is $S_2O_8^{2-}$, owing to its high stability in subsurface and ability to produce sulfate radicals [27]. ERH/$S_2O_8^{2-}$ is effective for the treatment of soil-sorbed hydrophobic organic contaminants and DNAPL. However, the treatment performance is more limited than equivalent treatments in water matrices because of the slow transport of the organic contaminants to the aqueous phase, where the oxidation occurs. The high temperatures that can be reached in ERH are favorable, as compared to EKR/$S_2O_8^{2-}$, because they facilitate the desorption/dissolution of organic compounds and activation of persulfate. However, excessive heating should be avoided to minimize the exponential decomposition of persulfate at high temperature. Based on this, low temperature (lower than typical sub-boiling) ERH is preferred in this version of the electrothermal process.

Li et al. (2020) developed a combo ISCO system capable of $S_2O_8^{2-}$ recirculation as the ERH was applied, aiming to promote even volumetric heating [27]. The injection of chemicals with a recirculation system is typical in ISCO to ensure their homogeneous delivery and dispersion through soil. Moreover, recirculation of such synthetic solutions promotes forced convective heat transfer, enhancing the heat distribution. An additional benefit comes from the rise in conductivity as persulfate salts enter into the soil. The system consisted of three injection/recirculation-electrode wells and one extraction-electrode well, and three-phase electric power was supplied. The electrodes were arranged in a star (Y) or the delta (Δ) configuration. In the former (Fig. 3), each injection/recirculation-electrode wells was connected to a hot wire, whereas the extraction-electrode well was connected to the neutral wire. In the delta configuration, the three injection/recirculation-electrode wells were connected as before, but the extraction-electrode only worked as an extraction well. The authors demonstrated that simultaneous $S_2O_8^{2-}$ recirculation and ERH can achieve an even volumetric heating, eventually improving the degradation of contaminants.

Another important AC technique with wide implantation at field scale, very closely related to ERH, is the electro-thermal dynamic stripping process (ET-DSP), developed in the mid-nineties [28]. Similarly to technologies previously discussed, ET-DSP has been primarily employed to remove VOCs, as well as non-volatile organic matter, from contaminated soils. Three-phase line power supply and water injection at electrode wells placed following a pattern are required for in situ soil heating. As mentioned in ERH, the heating tends to dry the soil, thus producing steam, which results in an increase in the permeability and dynamic stripping of the soil [28, 29]. Numerical results of the temperature distribution during extraction and electrical heating operations were shown in that study, but additional information on this will be given in Sect. 2.

ET-DSP combines the features of conventional ERH (i.e., ERH without water injection) with heat transfer by convection upon water addition into the ends of the electrode where the power density is most intense. The injected water carries heat away from the electrode into the soil. Convective heat transfer favors a rapid and uniform heating. A limiting feature of ET-DSP is the large number of electrode and extraction wells needed (spaced approximately 16 m apart, which results in 18 electrodes and 8 extraction wells per ~4000 $m^2$) [30]. On the other hand, the principal benefit is its high recovery factor within a short period of time.

In 2004, E-T Energy was created to develop and implement the ET-DSP process to extract bitumen. The company led a proof-of-concept pilot-scale project in 2006 and 2007. In the case of oil sands, the conducting path that the electrical current follows is through water that envelops the sand particles. Electrical energy is thus

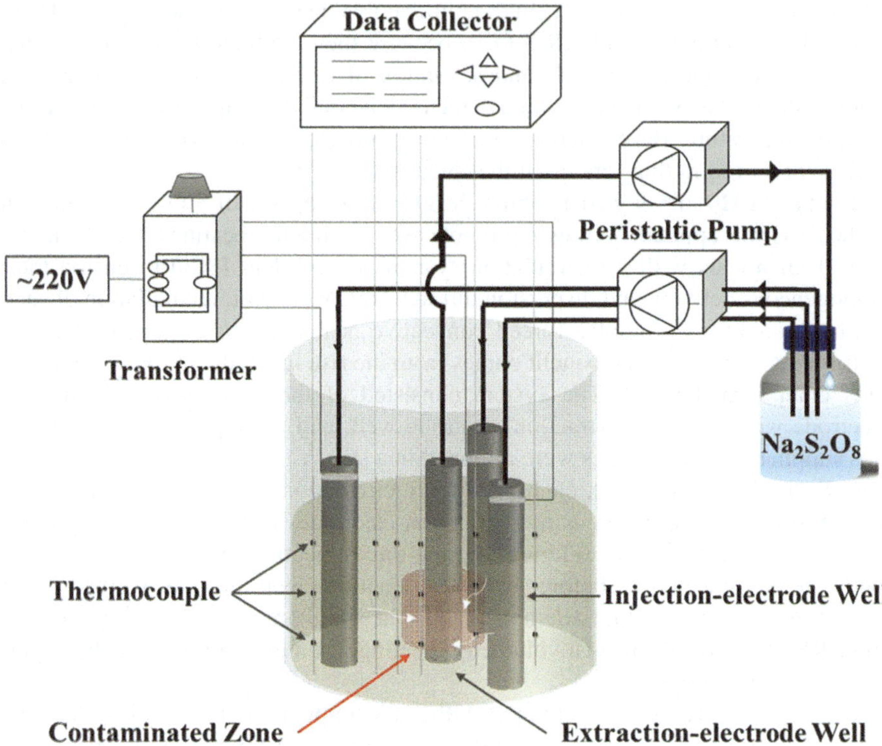


**Fig. 3** Conceptual operation scheme of the combo system. (Reproduced with permission from [27]. Copyright 2020 Elsevier)

converted to heat, which is transferred to the oil and sand particles by conduction. Due to the large contact area between the water film, the oil and the sand particles, heat transfers quickly.

A last comment to end this section refers to the potential use of electro-thermal ISTR technologies for biosolids dewatering (i.e., electro-dewatering, ED) [31]. The application of an electric field to mechanically dewatered biosolids removes 35 wt% water content, using less than 25% of the energy required for thermal drying. During ED, two main phenomena contribute to the drying process: electroosmosis and Joule heating, similarly to EKR. High temperatures may facilitate dewatering by reducing the water viscosity and enhancing evaporation.

## 2 Additional Fundamentals of Electrothermal Techniques Reviewed and Mathematical Considerations

As stated in Sect. 1, the background of the electro-thermal technologies under consideration is the promotion of Joule heating effect, reason why these methods are also classified within the so-called Joule heating soil remediation.

The resistivity and heat transfer characteristics of different types of soil have been assessed by many researchers [32]. The relationship between soil texture and its thermal characteristics has also been considered. Within this context, it has been reported that the soil resistivity decreases when temperature increases above 0 °C, but the tendency is the opposite above 50 °C. The enhanced water evaporation gradually dries the soil, having a negative impact on ion conductivity. As a result, an exponential increase in soil resistivity is observed. Soil composition plays an important role, since it is accepted that soil volume heat capacity is determined by soil bulk density and moisture content. Moreover, the soil thermal conductivity is mainly determined by volumetric water content and clay content.

On the other hand, water properties are also seriously affected by temperature changes. The behavior of fluid density ($\rho$) and viscosity ($\eta$) is well understood, and equations describing those effects are available. In general, a temperature rise causes a decrease in both, fluid $\rho$ and $\eta$, leading to faster buoyant groundwater flow. Note that the buoyancy phenomenon, so-called free convection, is promoted by $\rho$ changes linked to temperature variation. Convective water flow has been detected in regions of high permeability, as a result of $\rho$ variation caused by geothermal heating [33]. Consequently, this has impact on mass transport of contaminants. However, when these nonlinear effects (i.e., fluid flow and mass transport) are coupled during flow through porous media like soil, the implications for flow and transport are not so evident [25]. Temperature-dependent $\rho$ and $\eta$ should be included in computational models to accurately model heat transport, particularly when temperature differences are above 15 °C [34].

Some authors agree to criticize that there is a gap between field applications of electrochemical ISTR and the set of supporting equations, system design and process optimization. In response, several scholars have attempted the mathematical modelling of in situ electro-thermal technologies, with predominance of EKR and ERH, to simulate and predict the impact of Joule heating on both, temperature and the hydrodynamic velocity profiles of the system. This is certainly required to unveil a rigorous up-scaling and technology design. Unfortunately, few studies have simulated the hydraulics associated with heated soil and groundwater.

Mathematical studies available in the literature for DC techniques refer to some EKR applications. Laboratory-scale experiments and models developed for EKR generally assumed isothermal conditions at room temperature, since the applied current density ($j$) tends to be lower than 5 mA/cm$^2$ and the trials are short enough to neglect thermal effects. However, even small temperature changes can affect conductivities, ion mobilities (i.e., electromigration), water transport (i.e., electroosmosis), and sorption processes, eventually accelerating or delaying the removal of pollutants.

Baraud et al. [19] investigated the effect of temperature on the transport of a model cation and anion present in the pore solution when an electric field was applied through a kaolinite soil (a low permeability soil). The experiments were run at 20 and 40 °C. Ion diffusion, which results when concentration gradients exist or appear, was neglected, as well as hydraulic transport (so called advective flow). Both mechanisms were considered much less relevant than electromigration, resulting from the electrical potential gradient, and electroosmosis. The corresponding migration velocity of a species in the pore solution is known as electrokinetic velocity, which depends on the mobility and the electric field. The electroosmotic velocity was also established and hence, the total velocity of a given ion was the sum of both terms. It was observed that a rise in temperature mainly caused an increase of the electrokinetic velocities, for the two ions. The primary effect of temperature is through the change of the fluid viscosity, which can be approximated as 2% per degree centigrade. Since the ion mobility is inversely proportional to viscosity according to Eq. (1), where $z_i$ is the charge of the species, $e_0$ is the elementary charge and $r_i$ is the Stokes radius, then an increase of 2% per degree centigrade is observed in electromigration velocity.

$$u_i = \frac{z_i e_0}{6\pi\eta r_i} \tag{1}$$

Regarding the electroosmotic flow (and velocity), the influence of temperature is not so well known, since many parameters appear according to the Helmholtz-Smoluchowski theory. Some authors report that this flow increases with temperature, but actually the effect cannot be precisely predicted.

In conclusion, this simplified theoretical analysis predicted the simultaneous increase of electromigration and electroosmosis with temperature rising. For the cation, these two mechanisms are additive, meaning that the removal of cationic species should be enhanced upon temperature rise. For the anion, the temperature effect would depend on the relative contribution of each mechanisms, as the transport of anionic species is delayed due to the opposite electroosmotic flow.

Oyanader and Arce [35] also dedicated some efforts to Joule heating modelling for EKR. Temperature changes and Joule effect were studied in the zone near a cylindrical electrode (i.e., inside the boundary layer), aiming to provide a numerical solution. The temperature difference between the wall surface of the electrode and the fluid outside the boundary layer region yielded a $\rho$ gradient. This caused buoyancy-driven flows to occur as a thin boundary layer moving tangential to the vertical cylinder. A key idea in this and other works is that the Joule heating can modify the hydrodynamics of the system. In this particular example, hydrodynamics varied inside the boundary layer, near the cylindrical electrode wall.

Energy (heat) transfer and mass transport (addressed on the basis of hydrodynamic models) equations were coupled. Regarding the former, the heat transfer model was considered, by including a heat generation term associated to the Joule heating effect ($Q$) in the energy conservation equation, as follows:

$$\rho C_p \frac{DT}{Dt} = \nabla\left(k\nabla T\right) + Q \tag{2}$$

where $D$ accounts for the convective derivative (so-called advective derivative, taken with respect to a moving coordinate system), $C_p$ is the specific heat (or heat capacity of the soil), and $k$ is the bulk thermal conductivity of the fluid. Subsequently, the expression was converted to dimensionless variables, reaching a non-dimensional differential equation for the energy balance.

In conclusion, temperature profiles were modelled and the Joule heating was responsible for an increase in axial velocity.

In a more recent work, Torres et al. simulated the effect of Joule heating on heat transfer and hydrodynamics (i.e., on temperature and hydrodynamic profiles, respectively) [36]. The model, applied to a rectangular capillary (soils can be considered porous media and can be modelled as a set of capillaries), assumes the presence of different Nusselt numbers along the walls of the capillary in order to investigate the role of the non-uniform heat transfer properties. This allowed determining the temperature profiles upon variation of the heat transfer parameters, as suggested by different Nusselt numbers. In the end, a range of heat transfer "regimes" were identified, which had influence on the flow regimes appearing within the capillary (each with a characteristic velocity profile).

Some important considerations must also be made on the AC techniques. The overall heating pattern in ERH is said to be remarkably even throughout the treated volume [24]. However, the electricity takes preferential pathways (of lower resistance) when moving between electrodes, and heating is slightly faster in such pathways. Cases with low resistance pathways in the subsurface include silt or clay lenses and areas of higher free ion content. A simple way to ensure more uniform heating consists in the injection of water in the electrode wells to promote convection.

A remarkable work was made by Krol et al. to model the effect of temperature on flow and contaminant transport in ERH, with the intention to unravel the dominant mechanisms affecting subsurface flow and transport [25]. The effects of sub-boiling heating (temperatures reaching 50 °C) were assessed in a series of 2D tank experiments. The great interest of this investigation is that the authors developed a fully coupled 2D finite difference electro-thermal model, which means that it included the temperature-dependent fluid flow and mass transport arising from current supply. Temperature-dependent equations for $\rho$, $\eta$, diffusion coefficient ($D$) and electrical conductivity of the medium were employed in order to characterize the nonisothermal processes that prevail in the subsurface. In addition, the model was validated with laboratory-scale experiments, which is an uncommon but very important feature. Images of the tracer (Rhodamine WT) were captured using a Nikon D90 camera interfaced to a computer for time-lapse image capture.

As a reminder, the Joule effect is based on the accomplishment of Ohm's law, which relates $j$ to the electric voltage ($V$) and $\sigma$, as:

$$j = -\sigma V \tag{3}$$

On the one hand, note that $\sigma$ is an important parameter in ERH. It depends on the fluid electrical conductivity, indirectly depending on mineral composition, temperature and dissolved ions. It can vary in time and space. The effective conductivity in nonconductive media can be approximated by:

$$\sigma = \sigma_{\mathrm{w}} \frac{\varphi^m}{a} S_{\mathrm{w}}^n f\left(T, T_0\right) \tag{4}$$

where $\sigma_w$ is the electrical conductivity of the pore water, $S_w$ is the water saturation, $m$ is the cementation exponent, $a$ is the tortuosity factor, $n$ is the saturation exponent and $f(T,T_0)$ is a function for temperature ($T$) increase from initial temperature ($T_0$). This equation describes the temperature dependence of $\sigma$.

Regarding the electric voltage, it is time-dependent because alternating current is employed in ERH, giving rise to a phase-shifted potential distribution:

$$V = V_0 \cos\left(\omega t + \varphi\right) \tag{5}$$

where $V_0$ is the voltage amplitude, $\omega$ is the angular frequency, and $\phi$ is the phase angle.

The energy (heat) transfer equation yields the subsurface temperature distribution:

$$\frac{\partial}{\partial t}\left[\rho_{\mathrm{w}} \varphi c_p T + (1-\varphi)\rho_{\mathrm{b}} c T\right] + c_p \rho_{\mathrm{w}} \nabla \bullet \left[\vec{q} T\right] - k \nabla^2 T - U = 0 \tag{6}$$

which is analogous to Eq. (2) used for heat transfer in EKR, but more detailed. Here, $\rho_w$ is the water density, $\varphi$ is soil porosity, $\rho_b$ is the bulk density of the soil, $c$ is the heat capacity of the soil, $\vec{q}$ is the Darcy velocity vector and $U$ accounts for the Joule heating. The Darcy's law was further modified taking into account the varying density.

The decrease of liquid viscosity with increased temperature was considered from the Andrade equation, whereas the concomitant decrease of liquid density was approximated by polynomial interpolation of a linear relationship. Heat and flow equations were coupled through the temperature dependence of water properties.

Figure 4a, b show the experimental and modelled temperature distribution, respectively, for the small tank 1.5 h after injecting the tracer (this corresponds to 3.5 h of heating). The simulated temperatures in the heated zone show good correspondence with the experimental results. The largest disagreement between measured and simulated temperature values was seen in the upper part of the tank, where the model underestimated the temperatures by approximately 9 °C.

Two additional main conclusions were drawn from this accurate study: (a) Temperature-induced buoyant flow and contaminant transport in the subsurface can be certainly significant in ERH, even at 50 °C (water $\rho$ changes by 1.3%); and (b) the dependence of $\sigma$ on temperature change has a direct impact on ERH power consumption.

Another remarkable consideration that has been exposed by some authors during ERH trials refers to gas production and transport [37]. Sometimes, as in the case of VOCs, gas release is the predominant mechanism of removal and hence, successful capture of gas produced (i.e., vaporized molecules) is mandatory. In such cases of volatile pollutants, inefficient gas production or removal from subsurface may end

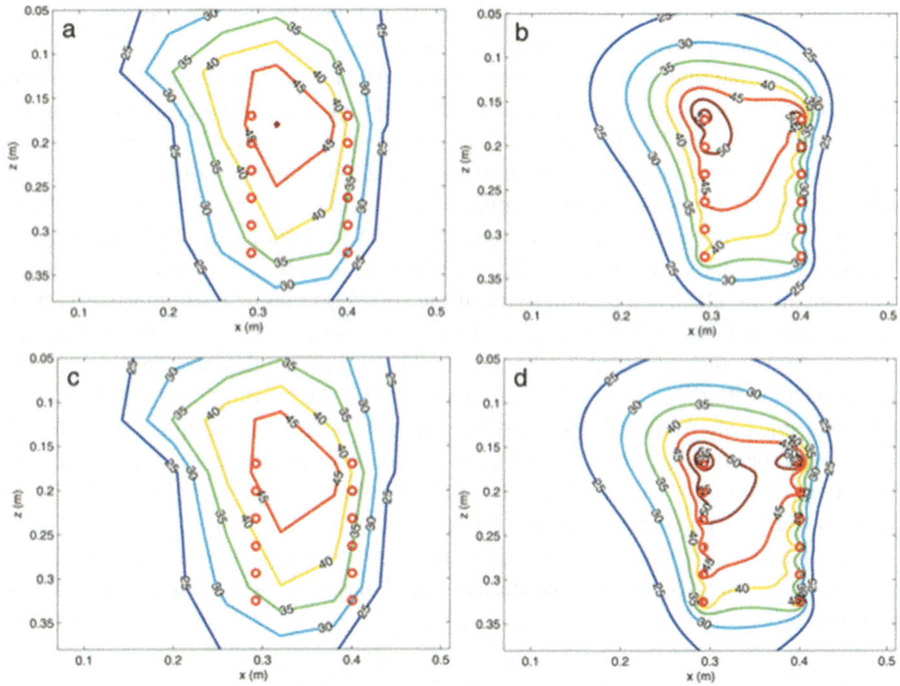

**Fig. 4** Comparison of experimental and predicted temperature distributions: (**a**) experiment in the small ERH tank, after 1.5 h; (**b**) model, 1.5 h. Red circles indicate electrode positions. (Reproduced with permission from [25]. Copyright 2011 Wiley)

in poor soil remediation performance. A typical failure occurs when lateral gas migration is feasible due to the presence of low permeability soil layers and, as a result, contaminant vapor is transported and condensed to form a DNAPL zone outside from the treatment area [38]. Nevertheless, problems may also arise in permeable soils. The quick groundwater flux may cause convection heat loss out of the treatment volume and, as a result, it becomes complicated to attain water boiling temperature [39]. This is especially easy in areas far from the electrodes, where gas condensation and further pollutant redissolution can occur.

In the event of gas production, gas bubbles nucleate and grow and, ideally, they give rise to a connected gas phase that can be extracted to be treated ex situ [15]. Numerical models have shown that the highest subsurface resistive heating (i.e., Joule effect) resulting in gas production occurs in the vicinity of the electrodes, which must be monitored to avoid soil dry-out and reduction in $\sigma$. Loss of electrical continuity between electrodes long before sufficient heating of the far-electrode soil region has been achieved is a serious consequence. The electrode geometry, dimensions and arrangement and the magnitude of the applied voltage have a major impact on the excessive heating in specific zones. It has been reported that conductive and advective (convective) heat transfer, as well as the superposition of electric fields from multiple electrodes, mitigates this effect and ensures a more uniform subsurface temperature distribution in space and time.

As mentioned above in this section, water can be injected in the wells to avoid dry-out, thus enhancing convection. However, the typically low permeability of the Joule heating target layer, such as clay, makes it very difficult to mitigate such effects by simply pumping water into the wells. Within this context, a different alternative to solve non-uniform heating of soil layers near the electrodes has been proposed by some authors [40]. It consists in the incorporation of electroosmotic infusion of groundwater (or an ad hoc electrolyte) to enhance, maintain, or restore the value of the electrical conductivity of the soil. This can be particularly effective for low permeability soil. In brief, electroosmotic infusion is requires the incorporation of one or more electrodes adjacent the heating electrodes and applying a distinctive DC voltage between two or more electrodes. Depending on the polarities of the electrodes, the induced flow will be directed toward the heating electrodes or away from them. Additionally, these extra electrodes may be located throughout the target area to modify the conductivity of the whole area. Periodic polarity reversal prevents large pH changes at the electrodes.

Finally, specific information on heat (and mass) transfer found in ES-TDP applications was evaluated through numerical simulations, showing temperature distribution maps [30].

## 3 Applications and Setups

Based on examples given in the literature, ERH is the process in which the influence of thermal heating is more evident among pure in situ electro-thermal technologies. The first subsection below is thus dedicated to the description of some selected experimental trials, with emphasis on the quantification of removal performance.

Important considerations on the scale-up of electrothermal technologies are addressed in the second subsection. In particular, it will be shown that, at large scale, the predominant factor that determines the mechanisms for pollutant removal is the temperature change (i.e., heating). This means that, in electrochemical thermal-based soil remediation technology, size (of the experimental setup) really matters, as has been especially evidenced in EKR.

### 3.1 *Selected Applications Showing the Impact of Electro-Thermal Effect*

Indeed, ERH stands out as one of the most prominent methods amongst those addressed in this chapter, as reflected by the fact that it accounts for over half of the ISTR applications reviewed by Tripplett Kingston to remediate DNAPL source zones [7].

For example, Hegele and Mumford investigated the gas production and transport during ERH treatment of a mixture of water and trichloroethene (TCE) [37]. When a polluted sited contains volatile and semi-volatile compounds, temperature must be high enough to vaporize them and allow gas transport toward extraction wells. In such situations, subsurface temperature must be increased to the boiling point of groundwater, resulting in the appearance of steam conduits. In that work, the authors employed a bench-scale system in which water was boiled alone or co-boiled with pooled DNAPL TCE in silica sand. Note that azeotropic mixtures are typically formed when such mixtures occur, allowing boiling at temperatures below the boiling points of the single components. The boiling points of most common VOCs in air are either below or just slightly above the boiling point of water (100 °C). However, when a VOC is immersed or dissolved in water, its boiling point is depressed, as described by Dalton's law of partial pressures. For instance, perchloroethylene has a boiling point in air of 121 °C, which decreases to 88 °C in contact with water or moist soil. In the TCE/water trials, co-boiling gave rise to growth of gas phase and its discontinuous transport above the DNAPL pool. It was then concluded that the subsurface should be heated to water boiling temperatures to facilitate gas transport following continuous pathways from DNAPL location to extraction points.

The performance of ERH to treat DNAPL sources areas at different scales has been carefully elsewhere, emphasizing the decontamination degree as compared to that reached via other ISTR technologies [7]. In a successive work, the same Mumford's group studied the gas production during DNAPL remediation, using TCE and chloroform DNAPL pools in silica sand. The spatial and temporal temperature distribution was determined, also employing image capture to evaluate the gas production [38]. In earlier sections of our chapter, the benefits derived from water addition to the electrode wells have been explained on the basis of conductivity increase and convection promotion. Conversely, in this specific study it was shown that high natural groundwater flow rates limit the subsurface heating rate. This can be accounted for by the transport of warm water away from the target heated zone, in concomitance with cool water arrival. This suggests that accurate control of hydraulics and electrical power supply is mandatory [8].

The progression of the experiment with the TCE/water mixture is shown in Fig. 5, with particular attention to DNAPL removal (Fig. 5a–c), gas accumulation in the coarse sand layer (Fig. 5d–f), and temperature increase (Fig. 5g–i).

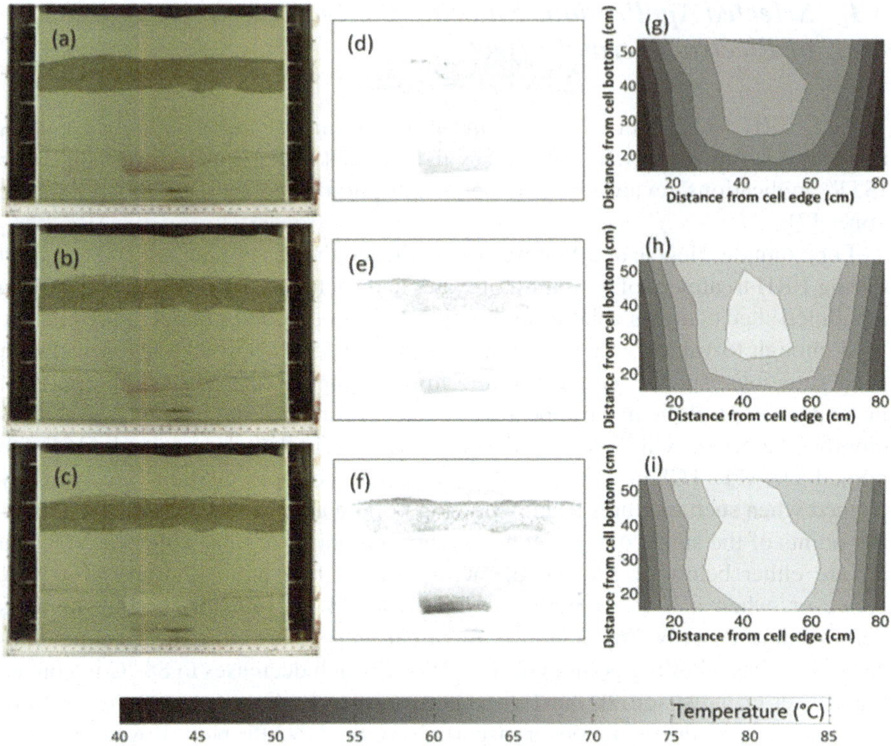


**Fig. 5** Results obtained upon ERH treatment of a TCE DNAPL pool for (**a**, **d**, **g**) 6.33 h, (**b**, **e**, **h**) 11.40 h and (**c**, **f**, **i**) 14.78 h after starting the heating. (**a–c**) Digital images of the flow cell's front face; (**d–f**) processed images showing higher gas saturations or decreased DNAPL saturations as darker shaded pixels; and (**g–i**) temperature distributions based on measurements at each of the 31 thermocouple locations. (Reproduced with permission from [38]. Copyright 2016 Elsevier)

Later, aiming at completing this series of focused studies, the same Mumford's group addressed the gas production upon clay heating during ERH. They studied the threshold pressure above which a connected gas pathway was formed [15].

Another interesting example concerns the application of ERH to 1,4-dioxane removal [41]. This chemical is highly soluble in water and has a low Henry's law constant, which complicates its removal from soils. Worth noting, 1,4-dioxane concentration reductions as high as 99.8% (from average initial content of 25,000 μg/L) have been observed in field tests after 186 days of operation, thanks to effective steam stripping. The wells also contained TCE and 1,1,1-trichloroethane. The authors estimated that the treatment cost would be between $150 and $300 per cubic yard, depending on the size and geometry of the site and the degree of cleanup required. For comparison, the cost to treat a common site by ERH typically ranges from $150 to $250 per cubic yard.

The ability of ERH to remove 16 common PAHs from soil at laboratory scale has also been investigated, further comparing the results with those from contact heating (CH) tests in which a quartz tube furnace was employed [42]. As expected, moisture and salinity were key factors, since they affect the electrical conductivity. The PAHs removal efficiency achieved in ERH was higher than that observed in CH, owing to the higher water evaporation and co-boiling of the pollutants with water (proven to be the main mechanism). In ERH, the removal was greater than 40% at 90 min.

On the other hand, the performance of the more advanced ERH/$S_2O_8^{2-}$ technology was assessed by treating phenanthrene (PHE) in soil [27]. The recirculation of $S_2O_8^{2-}$ was very favorable, as demonstrated by the exponential volumetric heating rate. This was mainly due to the inherently higher conductivity resulting from the presence of the added anions. Thus, the temperature difference between two selected points was 7.2 °C in the absence of recirculation after 2 h, but it was reduced to only 0.4 °C at a recirculation rate of 30 mL/min. In terms of decontamination, in the absence of recirculation, the residual PHE percentage at the sampling points at a 3.5 cm layer was 23.7–35.1% after 7 days, which increased to 49.1–60.9% at the 8.5 cm layer. The implementation of recirculation yielded residual contents of 19.3–24.2% and 33.2–38.1%, respectively, which confirms the enhancement. Worth highlighting, the application of the more efficient ERH/$S_2O_8^{2-}$ process allows diminishing the temperatures below the boiling point of water, which eventually reduces the complexity of the setup and the treatment cost.

As explained in Sect. 1, $S_2O_8^{2-}$ also has a positive contribution in EKR [21, 23]. This method was first employed to treat 50 mg/kg tetrachlorobiphenyl in kaolin and glacial till soil. The authors saw that the presence of $S_2O_8^{2-}$ was beneficial to remediate contaminated kaolin, but the effect in the glacial till was insignificant. These results were justified by the pH evolution (final pH of 1.0 and 7.0 in kaolin and glacial till, respectively, due to the high buffering capacity of the latter). Therefore, in kaolin, a double activation route was available for $S_2O_8^{2-}$: high temperature and low pH. As a result, the pollutant removal was 77.9% at 7th day, whereas it only reached 14.4% in glacial till. On the other hand, low permeability clayey soils were also investigated, finding that the optimal $Na_2S_2O_8$ dosage is around 30%.

Finally, ERH/$S_2O_8^{2-}$ and EKR/$S_2O_8^{2-}$ methods have been combined in a single lab-scale unit by some authors to remove chlorinated solvents from low permeability soil [43]. The idea is to use EKR to deliver $S_2O_8^{2}$, followed by ERH for heat activation, employing the same pair of electrodes. The time course of persulfate concentration and temperature, as well as the contents of sulfate and the chlorinated pollutant during the 91 days is shown in Fig. 6.

ERH could effectively sustained the target temperature to activate the persulfate As a result, the pollutant concentrations clearly decreased to below detection limit within some few weeks (Fig. 6d). Moreover, it was found that the activation at ~36 °C resulted in a larger removal as compared to the activation at >41 °C.

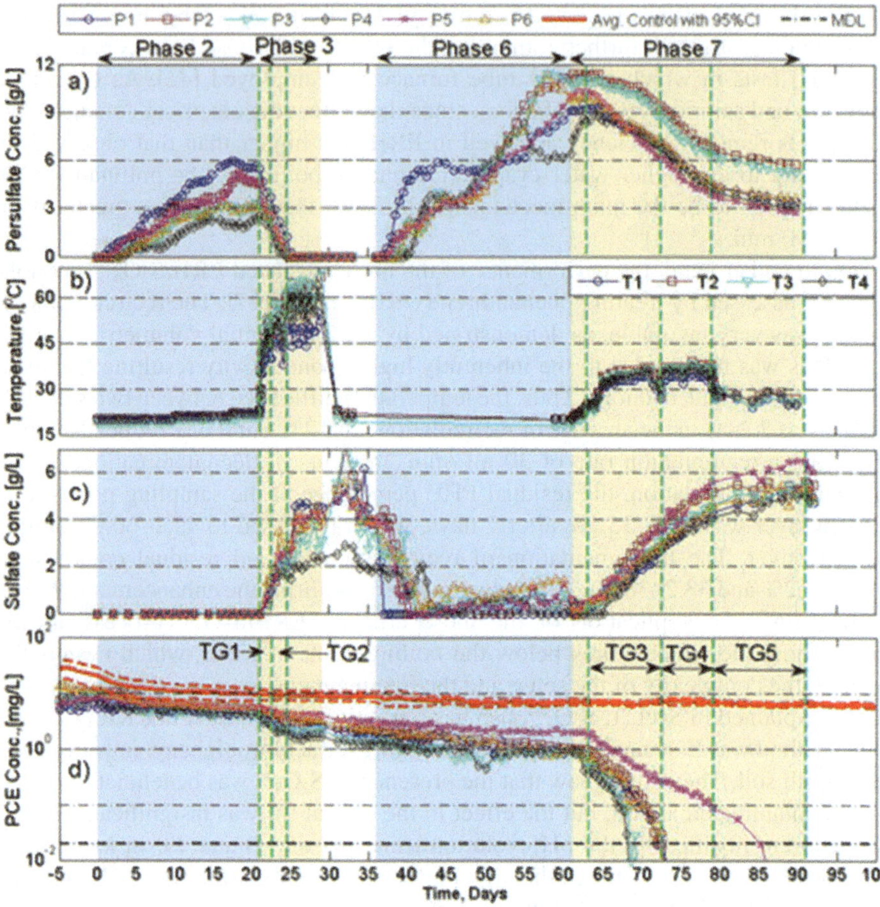


**Fig. 6** Time course of (**a**) aqueous persulfate concentration, (**b**) measured temperatures, (**c**) sulfate concentrations, and (**d**) PCE concentrations at different locations. The "blue boxes" represent the EKR/$S_2O_8^{2-}$ step, whereas the "orange boxes" represent the ERH/$S_2O_8^{2-}$. The region bounded between vertical green lines represent the different temperature regimes (TG1 to TG5) from left to right. (Reproduced with permission from [43]. Copyright 2017 American Chemical Society)

## 3.2 *Observation of the Scale Effect*

In previous sections, the occurrence of the Joule heating in the electro-thermal systems under review has been mathematically demonstrated, and its effect on the process performance has been directly assumed or, in some cases, determined through temperature measurements. However, the impact of the Joule effect in the aforementioned technologies upon gradual scale-up has not been quantified, which constitutes a serious gap because the largest increases of temperature during applications based on electric current or potential supply have been reported for pilot- or field-scale experiments [44].

The considerations and corrections that are necessary to be implemented to correctly operate full-size setups being based on models made or measurements obtained from small-scale systems give rise to the so-called scale effect. This subsection addresses those studies, most of which have been based on EKR trials, being especially highlighting the unrivalled leap ahead by Professors Rodrigo and Sáez from the Universidad de Castilla-La-Mancha in Spain. Their great engineering efforts allow obtaining quantitative data and drawing sound conclusions on the real impact of the electrothermal phenomena.

Temperature change was been barely observed at small scale. For example, Vieira dos Santos et al. [45] investigated the removal of four pesticides spiked into 3 kg of soil, by EK soil flushing (similar to EKR described above, but injecting a flushing fluid near the anodes to make collection near the cathodes). They employed a bench-scale setup operated at 1.0 V/cm. After 15 days, more than 80% pesticide removal was observed. Temperature remained almost constant at room T and hence, volatilization observed was simply due to a natural phenomenon rather than to Joule heating.

Similarly, Risco et al. [46] assessed the removal of oxyfluorfen by EK soil flushing. They simulated a spill in a soil mockup (silty loam soil, 175 L), using a row of anodes facing a row of cathodes, and studied the change of different parameters for 34 days upon the application of a constant electric field (i.e., voltage gradient) of 1.0 V/cm (total cell voltage of 38 V). According to the post-mortem soil analysis, a 3D map of pollutant distribution showed that removal was due to: (a) transport toward the anode and cathode wells, (b) gravity fluxes, and (c) natural volatilization. As in the previous study, heating was insignificant because temperature was almost constant never exceeding the ambient temperature. It can thus be deduced that, at such small scale, water acts as a cooling agent since it can either absorb the generated heat.

In contrast, temperature increased quite significantly at pilot scale and caused volatilization during EKR far beyond the natural phenomenon [47]. The group of Rodrigo and Sáez employed an unprecedented reactor of 32 $m^3$ (2 m in height and a square plant of 4 m × 4 m), also used in another parallel study [48]. The engineering and technological parameters when up-scaling to such dimensions must be carefully controlled and, to this aim, an analogous reactor of 16 $m^3$ with two linear rows of cathodes and anodes facing each other had been previously employed for EKR [49]. Based on the experience gained with that reactor, in the 32-$m^3$ reactor several graphite cylinders (15 cm in diameter, 100 cm in length) were positioned as electrodes in semipermeable perforated cylindrical electrolyte wells (31.5 cm in diameter, 140 cm in length), distributed in an hexagonal arrangement. This electrode configuration corresponds to an EK fence composed of six alternating electrodes (i.e., three cathodes and three anodes). Thermocouples were located in different positions to monitor the temperature during the test. Within the context of this chapter, the most relevant finding was the huge temperature increase determined at the end of the test: threefold for soil and fourfold in the wells, starting the experiment at

10 °C. Electric soil heating was thus confirmed. As evidenced in 2D temperature profiles, the more active areas were in the vicinity of the electrodes, not finding any difference between anodes and cathodes. Since the electrode wells were connected to a gas extraction system, pesticide volatilization was a major route for soil remediation in the prototype.

When comparing the thermal data of this work [47] to those obtained by Risco et al. at the smaller scales [46], it can be observed that the average temperature using the 32-$m^3$ prototype was 29.4 °C, being more than 10 °C above the average temperature typically measured in the 175-L setup (i.e., 18.2 °C for the oxyfluorfen tests). The scarce Joule effect ($P = IR^2$) at small scale was partly explained by the lower current density (i.e., lower $I$, in the range of 10–20 mA), but also by the smaller inter-electrode distance, which decreased the ohmic losses (i.e., lower $R$). Therefore, any possible heating was immediately counterbalanced by evaporative (water) cooling and heat exchange (convection) with the surroundings. In addition, the researchers also appreciated a different temperature distribution map when comparing both scales. In the mockup, the temperature of the soil near the electrodes (i.e., the electrokinetic zone) was almost 4 °C higher than that farther away. However, such temperature gradient was insufficient to force the pollutant volatilization. Conversely, in the prototype, a lower the temperature was found in the near-electrode area. A clear explanation was not offered, but it is worth noting that despite the analogous heat transfer and mass transport principles supposed at both scales, their effect in practice is quite different. Obviously, the dissimilar temperature distribution in the prototype is expected to have influence on the pesticide distribution, specially taking into account the high volatility of oxyfluorfen. Figure 7 compares the contribution of electrokinetic and volatilization mechanisms for the removal of this pesticide, as well for 2,4-dichlorophenoxyacetic acid (2,4-D) that was also tested. The

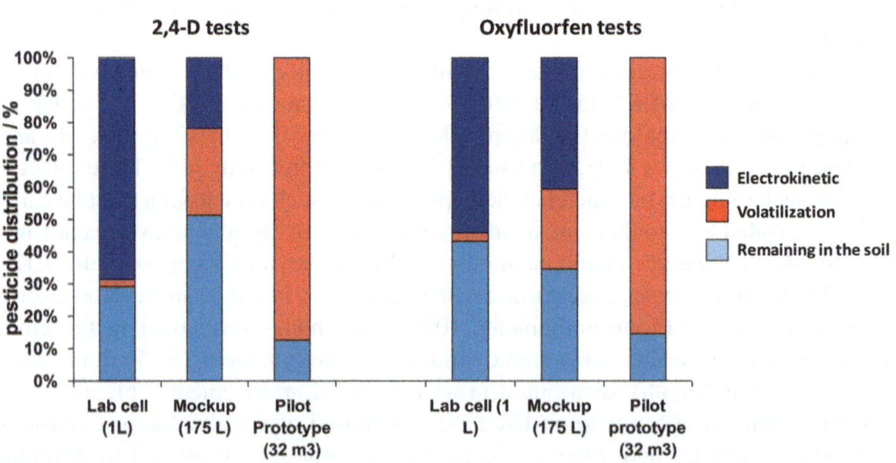


**Fig. 7** Contribution of different mechanisms (EK and volatilization) to pesticide removal (in %) in EKR trials carried out at three different scales. (Reproduced with permission from [47]. Copyright 2017 Elsevier)

performance of the prototype, the mockup and a lab-scale cell (1 L) is shown. The main conclusion is that a larger scale promotes the volatilization over the EK, mechanism, as a consequence of the higher temperature derived from Joule heating.

In a similar comparison between small and large scales, a 25% rise in electric heating was determined when up-scaling the EKR from the 175-L system to the 16-$m^3$ EKR reactor [49]. In other words, the temperature increase observed at small scale was 2 °C, whereas at pilot scale it was 8 °C. Furthermore, it must be taken into account that the pilot was exposed to the open atmosphere, at outside temperature of 11.2 °C, which cooled down the pilot to some extent. Otherwise, a larger difference in Joule heating could probably have been found. From a practical point of view, such a significant temperature increase in the pilot can cause shrinkage cracks on the soil surface, thus altering the expected evolution of the measured parameters during the EKR trial.

In a subsequent study [18], the same research group used the 32-$m^3$ reactor (Fig. 8a) but modified to couple a biological treatment in situ. The central well, designed for the initial herbicide discharge in the previous EK flushing trial, was replaced by a well housing a biobarrier (see BB in Fig. 8b). In fact, the central position contained concentrated activated sludge, aiming to stimulate the electro-bioremediation (EBR) of a mixture of oxyfluorfen and 2,4-D pesticides. Volatilization was identified as the main remediation mechanism. As expected from the above-mentioned EKR test with a closely related prototype, temperature increased as a consequence of the ohmic soil heating, reaching a steady-state value. Nonetheless, an important observation was that heating was greater at the cathodes (see Fig. 8c), despite the higher conductivity of the liquid contained in cathodic wells. This could be accounted for the passivating carbonate layer formed on the electrode surface, increasing the electric resistance.

The scale effect was also observed by other groups. For instance, Da Silva et al. applied the EKR to the remediation of petroleum-contaminated soil at a 2.55-L scale [50]. Graphite electrodes were employed for 360 h. The authors reported that the increase in scale directly influenced the amount of energy supplied to the soil, resulting in a more intense electric heating as compared to a previous study at a smaller-scale. However, no specific measurements were made.

Finally, one of the largest EKR setups reported in scientific journals was employed to remove As, Cu and Pb from paddy rice field soil using a hexagonal electrode configuration at field scale (17 m in width, 12.2 m in length, 1.6 m in depth) [51]. The arrangement, however, differed from the ones shown and described above because the anode was located at the center of the hexagon, whereas six cathodes were installed at the apex of the hexagon. The inter-electrode gap (either anode–cathode or cathode–cathode) was 2 m, and the total number of anodes and cathodes was 24 and 48, respectively. A constant voltage of 100 V was applied (i.e., electric field of 0.5 V/cm). The variation in the soil temperature was monitored through four thermometers placed at a soil depth of 50 cm. The removal was 44.4% for As, 40.3% for Cu, and 46.6% for Pb after 24 weeks of operation. Surprisingly, the energy consumption was very low when compared to that found at a small scale because, according to the authors, there was less energy consumption due to Joule

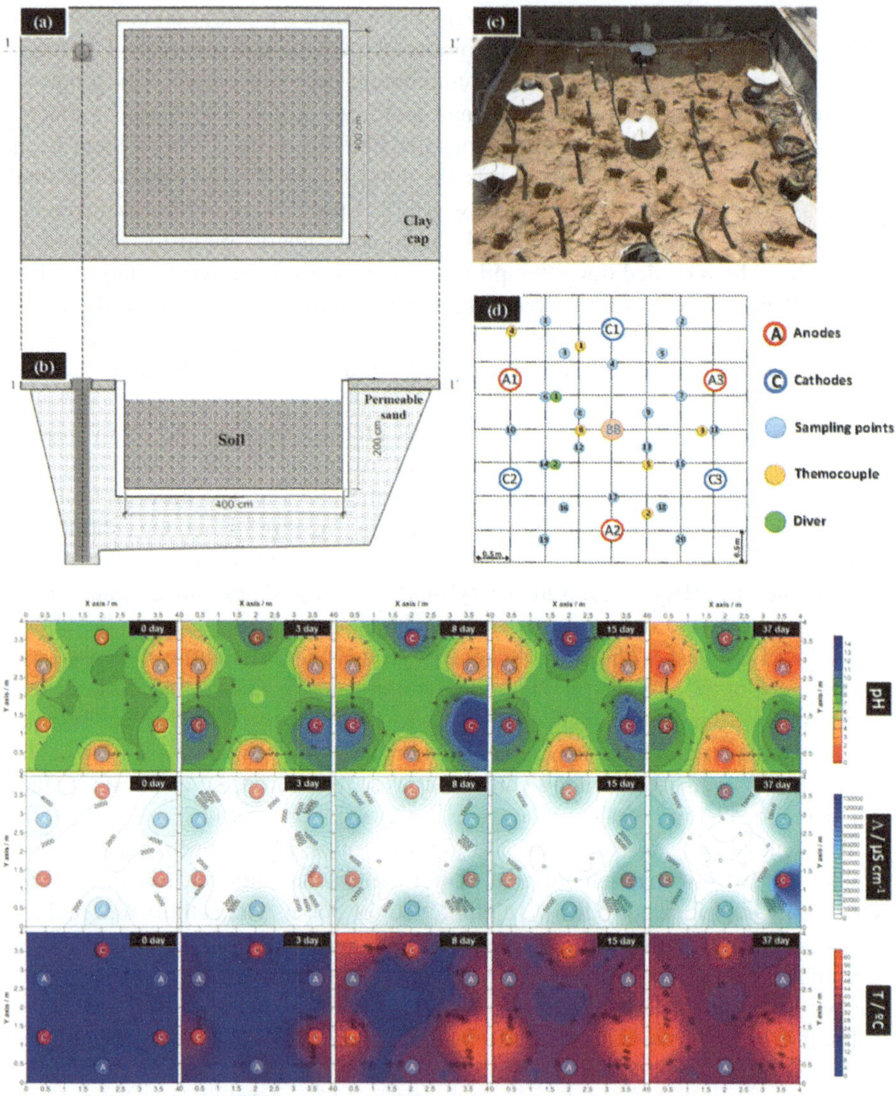


**Fig. 8** (**a**) Picture of the 32-m$^3$ EBR prototype, (**b**) scheme of the location of electrode wells and the main instrumentation (BB, biobarrier), and (**c**) 2D maps showing changes soil and well temperature over time. (Reproduced with permission from [18]. Copyright 2018 Elsevier)

heating. Further discussion was not made. We believe that this finding could arise from the small depth at which temperature was measured, especially taking into account the large length of the electrodes. In fact, the pattern of soil temperature was very similar to that of the atmosphere.

In general, based on most of these studies, we can conclude that it is a good idea to include capillary barriers or an impermeable tarp on the soil surface if volatilization has to be minimized. This will maximize the moisture content in the atmosphere right above the soil, significantly reducing further evaporation.

# 4 Real Case Studies and Companies in the Sector

In previous sections, most of the examples discussed to highlight the particularities and performance of the electro-thermal processes were referred to trials carried out either at laboratory or pilot scale. Now, the aim is to show the operation at real contaminated sites and results achieved. As could be presumed, this kind of studies have been mainly undertaken by companies, some of which have acquired great expertise in recent years and have made reality the expectations that were born as the initially smaller setups were successfully tested by a reduced group of pioneers.

TerraTherm, Inc. [16], a subsidiary of Cascade Environmental [52], is the worldwide leader in the development and implementation of ISTR of contaminated soil and groundwater. In 2016, the company acquired the recognized ISTR technology provider Current Environmental Solutions (CES, founded in 1997 by Battelle Memorial Institute (BMI) and Terra Vac as the first private company to commercialize ERH), which nowadays remains as the most experienced ERH vendor in the world. Nocon Remediation BV, licensed by CES, is the thermal remediation contractor with the longest track record in Europe [53]. Since 1997, Nocon has performed many successful projects in conjunction with different partners and clients. Since 1999, they have been applying ERH and, in 2002, they initiated the first ERH project in Europe.

One of the first Nocon's projects was made in collaboration with Terra Vac UK in a 2-ha site in Sheffield (UK). At that time, it was a densely populated residential area with two main contamination hotspots rich in chlorinated hydrocarbons like TCE. The target volume to be treated was ~3500 $m^3$, extending to 7 m below grade surface (bgs), which had a major impact on groundwater quality (630 $m^3$). Six-phase ERH was applied, causing water boiling that produced steam, which could then evacuate the volatile pollutants. The three-phase electricity distributed by the national grid was split into six phases by means of sophisticated electrical hardware. As stated by the company, the six-phase current is suitable to operate in hexagonal electrode arrays because it distributes proportionately more electric power and provides good geometric coverage. Final reductions of adsorbed and dissolved TCE concentrations were greater than 98% and 99% respectively, after 20 weeks.

In another project in France, Nocon tested the ERH/$S_2O_8^{2-}$ technology. The results showed tremendous VOCs reduction in a highly polluted aquifer.

Another big and experienced competitor of the previous companies is Thermal Remediation Services, Inc. (TRS Group, Inc.), one of the leaders in ERH applications [24, 54, 55]. TRS has completed more than 160 projects in the USA and has joint ventures in Brazil, China and Europe. In 1993, ERH was used at the Department of Defense Savannah River (South Carolina) facility to clean tetrachloroethene from a 10-ft-thick clay lens located 30 ft below the surface. This application was part of an integrated demonstration to remove VOCs in non-arid soils. More than 99% of the contaminants in the treatment area were removed within 25 days. In 1996, ERH was deployed at the US Air Force Reserve's Niagara Falls International Airport. In 3 months, ERH removed four times the mass of TCE initially estimated to be present in the area.

TRS addressed another case study on TCE remediation in low permeability soil (silty clay) in Illinois [55]. The groundwater table was approximately 20-ft bgs and the combined target volume was ~382 $m^3$. TRS was able to achieve a 99.9999% reduction in TCE concentration after 56 days, being 100,000 times lower than the required standards.

In this 2006, TRS worked with Shaw Environmental & Infrastructure to develop a site remediation program to remove chlorinated solvents in Annapolis (Maryland) [56]. A release of 1,1,2,2-tetrachloroethane (TeCA) had occurred in the 1940s from a small landfill. The site was composed of dense fine sand with silt layers, which extended to 40 ft bgs, where it became cemented. The remediation area had an elliptical shape and measured ~790 $m^2$ (target volume ~10,000 $m^3$). As the ERH treatment was run, the released chloride ions increased the conductivity, resulting in slightly faster heating. After 125 days, VOCs concentrations had been reduced by 99.9%.

More recently, TRS has been involved in a quite new ISCO-ERH application, in which zero-valent iron (ZVI) was employed instead of $S_2O_8^{2-}$ as chemical [57]. Initial groundwater TCE concentration was 3800 μg/L, accompanied by 20% of 1,2-dichloroethene. The field test was carried out for 345 days. Increased degradation of the pollutant was observed as temperature became higher (always up to moderate values lower than 50 °C), with minimal TCE volatilization. The results suggest that ZVI-based treatments can be certainly enhanced if operated with electrothermal methods.

Alternative Restoration Technology Team (ARTT), a US Naval Facilities Engineering Command (NAVFAC) workgroup, was established in 1842 to promote innovative technologies like ERH within the Navy Installation Restoration program. According to their report in spring 2006 [58], studies funded by ARTT Navy were conducted to study the performance of ERH at several US Navy and NASA sites. They concluded that drying around the electrodes, depression of the water table, and inefficient heating are key factors that make heating difficult in low permeability sites. In those places, the performance can be enhanced by: (a) including a higher electrode density, (b) installing deeper electrodes, (c) implementing a slower heating to allow recharge and prevent drying, (d) using ground rods to distribute heat and (e) adding water to the electrode wells. Preferably, temperatures should reach the boiling point of water, which is not easy with increasing depth because of the higher pressure and the presence of non-volatile solutes (ions). They also suggested that ERH design is better made considering the volume of aquifer to treat, not the VOC mass.

All these studies allow concluding that, at sites where ERH is applicable, the time spent in remediation is typically reduced from years to months when compared with other ISTR methods. Larger ERH sites are cleaned within 6–12 months and smaller sites can be cleaned in half that time. For large sites, the cost of ERH is about half that of excavation and disposal [55].

## 5 Conclusions

A nice set of technologies is currently at clients' disposal to ensure the effective removal of soil contaminants. Considering the case of volatile pollutants, ISTR methods are particularly suitable because temperature has a direct impact on soil and water physical properties. In turn, this causes many simultaneous or consecutive phenomena to such organic chemicals, including desorption, dissolution, transport or volatilization. An interesting feature of the dominant electrothermal technologies like EKR (using DC) and ERH (using AC) is that they have been already deployed at full scale by strong companies, once the theoretical fundamentals and mathematical basis have been appropriately addressed in past years. Note that the intrinsic aim of EKR is not heating-up soil and water, but this is rather a collateral effect that arises from the Joule effect. In contrast, in ERH, the actual goal is heating the subsurface soil and water, reaching sub-boiling or boiling temperatures.

Nowadays, numerical simulations make it possible to have temperature distribution maps throughout the target contaminated volume, which gives information on heat transfer efficiency. When establishing performance objectives for the ISTR technologies reviewed, it is not always necessary to reach total cleanup standards as the electro-thermal treatment finishes. Instead, remediation may be completed via enhanced bioremediation and natural attenuation, depending on the urgency to occupy or use the remediated soil. It is important to include off-gas collection and treatment units combined with the electrothermal systems, especially in terms of environment conservation. To our knowledge, there is still a lack of studies that rigorously address the sizing of such collectors and the energy and water requirements of the different technologies.

Similarly, the alteration of soil properties as part of the electro-thermal remediation, potentially leading to changes in soil fertility and ecosystem dynamics, has not been investigated in detail. In addition, most of the thermal-based technologies cause soil desiccation, which requires post-treatment with water. Based on these considerations, the lowest effective treatment temperature should be employed. This is also important to keep energy usage and associated costs low. As a corollary, high temperatures should be only considered to treat highly refractory pollutants.

A final remark on ISTR is that the injection of potentially hazardous chemicals is unnecessary. However, it has been demonstrated that some chemicals like $S_2O_8^{2-}$ and ZVI may have a positive impact, giving rise to highly effective electrochemical ISCO-ISTR processes. The exploration of non-toxic and/or biodegradable compounds that can be thermally activated may be an interesting approach in this regard.

**Acknowledgments** The authors kindly acknowledge support from project CTQ2016-78616-R (AEI/FEDER, EU).

## References

1. https://waste-management-world.com/a/340-000-reasons-to-choose-in-situ-remediation-instead-of-ex-situ-and-mass-transfer
2. M.T. Ricart, M. Pazos, S. Gouveia, C. Cameselle, M.A. Sanromán, Removal of organic pollutants and heavy metals in soils by electrokinetic remediation. J. Environ. Sci. Health A **43**, 871–875 (2008)
3. M.A. Rodrigo, N. Oturan, M.A. Oturan, Electrochemically assisted remediation of pesticides in soils and water: a review. Chem. Rev. **114**, 8720–8745 (2014)
4. O. Cuevas, R.A. Herrada, J.L. Corona, M.G. Olvera, S. Sepúlveda-Guzmán, I. Sirés, E. Bustos, Assessment of $IrO_2$-$Ta_2O_5$|Ti electrodes for the electrokinetic treatment of hydrocarbon-contaminated soil using different electrode arrays. Electrochim. Acta **208**, 282–287 (2016)
5. https://frtr.gov/matrix2/section3/sec3_int.html
6. J.E. Vidonish, K. Zygourakis, C.A. Masiello, G. Sabadell, P.J.J. Alvarez, Thermal treatment of hydrocarbon-impacted soils: a review of technology innovation for sustainable remediation. Engineering **2**, 426–437 (2016)
7. J.L. Triplett Kingston, P.R. Dahlen, P.C. Johnson, State-of-the-practice review of in-situ thermal technologies. Groundw. Monit. Remediat. **30**, 64–72 (2010)
8. H.F. Stroo, A. Leeson, J.A. Marqusee, P.C. Johnson, H. Ward, M.C. Kavanaugh, T.C. Sale, C.J. Newell, K.D. Pennell, C.A. Lebrón, M. Unger, Chlorinated ethane source remediation: lessons learned. Environ. Sci. Technol. **46**, 6438–6447 (2012)
9. I. Ross, J. McDonough, J. Miles, P. Storch, P.T. Kochunarayanan, E. Kalve, J. Hurst, S.S. Dasgupta, J. Burdick, A review of emerging technologies for remediation of PFASs. Remediation **28**, 101–126 (2018)
10. S. Kuppusamy, P. Rhavamani, K. Venkateswarlu, Y.B. Lee, R. Naidu, M. Megharaj, Remediation approaches for polycyclic aromatic hydrocarbons (PAHs) contaminated soils: technological constraints, emerging trends and future directions. Chemosphere **168**, 944–968 (2017)
11. M. Fingas, An overview of in-situ burning, in *Oil Spill Science and Technology*, ed. by M. Fingas, (Gulf Professional Publishing, Burlington, USA, 2010), pp. 737–903
12. U. Roland, F. Holzer, D. Buchenhorst, F.-D. Kopinke, Thermisch unterstützte Bodenreinigung durch direkte Erwärmung mittels Radiowellen - Teil 1: Grundlagen und verfahrenstechnische Aspekte. Chem. Ing. Tech. **79**, 1667–1678 (2007)
13. U. Roland, D. Buchenhorst, F. Holzer, F.-D. Kopinke, Engineering aspects of radio-wave heating for soil remediation and compatibility with biodegradation. Environ. Sci. Technol. **42**, 1232–1237 (2008)
14. U. Roland, F. Holzer, F.-D. Kopinke, Combining different frequencies for electrical heating of saturated and unsaturated soil zones. Chem. Eng. Technol. **34**, 1645–1651 (2011)
15. E.J. Martin, K.G. Mumford, B.H. Kueper, G.A. Siemens, Gas formation in sand and clay during electrical resistance heating. Int. J. Heat Mass Transf. **110**, 855–862 (2017)
16. http://www.terratherm.com/
17. A.T. Yeung, Contaminant extractability by electrokinetics. Environ. Eng. Sci. **23**, 202–224 (2006)
18. S. Barba, R. López-Vizcaíno, C. Saez, J. Villaseñor, P. Cañizares, V. Navarro, M.A. Rodrigo, Electro-bioremediation at the prototype scale: what it should be learned for the scale-up. Chem. Eng. J. **334**, 2030–2038 (2018)
19. F. Baraud, S. Tellier, M. Astruc, Temperature effect on ionic transport during soil electrokinetic treatment at constant pH. J. Hazard. Mater. **B64**, 263–281 (1999)
20. Z. Zhou, X. Liu, K. Sun, C. Lin, J. Ma, M. He, W. Ouyang, Persulfate-based advanced oxidation processes (AOPs) for organic-contaminated soil remediation: a review. Chem. Eng. J. **372**, 836–851 (2019)
21. Y. Yukselen-Aksoy, K.R. Reddy, Effect of soil composition on electrokinetically enhanced persulfate oxidation of polychlorobiphenyls. Electrochim. Acta **86**, 164–169 (2012a)

22. D.A. Reynolds, In situ remediation of soils and ground water containing organic contaminants, US patent 9004816 B2 (2015)
23. Y. Yukselen-Aksoy, K.R. Reddy, Electrokinetic delivery and activation of persulfate for oxidation of PCBs in clayey soils. J. Geotech. Geoenviron. Eng. **139**, 175–184 (2012b)
24. https://www.thermalrs.com/
25. M.M. Krol, B.E. Sleep, R.L. Johnson, Impact of low-temperature electrical resistance heating on subsurface flow and transport. Water Resour. Res. **47**, W05546 (2011)
26. H.M. Buettner, W.D. Daily, Cleaning contaminated soil using electrical heating and air stripping. J. Environ. Eng. **121**, 580–589 (1995)
27. J. Li, L. Wang, L. Peng, Y. Deng, D. Deng, A combo system consisting of simultaneous persulfate recirculation and alternating current electrical resistance heating for the implementation of heat activated persulfate ISCO. Chem. Eng. J. **385**, 123803 (2020)
28. B.C.W. McGee, C.W. McDonald, L. Little, Comparative proof of concept results for electrothermal dynamic stripping process: integrating environmentalism with bitumen production. SPE Projects, Facilities & Construction (2009)
29. B.C.W. McGee, Electro-thermal dynamic stripping process for in situ remediation under an occupied apartment building. Remediation **13**, 67–79 (2003)
30. B.C.W. McGee, F.E. Vermeulen, The mechanisms of electrical heating for the recovery of bitumen from oil sands. J. Can. Petrol. Technol. **46**, 28–34 (2007)
31. T. Navab-Daneshmand, R. Beton, R.J. Hill, D. Frigon, Impact of Joule heating and pH on biosolids electro-dewatering. Environ. Sci. Technol. **49**, 5417–5424 (2015)
32. X. Wen, M. Jing, H. Cai, Y. Zhang, S. Hu, Y. Teng, G. Liu, L. Lan, H. Lu, Temperature characteristics and influence of water-saturated soil resistivity on the HVDC grounding electrode temperature rise. Electr. Power Energy Syst. **118**, 105720 (2020)
33. H.-J. Diersch, O. Kolditz, Variable-density flow and transport in porous media: approaches and challenges. Adv. Water Resour. **25**, 899–944 (2002)
34. R. Ma, C. Zheng, Effects of density and viscosity in modeling heat as a groundwater tracer. Ground Water **48**, 380–389 (2010)
35. M.A. Oyanader, P.E. Arce, Role of aspect ratio and Joule heating within the fluid region near a cylindrical electrode in electrokinetic remediation: a numerical solution based on the boundary layer model. Int. J. Chem. Reactor Eng. **11**, 687–699 (2013)
36. C.M. Torres, P.E. Arce, F.J. Justel, L. Romero, Y. Ghorbani, Joule heating effects in electrokinetic remediation: role of non-uniform soil environments: temperature profile behavior and hydrodynamics. Environments **5**, 92 (2018)
37. P.R. Hegele, K.G. Mumford, Gas production and transport during bench-scale electrical resistance heating of water and trichloroethene. J. Contam. Hydrol. **165**, 24–36 (2014)
38. J.L. Munholland, K.G. Mumford, B.H. Kueper, Factors affecting gas migration and contaminant redistribution in heterogeneous porous media subject to electrical resistance heating. J. Contam. Hydrol. **184**, 14–24 (2016)
39. J.L. Triplett Kingston, P.R. Dahlen, P.C. Johnson, Assessment of groundwater quality improvements and mass discharge reductions at five in situ electrical resistance heating remediation sites. Groundw. Monit. Remediat. **32**, 41–51 (2012)
40. C.R. Carrigan, J.J. Nitao, Electro-osmotic infusion for Joule heating soil remediation techniques, US patent 5975799 (1999)
41. D. Oberle, E. Crownover, M. Kluger, *In situ* remediation of 1,4-dioxane using electrical resistance heating. Remediation **25**, 35–42 (2015)
42. Z. Han, W. Jiao, Y. Tian, J. Hu, D. Han, Lab-scale removal of PAHs in contaminated soil using electrical resistance heating: removal efficiency and alteration of soil properties. Chemosphere **239**, 124496 (2020)
43. A.I.A. Chowdhury, J.I. Gerhard, D. Reynolds, D.M. O'Carroll, Low permeability zone remediation via oxidant delivered by electrokinetics and activated by electrical resistance heating: proof of concept. Environ. Sci. Technol. **51**, 13295–13303 (2017)

44. R. Lageman, Electroreclamation: applications in The Netherlands. Environ. Sci. Technol. **27**, 2648–2650 (1993)
45. E. Vieira dos Santos, F. Souza, C. Saez, P. Cañizares, M.R.V. Lanza, C.A. Martínez-Huitle, M.A. Rodrigo, Application of electrokinetic soil flushing to four herbicides: a comparison. Chemosphere **153**, 205–211 (2016)
46. C. Risco, S. Rodrigo, R. López Vizcaíno, A. Yustres, C. Saez, P. Cañizares, V. Navarro, M.A. Rodrigo, Removal of oxyfluorfen from spiked soils using electrokinetic soil flushing with linear rows of electrodes. Chem. Eng. J. **294**, 65–72 (2016)
47. R. López-Vizcaíno, C. Risco, J. Isidro, S. Rodrigo, C. Saez, P. Cañizares, V. Navarro, M.A. Rodrigo, Scale-up of the electrokinetic fence technology for the removal of pesticides. Part II: does size matter for removal of herbicides? Chemosphere **166**, 549–555 (2017)
48. R. López-Vizcaíno, C. Risco, J. Isidro, S. Rodrigo, C. Saez, P. Cañizares, V. Navarro, M.A. Rodrigo, Scale-up of the electrokinetic fence technology for the removal of pesticides. Part I: Some notes about the transport of inorganic species. Chemosphere **166**, 540–548 (2017)
49. R. López-Vizcaíno, V. Navarro, M.J. León, C. Risco, M.A. Rodrigo, C. Sáez, P. Cañizares, Scale-up on electrokinetic remediation: engineering and technological parameters. J. Hazard. Mater. **315**, 135–143 (2016)
50. E.B.S. Da Silva, N.S. Fernandes, E.C.T. de A. Costa, D.R. da Silva, D.M. de Araújo, C.A. Martínez-Huitle, Scale-up of electrokinetic treatment of polluted soil with petroleum: effect of operating conditions. Int. J. Electrochem. Sci. **12**, 4001–4015 (2017)
51. E.-K. Jeon, J.-M. Jung, W.-S. Kim, S.-H. Ko, K. Baek, In situ electrokinetic remediation of As-, Cu-, and Pb-contaminated paddy soil using hexagonal electrode configuration: a full scale study. Environ. Sci. Pollut. Res. **22**, 711–720 (2015)
52. https://www.cascade-env.com/
53. http://www.nocon.eu/Electric-Resistance-Heating/
54. http://www.thermalrs.com/TRS_docs/EPA%20Skokie%20Cost%20Perform%20Rpt.pdf
55. http://www.thermalrs.com/TRS_docs/epa542r04010.pdf
56. M. Kluger, G.L. Beyke, Electrical resistance heating of volatile organic compounds in sedimentary rock. Remediation **20**, 69–82 (2010)
57. M.J. Truex, T.W. Macbeth, V.R. Vermeul, B.G. Fritz, D.P. Mendoza, R.D. Mackley, T.W. Wietsma, G. Sandberg, T. Powell, J. Powers, E. Pitre, M. Michalsen, S.J. Ballock-Dixon, L. Zhong, M. Oostrom, Demonstration of combined zero-valent iron and electrical resistance heating for in situ trichloroethene remediation. Environ. Sci. Technol. **45**, 5346–5351 (2011)
58. https://www.ready.navy.mil/search.html?q=rits+2006

# Electrochemically Assisted Dewatering

**Maria Villen-Guzman and Jose M. Rodriguez-Maroto**

## 1 Introduction

Electrochemically assisted dewatering, also called electro-dewatering (EDW), electrokinetic dewatering (EKD) and field-assisted dewatering is a technology widely used in soil engineering. Under the application of an electrical potential across a soil sample, species positively charged are transported toward the cathode and those negatively charged toward the anode. This movement entails the transport of water, usually, toward the cathode. However, the direction of this net flow depends on the sign of zeta potential on soil particles. This flow is known as electro-osmosis and depends mainly on the coefficient of electro-osmotic hydraulic conductivity ($k_e$) and the voltage gradient [1, 2].

EDW is usually applied to porous matrices with low hydraulic conductivity since the application of an electrical current has been widely demonstrated to be more efficient to produce water transport in fine-grained than hydraulic gradient. It has been proved that the grain size influences directly on hydraulic conductivity but not on the electro-osmotic permeability. Jones et al. estimated that electro-osmotic flow rates are 100–10,000 times greater than hydraulic conductivity in fine-grained materials [3].

The technology presented in this chapter has been largely used for the consolidation and strengthening of soils due to the influence of the pore water content on the stability of soils. Reduction of the pore water pressure as a consequence of the application of electric current entails the improvement of the soil behavior [4]. Comparing with conventional techniques, the consolidation of the soil by electro-osmosis allows accelerating the transport of water. The technique has been applied not only to soil stabilization but also to remediation for porous matrices,

M. Villen-Guzman (✉) · J. M. Rodriguez-Maroto
Department of Chemical Engineering, University of Malaga, Málaga, Spain
e-mail: mvillen@uma.es; maroto@uma.es



M. A. Rodrigo, E. V. Dos Santos (eds.), *Electrochemically Assisted Remediation of Contaminated Soils*, Environmental Pollution 30,
https://doi.org/10.1007/978-3-030-68140-1_16

stabilization of mine tailings, and many other geotechnical engineering applications [5]. The application of an electric current entails, in addition to water movement, drawing other species such as contaminant. For this reason, a large number of studies using electro-osmosis as a remediation technology have been reported.

The EDW applied to consolidate soft silty clays dated back to 1930s [6] and several patents on the removal of water from soil were issued before World War II [7]. The starting development of the technology was significantly due to a continuous research carried out by L. Casagrande. The EDW is especially suitable for projects requiring a rapid treatment to improve soils. Bjerrum et al. applied successfully the technique to a quick clay soil in Ås, a municipality located about 30 km south of Oslo, Norway in 1967 [8]. Also in 1967, the technique was investigated to stabilize a dam located on Mahoning River in Northeastern Ohio, USA. After feasibility studies carried out by L. Casagrande, it was decided to stabilize the dam by electro-osmotic process. Thereafter, other successful studies which will be discussed in this chapter has been carried out [9–14].

This chapter will cover the fundamental principles of electro-osmotic transport and the influence of operational conditions on the efficiency of EDW. Several applications are reviewed through the discussion of real cases, from traditional application to consolidation of soils to new approaches of dewatering of wastewater sludge.

## 2 Discussion

### *2.1 Electrochemical Transport Processes*

Upon electric field application, the transport mechanism of contaminants through the porous media are electromigration, electro-osmosis and electrophoresis. Electromigration is the movement of ions under an applied electric field toward the electrodes. It is the predominant transport mechanism for ionic metals, ionic micelles, polar organic molecules and colloidal electrolytes [15]. Electro-osmosis is the movement of liquid containing dissolved ionic and non-ionic species induced under electric gradient. This mechanism is the most important for organic and inorganic non-ionic contaminants that are dissolved, suspended, emulsified or such similar forms. A more detailed discussion about this mechanism will be presented in the follow section. Finally, electrophoresis is the transport of colloids or charged species under an electric field. For soils with low hydraulic permeability, mass transport by this mechanism is negligible [16].

In addition to these transport processes, electrolysis reactions occur at the electrodes under an electric potential gradient, as follows:

$$\text{Anode}\left(\text{oxidation}\right): H_2O \rightarrow \%O_{2(g)} + 2H^+ + 2e^- \quad (1)$$

$$\text{Cathode}\left(\text{reduction}\right): 2H_2O + 2e^- \rightarrow H_{2(g)} + 2OH^- \quad (2)$$

The decomposition of water at the electrodes generates an acidic medium at the anode and an alkaline medium at the cathode [15]. The influence of these reactions on transport, transformation and degradation processes of contaminants is of great importance as can be concluded from Sect. 2.3.

## 2.2 Principles of Electro-osmosis

The application of a direct current to generate flow of water through clayey soil was firstly reported by Reuss in 1809 [17], with his classical experiment reproducing the study carried out by Nicholson and Carlisle in 1800. The test consisted on the decomposition of water by an electric current. The experimental system, schematically presented in Fig. 1, consisted of a U-tube filled with water in which the powder was placed at the bottom, capillarity tubes connected with the U-tube and two platinum wires acting as electrodes. After applying an electrical potential, the water level in the arm containing the negative electrode increase significantly. Thus, Reuss found that under an electrical current application movement of water through a porous media occurs from the anode to the cathode against gravity.

The first quantitative measurement of the electro-osmosis phenomenon dated from 1852 to 1856 performed by Wiedemann. He observed two important facts: (a) the hydraulic pressure difference between the two sides of the plug was not only proportional to the electrical potential applied across the plug but also independent

**Fig. 1** Reuss experiment about electro-osmosis

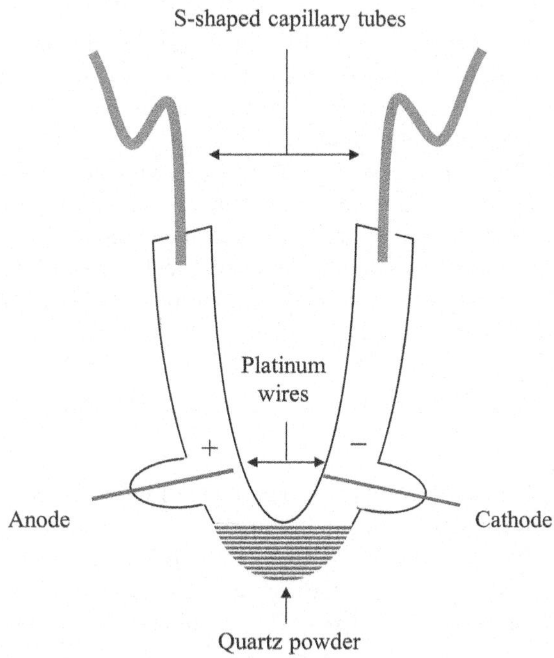

of the plug dimensions; (b) maintaining the same liquid level on both arms of U-tube, the flow of liquid was proportional to the electric current applied and not dependent on the plug dimension [18]. Until research carried out in 1861 by Quincke, it was supposed that the flow liquid during electro-osmosis would be always from the anode to the cathode as the electric current induced. Quincke introduced the concept of electrical double layer (EDL), discussed in detail in Sect. 2.2.1 [19]. This idea was fundamental for the model of electro-osmotic phenomena developed by Helmholtz in 1879, which assumed the EDL as a simple capacitor. Smoluchowski later modified this theory in 1921 to be also applied to electrophoretic velocity. This theory will be discussed in the next section.

The electro-osmotic volume flow rate for unidirectional flow can be described as:

$$Q_e = -k_e \frac{\delta V}{\delta x} A \tag{3}$$

where $k_e$ ($m^2\ V^{-1}\ s^{-1}$) is the electro-osmotic permeability, $\delta V/\delta x$ ($V\ m^{-1}$) is the gradient of the electrical potential at distance $x$ (m) to the cathode and $A$ ($m^2$) is the area perpendicular to the direction of the flow ($x$). Determining the flow rate of water through a soil of length and cross section known under an applied electrical gradient allows measuring the electro-osmotic permeability $k_e$.

On the other hand, the flow of water associated with the hydraulic gradient in saturated soils according to the Darcy's law is:

$$Q_h = -k_h \frac{\delta H}{\delta x} A \tag{4}$$

where $k_h$ ($m\ s^{-1}$) is the hydraulic permeability and $\delta H/\ \delta x$ (–) is the hydraulic gradient along the direction of flow. This flow contribution is relevant for permeable soils, such as sand. However, it is negligible for soils characterized by a low hydraulic conductivity, such as clay.

For applying equations to hydrodynamic consolidation, the drainage conditions should be considered. The setup configuration for dewatering consists of an anode closed and the drainage open at the cathode. Thus, the coupling of both transport phenomena, hydraulic and electrical, should be considered. The water flow is obtained as a sum of the contributions associated with electro-osmotic and hydraulic gradient flows, as following:

$$Q = -k_e \frac{\delta V}{\delta x} A \left[ 1 + \frac{k_h}{k_e} u \right] \tag{5}$$

where $u = \delta H/\ \delta V$ is the increment of interstitial pressure due to the electrical gradient applied.

The ratio of the coefficients of hydraulic conductivity to electro-osmotic permeability ($k_h/k_e$) is an indicative of the relative importance of hydraulic flow versus electro-osmotic flow. The smallest the ratio, the more important is the

**Table 1** Values for the coefficient of hydraulic and electro-osmotic permeability [2]

| Material | Water content (%) | $k_e \times 10^5$ ($cm^2\ s^{-1}\ V^{-1}$) | $k_h \times 10^8$ ($cm\ s^{-1}$) | $k_h/k_e \times 10^3$ ($V\ cm^{-1}$) |
|---|---|---|---|---|
| London clay | 52.3 | 5.8 | 1 | 0.2 |
| Boston blue clay | 50.8 | 5.1 | 1 | 0.2 |
| kaolin | 67.7 | 5.7 | 10 | 1.8 |
| Clayey silt | 31.7 | 5.0 | 100 | 20 |
| Rock flour | 27.2 | 4.5 | 10 | 2.2 |
| Mica powder | 49.7 | 6.9 | $10^3$ | 145 |
| Fine sand | 26.0 | 4.1 | $10^4$ | 2439 |
| Quartz powder | 23.5 | 4.3 | $10^4$ | 2326 |
| Ås quick clay | 31.0 | 2.5 | 2.0 | 0.1 |
| Bootlegger Cove clay | 30.0 | 2.4–5.0 | 2.0 | 0.4–0.8 |
| Silty clay, West Branch Dam | 32.0 | 3.0–6.0 | 1.2–6.5 | 0.2–2.2 |
| Clayey silt, Ontario | 26.0 | 1.5 | 1.2 –6.5 | 0.8–4.3 |

electro-osmotic flow. The dewatering technology is more efficient for soils with $k_h$ equal or smaller than $k_e$, such as the clayey soils. Values of this coefficient obtained from experimental test are given in Table 1.

### 2.2.1 Double Diffuse Layer

The electric double layer (EDL) term refers to the interface between a solid surface and an electrolyte solution. These interfaces are common in soil or clay suspensions due to the presence of negative charge in the solid surface attracting cations. The understanding of the processes taking place in the EDL is essential for the application of electro-osmotic phenomena to consolidate soils or remediate polluted soils, among other EDW applications. In this section, the most important models describing quantitatively the EDL concept will be discussed. The earliest model was developed by Helmholtz in 1879 which has been completed by Gouy [20], Chapman [21] and Stern [22]. The most applied theory for the description of EDL was the proposed by Gouy [20] and Chapman [21]. The electrochemical potential distribution for the different models described is presented in Fig. 2 in which: $\psi$ represents the total potential and $x$ distance.

The simplest theory describing the EDL is probably the proposed by Helmholtz. As discussed in the next section, the theory considers the EDL as a simple capacitor. The negative charge is considered to be distributed over the surface and the countercharge is concentrated in a plane parallel to the surface located at a very small distance ($x$), called the outer Helmholtz plane [23]. This distribution results in a linear potential drop as presented in Fig. 2a. The theory described by Gouy and Chapman (GC) was pioneer in describing the behavior of double layer at low ionic concentration. They considered, as Helmholtz theory, the negative charge distributed over the

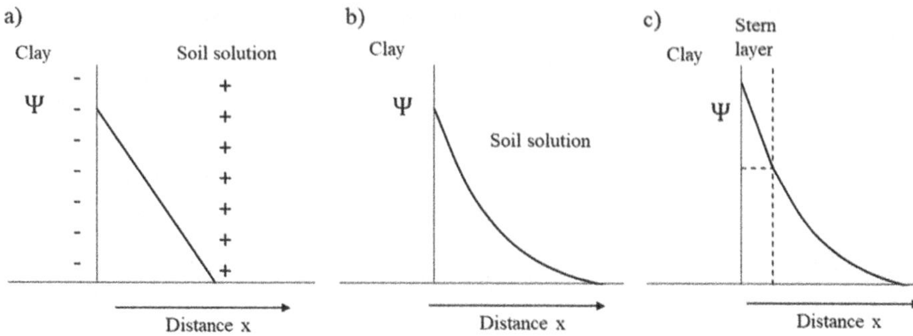


**Fig. 2** Electrochemical drop potential distribution in an EDL according to (**a**) Helmholtz, (**b**) Gouy-Chapman and (**c**) Gouy-Chapman-Stern models

surface. However, the counterions are supposed to be dispersed in the liquid layer. According to this model, the electric potential is maximum at the solid surface and decreases exponentially with distance being electrically neutral in bulk solution (Fig. 2b). The GC theory assumed the hydrated ions as point charges [23]. Results obtained from the GC theory shows some differences from experimental results. According to Stern [22], the modeling chemical species as dimensionless points was not valid at the immediate vicinities of the charged surfaces. He corrected this effect by taking into consideration the influence of ionic dimension. The layer of surface charge is defined in a similar way to that of previous theories. However, the layer in the liquid adjacent to the surface is divided as a compact layer close to the colloidal surface, known as Stern layer and with a thickness approximately equal to the radius of one hydrated ion and a diffuse layer. The theory called Gouy-Chapman-Stern (GCS) describes the potential distribution in the EDL as a combination of a linear (Helmholtz) and exponential (Gouy-Chapman) contributions (Fig. 2c) [23].

The GCS model still presents some limitations. Regarding the interaction of electrode/electrolyte, the assumption of the electrode as a perfect conductor neglects the oxide formation on the metal surface [24, 25] which has been proved to have an important influence on the potential distribution. Thus, researchers are currently dealing with new approaches to the EDL description. Paz-Garcia et al. proposed a model to reproduce the transient formation of an EDL between an electrolyte assuming chemical equilibrium and a chemically active flat surface (Fig. 3). A rate-controlled complexation model was used to describe the reactivity of the flat surface. The protonation/deprotonation reactions make uncharged actives sites, XH, form $XH_2^+$ or $X^-$, respectively. The thickness of the diffuse layer is characterized by the Debye length ($\kappa^{-1/2}$, m). Comparing a thin compact layer with the diffuse layer, the surface potential, $\phi_s$, is approximate by the potential at the shear plane (zeta potential, $\zeta$) in many cases. This model assumes that only surface species exist and the potential at the Helmholtz plane ($\phi_d$) is equal to the potential at the surface ($\phi_s$). With the aim of computing the EDL distribution in complex systems, authors proposed an unsaturated solution of $CaCO_3$ in equilibrium with atmospheric $CO_2$ (g).

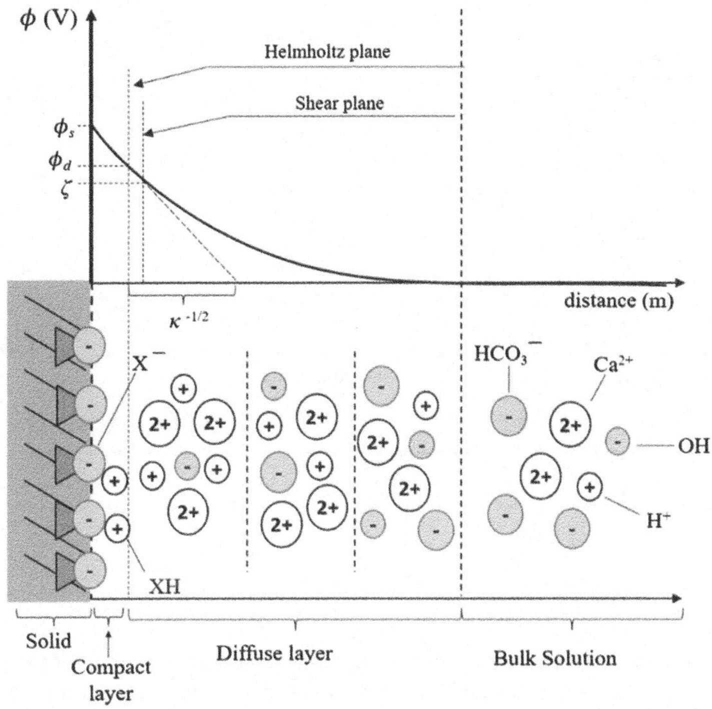


**Fig. 3** Modeled system of EDL considering chemical reaction effects [26]

Instantaneous local chemical equilibrium for the aqueous reaction was assumed while chemical kinetics for the heterogeneous surface reactions were considered. It is worth to point out that the proposed model has not limit to any range of surface potential and can be applied to different chemical system in presence of different electrolytes, surface interactions and geometries [26].

### 2.2.2 Theory of Helmholtz–Smoluchowski

The Helmholtz–Smoluchowski model assumed, as shown in Fig. 4, a liquid-filled capillary as an electrical condenser containing co-ions close to or on the surface and mobile ions of opposite sign, counterions, at a very small distance from surface [18].

The theory supposed that the layer thickness of charged fluid next to the pore wall, where the counterions are concentrated, is negligible comparing with the diameter of pores in the medium. This layer is known as aforementioned: electrical double layer (EDL). For surfaces negatively charged, cations in the liquid constitute the mobile phase. The charge on the soil in contact with an aqueous solution depends on several effects, such as, chemical and physical adsorption and lattice imperfections [27]. Under electrical field, the mobile cation phase is transported toward the

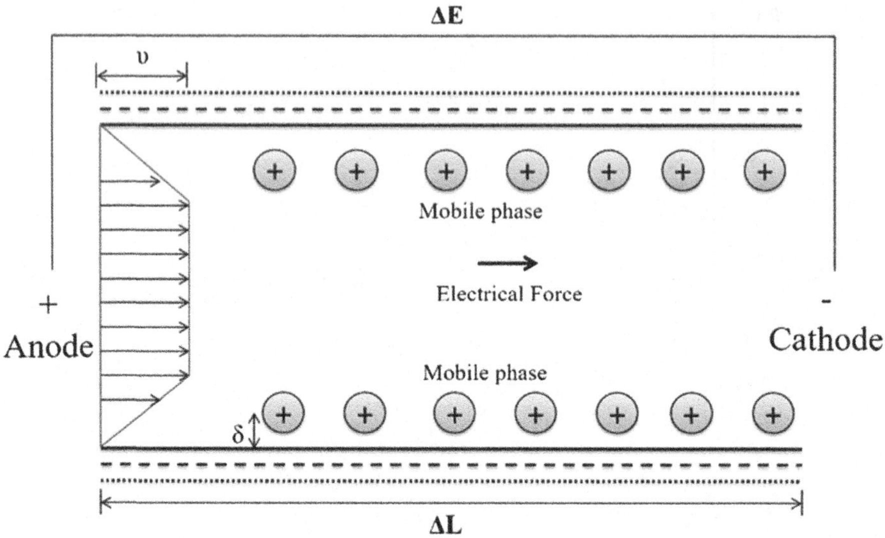


**Fig. 4** Model of Helmholtz–Smoluchowski [18]

cathode dragging the surrounding fluid molecules. This movement is controlled by the equilibrium of the electrical and frictional forces between liquid and solid phases. Assuming that the average velocity of the mobile phase is $\upsilon_{eo}$ (m $s^{-1}$), the distance between the mobile phase and the solid surface is $\partial$ (m) and the flow velocity at the surface is zero, the gradient is $\upsilon_{eo}/\partial$ ($s^{-1}$). At equilibrium, the frictional and electrical forces per unit area are equalled:

$$\frac{\eta \upsilon_{eo}}{\delta} = -\sigma \frac{''E}{''L} \tag{6}$$

where $\sigma$ (C $m^{-2}$) is the surface charge density, $\Delta E/\ \Delta L$ (V $m^{-1}$) is the electric field applied, $\eta$ (Ns $m^{-2}$) is the liquid viscosity and the negative sign represents that the electrical force on mobile phase are acting in the direction of decreasing electrical potential. Defining $\varepsilon_s$ as permittivity of the pore fluid (F $m^{-1}$), the difference between electrical potential (V) in the liquid $\phi_0$ and at the surface $\phi_s$ and could be written as:

$$\phi_0 - \phi_s = -\zeta = \frac{\sigma \delta}{\varepsilon_s} \tag{7}$$

Assuming the potential in the liquid as zero potential, the potential in the surface is zeta potential, $\zeta$ (V). Combining Eqs. (6) and (7):

$$\upsilon_{eo} = \frac{\varepsilon_s \zeta}{\eta} \cdot \frac{''E}{''L} = -k_e \cdot \frac{''E}{''L} \tag{8}$$

Generally, the value of the zeta potential for soils is negative if the mobile phase ions are cations and the net particle surface charge is negative, consequently, the electro-osmotic flow direction is from anode toward cathode. The potential zeta is dependent on many parameters, such as, solution pH, ionic concentration of the pore fluid, the electric field and temperature [27]. Thus, the sign and magnitude of $\zeta$ potential is highly influenced on pore fluid chemistry. This parameter plays an important role in the application of EDW. Regarding the dependence of potential zeta on the pH, Lorenz et al. evaluated in 1969 this relationship for sodium kaolinite [28]. The data obtained from this study was regressed by Eykholt and Daniel [29] which is presented in Fig. 5.

The concentration of $H^+$ and $OH^-$ has a significant effect on the soil zeta potential. Decreasing in pH values entails an increment of zeta potential, in other words, the zeta potential become less negative reaching positive values at low enough pH values. The pH value at which the net total particle charge of soil is zero is known as point of zero charge (PCZ), one of the most important parameters used to describe the variation of charges in surfaces [30]. If the pH of a soil is above its PZC, the soil surface will have a net negative charge and the direction of electro-osmotic flow would be from anode toward the cathode. The soil will mainly retain electrostatically anions if its pH is below its PZC changing the direction of electro-osmotic flow [1, 15].

According to Yeung, normal values of zeta potential are ranged from −50 to 50 mV, associating the highest values with high salt concentration in the liquid [18]. When values of zeta potential are between 0 and −50 mV, for most cases in soils, the charge of the mobile phase is positive, and the direction of electro-osmotic flow is the same as those of decreasing electrical potential. The concentration and type of salt in the pore fluid directly affects the zeta potential of porous media matrices such as clay minerals [31–34]. Yukselen et al. determined the main factors affecting the zeta potential of kaolinite using electrophoretic mobility as function of pH and concentration. The presence of salts, such as NaCl and LiCl, modified the values of zeta potential compared to water. The zeta potential of kaolinite in presence of NaCl and

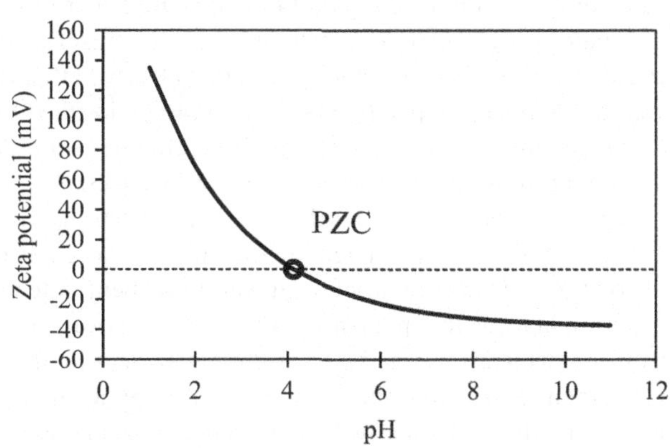


**Fig. 5** Empirical relationship between zeta potential and pH [28, 29]

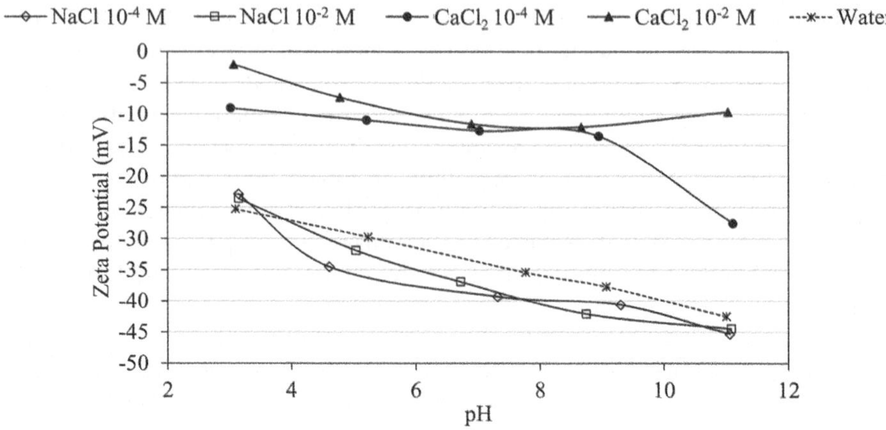


**Fig. 6** Zeta potential values of kaolinite in presence of dissolutions of NaCl and $CaCl_2$ at different pH values obtained by Yukselen et al. [31]

$CaCl_2$ at different concentrations experimentally obtained by Yukselen et al. are presented in Fig. 6.

Comparing values of zeta potential for monovalent cations and water, the more negative value obtained were related to the presence of monovalent cations. This fact was associated with the exchange of these cations with $H^+$ present in the system entailing the increase in zeta potential value. According to the Gouy-Chapman model, the higher cation valence, as for $CaCl_2$, the lower the thickness of diffuse electrical double layer entailing a more positive zeta potential. The same study also evaluated the zeta potential in presence of $CoCl_2$, $Pb(NO_3)_2$ and $CuCl_2$ solutions with different concentrations. The dependence of pH and zeta potential values was similar in presence of heavy metals: the higher concentration of ions, the more positive zeta potential values at acidic or neutral pH values. However, two apparent PCZs were observed in the evaluation of zeta potential values, as observed in Fig. 7 for $CoCl_2$. As mineral can only give one PZC, the precipitation of heavy metals as hydroxides at higher pH values explained the presence of the second PZC. As indicated in Fig. 7, the PZC of kaolinite appeared near to neutral pH value [31].

Hunters et al. also demonstrated that the presence of hydrolyzed metal cations could cause reversals for kaolinite. They observed that the higher the metal ion concentration, the more positive becomes potential zeta at low pH. This fact was related to the accumulation of a large positive charge in the electrical double layer compared with those present on the solid surface [33].

The soil properties also influence on the relationship between pH and zeta potential values. Vane et al. carried out electrophoresis experiments to evaluate the effect of soil type on zeta potential using kaolinite, bentonite and silty-clay soil [34]. From experimental results, presented in Fig. 8, the stronger dependence of zeta potential value on pH was observed for kaolinite. Changes in pH did not alter significantly the zeta potential of bentonite, which had a more negative value than that of kaolinite. Regarding the silty-clay soil under study, the zeta potential value was more sensitive to pH changes at lower pH values than the other materials.

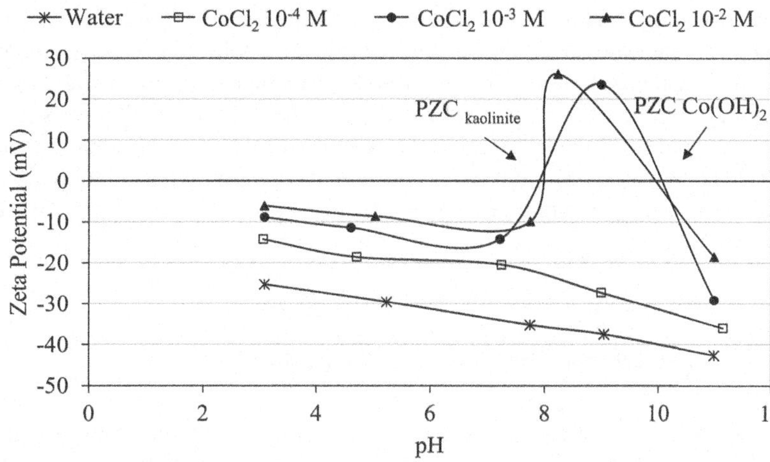


**Fig. 7** Zeta potential values of kaolinite in presence of solutions of $CoCl_2$ of different concentrations obtained by Yukselen et al. [31]

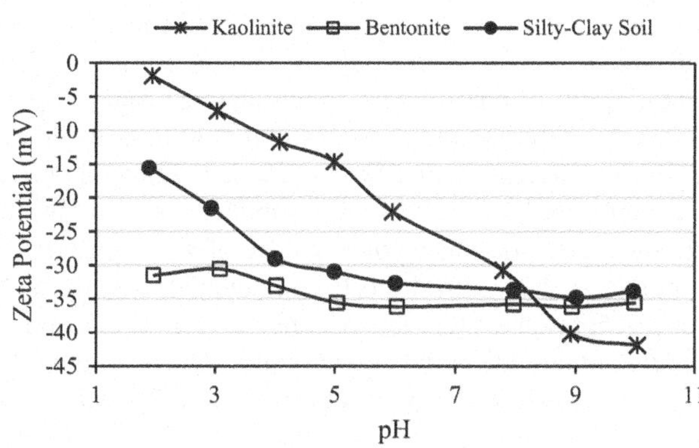


**Fig. 8** Zeta potential values at different pH values for kaolinite, bentonite and silty-clay soil obtained by Vane and Zang [34]

Regarding methods to determine the point zero charge in soils, the potentiometric titration has been widely used. This method estimates changes in surface potential from changes in the activities of $H^+$ and $OH^-$ [35–37]. Other methods, such as non-specific ion adsorption, based on electrostatic adsorption of a cation and anion caused by changes in the activities of $H^+$ and $OH^-$, has been also applied to soils.

In addition to Helmholtz–Smoluchowski theory, others have been developed with the same aim of describing electro-osmosis. The most known are: Schmid theory, Spiegler friction model and ion hydration approach. Schmid theory considered the extension of the counterion layer into the pore as opposed to Helmholtz–Smoluchowski theory, which is considered a large-pore theory. The presence of excess of ions over the balance charge was also considered in this theory. According to the mathematical developed of this theory, the $k_e$ depend on the squared pore size whereas according to Helmholtz–Smoluchowski theory (Eq. 8) the coefficient was not dependent on pore size. When both theories are applied to soils, better results are obtained by the Helmholtz–Smoluchowski. This better approach could be associated with presence of particles characterized by large pores in clay soils [2]. The model developed for Spiegler suggested a new description of the interaction among the mobile phases and with pore walls and, finally, the ion hydration model considered the use of transport numbers to describe the transport of water in a direct current electric field.

## *2.3 Electro-osmosis Setup in Field Applications*

The electro-osmosis system for field applications (Fig. 9) is based on the application of a DC electric field between a set of anode and cathode electrodes inserted into the porous matrix. Consequently, the liquid present in the porous is transported toward one set of electrodes, typically, as aforementioned, the cathodes. With the aim of improving electro-osmosis effects in the porous matrix, some purge solution can be used into the soil during the treatment. The addition of chemical solutions pursues washing, treating and also preventing cracks formation in soil [38, 39]. The effluent is then removed by means of pumping or siphoning to a proper disposal. The use of electro-osmotic for the stabilization of porous matrix entails also the installation of drainage in the cathode together with a sealed anode to obtain a porous matrix with a lower water content, higher shear strength and lower compressibility [40]. To prevent possible short circuiting as a consequence of water presence in the ground surface, the upper part of the electrodes is usually insulated with a dielectric coating [41]. The power distribution system consists usually of several units connected in series or in parallel to obtain the target power output. The operation of power system could be under constant voltage or current in continuous or pulsed mode. Changes

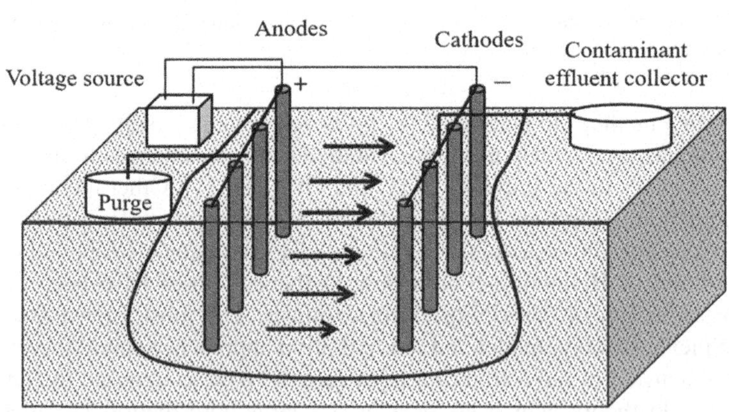


**Fig. 9** Schematic of a field system of electro-osmosis remediation [38]

in physical and chemical properties of the soil along the treatment time entail variation of soil resistance, the major electrical resistance of the system [16].

The electrode material is a crucial aspect in the application of electro-osmotic technology with high influence on their efficiency, corrosion and lifespan [42]. Extensive research has been carried out dealing with the study of the used electrode materials. In 1983, Lockhart et al. evaluated the comparison of the most used electrodes materials: mild steel, copper, graphite and aluminum. Experiments carried out for Cu kaolinite showed a better performance of copper electrodes than mild steel and carbon electrodes [43]. However, these differences were not observed in previous works using Na kaolinite [44]. The best performance of copper electrode for Cu kaolinite was associated with the reversible electrochemical half reaction Cu/$Cu^{2+}$–kaolinite–$Cu^{2+}$/Cu together with the cathode reaction $Cu^{2+} + 2e^- \rightarrow Cu$. The most appropriate option for field application should be evaluated taking into account not only the treatment efficiency but also the cost of electrodes. Wu et al. studied the electro-osmosis treatment of a sodium bentonite with electrodes of different materials: copper, iron, graphite and stainless steel. The use of reactive electrodes (copper and iron) resulted in a better improvement effect in the porous matrix than the inert electrodes (graphite and stainless) [45]. However, the metallic electrodes tend to be corroded by the acidic conditions generated at the anode because of water electrolysis. It is an important limitation for the field application of the EDW since the electrode corrosion entails poor electrical contact between electrodes and the porous media, difficulties in removing gasses and high cost of electrodes [46]. With the aim of mitigating corrosion problems, novel electrodes have been proposed. Jones et al. introduced the concept of electrically conducting geosynthetics materials (EKG) consisting of conducting elements coated by a corrosion-resistant material incorporated into a geosynthetic material. These materials do not only provide filtration, drainage and reinforcement as conventional geosynthetics but also can be improved by electrokinetic techniques to obtain an effective transport of water and chemical species within fine-grained low permeability soils. In 1996, Jones et al. evaluated the use of electrodes made of conductive geotextiles concluding that the behavior of this materials was similar to those observed for a conventional copper electrode [47]. Although these materials avoid the problems associated with electrode degradation and enhanced electro-osmotic effects, further studies should be carried out [48]. The use of metal or carbon inserted into prefabricated vertical drains (PVD) [49] and electrical vertical drains (EVD) [14] has been also reported as an alternative to conventional metal electrodes.

In addition to material, other operation parameters related to the electrode as configuration and spacing between electrodes plays an important role in the application of the technology [42]. According to Casagrande, the optimal spacing between electrode sets of the same polarity should be much less than spacing of the opposite polarity which typically ranged from 1 to 3 m [6].

### 2.3.1 Experimental Operation Conditions

Geochemical processes with high influence on the solid/liquid interface are of great relevance for the feasibility of EDW technology. The generation of pH gradient modifies the physical and chemical properties of soil. Thus, understanding pH-buffering processes is of maximum importance to control, among other aspects, the direction of electro-osmotic flow, the processes of sorption and desorption of contaminants onto the soil surface, the mechanisms of complexes formation and the oxidation-reduction reactions.

The change in direction of electro-osmotic flow could be controlled by the addition of enhancements agents. Yeung et al. studied the influence of chemistry in electrolyte solutions on the direction of electro-osmotic flow. An experiment was carried out on a Milwhite kaolinite specimen adding different agents into the anode or cathode compartments in eight stages. Details of the experimental conditions and results are presented in Fig. 10. During the first step, $NaHCO_3$ solution (pH = 9) was added into the anode compartment with the aim of producing an electro-osmotic flow from the anode toward the cathode (positive electro-osmotic volume flow rate). In the second stage, the $NaHCO_3$ solution was replaced by deionized water entailing reverse electro-osmotic flow (negative electro-osmotic volume flow rate) after more than 73 h. This result was associated with the required time to the transport of hydrogen ions produced by electrolysis toward the vicinity of anode. In stage 4, after 43 h of reverse electro-osmotic flow (stage 3), $NaHCO_3$ solution was introduced again into the anode compartment to change the direction of electro-osmotic flow from anode to cathode without success. These results indicate that the addition

| Reservoir solution/Stage | 1 | 2 | 3 | 4 | 5 | 6 | 7 | 8 |
|---|---|---|---|---|---|---|---|---|
| Anode | $NaHCO_3$ | $H_2O$ | $H_2O$ | $NaHCO_3$ | $NaHCO_3$ | $NaHCO_3$ | $CH_3COOH$ | $NaHCO_3$ |
| Catode | $H_2O$ | $H_2O$ | $H_2O$ | $H_2O$ | $NaHCO_3$ | $NaHCO_3$ | $NaHCO_3$ | $NaHCO_3$ |

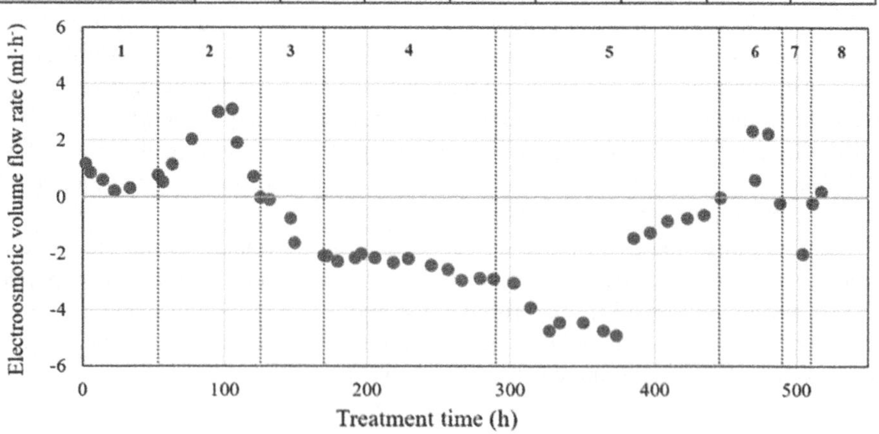


**Fig. 10** Study of the electro-osmotic flow direction through Milwhite kaolinite under the influence of several chemistry of reservoir fluids carried out by Yeung [50]

of a high pH solution to the anode is not a good procedure to change the electro-osmotic flow from reverse to forward. After the addition of $NaHCO_3$ into the cathode compartment in stage 5, the flow direction was change to the forward direction for 42 h (stage 6). With the aim of evaluating the impact of soil pH on flow direction, the anode solution was replaced by acetic acid in stage 7. As can be observed, the flow direction changes immediately. During the last stage, a $NaHCO_3$ solution was added to the anode compartment causing the change of flow direction to forward. It should be noted that during stage 8 the soil pH was higher than for stage 4 in which the $NaHCO_3$ solution was not able to change the reverse electro-osmotic flow to forward. These results corroborate the importance of soil pH to control the direction of electro-osmotic flow. Authors concluded that the change of electro-osmotic flow direction from reverse to forward takes higher time due to the difficulty of the high pH solution transport from the cathode toward the anode. In practice, it has been observed that the direction of electro-osmotic flow is mainly determined by the pH soil close to the anode compartment [50].

The ionic migration of electrolysis products entails important changes not only in the electro-osmosis flow direction but also in the chemistry of the solid matrix [16]. The precipitation of contaminants represents a problem for the recovery by electro-osmosis. The products generated at the anode migrate toward the cathode and the hydroxides ions produced at the cathode migrate toward the anode. The transport of these ions through the soil affects the contaminant migration and the removal during the remediation treatment. Thus, the reagent addition is the most typical approach to control the pH and, subsequently, to avoid the entrance of the basic front generated in the cathode compartment into the soil. From electrokinetic remediation studies, Reddy et al. in 2004 found that the recovery of heavy metals from a glacial till soil contaminated with Cr, Ni and Cd, was not achieved due to their precipitation under high pH conditions. They proposed the addition of different reagents to the electrolyte, such as EDTA, citric acid and acetic acid, to avoid the precipitation and to promote the formation of soluble metal complexes. Results showed that an adequate selection of the enhancements agents results beneficial and, even, necessary for soils with high buffering capacity [51]. Interesting studies using different acids solutions to enhance the technique has been reported [52–55]. Villen et al. compared the use of strong and weak acids for a real contaminated soil observing larger removal yield when a weak acid, as acetic acid, was added to the cathode. These results were associated with differences in the fraction of the electric current carried out by ions traveling from the anode to the cathode [56]. Surfactants and complexing agents have been also widely studied to enhance the mobility of contaminants. Surfactants are usually used to deal with organic contaminants [57, 58]. For soils contaminated with heavy metals, the use of chelating agents containing ligands to form complexes has been also reported [59–63]. Chelating agents have been proved to be excellent solubilizing agents for many metals such as Pb [64, 65].

The EDW to improve the mechanical properties of soils is also enhanced by the injection of chemical solutions. Some works demonstrated that the use of calcium chloride solution before the injection of sodium silicate solution during the

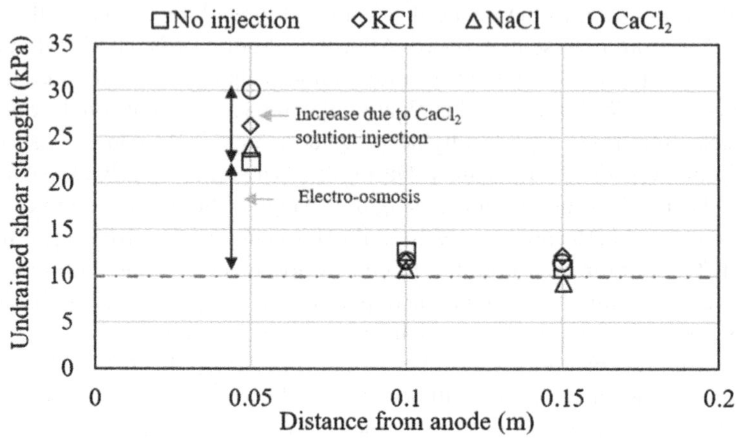


**Fig. 11** Influence of saline solution injection in soil stabilization experiments carried out by Ou et al. [67]

treatment can entail the soil cementation close to the electrode compartments [66]. Ou et al. studied the effects of electro-osmosis with injection of saline solutions on a silty clay. The saline solutions, consisting of solutions of KCl, NaCl and $CaCl_2$ with different concentration, were injected in the anode compartment. The treatment time was a 40% lower for the experiment with the use of the $CaCl_2$ solution comparing with experiments without solution injection. The average undrained shear strength for experiments with the injection of the solutions of higher valence, such as $CaCl_2$, was also improved in five times with respect to the untreated soil (Fig. 11). The improvement of the soil properties depended also on the distance from anode [67, 68].

With the aim of verifying the suitable operation process at larger scales, Chien et al. carried out tests at field scale applying electro-osmotic chemical treatments. The injection of calcium chloride solution followed by the injection of sodium silicate solution allowed the current decreased and the decrease of the drainage from the cathode. From results, it was confirmed the applicability of the electro-osmosis technique enhanced with chemical treatments for further field tests [69].

## 2.4 Applications

This section reviews real cases of EDW application pursuing different goals. Successful application reported soil stabilization, recovery of contaminant soils, dewatering of numerous matrixes such as soil, sludge and mine tailing.

### 2.4.1 Consolidation of Fine-Grained Matrices

The consolidation of saturated soils applying electro-osmosis is reached because of lower soil water content entails higher shear strength and lower compressibility. The changes in physical and chemical properties of the soil have also a direct influence on the soil stabilization. After first applications of the electro-osmosis by Casagrande, many successful trials applied to soils requiring an effective and rapid improvement have been reported over the years. This section reviews some of the most interesting field application whose main characteristics are summarized in Table 2.

The stabilization of an excavation in Norway carried out by Bjerrum et al. in the 1960s is one of the first reported successful cases. The electro-osmosis consolidation of the excavation was the most appropriate technique due to security reasons. The treatment was applied during 120 days with a total energy consumption of 30,000 kWh. The electrode installation consisted of reinforcing steel bars of 19 mm diameter and 10 m long covering a total area of 200 $m^2$. The electrodes suffered an important process of degradation with a 37% of steel anode corroded. The soil after treatment presented a higher plasticity due to changes in their mineralogy. Also, the addition of iron ions from the anode electrode entailed the formation of iron-saturated clay with higher plasticity [8].

The slope stabilization in southern British Columbia, Canada, is other historical example of electro-osmosis application. The treatment lasted 9 months during 1972–1973 with a volume of soil treated of 400,000 $m^3$. For safety reasons, an excavation of the slope was required to increase the soil stability. Before the sloped were excavated, it was decided to apply electro-osmosis technique to increase the shear strength and to reduce the water content in soil pore. A sequential application of the electro-osmosis treatment was carried out dividing the slope in several zones, as schematically presented in Fig. 12.

Once the higher part of the slope was treated, the zone was excavated until the next level with the aim of installing the electrodes and pumping wells. This method

**Table 2** Summary of real cases of electrochemically assisted dewatering to consolidate soils.

| Case | Reference | Treatment time | Treated zone | Power consumption |
|---|---|---|---|---|
| Stabilization of an excavation in As, Norway | Bjerrum et al. [8] | 120 days | 1765 $m^3$ | 17 kWh $m^{-3}$ of soil |
| Stabilization of a slope in British Columbia, Canada | Wade [11] | 9 months | 400,000 $m^3$ | 6.7 kWh $m^{-3}$ of soil |
| Stabilization of an embankment in Singapore | Chappell and Burton [10] | 9 days | 2400 $m^3$ | 0.5 kWh $m^{-3}$ of soil |
| Strengthening of soft sensitive clay | Lo et al. [12, 41] | 32 days | 1205 $m^2$ | 2136 kWh |
| Stabilization of a marine clay in Singapore | Chew et al. [70] | 13 days | 2500 $m^2$ | 1.8 kWh $m^{-3}$ of soil |
| Stabilization of a clay silt in Malaysia | Rittirong et al. [71] | 14 days | 2240 $m^2$ | 0.7 kWh $m^{-3}$ of soil |

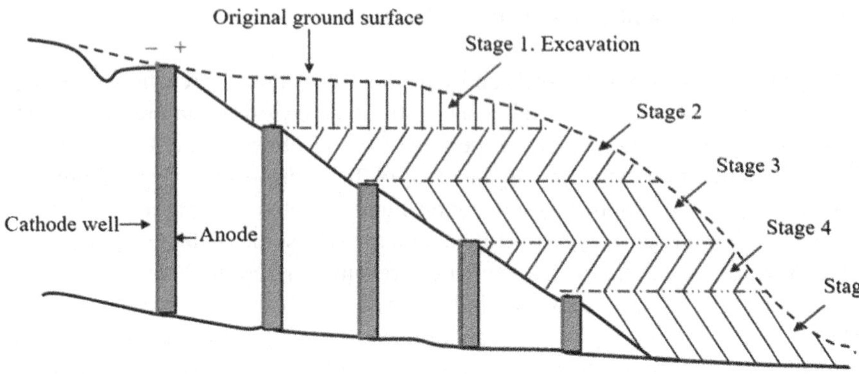


**Fig. 12** Sequential excavation for the slope stabilization carried out by Wade in Canada (1976) [11, 40]

was applied to the bottom of the slope. During the treatment, no slope movements were registered. Results put forward the success of electro-osmosis technology to consolidate the soil [11].

Chappel and Burton also applied electro-osmosis technology to the consolidation of an unstable embankment in Singapore in 1975. The soil causing the instability was mainly silt slurry, characterized by low permeability. Before deciding to apply electro-osmosis treatment, other attempts of stabilizing the embankment had not success. The effectiveness of the technique was proved with a reduction of the movement of the embankment from 1 m day$^{-1}$ to 1 cm day$^{-1}$ after only 9 days of treatment. The energy consumption was 0.7 kWh m$^{-3}$ of soil, significantly lower than other historical cases (Table 2). This fact was associated with the higher electrical conductivity of the soil in this zone as a consequence of higher contents of salts [10].

The historical case of strengthening of soft sensitive clay carried out in Ottawa Valley, Canada, in 1989, is characterized by the application of pioneering operational conditions. The perforated copper electrodes were designed to allow the transport of water into the cathode and out toward the surface by electro-osmosis without pumping. The electrode designed for the field test (Fig. 13) consisted of a cone-shaped steel shoe to ease pushing the electrode into the ground. With the aim of avoiding short circuiting of the system, the electrode part in contact with high conductivity layers, such as crush, fine sand and silt, was insulated with several layers of varnish. The purpose of the weep holes below the top of electrode was allowing water flow out of the pipe.

During the application of electro-osmosis, the polarity was changed after 17 days of treatment. Then, the treatment continued for a further 15 days. This polarity reversed was proved to achieve some improvement in the soil, such as an increment of shear strength, comparing with other real cases. The total energy consumption was low, about 1% of the total cost of the treatment. Tests carried out after treatment verified that the improvements observed in soil were permanent [12].

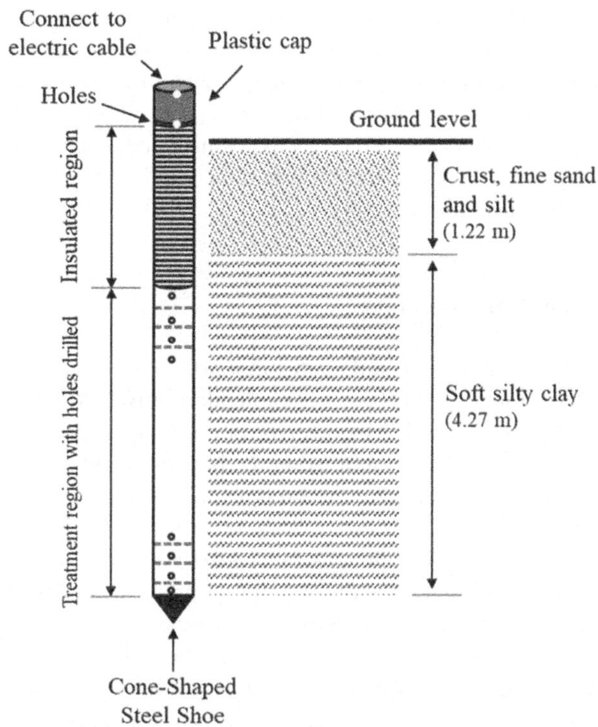


**Fig. 13** Copper electrode designed for the field test of electro-osmotic strengthening of soft sensitive clay [12, 40]

In a more recent studied, Chew et al. [70] carried out tests to probe the efficiency of electro-osmosis applied to a marine clay in Singapore at laboratory and field scales. As metal electrodes are expensive when applying the technique to marine clay due to anode corrosion, a patented electrically conducting polymer drain was used in this research. They used EVDs and PVDs as electrode and vertical drains, respectively. The polarity reversal was also applied to this study case. The total energy consumption was ranged from 14 to 28.8 kWh $m^{-3}$ of soil for laboratory scale application while the electrical used was of 1.8 kWh $m^{-3}$ for field scale. The field trial was carried out in an area of 2500 $m^2$ for 314.8 h. The polarity reversal started after the first 25 h and it was applied at time intervals of 7–10 h. From results, it was concluded that the application of electro-osmosis was effective using EVD even when the soft clay to treat was under 18 m of sand fill. Conditions required for a success application were: high conductivity of the EVD installed, good insulation of the metallic shoe and reversed polarity. Studies of pore pressure response suggested the improvement of shear strength in both scales. The used of electro-osmosis was proved to be about ten times faster than conventional prefabricated vertical drains (PVD) only using hydraulic gradient by external loading [70].

Rittirong et al. [71] applied electro-osmosis technology using EVDs to two natural soils at laboratory and field scale. This work aimed at a better knowledge about the improvement in soils using EVDs. They found that even when the electro-osmosis flow stopped, the soil strength increased. Therefore, the improvement of soil was not only associated with the electro-osmotic consolidation but also with electrochemical effects. From results, it was concluded that the use of EVDs was highly efficient to soil consolidation [71].

### 2.4.2 Removal of Contaminants

The recovery of contaminated soil using electro-osmosis technology has been widely studied over the years. This is a feasible alternative for heterogeneous saturated porous media such as soils characterized by low hydraulic permeability. The EDW is focus on contaminants contained in the porous liquid, such as organic species. For ionic species, the electromigration transport can contribute to the recovery of contaminants.

The removal of organic contaminants by electro-osmosis requires their solubilization in the pore fluid. Therefore, the adsorptive properties of the porous matrix, such as distribution coefficient, solubility and octanol/water partition coefficient, play an important role. The organic pollutant can be classified according to their solubility in water. Thus, it could be distinguished insoluble organics, such as heavy hydrocarbons and soluble organics, such as benzene, toluene, xylene, chlorinate solvents and phenolic compounds [27]. Since the hydrophobic organic compounds remain adsorbed onto the organic matter in the soil and the soil particles, the use of solubilizing agents is required in this case. To promote the dissolution/solubilization of organic compounds, it has been widely used surfactants, bio-surfactants, co-solvents and cyclodextrins [72].

This section will report some practical cases of soil decontamination by electro-osmosis technology paying special attention to the technique evolution. With this aim, earliest experimental studies using spiked model soils were designed to a better understanding of the fundamentals of the technique. However, as it is widely accepted, researches with real soils are also needed for the prediction of the electro-osmosis performance at larger scales.

Clifford et al. studied the application of electro-osmosis technology with the aim of removing hydrocarbons from a fine-grained soil in 1992. They evaluated the mobilization of dissolved benzene, toluene, *m*-xylene, hexane, isooctane and trichloroethlylene (TCE) from a spiked clay soil. Experiments were carried out in soil columns under an electrical gradient of 0.4 V $cm^{-1}$ of soil length at laboratory scale. The higher degree of mobility was obtained for the organic compounds with higher water solubility, as can be observed in Table 3. The estimation of distribution coefficient value was found to be a potential tool for studying the mobility of organic contaminants by electro-osmosis. From experimental results obtained in laboratory test, field predictions suggested the electro-osmosis technology as a viable tool to remediate fine-grained soils containing gasoline hydrocarbons and chlorinated

**Table 3** Experimental results obtained by Clifford et al. [73]

| Pollutant | Treatment time (day) | Contaminant removal (%) | Aqueous solubility at 20 °C (mg $L^{-1}$) |
|---|---|---|---|
| Benzene | 5 | 27 | 1780 |
| TCE | 5 | 25 | 1100 |
| *m*-Xylene | 5 | 19 | 146 |
| Isooctane | 25 | 7 | 2.4 |

solvents. However, further studies at bench-scale were needed to a better understanding of the feasibility of the technique at larger scales [73].

In the same year, 1992, Acar et al. evaluated the recovery of phenol from a kaolinite applying an external electric field. First, adsorption tests were carried out to evaluate the adsorption kinetic of phenol onto the kaolinite. Results suggested that the organic compound was adsorbed onto the porous matrix in a short period. The generation of an acid front at the anode as result of water electrolysis was found to be an improvement for the contaminant desorption. This work evaluated energetic aspects finding that the energy consumed for removal the phenol was from 18 to 39 kWh $m^{-3}$ of soil. Authors also highlighted the need of pilot-scales studies to develop the process at real scale [74].

Shapiro et al. carried out other relevant work about the feasibility of using electro-osmosis to recover soils contaminated with organic species in 1993. Experimental results were compared with those obtained from a theoretical model. The soil sample treated at two different scales was a kaolinite spiked with a known concentration of phenol or acetic. The tests at small scale were designed to allow the identification of potential limitations obtained at large scale. An efficient decontamination rate was obtained after applying electro-osmosis. These results were associated with the capacity of controlling the flow direction and uniformity when electro-osmosis technology was applied to a fine-grained soil. The important dependence of the electro-osmotic flow on the soil pH values was also evaluated. After analyzing results, it was clear that even for simplified cases, as the experiments carried out with spiked soils, the correlation between transport mechanisms should be considered to a better understanding [38].

Ribeiro et al. studied the atrazine behavior in soils when an electric field was applied. A one-dimensional mathematical model was developed not only to describe the behavior of the contaminant-soil system, but also to predict the generalized removal of other uncharged pollutants from soils. This model considered the reversed electro-osmosis, from cathode toward the anode, due to the low pH of the soil. From results, the mobilization of atrazine was mainly associated with the reversed electro-osmosis for natural and spiked soils studied [75]. Other mathematical model for the removal of organic compounds, molinate and bentazone, in soils was developed in 2011. Experimental results were obtained from the application of electrokinetic technique to three real soils from Portugal. From comparison of experimental and theoretical results, the high dependence of electro-osmotic flow on the soil pH was proved. From varying the intensity conditions, the importance of electromigration versus electro-osmostic transports was compared. The application of higher current intensity entailed the mobilization of bentazone toward the anode because of the electromigration transport of enolate, counteracting the expected electro-osmotic transport of bentazone toward the cathode. For lower intensity conditions, the opposite movements took place. This fact was associated with the higher dependence of electromigration flow on the current intensity comparing with electro-osmotic transport [76].

Cameselle and Reddy carried out the evaluation of the influence of some important variables on electro-osmotic flow, such as soil composition. The soils used in this study were kaolin and glacial till. As a low permeability soil with a negligible organic matter content and a low buffering capacity and cation exchange capacity (CEC), kaolin was selected. Glacial was used as a low permeability soil with low organic content, high pH and a cation exchange capacity higher than kaolin. The contaminants added artificially in the laboratory were heavy metals (Mn, Ni, Cr, Hg and Cd) and a hydrophobic organic pollutant, phenanthrene. With the aim of promoting the contaminant dissolution, different chemical solutions were added. To evaluate the influence of the chemical reagents, a wide variety of solutions were tested: acid and alkaline agents ($H_2SO_4$, NaOH), complexing agents (citric acid, EDTA, oxalic acid, potassium iodide), surfactants (Tween 80, Igepal CA-720, Witco 207 and Witconol 2722), co-solvent (ethanol and butylamine) and cyclodextrins (hydroxypropy-β-cyclodextrin). The experimental setup used in this research is schematically presented in Fig. 14.

Although the removal of heavy metals was mainly associated with electromigration of charged species, electro-osmosis was proved to have a direct influence on the behavior of these contaminants. Depending on the transport direction of contaminants, electro-osmosis enhanced or impeded the removal. It is worth mentioning that the soil composition played an important role in the heavy metal recovery. For glacial till containing heavy metals, the pH changes associated with high content of

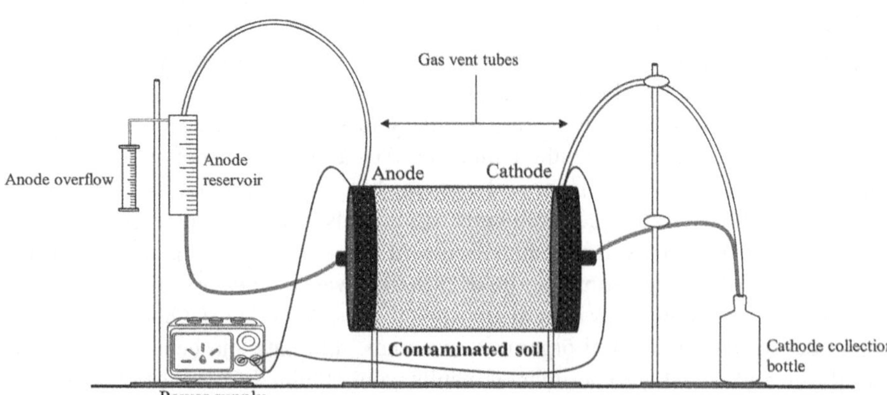


**Fig. 14** Experimental setup used for the study of the enhancement of electro-osmotic flow for removal of pollutants from soils [72]

carbonates and organic matter during the treatment entailed important changes in the direction and magnitude of electro-osmosis flow. In addition, the use of enhanced agents, such as citric acid, influences over the value of zeta potential resulting in a higher electro-osmotic flow. Regarding soil polluted with phenanthrene, the capacity of the enhanced agent (surfactants and co-solvents) to solubilize the contaminants was a key point to the soil recovery. Although the high acid buffering capacity of glacial till promoted a higher electro-osmotic flow, the phenanthrene was more difficult to remove from this soil than from kaolin. These results were related to the strong association of phenanthrene with the organic matter in glacial till. Authors concluded that the electro-osmotic flow was highly affected by experimental conditions, such as pH changes in soil, ionic strength, electric field intensity and soil type. They recommended the development of theoretical models for electro-osmosis together with laboratory studies as an approach to optimize experimental conditions for field scale [72].

The use of a wide variety of soil types in the application of electrokinetic remediation studies has proved that the mineralogical composition of soil and the presence of natural substances play an important role in the application of the technique at field scale [62, 77, 78]. So, although early studies on electrokinetic remediation performed at lab scale were aimed at a better understanding of the fundamentals of the processes involved, the extrapolation of results obtained in this studies to real soils is not possible to predict the feasibility technique at larger scale [79]. López-Vizcaíno et al. (2014) studied the electrokinetic treatment of a real soil contaminated with phenanthrene at a pilot plant [58]. As aforementioned, the high persistence of phenanthrene in soils entails the necessity of using special solutions of surfactants and co-solvents. Authors based on previous washing assessment studies selected the sodium dodecyl sulfate as the flushing fluid [80]. Although the removal mechanism of phenanthrene desorption as a consequence of electric heating was insignificant in laboratory experiments, this phenomenon became relevant when experiments were performed at pilot scale. Thus, the main mechanism involved in the removal of phenanthrene was not only the electro-osmosis and electrophoresis but also the desorption produced by electric heating of the soil. The experimental setup at pilot scale was open to atmosphere, as shown in Fig. 15, which entailed an evaporation fluxes during the treatment. Contrary to results obtained from laboratory scale, the gravity and evaporation fluxes were demonstrated more important even than electrokinetic fluxes at pilot scale.

The main advantage of using EDW for the environmental prevention of soil pollution is its suitability for in situ treatment of polluted soils with low hydraulic conductivity, which cannot be treated by other conventional techniques. Other important advantages of EDW are the treatment of inaccessible areas, the control of the water flow and contaminants and the treatment of soil containing multiple contaminants. Regarding limitations of the technique, the most important are the problems due to electrode corrosion together with the possible soil heating. In this point, it is worth noting that new studies in EDW are dealing with these limitations.

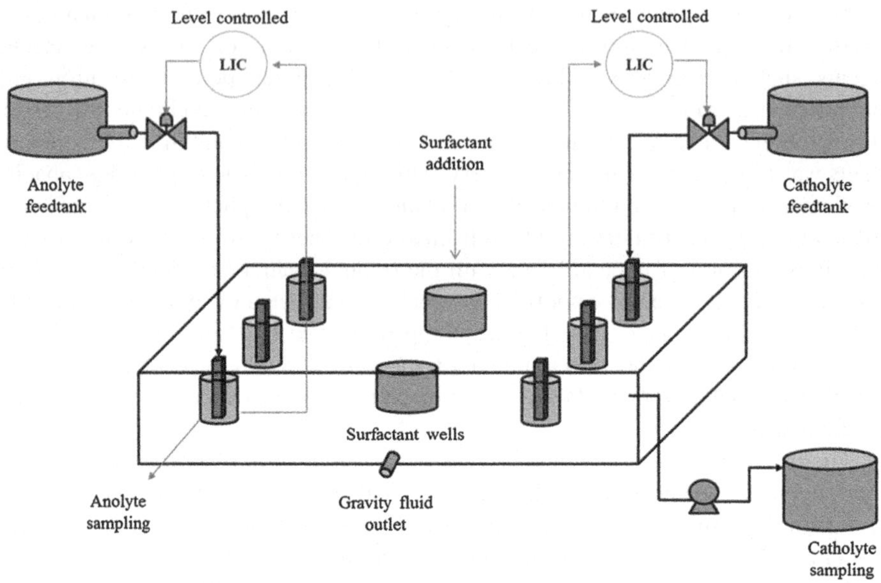


**Fig. 15** Electrokinetic remediation pilot plant for a soil polluted with phenantrene [58]

### 2.4.3 Other Matrices

The EDW has been also successfully applied to other porous matrices different from soils, such as sludge sewage, mine tailing, oily sludge and marine clay. In general, EDW has been proved suitable for materials that are not treatable by conventional methods for the water reduction. The dewatering of wastewater sludge is considered the major difficulty in the sludge disposal [81]. The problems associated with this process are related to the very fine particles of this matrix. Therefore, improvements in the conventional mechanical dewatering processes for sludge are needed [82]. To a better understanding of the sludge dewatering applying an electric field, Vesiling et al. differentiate four categories of water in sludge, from more to less mobilizable: free water not linked to particles, interstitial water trapped within the flocs, surface water adsorbed or adhered on the surface solid particle and intercellular and chemically bound water [83]. The quantities magnitude of each different water fractions results essential to the performance of the dewatering systems [84]. The water associated with free and interstitial phases is easily removed using conventional processes. However, the water adsorbed on the surface solid particle, consisted of layered and ordered water molecules, is not removed mechanically [85]. Thus, electro-dewatering is a useful tool when sludge is not sufficiently dewatered using mechanical techniques. The most used electro-dewatering technique is known as pressurized electro-dewatering [86].

Concerning the electro-osmosis phenomenon, the understanding of processes taking place in EDL is essential, as described in detail in Sect. 2.2. The sludge particles are characterized by large surface area negatively charged, which facilitates the water adsorption. The water contained in the EDL is associated with interstitial water and surface water adsorbed [87]. Successful experiences for electrical treatment of sludge has been reported. Smollen and Kafaar [87] carried out experiments of sludge dewatering assisted with electro-osmosis in a pilot plant. Based on previous laboratory studies, they designed a pilot plant characterized by incorporating electro-osmosis technology into filter-best press. With this study, they pretended not only to develop a further complete pilot plant but also to contribute with relevant results about difficult problems typically found in sludge dewatering. The experimental setup was divided into: a feed box, horizontal belt for gravity dewater and electro-osmotic section. The electro-osmostic section consisted of two belts made of an electrically conductive material for squeezing the sludge and connected to the anode and cathode electrodes. Thus, the water contained in the sludge would be transported toward the cathode while the negatively charged species in the sludge would move toward the anode electrode. The experimental tests were carried out in two stages passing the sludge two times through the system; the dewatering in the first stage was carried out by gravity and belt pressing and in the second stage applying electro-osmotic technology. With the aim of comparing the current efficiency and energy requirements, experiments were run at different voltage. It should be noted that studies were carried out applying the direct current maintaining constant the voltage or the electric current. Results demonstrated that decreasing in current intensity was observed when experiments were carried out at constant voltage. These results are related to the increase of electric resistance because of lower water content. The addition of polyelectrolyte did not influence significantly on the electro-osmotic processes. The use of electro-osmosis technology for sludge dewatering resulted in an important improvement, specially, for gelatinous and fine particles sludge. The use of electro-dewatering demonstrated important advantages in a wide recognized drawback of processes filtration processes: the blinding of the filter media [87].

The application of a low electric field to the wastewater sludge is not only used to improve the liquid-solid separation but also to remove contaminants. This remediation is desired for reusing of waste sludge as composting, fertilizer or even incineration and disposal to landfill [88]. Hwang et al. demonstrated that the combination of mechanical and electro-osmosis processes improved the dewatering of sludge and reduced the pollutants. They associated the reduction concentration of heavy metals with the accelerated removal of water because of electro-osmotic flow. In addition, the cake heating value, a relevant parameter for incineration disposal, was proved higher for sludge treated with electro-osmotic technique. Thus, the required fuel for incineration would be reduced from two to three times. Authors also highlighted that the application of a low electric field facilitates the control of proper water content of sludge, depending on the reuse proposal, by controlling the current density [89].

The development of efficient dewatering processes also results essential to treat mine wastes with high contents of water, which entails high risks of instability. The most used approach for in-situ dewatering consists of the installation of prefabricated vertical drains. The US Bureau of Mines was pioneer in the application of electro-dewatering of mine tailing in the 1960s. Shang et al. studied the electro-dewatering of a phosphate clay in an experimental program. The experimental setup designed for the electrokinetic experiments (Fig. 16) consisting of horizontal electrodes allowed controlling the drainage at the bottom of the cell.

The sedimentation velocity was directly associated with the current density. In addition, the polarity reversal and the application of intermittent current improved the dewatering process due to not only the reduction of power consumption but also the neutralization of adverse effects of electrode reactions [42]. The energy consumption for the dewatering application to mine tailing reported in [90] varies from 0.6 to 880 kWh $dry\_tonne^{-1}$. Fourie et al. proposed the use of electrokinetic geosynthetics (EKG) not only to optimize costs but also to avoid problems associated with the anode corrosion, as reported in Sect. 2.3. The application of a low voltage gradient together with the optimization of other operational conditions allowed obtaining an energy consumption rate of 0.9 kWh $dry\_tonne^{-1}$. Regarding the use of EKGs electrodes, no corrosion was detected after 2 months of use in an outdoor experiment. The application of this technique to different tailing materials should be previously evaluated due to the important role of the mine tailing properties on the electro-osmosis processes [90].

Semi-continuous and continuous electrokinetic dewatering processes have been designed with the aim of being applied to the reduction of water content of phosphate clay suspensions. Kong et al. proposed the semi-continuous system schematically presented in Fig. 17a as a previous prototype of a fully continuous process. The dilute particulate suspension was continuously fed while the supernatant liquid was continuously removed. The effect of electric field over the fine particles moving toward the anode causes the formation of a cake. The solid was accumulated at the tank bottom by gravity generating a solids-rich suspension. The experiments were

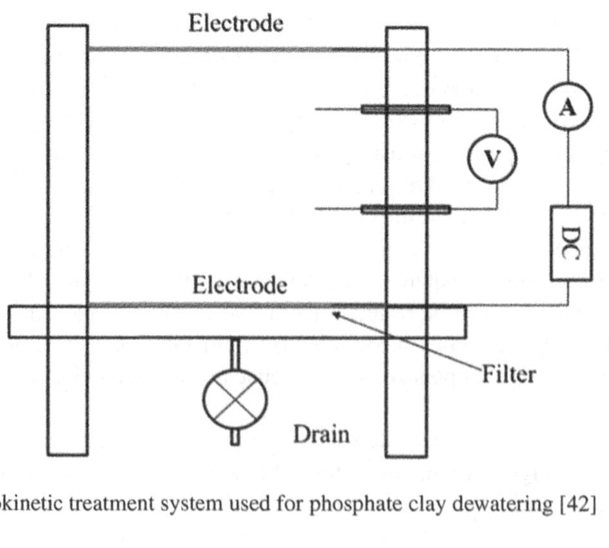


**Fig. 16** Electrokinetic treatment system used for phosphate clay dewatering [42]

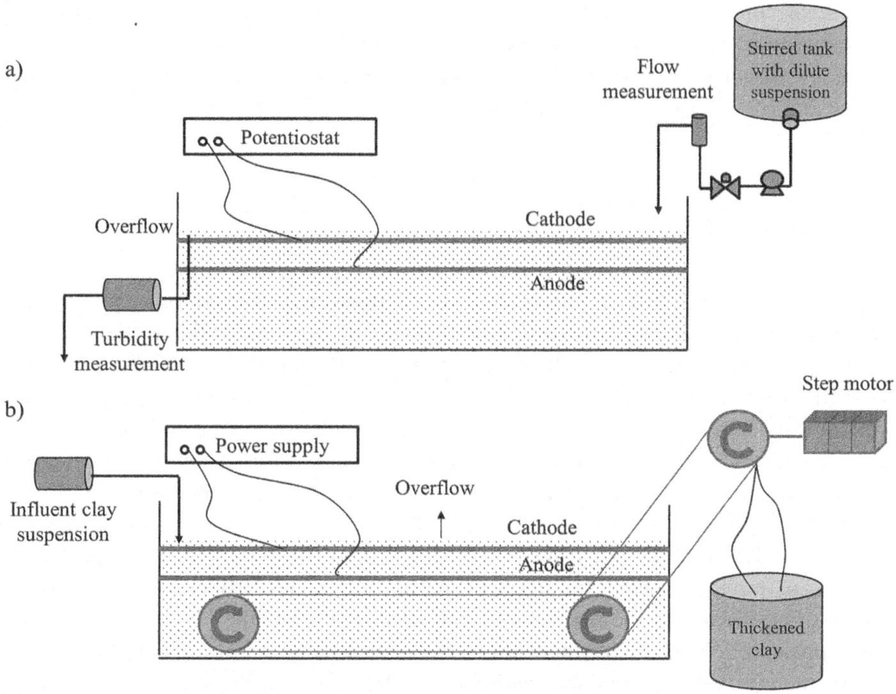


**Fig. 17** Schematic representation of (**a**) semi-continuous and (**b**) fully continuous systems for electro-dewatering of phosphatic clay suspension designed by Kong et al. [91, 92]

finished once the accumulated solid volume exceeded the capacity of the store tank. From results, authors concluded that fully continuous separation should provide better results not only by obtaining a steady stream of low-turbidity water but also by improving the efficiency of the process [91]. The design of a fully continuous prototype combined the idea of previously designed systems for water and solids recovery. The system schematically represented in Fig. 17b, consisted of two parts: the horizontal part where the cake is formed and the angled part where the cake is dewatered. The electrodes configuration was parallel in each zone. The electro-osmosis was proved the major transport mechanism in the angled zone, knowing this region of the experimental system as the cake-dewatering zone. The final solid content obtained was of 35% under an applied electric field of 1 V $cm^{-1}$ with an energy consumption of 4.7 Wh/kg of solid. Compared with static electrokinetic separation, the fully continuous prototype was proved to consume significantly less energy and to improve solid content [92]. Among other limitations associated with a poor contact in the cake-dewatering zone and the difficulty of collecting the low-turbidity supernatant water, further studies about the clay cake microstructure are desirable to a better understanding of the electrokinetic process [93].

In addition to the matrices aforementioned, the electro-dewatering process could be applied to oily sludge, one of the most significant solid wastes generated in the petroleum industry. The oily sludge consists of emulsion of various petroleum hydrocarbons, water, heavy metals and solid particles. The application of a low-intensity direct current to this matrix has successfully allowed the separation of water and oil phases because of electro-osmosis transport [94, 95].

## 3 Conclusion

The application of EDW requires a well understanding of the coupling phenomena transport. Thus, the combination of mechanisms taking place under an electric field, such as electromigration, electrophoresis, diffusion and electro-osmosis, plays an essential role in the effectiveness of the technique. The understanding of some theoretical concepts, such as zeta potential, electrical double layer and point of zero charge, assists the design of experimental setup for different applications. From theoretical and experimental studies, the use of enhancement methods to optimize the technology has been discussed. In this chapter, from the report of numerous real cases of EDW application, it has been highlighted the importance of operational conditions paying especial attention to the porous matrix properties. From a wide range of applications, the main factors affecting the efficiency of technology have been presented. This chapter has not only discussed the traditional applications of EDW, the civil engineering applications, but also others such as recovery of pollutant solid matrices. An especial attention has been paid to the recovery of contaminated soils through the study of the most important properties affecting the electro-osmotic transport. Currently, the development of promising experimental setups from studies dealing with traditional and novel applications reflects the versatility of the EDW to solve important environmental problems. Further experimental and theoretical studies will allow the optimization of economic and operational aspects required to be widely applied at field scale.

**Acknowledgments** Villen-Guzman acknowledges the "Contrato Postdoctoral" obtained from University of Malaga.

## References

1. A. Ribeiro, J. Rodríguez Maroto, Electroremediation of heavy metal-contaminated soils-processes and applications, in *Trace Elements in the Environment*, (CRC Press, Taylor & Francis, Boca Raton, 2006), pp. 325–340
2. J.K. Mitchell, K. Soga, *Fundamentals of Soil Behavior* (Wiley, Hoboken, 2005)
3. C.J.F.P. Jones, J. Lamont-Black, S. Glendinning, Electrokinetic geosynthetics in hydraulic applications. Geosynth. Hydraul. Appl. **29**, 381–390 (2011). https://doi.org/10.1016/j.geotexmem.2010.11.011
4. A.R. Estabragh, M. Naseh, A.A. Javadi, Improvement of clay soil by electro-osmosis technique. Appl. Clay Sci. **95**, 32–36 (2014). https://doi.org/10.1016/j.clay.2014.03.019

5. P.G. Nicholson, Chapter 10—Electro-osmosis (electrokinetic dewatering), in *Soil Improvement and Ground Modification Methods*, (Butterworth-Heinemann, Boston, 2015), pp. 221–228
6. L. Casagrande, Stabilization of soils by means of electro-osmosis—state-of-the-art. J. Boston Soc. Civ. Eng. Sect. Am. Soc. Civ. Eng. **69**, 255–302 (1983)
7. L. Casagrande, Hardening soil (1937) United States Patent US2099328
8. L. Bjerrum, J. Moum, O. Eide, Application of electro-osmosis to a foundation problem in a Norwegian quick clay. Geotechnique **17**, 214–235 (1967). https://doi.org/10.1680/geot.1967.17.3.214
9. L.G. Soderman, V. Milligan, Capacity of friction piles in varved clay increased by electro-osmosis, in *Proceedings of the 5th International Conference on Soil Mechanics and Foundation Engineering*, vol. 2, (1961), pp. 143–147
10. B.A. Chappell, P.L. Burton, Electro-osmosis applied to unstable embankment. J. Geotech. Eng. Div. **101**, 733–740 (1975)
11. N.H. Wade, Slope stability by electro-osmosis, in *Proceedings, 29th Canadian Geotechnical Conference*, Vancouver, (1976), pp. 44–66
12. K.Y. Lo, K.S. Ho, I.I. Inculet, Field test of electroosmotic strengthening of soft sensitive clay. Can. Geotech. J. **28**, 74–83 (1991). https://doi.org/10.1139/t91-008
13. V. Milligan, First application of electro-osmosis to improve friction pile capacity—three decades later. Proc. ICE Geotech. Eng. **113**, 112–116 (1995)
14. G.P. Karunaratne, Prefabricated and electrical vertical drains for consolidation of soft clay. Geotext. Geomembr. **29**, 391–401 (2011). https://doi.org/10.1016/j.geotexmem.2010.12.005
15. Y.B. Acar, Principles of electrokinetic remediation. Environ. Sci. Technol. **27**, 2638–2647 (1993)
16. K.R. Reddy, C. Cameselle, Overview of electrochemical remediation technologies, in *Electrochemical Remediation Technologies for Polluted Soils, Sediments and Groundwater*, (2009), pp. 1–28
17. F.F. Reuss, Sur un nouvel effet de l'électricité galvanique. Mém. Soc. Impériale Dês Nat. Moscou. **2**, 327–337 (1809)
18. A. Yeung: Electrokinetic flow processes in porous media and their applications. In: *Advances in Porous Media*. pp. 309–395. Elsevier Science (1994)
19. G. Quincke, Pogg. Ann. Physik. **113**, 513 (1861)
20. L. Gouy, Sur Const. Charge Electrique Surf. D'un Electrolyte. **9**, 457 (1910)
21. D.L. Chapman, LI. A contribution to the theory of electrocapillarity. Lond. Edinb. Dublin Philos. Mag. J. Sci. **25**, 475–481 (1913). https://doi.org/10.1080/14786440408634187
22. O. Stern, Zur theorie der elektrolytischen doppelschicht. Z. Elektrochem. Angew. Phys. Chem. **30**, 508–516 (1924)
23. W. Schmickler, Electrochemical theory: double layer, in *Reference Module in Chemistry, Molecular Sciences and Chemical Engineering*, (Elsevier, Amsterdam, 2014)
24. S. Trasatti, Structure of the metal/electrolyte solution interface: new data for theory. Electrochim. Acta **36**, 1659–1667 (1991). https://doi.org/10.1016/0013-4686(91)85023-Z
25. J. Huang, A. Malek, J. Zhang, M.H. Eikerling, Non-monotonic surface charging behavior of platinum: a paradigm change. J. Phys. Chem. C **120**, 13587–13595 (2016). https://doi.org/10.1021/acs.jpcc.6b03930
26. J.M. Paz-Garcia, B. Johannesson, L.M. Ottosen, A.B. Ribeiro, J.M. Rodriguez-Maroto, Modeling of electric double-layers including chemical reaction effects. Electrochim. Acta **150**, 263–268 (2014). https://doi.org/10.1016/j.electacta.2014.10.056
27. R.F. Probstein, R.E. Hicks, Removal of contaminants from soils by electric fields. Science **260**, 498–503 (1993). https://doi.org/10.1126/science.260.5107.498
28. P.B. Lorenz, Surface conductance and electrokinetic properties of kaolinite beds. Clays Clay Miner. **17**, 223–231 (1969)
29. G.R. Eykholt, D.E. Daniel, Impact of system chemistry on electroosmosis in contaminated soil. J. Geotech. Eng. **120**, 797–815 (1994). https://doi.org/10.1061/(ASCE)0733-9410(1994)120:5(797)
30. G. Sposito, *The Chemistry of Soils* (Oxford University Press, New York, 1989)

31. Y. Yukselen, A. Kaya, Zeta potential of kaolinite in the presence of alkali, alkaline earth and hydrolyzable metal ions. Water Air Soil Pollut. **145**, 155–168 (2003). https://doi.org/10.1023/A:1023684213383
32. Y. Kitagawa, S. Yamaguchi, Y. Yoruzo, Effects of sodium chloride concentrations on zeta potentials of clay minerals estimated by an electrokinetic sonic amplitude method. Clay Sci. **12**, 91–96 (2003). https://doi.org/10.11362/jcssjclayscience1960.12.91
33. R.J. Hunter, M. James, Charge reversal of kaolinite by hydrolyzable metal ions: an electroacoustic study. Clays Clay Miner. **40**, 644–649 (1992)
34. L.M. Vane, G.M. Zang, Effect of aqueous phase properties on clay particle zeta potential and electro-osmotic permeability: implications for electro-kinetic soil remediation processes. Electrochem. Decontam. Soil Water. **55**, 1–22 (1997). https://doi.org/10.1016/S0304-3894(97)00010-1
35. B. Van Raij, M. Peech, Electrochemical properties of some oxisols and alfisols of the tropics. Soil Sci. Soc. Am. J. **36**, 587–593 (1972). https://doi.org/10.2136/sssaj1972.03615995003600040027x
36. E. Marcano-Martinez, M. Mcbride, Comparison of the titration and ion adsorption methods for surface charge measurement in oxisols. Soil Sci. Soc. Am. J. **53**, 1040–1045 (1989). https://doi.org/10.2136/sssaj1989.03615995005300040009x
37. C. Appel, L.Q. Ma, R. Dean Rhue, E. Kennelley, Point of zero charge determination in soils and minerals via traditional methods and detection of electroacoustic mobility. Geoderma **113**, 77–93 (2003). https://doi.org/10.1016/S0016-7061(02)00316-6
38. A.P. Shapiro, R.F. Probstein, Removal of contaminants from saturated clay by electroosmosis. Environ. Sci. Technol. **27**, 283–291 (1993). https://doi.org/10.1021/es00039a007
39. A.N. Alshawabkeh, T.C. Sheahan, X. Wu, Coupling of electrochemical and mechanical processes in soils under DC fields. Coupled Chemo-Mech. Phenom. **36**, 453–465 (2004). https://doi.org/10.1016/S0167-6636(03)00071-1
40. A. Rittirong, J. Shang, Chapter 34: Electro-osmotic stabilization (2005). In: Ground Improvement. Case Histories. pp. 967–996. Elsevier (2005). https://doi.org/10.1016/S1571-9960(05)80037-0
41. K.Y. Lo, K.S. Ho, The effects of electroosmotic field treatment on the soil properties of a soft sensitive clay. Can. Geotech. J. **28**, 763–770 (1991). https://doi.org/10.1139/t91-093
42. J.Q. Shang, K.Y. Lo, Electrokinetic dewatering of a phosphate clay. J. Hazard. Mater. **55**, 117–133 (1997). https://doi.org/10.1016/S0304-3894(97)00018-6
43. N.C. Lockhart, Electroosmotic dewatering of clays, III. Influence of clay type, exchangeable cations, and electrode materials. Colloids Surf. **6**, 253–269 (1983). https://doi.org/10.1016/0166-6622(83)80017-1
44. R.H. Sprute, D.J. Kelsh, Dewatering fine-particle suspensions with direct current. Fines Part. Process Symp. **2**, 1828–1844 (1980)
45. H. Wu, L. Hu, Q. Wen, Electro-osmotic enhancement of bentonite with reactive and inert electrodes. Appl. Clay Sci. **111**, 76–82 (2015). https://doi.org/10.1016/j.clay.2015.04.006
46. D. Kalumba, S. Glendinning, C. Rogers, M. Tyrer, D.I. Boardman, Dewatering of tunneling slurry waste using electrokinetic geosynthetics. J. Environ. Eng. **135** (2009). https://doi.org/10.1061/(ASCE)0733-9372(2009)135:11(1227)
47. S. Glendinning, C.J. Jones, J. Lamont-Black, Chapter 35: The use of electrokinetic geosynthetics (EKG) to improve soft soils, in *Elsevier Geo-Engineering Book Series*, ed. by B. Indraratna, J. Chu, (Elsevier, Amsterdam, 2005), pp. 997–1043
48. S. Glendinning, J. Lamont-Black, C.J.F.P. Jones, Treatment of sewage sludge using electrokinetic geosynthetics. J. Hazard. Mater. **139**, 491–499 (2007). https://doi.org/10.1016/j.jhazmat.2006.02.046
49. H.O. Abiera, N. Miura, D.T. Bergado, T. Nomura, Effects of using electro-conductive PVD in the consolidation of reconstituted Ariake clay. Geotech. Eng. **30**, 67-83 (1999)
50. A.T. Yeung, Fundamental aspects of prolonged electrokinetic flows in kaolinites. Geomech. Geoeng. **1**, 13–25 (2006). https://doi.org/10.1080/17486020500528026

51. K.R. Reddy, S. Chinthamreddy, Enhanced electrokinetic remediation of heavy metals in glacial till soils using different electrolyte solutions. J. Environ. Eng. **130**, 442–455 (2004). https://doi.org/10.1061/(ASCE)0733-9372(2004)130:4(442)
52. S.S. Al-Shahrani, E.P.L. Roberts, Electrokinetic removal of caesium from kaolin. J. Hazard. Mater. **122**, 91–101 (2005). https://doi.org/10.1016/j.jhazmat.2005.03.018
53. M. Pazos, M.T. Alcántara, C. Cameselle, M.A. Sanromán, Evaluation of electrokinetic technique for industrial waste decontamination. Sep. Sci. Technol. **44**, 2304–2321 (2009). https://doi.org/10.1080/01496390902979867
54. B.-G. Ryu, G.-Y. Park, J.-W. Yang, K. Baek, Electrolyte conditioning for electrokinetic remediation of As, Cu, and Pb-contaminated soil. Sci. Adv. Innov. Appl. Electrokinet. Remediat. **79**, 170–176 (2011). https://doi.org/10.1016/j.seppur.2011.02.025
55. C. Cameselle, A. Pena, Enhanced electromigration and electro-osmosis for the remediation of an agricultural soil contaminated with multiple heavy metals. Process Saf. Environ. Prot. **104**, 209–217 (2016). https://doi.org/10.1016/j.psep.2016.09.002
56. M. Villen-Guzman, J.M. Paz-Garcia, J.M. Rodriguez-Maroto, C. Gomez-Lahoz, F. Garcia-Herruzo, Acid enhanced electrokinetic remediation of a contaminated soil using constant current density: strong vs. weak acid. Sep. Sci. Technol. **49**, 1461–1468 (2014). https://doi.org/10.1080/01496395.2014.898306
57. C.C. West, J.H. Harwell, Surfactants and subsurface remediation. Environ. Sci. Technol. **26**, 2324–2330 (1992). https://doi.org/10.1021/es00036a002
58. R. López-Vizcaíno, J. Alonso, P. Cañizares, M.J. León, V. Navarro, M.A. Rodrigo, C. Sáez, Electroremediation of a natural soil polluted with phenanthrene in a pilot plant. J. Hazard. Mater. **265**, 142–150 (2014). https://doi.org/10.1016/j.jhazmat.2013.11.048
59. D.-M. Zhou, C.-F. Deng, L. Cang, Electrokinetic remediation of a Cu contaminated red soil by conditioning catholyte pH with different enhancing chemical reagents. Chemosphere **56**, 265–273 (2004). https://doi.org/10.1016/j.chemosphere.2004.02.033
60. S. Amrate, D.E. Akretche, Modeling EDTA enhanced electrokinetic remediation of lead contaminated soils. Chemosphere **60**, 1376–1383 (2005). https://doi.org/10.1016/j.chemosphere.2005.02.021
61. M.G. Nogueira, M. Pazos, M.A. Sanromán, C. Cameselle, Improving on electrokinetic remediation in spiked Mn kaolinite by addition of complexing agents. Electrochim. Acta **52**, 3349–3354 (2007). https://doi.org/10.1016/j.electacta.2006.03.115
62. A. García-Rubio, J.M. Rodríguez-Maroto, C. Gómez-Lahoz, F. García-Herruzo, C. Vereda-Alonso, Electrokinetic remediation: the use of mercury speciation for feasibility studies applied to a contaminated soil from Almadén. Electrochim. Acta **56**, 9303–9310 (2011)
63. J.D. Subirés-Muñoz, A. García-Rubio, C. Vereda-Alonso, C. Gómez-Lahoz, J.M. Rodríguez-Maroto, F. García-Herruzo, J.M. Paz-García, Feasibility study of the use of different extractant agents in the remediation of a mercury contaminated soil from Almaden. Sep. Purif. Technol. **79**, 151–156 (2011)
64. J.S.H. Wong, R.E. Hicks, R.F. Probstein, EDTA-enhanced electroremediation of metal-contaminated soils. Electrochem. Decontam. Soil Water. **55**, 61–79 (1997). https://doi.org/10.1016/S0304-3894(97)00008-3
65. M. Villen-Guzman, A. Garcia-Rubio, J.M. Paz-Garcia, J.M. Rodriguez-Maroto, F. Garcia-Herruzo, C. Vereda-Alonso, C. Gomez-Lahoz, The use of ethylenediaminetetraacetic acid as enhancing agent for the remediation of a lead polluted soil. Electrochim. Acta **181**, 82–89 (2015). https://doi.org/10.1016/j.electacta.2015.03.061
66. S.-C. Chien, C.-Y. Ou, Y.-C. Lee, A novel electroosmotic chemical treatment technique for soil improvement. Appl. Clay Sci. **50**, 481–492 (2010). https://doi.org/10.1016/j.clay.2010.09.014
67. C.-Y. Ou, S.-C. Chien, Y.-G. Wang, On the enhancement of electroosmotic soil improvement by the injection of saline solutions. Appl. Clay Sci. **44**, 130–136 (2009). https://doi.org/10.1016/j.clay.2008.12.014

68. C.-Y. Ou, S.-C. Chien, R.-H. Liu, A study of the effects of electrode spacing on the cementation region for electro-osmotic chemical treatment. Appl. Clay Sci. **104**, 168–181 (2015). https://doi.org/10.1016/j.clay.2014.11.027
69. S.-C. Chien, F.-C. Teng, C.-Y. Ou, Soil improvement of electroosmosis with the chemical treatment using the suitable operation process. Acta Geotech. **10**, 813–820 (2015). https://doi.org/10.1007/s11440-014-0319-y
70. S. Chew, G. Karunaratne, V. Kuma, L. Lim, M. Toh, A. Hee, A field trial for soft clay consolidation using electric vertical drains. Spec. Issue Prefabr. Vert. Drains. **22**, 17–35 (2004). https://doi.org/10.1016/S0266-1144(03)00049-9
71. A. Rittirong, R. Douglas, J. Shang, E. Lee, Electrokinetic improvement of soft clay using electrical vertical drains. Geosynth. Int. **15**, 369–381 (2008)
72. C. Cameselle, K.R. Reddy, Development and enhancement of electro-osmotic flow for the removal of contaminants from soils. Electrochim. Acta **86**, 10–22 (2012). https://doi.org/10.1016/j.electacta.2012.06.121
73. C.J. Bruell, B.A. Segall, M.T. Walsh, Electroosomotic removal of gasoline hydrocarbons and TCE from clay. J. Environ. Eng. **118**, 68–83 (1992). https://doi.org/10.1061/(ASCE)0733-9372(1992)118:1(68)
74. Y.B. Acar, H. Li, R.J. Gale, Phenol removal from kaolinite by electrokinetics. J. Geotech. Eng. **118**, 1837–1852 (1992)
75. A.B. Ribeiro, J.M. Rodriguez-Maroto, E.P. Mateus, H. Gomes, Removal of organic contaminants from soils by an electrokinetic process: the case of atrazine. Experimental and modeling. Chemosphere **59**, 1229–1239 (2005). https://doi.org/10.1016/j.chemosphere.2004.11.054
76. A.B. Ribeiro, E.P. Mateus, J.-M. Rodríguez-Maroto, Removal of organic contaminants from soils by an electrokinetic process: the case of molinate and bentazone. Experimental and modeling. Sep. Purif. Technol. **79**, 193–203 (2011). https://doi.org/10.1016/j.seppur.2011.01.045
77. K.R. Reddy, U.S. Parupudi, S.N. Devulapalli, C.Y. Xu, Effects of soil composition on the removal of chromium by electrokinetics. J. Hazard. Mater. **55**, 135–158 (1997)
78. M. Villen-Guzman, A. Garcia-Rubio, J.M. Paz-Garcia, C. Vereda-Alonso, C. Gomez-Lahoz, J.M. Rodriguez-Maroto, Aging effects on the mobility of Pb in soil: influence on the energy requirements in electroremediation. Chemosphere **213**, 351–357 (2018). https://doi.org/10.1016/j.chemosphere.2018.09.039
79. L.M. Ottosen, K. Lepkova, M. Kubal, Comparison of electrodialytic removal of Cu from spiked kaolinite, spiked soil and industrially polluted soil. J. Hazard. Mater. **137**, 113–120 (2006)
80. R. López-Vizcaíno, C. Sáez, P. Cañizares, M.A. Rodrigo, The use of a combined process of surfactant-aided soil washing and coagulation for PAH-contaminated soils treatment. Sep. Purif. Technol. **88**, 46–51 (2012). https://doi.org/10.1016/j.seppur.2011.11.038
81. S.A. Clayton, O.N. Scholes, A.F.A. Hoadley, R.A. Wheeler, M.J. McIntosh, D.Q. Huynh, Dewatering of biomaterials by mechanical thermal expression. Drying Technol. **24**, 819–834 (2006). https://doi.org/10.1080/07373930600733093
82. J. Vaxelaire, P. Cézac, Moisture distribution in activated sludges: a review. Water Res. **38**, 2214–2229 (2004). https://doi.org/10.1016/j.watres.2004.02.021
83. K.R. Tsang, P.A. Vesilind, Moisture distribution in sludges. Water Sci. Technol. **22**, 135–142 (1990). https://doi.org/10.2166/wst.1990.0108
84. M. Smollen, Moisture retention characteristics and volume reduction of municipal sludges. Water SA **14**, 25–28 (1988)
85. P.A. Vesilind, The role of water in sludge dewatering. Water Environ. Res. **66**, 4–11 (1994)
86. A. Mahmoud, J. Olivier, J. Vaxelaire, A.F.A. Hoadley, Electrical field: A historical review of its application and contributions in wastewater sludge dewatering. Water Res. **44**, 2381–2407 (2010). https://doi.org/10.1016/j.watres.2010.01.033
87. M. Smollen, A. Kafaar, Electroosmotically enhanced sludge dewatering: pilot-plant study. Water Sci. Technol. **30**, 159–168 (1994)

88. P. Mantovi, G. Baldoni, G. Toderi, Reuse of liquid, dewatered, and composted sewage sludge on agricultural land: effects of long-term application on soil and crop. Water Res. **39**, 289–296 (2005). https://doi.org/10.1016/j.watres.2004.10.003
89. S. Hwang, K.-S. Min, Improved sludge dewatering by addition of electro-osmosis to belt filter press. J. Environ. Eng. Sci. **2**, 149–153 (2003). https://doi.org/10.1139/s03-013
90. A. Fourie, D. Johns, C.F. Jones, Dewatering of mine tailings using electrokinetic geosynthetics. Can. Geotech. J. **44**, 160–172 (2007)
91. R. Kong, M.E. Orazem, Semi-continuous electrokinetic dewatering of phosphatic clay suspensions. Electrochem. New Era **140**, 438–446 (2014). https://doi.org/10.1016/j.electacta.2014.07.016
92. R. Kong, Continuous electrokinetic dewatering of phosphatic clay suspensions. University of Florida (2015)
93. C. Peng, H. Lai, M.E. Orazem, S. Moghaddam, Microstructure of clay fabric in electrokinetic dewatering of phosphatic clay dispersions. Appl. Clay Sci. **158**, 94–101 (2018). https://doi.org/10.1016/j.clay.2018.03.020
94. L. Yang, G. Nakhla, A. Bassi, Electro-kinetic dewatering of oily sludges. J. Hazard. Mater. **125**, 130–140 (2005). https://doi.org/10.1016/j.jhazmat.2005.05.040
95. G. Hu, J. Li, G. Zeng, Recent development in the treatment of oily sludge from petroleum industry: a review. J. Hazard. Mater. **261**, 470–490 (2013). https://doi.org/10.1016/j.jhazmat.2013.07.069

# Part III
# Scale-up

# Fundamentals of the Scale-Up of the Electrochemically Assisted Soil Remediation Processes

**Cristina Sáez**

## 1 Introduction

In the last decades, much of the works focussed on electrochemically assisted soil remediation processes were limited to research conducted on laboratory [1–9], in which hermetically sealed cylindrical or prismatic cells were used as electrokinetic reactors. Working at this scale the soil is isolated from the surrounding environment, greatly reducing the variables that must be controlled during processing. This reduce size facilitates the analysis of the different electrokinetic and electrochemical processes that occur in the soil as well as the estimation of the most relevant variables. Nevertheless, although the use of these small and well-controlled units is very useful to obtain high accuracy value data [10], they do not reproduce accurately the behaviour of the full application due to the experimental working conditions are far from resembling the real field.

The idea of going from a laboratory scale to a full application directly is very tempting from the point of view economic and time-consuming. However, following this direct path requires either an information fund that is often inaccessible or a series of technical evaluations beyond what is normally considered possible or desirable. In this way, it is necessary to go through a path that includes going through one or more intermediate scales before achieving the final objective. However, it is necessary to verify that the hypotheses obtained through the first experiments in the laboratory continue to be valid as one moves from one scale to another.

In this point, it is worth to point out that in disciplines such as chemical and environmental engineering scaling-up involves a deeper understanding of the process. This entails to pay attention to the basis and fundamentals of remediation

C. Sáez (✉)
Chemical Engineering Department, Faculty of Chemical Sciences & Technologies, University of Castilla-La Mancha, Ciudad Real, Spain
e-mail: cristina.saez@uclm.es



M. A. Rodrigo, E. V. Dos Santos (eds.), *Electrochemically Assisted Remediation of Contaminated Soils*, Environmental Pollution 30,
https://doi.org/10.1007/978-3-030-68140-1_17

process and, even more important, to the control mechanisms. In the case of soil remediation, three electrical drive processes (electrokinetic, electrochemical, and electrical heating processes) coexist. Additionally, other chemical processes, such as precipitation or exchange reactions, and physical processes, such as hydraulic flux or evaporation, may occur simultaneously. Any one of these mechanisms is more ubiquitous than others and may result in the performance of the technology being very different. If the results of the small scale are intended to be fully inferred [10, 11], it is worth evaluating how the size of the experimental device affects these mechanisms.

From the technological viewpoint, few research groups have focussed their works on the scale-up of electroremediation process [10, 12–17]. In contrast, many companies have the experience and know-how in conducting large-scale research but generally they are not interested in sharing it. Among these companies, there are R + D sections of big chemical or environmental companies (such as DuPont or Monsanto) and also small spin-off companies (such as the relevant geokinetics or its actual matrix company Hak Milieutechniek BV). This limits markedly the knowledge because they are not interested in spreading their knowledge, but in applying it in the remediation of particular polluted sites: (1) the company Electrokinetics developed soil remediation processes based on the use of perforated electrodes through which water retained in the soil is mobilized, or through which improving fluids are introduced into the in-treatment soil; (2) the company Geokinetics International, Inc. worked on an electroremediation process patent in which the main claim was the use of ion-permeable electrolyte compartments (wells) inside which the electrodes are placed (both anodes and cathodes); (3) the company Hak Milieutechniek BV developed a process in which the electrodes heat the ground to favour the volatilization of organic species and the fuel movement; (4) the company Isotron Corporation developed, together with the Department of Energy of the United States, electrochemical decontamination processes for soils and groundwater polluted with heavy metal, especially radionuclides, being particularly interested on the development of special electrode materials that promote removal processes such as adsorption or ionic exchange [18]; (5) the company SRI International in collaboration with USAF created an electrochemical reactor to treat groundwater by means of advanced oxidation techniques; (6) a consortium of companies such as Monsanto, E.I. du Pont de Nemours and Company (DuPont) and General Electric (GE) designed the Lasagna process, that combines electroosmosis together with reactive permeable barriers [12].

The larger numbers of processes involved in the electrochemically assisted soil remediation processes, and the strong interactions of parameters in those processes, makes every application a unique case from which a direct application to other case cannot be expected but from which important lessons can be learned and applied to many other cases. In this context, this chapter reports information about the scaling-up of electrochemically assisted soil remediation processes, trying to highlight the lessons learned on this field and the main inputs that should be accounted for a proper scale-up study.

## 2 From Lab Scale to Full Application: Main Aims and Cell Design

In the field of environmental and chemical engineering, the scaling-up of any process is, generally, classified into five levels or scales: laboratory, bench or semi-pilot, pilot, prototype or semi-industrial and industrial. In each step, the size increases and the magnitudes which marks the type of scale used are the size and/or the volume. However, in terms of engineering, this concept is too basic, and a multitude of factors, some more important than others, must be evaluated. But if a criterion must be established to discern between the different scales, it could be said that they can be sort based on the purpose for which they were created. Therefore, each scale provides the information to be used in the design of the next scale, and all this information must be collected to establish a successful strategy and achieve the ultimate goal of scaling-up: to obtain comprehensive knowledge of the remediation process. Finally, developed technologies should be validated in a relevant environment.

***Laboratory scale.*** The purpose of this scale is to verify that the theory is fulfilled, as well as to constitute a source of obtaining data for the formulation of theoretical models. Thus, it allows confirming or rejecting the hypotheses obtained from prior knowledge and from the literature and it is also a fundamental piece for obtaining of the systematized information that will form the basis for higher scales. Additionally, this scale allows to evaluate in the most efficient way the effects that the different disturbance variables have on the process.

In soil remediation, the laboratory scale trials are carried out in hermetically sealed systems that are easy to operate and reproduce. These trials require limited economic and time investment, and, thus, they can be easily replicate. However, as these systems are isolated, the information obtained cannot be used directly to perform any kind of inference, as all processes are studied under extremely controlled conditions. On contrary, this extreme isolation from external agents is useful to estimate the thermodynamic, transport and chemical kinetic parameters.

Figure 1 shows a schematic representation of a typical lab-scale cell, proposed by Sandu et al. [9]. This lab-scale cell has a capacity of 150 mL (within the typical range, from 50 to 200 mL) and is divided into three compartments: two electrode compartments and a central one where the soil is placed. These compartments are generally separated using cellulose filters or plastic meshes. Carbonaceous or metallic fine bars are used as electrodes which are connected to a power supply to maintain a constant DC. The electrode chambers are filled with the selected processing fluid that is recirculated with a pump. Additionally, electrolytes are transferred to collector flask.

***Bench or semi-pilot scale.*** This scale has more technical characteristics. Contrary to the first, the scale already has equipment whose geometry is more comparable to industrial scale and requires a higher level of instrumentation and control. Bench-scale trials allow to have a complete preliminary understanding of the performance of the process and pay attention to other aspects related to safety, waste generation, and control and instrument requirements. In addition, this scale mainly

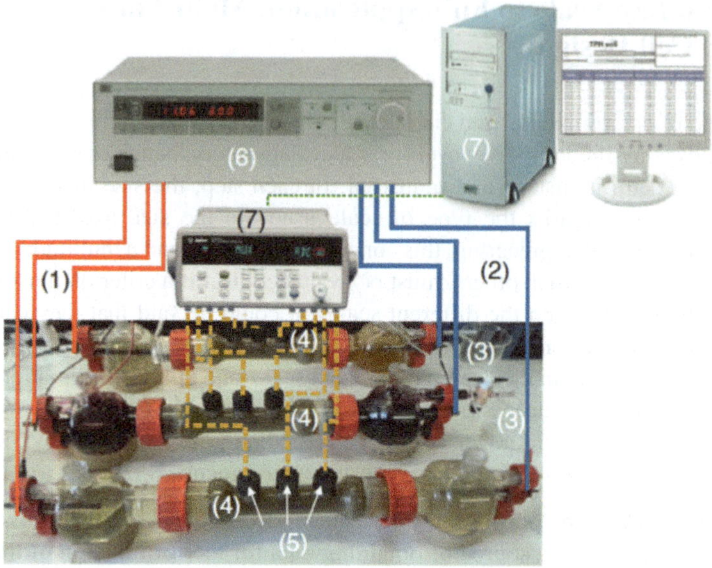


**Fig. 1** Experimental set-up for the simulated bench-scale in situ experiments: (*1*) anode, (*2*) cathode, (*3*) electroosmotic flow reservoir, (*4*) soil sample, (*5*) auxiliary electrodes, (*6*) power supply, (*7*) data acquisition system [9]

confirms the mathematical model obtained in the laboratory stage and specifies and expands the information obtained from the laboratory. However, the information obtained is still related to the main direction of change (one-dimensional) and the extrapolation to the full range is still very questionable. In any case, results obtained at this scale will have an impact on the decision to carry out the next highest scale (pilot scale) or not.

Figure 2 shows a schematic representation of a bench-scale electrokinetic reactor used in literature [19] and a picture of the bench-scale cell used by Vieira do Santos et al. [7]. Generally, this type of bench-scale cell has a capacity ranging from 0.2 to 2 L g of soil and it is divided into three compartments, as in lab-scale cells, and where electrode compartment are not immersed in the soil but located in the extreme of the cell (immersed in electrolyte compartments). In this case, bars (with 2–3 cm of diameter and 4–5 m of length) or plates (with dimensions around 10.0 cm × 10.0 cm × 1.0 cm) of carbonaceous or metallic materials are generally used as electrodes. As before, electrolytes are recirculated by means of a pump and collected in reservoir tanks.

***Pilot scale.*** The operation at this scale is more expensive than the previous ones, but its use is crucial by the level of information obtained to size the industrial unit. Thus, the pilot scale is recognized as a key issue in scaling-up that requires the information of the previous stages (research and development). With this previous information, the purpose of the pilot scale is the verification of the results in a three-

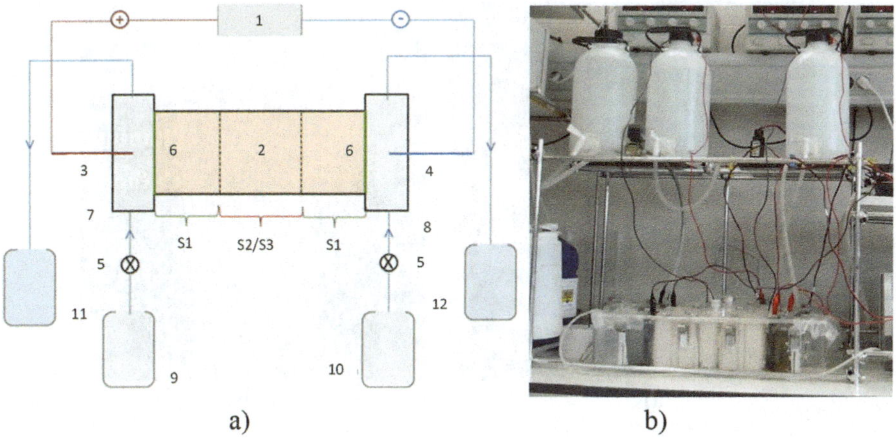


**Fig. 2** (**a**) Schematic of bench-scale electrokinetic reactor: (*1*) DC power supply; (*2*) soil cell; (*3*) anode; (*4*) cathode; (*5*) peristaltic pump; (*6*) cellulose filters; (*7*) anode chamber; (*8*) cathode chamber; (*9*) anolyte reservoir; (*10*) catholyte reservoir; (*11*) anolyte receiving reservoir; (*12*) catholyte receiving reservoir; (*S1*) non-contaminated soil S1; (*S2/S3*) contaminated soils. (Adapted from [19]), (**b**) Picture of bench-scale electrokinetic reactor used by [7]

dimension (3D) configuration. Therefore, this entails changes in the design of cells at pilot scale, with the use of open systems with a capacity ranging around 150–300 L generalized. That is, prismatic cells open to atmosphere where electrodes can be directly inserted in the soil or immersed in electrolyte wells are used. At this size, the auxiliary service and maintenance requirements can be estimated, as well as a first evaluation of the typical operational problems.

Figure 3 shows pictures of different pilot-scale electrokinetic reactors reported in literature [20, 21]. The reactor proposed by Risco et al. [21] consists of a methacrylate prism with a soil capacity of 175 L ($70 \times 50 \times 50$ cm$^3$). The electrodes used for both the anodes and the cathodes are graphite rods with dimensions of $1 \times 1 \times 10$ cm$^3$ and positioned in semipermeable electrolyte wells of 100 cm$^3$ of capacity. On the other hand, the pilot-scale equipment proposed by Kim et al. [20] consists of a cell of 50 L of capacity ($50 \times 50 \times 20$ cm$^3$), with anode and cathode compartments with dimensions of $50 \times 50 \times 12$ cm$^3$ and $50 \times 50 \times 15$ cm$^3$, respectively. In this case, a mixed metal oxide ($50 \times 50$ cm) is used as anode and titanium as cathode.

The large volume of soil treated at this scale and its typical heterogeneous make important to control soil conditions to reproduce conditions similar to field, to ensure the homogeneity of the soil between tests and to avoid the formation of preferential flow paths. This is a relevant point (that is critical in the next level, prototype) due to the macroscopic structure of the soil becomes altered during extraction, transport, and/or storage before the experiment. Therefore, to use soils with hydromechanical properties analogous to those in the field, some authors [10, 22] inform about the necessity of the reestablishment of initial properties. To do this, it is proposed the installation of a granular material (double layer of gravel and sand) that

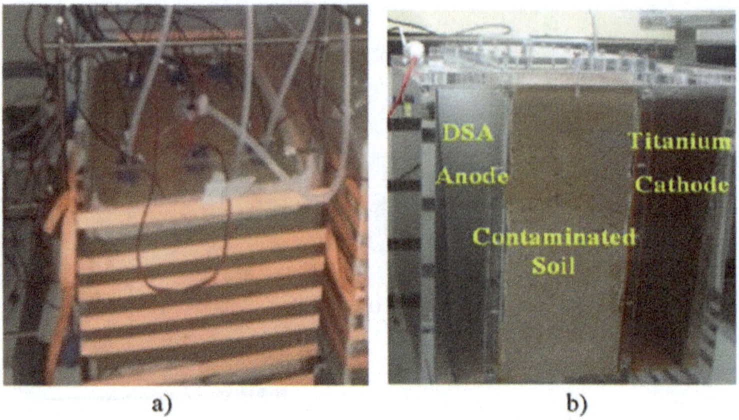


**Fig. 3** Pilot electrokinetic reactors reported in literature: (**a**) 175 L [21], (**b**) 50 L [20]

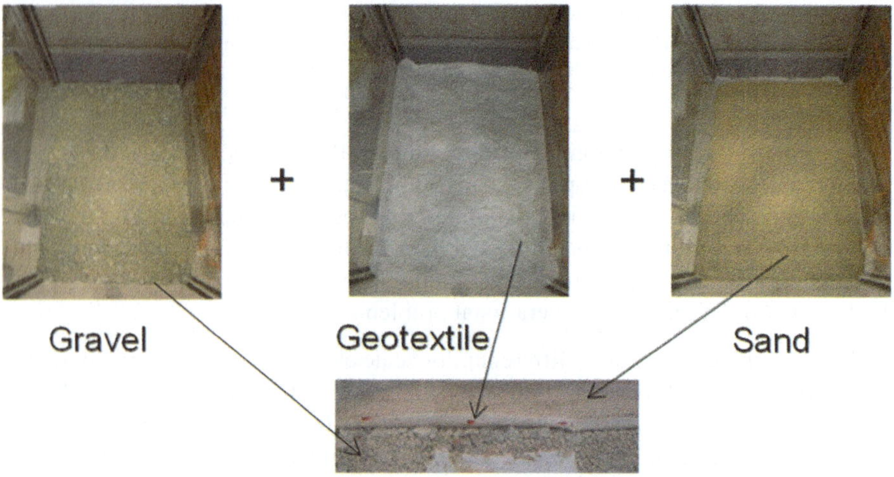


**Fig. 4** Pictures and scheme of the drain support recommended by Lopez-Vizcaíno et al. [10] for pilot scale

acts as a drain support (Fig. 4). Then, the soil, with controlled humidity conditions, is deposited into the reactor by compaction of layers of controlled height. Finally, the installation of a layer of sand is deposited on the top to act as a capillary barrier to minimize evaporation losses. The compaction necessary depend on soil properties and the optimal compaction energy must be evaluated in each case [10, 22].

Unlike previous scales, its higher size makes possible the evaluation of different electrodic configuration. In fact, the position of the electrodes placed in the soil is a key parameter in the efficiency of the remediation, because this determines the direction and magnitude of the electrokinetic processes, and consequently, the significance of the transport of pollutants. The number of the electrodes and the electrode spacing used has to be directly related to the mass of treated soil, and their configuration depends on the aim of the treatment. From a viewpoint of transport of pollutant, the electrode configuration more effective is the use of facing rows of electrodes or the use of a group of anodes surrounding a central cathode or vice versa, depending on the transport characteristic of the pollutant. On the other hand, to minimize the pollution dispersion an electrokinetic fence is recommended. It consists of a sequence of alternate electrodes facing the groundwater plume with the aim to capture pollutants contained in the soil and avoid their dispersion by diffusion or advention. Figure 5a shows a scheme of some of electrode configuration used in literature.

***Prototype or full-scale.*** The purpose of this scale is to reproduce a real application using cells built with the same concept but with soil-treatment capacities in the range of few cubic meter. Trials carried out at this scale contribute to attain the complete understanding of involved remediation processes and coupled phenomena. They also take into consideration the interaction between the system and the surrounding environment, in terms of lateral flow and atmospheric interaction processes, which is not relevant in downsizing scales. In addition, prototype trials give information to overcome practical problems and to establish protocols for waste management and control and instrumentation. As negative aspects, the execution of

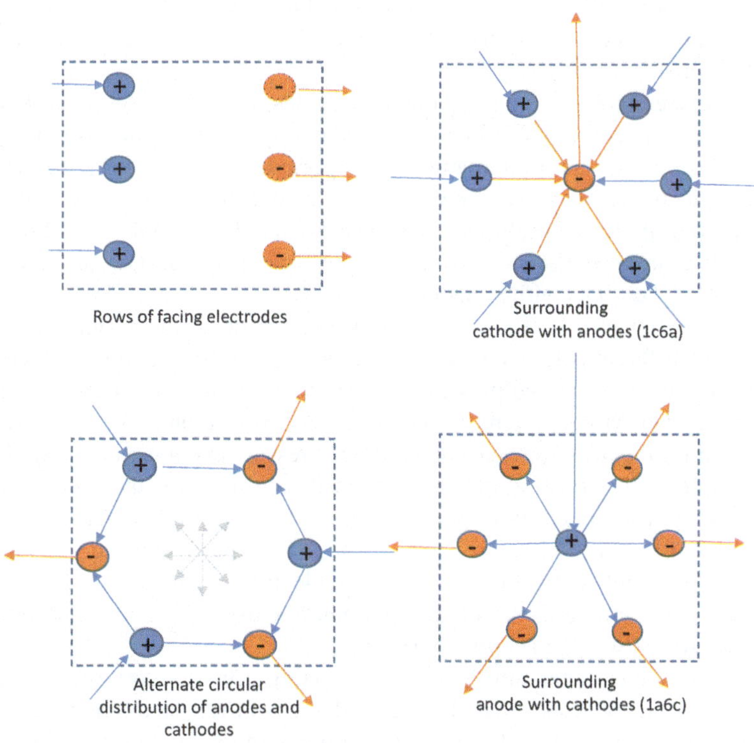


**Fig. 5** Scheme of some of electrode configuration used in literature

these trials implicate a large consumption of time and financial resources, and this has limited the number of works carried out at this scale.

Figure 6 shows prototype electrokinetic reactors used in literature, ranging from 3.3 to 90 $m^3$ of capacity. As can be observed, they consist of high size prismatic cells. The prototype proposes by Ochoa et al. [23] has a capacity of 3.3 $m^3$ and is built with a circular arrangement of electrodes: a central Ti cathode and six $IrO_2$-$Ta_2O_5$ Ti surrounding anodes. Each electrode has a diameter of 24 cm and is 0.6 m in length, buried only 0.5 m into the ground, and the distance between the electrodes is 117 cm (Fig. 6a). In the prototype of 4 $m^3$ (Fig. 6b) [24] a pair of machined cast iron electrodes (20 cm long and 1 cm in diameter) are centred and inserted vertically into sand wells (using fine acid washed sand, to facilitate drainage of water and prevent sample desiccation) at the edges of the soil mass in each container at an electrode separation of 16 cm. In prototypes shown in Fig. 6c [17] and Fig. 6d [16], electrodes are positioned in semipermeable electrolyte wells and separated between 1.5 and 2 m, depending on the electrode configuration: facing rows in Fig. 6c and circular with alternate anodes and cathodes in Fig. 6d. In this case, the electrodes used are cylinders of graphite (15 cm in diameter and 100 cm in length) and the wells used are PVC cylinders with lateral perforations (31.5 cm in diameter and 140 cm in depth) to facilitate the flow and transport of electrolytes. The prototype shown in Fig. 6e [25] has a total capacity of 90 $m^3$. It is equipped with steel bars in polyvinyl chloride (PVC) casings as anodes and hollow stainless steel as cathode. It allows different electrode configuration: parallel (24 $m^2$), square (20 $m^2$), and hexagonal (16 $m^2$).

As a differentiating element of previous scales, the design of electrolyte wells and the size and geometry of electrodes play an important role at this point, and, as shown in Fig. 6, different alternatives are reported. Figure 7 shows the schematic design of electrolyte well used by Woo-Seung et al. [26]. In this work, authors propose the use of perforated cathodes that consist of a hollow stainless steel with many holes to facilitate the transport of electrolytes, while the anodes are a steel bars located in PVC casings with many holes.

On the other hand, at this scale the instrumentation of the system is crucial to monitor the different parameters during the test. Generally, positioning thermocouples, tensiometers water and pressure sensors conveniently distributed into the soil is recommended. Moreover, micro-boreholes can dispose into the soil for sampling and to monitor the pH, electrical conductivity, pressure and temperature of the pore. Figure 8 shows the location of the instrumentation used for the operation of the 16 $m^3$ prototype operated by Lopez-Vizcaíno et al. [10]. In this work, authors also propose the use of a level-control system coupled to the electrolyte wells and connected to a feed tank to regulate the volume of electrolyte added to the soil. All these variables can be remotely controlled and monitored using a supervisory control and data acquisition (SCADA) software package.

As will be shown later, at this scale soil heating becomes very relevant and this has an important consequence in the overall design of the system: the need to implement gas treatment system. In this point, some authors have proposed the implementation of granular activated carbon beds or scrubber to retain volatilized compounds carried out by gas flow [16, 17].

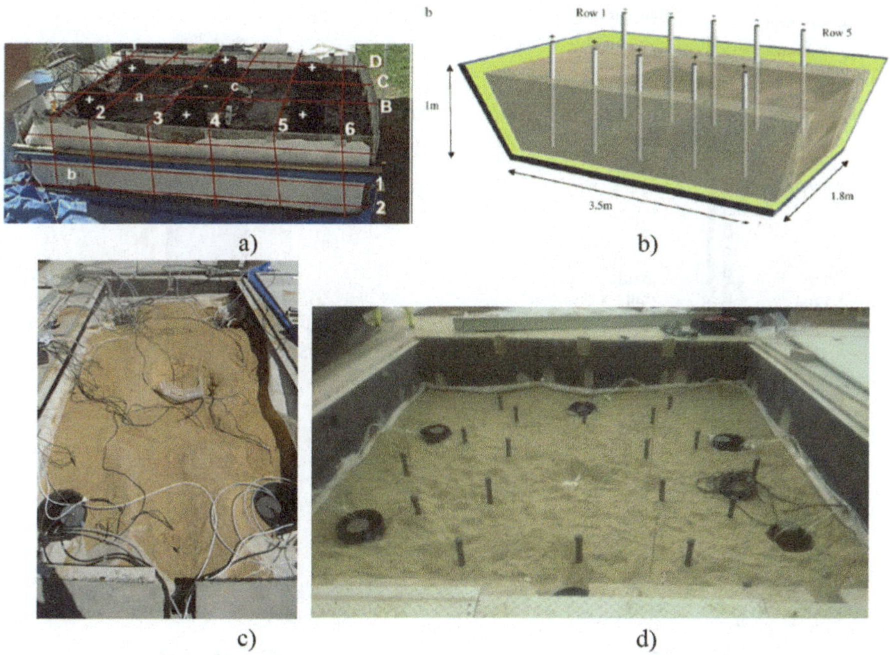


**Fig. 6** Prototype electrokinetic reactors. (**a**) 3.3 m$^3$ [23], (**b**) 4 m$^3$ [24], (**c**) 16 m$^3$ [17], (**d**) 32 m$^3$ [16], (**e**) 90 m$^3$ [26]

***Industrial scale.*** The design and dimensions of the electrochemically assisted soil remediation system to be deployed at a site are based on and derived from the data collected during the foregoing scales. The objective of the field demonstration is to evaluate the electrochemically assisted soil remediation in a relevant environment, and to identify, collect, and verify the economic, operational, and performance data that would be used to validate and transfer this technology to potential users. Engineering and technological considerations (electrode material, well design, instrumentation, auxiliary systems, etc.) are like those proposed at prototype level.

## 3 Technological Outcomes

As mentioned before, the electrochemically assisted soil remediation processes are the sum of many contributing processes started by the application of an electric field between electrodes placed in the polluted soil: electrokinetic (electroosmosis, electrophoresis, electromigration), electrochemical (water oxidation and reduction), chemical (ionic exchange, precipitation) and physical (heating). From a technical point of view, the most important scale-up effect is related to the significance of these processes and to the controlling transport mechanism [10, 17]. This

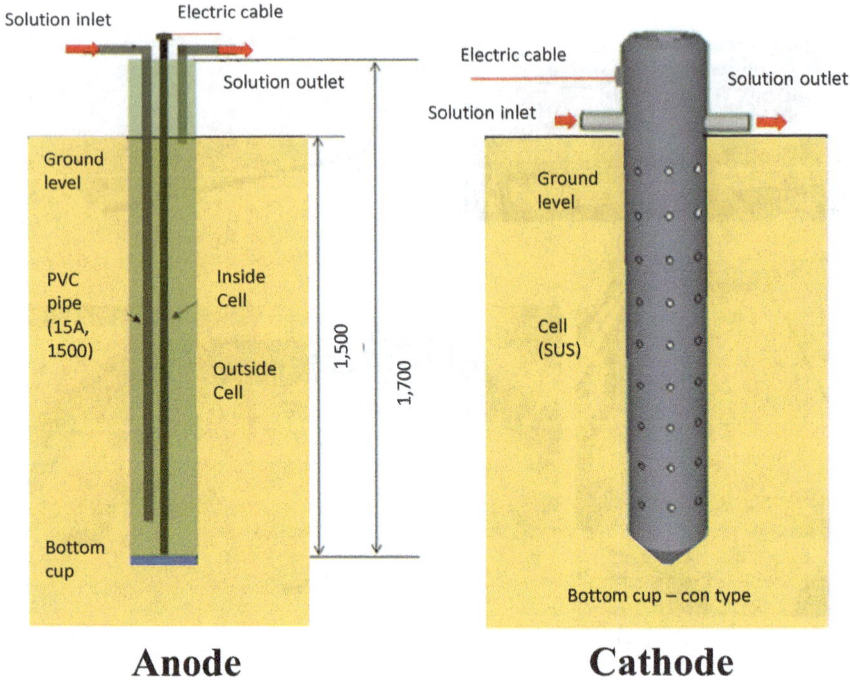


**Fig. 7** Schematic diagram of electrode system proposed by Woo-Seung et al. [26]

information can be discerned from the analysis of the main operation parameters and of the main output variables, such as current intensity, pH, soil temperature and electrokinetic flows.

Unfortunately, very few works have evaluated the comparison at different scales of the same electrochemically assisted soil remediation process, and the information available in the literature corresponds, mainly, to non-connected works where the mineralogical composition and geochemical properties of the soil, the electrode material and electrode configurations and the target pollutant vary from one to the other. In addition, there is a lack of coherence on how to report data on voltage distribution and transport rates, which complicate the comparison of studies [27]. Such varied conditions lead to varied information whose lack of uniformity hinder the delivery of very valuable data. Hence, the formulation of general conclusions and recommendations is only possible.

One of the key parameters of electrochemically assisted soil remediation is the current intensity which results for the electric field applied between the electrode. In this point, it is important to remember that electrokinetic system generally operates in potentiostatic mode and that the resulting current intensity depends on the electric potential but also on the ionic resistance of the soil sited between electrodes. In soil remediation, the reference parameter is the electric potential gradient ($E_z$,

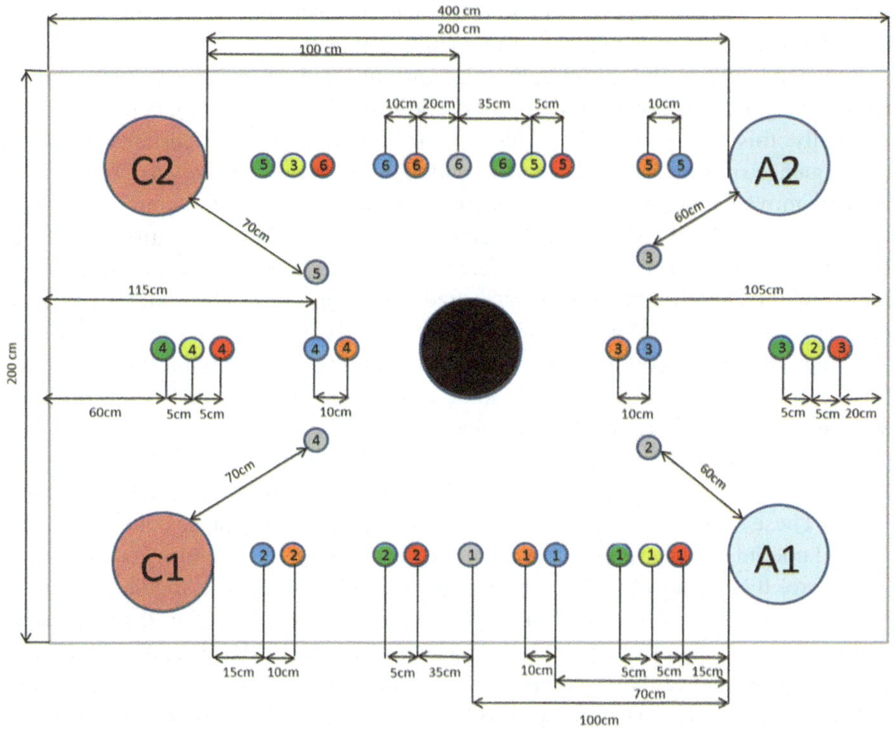


**Fig. 8** Schematic of the instrumentation distribution in 16 $m^3$ prototype [10]: (red circle) thermocouples (1 m from soil surface), (yellow circle) thermocouples (0.1 m from soil surface), (green circle) moisture sensors (1 m from soil surface), (orange circle) conductimeters (0.1 m from soil surface), (gray circle) tensiometers (0.1 m from soil surface), (blue circle) pH meters (0.1 m from soil surface)

V $cm^{-1}$). This means that the applied electric potential required to reach the desired $E_z$ increases with the interelectrode distance. Therefore, the electric potential is affected by the scale due to the distance between anode and cathode increases from few centimetres (4–6 cm) at lab scale to 40–50 cm at pilot scale or to 200–300 cm at prototype. So, assuming a typical $E_z$ of 1 V $cm^{-1}$, the applied voltage can vary from 4 V at lab scale to 300 V at prototype. This high electric potential at prototype requires the use of high-performance power source but, more importantly, can implicate the promotion of non electrokinetic processes because the energy supplied is used in electrode reaction and generation of electric field to transport contaminants through soil, but also part of this energy might be used to heat the soil.

Regarding the resulting current intensity, it may differ between trials carried out at the same scale and electric potential, due to the effect of other factors such as the ionic conductivity of the flushing fluid used, soil moisture or ionic species contained in the soil. To illustrate it, Fig. 9 shows the relation between electric potential and electric current using acetic acid and nitric acid as flushing fluids during the operation of a pilot-scale to remediate a radioactive soil [19].

Independently of the initial value, the current intensity generally varies during the first stages of the remediation process and it stabilizes in few hours of operation. Typically, this rise is related to changes in the conductivity of the electrolyte as consequence of the generation of $OH^-$ and $H^+$ from water electrolysis. Then, these ions are moved through the soil by electromigration and this can favour the precipitation or redissolution of chemical species, and the fixation or liberation of species by ionic exchange. The processes affecting inorganic species generally occur very rapid and the operation parameters stabilize in few days or in few hours (in the case of pH). The magnitude of these chemical processes depends on the nature of the pollutant and on the composition of the soil, and this could explain the fluctuation observed in the current intensity in some cases. However, the scale used does not seem to be the key factor of these changes.

As mentioned before, the electric current is the driving force of electrochemical and electrokinetic processes, but also of the electrical heating caused by ohmic losses. These ohmic losses are initiated by the large ionic resistances of the soil, that depend not only on the soil ionic conductivity but also on the interelectrode distance. Therefore, the higher electric current applied at high size scale together with the higher interelectrode distance promote the electrical heating, and thus the increase of soil temperature. At this respect, Lopez-Vizcaíno et al. [17] have reported that temperature at the end of the prototype trial multiplies by three or four the initial value in the soil and wells, respectively. Additionally, this rise is not homogeneous, and it generates a clear profile of temperatures in the soil, with hot areas in the nearness of the electrodes (both anodes and cathodes). Part b of Fig. 9 shows the 2D maps of soil temperature registered before and after electroremediation trial carried out at prototype [17]. On the other hand, comparing temperature profiles at different scales [10], no significant changes are observed over the lab-scale tests due to, in this case, the very low resulting current intensity does not lead to relevant heating. In addition, this small temperature rise can be compensated by the evaporative cooling and also by the exchange of energy with the environment, because of the high relationship surface/volume of the lab-scale cell. On contrary, the average temperature during pilot scale trials increases, although this increment is less relevant than at prototype, due to the lower ohmic losses associated to the closer position between electrodes in the pilot scale. Therefore, the observed discrepancies between scales lead to reasonable doubts regarding the validity of making predictions about the effectiveness of a process using results obtained from reduced-scale tests. Part a of Fig. 9 illustrates the changes in the average temperature during electrochemically assisted remediation trials performed at the pilot scale and prototype [17].

The main effect of temperature rise is related to the transport mechanisms and the mobility of species during electroremediation treatment. Additionally, it can also favour the intensification of various processes, such as the evaporation of soil water or the desorption of volatile or semivolatile species. Among them, the evaporation of water is of great importance in the development of electrokinetic remediation because it can cause contraction defects on the surface of the soil, which considerable disturb the development of electrokinetic processes. Nevertheless, this

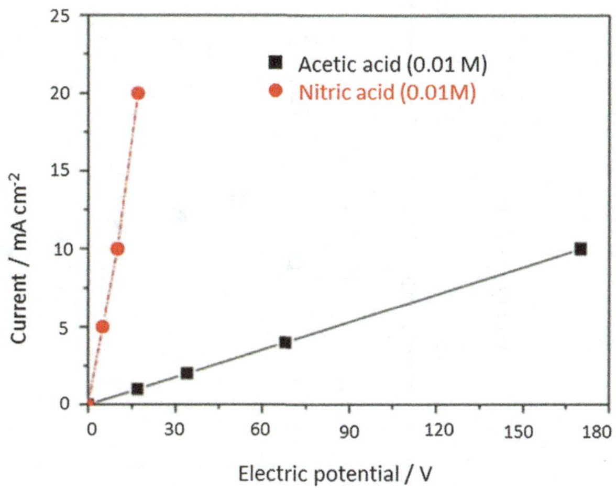


**Fig. 9** Relation between electric potential and electric current with acetic acid and nitric acid (as flushing fluids) in the pilot-scale electrokinetic remediation of a radioactive soil. Interelectrode distance: 20 cm. Anode: metal mixed oxide. Cathode: titanium [20]

evaporation can be minimized including capillary barriers and/or impermeable cover on the surface of the soil [10]. Regarding desorption of volatile compounds, this is not observed in lab scale due to the lower electric current applied and also to the peculiarities of the lab system (hermetically sealed cell). However, the scaling-up (higher size and higher electric current) and the use of open to atmosphere system highlight the relevance of volatilization transport mechanisms, mainly in the treatment of hydrocarbon polluted soils (Fig. 10). To illustrate this, Table 1 shows overall remediation percentage attained in different prototype trials and Fig. 11 shows the percentages of pesticide remaining in soil and removed by electrokinetic and desorption after tests carried out at three different scales.

Therefore, from a technical point of view, the most important scale effect does not seem to be related to the mechanisms involved in the electrochemically assisted soil remediation processes (they are the same in the different scales) but to the their relevance and magnitude. Table 2 summarizes the main outcomes of each scale.

## 4 Conclusions

In the scaling of an electrochemically assisted soil remediation, each scale contributes with relevant and non-overlapping information. All scales are useful and the missing of some of them can lead to misinterpretation of the process and can difficult the design of the real application. Laboratory and bench scales contribute to the assess the technical feasibility of the process, while pilot and prototype contribute information of use in the operation of the process. The mechanisms of transport are the same in the different scales but the higher the scale, the lower is the magnitude of the electrokinetic processes and the higher is the relevance of the volatilization. This outcome is related to the higher temperature reached in the larger scale facilities, which in turn is related to the higher ohmic losses in energy.

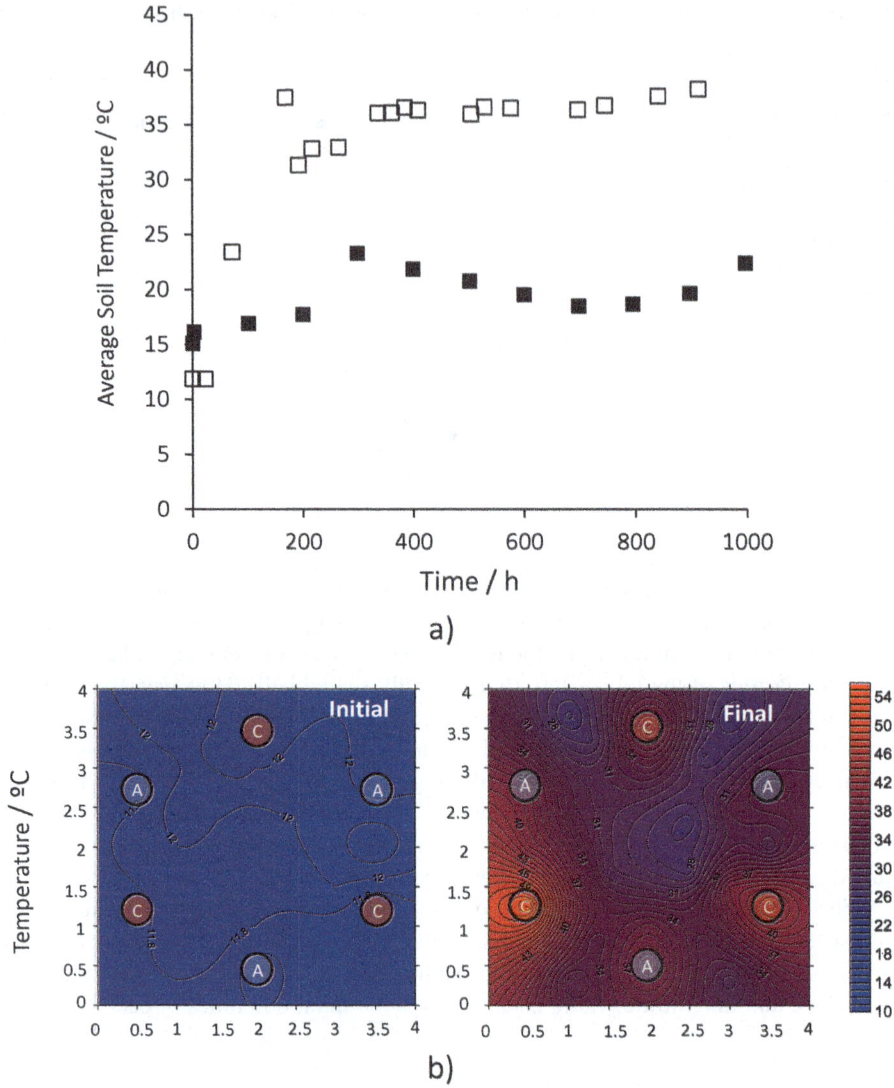


**Fig. 10** (**a**) Changes in the average temperature of soil during pilot scale (filled square) and prototype (open square) trials. (**b**) Initial and final 2D maps of temperature in soil during prototype trial. $E_z$: 1 V cm$^{-1}$. Pilot scale: 173 dm$^3$. Prototype: 32 m$^3$ [16]

**Table 1** Some results of electrochemically assisted soil remediation processes carried out at prototype

| Pollutant | Concentration, mg kg$^{-1}$ | Soil capacity, m$^3$ | $E_z$, V cm$^{-1}$ | Duration, days | Percentage removal, % | Reference |
|---|---|---|---|---|---|---|
| Gasoline | 1126 | 3.3 | 1 | 21 | 80 | 23 |
| As/Cu/Pb | 55/80/170 | 4 m$^2$ (surface) | 1 | 30 | 39.8/17.2/19.4 | 26 |
| 2,4-D/ oxyfluorfen | 20/20 | 32 | 1 | 30 | 90/85 | 17 |

**Acknowledgments** Financial support from the Spanish Ministry of Economy, Industry and Competitiveness and European Union through project CTM2016-76197-R (AEI/FEDER, UE) is gratefully acknowledged.

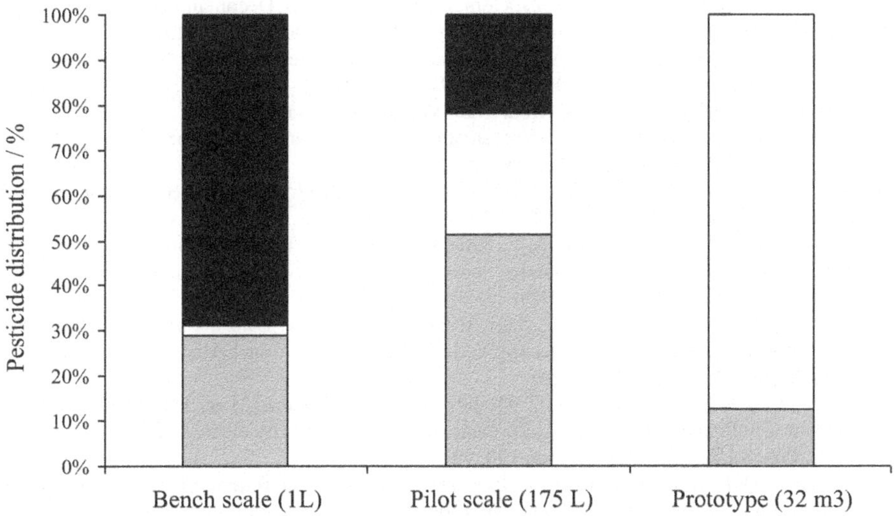


**Fig. 11** Percentage of 2,4-D (model pesticide) remaining in soil (gray shaded square) and removed by electrokinetic (filled square) and volatilization (open square) after the tests carried out at three different scales. $E_z$: 1 V cm$^{-1}$ [16]

**Table 2** Summary of the role of the scale on the relevance of remediation processes

| Scale | Main outcomes |
|---|---|
| Laboratory | Electrokinetic processes are predominant. No other side transport processes are observed. The extreme isolation from external agents is useful to estimate electroosmotic flux, diffusivity and electromigration parameters, or the ion exchange capacity of the soil and its chemical and electrochemical reactivity. |
| Bench | Reduction in the relative importance of electrokinetic and reactive processes, although they are still the main process. Very valuable information about the electroosmotic flux and the acidic and basic fronts is obtained, but extrapolation to the full range is still not accurate. |
| Pilot | The size of the device is large enough to begin to show the important difference between the electrokinetic and reaction processes and the soil heating suffered by the soil. The higher resistance associated with the larger distance between the electrodes begins to indicate that the thermal process is not negligible, but a key process in electrochemically assisted soil remediation processes. |
| Prototype | The observations related to the main mechanisms in the pilot plant became more important in the prototype: the electrokinetic processes became almost negligible compared with soil heating. This has an important consequence in the design of the system: the need to implement gas treatment system. |

## References

1. J. Gomez, M.T. Alcantara, M. Pazos, M.A. Sanroman, Remediation of polluted soil by a two-stage treatment system: desorption of phenanthrene in soil and electrochemical treatment to recover the extraction agent. J. Hazard. Mater. **173**, 794–798 (2010)
2. M. Pazos, E. Rosales, T. Alcantara, J. Gomez, M.A. Sanroman, Decontamination of soils containing PAHs by electroremediation: a review. J. Hazard. Mater. **177**, 1–11 (2010)
3. K.R. Reddy, K. Darko-Kagya, A.Z. Al-Hamdan, Electrokinetic remediation of pentachlorophenol contaminated clay soil. Water Air Soil Pollut. **221**, 35–44 (2011)
4. A.B. Ribeiro, E.P. Mateus, J.-M. Rodriguez-Maroto, Removal of organic contaminants from soils by an electrokinetic process: the case of molinate and bentazone. Experimental and modeling. Sep. Purif. Technol. **79**, 193–203 (2011)
5. M.T. Alcantara, J. Gomez, M. Pazos, M.A. Sanroman, Electrokinetic remediation of lead and phenanthrene polluted soils. Geoderma **173**, 128–133 (2012)
6. H.I. Gomes, C. Dias-Ferreira, A.B. Ribeiro, Electrokinetic remediation of organochlorines in soil: enhancement techniques and integration with other remediation technologies. Chemosphere **87**, 1077–1090 (2012)
7. E. Vieira dos Santos, F. Souza, C. Saez, P. Cañizares, M.R.V. Lanza, C.A. Martinez-Huitle, M.A. Rodrigo, Application of electrokinetic soil flushing to four herbicides: a comparison. Chemosphere **153**, 205–211 (2016)
8. M. Popescu, E. Rosales, C. Sandu, J. Meijide, M. Pazos, G. Lazar, M.A. Sanromán, Soil flushing and simultaneous degradation of organic pollutants in soils by electrokinetic-Fenton treatment. Process Saf. Environ. Protect. **108**, 99–107 (2017)
9. C. Sandu, M. Popescu, E. Rosales, M. Pazos, G. Lazar, M.A. Sanromán, Electrokinetic oxidant soil flushing: a solution for in situ remediation of hydrocarbons polluted soils. J. Electroanal. Chem. **79915**, 1–8 (2017)
10. R. Lopez-Vizcaino, V. Navarro, M.J. Leon, C. Risco, M.A. Rodrigo, C. Saez, P. Canizares, Scale-up on electrokinetic remediation: engineering and technological parameters. J. Hazard. Mater. **315**, 135–143 (2016)
11. D. Li, D. Sun, S. Hu, J. Hu, X. Yuan, Conceptual design and experiments of electrochemistry-flushing technology for the remediation of historically Cr(VI)-contaminated soil. Chemosphere **144**, 1823–1830 (2016)
12. S.-V. Ho, C.J. Athmer, P.W. Sheridan, A.W. Shapiro, Scale-up aspects of the LasgnaTM process for in situ soil remediation. J. Hazard. Mater. **55**, 69–60 (1997)
13. B.-K. Kim, K. Baek, S.-H. Ko, J.-W. Yang, Research and field experiences on electrokinetic remediation in South Korea. Sep. Purif. Technol. **79**, 116–123 (2011)
14. M. Villen-Guzman, J.M. Paz-Garcia, J.M. Rodriguez-Maroto, F. Garcia-Herruzo, G. Amaya-Santos, C. Gomez-Lahoz, C. Vereda-Alonso, Scaling-up the acid-enhanced electrokinetic remediation of a real contaminated soil. Electrochim. Acta **181**, 139–145 (2015)
15. R. Lopez-Vizcaino, C. Risco, J. Isidro, S. Rodrigo, C. Saez, P. Canizares, V. Navarro, M.A. Rodrigo, Scale-up of the electrokinetic fence technology for the removal of pesticides. Part I: some notes about the transport of inorganic species. Chemosphere **166**, 540–548 (2017)
16. R. Lopez-Vizcaino, C. Risco, J. Isidro, S. Rodrigo, C. Saez, P. Canizares, V. Navarro, M.A. Rodrigo, Scale-up of the electrokinetic fence technology for the removal of pesticides. Part II: does size matter for removal of herbicides? Chemosphere **166**, 549–555 (2017)
17. S. Barba, R. Lopez-Vizcaino, C. Saez, J. Villasenor, P. Canizares, V. Navarro, M.A. Rodrigo, Electro-bioremediation at the prototype scale: what it should be learned for the scale-up. Chem. Eng. J. **334**, 2030–2038 (2018)
18. D.B. Gent, R.M. Bricka, A.N. Alshawabkeh, S.L. Larson, G. Fabian, S. Steve Granade, Bench- and field-scale evaluation of chromium and cadmium extraction by electrokinetics. J. Hazard. Mater. **110**, 53 (2004)
19. P. Guedes, E.P. Mateus, N. Couto, Y. Rodríguez, A.B. Ribeiro, Electrokinetic remediation of six emerging organic contaminants from soil. Chemosphere **117**, 124–131 (2014)

20. G.-N. Kim, S.-S. Lee, D.-B. Shon, K.-W. Lee, U.-S. Chung, Development of pilot-scale electrokinetic remediation technology to remove $^{60}Co$ and $^{137}Cs$ from soil. J. Ind. Eng. Chem. **16**, 986–991 (2010)
21. C. Risco, R. López-Vizcaíno, C. Sáez, A. Yustres, P. Cañizares, V. Navarro, M.A. Rodrigo, Remediation of soils polluted with 2,4-D by electrokinetic soil flushing with facing rows of electrodes: a case study in a pilot plant. Chem. Eng. J. **285**, 128–136 (2016)
22. R. Lopez-Vizcaino, V. Navarro, J. Alonso, A. Yustres, P. Canizares, M.A. Rodrigo, C. Saez, Geotechnical behaviour of low-permeability soils in surfactant-enhanced electrokinetic remediation. J. Environ. Sci. Health Part A Toxic/Hazard. Subst. Environ. Eng. **51**, 44–51 (2016)
23. B. Ochoa, L. Ramos, A. Garibay, M. Pérez-Corona, M.C. Cuevas, J. Cárdenas, M. Teutli, E. Bustos, Electrokinetic treatment of polluted soil at pilot level coupled to an advanced oxidation process of its wastewater. Phys. Chem. Earth **91**, 68–76 (2016)
24. K. Agnew, A.B. Cundy, L. Hopkinson, I.W. Croudace, P.E. Warwick, P. Purdie, Electrokinetic remediation of plutonium-contaminated nuclear site wastes: results from a pilot-scale on-site trial. J. Hazard. Mater. **186**, 1405–1414 (2011)
25. D.-H. Kim, J.-C. Yoo, B.-R. Hwang, J.-S. Yang, K. Baek, Environmental assessment on electrokinetic remediation of multimetal-contaminated site: a case study. Environ. Sci. Pollut. Res. **21**, 6751–6758 (2014)
26. W.-S. Kim, G.-Y. Park, D.-H. Kim, H.-B. Jung, S.-H. Ko, K. Baek, In situ field scale electrokinetic remediation of multi-metals contaminated paddy soil: influence of electrode configuration. Electrochim. Acta **86**, 89–95 (2012)
27. L.M. Ottosen, T.H. Larsen, P.E. Jensen, G.M. Kirkelund, H. Kerrn-Jespersen, N. Tuxen, B.H. Hyldegaard, Electrokinetics applied in remediation of subsurface soil contaminated with chlorinated ethenes: a review. Chemosphere **235**, 113–125 (2019)

# Electrochemical Technologies for Petroleum Contaminated Soils

**Sibel Pamukcu**

## 1 Introduction and Background

Contamination of soils and sediments with petroleum liquids is a major challenge to groundwater resources [1–3]. The filtration and retention properties of the soil often dictate the extent of contamination by petroleum liquids in subsurface. When the contaminants, particularly the immiscible liquids, are adsorbed onto soil components (i.e., iron-oxide surfaces, soil colloids and organic matter), or retained in clay interstices and pore fluid in the form of immobile products, the available technologies may not be able to treat the entire site effectively. Electrokinetic remediation may decrease the petroleum liquids contamination by transporting and/or transforming these contaminants. Electrokinetic processing is a resource recovery method as it may recover trapped oils and petroleum, which may not be extractable feasibly by other means from porous media. In these regards, electrokinetically assisted in situ processes has been most viable for low permeability clayey deposits containing compounds difficult to treat [3–18].

Originally used for consolidating clays [19], electrokinetic remediation was mostly studied and applied in pilot-scale tests between 1980s and early 2000s. These studies and applications included extracting heavy metals [20–44], extracting and treating some organic compounds [45–52], and dewatering of hazardous slurries [53–56]. The field-scale applications of electrokinetic remediation methods have increased since 1980s [57–64]. Currently, electrokinetic processing has been shown to successfully treat soils containing radioactive materials [25, 65–67] and a variety of organic and mixed compounds [68–76], and wastewater sludges and biomass [77–83]. In many applications, electrokinetics have been used to infiltrate

S. Pamukcu (✉)
Department of Civil and Environmental Engineering, Lehigh University,
Bethlehem, PA, USA
e-mail: sp01@lehigh.edu



M. A. Rodrigo, E. V. Dos Santos (eds.), *Electrochemically Assisted Remediation of Contaminated Soils*, Environmental Pollution 30,
https://doi.org/10.1007/978-3-030-68140-1_18

active agents to enhance solubilization, desorption, transport, transformation or degradation of the contaminants [84–97]. Electrokinetics have also been shown to aid in stripping and transporting non-polar compounds including heavy oils, from soils and sediments [98–100]. Furthermore, electrokinetics have been used in tandem with other in situ techniques, including bioremediation [101–110], phytoremediation [111–113], nano-scale zero valent iron (nZVI) treatment [87–92, 114–116], oxidation-reduction and Fenton-like methods [117–121], permeable reactive barriers or electric fences [122–127] and electrodialysis that manage electrode products [6, 7, 43, 74, 80, 128–131]. The technology still faces engineering challenges in the field due to process generated or site-specific side effects [9, 11, 17, 132–135]. For example, temporally and spatially changing process parameters, such as the voltage gradients, pore water pressures (PWPs), and migrating fronts of ion concentrations can influence the transport and/or recovery process [136, 137].

Some innovative applications of electrokinetics technology in geological media include use of pulsed/sinusoidal electric field [138, 139], DC/AC coupled electric field [64], photovoltaic solar panels [140], and oil recovery by mass displacement in sediment and rock formations [98–100]. Electrokinetics has been demonstrated to enhance transformation reactions through double layer polarization of the clay surfaces leading to Faradaic processes under the applied DC electric field [4, 141, 142]. The low direct current applied during electrokinetics was shown to aid in separation and demulsification processes in oily sediments [77, 78, 143]. Furthermore, electrokinetic treatment of TCE contaminated soils were shown to result in electrochemical degradation of chlorinated solvents near the electrodes [144–146].

This chapter provides an overview of the use of electrokinetics in environmental mitigation of mainly petroleum contaminated soils, for the purpose of extracting/recovering or transforming the petroleum hydrocarbons.

## 2 Electrokinetic Extraction and Recovery

In electrokinetic extraction, the contaminant is extracted from the soil by advective flow of water toward the negative electrode site. Additionally, compounds that remain ionic under changing pH and redox conditions within the electric field electromigrate toward the oppositely charged electrode site. As the ions separate, the electro-neutrality is maintained by the hydrogen and the hydroxide ions produced by the electrode reactions. The extraction of organic compounds, unless they are driven by colloidal transport (i.e., sorbed onto electrophoretically moving colloids), and/or maintain a surface charge, depends on the advective action of electroosmosis. In most cases, surface tension reducing agents (i.e., surfactants) are used to make the organic compounds available for electroosmotic transport.

Several selected laboratory experiments using field retrieved soil samples are described in the following sections to exemplify the variety of electrokinetic extraction and separation processes applicable to petroleum hydrocarbon contaminated soils.

## 2.1 Extraction/Treatment of VOCs, PAHs, and Organic Acids

Despite recent advances, there is still much work to be done in investigation of electrokinetic transport of NAPLs in contaminated subsurface. Early examples of work include those by Bruell et al. [45] who showed transport of TCE and several gasoline hydrocarbons by electroosmosis, and Acar et al. [147] who showed transport of phenol by electroosmosis and electromigration. Ho et al. [59–61] conducted field experiments demonstrating that the dissolved fraction of TCE could be transported by electroosmosis, through the Lasagna™ [1] process. The horizontally layered configuration of treatment zones had inspired its name for this process. In their work, horizontal treatment zones containing zero valent iron were installed in the subsurface along with electrodes to generate electroosmotic flow of the TCE into these zones. The materials used for the electrodes and the treatment zones were compatible with the environment and designed to be left in place at the completion of the project. In these field experiments, the layered configuration of the treatment zones allowed for sequential de-chlorination of TCE and its daughter products (i.e., cis-dichloroethylene, vinyl chloride) into non-toxic compounds (i.e., ethylene and ethane) using zero valent iron.

In 1994, Pamukcu [48] conducted extensive laboratory treatments of field retrieved specimens of coal tar contaminated soil from a manufactured gas plant (MGP) site in Champaign-Urbana, Illinois. The coal tar compounds evaluated were: naphthalene, acenaphthylene, acenaphthene, fluorene, phenanthrene, anthracene, fluoranthene, pyrene, chrysene, benz(a)anthracene, benzo(b)fluoranthene, benzo(k)fluoranthene, benzo(a)pyrene, dibenzo(a,h)anthracene, benzo(g,h,i)perylene, and indeno(1,2,3-c,d)pyrene. The results indicated that the water-soluble PAH compounds with lower molecular weight (i.e., naphthalene) could be transported electroosmotically in the absence of surface tension reducing agents. The posttreatment average distributions of the total PAHs are shown in Fig. 1. The electroosmotic advection, referred to as "EO," transported the PAHs from the anode toward the cathode side of the test sample, but did not completely remove them from the soil. The PAHs were accumulated adjacent to the cathode electrode interface at near original concentration. When an anionic surfactant (sodium dodecyl sulfate (SDS)) and a co-surfactant (butanol) was injected into the system while along with electroosmosis, the soil PAH mass was reduced significantly, to below 20% the original in the case of surfactant only.

In 1998, Pamukcu and Pervizpour [148], conducted laboratory treatments of field retrieved specimens of TCE (trichloroethylene) contaminated soil from a DOE site in California. Due to possible loss of the residing TCE by volatilization during sampling, the retrieved specimens were re-injected with TCE solution below its solubility limit. The TCE injection was accomplished by electroosmotic advection of the solution from the anode (influent chamber) to the cathode (effluent chamber), while keeping constant the concentration of the solution in the anode chamber. The

[1] www.terrancorp.com/remediation/electro.htm.

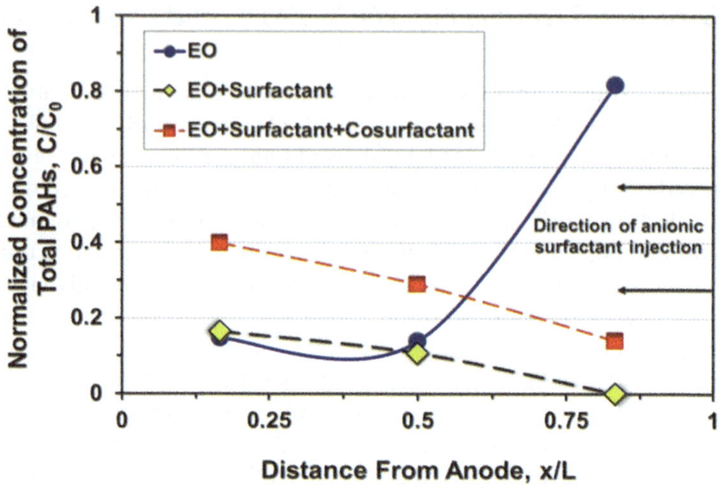


**Fig. 1** Post-EK distribution of the normalized concentration of total PAHs in coal tar contaminated specimens from an MGP site in Illinois [48]

resulting breakthrough of TCE into the cathode chamber is shown in Fig. 2. The breakthrough at cathode occurred at about 15,000 min of EO flow, which corresponded to approximately 1.5 pore volumes. The hydraulic conductivity of the core was measured as $k_h = 10^{-8}$ cm/s, and its electroosmotic conductivity was determined to be $k_{eo} = 2 \times 10^{-6}$ cm$^2$/s V. The equivalent hydraulic conductivity, described as $k_{eh} = k_{eo} \cdot i_{eo}$, is plotted with time in Fig. 2. The $k_{eh}$ decreased initially but attained an average constant value of $5 \times 10^{-6}$ cm/s in time. The breakthrough curve indicates retardation of TCE on the solid phase as expected. The results showed that the soluble fraction of TCE could be transported by electroosmosis, in agreement with the earlier Lasagna™ experiments in field [59–61].

Samples of drilling mud fluid containing potassium formate (i.e., potassium salt of formic acid) was treated electrokinetically to extract the potassium content of the mixture for recycling [149]. In this process, potassium was intended to separate from the formic acid as shown in the conceptual schemes in the single and double-cell test configurations shown in Fig. 3. The potassium formate solution was contained in the center or the anode chambers, with soil packs acting as filters in between. In either case, potassium was expected to filter through the soil into the oppositely charged electrode chamber. Figure 4 shows the temporal distribution of potassium and formic acid in the electrode chambers of a single cell unit during electrokinetic treatment. In all the experiments, the potassium concentration increased at the cathode, while it decreased at the anode and the center chambers. Formic acid decreased in the anode chamber also, indicating diffusion of the compound into the soil filter. The potassium separation and recovery appeared to be most successful in the single cell test, as majority of the potassium (~80%) moved into the cathode chamber and its concentration was reduced to almost zero in the anode chamber, as shown in Fig. 4.

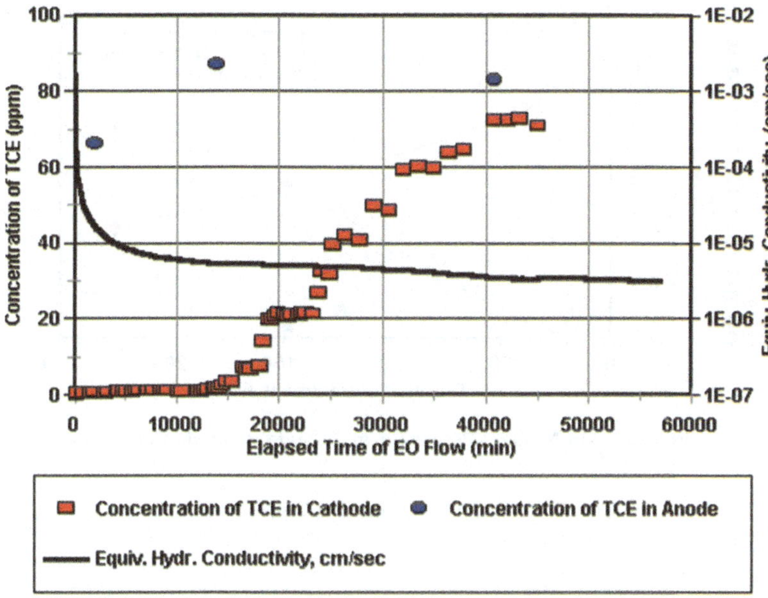


**Fig. 2** Cumulative mass transport of TCE from anode to cathode electrode chamber by EO in soil samples from LLNL DOE Complex site [148]

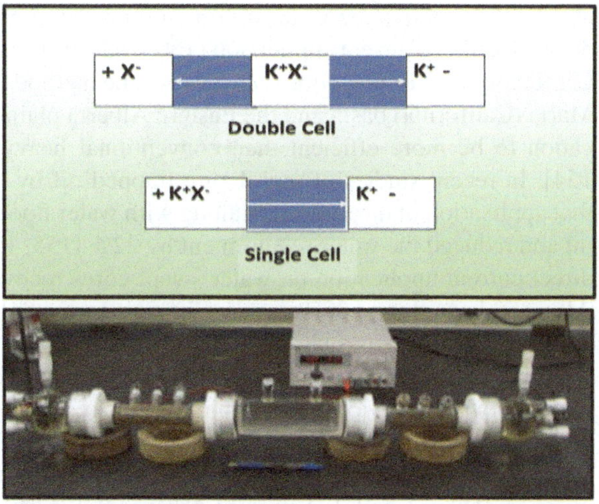


**Fig. 3** Single and double-cell EK test configurations and photo of a test cell for extracting $K^+$ from drilling mud [149]

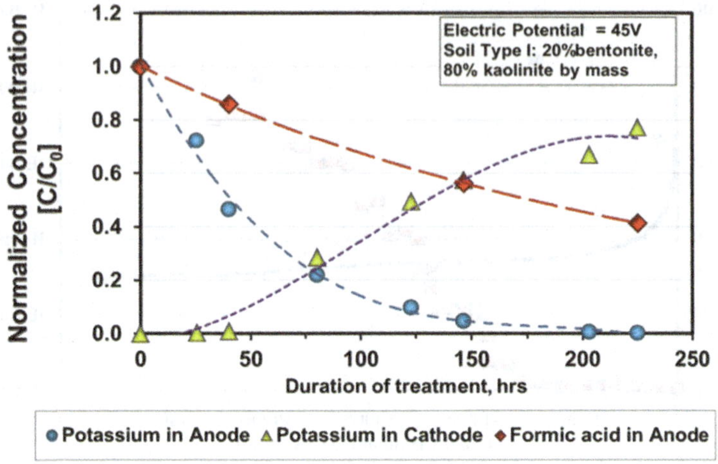


**Fig. 4** Evolution of potassium and formic acid concentrations in the electrode sites in *single cell* EK test [149]

## 2.2 Extraction of Crude Oil

In early 1960s, researchers in petroleum engineering field found that direct electric current application enhanced the flow rate of oil and water in rock formations over that of water flooding only (i.e., by up to 34-fold). Additionally, the effective permeability of the entrapped water in formation pores containing kerosene increased by 50% with direct current application [150–152]. Referred to as Electrically Enhanced Oil Recovery (EEOR), pilot-scale tests of the method in heavy oil fields in the Santa Maria (California) basin and the Eastern Alberta plains showed direct current application to be more efficient than conventional heavy oil recovery methods [153, 154]. In recent studies of mobilizing trapped oil by core-flooding it was reported that application of direct current along with water flooding recovered up to 4% more oil and reduced the water requirement by 22% [155, 156]. Furthermore, subsequent direct current application on water swept cores recovered up to 9% additional oil. All tests showed upto 30% increase in core permeability, potentially due to colloid dislodgement by electrophoretic transport [99, 157].

Similarly, a series of electrokinetic oil extractability tests on oil-bearing rock cores from a shallow oil formation in Kentucky showed marked change in the rock-core permeability upon direct current application [158]. The physical properties of the rock cores and of the natural oil and surrogate formation water for the tests are given in Tables 1 and 2. A current density of 1 A/m$^2$, similar to that achievable in field applications (1.0–1.5 A/m$^2$), was applied to the cores for about 120 h. Most of the oil recovered was produced at the catholyte, in the range of 5–11% by mass of the initial oil content of the cores. The results suggested a direct relation between the initial oil content and the achieved oil recovery. Modeling studies of two-phase

**Table 1** Physical properties and oil recovery in the natural sand stone cores from Kentucky [158]

| Core | Porosity (%) | Bulk density ($g/cm^3$) | Permeability (mD) | Residual oil (%) | Residual water (%) | Recovered oil at the anolyte (ml) | Recovered oil at catholyte (ml) | Total recovered oil (%) |
|---|---|---|---|---|---|---|---|---|
| KY1 | 10.2 | 2.48 | 2.67 | 12.9 | 61 | 0.000 | 0.021 | 5.1 |
| KY2 | 13.6 | 2.39 | 11.59 | 28.1 | 43.9 | 0.005 | 0.104 | 9.9 |
| KY3 | 14.7 | 2.37 | 41.88 | 30.2 | 33.9 | 0.002 | 0.157 | 11.4 |

**Table 2** Physical properties of the Kentucky formation oil and surrogate water samples [158]

| Fluid | Properties |
|---|---|
| Natural formation oil | API 22; Dynamic viscosity = 66.5 cP |
| | Specific gravity = 0.92 at 20 °C |
| Electrolyte solution (Surrogate formation water) | Salinity = 33,000 ppm; Electrical conductivity = 45,000 μS |
| | pH = 7.50; Major elements: Na, Cl, Mg, S, K, Ca, Br |

EO flow showed that the oil production decreased when water saturation was above 50% and the oil saturation dropped below 50% [158, 159].

Electrokinetic oil extractability tests on oil-bearing sandstone cores from a deep oil formation in Pennsylvania were conducted. These cores were spiked with oil of a known composition (API 47; Dynamic viscosity = 38.7 cP) and formation water prior to testing. The cores were impregnated to a desired initial oil-water saturation using a high-pressure injection pump. They were tested under a constant current density of 1 $A/m^2$ for 130 h. Table 3 presents the properties and the test results of the four Pennsylvania cores. Once again, most of the oil recovered was produced at the catholyte, in the range of 0.5–6% by mass of the initial oil content of the cores. The highest oil recovery was achieved in those cores with the highest initial oil content and highest hydraulic permeability.

Cumulative oil and water production were monitored during an EEOR application in the Pennsylvania field for 180 days. In addition to oil and water, paraffin was also observed in the production well. It was postulated that during EEOR, as pH and dissolved minerals (e.g., calcite) concentration increase at the production well (cathode), it sets the ideal conditions for a portion of the oil to transform into paraffin. Figure 5 presents the cumulative production of the three fluids over time. The cumulative oil was about 5% of the water by volume, while the cumulative paraffin was about 4% of the oil by volume. There were several power-off and power-on cycles during the treatment. It was observed that the oil production was not affected during the power-off cycles. The continuity of the oil production was attributed to the initial build-up of the electroosmotic seepage and suction pressures, which would require time to dissipate and adjust in low permeability formations when the driving gradient suddenly subsides [136]. The paraffin production increased with power and temperature, and ceased with reversal of current, as shown in Fig. 6. Since reversal of current would decrease the pH and oil production, these conditions

**Table 3** Physical properties and oil recovery in the oil impregnated cores from Pennsylvania [158]

| Core | Porosity (%) | Bulk density (g/cm$^3$) | Perme-ability (mD) | Oil saturation (%) | Water saturation (%) | Recovered oil at the anode side (ml) | Recovered oil at cathode side (ml) | Total recovered oil (%) |
|---|---|---|---|---|---|---|---|---|
| PA1 | 14.5 | 2.49 | 3.6 | 42 | 58 | 0.0056 | 0.0457 | 1.1 |
| PA2 | 12.8 | 2.43 | 2.3 | 48 | 52 | 0.014 | 0.23 | 3.8 |
| PA3 | 14.4 | 2.47 | 8.3 | 53 | 47 | 0.020 | 0.377 | 6.3 |
| PA4 | 11.5 | 2.45 | 0.5 | 40 | 60 | 0.0024 | 0.0198 | 0.6 |

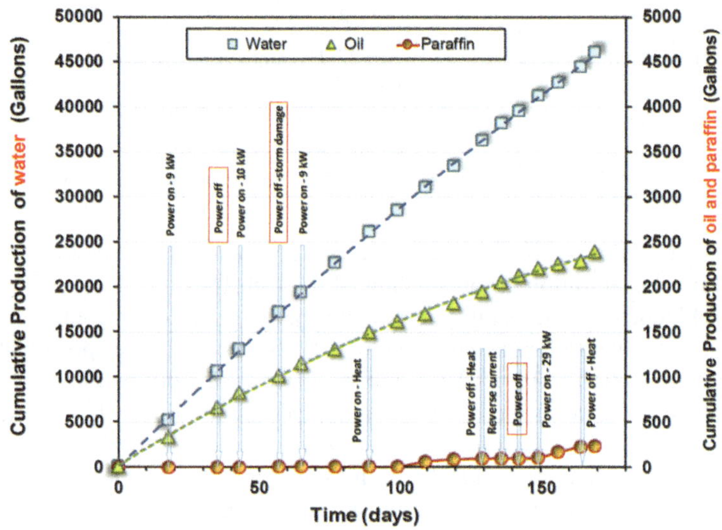


**Fig. 5** Water, oil, and paraffin production (Pennsylvania field) [158]

would also decrease paraffin production. It is also noted from Fig. 6 that at times of increased paraffin production, the oil production did not follow the trend. This incidence may be attributed to increased availability of dissolved minerals at elevated temperature for paraffin production.

## 2.3 *Laboratory Evidence of Crude Oil Transport by Water Displacement*

Electroosmotic advection can assist mass transport of immiscible hydrocarbon compounds in natural porous media. The moving water and its ionic constituents can drag the nearby hydrocarbon compounds and their micelles along. Viscous coupling between the two liquid phases (i.e., pore water and oil) and the resulting drag action is one of the mechanisms that can transport hydrocarbon compounds in

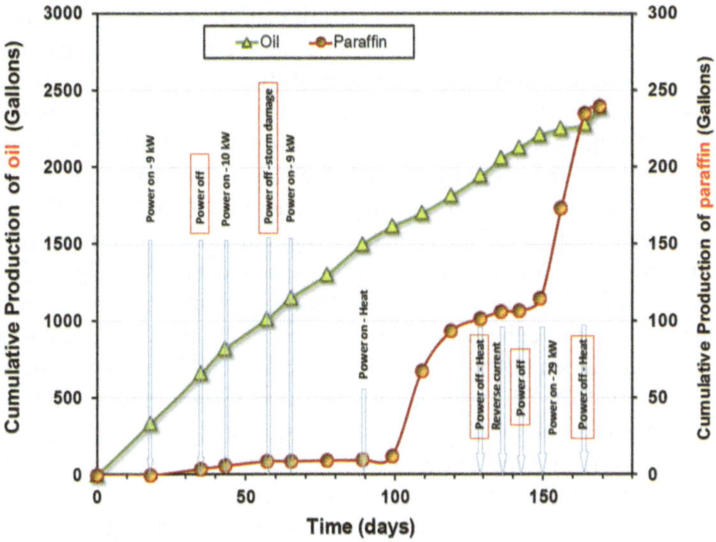


**Fig. 6** Oil and paraffin production (Pennsylvania field) [158]

water-wet porous media. The other mechanism stem from electrochemical reactions that can reduce the viscosity of the oil and the interfacial tension between water and oil, making the oil more mobile. It is possible that both mechanisms occur simultaneously leading to increased mobility of oil by electroosmotic advection. The reduction in interfacial tension occur as a result of the reactions between electrolysis products of formation water (hydroxyl ions) and the acid impurities of oil (carboxylic acids) [160]. The reaction products are low solubility surface-active carboxylates, analogous to a surfactant [161]. The magnitude of interfacial tension reduction is reported to be proportional to the applied electric potential [162].

The viscous dragging of oil requires coupling of the oil mass with the water through an interface, similar to the diffuse double layer (DDL) of clay that gives rise to electrokinetic potential. A similar interface and potential arise between two adjacent immiscible liquids (i.e., oil and water) based on the degree of their polarity, the of salinity of the water, and the presence of functional groups and acidic impurities (i.e., O, N, S compounds, carboxylic acids, amides) in the oil [163]. When one liquid is notably more conductive than the other, and the dipolar coupling force between them is weak [164], the interface does not develop and viscous coupling may not take place. Instead, when put into motion by electroosmosis, the conductive liquid (i.e., pore water) flows pass-by the non-conductive mass (i.e., oil ganglion), as long as the pore space is open for flow. In compact clay sediments where the pore space is tight and water connectivity is poor, and/or the chemistry of the two liquids do not support dispersion or dissolution of one in the other, the net result of a strong electroosmotic (EO) drive may be the separation of two liquids by “displacement” of one with the other. In other words, when water rushes into a confined space (i.e., tight pores of clay sediments) with a high EO driver (i.e., high electric field) and potentially with high seepage pressure [136], it can displace and force the oil mass to move in the opposite direction rather than dragging it forward in the direction of flow. Water moving forward in the direction of the electric field can progressively displace the oil mass toward the opposite direction, causing it to accumulate into an "oil bank" within the porous medium. This phenomenon of aided oil transport in a confined water-wet pore space is referred to as *electrically induced displacement*, particularly relevant for nonpolar and non-conductive oil phases [100].

The electrically induced displacement of non-conducting non-polar oil phase in saline sediments was examined in floor-scale experiments [100]. Figure 7 shows the visuals of post-test accumulation of clay colloids and oil ganglion at various positions along the test soil core. The "clay bank" is the light tan colored zone of fine material at the anode end of the soil core, while the "oil bank" is the dark colored zone across interval E4-E5, particularly under port P3 in Fig. 7. Clay bank formation results from accumulation of colloidal constituents in pore space where electrophoretic transport deposit them at locations where the enabling electrical gradient is weakened and the permeability is reduced by progressive accumulation [100, 136]. Oil bank forms when the pore space is occupied with agglomerated oil ganglia and/or oily micelles that may become too large in size to permeate through the soil pores. Water flow restriction at the regions of "oil banks" would give rise in pore water pressure (PWP) when the EO drive or the electric field is high. This behavior is observed in Fig. 7 from the voltage and PWP distribution at port P3, which is unlike the trends observed at P1, P2, and P4 where flow is not restricted.

The post-test distribution of normalized volumetric oil and water fractions along the specimen ports of P1 through P4 are shown in Fig. 8 [100]. The graph also shows the distribution of volumetric ratio of oil/water (O/W). The O/W ratio is up by about 50% over the anode side half of the core, while it is down by about 30% over the cathode side half of the core. The largest increase in O/W appears at P3 location, where potentially EO flow was restricted due to formation of an oil bank there. The increase in voltage and PWP at P3 also confirms restriction of flow. The significant increase in O/W at this location is attributed to increase in oil content since the water content change was only within standard deviation. Potential "displacement" of the resident oil by electroosmotically advanced water volume toward cathode explains the substantial decrease in O/W ratio over the cathode side half of the core. In the absence of potential viscous coupling between non-polar oil phase and the saline pore water, *electrically induced displacement* emerges as the most plausible mechanism for the observed transport and separation of the immiscible fluids.

## 3 Electrokinetically Enhanced Transformations

In situ transformation of the soil contaminants through beneficial electrochemical reactions have been explored. Electrochemical transformations do take place in natural systems either through intervention of redox active compounds or by microbial activity [165, 166]. Most of the electrochemical transformations take place in

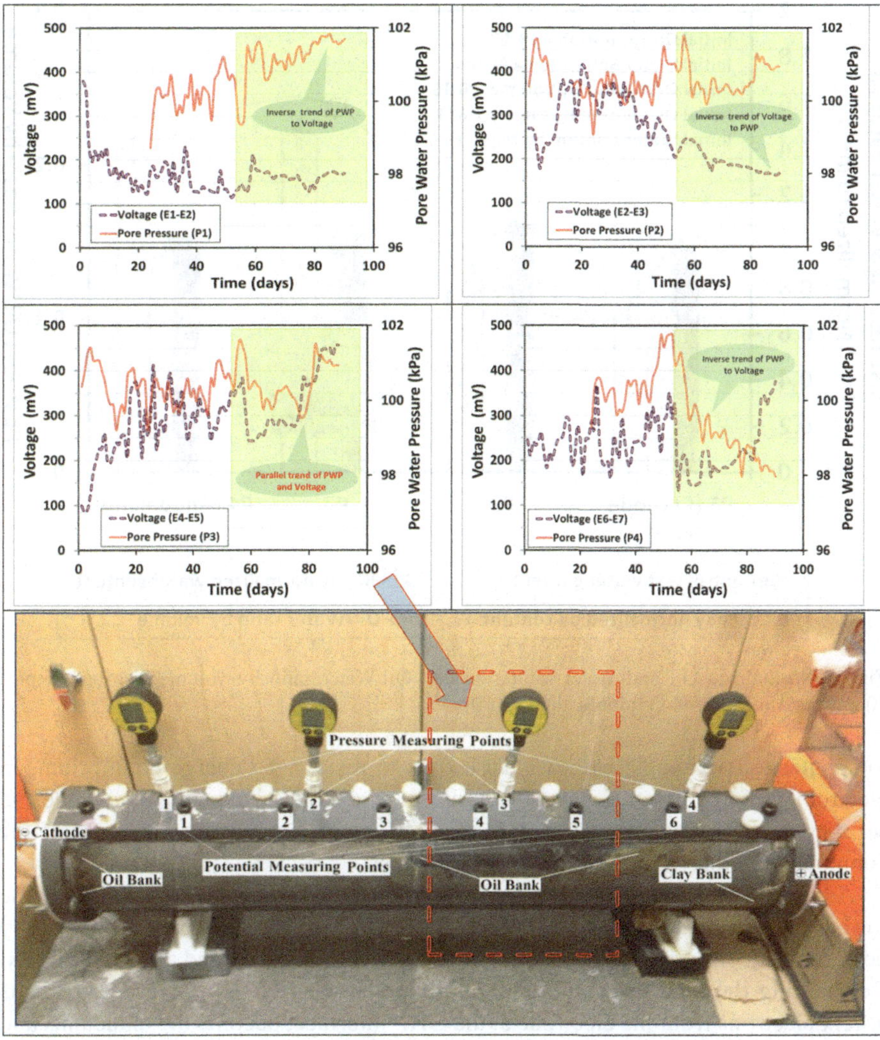


**Fig. 7** Transient distribution of Pore Water Pressure (PWP) and corresponding Voltage at four measurement ports (intervals) of test cell P1: (E1-E2); P2: (E2-E3); P3: (E4-E5); P4: (E6-E7) and Photograph of the test cell at the end of 90 days; *note oil bank formation between E4-E5* [100]

the bulk fluid, and the clay DDL contribute to the process in ion-exchange and sorption reactions.

Diffuse double layer (DDL) processes of clay have been suggested to influence in situ conversion of soil contaminants when a direct current is applied to contaminated clay-electrolyte system [4, 141, 142]. In the event of lowered pH and/or increased ionic concentration in the pore fluid, DDL shrinks due to accumulation of charges, reducing the electrokinetic potential (i.e., zeta potential) of the clay-electrolyte system [167]. Therefore, as the electrokinetic potential changes under varying pH or ionic concentration, the clay-electrolyte interface may become prone

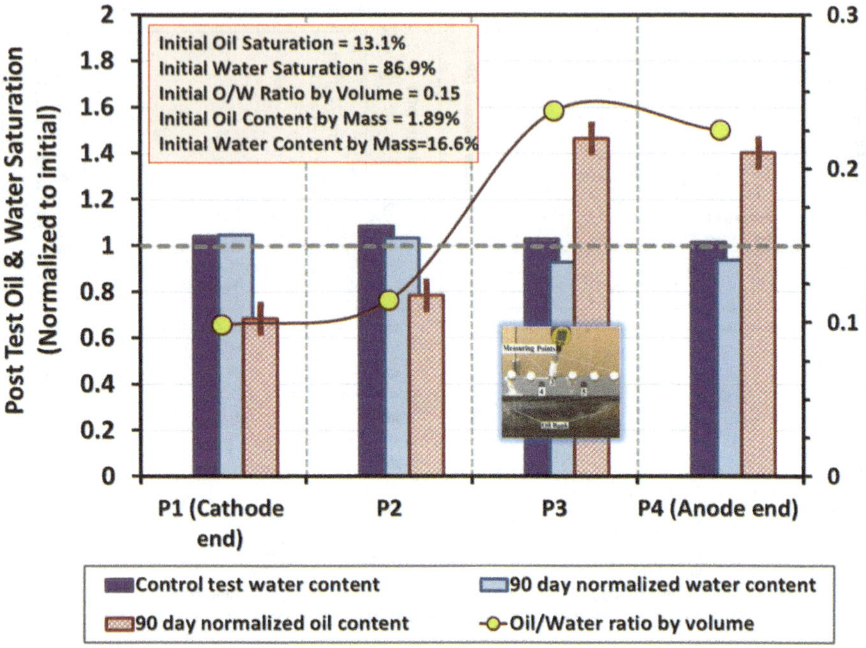


**Fig. 8** Normalized Oil and Water saturation and Oil/Water ratio by volume distribution post 90-day test (normalized to initial values) [100]

to spontaneous Faradaic reactions as it tries to restore the charge equilibrium. In electrokinetic treatment of clays, the transient acid front from anode to cathode can set the stage for triggering Faradaic reactions resulting in transformation of the compounds in the vicinity of the DDL interfaces.

A charged electrode surface, where the oxidants with a positive charge diffuse toward the negative electrode is analogous to surface of a clay particle with fully developed electric double layer (DDL), as shown in Figs. 9 and 10, respectively. Considering the inner and outer Helmholtz planes of IHP and OHP, the oxidants accept electrons from the electrode at the OHP interface, become the reductants and diffuse back toward the bulk solution. The equivalent circuit model parameters in Fig. 9 can be mapped to the clay model as $C_{ddl}$ depicts the double layer capacitor, $R_p$ the polarization resistor between IHP-OHP, $Z_W$ diffusion-controlled impedance of the diffused layer, and $R_b$ is the bulk solution resistance [168]. Therefore, active interfaces of clay particles (i.e., IHP and OHP) can be considered to "act" similar to electrode interfaces when an external electric field is supplied to them [115].

In a clay-electrolyte system, two layers of electrical conductivity can be identified. These layers are: (1) the DDL of clay particles with low conductivity ($\sigma_s$), and (2) the surrounding pore fluid with high conductivity ($\sigma_b$). Given the incompatibility between the conductivity of these two conducting layers, a large electrical potential develops across the DDL under an external electric field. As a result, DDL compresses, rendering the DDL a "capacitor" with higher charge density [169]. The compression of DDL causes the electric potential distribution shift backward (i.e., toward the solid),

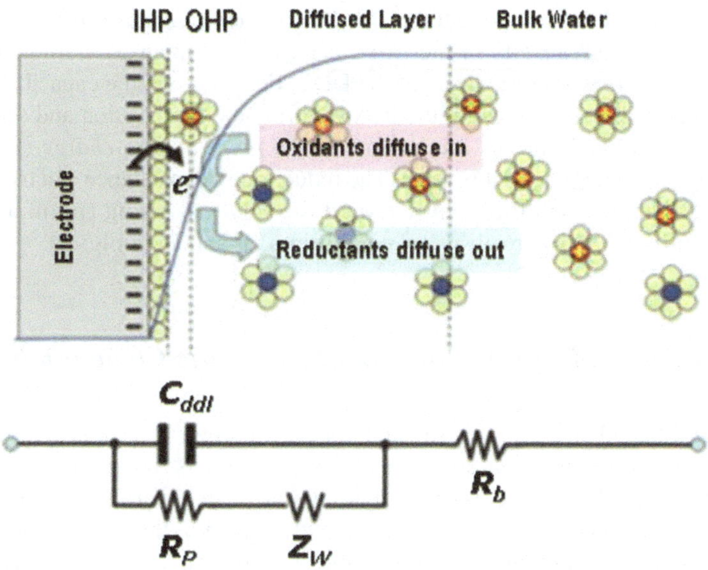


**Fig. 9** Schematic of a DDL interface of an electrode surface with equivalent circuit representation in the usual *Grahame* notation: $C_{ddl}$; $R_p$; $Z_W$; $R_b$ [168]

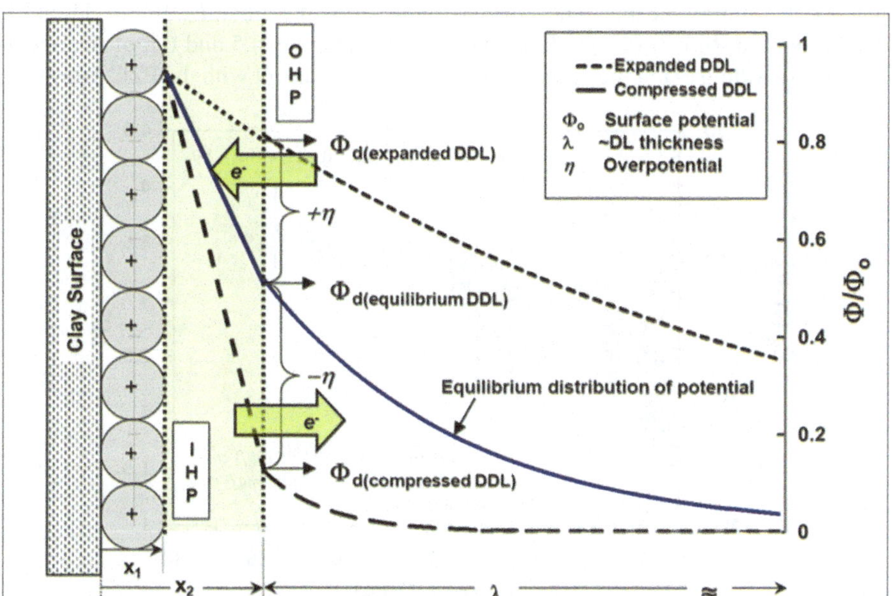


**Fig. 10** The concept of Faradaic current development in GCSG (Gouy-Chapman-Stern-Grahme) model of clay DDL [115, 142]

lowering the electrokinetic potential at the shear plane ($\varsigma$, *zeta potential*). The potential distribution shift results in increased electric field intensity within IHP-OHP interval, and decreased capacitance of the DDL. Furthermore, the capacitance of the double layer is directly related to the bulk electrolyte concentration and the electrokinetic potential, and reaches a minimum at the *point of zero charge* (PZC →pH where electrokinetic potential is zero). The reduction in capacitance can trigger electron transfer across the diffuse layer toward the solution, giving rise to a Faradaic cathodic current that restore electrical equilibrium, as shown in Fig. 10.

## 3.1 Evidence of Transformation of Inorganic Compounds

Brosky and Pamukcu [141] investigated apparent current discharge with compression of the DDL through measurements of nonspontaneous changes in the oxidation-reduction potential ($E_h$) of clay-electrolyte mixtures. In experiments of electrically enhanced reduction of Cu(II) by Fe(II) using clay slurries, they showed a upshift of the measured $E_h$ values with increasing clay concentration, as shown in Fig. 11. The data indicated that, with other parameters remaining constant, increased clay concentrations enhanced the electrochemical reduction of $Cu^{2+}$ to $Cu^{+}$ or $Cu^{0}$ in the system. During an electrokinetic processing, as the matrix pH dips below the "apparent" PZC, the clay surface potential can switch sign from negative to positive [167]. The PZC for the kaolinite-Cu(II)-Fe(II) test slurries were reached at pH 2.4, as shown in Fig. 12. At PZC, the DDL capacitor stores the maximum charge and therefore has reached its minimum capacitance, or maximum redox potential, $E_h$. In Fig. 11, at the time when the shift from base $E_h$ is maximized (in between 5 and 6 h of treatment) the corresponding pH of the system was about 2.4, consistent with the PZC measured.

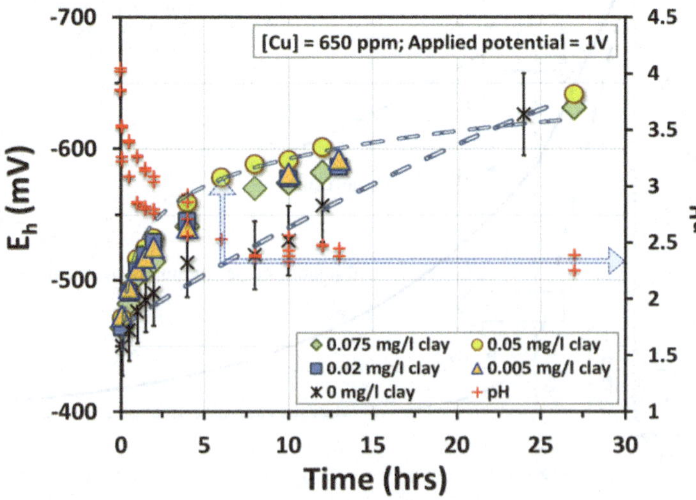


**Fig. 11** Temporal variation and shift of $E_h$ with direct current in Cu(II)-Fe(II) mixtures of clay where maximum $E_h$ shift up is 0.50 V [141]

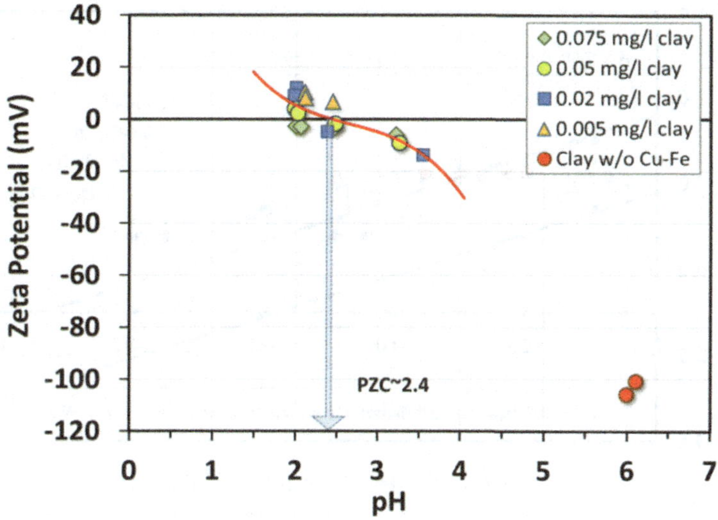


**Fig. 12** Zeta Potential with pH in Cu(II)-Fe(II) mixtures of clay [141]

In laboratory experiments Cr(VI) containing kaolinite clay was injected with Fe(II) using electrokinetic advection to promote reduction of Cr(VI) to Cr(III) [4]. Capacitive changes on the clay surfaces manifested by a positive upshift of the clay-electrolyte ORP (i.e., $E_h$) from the standard ORP showed that the applied electric field may have caused an additional "cathodic current" that drive forth the transformation reactions [6]. The ORP measurements are plotted against the reaction quotient of the Nernst relation in common pH range groups in Fig. 13. The low pH range (pH range 2→3) test data agreed best with the average linear fit to all data, which was shifted 0.37 V from the Nernst relation line of the aqueous system. At low pH, the DDL is compressed possessing higher ion concentration, higher electric field [167], hence higher degree of polarization, and lowered capacitance. At higher pH, the DDL is expanded with reduced degree of polarization, consequently higher capacitance. The ions in an expanded DDL are less restricted to move, therefore discernable fluctuations are likely to occur in the potential development across the diffuse layer. Therefore, the relative scatter of the data at higher pH ranges is attributed to poor polarization of the electric layer.

## 3.2 *Evidence of Transformation of Crude Oil*

The ambient pH and redox conditions influence the interactions between the pore electrolyte and the soil solids naturally. The pH and redox conditions change upon application of a DC electric field, which can drive electrochemical reactions. The electrochemical reactions may transform heavy molecular structure of crude oil into its lighter components with lower viscosity [153, 154]. Considering such a process

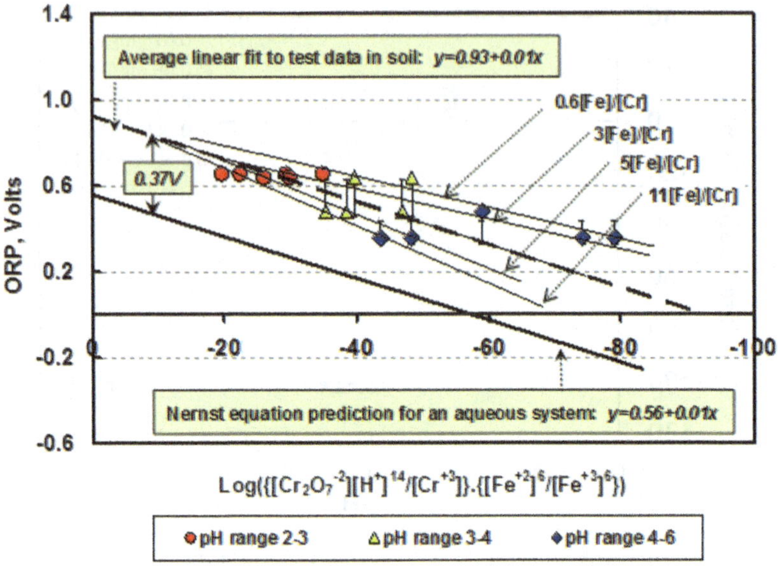


**Fig. 13** Measured and predicted redox potential variation with reaction quotient of species concentration in Nernst relation (average redox potential shift = 0.37 V) [4]

on the four functional group fractions of the crude oil, SARA (i.e., *S*aturates, *A*romatics, *R*esins, and *A*sphaltenes), it is expected that the asphaltene and resin content of the oil decrease and the aromatic content increase, as the viscosity of the oil decrease. Reduction in the oil viscosity under applied electric field has been reported in the literature [153, 154, 161, 162]. The less viscous oil has higher mobility in soil or rock formation pores, hence easier to extract.

Synthetic oil formation cores of 100 cm long and 16.2 cm diameter were tested to investigate heavy oil transformation to its lighter components under an applied DC electric field, as shown in Fig. 14 [158]. Three of the cores were subjected to constant low current density of about 0.1 A/m$^2$, and three of them to high current density of about 1.0 A/m$^2$. Three control cells were run with no electric field. Prepared with titanium electrodes, ion-exchange resin packs and hydraulic gradient controls, the cells were designed to minimize the pH front and electroosmotic water flow through the cores. Samples were retrieved through the corked ports on the cell walls. The solids composition of the cores simulated the formation characteristics of an oil field in Kern County, California, which included 52% sand ($D_{mean}$ = 0.23 mm), 40% silt ($D_{mean}$ = 0.025 mm), and 8% kaolinite clay ($D_{mean}$ = 0.0011 mm), by mass. The mix water used to prepare the cores simulated the electrolyte type and content of groundwater at the same oil field. Three different crude oils representing a range of dynamic viscosities were used in preparation of the cores. The crude oils were labeled as Pennsylvania (ARG field—low viscosity), California (E&B field—medium viscosity) and Canada (Wilkie field—high viscosity) field oils. The properties of these oils are presented in Table 4. The initial porosity, water, and oil

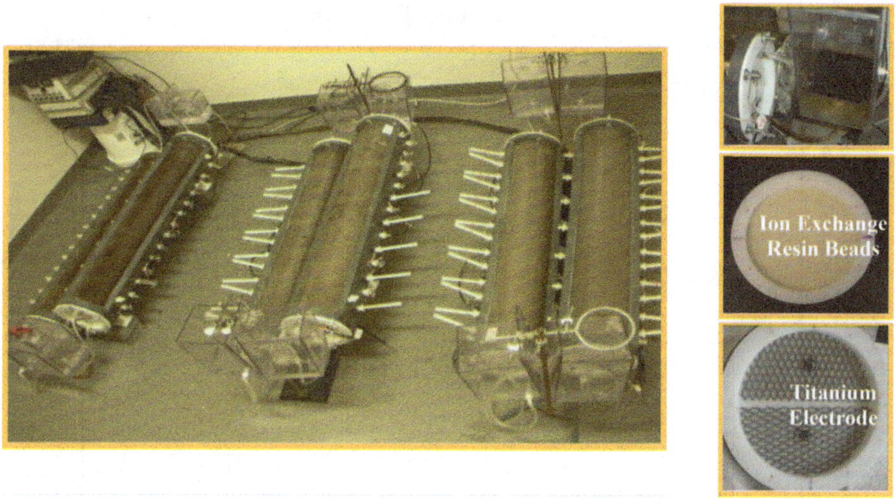


**Fig. 14** Floor-scale surrogate crude oil core test set-up [158]

**Table 4** Physical properties of the formation oils and surrogate water [158]

| Fluid | Properties |
|---|---|
| PA field oil | API 47; Dynamic viscosity = 38.7 cP |
| | Specific gravity = 0.79 at 20 °C |
| CA field oil | API 17.4; Dynamic viscosity = 2782 cP |
| | Specific gravity = 0.95 at 20 °C |
| Canadian field oil | API 13; Dynamic viscosity = 52,000 cP |
| | Specific gravity = 0.98 at 20 °C |
| Electrolyte solution (Surrogate formation water) | Salinity = 33,000 ppm; Electrical conductivity = 45,000 μS pH = 7.50; Major elements: Na, Cl, Mg, S, K, Ca, Br |

saturation of the three different oil-type synthetic cores were determined to be 30%, 46%, and 48%, respectively. The average core permeability was 510 mD (millidarcy),[2] placing them at the high end of the permeability spectrum for oil formations.

The DC electric field was applied for 107 days. Voltage, current, and ORP were measured at eight different sections along the center line ports of the cells. Figure 15 shows the voltage gradient and current density traces recorded for one of the cores. Vacuum extracted samples were taken for gravimetric, UV and Fourier transfer infrared (FTIR) analysis following any marked changes observed in the ORP. At the end of 107 days of treatment, the cells were dismantled and formation oil was

[2] 1 Darcy permits a flow of 1 $cm^3/s$ of a fluid with viscosity 1 cP (1 mPa s) under a pressure gradient of 1 atm/cm acting across an area of 1 $cm^2$.

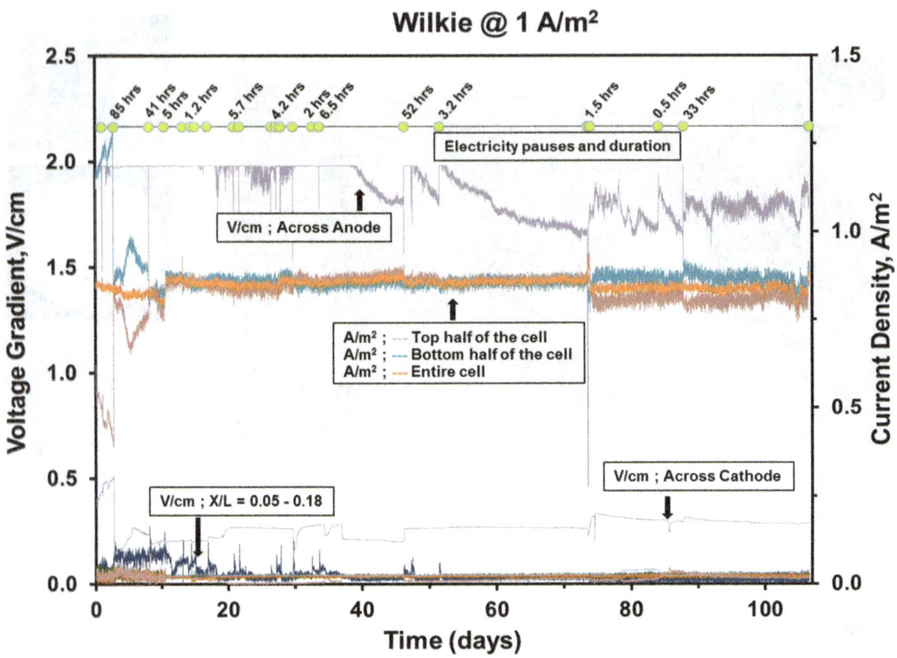


**Fig. 15** The voltage gradient and current density evolution of Wilkie field oil core (Canada) run at 1 A/cm² [158]

extracted from the sampling locations to run the FTIR, SARA, and viscosity analysis.

Figures 16 and 17 present the post-test distributions of the viscosity of the oil specimens extracted from ARG and Wilkie oil cores run at high current density. While they all show marked reductions from the control values, the 35% and the 69% drop in viscosity at the anode sections of the cores is noteworthy. Figures 18 and 19 show the post-test distributions of the SARA content in E&B and Wilkie oil cores run at high current density. The Wilkie specimens extracted from the anode and cathode regions showed marked decrease in the asphaltene content, and marked increase in the saturate content. The post-test viscosity and SARA content results of the Wilkie oil core are complimentary and consistent, supporting the transformation effect of the applied DC field on the heavy crude oil. Conversely, the SARA analysis of the E&B oil showed a decrease in saturates and aromatics, while a substantial increase of asphaltenes took place in the anode region. These changes were in concert with post-test viscosity increase for the E&B oil in general.

The temporal and spatial distributions of ORP (in color-coded regions) and the pH (in labeled line contours) for the high current density (1 A/m²) tests of the three different viscosity oil cells are shown in Fig. 20a–c. A constant pH of about 6 was maintained within the formations for about 30 days after which the ion-exchange resins were depleted and had to be replaced. This process introduced a sharp change

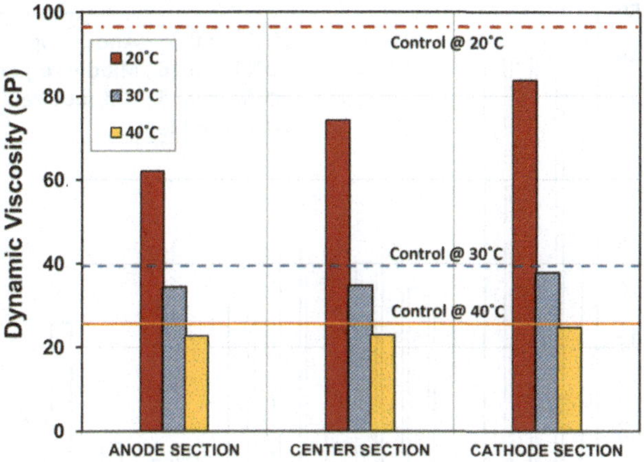


**Fig. 16** The post-test viscosity distribution of ARG field oil core (Pennsylvania) compared to its control [158]

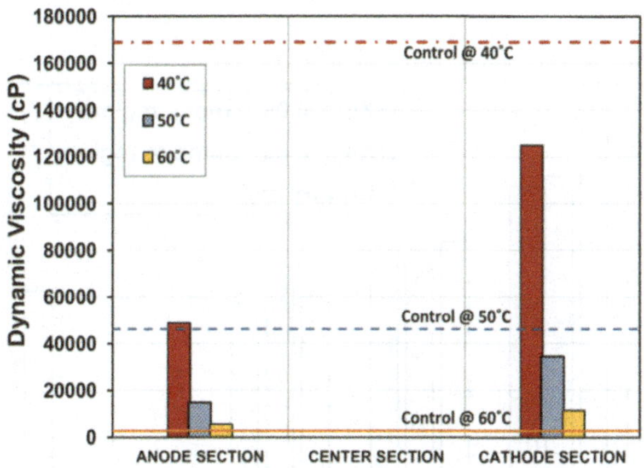


**Fig. 17** The post-test viscosity distribution of Wilkie field oil core (Canada) compared to its control [158]

in the pH of the formations close to the anode and cathode but did not propagate deep into the formations for the next 60 days of treatment. The pH of the central sections of the cores remained constant at about 6 at all times. After the initial 60 days, the ORP at regions close to the anode ($X/L = 0$ to $X/L = 0.2$) reached oxidative state and remained at that state for the test duration. The ORP values did not change appreciably in the middle regions ($X/L = 0.2$ to $X/L = 0.80$), but fluctuated within the reductive state. The ORP values shifted to reductive state close to the cathode region ($X/L = 0.80$ to $X/L = 1.0$) early on and remained there for the test duration. Overall, the reductive trend prevailed over time in larger portions of the cells.

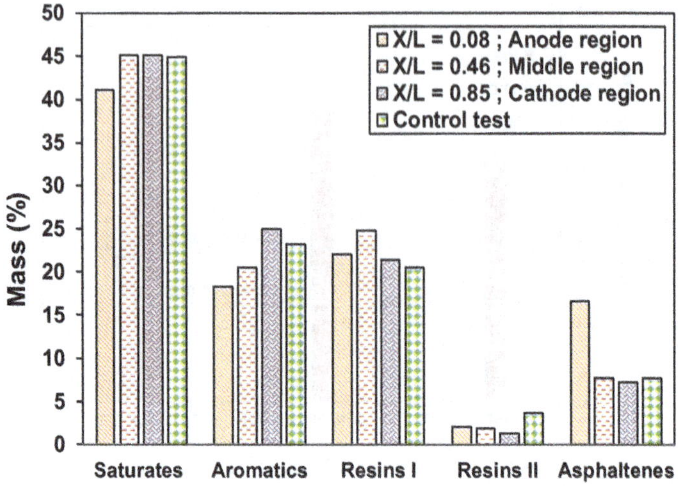


**Fig. 18** The post-test viscosity distribution of E&B field oil core (California) compared to its control [158]

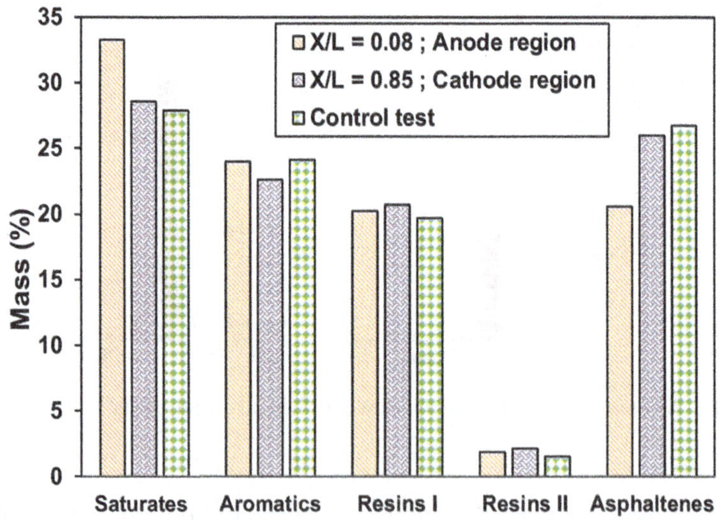


**Fig. 19** The post-test SARA distribution of Wilkie field oil core (Canada) compared to its control [158]

As discussed earlier, the general composition of crude oils can be classified in terms of Saturates, Aromatics, Resins, and Asphaltenes (SARA). Saturates are iso- and cyclo-paraffins, while aromatics, resins, and asphaltenes form a continuum of molecules with increasing molecular weight, aromaticity, and heteroatom contents. The aromatics are the single ring or condensed ring compounds $(C_6H_5)_n$; and

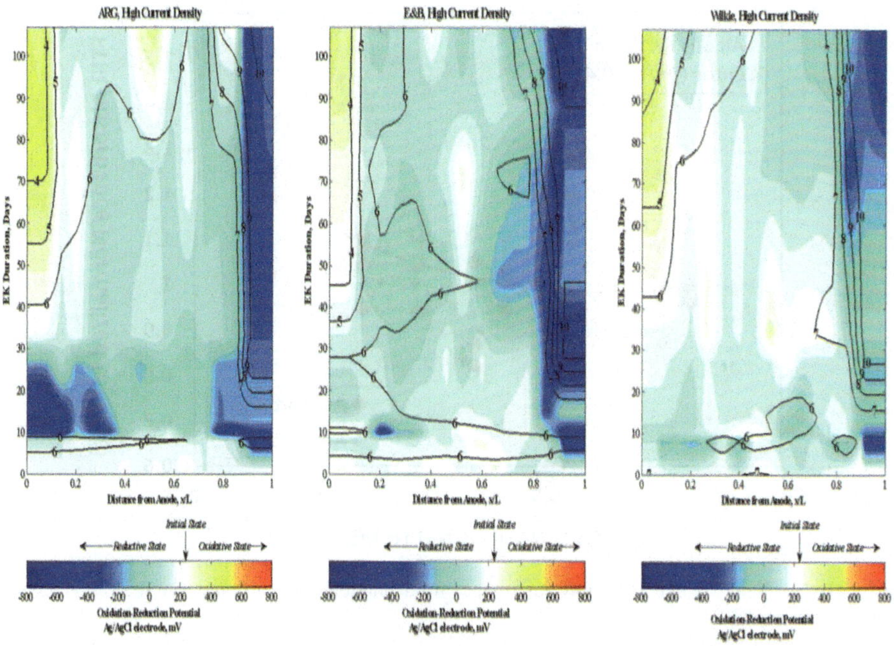


**a**. ORP and pH for ARG field oil core at 1 A/m²

**b**. ORP and pH for E&B field oil core at 1 A/m²

**c**. ORP and pH for at Wilkie field oil core at 1 A/m²

**Fig. 20** Evolution of ORP and pH distribution for three oil field cores during high current density EK test [158]. (**a**) ORP and pH for ARG field oil core at 1 A/m$^2$. (**b**) ORP and pH for E&B field oil core at 1 A/m$^2$. (**c**) ORP and pH for at Wilkie field oil core at 1 A/m$^2$

asphaltenes consist primarily of nitrogen, oxygen, and sulfur containing compounds. Resins are similar to asphaltenes. In FTIR, increase or decrease of the asphaltene content can be determined by the heteroatoms (i.e., N, O, S). Increase in those elements shows rise in polarity of the compounds. In the FTIR analysis, percent change of absorbance of C-O, C=O, OH, C-N functional groups were considered to determine the change in the asphaltene content compared to the baseline FTIR spectra of the particular virgin oil. Complete FTIR analysis was conducted only on the 1 A/m$^2$ run core samples since the low current density cores showed very little or no change in their FTIR spectra and ORP distribution over time.

Oil samples were extracted from the three regions (i.e., anode, central, and cathode) of the test and control cores at time intervals of 10, 30, 45, 75, and 107 days for the ORP and FTIR analysis. The normalized ORP and the average normalized FTIR absorbance (i.e., normalized to their respective initial values) at the center region of each core are plotted for all three oils in Fig. 21. The change in these values were examined at the center of the cores where pH remained constant at about 6, therefore allowed to rule out pH as a factor for potential changes in ORP and FTIR absorbance. As observed, those samples for which ORP increased above the initial, the asphaltene content increased also. The asphaltene content decreased when the

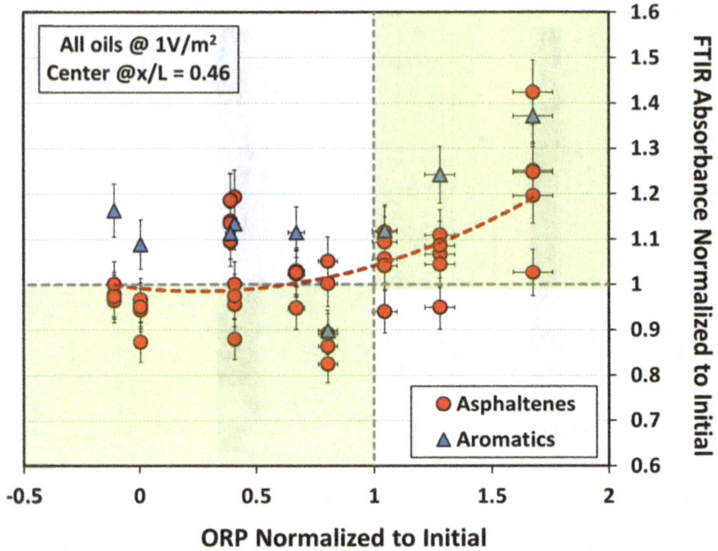


**Fig. 21** Normalized ORP and FTIR absorbance for asphaltenes and aromatics at the *center region* of all cores [158]

ORP decreased below the initial value. Consequently, at constant pH conditions, a reductive environment promoted a drop in asphaltene content while an oxidative environment helped to increase it. The same is not observed with aromatics, as their trend appeared to be less dependent on the state of ORP. The ORP versus the normalized FTIR absorbance for all specimens (i.e., all regions) are shown in Fig. 22. There are three discernible regions on the graph. The first is the cathode region in which ORP is negative and the FTIR absorbance of the asphaltenes has reduced compared to their initial value. The second is the central region where the ORP is positive and the FTIR absorbance appears to be constant compared to initial value, yet it remains a transition region. The third is the anode region where ORP is positive and well above the initial and the FTIR absorptions are all above the initial values. The aromatic contents are almost all above the original for all ORP distribution, showing no discernible relation between the two parameters.

The data presented in Figs. 16, 17, 18, 19, 20, 21, and 22 shows clearly that electrochemical reactions do take place which affect the transformation of heavy crude oil into its different components by passage of current through the formation. This process appeared to take place in the central regions of the cores where pH remained constant independent of the sharp changes at the electrode sites. As the lighter components (aromatics) of the oil increased and the heavier components (asphaltenes) decreased, and the viscosity of the oil decreased. As the lighter components (aromatics) of the oil increase and the heavier components (asphaltenes) decrease, the viscosity of the oil is found to decrease as well. These changes are also signaled by a drop in ORP at neutral or slightly alkaline conditions. The process ultimately

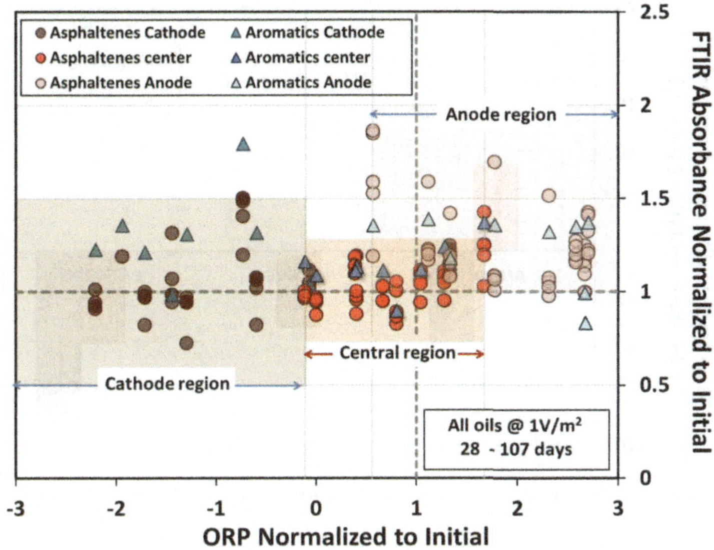


**Fig. 22** Regional distribution of ORP versus FTIR absorbance for asphaltenes and aromatics for all cores [158]

increases the mobility of the crude oil as a result of reduced surface tension between the oil and water.

The post-test changes in the Carbon fractions from low to high Carbon Numbers in the three oils were compared with respect to their controls. In the Pennsylvania (ARG) and Canada (Wilkie) oils, there was an increase ("+" positive change from control) in the smaller Carbon numbers, where there was a decrease ("–" negative change from control) in the higher numbers. This indicated increase in light components of oil (i.e., gasoline) and decrease in heavier components. There appeared an opposite trend in the California (E&B) oil regarding the lower C number components. All showed a substantial decrease in the higher carbon number (C100+) fraction at the cathode section as presented in Fig. 23. The reduction in C100+ appears to be more prevalent in higher viscosity crude oil (i.e., Wilkie) than the lower viscosity crude oil (i.e., ARG). The E&B oil, with mid-level viscosity among the three, represents an average behavior with significant reduction of C100+ in the middle section of the core as well.

Although the tests were not designed to monitor and measure oil production at the anode and cathode water reservoirs, some oil flow was observed in the cells. In the Pennsylvania oil case (ARG cells), oil was visually observed to seep into the anode and cathode reservoirs. This was attributed to low viscosity of the ARG oil, and higher tendency for it to be transported. In the California oil case (E&B cells) with 1.0 $A/m^2$ current density application, oil transport into the cathode side water reservoir occurred at about 45 days. No visual oil transport was observed to either of the water reservoirs in the Canada oil case (Wilkie cells) which were attributed to

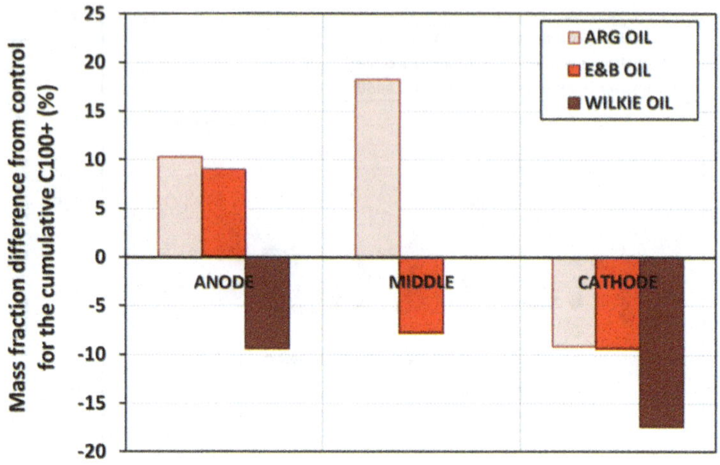


**Fig. 23** Spatial distribution of post-test C100+ fraction change in all core oils compared to controls [158]

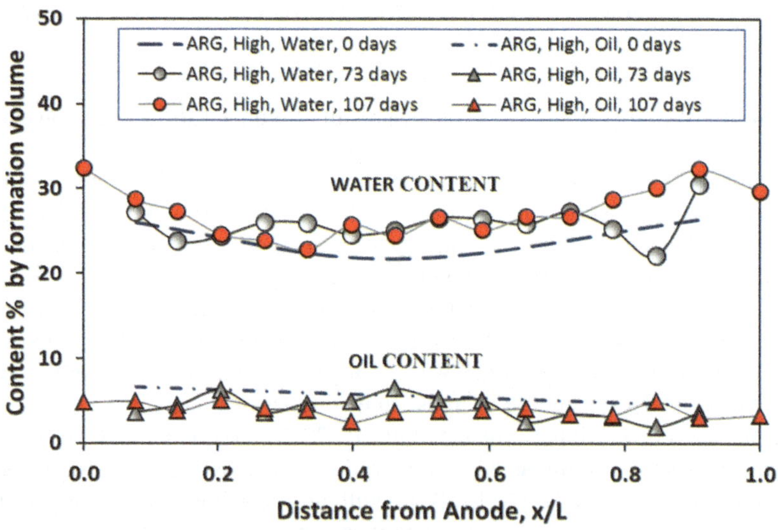


**Fig. 24** Oil-water content measured in ARG field oil core at 1.0 A/m$^2$ [158]

the high viscosity of the Wilkie oil. Despite the visual transport and diffusion of oil in some of the cells, the measured oil and water contents did not change appreciably throughout the entire test period. Figure 24 shows the variation of the oil and water content over time for the Pennsylvania (ARG) oil core. This was the core for which the highest oil diffusion and transport was observed. As a result, slight reduction of oil and slight increase of water had occurred, but the overall oil and water distribution remained uniform in the core throughout the test period. Similar trends were observed in the other cores' oil and water content distributions.

Finally, for low current density (i.e., 0.1 mA/m$^2$) cells, most of the measurements including the FTIR absorbance values and the ORP measurements indicated negligible changes in the formation. This observation indicated that the current density had to be above a threshold level in order for electrochemical reactions to be effective in oil transformation.

## 4 Summary and Conclusions

Electrochemical technologies have been shown to transport, recover or transform inorganic, organic and mixed substances that reside in wet soils, sediments or sludges. Electrokinetics particularly is a versatile technology that can be applied alone or in tandem with other technologies to accomplish in situ soil remediation and/or recover resources, such as oil. The various results presented in this chapter provide evidence that a well-engineered electrokinetic system may serve well the goals of a remediation/recovery project in a petroleum contaminated field or an oil formation.

## References


1. A. McBratney, D.J. Field, A. Koch, The dimensions of soil security. Geoderma **213**, 203–213 (2014). https://doi.org/10.1016/j.geoderma.2013.08.013
2. K.R. Reddy, Technical challenges to in-situ remediation of polluted sites. Geotech. Geol. Eng. **28**, 211–221 (2010)
3. A.T. Lima, P.J. Kleingeld, K. Heister, J.P.G. Loch, In situ electro-osmotic cleanup of tar contaminated soil - removal of PAHs. Electrochim. Acta **86**, 142–147 (2012)
4. S. Pamukcu, A. Weeks, J.K. Wittle, Enhanced reduction of Cr(VI) by direct electric current in a contaminated clay. Environ. Sci. Technol. **38**(4), 1236–1241 (2004). https://doi.org/10.1021/es034578v
5. D.B. Gent, R.M. Bricka, A.N. Alshawabkeh, S.L. Larson, G. Fabian, S. Granade, Bench- and field-scale evaluation of chromium and cadmium extraction by electrokinetics. J. Hazard. Mater. **110**, 53–62 (2004). https://doi.org/10.1016/j.jhazmat.2004.02.036
6. G.M. Nystroem, L.M. Ottosen, A. Villumsen, Electrodialytic removal of Cu, Zn, Pb, and Cd from harbor sediment: influence of changing experimental conditions. Environ. Sci. Technol. **39**, 2906–2911 (2005)
7. P.E. Jensen, L.M. Ottosen, C. Ferreira, Electrodialytic remediation of soil fines (<63μm) in suspension—influence of current strength and L/S. Electrochim. Acta **52**, 3412–3419 (2007). https://doi.org/10.1016/j.electacta.2006.03.116
8. M. Elektorowicz, Electrokinetic remediation of mixed metals and organic contaminants, in *Electrochemical Remediation Technologies for Polluted Soils, Sediments and Groundwater*, ed. by K. Reddy, C. Cameselle, (John Wiley & Son, Ltd, New York, NY, 2009), pp. 315–331. Chapter 15
9. A.N. Alshawabkeh, Electrokinetic soil remediation: challenges and opportunities. Sep. Sci. Technol. **44**, 2171–2187 (2009)

10. G. Traina, Electrokinetic stabilization as a reclamation tool for waste materials polluted by both salts and heavy metals. Chemosphere **75**, 819–824 (2009)
11. A.T. Yeung, Milestone developments, myths, and future directions of electrokinetic remediation. Sep. Purif. Technol. **79**, 124–132 (2011). https://doi.org/10.1016/j.seppur.2011.01.022
12. A. Nieto Castillo, R.A. García-Delgado, V. Cala Rivero, Electrokinetic treatment of soils contaminated by tannery waste. Electrochim. Acta **86**, 110–114 (2012). https://doi.org/10.1016/j.electacta.2012.04.132
13. O.K. Merkx, J.P.G. Loch, A.T. Lima, J.A. Dijk, J.F. Kreuk, P.J. Kleingeld, The effectiveness of electro-remediation of aged, metal-contaminated sediment in relation to sequential extraction of metals. Water Air Soil Pollut. **224**, 1–12 (2013). https://doi.org/10.1007/s11270-013-1667-1
14. A. Demir, S. Pamukcu, R.A. Shrestha, Simultaneous removal of Pb, Cd and Zn from heavily contaminated mine tailing soil using enhanced electrochemical process. Environ. Eng. Sci. **32**(5), 416–424 (2014). https://doi.org/10.1089/ees.2014.0384
15. H.I. Gomes, L.M. Ottosen, A.B. Ribeiro, C. Dias-Ferreira, Treatment of a suspension of PCB contaminated soil using iron nanoparticles and electric current. Environ. Manag. **151**, 550–555 (2015). https://doi.org/10.1016/j.jenvman.2015.01.015
16. V. Valdovinos, F. Monroy-Guzmán, E. Bustos, Electrokinetic removal of radionuclides contained in scintillation liquids absorbed in soil type Phaeozem. J. Environ. Radioact. **162**, 80–86 (2016). https://doi.org/10.1016/j.jenvrad.2016.05.017
17. A.T. Lima, A. Hofmann, D. Reynolds, C. Ptacek, P. Van Cappellen, L. Ottosen, S. Pamukcu, A. Alshawabekh, D. O'Carroll, C. Riis, E. Cox, D. Gent, R. Landis, J. Wang, A. Chowdhury, E. Secord, A. Sanchez-Hachair, Environmental electrokinetics for a sustainable subsurface. Chemosphere **181**, 122–133 (2017). https://doi.org/10.1016/j.chemosphere.2017.03.143
18. E. Bustos, E. Méndez, F. Prieto-Garcia, G. Hernández, J. García, S. Solís, S. Pamukcu, Study of electrochemical removal of phenantrene in bentonite clay by physicochemical indicators. Sep. Purif. Technol. **12**, 92–99 (2019). https://doi.org/10.1016/j.seppur.2018.04.078
19. L. Casagrande, Electro-osmosis in soils. Geotechnique **1**, 159–177 (1949)
20. C. Ray, R.H. Ramsey, Removal of heavy metals in wastewater in a clay soil matrix using electro-osmosis. Environ. Prog. **6**(3), 145–149 (1987)
21. L.I. Khan, S. Pamukcu, Validity of electroosmosis for groundwater decontamination, in *Proc. of Environmental Engineering Conf., ASCE*, (1989), pp. 563–570
22. S. Pamukcu, L.I. Khan, H.Y. Fang, Zinc detoxification of soils by electro- osmosis. Trans. Res. Rec. **1289**, 41–46 (1990)
23. J. Hamed, Y.B. Acar, R.J. Gale, Pb(II) removal from kaolinite by electrokinetics. J. Geotech. Eng. **117**(2), 241–271 (1991)
24. A.N. Alshawabkeh, Y.B. Acar, Removal of contaminants from soils by electrokinetics: a theoretical treatise. J. Environ. Sci. Heal. Part A Environ. Sci. Eng. Toxicol. **27**, 1835–1861 (1992). https://doi.org/10.1080/10934529209375828
25. S. Pamukcu, J.K. Wittle, Electrokinetics for removal of low-level radioactivity from soil, in *The 14th Annual DOE Low-Level Radioactive Waste Management Conference, Phoenix, AZ, November 18–20*, (1992), pp. 256–278. Conf-921137-Proc
26. S. Pamukcu, J.K. Wittle, Electrokinetic removal of selected heavy metals from soil. Environ. Prog. **11**(3), 241–250 (1992)
27. Y.B. Acar, A.N. Alashawabkeh, Principles of electrokinetic remediation. Environ. Sci. Technol. **27**(13), 2638–2647 (1993)
28. Y.B. Acar, A.N. Alshawabkeh, Electrokinetic remediation I: pilot scale tests with lead spiked kaolinite. Geotech. Geoenviron. Eng. **122**(3), 173–185 (1996)
29. R. Lageman, Electroreclamation: applications in The Netherlands. Environ. Sci. Technol. **27**, 2648–2650 (1993)
30. R.F. Probstein, R.E. Hicks, Removal of contaminants from soils by electric fields. Science **260**, 498–503 (1993)

31. J.K. Wittle, S. Pamukcu, *Electrokinetic Treatment of Contaminated Soils, Sludges and Lagoons, DOE/CH-9206, No. 02112406* (US Department of Energy, Argonne National Laboratories, Argonne, IL, 1993) 65 p
32. A.P. Shapiro, R.F. Probstein, Removal of contaminants from saturated clay by electroosmosis. Environ. Sci. Technol. **27**, 283–291 (1993)
33. G.R. Eykholt, D.E. Daniel, Impact of system chemistry on electroosmosis in contaminated soil. Geotech. Geoenviron. Eng. **120**(5), 797–815 (1994)
34. Y.B. Acar, J. Hamed, A.N. Alshawabkeh, R. Gale, Cd(II) removal from saturated kaolinite by application of electrical current. Geotechnique **44**(3), 239–254 (1994)
35. Y.B. Acar, R.J. Gale, A.N. Alshawabkeh, R.E. Marks, S. Puppala, M. Bricka, R. Parker, Electrokinetic remediation: basics and technology status. J. Hazard. Mater. **40**, 117–137 (1995). https://doi.org/10.1016/0304-3894(94)00066-P
36. S. Pamukcu, A. Weeks, J.K. Wittle, Electrochemical extraction and stabilization of selected inorganic species in porous media. J. Hazard. Mater. **55**, 305–318 (1997). https://doi.org/10.1016/S0304-3894(97)00025-3
37. B.S. Haran, G. Zheng, B.N. Popov, R.E. White, Electrochemical decontamination of soils: development of new electrochemical method for decontamination of hexavalent chromium from sand, in *Electrochemical Society Proc., V95-12*, (1995), pp. 227–251
38. A.P. Shapiro, R.F. Probstein, R.E. Hicks, Removal of cadmium (II) from saturated kaolinite by application of electric current. Geotechnique **45**, 355–359 (1995)
39. G. Denisov, R.E. Hicks, R.F. Probstein, On the kinetics of charged contaminant removal from soils using electric fields. J. Colloid Interface Sci. **178**, 309–323 (1996)
40. J.K. Wittle, S. Pamukcu, *Electrochemical System and Method for the Removal of Charged Species from Contaminated Liquid and Solid Wastes*. US 5614077 A (1997)
41. K.R. Reddy, U. Parupudi, Removal of chromium, nickel, and cadmium from clays by in-situ electrokinetic remediation. Soil Contam. **6**(4), 391–407 (1997)
42. A.B. Ribeiro, J.T. Mexia, A dynamic model for the electrokinetic removal of copper from a polluted soil. J. Hazard. Mater. **56**(3), 257–271 (1997)
43. H.K. Hansen, L.M. Ottosen, B.K. Kliem, A. Villumsen, Electrodialytic remediation of soils polluted with Cu, Cr, Hg, Pb and Zn. J. Chem. Technol. Biotechnol. **70**, 67–73 (1997)
44. A.T. Yeung, C.N. Hsu, R. Menon, EDTA-enhanced electrokinetic extraction of lead. Geotech. Eng. **122**(8), 666–673 (1996)
45. C.J. Bruell, B.A. Segal, M.T. Walsh, Electroosmotic removal of gasoline hydrocarbons and TCE from clay. J. Environ. Eng. **118**(1), 63–68 (1992)
46. E.E. Hicks, S. Tondorf, Electrorestoration of metal contaminated soils. Environ. Sci. Technol. **28**(12), 2203–2210 (1994)
47. C. Pulgarin, N. Adler, P. Peringer, C. Comninellis, Electrochemical detoxification of a 1,4-benzoquinone solution in wastewater treatment. Water Res. **28**, 887–893 (1994)
48. S. Pamukcu, *Electrokinetic Removal of Coal Tar Constituents from Contaminated Soils. EPRI TR-103320* (Electric Power Research Institute, Palo Alto, CA, 1994)
49. S. Pamukcu, I. Filipova, J.K. Wittle, The role of electroosmosis in transporting PAH compounds in contaminated soils, in *Electrochemical Techn. Applied to Environmental Problems, Electrochemical Society Trans., PV95-12*, (1995), pp. 252–266
50. M. Elektorowicz, V. Boeva, Electrokinetic supply of nutrients in soil bioremediation. Environ. Sci. Technol. **17**, 1339–1349 (1996)
51. G.V. Chilingar, W.W. Loo, L.F. Khilyuk, S.A. Katz, Electrobioremediation of soils contaminated with hydrocarbons and metals. Energy Sources **19**, 129–146 (1997)
52. R.E. Saichek, K.R. Reddy, Effect of pH control at the anode for the electrokinetic removal of phenanthrene from kaolin soil. Chemosphere **51**, 273–287 (2003)
53. N.C. Lockhart, R.E. Stickland, Dewatering coal washery tailings ponds by electroosmosis. Powder Technol. **40**, 215–221 (1984)

54. J. Huang, M. Elektorowicz, J. Oleszkiewicz, Dewatering and disinfection of aerobic and anaerobic sludge using an electrokinetic (EK) system. Water Sci.Tech. **57**(2), 231–236 (2008). https://doi.org/10.2166/wst.2008.009
55. L. Yang, G. Nakhla, A. Bassi, Electro-kinetic dewatering of oily sludges. J. Hazard. Mater. **125**, 130–140 (2005)
56. G.C.C. Yang, M.C. Chen, C.F. Yeh, Dewatering of a biological industrial sludge by electrokinetics-assisted filter press. Sep. Purif. Technol. **79**, 177–182 (2011). https://doi.org/10.1016/j.seppur.2011.02.012
57. S. Banerjee, J. Horng, J.F. Ferguson, P.O. Nelson, *Field Scale Feasibility Study of Electrokinetic Remediation. CR811762-01* (Risk Reduction Engineering Laboratory, Office of Research and Development, US EPA, Cincinnati, OH, 1987) 129 p
58. E.R. Lindgren, *Electrokinetic Demonstration at the Unlined Chromic Acid Pit* (Sandia National Laboratories, DOE-Office of Scientific and Technical Information, Springfield, VA, 1998) 144 p
59. S.V. Ho, C. Athmer, P.W. Sheridan, B.M. Hughes, R. Orth, D. McKenzie, P.H. Brodsky, A.M. Shapiro, T.M. Sivavec, J. Salvo, D. Schultz, R. Landis, R. Griffith, S. Shoemaker, The Lasagna Technology for in situ soil remediation. 1. Small field test. Environ. Sci. Technol. **337**, 1086–1091 (1999). https://doi.org/10.1021/ES980332S
60. S.V. Ho, P.W. Sheridan, C.J. Athmer, M.A. Heitkamp, J.M. Brackin, D. Weber, P.H. Brodsky, Integrated in situ soil remediation technology: the Lasagna process. Environ. Sci. Technol. **29**, 2528–2534 (1995)
61. S.V. Ho, C. Athmer, P.W. Sheridan, B.M. Hughes, R. Orth, D. McKenzie, P.H. Brodsky, A.M. Shapiro, T.M. Sivavec, J. Salvo, D. Schultz, R. Landis, R. Griffith, S. Shoemaker, The Lasagna technology for in situ soil remediation. 2. Large field test. Environ. Sci. Technol. **337**, 1092–1099 (1999). https://doi.org/10.1021/ES980414G
62. J.K. Wittle, S. Pamukcu, D. Bowman, L.M. Zanko, F. Doering, Field studies on sediment remediation, in *Electrochemical Remediation Technologies for Polluted Soils, Sediments and Groundwater*, ed. by K. Reddy, C. Camaselle, (John Wiley & Sons, New York, NY, 2009), pp. 661–696. https://doi.org/10.1002/9780470523650.ch32
63. R. Lageman, M.S. Godschalk, Electro-bioreclamation. A combination of in situ remediation techniques proves successful at a site in Zeist, the Netherlands. Electrochim. Acta **52**, 3449–3453 (2007)
64. L.M. Zanko, J.K. Wittle, S. Pamukcu, Case study: electrochemical geo-oxidation (ECGO) treatment of Massachusetts New Bedford Harbor Sediment PCBs. Electrochim. Acta **354**, 136690 (2020). https://doi.org/10.1016/j.electacta.2020.136690
65. N. Maes, H. Moors, A. Dierckx, P. De Cannière, M. Put, The assessment of electromigration as a new technique to study diffusion of radionuclides in clayey soils. J. Contam. Hydrol. **36**, 231–247 (1999). https://doi.org/10.1016/S0169-7722(98)00146-6
66. Y. Roh, S.Y. Lee, M.P. Elless, K.S. Cho, Electro-enhanced remediation of radionuclide-contaminated groundwater using zero-valent iron. J. Environ. Sci. Health A Tox. Hazard. Subst. Environ. Eng. **35**, 1043–1059 (2000)
67. K.H. Kim, S.O. Kim, C.W. Lee, M.H. Lee, K.W. Kim, Electrokinetic processing for the removal of radionuclides in soils. Sep. Sci. Technol. **38**, 2137–2163 (2003). https://doi.org/10.1081/SS-120021617
68. R.E. Saichek, K.R. Reddy, Electrokinetically enhanced remediation of hydrophobic organic compounds in soils: a review. Crit. Rev. Environ. Sci. Technol. **35**, 115–192 (2005)
69. M.T. Alcántara, J. Gómez, M. Pazos, M.A. Sanromán, Electrokinetic remediation of lead and phenanthrene polluted soils. Geoderma **173–174**, 128–133 (2012). https://doi.org/10.1016/j.geoderma.2011.12.009
70. A.B. Ribeiro, J.M. Rodriíguez-Maroto, E.P. Mateus, H. Gomes, Removal of organic contaminants from soils by an electrokinetic process: the case of atrazine. Experimental and modeling. Chemosphere **59**, 1229–1239 (2005)

71. A.B. Ribeiro, E.P. Mateus, J.M. Rodríguez-Maroto, Removal of organic contaminants from soils by an electrokinetic process: the case of molinate and bentazone: experimental and modeling. Sep. Purif. Technol. **79**(2), 193–203 (2011)
72. V.A. Korolev, O.V. Romanyukha, A.M. Abyzova, Electrokinetic remediation of oil-contaminated soils. J. Environ. Sci. Health A Tox. Hazard. Subst. Environ. Eng. **43**(8), 876–880 (2008)
73. A.T. Lima, P.J. Kleingeld, K. Heister, J.P.G. Loch, Removal of PAHs from contaminated clayey soil by means of electro-osmosis. Sep. Purif. Technol. **79**, 221–229 (2011)
74. A.T. Lima, L.M. Ottosen, K. Heister, J.P.G. Loch, Assessing PAH removal from clayey soil by means of electro-osmosis and electrodialysis. Sci. Total Environ. **435–436**, 1–6 (2012). https://doi.org/10.1016/j.scitotenv.2012.07.010
75. K.R. Reddy, K. Darko-Kagya, A.Z. Al-Hamdan, Electrokinetic remediation of pentachlorophenol contaminated clay soil. Water Air Soil Pollut. **221**, 35–44 (2011)
76. P. Guedes, E.P. Mateus, N. Couto, Y. Rodríguez, A.B. Ribeiro, Electrokinetic remediation of six emerging organic contaminants from soil. Chemosphere **117**, 124–131 (2014)
77. M. Elektorowicz, S. Habibi, R. Chifrina, Effect of electrical potential on the electrodemulsification of oily sludge. J. Colloid Interface Sci. **295**(2), 535–541 (2006). https://doi.org/10.1016/j.jcis.2005.08.042
78. S. Taslimi Taleghani, A. Fellah Jahromi, M. Elektorowicz, Electro-demulsification of water-in-oil suspensions enhanced with implementing various additives. Chemosphere **233**, 157–163 (2019). https://doi.org/10.1016/j.chemosphere.2019.05.161
79. H. Lei, K. Chen, Y. Li, Electrokinetic recovery of copper, nickel and zinc from wastewater sludge: effects of electrical potentials. Environ. Eng. Sci. **29**, 472–478 (2012)
80. P. Guedes, N. Couto, L.M. Ottosen, A.B. Ribeiro, Phosphorus recovery from sewage sludge ash through an electrodialytic process. Waste Manag. **34**, 886–892 (2014). https://doi.org/10.1016/j.wasman.2014.02.021
81. S. Hasan, M. Elektorowicz, J. Oleszkiewicz, Start-up period investigation of pilot-scale submerged membrane electro-bioreactor (SMEBR) treating raw municipal wastewater. Chemosphere **97**, 71–77 (2014). https://doi.org/10.1016/j.chemosphere.2013.11.009
82. K. Bani-Melhem, M. Elektorowicz, Development of a novel submerged membrane electro-bioreactor (SMEBR): performance for fouling reduction. Environ. Sci. Technol. **44**, 3298–3304 (2010)
83. K. Bani-Melhem, M. Elektorowicz, Performance of the submerged membrane electro-bioreactor (SMEBR) with iron electrodes for wastewater treatment and fouling reduction. J Mem. Sci. **379**(1–2), 434–439 (2011)
84. J.H. Chang, Y.H. Shi, C.H. Tung, Stepwise addition of chemical reagents for enhancing electrokinetic removal of Cu from real site contaminated soil. J. Appl. Electrochem. **40**, 1153–1160 (2010)
85. J.L. Chen, S.F. Yang, C.C. Wu, S. Ton, Effect of ammonia as a complexing agent in electrokinetic remediation of copper-contaminated soil. Sep. Purif. Technol. **79**, 157–163 (2011)
86. L. Hannum, S. Pamukcu, Aided transport of nano-iron in clay soils using direct electric field, in *NSTI-Nanotech2007*, vol. V2, (2007), pp. 635–638
87. A.I.A. Chowdhury, D.M. O'Carroll, Y. Xu, B.E. Sleep, Electrophoresis enhanced transport of nano-scale zero valent iron. Adv. Water Resour. **40**, 71–82 (2012). https://doi.org/10.1016/j.advwatres.2012.01.014
88. Y.C. Huang, Y.W. Cheng, Electrokinetic-enhanced nanoscale iron reactive barrier of trichloroethylene solubilized by Triton X-100 from groundwater. Electrochim. Acta **86**, 177–184 (2012). https://doi.org/10.1016/j.electacta.2012.03.048
89. H.I. Gomes, C. Dias-Ferreira, A.B. Ribeiro, S. Pamukcu, Electrokinetic enhanced transport of zero-valent iron nanoparticles for chromium (VI) reduction in soils. Chem. Eng. Trans. **28**, 139–144 (2012)

90. H.I. Gomes, C. Dias-Ferreira, A.B. Ribeiro, S. Pamukcu, Enhanced transport and transformation of zerovalent nanoiron in clay using direct electric current. Water Air Soil Pollut. **224**, 1710 (2013). https://doi.org/10.1007/s11270-013-1710-2
91. H.I. Gomes, C. Dias-Ferreira, A.B. Ribeiro, S. Pamukcu, Influence of electrolyte and voltage on the direct current enhanced transport of iron nanoparticles in low permeability media. Chemosphere **99**, 171–179 (2014). https://doi.org/10.1016/j.chemosphere.2013.10.06
92. H.I. Gomes, J.M. Rodríguez-Maroto, C. Dias-Ferreira, S. Pamukcu, A.B. Ribeiro, Numerical prediction of diffusion and electric field-induced iron nanoparticle transport. Electrochim. Acta **18**, 5–12 (2015). https://doi.org/10.1016/j.electacta.2014.11.157
93. R.E. Saichek, K.R. Reddy, Evaluation of surfactants/cosolvents for desorption/solubilization of phenanthrene in clayey soils. Int. J. Environ. Stud. **61**, 587–604 (2004)
94. R.E. Saichek, K.R. Reddy, Surfactant-enhanced electrokinetic remediation of polycyclic aromatic hydrocarbons in heterogeneous subsurface environments. J. Environ. Eng. Sci. **4**(5), 327–339 (2005)
95. B.G. Ryu, G.Y. Park, J.W. Yang, K. Baek, Electrolyte conditioning for electrokinetic remediation of As, Cu and Pb-contaminated soil. Sep. Purif. Technol. **79**(2), 170–176 (2011)
96. G. Fan, L. Cang, G. Fang, D. Zhou, Surfactant and oxidant enhanced electrokinetic remediation of a PCBs polluted soil. Sep. Purif. Technol. **123**, 106–113 (2014)
97. N. Roach, K.R. Reddy, Electrokinetic delivery of permanganate into low-permeability soils. Int. J. Environ. Waste Manag. **1**, 4–19 (2006)
98. E. Ghazanfari, R.A. Shrestha, A. Miroshnik, S. Pamukcu, Electrically assisted liquid hydrocarbon transport in porous media. Electrochim. Acta **86**, 185–191 (2012). https://doi.org/10.1016/j.electacta.2012.04.077
99. E. Ghazanfari, S. Pamukcu, M. Pervizpour, Z. Karpyn, Investigation of generalized relative permeability coefficients for electrically assisted oil recovery in oil formations. Transp. Porous Media **105**, 235–253 (2014). https://doi.org/10.1007/s11242-014-0368-6
100. S. Pamukcu, R.A. Shrestha, A.B. Ribeiro, E.P. Mateus, Electrically induced displacement transport of immiscible oil in saline sediments. J. Hazard. Mater. **313**, 185–192 (2016). https://doi.org/10.1016/j.jhazmat.2016.04.005
101. Z.-Y. Dong, W.-H. Huang, D.-F. Xing, H.-F. Zhang, Remediation of soil co-contaminated with petroleum and heavy metals by the integration of electrokinetics and bio-stimulation. J. Hazard. Mater. **260**, 399–408 (2013)
102. M.F. DeFlaun, C.W. Condee, Electrokinetic transport of bacteria. J. Hazard. Mater. **55**, 263–277 (1997). https://doi.org/10.1016/S0304-3894(97)00023-X
103. L.Y. Wick, P.A. Mattle, P. Wattiau, H. Harms, Electrokinetic transport of PAH-degrading bacteria in model aquifers and soil. Environ. Sci. Technol. **38**, 4596–4602 (2004). https://doi.org/10.1021/es0354420
104. Q. Luo, H. Wang, X. Zhang, X. Fan, Y. Qian, In situ bioelectrokinetic remediation of phenol-contaminated soil by use of an electrode matrix and a rotational operation mode. Chemosphere **64**, 415–422 (2006). https://doi.org/10.1016/j.chemosphere.2005.11.064
105. J.-L. Niqui-Arroyo, M. Bueno-Montes, R. Posada-Baquero, J.-J. Ortega-Calvo, Electrokinetic enhancement of phenanthrene biodegradation in creosote-polluted clay soil. Environ. Pollut. **142**, 326–332 (2006). https://doi.org/10.1016/j.envpol.2005.10.007
106. J.-L. Niqui-Arroyo, J.-J. Ortega-Calvo, Integrating biodegradation and electroosmosis for the enhanced removal of polycyclic aromatic hydrocarbons from creosote-polluted soils. J. Environ. Qual. **36**, 1444–1451 (2007)
107. X. Mao, J. Wang, A. Ciblak, E.E. Cox, C. Riis, M. Terkelsen, D.B. Gent, A.N. Alshawabkeh, Electrokinetic-enhanced bioaugmentation for remediation of chlorinated solvents contaminated clay. J. Hazard. Mater. **213–214**, 311–317 (2012). https://doi.org/10.1016/j.jhazmat.2012.02.001
108. M. Elektorowicz, S. Hasan, J. Oleszkiewicz, Charge it: a submerged membrane bioreactor achieves high removal efficiencies. Water Environ. Technol. **24**(1), 60–63 (2012)

109. C. Riis, M. Bymose, E. Cox, J. Wang, D. Gent, M. Terkelsen, Succesfull pilot test of electrokinetic enhanced bioremediation (EK-BIO) as an innovative remedial approach for PCE DNAPL source area, in *NORDROCS 2012: 4th Nordic Joint Meeting on Remediation of Contaminated Sites, International Conference, Oslo, Norway*, (2012) 4 p
110. R.T. Gill, M.J. Harbottle, J.W.N. Smith, S.F. Thornton, Electrokinetic-enhanced bioremediation of organic contaminants: a review of processes and environmental applications. Chemosphere **107**, 31–42 (2014). https://doi.org/10.1016/j.chemosphere.2014.03.019
111. H. Aboughalma, R. Bi, M. Schlaak, Electrokinetic enhancement on phytoremediation in Zn, Pb, Cu and Cd contaminated soil using potato plants. J. Environ. Sci. Health A Tox. Hazard. Subst. Environ. Eng. **43**, 926–933 (2008). https://doi.org/10.1080/10934520801974459
112. J.J. Kubiak, P.J. Khankhane, P.J. Kleingeld, A.T. Lima, An attempt to electrically enhance phytoremediation of arsenic contaminated water. Chemosphere **87**, 259–264 (2012)
113. C. Cameselle, R.A. Chirakkara, K.R. Reddy, Electrokinetic-enhanced phytoremediation of soils: status and opportunities. Chemosphere **93**, 626–636 (2013). https://doi.org/10.1016/j.chemosphere.2013.06.029
114. J.W. Moon, H.S. Moon, H. Kim, Y. Roh, Remediation of TCE-contaminated groundwater using zero valent iron and direct current: experimental results and electron competition model. Environ. Geol. **48**, 805–817 (2005). https://doi.org/10.1007/s00254-005-0023-1
115. S. Pamukcu, L. Hannum, J.K. Wittle, Delivery and activation of nano-iron by DC electric field. J. Environ. Sci. Health A Tox. Hazard. Subst. Environ. Eng. **43**(8), 934–944 (2008). https://doi.org/10.1080/10934520801974483
116. C.H. Weng, Y.T. Lin, T.Y. Lin, C.M. Kao, Enhancement of electrokinetic remediation of hyper-Cr(VI) contaminated clay by zero-valent iron. J. Hazard. Mater. **149**, 292–302 (2007). https://doi.org/10.1016/j.jhazmat.2007.03.076
117. M.M. O'Mahony, A.D.W. Dobson, J.D. Barnes, I. Singleton, The use of ozone in the remediation of polycyclic aromatic hydrocarbon contaminated soil. Chemosphere **63**, 307–314 (2006). https://doi.org/10.1016/j.chemosphere.2005.07.018
118. P. Isosaari, R. Piskonen, P. Ojala, S. Voipio, K. Eilola, E. Lehmus, M. Itävaara, Integration of electrokinetics and chemical oxidation for the remediation of creosote-contaminated clay. J. Hazard. Mater. **144**, 538–548 (2007). https://doi.org/10.1016/j.jhazmat.2006.10.068
119. B.G. Petri, R.J. Watts, A.L. Teel, S.G. Huling, R.A. Brown, Fundamentals of ISCO using hydrogen peroxide, in *In Situ Chemical Oxidation for Groundwater Remediation*, ed. by R. L. Siegrist, (Springer, Berlin, 2011), pp. 33–88. https://doi.org/10.1007/978-1-4419-7826-4_2
120. Y. Yukselen-Aksoy, K.R. Reddy, Effect of soil composition on electrokinetically enhanced persulfate oxidation of polychlorobiphenyls. Electrochim. Acta **86**, 164–169 (2012). https://doi.org/10.1016/j.electacta.2012.03.049
121. I. Innocenti, I. Verginelli, F. Massetti, D. Piscitelli, R. Gavasci, R. Baciocchi, Pilot-scale ISCO treatment of a MtBE contaminated site using a Fenton-like process. Sci. Total Environ. **485–486**, 726–738 (2014). https://doi.org/10.1016/j.scitotenv.2014.01.062
122. H.I. Chung, M. Lee, A new method for remedial treatment of contaminated clayey soils by electrokinetics coupled with permeable reactive barriers. Electrochim. Acta **52**, 3427–3431 (2007). https://doi.org/10.1016/j.electacta.2006.08.074
123. E.M. Ramírez, C.S. Jiménez, J.V. Camacho, M.A.R. Rodrigo, P. Cañizares, Feasibility of coupling permeable bio-barriers and electrokinetics for the treatment of diesel hydrocarbons polluted soils. Electrochim. Acta **181**, 192 (2015). https://doi.org/10.1016/j.electacta.2015.02.201
124. E. Mena, J. Villaseñor, M.A. Rodrigo, P. Cañizares, Electrokinetic remediation of soil polluted with insoluble organics using biological permeable reactive barriers: effect of periodic polarity reversal and voltage gradient. Chem. Eng. J. **299**, 30–36 (2016). https://doi.org/10.1016/j.cej.2016.04.049

125. M. Zhou, S. Zhu, Y. Yi, T. Zhang, An electrokinetic/activated alumina permeable reactive barrier-system for the treatment of fluorine-contaminated soil. Clean Techn. Environ. Policy **18**, 2691–2699 (2016). https://doi.org/10.1007/s10098-016-1156-5
126. E. Vieira dos Santos, C. Sáez, P. Cañizares, C.A. Martínez-Huitle, M.A. Rodrigo, Reversible electrokinetic adsorption barriers for the removal of atrazine and oxyfluorfen from spiked soils. J. Hazard. Mater. **322**, 413–420 (2017). https://doi.org/10.1016/j.jhazmat.2016.10.032
127. R. López-Vizcaíno, C. Risco, J. Isidro, S. Rodrigo, C. Saez, P. Cañizares, V. Navarro, M.A. Rodrigo, Scale-up of the electrokinetic fence technology for the removal of pesticides. Part II: Does size matter for removal of herbicides? Chemosphere **166**, 549–555 (2017). https://doi.org/10.1016/j.chemosphere.2016.09.114
128. L.M. Ottosen, P.E. Jensen, G.M. Kirkelund, C. Dias-Ferreira, H.K. Hansen, Electrodialytic remediation of heavy metal polluted soil - Treatment of water saturated or suspended soil, in *Chemical Engineering Transactions. Italian Association of Chemical Engineering - AIDIC*, (2012), pp. 103–108. https://doi.org/10.3303/CET1228018
129. H.K. Hansen, A. Rojo, L.M. Ottosen, Electrodialytic remediation of copper mine tailings. J. Hazard. Mater. **117**, 179–183 (2005)
130. L.M. Ottosen, H.K. Hansen, C.B. Hansen, Water splitting at ion-exchange membranes and potential differences in soil during electrodialytic soil remediation. J. Appl. Electrochem. **30**, 1199–1207 (2000)
131. T.R. Sun, L.M. Ottosen, P.E. Jensen, G.M. Kirkelund, Electrodialytic remediation of suspended soil--comparison of two different soil fractions. J. Hazard. Mater. **203–204**, 229–235 (2012). https://doi.org/10.1016/j.jhazmat.2011.12.006
132. A.N. Alshawabkeh, H. Sarahney, Effect of current density on enhanced transformation of naphthalene. Environ. Sci. Technol. **39**, 5837–5843 (2005)
133. C. Peng, J.O. Almeira, Q. Gu, Effect of electrode configuration on pH distribution and heavy metal ions migration during soil electrokinetic remediation. Environ. Earth Sci. **69**, 257–265 (2013)
134. E. Mattson, D. Bowman, S. Robert, E.R. Lindgren, Electrokinetic ion transport through unsaturated soil: 1. Theory, model development, and testing. Contam. Hydrol. **54**(1–2), 99–120 (2002)
135. A.J. Franz, J.W. Rucker, J.R.V. Flora, M.E. Meadows, W.G. Irwin, Electrolytic oxygen generation for subsurface delivery: effects of precipitation at the cathode and an assessment of side reactions. Water Res. **36**(9), 2243–2254 (2002). https://doi.org/10.1016/S0043-1354(01)00443-2
136. T. Muraoka, E. Ghazanfari, R.A. Shrestha, S. Pamukcu, Electrically induced pore pressures in high salt content clay slurry. Sep. Purif. Technol. **79**, 133–138 (2011)
137. S. Pamukcu, Electrochemical transport and transformations, in *Electrochemical Remediation Technologies for Polluted Soils, Sediments and Groundwater*, ed. by K. R. Reddy, C. Cameselle, (John Wiley & Sons, Inc., Hoboken, NJ, 2009), pp. 29–65. https://doi.org/10.1002/9780470523650
138. A. Rojo, H.K. Hansen, M. Cubillos, Electrokinetic remediation using pulsed sinusoidal electrical field. Electrochim. Acta **86**, 124–129 (2012)
139. E. Kariminezhad, M. Elektorowicz, Comparison of constant, pulsed, incremental and decremental direct current applications on solid-liquid phase separation in oil sediments. J. Hazard. Mater. **358**(15), 475–483 (2018). https://doi.org/10.1016/j.jhazmat.2018.04.002
140. S. Zhang, J. Zhang, X. Cheng, Y. Mei, C. Hu, M. Wang, J. Li, Electrokinetic remediation of soil containing Cr(VI) by photovoltaic solar panels and a DC-DC converter. J. Chem. Technol. Biotechnol. **90**, 693–700 (2015). https://doi.org/10.1002/jctb.4359
141. R.T. Brosky, S. Pamukcu, Role of DDL processes during electrolytic reduction of Cu(II) in a low oxygen environment. J. Hazard. Mater. **262**, 878–882 (2013). https://doi.org/10.1016/j.jhazmat.2013.09.032
142. T.R. Sun, S. Pamukcu, L.M. Ottosen, F. Wang, Electrochemically enhanced reduction of hexavalent chromium in contaminated clay: kinetics, energy consumption, and applica-

tion of pulse current. Chem. Eng. J. **262**, 1099–1107 (2015). https://doi.org/10.1016/j.cej.2014.10.081
143. A. Fellah Jahromi, M. Elektorowicz, Electrokinetically assisted oil-water phase separation in oily sludge with implementing novel controller system. J. Hazard. Mater. **358**, 434–440 (2018). https://doi.org/10.1016/j.jhazmat.2018.07.032
144. S. Yuan, X. Mao, A.N. Alshawabkeh, Efficient degradation of TCE in groundwater using Pd and electro-generated H2 and O2: a shift in pathway from hydrodechlorination to oxidation in the presence of ferrous ions. Environ. Sci. Technol. **46**, 3398–3405 (2012). https://doi.org/10.1021/es204546u
145. L. Rajic, N. Fallahpour, S. Yuan, A.N. Alshawabkeh, Electrochemical transformation of trichloroethylene in aqueous solution by electrode polarity reversal. Water Res. **67**, 267–275 (2014). https://doi.org/10.1016/j.watres.2014.09.017
146. L. Rajic, N. Fallahpour, E. Podlaha, A. Alshawabkeh, The influence of cathode material on electrochemical degradation of trichloroethylene in aqueous solution. Chemosphere **147**, 98–104 (2016). https://doi.org/10.1016/j.chemosphere.2015.12.095
147. Y.B. Acar, H. Li, R.J. Gale, Phenol removal from kaolinite by electrokinetics. J. Geotech. Eng. **118**, 1837–1852 (1992)
148. S. Pamukcu, M. Pervizpour, *Electroosmotically Aided Restoration of TCE Contaminated Soil. Final Report, Contract No. B3460123* (Lawrence Livermore National Laboratory, Environmental Restoration Division, University of California, Livermore, CA, 1998) 52 p
149. S. Pamukcu, In-situ soil and sediment remediation: electrokinetic and electrochemical methods, in *Handbook of Environmental Engineering*, ed. by M. Kutz, (Wiley, New York, NY, 2018). https://doi.org/10.1002/9781119304418.ch8. ISBN-13: 978-1118712948
150. S.A. Amba, G.V. Chilingar, C.M. Beeson, Use of direct electrical current for increasing the flow rate of reservoir fluids during petroleum recovery. J. Can. Pet. Technol. **3**, 8–14 (1964). https://doi.org/10.2118/64-01-02
151. G.V. Chilingar, A. El-Nassir, R.G. Stevens, Effect of direct electrical current on permeability of sandstone cores. Pet. Technol. **22**, 830–836 (1968). https://doi.org/10.2118/2332-PA
152. G.V. Chilingar, C.K. Sang, J.E. Davis, H. Farhangi, L.G. Adamson, S. Sawabini, Possible use of direct electric current for augmenting reservoir energy during petroleum production. Pet. Technol. **4**, 272–285 (1968)
153. J.K. Wittle, D.C. Hill, G.V. Chilingar, Direct current electrical enhanced oil recovery in heavy-oil reservoirs to improve recovery, reduce water cut, and reduce H2S production while increasing API gravity, in *SPE Western Regional and Pacific Section AAPG Joint Meeting*, (2008) 114012
154. J.K. Wittle, D.G. Hill, G.V. Chilingar, Direct electric current oil recovery (EEOR)—a new approach to enhancing oil production. Energy Sources Part A Recover. Util. Environ. Eff. **33**, 805–822 (2011). https://doi.org/10.1080/15567036.2010.514843
155. E.W. Al Shalabi, B. Ghosh, M. Haroun, S. Pamukcu, The application of direct current potential to enhancing waterflood recovery efficiency. Pet. Sci. Technol. **30**(20), 2160–2168 (2012). https://doi.org/10.1080/10916466.2010.547902
156. E.W. Al Shalabi, B. Ghosh, M. Haroun, S. Pamukcu, Stimulation of sandstone reservoirs using DC potential. J. Petrol. Sci. Tech. **30**(20), 2137–2147 (2012)
157. M.R. Haroun, G.V. Chilingar, S. Pamukcu, J.K. Wittle, H. Belhaj, M.N. Al Bloushi, Optimizing electroosmotic flow potential for electrically enhanced oil recovery (EEORTM) in carbonate rock formations of Abu Dhabi based on rock properties and composition, in *International Petroleum Technology Conference. International Petroleum Technology Conference*, (2013). https://doi.org/10.2523/IPTC-13812-MS
158. E. Ghazanfari, *Development of a Mathematical Model for Electrically Assisted Oil Transport in Porous Media*. PhD Dissertation, Department of CEE, Lehigh University, Bethlehem, PA (2013)
159. E. Ghazanfari, S. Pamukcu, Mathematical modeling of electrokinetic transport and enhanced oil recovery in porous geo-media, in *Electrokinetics for Petroleum and Environmental*

*Engineers*, ed. by G. V. Chilingar, M. Haroun, (John Wiley & Sons, Inc., Hoboken, NJ, 2014), pp. 177–236. https://doi.org/10.1002/9781118842805
160. J.M. Fleureau, M. Dupeyart, Influence of an electric field on the interfacial parameters of a water/oil/rock system: application to oil enhanced recovery. J. Colloid Interface Sci. **123**(1), 249–258 (1988)
161. M. Sato, N. Kudo, M. Saito, Surface tension reduction of liquid by applied electric field using vibrating jet method. IEEE Trans. Ind. Appl. **34**(2), 415–416 (1998)
162. R. Tao, X. Xu, Reducing the viscosity of crude oil by pulsed electric or magnetic field. Energy Fuel **20**, 2046–2051 (2006)
163. E.M. Johansen, The interfacial tension between petroleum products and water. Ind. Eng. Chem. **16**, 2 (1923)
164. V.B. Warshavsky, T.V. Bykov, X.C. Zeng, Effects of external electric field on the interfacial properties of weakly dipolar fluid. J. Chem. Phys. **114**, 504–512 (2001)
165. A. Kappler, S.B. Haderlein, Natural organic matter as reductant for chlorinated aliphatic pollutants. Environ. Sci. Technol. **37**(12), 2714–2719 (2003)
166. J.P. Busalmen, A. Esteve-Nuñez, J.M. Feliu, Whole cell electrochemistry of electricity-producing microorganisms evidence an adaptation for optimal exocellular electron transport. Environ. Sci. Technol. **42**(7), 2445–2450 (2008)
167. R.J. Hunter, *Zeta Potential in Colloidal Science Principles and Applications* (Academic Press, New York, NY, 1981)
168. S.-M. Park, J.-S. Yoo, Electrochemical impedance spectroscopy for better electrochemical measurements with impedance data, a complete description of an electrochemical system is possible. Anal. Chem. **75**(21), 455A–461A (2003). https://doi.org/10.1021/ac0313973
169. A.J. Bard, L.R. Faulkner, *Electrochemical Methods: Fundamentals and Applications* (John Wiley & Sons, New York, NY, 1980)

# Treatment of Gaseous Effluents Produced During Electrochemically Assisted Soil Remediation Processes

**I. Ferrara and O. Scialdone**

## 1 Introduction

As discussed in previous chapters, various electrochemical approaches can be used in order to achieve the remediation of soils contaminated by numerous kinds of pollutants. It is important to highlight that, in many cases, electrochemically assisted soil remediation processes, like electrokinetic or electroheating ones, are likely to generate gaseous effluents that need to be properly collected and treated. The composition of gaseous stream depends on the nature of soil or groundwater contaminants and on the adopted remediation process. As an example, in electrokinetic processes, the gas obtained at the electrodes can be contaminated by chlorine or ammonia while various kinds of volatile pollutants can be present in gas collected in the frame of electroheating. The choice of the treatment of gaseous effluents, produced during electrochemically assisted soil remediation processes, depends mainly on the nature and the concentration of the pollutant in gaseous stream. When the pollutants present a high solubility in water (or in water solution), absorption processes can be evaluated. For volatile organic compounds (VOCs), that present a poor solubility in aqueous media, thermal or catalytic oxidation can be used. However, if the concentrations of VOCs are rather low, adsorption processes are often preferred. Furthermore, adsorption processes can be used also for many other pollutants.

In this chapter, some examples of generation of gaseous effluents by electrochemically assisted soil remediation processes will be briefly presented. After,

I. Ferrara
Dipartimento Regionale dell'Ambiente, Regione Siciliana, Palermo, Italy

O. Scialdone (✉)
Dipartimento di Ingegneria, Università degli Studi di Palermo, Palermo, Italy
e-mail: onofrio.scialdone@unipa.it



M. A. Rodrigo, E. V. Dos Santos (eds.), *Electrochemically Assisted Remediation of Contaminated Soils*, Environmental Pollution 30,
https://doi.org/10.1007/978-3-030-68140-1_19

various kinds of conventional and innovative processes for the treatment of gaseous effluents will be described more in detail.

## 2 Generation of Gaseous Effluents by Electrochemically Assisted Soil Remediation Processes

In electrokinetic remediation processes, the application of a direct electric current via electrodes immersed in a moisture-saturated soil results in water oxidation at the anode with production of oxygen gas and $H^+$ (Eq. 1), and in water reduction at the cathode with hydrogen gas and hydroxyl anions formation (Eq. 2).

$$2H_2O \rightarrow O_2(g) + 4H^+ + 4e^- \tag{1}$$

$$2H_2O + 2e^- \rightarrow 2OH^- + H_2(g) \tag{2}$$

Gas produced by electrode reactions is extracted by vents from cathode and anode compartments in order to limit the increase of the resistance at the electrodes due to the low conductivity of gas bubbles attached to the electrodes and in order to prevent gas accumulation, which may lead to a pressure in opposition to the electro-osmotic flow [1]. From reduction and oxidation process of contaminants present in saturated soil, gas compounds can be generated or volatile compounds can be extracted from venting. If the gaseous stream contains pollutants that need to be treated, suitable abatement processes have to be selected. As an example, in electrokinetic processes, highly soluble ionized inorganic species, like chlorides, nitrates, and phosphates, are transported to the electrodes of opposite charge [2, 3], and some of them can react forming gaseous products. In particular, chlorides are expected to be oxidized at the anode [3, 4] with formation of gaseous chlorine (Eq. 3).

$$2Cl^- \rightarrow Cl_2(g) + 2e^- \tag{3}$$

The liberation of chlorine from an aqueous solution would be impossible on a pure thermodynamic basis all over the pH range, since the oxygen evolution is thermodynamically preferred (Fig. 1). However, chlorides oxidation can take place in acidic conditions, because oxygen evolution is reduced at low pH and it shows a much higher overpotential [5] at various electrodes including graphite and DSA ones (Fig. 1), often used in electrokinetic processes. In particular, chlorine evolution is helped by the fact that oxygen evolution gives rise to local acidification of the solution (Eq. 1) with consequent shift of this reaction to more anodic potential.

In the presence of nitrate, at the cathode the evolution of nitrogen and, in limited amounts, the formation of ammonia (Eq. 4) may take place.

$$NO_3^- + 9H^+ + 8e^- \leftrightarrow NH_3 + 3H_2O \tag{4}$$

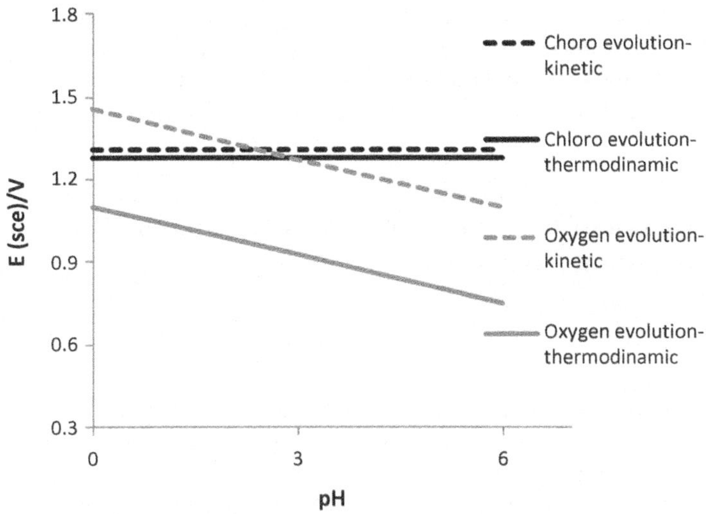


**Fig. 1** Electrode potential-pH diagram for oxygen and chlorine reactions in acidic solutions: (–) Thermodynamic equilibrium conditions; (---) kinetic conditions of anodic gas evolution at 1 mA cm$^{-2}$. The pH dependence is based on the form of the overall reaction in both cases [5]

As an example, Chew and coauthors [6] found that a part of nitrate was converted to ammonia at graphite cathodes in the case of a nitrate-contaminated soils by electrokinetic process assisted by iron walls. Hence, in this case, gas produced at the cathode can contain hydrogen, nitrogen, and ammonia that needs to be treated.

Hydrogen sulfide evolution was also found in cathode wells by some authors [7, 8] and it was attributed to the presence of metal sulfides in the soil [8]. According to some authors [8], $H_2S$ may be generated as the hydrogen ion front (generated at the anolyte) solubilizes iron sulfides in the soil as it moves from the anode to the cathode array (Eq. 5).

$$2Fe_2S_3 + 12H^+ \rightarrow 6H_2S_{(g)} + 4Fe^{2+} \quad (5)$$

Furthermore, the heat produced during electrokinetic processes can allow to volatilize some of the contaminants from the water that are pumped out near the electrodes. Volatile and semi-volatile organics from soil can be removed by various assisted electrochemical approaches. As an example, since soil is not a good electric conductor, the passage of an electric current generates heat. In electrokinetic remediation, a continuous electric current is used because the objective is to transport ionic and nonionic contaminants out of the soil. In the case of electroheating, the electric field is often used as a source of energy that is transformed from electric energy into heat. That is why in electroheating the continuous electric field can be substituted by an alternate current that supplies the energy but does not induce transportation. In all cases, when organic compounds are generated, the gaseous effluents need to be captured and treated. As an example, Lageman and coauthors

[9] used a combination of in situ techniques: heating of soil and groundwater in the source areas, combined with soil vapor extraction to remove various organics like perchloroethylene (PCE) and trichloroethylene (TCE) and their volatile degradation products cis-1,2-dichloroethene (C-DCE) and vinyl chloride.

# 3 Treatment of Gaseous Effluents Generated by Soil Remediation Processes

## *3.1 Adsorption*

In the adsorption process, gas pollutants are retained on a solid surface that prefers specific compounds to others removing them from gaseous effluents. When the surface is saturated, in most of cases the adsorbent is regenerated while in few cases it is disposed and replaced. When desorbed, the contaminants are present at high concentration and they can be either recovered if they present an economic value or disposed.

Adsorption is traditionally used for treating off-gases containing VOCs and BTEX compounds both in industrial processes and soil remediation applications.

Major types of adsorption systems are:

- fixed-bed adsorption
- fluidized-bed adsorption
- continuous moving-bed adsorption
- pressure swing adsorption (PSA)

The process is exothermic and favored by low temperatures and high pressures (which are not usually necessary). The pollutant is often physically adsorbed and, usually, it can be desorbed relatively easily by heat at relatively low temperatures and/or vacuum processes. However, in various cases a chemical adsorption is needed often to convert the pollutant in a less toxic substance. Table 1 reports various examples of abatement efficiencies reported in literature for the treatment of some pollutants.

The most used adsorbent for the treatment of off-gas from soil remediation is activated carbon, but also other adsorbents such as aluminum-silicate zeolites and synthetic polymers are adopted. In adsorption, the pollutant is collected on internal and external surface of adsorbent. The adsorption is described from a thermodynamic point of view by adsorption isotherms. Adsorption isotherms allow to select the adsorbents that give the best performances from a thermodynamic point of view. As an example, it is possible to consider the case of a gas containing toluene and water. Adsorption isotherms show that activated carbon presents low adsorption of water for low humidity and high adsorptions of toluene also for low partial pressures of toluene. Hence, activated carbon can be selected for this specific case.

**Table 1** Abatement efficiencies and emission levels associated with adsorption

| | Abatement efficiency (%) | | | |
|---|---|---|---|---|
| Substance | Granulated Activated Carbon | Powdered Activated Carbon | Zeolite | Notes and reference |
| Organic Hg | 80 | | | Low feed [10, 11] |
| Inorganic Hg | 80 | | | Feed 29µg $L^{-1}$ [10] |
| Hexachlorobenzene | 93 | | | Feed 7.7µg $L^{-1}$ |
| | | 99 | | Feed 10µg $L^{-1}$ [10, 11] |
| Atrazine | 84 | | | Feed 0.61µg $L^{-1}$ [10, 11] |
| Trichlorobenzene | | 70–93 | | Feed 200µg $L^{-1}$ [10, 11] |
| Ammonia | | | 98 | Feed 25 mg $L^{-1}$ [10, 11] |
| COD | 50 | | | Moving bed, feed 5 mg $L^{-1}$ [10, 12] |
| | 65–75 | | | Feed 300–400 mg $L^{-1}$ [10] |
| Phenols | 75 | | | Moving bed, feed 70 mg $L^{-1}$ [10, 13] |
| | 60–80 | | | Feed 5 mg $L^{-1}$[10,13] |
| AOX | >90 | | | Moving bed, feed 60 mg $L^{-1}$ [10, 12] |

The conventional adsorption process is quite simple: The gas containing the contaminants flows through a packed bed of the adsorbing material and the pollutants are adsorbed onto its surface; the process is stopped when the concentration of the pollutants in the gas that goes out from the bed exceeds acceptable levels. As an example, when the humidity content of the gas is low, activated carbon can sorb 10–20% of its weight, due to its hydrophobic character; on the other hand, when relative humidity is high (as an example, above 50%), the capacity is inevitably reduced by sorption of water.

Adsorption materials such as activated carbons are often regenerated once they have reached their adsorption capacity. In regenerable systems, when the adsorption material is saturated by pollutants, it is treated to remove the adsorbed chemicals in order to be used again. Regeneration is performed by increasing the temperature using hot air or steam and/or decreasing the partial pressure. In some cases, two beds are used in parallel in order to have in the same time both the adsorption and the desorption stages (Fig. 2).

After the regeneration step, a small part of contaminants can be still adsorbed, thus resulting in a lower adsorption capacity of the adsorbent in the next cycle. After a certain number of cycles, the adsorption capacity becomes too low and the adsorbent requires replacement [14]. Adsorption systems with activated carbons are largely used since they are effective for most volatile organic compounds and chlorinated VOC and when the concentration of pollutants changes rapidly, although some organics such as vinyl chloride, methanol, and formaldehyde are not adsorbed in an adequate way. Furthermore, activated carbons are often added/impregnated with various compounds to enhance the removal of other contaminants, such as $H_2S$, Hg, and $NH_3$.

Adsorption with activated carbon is rather effective, flexible and present relatively low costs. However, the replacement of the adsorbent or its regeneration leads to relatively high operative costs in the presence of high concentrations of pollutants.

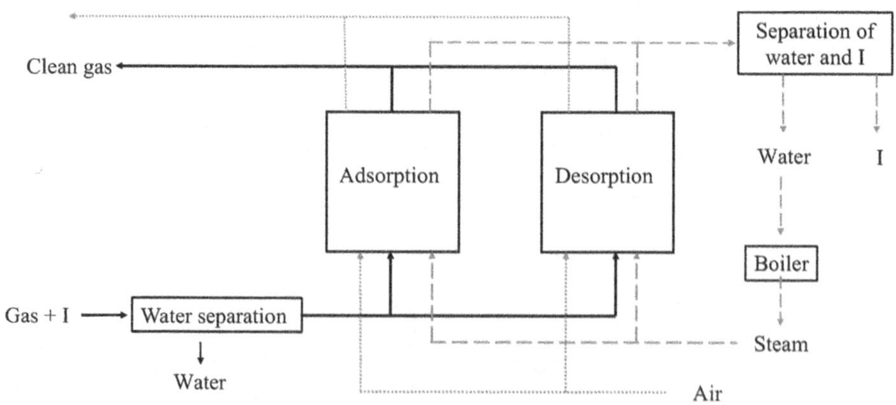


**Fig. 2** Scheme of process with two beds that work in parallel in adsorption and desorption (I indicates the pollutant)

### 3.2 Thermal and Catalytic Oxidation Processes

Oxidation systems are used to treat gas containing volatile organic compounds generated from soil and groundwater remediation systems and from gas streams produced from many industrial sources. In 1997, approximately 6000 oxidation systems were active in the world and about 20% of them were used for remediation [14]. Oxidation systems are often used for organic compounds, but they can be potentially involved also for some inorganic substances (see as an example, Table 2).

In a complete oxidation reaction, organics are oxidized to carbon dioxide and water. Oxidation can be used to treat many kinds of volatile organic compounds in a large range of concentrations with abatement removals often higher than 99%. Examples of compounds easily oxidized include alcohols, aliphatics, aromatics, esters, and ketones.

If halogenated compounds are present, the resulting combustion products can include acid gases (i.e., hydrochloric acid) and further treatments with a scrubber are needed (see Sect. 3.3). An oxidation unit usually consists of a combustion unit (reaction chamber) with one or more burners and a heat exchanger and, if necessary, a post-oxidation treatment (Fig. 3).

For small concentrations of volatile organic compounds, an auxiliary fuel must be added (like natural gas) to sustain the combustion process. On the other hand, if the concentrations of VOC are very high, thus leading to safety problems, ambient air must be added to the gaseous effluents to dilute them. In order to remove VOC emissions, two kinds of systems can be used:

- thermal oxidation (with direct-flame or flameless thermal oxidizers)
- catalytic oxidization

Many kinds of catalyst can be used, involving both metals and metal oxides. Palladium, platinum, rhodium, chromium, nickel, cobalt, and manganese are often adopted. In most of cases, ceramic or metal supports are used to decrease the amount of precious catalyst and to increase the mechanical resistance. Platinum-based catalysts can be effectively adopted for the oxidation of a large number of VOCs, but they are deactivated in short times if the effluents contain chlorine or chlorinated compounds. In this case, cobalt oxide, chromia/alumina, and copper oxide/manganese oxide are often adopted. Most of catalysts work usually between 300 and 700 °C, but some of them can work even at lower temperatures (200–250 °C) [10].

**Table 2** Examples of oxidation reactions

| Compound | Oxidation reaction |
|---|---|
| $C_7H_8$ (Toluene) | $C_7H_8 + 9O_2 \rightarrow 7CO_2 + 4H_2O$ |
| $CH_3OH$ | $CH_3OH + 1.5O_2 \rightarrow CO_2 + 2H_2O$ |
| $C_3H_6O$ (acetone) | $C_3H_6O + 4O_2 \rightarrow 3CO_2 + 3H_2O$ |
| $C_4H_8O_2$ (ethyl acetate) | $C_4H_8O_2 + 5O_2 \rightarrow 4CO_2 + 4H_2O$ |
| $NH_3$ | $NH_3 + 0.75O_2 \rightarrow 0.5N_2 + 1.5H_2O$ |
| $H_2S$ | $H_2S + 1.5O_2 \rightarrow SO_2 + H_2O$ |

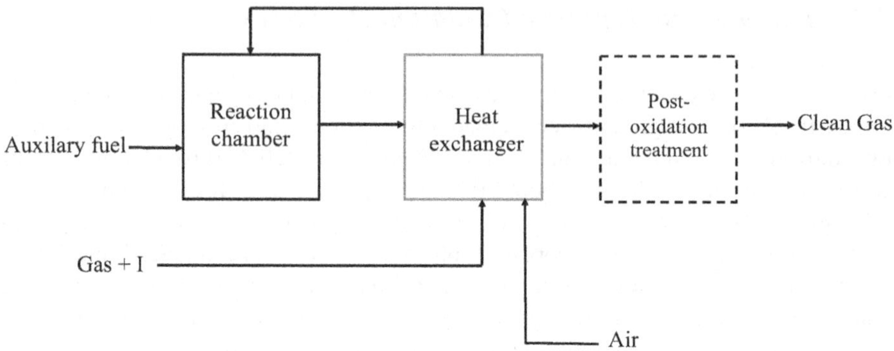


**Fig. 3** Generalized flow diagram of thermal oxidation system (I indicates the pollutant)

The selection of the oxidation system depends on many parameters including the abatement required for the VOC present in the effluents, their nature and concentration, the gas extraction flow rate, and the presence of other substances that can affect the performances of the process and of the catalyst. For high concentrations of volatile organic compounds, a high energy content is produced during their combustion, and thermal oxidation systems can be used; on the other hand, for low VOC concentrations, catalytic systems are often used to decrease the amount of auxiliary fuel necessary to sustain the oxidation at the required temperature, and operative costs are often reduced with respect to thermal processes.

However, catalytic systems require more maintenance than thermal ones and they are vulnerable to chemicals ("*poisons*") that can react with the catalysts and/or to particulate matter that can cover their surface. Poisons for catalysts include metals halides and phosphorus-, silicon- and sulfur-containing compounds. Catalysts can also be deactivated if the gas becomes too hot, altering the structure of the catalyst. Deactivated catalysts in some cases can be regenerated while in other cases they must be disposed and replaced.

Thermal and catalytic oxidation are consolidated technologies that have been effectively implemented in many sites. Relevant disadvantages of such methods are the high capital costs and the high cost for energy necessary to heat the gas when the concentrations of organic pollutants are not sufficiently high. For example, the cost of auxiliary fuel can often exceed the cost of active carbon replacement, making oxidation processes less convenient than adsorption ones. Furthermore, oxidation processes usually require more training of personnel and more maintenance with respect to adsorption ones. Hence, oxidation processes are often used when the concentration of organics is sufficiently high, thus limiting the operating costs for oxidation processes and increasing that of adsorption, or when organic pollutants contained in the off-gas are effectively destroyed by oxidation but not adequately removed by adsorption.

## 3.3 Absorption

Gas absorption technologies involve processes in which pollutants previously present in the gas phase are dissolved in a liquid solution. In some cases, the pollutant is simply physically dissolved (*physical absorption*) while in other cases it chemically reacts with a component of the solution (*chemical absorption*). If the solubility is sufficiently high, no chemical reactions are necessary. On the other hand, if the pollutant presents a low solubility, a component can be added to the solvent to react with the pollutant in order to increase its solubility and to reduce the volume of the absorption solution and of the scrubber. The process is exothermic and favored by low temperatures and high pressures. In few cases, the solution containing the pollutant has an economical value while in most of cases it has no value; in these last cases, the solvent can be separated by the pollutant and recirculated to the absorption tower, limiting the generation of liquid effluents and the need to replace the absorption solution. Figure 4 reports an example of an absorption scheme that involves two towers that work in parallel: The first tower is devoted to the absorption while the second allows a continuous regeneration of the solvent obtained by an increase of the temperature and/or by the utilization of vacuum.

Several aqueous solutions can be used including the following [10, 15].

- Water to remove substances that present a high solubility as HCl or $NH_3$.
- Oxidizing solutions.
- Alkaline solutions (containing chemicals such as sodium hydroxide and sodium carbonate) to remove acidic substances like hydrogen halides, phenols, sulfur dioxide, hydrogen sulfide, and chlorine. In most of cases, a pH between 8.5 and

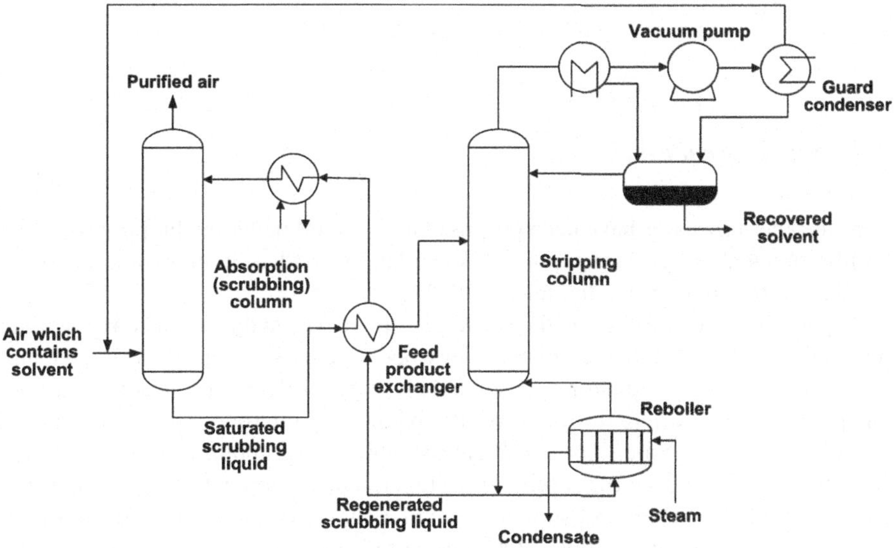


**Fig. 4** Absorption/desorption system

9.5 is used. However, the alkalinity of the solution depends on the nature of the pollutant and, as an example, a higher pH is used for $H_2S$.

- Alkaline solutions containing oxidants such as $H_2O_2$, NaOCl, $O_3$, or $ClO_2$.
- $NaHSO_3$ solutions used mainly to remove odorous compounds such as aldehydes.
- Acidic solutions used to absorb alkaline pollutants such as $NH_3$, esters, and amines. Usually, a pH between 3 and 6 is used. $H_2SO_4$ is often involved for its relatively low cost, even if other acids such as $HNO_3$ can be used for specific applications.
- Solutions containing monoethanolamine and diethanolamine, often used for $H_2S$.

Various kinds of scrubbers are used as [10, 14, 16] jet rotation, venturi, dry tower, spray and packed tower scrubbers.

The most adopted absorption solvent is water, which is suitable for polar compounds. Non-polar organic solvents can be used for apolar pollutants. As an example, organic solvents can be used for the treatment of small volatile organic compounds such as butanes and pentanes.

In the case of the treatment of HCl, the absorption solution is strongly acidic (pH 0–1) due to acid presence. In the case of chlorine, absorption is usually carried out using solutions of NaOH. The absorption takes place with the formation of HOCl (Eqs. 6 and 7) that is enhanced in the presence of high pH [17]. Furthermore, at high pH the reaction (Eq. 8) takes place.

$$Cl_2 + H_2O \leftrightarrow HOCl + H^+ + Cl^- \tag{6}$$

or

$$Cl_2 + OH^- \leftrightarrow HOCl + Cl^- \tag{7}$$

$$HOCl + OH^- \leftrightarrow OCl^- + H_2O \tag{8}$$

## *3.4 Other Processes*

Various other processes have been proposed in literature including bioprocesses like biofiltration and bioscrubbing, photo/UV oxidation, condensation, membrane separation, non-thermal plasma treatment [10, 14].

In particular, in biofiltration, the waste gas stream passes through a bed of organic material (like peat, heather, compost, root wood, tree bark, compost, softwood, and different kinds of combinations) or some inert material (like clay, activated carbon, and polyurethane), where it is biologically oxidized by microorganisms into carbon dioxide, water, inorganic salts, and biomass, while bioscrubbing combines absorption and biodegradation; the scrubbing water contains a population of microorganisms suspended in water and suitable to oxidize gaseous compounds. Biofiltration can be used to treat rather dilute VOC concentrations.

If optimum conditions are maintained, a properly designed biofilter may achieve greater than 90% and sometimes greater than 95% abatements. Specific classes of compounds readily biodegradable by biofilters include mono-aromatic hydrocarbons, alcohols, aldehydes, and ketones.

## 4 Comparison Between Treatment Processes and Final Considerations

Table 3 reports the process often used for the treatment of some pollutants [10]. It can be seen that in some cases, various alternative processes can be used. Hence, in these cases costs and operative considerations have to be used to choose the best option.

Adsorption with activated carbons and thermal or catalytic oxidation are the most common technologies used for assisted soil vapor extraction off-gas treatment. These two technologies are robust and mature vapor treatment methods that can address a variety of contaminants and concentrations [14]. According to EPA-542-R-05-028 [14], the general rule for selecting thermal/catalytic oxidation or carbon adsorption is that dilute off-gases are more cost-effectively treated by carbon adsorption, while thermal/catalytic oxidation becomes more cost-effective for off-gases that contain greater concentrations of vapor contaminants. Some sites have both thermal oxidation and adsorption systems. Thermal or catalytic oxidation systems have been used to treat higher initial concentrations and are replaced by adsorption systems once concentrations have decreased. A limited number of biofiltration systems are currently being used for soil vapor extraction applications. One

**Table 3** Examples of the waste gas treatment techniques with respect to the pollutants to be abated (♦ = the primary goal of the technique is the removal of specific pollutant; + = the primary goal of the technique is not the removal of specific pollutants, but these pollutants are also, in some cases partially, removed using the technique)

| | Pollutants | | | | |
|---|---|---|---|---|---|
| Process | VOC | $H_2S$ | $NH_3$ | HCl | $Cl_2$ |
| Thermal oxidation | ♦ | | | | |
| Catalytic oxidation | ♦ | | | | |
| Adsorption (general) | | | | + | |
| • Active coal | ♦ | ♦ | | | |
| • Zeolites | ♦ | | ♦ | | |
| • Polymeric | ♦ | | | | |
| Absorption (general) | ♦ | | ♦ | ♦ | |
| • Acid | | | ♦ | ♦ | |
| • Alkaline | | ♦ | | ♦ | ♦ |
| Bioprocess | ♦ | + | ♦ | | |

operative limitation is that this technology is rather sensitive to variations in operating parameters, like moisture content, temperature, pH, and nutrient levels.

Absorption processes are routinely used for the treatment of gases involving polar compounds, like HCl, chlorine, or ammonia, that are rather soluble in water solutions.

As an example, a gaseous pollutant produced by electrochemically assisted soil remediation processes is chlorine that is usually treated by chemical absorption in water solution by formation of HOCl and/or $ClO^-$, usually carried out in alkaline solutions.

## References

1. V. Ferri, S. Ferro, C.A. Martínez-Huitle, A. De Battisti, Electrokinetic extraction of surfactants and heavy metals from sewage sludge. Electrochim. Acta **54**, 2108–2118 (2009)
2. J. Virkutyt, M. Sillanpää, P. Latostenmaa, Electrokinetic soil remediation-critical overview. Sci. Total Environ. **289**, 97–121 (2002)
3. K.J. Kim, J.M. Cho, K. Baek, J.S. Yang, S.H. Ko, A review article: Electrokinetic bioremediation current knowledge and new prospects. J. Appl. Electrochem. **40**, 1139–1144 (2010)
4. R. Iannelli, M. Masi, A. Ceccarini, M.B. Ostuni, R. Lageman, A. Muntoni, D. Spiga, A. Polettini, A. Marini, R. Pomi, Electrokinetic remediation of metal-polluted marine sediments: experimental investigation for plant design. Electrochim. Acta **181**, 146–159 (2015)
5. S. Trasatti, Progress in the understanding of the mechanism of chlorine evolution at oxide electrodes. Electrochim. Acta **32**, 369–382 (1987)
6. C.F. Chew, T.C. Zhang, Water Sci. Technol. **38**, 135–142 (1998)
7. D.B. Gent, R.M. Bricka, A.N. Alshawabkeh, S.L. Larson, G. Fabian, S. Granade, Bench- and field-scale evaluation of chromium and cadmium extraction by electrokinetics. J. Hazard. Mater. **110**, 53–62 (2004)
8. *In-Situ Electrokinetic Remediation of Metal Contaminated Soils Technology Status Report*. US Army Environmental Center Report Number: SFIM-AEC-ET-CR-99022 (2000)
9. R. Lageman, M.S. Godschalk, Electro-bioreclamation: a combination of in situ remediation techniques proves successful at a site in Zeist, the Netherlands. Electrochim. Acta **52**, 3449–3453 (2007)
10. *Best Available Techniques (BAT) Reference Document for Common Waste Water and Waste Gas Treatment/Management Systems in the Chemical Sector*. Industrial Emissions Directive 2010/75/EU
11. Environment Agency (England and Wales), *Guidance for Operators and Inspectors of IPC Processes - Effluent Treatment Techniques* (Environment Agency (England and Wales), Bristol, 1997)
12. BMU/LAWA, *Chemische Industrie - Hinweise und Erläuterungen zu Anhang 22 der Abwasserverordnung* (Bundesanzeiger Verlag, Köln, 2000)
13. NOREC, *Contribution to the Drawing Up of the Original CWW BREF: Examples of Waste Water and Waste Gas Treatment in the Chemical Industry in Finland* (North Ostrobothnia Regional Environment Centre, Oulu, 2000)
14. EPA-542-R-05-028 (2006)
15. ETBPP, *Solvent Capture for Recovery and Re-Use from Solvent-Laden Gas Streams* (Environmental Technology Best Practice Programme, London, 1996)
16. F. Ullmann, Adsorption chapter, in *Ullmann's Encyclopedia of Industrial Chemistry*, 7th edn., (Wiley VCH, Weinheim, 2011)
17. K.S. Agrawal, Int. J. Innov. Res. Sci. Eng. Technol. **2**, 3201–3212 (2013)

# Solar-Powered Electrokinetic Remediation for Treatment to Soil Polluted with Organic Compounds

**Eduardo Expósito Rodríguez, Francisco Gallud Martínez, and Vicente Montiel Leguey**

## 1 Introduction

Throughout the preceding chapters of this book, different authors have shown us the basics and the conceptual and mathematical models of electrokinetic remediation processes. In addition, we have seen major possibilities and applications at soil decontamination, both applied individually as well as establishing synergies when in combination with other remediation technologies.

In this chapter, the possibilities of electroremediation (ERem) technology will be discussed; either used individually or combined with other remediation technologies, using a source of renewable energy such as solar photovoltaic. However, in order to provide the reader with the necessary judgement tools, the operation of a photovoltaic (PV) panel will be explained as well as the main peculiarities and characteristics of the power supply of electrochemical systems using a solar PV plant, emphasizing the current knowledge status regarding ERem processes supplied with PV modules.

A reflection on the different available options to power ERem processes—coupled or not to other remediation technologies— by means of a PV solar plant will be presented. The aim is to speculate— theorize about how it would affect to engage a solar plant in an ERem process and its influence on the efficiencies obtained in the treatment. Various thoughts will also be shared on:

- Characteristics of the solar plant of choice: use of batteries, direct connection, use of DC-DC converters, etc.

E. E. Rodríguez · F. G. Martínez · V. M. Leguey (✉)
Grupo de Electroquímica Aplicada, Instituto Universitario de Electroquímica, Departamento Química Física, Universidad de Alicante, Alicante, Spain
e-mail: vicente.montiel@ua.es



M. A. Rodrigo, E. V. Dos Santos (eds.), *Electrochemically Assisted Remediation of Contaminated Soils*, Environmental Pollution 30,
https://doi.org/10.1007/978-3-030-68140-1_20

- The need of solar plant modifications based on variations of experimental parameters such as pH, the soil resistance, etc. during the treatment.
- The need of solar plant modifications based on the number and disposition of electrodes (linear, hexagonal, etc.). We can speculate on how we could change the solar field configuration based on the electrode restructuration during the treatment.
- The influence of the use of PV modules on operating parameters that can be conditioned by significant decreases in irradiation or night periods, for example, soil pH control etc.

## 2 Basic Principles and Main Characteristics of the Electroremediation

Electrokinetic remediation is a soil decontamination technology, which has been proven useful in the elimination of a wide range of both organic and inorganic contaminants. In a very simple approach, we can say that this technology is based on the application of a continuous electric field across the soil to decontaminate using electrodes placed on the ground [1, 2]. The transport of the pollutants in the soil under the influx of a continuous electric field is mainly produced by three phenomena [3]:

- Electroosmosis, which involves the movement of the fluid present in the soil altogether with the pollutants dissolved in it.
- Electromigration, where the ions in the solution and electrically charged complexes move due to the action of the electric field towards the oppositely charged electrode.
- Electrophoresis, which relies on the advective transport of colloids, small particles, bacteria, etc., all of them electrically charged. Similarly, to the case of the electromigration, these species move towards the oppositely charged electrode.

The transport of contaminants in the soil is a very complex process. In addition to the physicochemical properties intrinsic to the ground—such as composition, porosity, etc.—and environmental—humidity, temperature, etc.—which determine the strategy and the success rate of the ERem treatment [4], there are a large number of different phenomena that can occur simultaneously during the treatment, such as pH gradient occurrence, desiccation due to the Joule's effect generated heat, salt precipitation, etc. These phenomena can change the extraction process dynamics, enhancing or inhibiting the extractability of pollutants.

Once contaminants have migrated to the proximity of the electrodes, there are different options for disposal. Among the most common are the electrodeposition in the case of metals, precipitation, adsorption using ion-exchange resins and the electrode reservoir solutions treatment.

The main operational parameters that must be defined when designing an experimental installation of ERem treatment are [3]:

- The manufacture material of the electrodes, their geometry and their relation with the immediate subsurface environment concerning the dissolution exchange. Electrodes with high electrical conductivity, chemically inert, low cost, easy to install in the field and those, which provide a good subsurface electrical contact, are pursued.
- The geometrical configuration of the electrode assembly to achieve the most efficient and effective elimination results based on the applied electric field. Zones without active electric field should be minimised.
- The use of enhancement techniques, among which we can highlight the addition of chemicals (surfactants, complexing agents, etc.) and the conditioning of electrode reservoir solutions. The first one allows increasing the solubility and mobility of pollutants and in the second case is used to prevent the negative impact of the electrode reactions in the treatment.

An application of the ERem on the rise in recent years is its joint implementation with other remediation technologies [5, 6] looking for a synergistic effect, in such a way that the results of the coordinated implementation exceed those obtained on an individual basis with each technology. The most studied combinations include the use of the ERem with:

- Fenton process [7]. This is an advanced oxidation technology based on the catalysed oxidation using hydrogen peroxide and $Fe^{2+}$ ions. The oxidation of $Fe^{2+}$ to $Fe^{3+}$ together with the decomposition of $H_2O_2$ and the generation of highly oxidizing hydroxyl radicals occur in a first stage. Then, organic pollutants are oxidised by the OH radicals.
- Permeable reactive barriers (PRBs) [1]. These are areas of 'in situ' treatment established in the passage of an underground-polluted fluid plume and are able to degrade the contaminants it contains. The most common case is the use of zero-valent iron granules (ZVI) mixed with soil.
- On the engaging of the electrokinetic remediation with the use of PRBs is not necessarily a natural advective transport of groundwater caused by a hydraulic gradient. The applied electric field provides the necessary driving force to transport the pollutants.
- Bioremediation [8]. This technology is based on the use of microorganisms with the ability to break down dangerous pollutants, which may be considered their 'nourishment'. This is a complex technology, where the success of its application depends on the transport of microorganisms, electron acceptor compounds and essential nutrients for the microorganisms to the polluted subsoil area. In a process of bioremediation coupled with ERem, the role of the latter is to force the transport of species: pollutants towards microorganisms or vice versa.

# 3 Use of Photovoltaic Panels in Electrochemical Processes

## 3.1 Performance of a Photovoltaic Panel

In normal conditions, the *V-I* curve of an electrochemical reactor resembles quite well to a straight line, where the voltage and intensity are proportional. In order to power supply an electrochemical system, it normally used a conventional power supply connected to the electrical grid. This configuration is the simplest because a power supply is characterized by maximum values of voltage (*V*) and current intensity (*I*) that are adjustable. If the experiment is conducted under controlled voltage, the operation mode is called potentiostatic. If the parameter that we control is the electric current that is applied to the electrodes, the mode of operation is called galvanostatic.

The figure below depicts the *V-I* characteristic curves of an electrochemical reactor—for this example, we have chosen a global system resistance of 11 Ω—and maximum power from a power supply with 80 V–5 A.

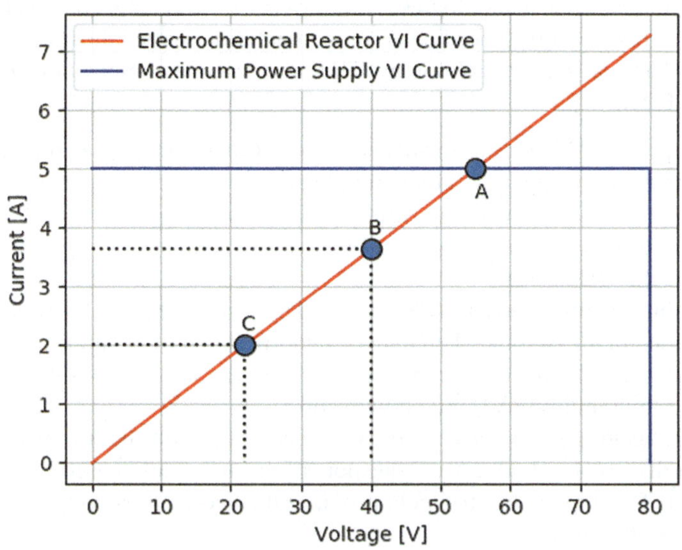


The figure also shows several working points of the system according to different applied voltage or intensity values. Thus, we can see how the system cannot reach a voltage value above 55 V, given that at this value, it has reached the maximum power supply's current, 5 A (point A). Other possible working points of the system are the couples 40 V–3.64 A and 22 V–2 A (points B and C of the graph, respectively). These pairs of *V-I* values are obtained irrespective of mode of operation—potentiostatic or galvanostatic—which only indicates what variable—voltage or intensity—is being controlled.

In the previous paragraphs, we have seen the electrical characteristics of an electrochemical system powered by a conventional power supply. Next, we will explain the workings of an electrochemical system powered by a PV solar plant.

The PV solar energy is one of the most widely used renewable sources of energy. In essence, a PV solar cell transforms solar energy directly into electrical energy. Among the main advantages of this technology, we can list that it is not polluting, noiseless, decentralized and modular (i.e. that we can have it in small increments), abundant and renewable. In addition, PV panels contain a series of solar cells and have no moving parts. They are manufactured in corrosion-resistant watertight structures, so its maintenance is minimal, and its lifespan is very long.

A PV cell consists of two or more thin layers of semiconductor material, typically silicon. When it is exposed to the light, electric charges are generated (electrons and holes), which travel to metal collectors arranged on the material's surface and lead to an external electric circuit. Low-voltage direct current (DC) is provided, so in practice arrays of PV cells are grouped and connected in series and parallel and encapsulated between glass covers in order to put together PV panels.

The characteristic curve of a solar PV panel—also known as current-voltage curve (abbreviated to *V-I* curve)—represents the values of voltage and current measured experimentally for a PV panel under certain consistent conditions of irradiance ($G$, W/m) and temperature ($T_{panel}$, °C) [9].

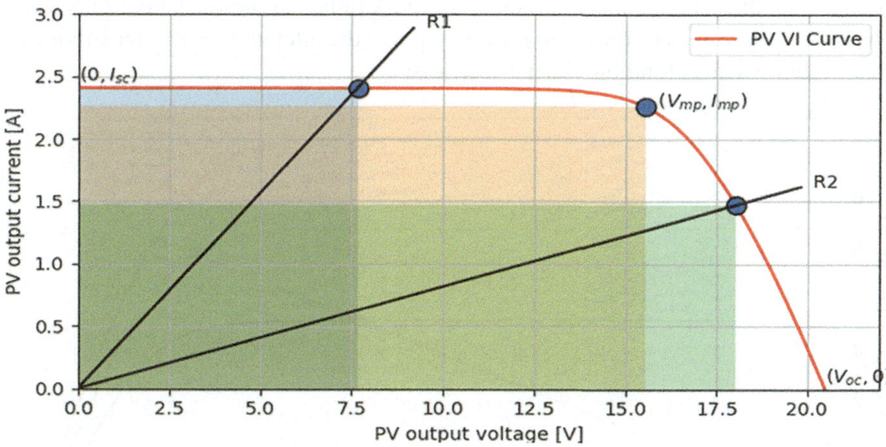


The shaded areas represent the power provided by the PV modules for different external charge (resistances $R1$ and $R2$). The parameters that define a PV module are:

- The short-circuit current ($I_{sc}$), which is the current intensity ($A$), which provides the PV module when the voltage between the terminals is zero. Its value depends mainly on solar irradiance.
- The open-circuit voltage ($V_{oc}$) is the maximum voltage value of the PV module, and it's obtained when it is not connected to an external charge.

- The maximum power ($P_m$) is the maximum electrical power that the PV module provides for a given irradiance and panel temperature conditions.

In the figure, we can see how at low and intermediate values of output voltage the *V-I* curve presents a plateau, where the current gently decreases when the voltage is increased. This behaviour remains until output voltage values approach to the maximum power point, where a sudden decrease in the intensity happens until it reaches zero at the open-circuit voltage.

The figure displays that a very important characteristic of the operation of a PV panel is that for an external electrical resistance (*R*, Ohm) connected to the terminals of the panel, the operating point of the system—the power provided by the PV module—is given by the intersection of the line $I = (1/R)V$ and the characteristic *V-I* curve of the PV module. It is of utmost importance to highlight that for the power (*W*) provided by the PV module, both the value of voltage and current are set ($W = VI$).

As we can see in the figure below, at a given electrical resistance of the electrochemical reactor, the main parameters that influence the operation point are the solar irradiance and the temperature of the panel, as they are the experimental parameters that modify the shape of the PV module's *V-I* characteristic curve in a more pronounced way. Judging by its practical application, this means that the voltage and intensity provided by the PV module will greatly vary throughout the day because of the daily solar irradiance curve. Another common cause of sudden changes in the values of provided power is due to the attenuation of solar irradiance that occurs when a cloud blocks the solar disk.

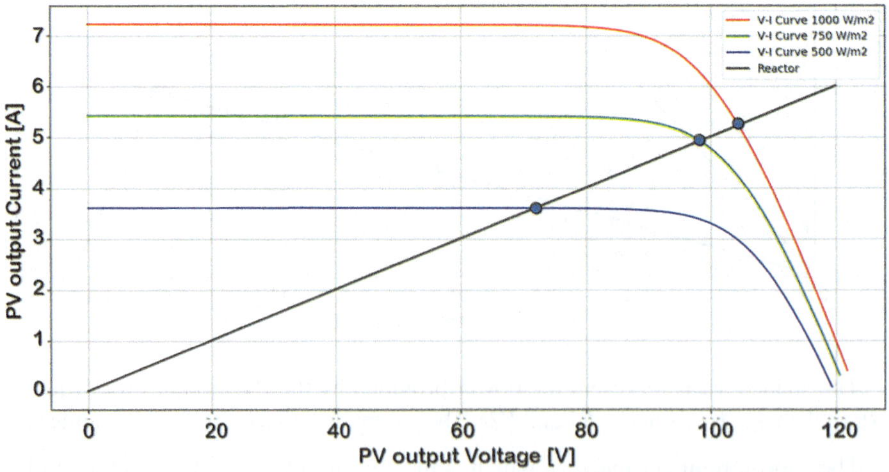


A PV panel has a curve similar to those of a conventional power supply. Differences are on the one hand that fixed voltage, and current areas in a PV panel are not as pronounced as in a conventional power supply. On the other hand, the maximum current provided by a PV panel depends on the solar irradiance that the panel receives.

In order to supply an electrochemical system with solar panels, one should bear in mind that the solar irradiance highly varies during the day. The figure below explains the operation of an electrochemical system connected directly to a PV generator throughout a day. In the figure, we can see the represented typical irradiance evolution on a sunny day (red curve), along with the voltage and current variation (curves in the colours green and blue, respectively) provided by a solar PV plant connected to an electrochemical system. In order to simplify the figure, the electrical resistance of the electrochemical system (black line) has been considered constant. To clarify the explanation, the figure also shows the characteristic *V-I* curves of the PV solar plant (curves in grey). Figure shows as the irradiance increases during the morning (point *G*1) up to the maximum towards the solar noon (point *G*2), to finally decrease during the afternoon. The curve shows how the clouds blocking the solar disk make the irradiance decrease very abruptly (point *G*3). This sharp decrease of irradiance is also reflected in the working current and voltage curves. For each irradiance value, there is a PV plant *V-I* curve. The intersection between the PV plant *V-I* curves and the electrochemical reactor $I = (1/R)V$ lines leads to the working points of the PV plant (VI1, VI2 and VI3 points).

In the morning and in the evening, the working point of the PV plant is located in the area of the *V-I* curve, where the current intensity is approximately constant. Therefore, the irradiance curve is similar to the current intensity one. In the middle of the day, where the irradiance is greater (point VI2), the figure shows how the system operates in the area, where the PV plant's current intensity strongly decreases with voltage, and so the irradiance and current intensity curves differ. In addition, it can be seen how in this time interval there is a plateau in the current curve.

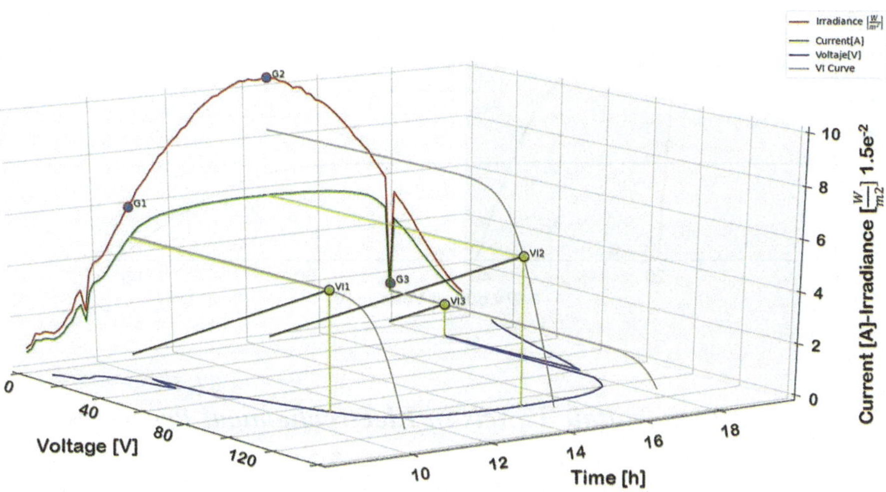


In the event of the electrochemical reactor needing more power than the provided by a single solar panel, several PV modules forming a solar plant are used. In this plant, the PV panels are grouped by series or parallel connections.

When an array of identical PV panels is connected in series, the characteristic curve of the generator changes, increasing the value of the open-circuit voltage ($V_{oc\ array}$) in proportion to the number of panels connected in series. Thus, we can state that $V_{oc\ array} \cong nV_{oc}$, where $n$ is the number of panels connected in series. The electrical power of a PV generator consisting of panels connected in series can be calculated using the expression $P_{array} = (nV_{mp})I_{mp}$, where $V_{mp}$ and $I_{mp}$ are the voltage and current corresponding to the maximum power point for a single panel on the same irradiance and temperature conditions.

Similarly, when a group of identical panels are connected in parallel, the short-circuit current ($I_{sc\ array}$) increases with the number of panels connected in parallel, being approximately $I_{sc\ array} \cong mI_{sc}$, where $m$ is the number of connected panels in parallel, and $I_{sc}$ is the current of short circuit for a single panel. The electrical power of a PV generator consisting of panels in series will come in this case, given by the expression $P_{array} = (mI_{mp})V_{mp}$.

The following figure shows the characteristic $V$-$I$ curves of various configurations of PV generators consisting of four PV modules (equal to those described previously) connected in parallel, four PV modules connected in series and two parallel lines of two PV panels connected in series for each line.

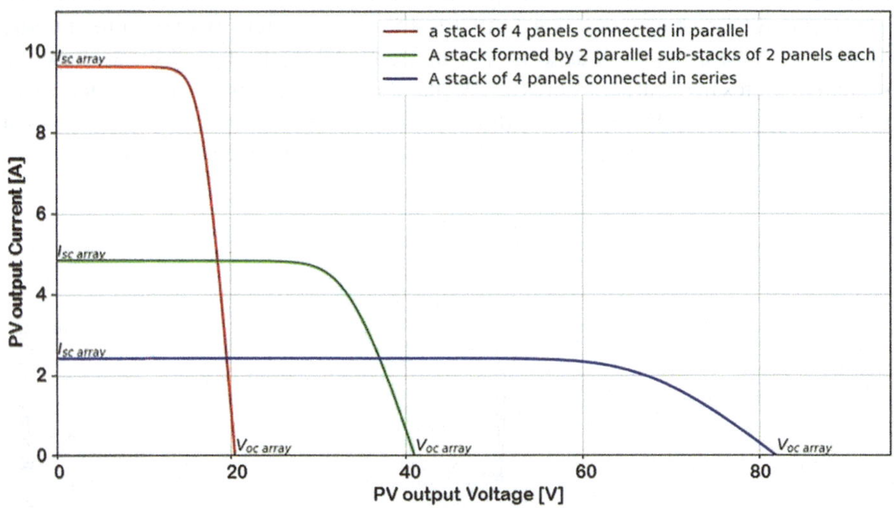


## 3.2 *Use of Photovoltaic Panels in Electrochemical Processes*

A series of studies focused on the use of PV modules for power supplying different electrochemical systems have emerged in the last two decades. The most frequent case are studies at laboratory scale, where only the electrochemical reactor is powered by one or several PV modules, either directly or through the use of batteries. In these works, the PV modules (pumps, sensors, etc.) do not power the rest of the equipment of the experimental system.

In this context, the works carried out by the Group of Applied Electrochemistry and Electrocatalysis at the University of Alicante deserve special mention. This research group has carried out a series of Industrial Pilot plant scale studies focused on the direct powering of different electrochemical technologies using PV plants.

Electrochemical technologies have been proved useful for the treatment of wastewater. In order to enhance their green characteristics, it seems interesting to use a green electric energy source such as PV cells.

In the first of these works [10], the use of an electrooxidation system (EO), directly powered by a PV plant for the treatment of a dye-containing solution, was carried out. An electrochemical filter press reactor of 0.33 $m^2$ of geometric electrode area and a 40-module PV array were used. The work proved that the PV array configuration has a strong influence on the efficiency of the use of the electric energy generated. Thus, the optimum PV array configuration depends on the solar irradiation, solution conductivity and the concentration of pollutant in the wastewater.

Electrodialysis (ED) is an electrochemical technology widely used for the desalination of brackish water coming from groundwater [9, 11, 12]. The desalination of brackish water by ED is a useful method for obtaining low-cost drinking water. PV energy can be used to power the ED system in remote areas in a reliable and autonomous way. Furthermore, the PV plant can be connected directly—without batteries—to the electrodialyzer. Throughout several works, the feasibility of the desalination of brackish water from aquifers (total dissolved solids = 2300–5100 g $m^{-3}$) by means of an ED system supplied directly with PV solar panels was shown. In addition, a mathematical model for the coupled electrodialysis-photovoltaic (ED-PV) system was developed. This model enables predicting the behaviour of the ED-PV system under different operational and weather conditions and the time required to reach a given final concentration. Lastly, the cost of ED-PV systems for reduced-scale applications in isolated locations off the grid has been estimated.

Finally, a last example of PV coupled with electrochemical wastewater treatment technologies in the treatment of wastewater generated in the Almond Manufacture Sector is presented. The approach used was the application of different electrochemical processes in an orderly manner to reuse water obtained from the wastewater treatment. This work was compiled by Valero on his doctoral thesis. Valero's work demonstrates that electrochemical processes: electrocoagulation (EC) [13] and EO [14] are able to be coupled with direct solar PV energy (without batteries) in order to carry out the electrochemical treatment of wastewater from the almond manufacture industry.

### 3.3 State of the Art of Photovoltaic Solar Powered Electroremediation

The features of the ERem technology make it and ideal candidate to be powered by a solar field, both for technical reasons as for logistical and operational reasons [15]. Like the rest of electrochemical technologies, the ERem works using DC as the PV modules provide. However, a great advantage of ERem is the simplicity both from a point of view of its control as well as its required equipment, not existing or being minimal the need of other auxiliary systems, which require to be supplied with electricity (either by DC or alternate current (AC)). Otherwise happens, for example, with EC. The latter technology works in a continuous mode of operation in a single step, requiring reagents addition to adjust pH and/or conductivity, solution pumping, filtering systems, etc. The simplicity of ERem systems avoids the use of DC-AC converters to supply the auxiliary systems, improves energy efficiency, reduces investment costs and minimizes the complexity of control systems, making it much more appealing for its application as an off-grid system. Furthermore, contaminated soils are often found in remote locations disconnected from the grid, which makes, particularly, interesting the use of a solar plant.

The use of PV modules in ERem processes is very recent, and most of the studies were performed at laboratory scale and in synthetic soils. Yuan [16] published the first work in this area, where a PV module directly connected to two electrodes was used to carry out the electrokinetic remediation of a cadmium-contaminated soil. Among the main conclusions of the work, the authors cited that the voltage provided by the module depended on weather conditions and confirmed that the removal efficiency was comparable to that obtained with a conventional power supply. Mohamed Elhassan [17] studied the feasibility at laboratory scale of ERem powered directly by solar panels for application as an electrokinetic barrier. The case under study prevented the contamination of a clayey soil by a hydraulic flow of cadmium from a raft of mine tailings. Solar panels provided energy and kept the Cd within the exclusion zone, even considering that the electrokinetic barrier worked only during periods of sunlight. Other examples of ERem-PV in similar works at laboratory scale are found by Zhou [18], Zhang [19], Hassan [20] and Jeon [21], who displayed the removal of fluorides, Cr(VI), copper and arsenic, respectively.

Souza and co-workers [22] carried out the removal of a model pesticide (2,4-Dichlorophenoxyacetic acid, 2,4-D) at pilot plant scale using electrokinetic soil flushing (EKSF) powered either by a DC power supply or by solar panels. The results in the last case showed changes in operating conditions during the solar test, which can be clearly related to the day-night cycle. The fluctuations in the current intensity supplied to the electrokinetic system led to softer pH and conductivity profiles between anode and cathode in the case of the PV-powered system. At the end of the tests, the authors observed that the PV-powered ERem was a less efficient system, and much higher amount of charge was required to remove an equal amount of pollutants in the soil. This fact was related to the significantly lower electroosmotic flux observed during the operation of the PV-powered system.

In the previous paragraphs, we have seen how works on ERem powered with PV solar panels are relatively recent and few in numbers. In this sense, if we focus on

ERem systems in combination with other soil decontamination technologies, we find that these are even scarcer.

Godschalk and colleague [23] worked on electrokinetic biofence at a chemical laundry in Netherlands, where the soil was polluted with volatile chlorinated hydrocarbons (VOCs). The fence-alternated electrodes in a row separated 5 m. Upstream of the electrodes infiltration wells for the addition of nutrients were installed. The aim of the work was to enhance the biodegradation of the VOCs in the groundwater by electrokinetic dispersion of the dissolved nutrients in the groundwater. The systems were operated for 2 years and the authors reported that: (a) the concentration of nutrients increased, (b) the chloride index decreased and (c) VOCs were dechlorinated by bioactivity. Electrical energy was supplied by solar panels on sunny days. However, both at night and on cloudy days, the electricity was taken from the grid.

Hassan et al. [24] carry out an electrokinetic bioremediation treatment on a phenanthrene-contaminated soil. *Mycobacterium pallens* was the microorganism selected to degrade the pollutant. The study reports result on the efficacy of this microorganism in the degradation of phenanthrene, the influence of electrokinetics for delivering nutrients and microorganisms to contaminated soil and the use of PV modules. The results showed that solar panels generated sufficient power for electrokinetic bioremediation and that the use of solar power with electrokinetic bioremediation provided a cost-effective approach to these technologies.

In the studies of ERem systems directly powered by PV modules shown so far, the authors concluded in all cases that the contaminated soil treatment using this combination of technologies (ERem-PV) was viable. On the other hand, its main features were the great daily variations in both the current intensity and the voltage gradient. These variations were due to the intrinsic performance of the PV modules and dependence of their power and characteristic *V-I* curves with the daily solar irradiance curve. Logically, no system with direct connection could work at night or on cloudy days. The other parameter that affects the characteristic *V-I* curve of the PV panel is the temperature of the panel. However, against the variations caused by solar irradiance, the influence of the former is less relevant.

The studies published so far were performed on laboratory-scale systems. These systems were of reduced size and required little power for its operation, so their supply could be performed with a single commercial PV panel. Likewise, authors focused their study on power supplying the electrodes, not dealing with in any case with any other of the auxiliary systems supply, such as the solution pumping system.

## 4 Design of a Real Electroremediation System Supplied with Photovoltaic Panels

### 4.1 General Considerations

The rest of this chapter consists in the exposing and developing of different connection options of a solar plant to an ERem system at field scale. At this point, we must define the characteristics—mainly the contaminated area to be treated and the power supply requirements—of the ERem system that we will use as a reference to carry out the dissertation. ERem systems that work in the field scale and can be found in literature can decontaminate surfaces from 10 to 100 $m^2$. All systems are multi-electrode, which can be structured in different geometric distributions—mainly square, rectangular or hexagonal—with an electrode separation that varies between 0.5 and 2 m. The most common potential gradient used in these systems is 1 V/cm. The power needed to feed these systems can vary between 1 and 10 kW, which enables working under hundreds of DC volts and tens of current amps.

It is necessary to highlight the interest of focusing our argument on the development of ERem-PV systems at a field scale. Thus, López Vizcaíno et al. [25, 26] analysed the effect of the scale up of electrokinetic remediation processes in natural soils. These authors proposed a procedure to prepare soils in order to obtain similar moisture content and density as real soils. The study was carried out in two facilities at different scales: a pilot scale (0.175 $m^3$) and a field-scale application (16 $m^3$). The authors demonstrated the influence of the work scale in electroosmotic and electromigration flows and electric heating.

On the other hand, the work behind the design of an ERem-PV system on field scale is also justified on the information that can be provided by the scale factor. To supply a field-scale ERem-PV system, the size of the solar plant must be increased, which enriches the discussion of the system's design. As we will see, we can consider different optimization strategies regarding the electrical power that must be generated in solar plants of a given power. These strategies are based on reconfigurations of the solar plant, the use of DC/DC converters and maximum power point monitoring systems, etc. Such strategies are meaningless in lab bench systems that use a single PV module to power supply the ERem system. On the other hand, in field-scale systems, we can consider extending the PV power supply to auxiliary systems such as the electrode solutions pumping, reagents addition, etc.

Regarding the main modalities of application of ERem technology and its different variants in field-scale systems, the most common situations are the followings [27, 28]:

- In situ remediation, where the electrodes are directly placed on the contaminated terrain and the pollutants are eliminated with minimum alterations on the environment. Within this type, we can find examples in remote places without grid access, as well as applications in urban areas or farms, where a conventional grid is available.
- Batch treatment, where the polluted soil is extracted and transported to a treatment facility, where it is decontaminated ex situ. In this case, the grid availability is not an issue.

The use of a solar plant powering system (ERem-PV) presents no interest in cases where conventional electricity grid is available, both for in situ and ex situ treatments. The limitations that we have previously seen regarding ERem-PV systems, together with the installation cost of a PV plant, make them only viable and interesting for the treatment of contaminated soils in isolated areas without access to the grid.

The next consideration to focus the ERem-PV system design refers to the use of support battery racks. Generally, a PV system using batteries consists of PV modules, the rack of batteries, a regulator or controller, an inverter (DC-AC) and loads (charges). The solar radiation incident on the surface of the PV array is transformed into electric energy (DC) by the PV modules. The generated electricity is passed to the regulator, which protects the batteries from overcharging or an excessive discharge. The batteries store energy that can be used as electrical back up during periods of low solar radiation, for example, during rainy, cloudy weather or at night. The inverter transforms the DC into AC for those devices that work with the latter.

An ERem-PV system with a support battery rack features a simple design, since it can work as an ERem system supplied with a conventional power supply. The configuration of the PV plant acquires much less relevance than in the case of being directly supplied, since all the energy generated by the solar plant passes through the rack of batteries, from where it is then provided to the ERem system. Regarding the sizing of the supply system—PV modules and batteries—this will depend on the number of effective working hours per day of the ERem system.

However, the use of the electrical energy supply by the PV array to the ERem system directly from the PV modules would substantially decrease the cost of investment and maintenance of these systems due to the high price of batteries, the short lifespan of the battery and to the higher maintenance costs compared to solar panels. Likewise, the environmental problems arising from the management of batteries waste by products after life cycle would be eliminated, increasing, therefore, the sustainability of the process. A particular case is the ERem systems that must work the whole day, as is the case of the electrokinetic barriers.

#### 4.1.1 Electroremediation System Directly Supplied with a Photovoltaic Solar Plant

At this point, the objective that we set ourselves is the design of an electrokinetic remediation system powered directly with a PV solar plant. In a first approach will address the simplest design: considering a direct connection between the PV modules' terminals and the ERem system electrodes. This has been the way used in all the papers published so far in this field, being the most intuitive in order to understand the interactions between the solar plant and the ERem system.

To support us in the process of design of the solar power plant, we will take as a guide a process of a real ERem of reference, where a treatment with conventional means takes place. Kim et al. [29, 30] have carried out a series of works focused on the in situ real-scale electrokinetic remediation of multi-metals contaminated paddy soil, which, by its nature, can be very useful as a reference. In these works, the authors evaluated the influence of electrode configuration on in situ electrokinetic remediation of As, Cu and Pb contaminated soil. The site used in the study was a rice field near a zinc refinery plant located at South Korea.

Among the features of these works, which make them interesting for our study, we can highlight: (a) they are a field-scale work with actual contaminated soil, (b) the duration of the experiments was very high, between 3 and 6 months, (c) the electrokinetic setup involves multi-electrode systems with different configurations, (d) parameters studied include the voltage gradient, (e) systems incorporate both the conditioning of electrolyte solutions and the addition of soil conditioning solutions to improve the extraction of pollutants and (f) during the experiments the authors performed a thorough monitoring of parameters such as current intensity and the temperature of the soil.

***Electrokinetic Set-up.*** The most important component of an ERem system is related to the number and arrangement of the electrodes in the contaminated soil to be treated. Thus, defining the solar plant necessary to supply the electrokinetic electrode setup is the first step when sizing the global solar plant. This electrokinetic setup is characterized by the separation between the electrodes of the same and opposite sign, along with the geometry of the electrode array—usually square, triangular, rectangular or hexagonal.

In the work of Kim et al. [29], the author installed four different electrode configurations that can be seen in the next figure. The anode-cathode separation was 1 m in A1, A2 y A4, and 2 m in A3. On the other hand, the separation between the electrodes with the same polarity was 0.5 m in A1, A2 y A3, and 1 m in A4. The area covered by electrode system of A1 and A2 was 4 m$^2$, but the area for A3 and A4 was 8 m$^2$. The potential gradients were 1 V/cm—most frequently voltage value used in processes of ERem—and 0.5 V/cm.

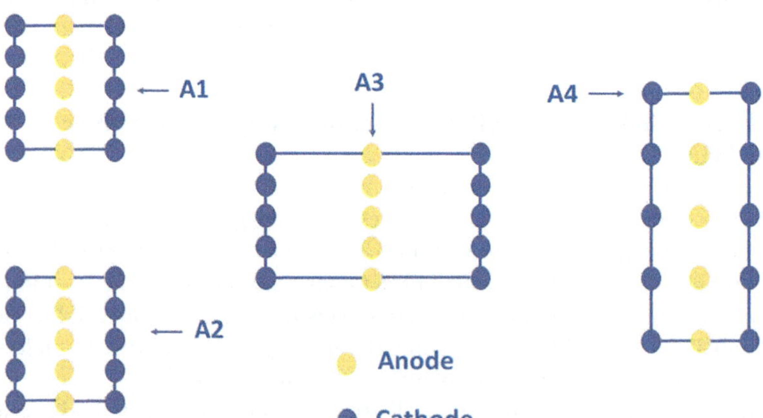


As we saw, when the performance of a PV panel was described, an ERem system *V-I* curve resembles a straight line, where voltage and intensity are proportional. When using a power supply in potentiostatic mode, we impose the voltage applied between electrodes, and the ERem process initial current is given by the electrical resistance of the system—basically the ohmic resistance of the soil. The power supply is selected in such a way that it can provide the voltage to be applied (100 V in the example of the work of Kim et al.), and its maximum current intensity value is widely superior to the working current of the ERem process. In fact, the current intensity is not a relevant experimental parameter in the ERem processes.

In the experiments conducted by Kim et al., the current intensity was different for each experiment, mainly depending on the number and arrangement of the electrodes as well as the applied voltage. Also, important variations of current intensity along the experiments were detected due to changes in the temperature of the soil and the pore conductivity. In all the experiments, the current increased at the beginning to a maximum around the 500 h of operation. Then, the intensity went down in all cases to reach values between 0.5 and 1.0 A around 1500 h. The experiment, where the highest current was registered, was A2 experiment with a maximum of approximately 10 A.

Given the peculiarities of the functioning of a solar PV generator, when designing a solar plant to a ***direct connection supply*** of the ERem system, we should mark as requirement that the operating point of the ERem-PV system within the characteristic *V-I* curve of the system takes place in voltage values slightly higher than those of the maximum power. It is important emphasize that the direct connection of the solar PV plant implies the supply of the electrodes connecting them directly to the PV modules' terminals, without any electronic instrumentation involved, such as converters, etc.

As we saw previously, the voltage that we can obtain from the solar plant is regulated based on the number of PV modules arranged in series. We have also seen that the characteristic *V-I* curve of a PV generator changes throughout the day, at the same time that the daily curve of incident solar irradiance does. Thus, we saw how the position of the working point of the system past the point of maximum power of the *V-I* curve minimizes the effect of daily variations of solar irradiance on the potential applied to the ERem system.

If the PV solar plant is designed in such a way that the ERem-PV system operation point was located in the current plateau area of the characteristic *V-I* curve, the irradiance variations would cause a large displacement of the position of the operating point at its voltage-current coordinates. This translates into some severe changes in the experimental working conditions, and their impact on the results obtained in the soil decontamination should be considered.

A greater number of PV modules arranged in parallel connection allows increasing the current provided by the solar plant. As the number of PV modules arranged in parallel increases, we minimize the effect of the daily variation in solar irradiance—both by the effect of the daily irradiance Gaussian distribution and sudden decreases of irradiance because of clouds. In practice, an increase in the number of PV modules placed in parallel translates into a greater number of daily work hours of the ERem-PV system. This happens because the solar plant is capable of providing the working voltage and current of the ERem system with lower values of solar irradiance (given at dawn and dusk of the day). Logically, the increase in daily working hours requires a greater economic investment in equipment.

We have seen an approximation to the sizing of the solar plant, which has allowed us to initially and simply define the settings to supply a real ERem system. There are a series of thoughts and interesting considerations that we must bear in mind to delve into the design and performance of the solar plant.

Thus, for example, it is very important to emphasize the great influence that the electrical conductivity of the soil has in the installed power of the solar plant used to supply the ERem. Operational lines with a high slope will characterize soils with high electrical conductivity due to factors such as a large amount of dissolved salts, high porosity, soils with a high degree of humidity, etc. This causes that in order to apply a standard potential gradient of 1 V/cm, the working point of the system lays at high current values, which are attained increasing the number of series of PV modules arrayed in parallel, involving high investment cost towards the treatment of a low conductivity soil.

On the other hand, we must consider that the PV array sizing has been carried out using the initial ERem system's line $I = (1/R)V$. However, although we can say in general that ERem processes are slow processes with slight experimental variations, several factors can influence the value of the electrical resistance of the soil during the treatment. These variations cause changes in the slope of the ERem operational line and its intersection with the characteristic *V-I* curve, to the point of making necessary a reconfiguration of the solar plant.

In order to a better understanding of the effect of these phenomena, we should explain that the electrical resistance of conductor (in our case, the soil to be decontaminated) is given by:

$$R = \rho \frac{l}{S},$$

where $R$ is the *electrical resistance* (Ω), which is the slope of the working function of the ERem system, $l$ (m) is the *distance between the electrodes* and $S$ ($m^2$) is the *soil section perpendicular to the electric field*. The parameter $\rho$ is the *electrical resistivity* (Ωm), and its value depends on the nature and concentration of the conductive species, as well as temperature.

- *Significant variations on the soil* change the conductivity of the solutions that are located in the soil's pores, changing the operational line's slope of the ERem process. Thus, for example, in the cited works of Kim et al., the system worked in the range of 10–65 °C. This pronounced variation was due to the change of season, as every experiment lasted for several months, as well as Joule's effect during operation.
- In this latter sense is very possible that we find qualitative differences between the behaviour of ERem-PV systems and conventional ERem systems about the soil temperature variation. Using a solar plant, the system workload is intermittent, and the electricity flows only during several hours per day when the solar irradiance is enough to supply the system. It is expected then that every day the heat generated by Joule's effect during the ERem-PV system's operating is dissipated during the night, so the soil attains temperature values comparable to its environment.

- *Addition of chemical agents* (surfactants, complexing agents, pH adjusters) with different purposes, for example, to improve the solubility and mobility of pollutants. If the added solutions have a high electrical conductivity, they will increase the electrical conductivity of the soil and the slope of the operational line of the ERem system. The working point will move to positions within the current plateau, characterized by lower voltage values (decrease of the applied voltage gradient), higher current and a greater variability against solar irradiance.
- In any case, it is accepted that contaminated soils improve their efficiencies of treatment by ERem when they present relatively low concentrations of dissolved ions [2, 3]. High concentrations of these disrupt electroosmosis phenomena and consume a disproportionate amount of power supply regarding the transport of pollutant ions. This reasoning applies, especially $H^+$ and $OH^-$ ions, due to their high ionic mobility.

It must be taken into account that we are just analysing the expected influence of experimental parameters on the position of the working point of the ERem-PV system in the characteristic *V-I* curve. Due to the complexity of the electromigration and electroosmosis processes and how changes in the latter affect removal efficiencies of different pollutants in soils in ERem processes, in any case, the above dissertation aims to extract conclusions about the influence of variations in the soil conductivity regarding pollutant removal efficiency.

From the reasoning explained in this section, we can outline some strategies to improve the use of the energy generated by the solar plant. In this sense, an interesting situation occurs in the initial period of the day when the solar irradiance is still not enough energetic to supply the ERem system, and the energy generated by the solar plant is not used. If we carry out the electrical connections between the solar plant and the electrodes setup in a sectioned way—so that only a part of the electrodes are connected in these periods of low irradiance—we could perform the treatment of a part of the terrain. The decrease of the terrain area to be treated increases the value of the total ERem system electrical resistance. This happens because anode-cathode electrodes arrays are connected in parallel. By eliminating part of the electrode groupings, we are decreasing the floor section available for the electric current to flow in (conductive medium section), with a consequent increase of $R$. In the characteristic *V-I* curve of an ERem-PV system, this originates a decrease of the ERem system's operational line slope, so the operational intersection point occurs at lower current values, which can be provided by the solar plant with lower irradiance values. This reasoning can be repeated analogously in the latest hours of the day when the irradiance energy cannot supply the ERem system.

Another different solar plant optimization strategy regarding low solar irradiance periods is based on the transitory decrease of the operating voltage gradient—that is, decreasing the usual value from 1 to 0.5 V/cm. To achieve this new operational voltage gradient value, we could reconfigure the solar plant as it follows: halving the PV modules connected in series and doubling those connected in parallel. The result is a solar plant with the same number of PV panels, but with double the $I_{sc}$ and half the $V_{oc}$ value. When the irradiance is energetic enough for the ERem-PV system to work at 1 V/cm voltage gradient conditions, the solar plant is reconfigured to its initial setup.

***Electrode Assembly.*** It is of utmost importance when designing a power supply with PV panels, where there are several aspects to take into account. With regard to the *manufacturing material* and *its geometry* electrodes with good electrical conductivity, chemically inert, low cost and easy to install on the ground are wanted. These features are independent of the power supply and are not influenced by the use of a solar plant.

It is relevant to study the specific relation of the electrode with the *immediate subsurface environment concerning the exchange of dissolution.* In its simplest configuration, the electrodes are directly introduced in the soil. This is the simplest case from the point of view of a solar plant supply, since the electrode is the only element involved.

However, in the majority of real applications, electrodes are immersed in more or less sophisticated wells containing an electrolyte solution. The reasons for choosing this arrangement are several and depend on the process and specific pollutants to eliminate. Sometimes, these solutions can be pumped between the electrode wells of the same polarity—to homogenize solutions—or opposite polarity—to neutralize $H^+$ and $OH^-$ generated in electrode reactions and avoid pH gradients in the soil. On other occasions, the electrode wells are used as injection point for additive pumping to improve the efficiency of the process. Examples of this operational strategy are the injection of surfactants, co-solvents, pH regulators, etc., depending on whether the decontamination process involves heavy metals, PAHs, petroleum by-products, etc. In the case of treatments for decontamination of heavy metals, sometimes electrode solutions—normally the catholyte—that accumulate extracted metals are pumped out to be treated externally.

In the works of Kim et al. taken as reference, electrolyte conditioning was carried out to enhance the removal of heavy metals. Authors used NaOH as an anolyte conditioning agent and EDTA as a catholyte processing fluid. A hollow stainless steel with many holes to facilitate the transport of electrolytes was used as cathode, and a steel bar into PVC casings with many holes was used as anode (see next figure). Anolyte was supplied through the bottom of the casing, and the overflowed anolyte was transported to the next anode. The electrodes were wrapped with a Gore-Tex and filtering textiles to prevent the leakage of electrolyte in anode and cathode.

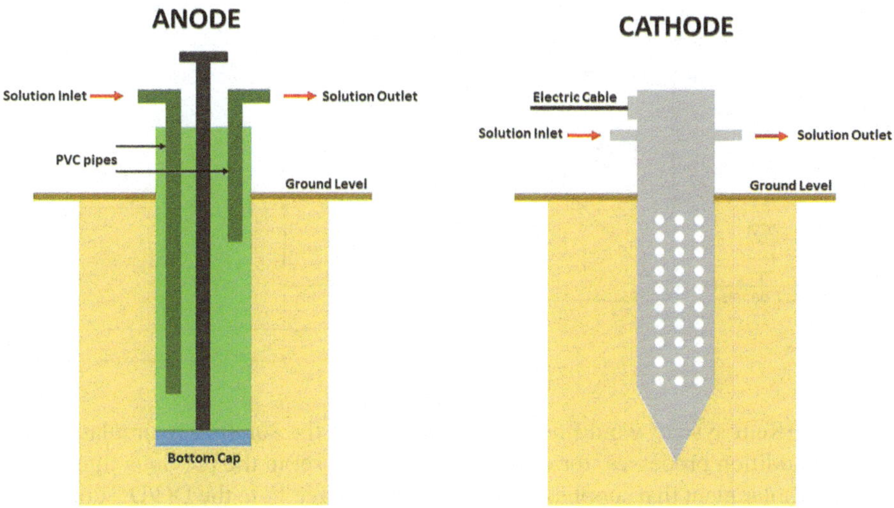


In many occasions, the ERem system incorporates a solution pumping system. The features of the pumping systems—pump power, flow rate, pressure, power on and off patterns, etc.—will be very different, involving different number of pumps with different functions (reagent addition, recirculation or extraction of solutions). Given that the solar plant's energy supply is discontinuous and can decrease dramatically in a very short time span—the clouds' shadow effect on irradiance is a clear example—it is important to evaluate the impact of the shutdown of every pumping system on parameters, such as decontamination efficiency or element stability, in order to define ERem-PV system's work strategies.

Fortunately, one of the advantages of the ERem, when it comes to be supplied directly with a solar plant, derives from the fact that they are very slow processes with large inertia. This fact, coupled with the absence of items particularly sensitive to sudden changes in their operating conditions—solutions' flow, applied current, etc., makes this technology sturdy against sudden changes in the solar irradiance.

Regarding the porous or semi-porous membranes used in electrode compartments, they have not been found studies about their behaviour in ERem-PV systems. Generally, problems related to system operation shutdowns during the night are not expected. However, in certain cases—for example, in case of heavy metals presence—it should not be dismissed the possibility of contamination by precipitation of very insoluble heavy metal salts or hydroxides during the overnight stops.

The simplest approach for the design of the solution pumping system's solar subplant power supply is to consider that all pumps used are DC and set a configuration type.

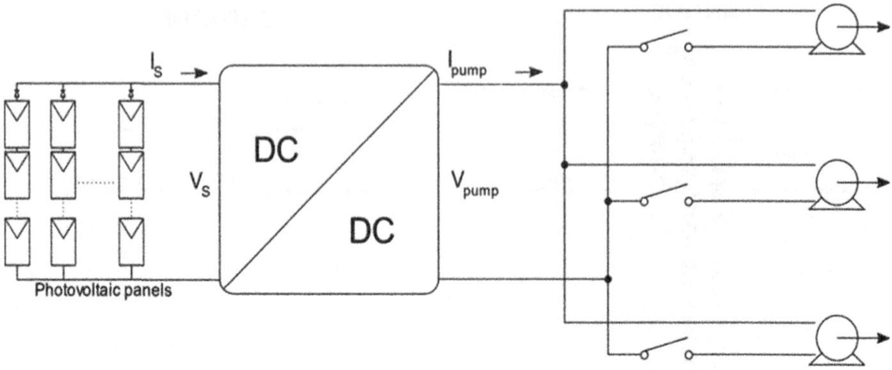


This ERem system would need three pumps for the solution recirculation and reagent addition processes, for example. The circuit set in the previous figure consists of a solar plant that supplies a current $I_S$ and voltage $V_S$ to the DC/DC converter, which converts the variable voltage generated by the solar field $V_S$ into the required voltage for the pumps to operate $V_{pump}$. In the circuit shown in the picture, three pumps are connected in parallel with switches that enable independent connections. The DC/DC converter ensures that the bombs are not damaged if the voltage in the solar plants increases. In addition, each pump must have fuses built-in in order to prevent the motor from burning down. The flow provided by the pumps will then be proportional to the current, $I_{pump}$, parameter that is given by the expression:

$$I_{pump} \leq \frac{I_S \times V_S}{V_{pump}}$$

$I_{pump}$ depends on the solar irradiance through the $I_S$ and $V_S$ supplied to the DC/DC converter. The DC/DC converter supplies the pumps the current required to work under open flow (as long as the irradiance is high enough). When more energy than required is generated (for example, when only one pump is working), this excess energy is wasted.

At this point, we must formulate the strategies to connect the solar plant designed to supply the electrokinetic setup with the solar plant required to supply the pumping systems. This can be achieved by applying the following design strategies.

In a first approximation, we can consider a system where the energy excess in the pumping system can be used by the ERem system. To achieve this goal, we have to work with a single solar plant according to the scheme shown in the following figure:

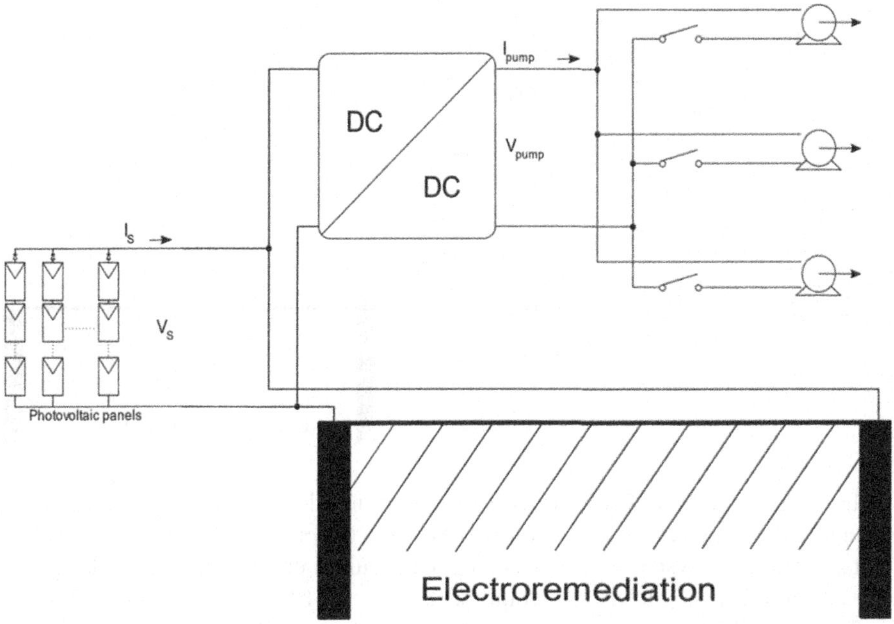


The pumping system takes the energy that it requires throughout the DC/DC converter, and the ERem process consumes the rest of the energy generated by the solar plant. The problem of this configuration is that changing the solar plant's configuration is a complex matter. It would require of a highly sophisticated DC/DC converter, losing this optimization prospect. It should be assessed if the energy obtained through the pumping is enough to compensate for the use of a DC/DC converter that enables the reconfiguration of the solar plant.

A more sophisticated modification implies that the system can work near the maximum power point at any time. As previously seen, at work voltages in the voltage drop zone of the characteristic *V-I* curve, the ERem system is less sensitive to irradiance variations. To do this, a configuration aimed to improve the solar plant is presented in the figure below:

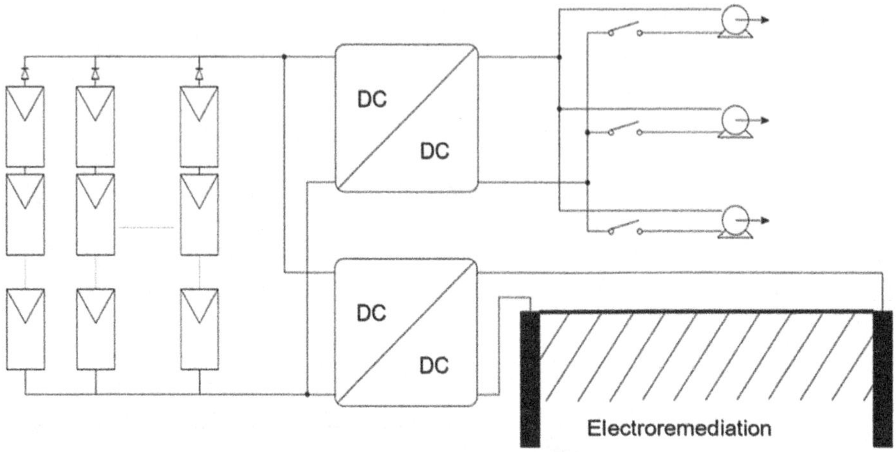


In this stage, a DC/DC converter is proposed in order to control the energy used to supply the ERem process. The circuit shown in the previous figure is designed so that the pumping system only consumes the required energy from the solar plant. The rest of the available energy would be used by the ERem system as it follows:

The solar plant configuration is fixed as in the first option, but in this case, electronic equipment is placed between the solar plant and the ERem system. This converter takes in the energy coming from the panels and converts it into the fixed voltage previously chosen for the ERem system (normally for a voltage gradient of 1 V/cm). The DC/DC converter modifies the voltage and current of the solar plant into the maximum power point. At the converter's output, the voltage and current are increased as high as possible regarding the available energy. Depending on the design, this allows to increase the voltage in the ERem in reasonable range or connecting more or less electrodes depending on the available power at each moment.

The *automation and control of the ERem system* must include the ERem process itself, as well as the electrode solution recirculation and reagent addition. As we will see, this automation should include the required actions on ERem's work conditions so they lead to an optimal solar plant energy exploiting. Regarding the *automation and control of the solar plant* that supplies the ERem system, the final objective is to design an ERem-PV prototype that should: (a) connect automatically each morning when the solar irradiance is enough to supply the pumping system and the electrodes, (b) carry out the soil decontamination and (c) shut down automatically when the solar irradiance at the end of the day is not enough to sustain a correct performance.

## 4.2 Design Choices for ERem-PV Systems. Examples Based on a Real Case

In this section, we will carry out the detailed design of several ERem systems supplied with a solar PV plant. Each design will be applied to a given conventional ERem system found in literature and which has been broadly described in the previous section: the works by Kim et al. on ERem treatment of heavy metal polluted rice crop fields due to a zinc processing plant located in South Korea. Among the experiments conducted by Kim et al., we will choose A2 experiment, characterized by a 100 V constant voltage with a 1 V/m gradient. The area covered by the system was 4 $m^2$, ten cathodes and four anodes were used in an electrode setup arranged in a square geometry and with a same polarity electrode separation of 0.5 m and an anode-cathode separation of 1 m.

### 4.2.1 ERem-PV System Without Batteries

The first step towards the design is the required solar plant sizing in order to supply the electrode setup and the pumping system. We should impose the following criteria. The first required data is *the latitude of the location in which the plant will be installed.* The longitude is not required as it has no effect on the solar time in the zone. In this case, the location is Janghang in South Korea, approximately 36°N.

The tilt and orientation of the solar plant are basic parameters to achieve the best performance of the PV generator. Since the Sun changes its position and height regarding the horizon, it is of utmost importance that the PV generator is oriented and tilted in the most appropriate way. In the temperate region of the Northern Hemisphere, solar PV modules are oriented to the south. With respect to inclination, the following design decision to make is to delimit the *degree of seasonality in the energy production* [31], that is, in which period of the year we want to achieve the maximum energy provided by the plant, that is, constant all the year, more in summer than in winter or vice versa, etc. Logically, this is a decision that is greatly influenced by the system's planned schedule. Usually the duration of treatment for ERem of a given contaminated soil does not cover a full year, so the seasonal nature of the production will take place in each case depending on the dates in which the treatment is going to be performed on and the estimated duration of it. Seasonality allows us to select the inclination of the field of panels with respect to the horizontal plane. For simplicity, in this example, we will consider that the power required is the same throughout the year. For these facilities the criteria is tilting the panels 10° more than the latitude value. This makes the power production not to drop in winter as much as it would otherwise. This choice sets the inclination of the panels plant to solar array tilt = 46°.

There are many types and models of PV panel available in the market. For this application, where moving the solar camp to another location from time to time could be required, high performance panels should be used; for this reason, the choice should be made seeking the highest peak power vs. area ratio possible. In this application, we will use monocrystalline silicon or either high efficiency polycrystalline silicon.

Next, we must estimate *the power requirements of the facility*. To do this, the list of elements that consume power in the system must be made, establishing their periods and daily use particularities. The consumed power, the number of daily working hours and the simultaneity in their use must be identified for each element. In this sense, we can distinguish between daily workload—as it is the case with the electrodes—or workload, which can be shifted hours—the case of the cathodic and anodic solutions' pumping—or even days—in cases where there is a periodical addition of some additive. An approach regarding the example that we are discussing:

- Electrode Assembly. In Kim et al. work, we can see that the system will work at 100 V of applied voltage. The initial current was 5 A. In this case, the nominal power would be 500 W.
- Normally, when the design of an ERem system is being studied, the current intensity data is not available. In these cases, there exist several options in order to estimate this parameter, which include from estimating or making experimental measurements of electrical resistance of the soil to be decontaminated, to conduct a laboratory scale ERem study (this procedure is recommended before implementing a real-scale system).
- In cases where it is expected that the electrical resistance of the soil may increase considerably during the ERem treatment—that is, seasonal changes that make the soil temperature decrease heavily, insoluble compounds precipitation processes, etc.—an estimate of the required power should be provided on the basis of the most unfavourable expected results.
- Regarding the number of hours of daily operation, this is an 'arbitrary' design parameter, which is directly related to the treatment capacity of the system and the required economic investment. The higher the number of daily operating hours, the faster the treatment of the contaminated soil will be and the greater the energy required. This implies a higher installed power of the PV generator and an increase in the associated investment. In this case, we can consider the example of the operation of the system during the five central hours of the day.
- Cathode and Anode Solution Pumping. Direct current pumps, which can be supplied directly from the PV generator with the lowest possible power, should be used. It is important to bear in mind that the pressure drop in the system is low and the pumping only purpose is the recirculation and homogenization of solutions. In this system two pumps would be required—for both catholyte and anolyte—and we could use pumps with an individual power of 100 W (24 V, 4 A).
- Given that ERem processes are very slow, it is not necessary that the pumps are operating permanently. Hence, in order to minimize the energy consumed by the pumping system, we set daily use patterns for this system. A feasible example would be the non-simultaneous operation of the pumps for a total period of 1 h a day, distributed in 5–10 min periods throughout the central hour of the day. Given that the pumps operating would not be simultaneous, the required installed power would be 100 W.

In order to decide the PV array power required, the criteria applied is that the system—during the worst day scenario (the winter solstice)—must have enough energy to:

1. Generate 100 V/5 A at least 5 h a day, so 500 W are required along the central 5 h of the day.

2. Recirculate the anolyte and catholyte solutions for 1 h per day while the soil is being treated with 100 V/5 A, so 600 W are required in the central hour of the day.

To find out which power criteria are more restrictive, the winter solstice registered irradiation curve is used. Ideally, the data should be registered in the place, where the solar plant will be installed. However, in this case, we do not have data of the zone, where the solar plant is intended to be installed, so a registered irradiation curve for a location in similar latitude (38°N) in Spain is used, and following the same criteria, the solar panel plant tilt is set. In this figure, the irradiance during central 5 h of that day—between 10:30 and 15:30—is shown.

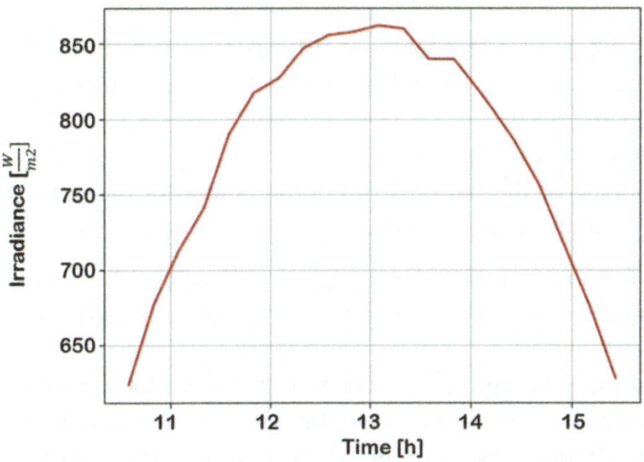


The figure shows that:

- The maximum irradiance was 862 W/m$^2$. The irradiance 1 h time around the maximum point was always higher than 842 W/m$^2$, and the minimum requirement is at least 600 W at that time zone.
- The irradiance 5 h around the maximum was always higher than 626 W/m$^2$ and the required is 500 W around that time range.

Therefore, the generation criterion of 500 W for 5 h a day is more demanding than the 600 W 1 h a day, as the power demand-irradiance ratio is higher. If we consider 15% losses in the DC/DC converter and cabling, then the output power of the panel plant is:

$$P_{\text{panels}} = \frac{P_{\text{Load}}}{1 - R},$$

where:

- $P_{\text{panels}}$ is the power generated by the solar panels.
- $P_{\text{Load}}$ is the available power for consume.

- $R$ is the power loss on cabling and converters. In this case 0.15 as mentioned before.

Therefore, the power generated in the panels under 626 W/m$^2$ of irradiance is 588 W. Nonetheless, PV panels' manufacturers provide their information on power production under 1000 W/m$^2$. Since the power generated by the PV panels is linear with the irradiation, the necessary solar field would be:

$$P_p = \frac{1000 \bullet P_{panels}}{I_{5h}},$$

where:

- $P_p$ is the required panels' peak power. The peak power is the power that the panels provide when the irradiance reaches 1000 W/m$^2$ and will be the value used to obtain the number of panels required.
- $P_{panels}$ is the power generated by the PV panels.
- $I_{5h}$ is the irradiance in the plane of the panels 2.5 h before and after the point at which the irradiance reaches its maximum value.

By applying the previous formula, we find that we need to install a PV plant of 940 Wp and an optimal tilt angle of 46°. This PV generator ensures at least 5 h of operation on sunny days.

Of course, the system allows the incorporation of all the energy exploiting strategies when the irradiance is not enough to supply the entire system, and which have been discussed throughout the chapter. For example, working only with part of the electrode setup or diminishing the voltage gradient. Similarly, in the days where the output is greater than the system's requirement and given that it cannot be stored, a possible solution involves making the system work under voltage gradients higher than 1 V/cm.

### 4.2.2 ERem-PV System with a Battery Pack

An ERem system powered by solar panels has the inconvenience of shutting down when the irradiance is lower than the required minimum. This occurs at cloudy days and at night. In order to solve this problem, a battery pack can be added to the system so it stores energy excess during daytime to use it on cloudy days and/or at night. Within this design strategy, the system can be sized according to the degree of reinforcement that we want the battery pack to provide. For example, using batteries only to avoid operation shutdown caused by clouds, but avoid their use at night. Another option is to extend the daily use of the system. Consequently, the more the number of daily operation hours of the ERem system—including periods without enough irradiance—the more the size of battery racks and the solar plant required. The electrical diagram of this configuration of ERem-PV system is shown in the following figure:

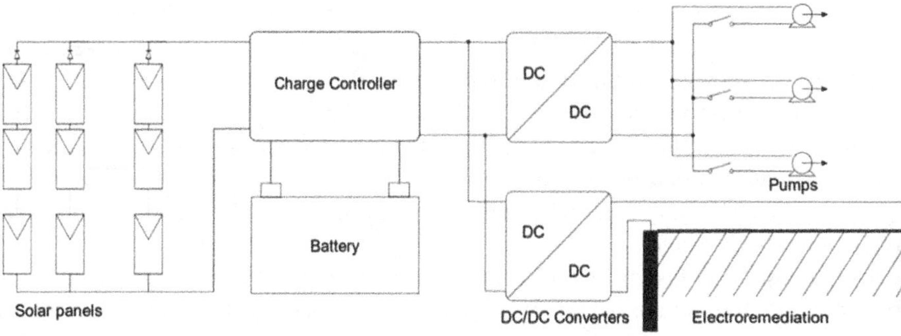


The scheme is similar to the one shown in the previous example, but two new elements have been incorporated: the charge controller and the battery. ERem-PV systems with battery racks use fixed output voltage DC/DC converters. They do not require configuration depending on the panels' power point since their supply is provided by a battery whose voltage is constant.

In this case, the PV plant supplies energy to the charge controller. This device is responsible for charging the battery and supply power to the rest of the system. Its functions are on the one hand to disconnect the panel plant if the battery reaches its maximum charge status. On the other hand, if the battery is discharged to its minimum charge status, it shuts down the ERem system.

A third function of the charge controller is to keep the PV modules at their maximum power point making use of maximum power point tracking (MPPT) techniques.

In this example, we will design an ERem-PV system capable of working 24 h a day. In addition, in case that several cloudy days happen consecutively, the setup will allow the battery to store enough energy to continue working without interruption for 3 days. This design strategy is particularly interesting in the case of electrokinetic barriers.

In the market, there are several rechargeable batteries technologies available, whose main features we can see in the following table:

| | Pb-acid | Ni-Cd | Ni-MH | Li-ion |
|---|---|---|---|---|
| Cell voltage (V) | 2 | 1.2 | 1.2 | 3.6 |
| Energetic density (Wh/kg) | 30–50 | 50–80 | 60–120 | 110–160 |
| Self-discharge (%/month) | <5 | 20 | 20 | 6 |
| N° cycles | Medium | High | Medium | Medium |
| Capacities (Ah) | 5–3000 | <6 | Medium | Low |
| Approximate cost (12 V, 100 Ah) | 250–600 | 2500 | 1200 | 4500 |

The ERem-PV system requires a high storage capacity to work several days with low irradiance. On the other hand, unlike start-stop applications that require high power but low energy, an ERem-PV system will operate with medium or low discharge currents and deep-discharge conditions. Based on the above table, good choices are valve-regulated lead-acid batteries (VRLA batteries), specifically a gel battery. These are lead-acid batteries with a gelified electrolyte. They are relatively economic and easier than liquid electrolyte batteries to maintain. On the other hand, this type of battery copes well with the adverse operation conditions in isolated PV facilities: variable operation rate, high temperatures, deep-cycle discharge and insufficient charge stages.

In this case, the design of the installation is performed following the conventional methodology in the design of grid isolated PV systems [31]. The first step to design the generation system is *to calculate the energy requirements of the equipment to be supplied.* In this case, it is decided that the ERem system will require, as in the previous case, 500 W to supply the electrodes with 100 V/5 A continuously for 24 h. Since the system must work 24 h a day, we will consider, in this case, a 100 W pumping system operating for 2 h per day. Making the total energy:

$$E_{\text{load}} = P_{\text{ER}} \bullet T_{\text{ER}} + P_{\text{PUMP}} \bullet T_{\text{PUMP,}}$$

where:

- $E_{\text{load}}$ is the energy that the whole system requires.
- $P_{\text{ER}}$ is the power required to carry out the ERem, which is 500 W.
- $P_{\text{PUMP}}$ is power the pumping system requires, that is, 100 W.
- $T_{\text{ER}}$ is the time that the ERem system must be operative each day, which is 24 h.
- $T_{\text{PUMP}}$ is the time the pumping system should be active each day, 2 h.

Therefore, we will need 12.2 kWh/day to power the system.

The battery capacity is given by:

$$C_{20} = \frac{A \bullet L_{\text{P}}}{PD_{\text{max}} \bullet \eta_{\text{DC}} \bullet \eta_{\text{BAT}}},$$

where:

- $C_{20}$ is the capacity of the battery in Ah if the discharge lasts 20 h.
- $A$ is the desired battery autonomy in days. In this case 3 days.
- $L_{\text{P}}$ is the daily intake in Ah.
- $\text{PD}_{\text{max}}$ is the maximum battery discharge status. In lead-acid batteries, this value is 0.5.
- $\eta_{\text{DC}}$ is the energetic efficiency of the DC/DC converter. This parameter's value for this kind of equipment is 0.9.
- $\eta_{\text{BAT}}$ is the energetic efficiency of the battery and its charging equipment. This parameter's value for this kind of equipment is 0.75.

In order to calculate the battery capacity, the only missing parameter is $L_{\text{P}}$. To obtain this value, the following expression is used:

$$L_P = \frac{E_{load}}{V_{BAT}},$$

where:

- $L_P$ is the daily intake in Ah.
- $E_{load}$ is the energy that the whole system requires, 12.2 kWh/day as seen before.
- $V_{BAT}$ is the battery's nominal voltage. In this case, 48 V batteries are used.

Therefore, the battery that the systems require is:

- Type: lead-acid.
- $C_{20} = 2259$ Ah.
- $V_{BAT} = 48$ V.

The energy provided by the PV panels would be $E_{load}$ if the battery, the charge regulator and the cabling were ideal. Unfortunately, it is not the case. Hence, the PV plant should provide daily energy following the next equation:

$$E_{min} = \frac{L_{load}}{\eta_w \bullet \eta_{DC} \bullet \eta_{BAT}},$$

where:

- $E_{min}$ is the energy that the PV plant should generate in a worst day scenario.
- $\eta_w$ is the energetic efficiency of the electric cabling. A typical value would be 0.9.
- $\eta_{DC}$ is the energetic efficiency of the DC/DC converters. A efficiency value for this equipment is 0.9.
- $\eta_{BAT}$ is the energetic efficiency of the battery and its charging equipment. A efficiency value for this type of equipment is 0.75.
- $L_{load}$ is the energy that the ERem system requires to operate for a day, 12.2 kWh/day as seen before.

Then, the PV plant should supply at least 20 kWh/day.

In order to design the system, we must obtain the irradiance on the panel's plane in the worst case scenario. The system is designed to provide the required power during the least productive day of the least productive month, December. To calculate the average irradiance in December on the panel's plane, we use experimental tables of average irradiance values around the world for each month of the year. In our case, in December at Janghang, South Korea, the average irradiance value is 1.94 kWh/m$^2$/day. Therefore, the panels are tilted in such a way that the production is increased in winter, the least productive month due to low irradiance. The panel plant tilt is then set 10° higher than the latitude.

Using the calculated tilt of the panels, a series of experimental data tables that relate the solar plant's tilt with the irradiance gain due to tilt is checked. In this case, in December and with a tilt 10° greater than the latitude, the irradiance is 1.7 times greater than the irradiance in the horizontal plane, which is 1.94 kWh/m$^2$/day as we calculated previously. Therefore, the average irradiance in the panel's plane is 3.3 kWh/m$^2$/day.

With this data and the generated energy required per day, we can calculate the power of the panels to be installed.

$$P_P = \frac{E_{min}}{E_{PLANE}},$$

where:

- $P_P$ is the power of the panels that should be installed (kWp).
- $E_{min}$ is the energy that the PV plant should generate in a worst case day, which is 20 kWh/day as it was calculated before.
- $E_{PLANE}$ is the daily energy that reaches the plane, where the panels have been placed. It is 3.3 kWh/m$^2$/day as it was calculated before.

Therefore, 6 kWp of panels are required at least to supply the system.

As a summary, the installation consists of:

- Panels: 6000 Wp power in a south orientation tilted 10° more than local latitude regarding the horizontal.
- 48 V and 2259 Ah 20 h discharge time battery.

This system could work non-stop all the year. Even in cloudy periods, the stops will be moderate as it can work 3 days in total darkness. Therefore, it could work normally during moderately cloudy periods.

It has to be taken in consideration which strategies can be followed in order to take advantage of the excess energy in months with highest irradiance. The exploitation strategy would be focused in the charge status of the battery. As the battery discharges, the pumping will decrease, reduce the number of the operating electrodes and/or lower the voltage of the electrodes. As the charge of the batteries reaches the maximum, the pumping would be increased, connect more electrodes and/or increase the ERem voltage within the possible values. Moreover, it would be very advisable that the control system could have access to the forecast of the area, so it could use strategies that are more aggressive if the several sunny days are to come, or on the contrary, save energy if it is cloudy.

As a summary, it is shown in the following table a comparison between the calculus methods for ERem systems with and without batteries.

| Electroremediation without batteries | Electroremediation with batteries |
|---|---|
| • Setting the power requirements in W. | • Establishing the necessary energy in Wh for a day. |
| • Obtaining the irradiation curve in the zone for the worst day of the period in which the operation will take place. | • Sizing the batteries so they work as long as required without supply from the panel plant. |
| • Establishing the power of the panels necessary to supply the system in the worst day intersecting the power needed with the irradiation curve for this day. | • Obtaining statistical data of how much energy is normally gathered during the period in which the soil will be treated. |
| | • Correcting this latter data with energy gathered if the panel was horizontal in order to take into account the panel's tilt. |
| | • Sizing the panel plant in order to obtain enough energy even in the worst month. |

### 4.2.3 Comparison Between ERem-PV Systems Without and with a Pack of Batteries

The decision whether to incorporate or not batteries depends largely on the costs and expectations of the system. In reality, there are two very different system types.

A system without batteries is small, portable and almost does not require maintenance. It can be transported to the place and left unattended until the treatment of the soil is completed. Once finished, the process can be removed, transported and installed in another location quickly. The system previously designed needs a plant consisting of 940 W panels, which occupies a panel area of 5 $m^2$. This type of system is appropriate to treat small soils spaced among themselves.

Systems with battery racks are much more expensive and voluminous system, but it operates much faster. For example, the previously designed system uses a plant consisting of 6000 Wp panels, which covers an area of 30 $m^2$ and includes a heavy battery rack that requires maintenance. This system may be more appropriate for treating larger areas, where the electrode assembly moves as small portions of contaminated soil are treated, while the solar field and the battery rack are fixed in one location.

## References


1. H.K. Hansen, L.M. Ottosen, Electrokinetic soil remediation: an overview, in *Electrokinetics Across Disciplines and Continents: New Strategies for Sustainable Development*, ed. by A. B. Ribeiro, E. P. Mateus, N. Couto, (Springer, Berlin, 2015), p. 472
2. J. Virkutyte, M. Sillanpää, P. Latostenmaa, Electrokinetic soil remediation—critical overview. Sci. Total Environ. **289**(1), 97–121 (2002). http://www.sciencedirect.com/science/article/pii/S0048969701010270
3. A.T. Yeung, Contaminant extractability by electrokinetics. Environ. Eng. Sci. **23**(1), 202–224 (2006). http://www.liebertonline.com/doi/abs/10.1089/ees.2006.23.202

4. J.M. Dzenitis, Soil chemistry effects and flow prediction in electroremediation of soil. Environ. Sci. Technol. **31**(4), 1191–1197 (1997)
5. A.T. Lima et al., Environmental electrokinetics for a sustainable subsurface. Chemosphere **181**, 122–133 (2017). https://doi.org/10.1016/j.chemosphere.2017.03.143
6. A.T. Yeung, Y.Y. Gu, A review on techniques to enhance electrochemical remediation of contaminated soils. J. Hazard. Mater. **195**, 11–29 (2011). https://doi.org/10.1016/j.jhazmat.2011.08.047
7. L. Ren et al., Enhanced electrokinetic technologies with oxidization-reduction for organically-contaminated soil remediation. Chem. Eng. J. **247**, 111–124 (2014)
8. R.T. Gill et al., Electrokinetic-enhanced bioremediation of organic contaminants: a review of processes and environmental applications. Chemosphere **107**, 31–42 (2014). https://doi.org/10.1016/j.chemosphere.2014.03.019
9. J.M. Ortiz et al., Electrodialysis of brackish water powered by photovoltaic energy without batteries: direct connection behaviour. Desalination **208**(1–3), 89–100 (2007). https://www.scopus.com/inward/record.uri?eid=2-s2.0-33947653318&doi=10.1016%252fj.desal.2006.05.026&partnerID=40&md5=f15ddc290a867b254af47cb0038eefdb
10. D. Valero et al., Electrochemical wastewater treatment directly powered by photovoltaic panels: electrooxidation of a dye-containing wastewater. Environ. Sci. Technol. **44**(13), 5182–5187 (2010). https://www.scopus.com/inward/record.uri?eid=2-s2.0-77954255453&doi=10.1021%252fes100555z&partnerID=40&md5=9f415cd84d9d1bfd49b9f4b057906c0f
11. J.M. Ortiz et al., Desalination of underground brackish waters using an electrodialysis system powered directly by photovoltaic energy. Sol. Energy Mater. Sol. Cells **92**(12), 1677–1688 (2008). https://www.scopus.com/inward/record.uri?eid=2-s2.0-55349122894&doi=10.1016%252fj.solmat.2008.07.020&partnerID=40&md5=92ec162bf3bbb78ef7649627e9c30a9e
12. J.M. Ortiz et al., Photovoltaic electrodialysis system for brackish water desalination: modeling of global process. J. Membr. Sci. **274**(1–2), 138–149 (2006). https://www.scopus.com/inward/record.uri?eid=2-s2.0-32644459581&doi=10.1016%252fj.memsci.2005.08.006&partnerID=40&md5=62c63919a1ba72f83dfddca3d2f419f8
13. D. Valero et al., Electrocoagulation of a synthetic textile effluent powered by photovoltaic energy without batteries: direct connection behaviour. Sol. Energy Mater. Sol. Cells **92**(3), 291–297 (2008). https://www.scopus.com/inward/record.uri?eid=2-s2.0-36849079908&doi=10.1016%252fj.solmat.2007.09.006&partnerID=40&md5=4d9176b6fc342243513087a2481ebafb
14. D. Valero et al., Electrochemical treatment of wastewater from almond industry using DSA-type anodes: direct connection to a PV generator. Sep. Purif. Technol. **123**, 15–22 (2014)
15. V. Montiel et al., Chapter 19—Prospective applications of renewable energy-based electrochemical systems in wastewater treatment, in *Electrochemical Water and Wastewater Treatment*, ed. by C. A. Martínez-Huitle, M. A. Rodrigo, O. Scialdone, (Butterworth-Heinemann, Oxford, 2018), pp. 513–541. http://www.sciencedirect.com/science/article/pii/B9780128131602000195
16. S. Yuan et al., Use of solar cell in electrokinetic remediation of cadmium-contaminated soil. J. Hazard. Mater. **162**(2–3), 1583–1587 (2009)
17. E. Mohamedelhassan, Laboratory study on integrated solar electrokinetic barrier for preventing cadmium contamination. Res. J. Environ. Earth Sci. **3**(5), 521–533 (2011)
18. M. Zhou et al., Removal of fluorine from contaminated soil by electrokinetic treatment driven by solar energy. Environ. Sci. Pollut. Res. **20**(8), 5806–5812 (2013). https://www.scopus.com/inward/record.uri?eid=2-s2.0-84880617583&doi=10.1007%252fs11356-013-1595-z&partnerID=40&md5=c93a8597a0624976b61127054224ce0f
19. S. Zhang et al., Electrokinetic remediation of soil containing Cr(VI) by photovoltaic solar panels and a DC-DC converter. J. Chem. Technol. Biotechnol. **90**(4), 693–700 (2015). https://www.scopus.com/inward/record.uri?eid=2-s2.0-84924283315&doi=10.1002%252fjctb.4359&partnerID=40&md5=d6d873d86876960b5f6bf63651aa876c
20. I. Hassan, E. Mohamedelhassan, E.K. Yanful, Solar powered electrokinetic remediation of Cu polluted soil using a novel anode configuration. Electrochim. Acta **181**, 58–67 (2015). https://

www.scopus.com/inward/record.uri?eid=2-s2.0-84942890996&doi=10.1016%252fj.electacta.2015.02.216&partnerID=40&md5=e8bad8a74bd126b708548c374dc2f259

21. E.-K. Jeon, S.-R. Ryu, K. Baek, Application of solar-cells in the electrokinetic remediation of As-contaminated soil. Electrochim. Acta **181**, 160–166 (2015). https://www.scopus.com/inward/record.uri?eid=2-s2.0-84945468450&doi=10.1016%252fj.electacta.2015.03.065&partnerID=40&md5=3a5dcae54ffb63279369ce11f99e5812
22. F.L. Souza et al., Solar-powered electrokinetic remediation for the treatment of soil polluted with the herbicide 2,4-D. Electrochim. Acta **190**, 371–377 (2016). https://doi.org/10.1016/j.electacta.2015.12.134
23. M.S. Godschalk, R. Lageman, Electrokinetic biofence, remediation of VOCs with solar energy and bacteria. Eng. Geol. **77**(3–4 spec. iss), 225–231 (2005)
24. I. Hassan et al., Solar power enhancement of electrokinetic bioremediation of phenanthrene by Mycobacterium pallens. Biorem. J. **21**(2), 53–70 (2017)
25. R. López-Vizcaíno et al., Scale-up on electrokinetic remediation: engineering and technological parameters. J. Hazard. Mater. **315**, 135–143 (2016)
26. R. López-Vizcaíno et al., Techno-economic analysis of the scale-up process of electrochemically-assisted soil remediation. J. Environ. Manage. **231**, 570–575 (2019). https://doi.org/10.1016/j.jenvman.2018.10.084
27. L. Van Cauwenberghe, Electrokinetics. Technology Overview Report TO-97-03. Electrokinetics: Technology Overview Report (1997), pp.1–17. https://clu-in.org/download/remed/elctro_o.pdf
28. R. Lageman (Lambda Consult) and W. Pool (Holland Environment), Electrokinetic remediation in practice experiences with field applications (2011). https://www.lambdaconsult.com/downloads/er-practice.pdf
29. W.S. Kim et al., Field application of electrokinetic remediation for multi-metal contaminated paddy soil using two-dimensional electrode configuration. Environ. Sci. Pollut. Res. **21**(6), 4482–4491 (2014)
30. W.S. Kim et al., In situ field scale electrokinetic remediation of multi-metals contaminated paddy soil: influence of electrode configuration. Electrochim. Acta **86**, 89–95 (2012). https://doi.org/10.1016/j.electacta.2012.02.078
31. J.M. Mendez Muñiz, R. Cuervo García, *Energía Solar Fotovoltaica 2º* (FC Editorial, Madrid, 2010)

# Electrokinetic Processes: Directions for Future Research and Constraints

**Karyn N. O. Silva, Suelya S. M. Paiva, and E. V. dos Santos**

## 1 Introduction

Currently, it is necessary to investigate the technical and economic feasibility of alternatives for the remediation of soils polluted by organic and inorganic compounds [1–4]. Since the industrial revolution, different chemical products derived from petroleum, agriculture, and other chemical activities have contaminated the environment in innumerable ways. Different technologies for soil remediation, including electrokinetic remediation (EKR) [5–9], have been investigated. As discussed in previous chapters, EKR has the potential to remove pollutants, sediments, and solid wastes from the contaminated soil when a gradient electric potential is applied between a set of electrodes [1].

Owing to the promising laboratory- and pilot-scale studies, EKR has attracted increased attention in the last decade [10–17]. These studies have shown that pollutants are transported direct or indirectly via electrolysis, electromigration, electroosmosis, electrophoresis, and electrical heating in EKR. During the electrolysis, water is oxidized and reduced at the anode and cathode, respectively. As a result, a high pH gradient is generated in the cathodic region, thereby resulting in the electrodeposition of metals. In electromigration, ions in the soil water move toward

K. N. O. Silva · S. S. M. Paiva
Federal University of Rio Grande do Norte, Natal, Brazil

E. V. dos Santos (✉)
School of Science and Technology, Federal University of Rio Grande do Norte, Natal, RN, Brazil
e-mail: elisamavieira@ect.ufrn.br



M. A. Rodrigo, E. V. Dos Santos (eds.), *Electrochemically Assisted Remediation of Contaminated Soils*, Environmental Pollution 30,
https://doi.org/10.1007/978-3-030-68140-1_21

electrodes depending on their charge, i.e., anions move toward the anode and cations move toward the cathode. Electroosmosis involves the transportation of water in the soil from the anode to the cathode in the presence of an electrical field. Electrophoresis involves the movement of charged particles in the soil under the influence of an electrical field. During pilot-scale studies, another important phenomenon that was observed is the significant increase in temperature near electrodes due to electrical heating. This can be attributed to the Joule–Thompson effect. The increase in temperature is proportional to the applied current intensity, distance between the anode and cathode, and soil resistance [1, 18–20].

In addition, the material and distribution of electrodes influence the performance of EKR [21]. Characteristics such as the corrosion resistance, chemical stability, current density, and mechanical stability are considered while choosing the electrode material [9, 21]. However, due to the variation of the soil pH, it is important to use inert electrodes, such as graphite, platinum, gold, and silver [1]. In contrast, in pilot studies, reliable and economically cheaper titanium, stainless steel, or plastic electrodes were used. In the case of inert electrodes, $H^+$ ions and oxygen gas are produced at the anode, whereas $OH^-$ ions and hydrogen gas are produced at the cathode. As a result, if the pH is not controlled, an acid front propagates into the soil pores from the anode, whereas the base front moves out from the cathode [22]. Moreover, electrodes can be placed either in the electrolyte or in direct contact with the soil [14, 20, 23].

Furthermore, the efficiency of removing soil contaminants using EKR also depends on the type, chemical properties, pH, and conductivity of the soil. EKR technology is found to be most suitable for remediation of low-conductive or fine-grained soil, which cannot be readily drained [15, 24, 25]. Moreover, sorption and buffering capacities are also affected by soil pH. Recent studies have shown that the soil pH should be maintained low to promote the dissolution of contaminants [25]. However, the removal of insoluble organic contaminants in the presence of an electric field is limited by their movement out of the soil. This can be achieved either using electroosmotic purging with surfactant that solubilizes the compounds [26].

Furthermore, heavy metals and solid wastes can be removed from soils under acidic conditions [15, 27]. In an acidic environment, heavy metals are extracted from the solid matrix and dissolved in the solution as cations, which migrate toward the cathode under the application of an electric field. In contrast, organic pollutants that are ionizable under prevailing soil conditions are transported by electroosmosis and electromigration [8, 28–30]. This chapter presents an overview of EKR, its fundamental applications, and the current research activities. Furthermore, the advantages and disadvantages of EKR when applied in the laboratory- and pilot-scale studies, as well as some recommendations and directions for future research, are discussed.

# 2 Factors Affecting EKR Technology

## 2.1 Soil Condition

The soil composition and its classification influence the EKR treatment process. For example, EKR is more effective on clay soils with particle sizes less than 2 μm and moderate plasticity, such as kaolinite. However, this method is not effective for soils containing high carbonate buffers, which results in high buffering capacity. Another important parameter to be considered is the soil water contents. This is because the soil moisture contributes to the conductivity required to promote electromigration [1]. To improve the removal efficiency of contaminants in EKR, different methods are often used to control the soil pH [5, 25, 31]. This is because variations in the soil pH affect the zeta potential, which in turn influences the surface charge of soils, thereby changing the direction of electroosmotic flow [18]. Therefore, a more negative soil surface zeta potential contributes more effectively to the electroosmotic flow. Full-scale studies have shown that electroosmotic flow is highly dependent on on-site geochemistry [20].

Moreover, the pH changes may also affect the soil chemistry. As a result, different chemical reactions, mainly precipitation or dissolution of salts and soil minerals, can occur. In EKR, the low pH near the anode causes the desorption and solubility of cationic metals, whereas the high pH near the cathode promotes the adsorption or precipitation of metals [15]. Meanwhile, the electroosmotic flow toward the cathode contributes to the improved elimination of cationic species either by precipitation or co-precipitation. In contrast, the high pH near the cathode promotes soil heating, especially near electrodes. This increase in temperature affects the transportation of pollutants through the volatilization of organic compounds [25].

## 2.2 Designing the Electrokinetic System

The design and implementation of an electrokinetic process to decontaminate soils is a complex task. Therefore, it is necessary to characterize interactions between the physical, chemical, and electrochemical processes that occur simultaneously, thereby simplifying the scalability of EKR technology [17, 18, 32].

**Voltage and current** The voltage and current intensities between two points in the medium are key parameters that influence the mobilization of pollutants in EKR. The electric field intensity across a contaminated site depends on system characteristics, such as the salt content, humidity, soil composition, and porosity. The applied voltage difference causes different oxidation and reduction reactions to occur at the anode and cathode, respectively. Despite providing the driving force for the transportation of pollutants, the electric field intensity also generates heat ($Q$) inside the soil owing to the Joule heating effect [1, 14, 23].

**Nature of electrodes** As described in the literature, electrodes can either be placed directly in contact with the soil to be treated or in compartments containing water or chemical agents (electrolytes and surfactants) [33]. When electrolytes are used to remediate the soil, the resistance between electrodes is decreased. Linear and radial electrode configurations with the active area located at the central region and extending to edges have been investigated [10].

## *2.3 Advantages and Disadvantages of EKR Technology*

The electrokinetic process has shown to be highly efficient in the remediation of inorganic, organic, and metalloid pollutants especially from the low-permeability soil [2, 24]. Compared to conventional soil remediation technologies, advantages of EKR technology are associated with the use of direct electric current to transport pollutants in fine-grained soils. Furthermore, during batch- and full-scale tests, EKR has been shown to generate excessive heat near electrodes. This induces higher volatilization of the organics present in the soil. Moreover, gases are generated at electrodes as a result of the electrolysis of water [19].

The electrodes can be designed to produce either a vertically uniform or non-uniform electric field, which facilitates horizontal or vertical migration of contaminants. Electrode configuration involving vertical nonuniform electric field, pollutants are transported upward to the surface of the treated soil [10]. A major disadvantage of EKR is the acidification that modifies the soil constituents, thereby affecting its zeta potential and decreasing the electroosmotic flow. Advantages and disadvantages are summarized in Table 1.

**Table 1** Summary of the main advantages and limitations of soil electrokinetic process [2, 10, 19]

| Advantages | Limitations |
|---|---|
| • One of the most effective in situ or ex situ approaches for removing pollutants from the soil, mainly in low-permeability soil. | • The solubility of pollutants in the aqueous phase is limited if the concentration of contaminants is low. Furthermore, the desorption of pollutants from the soil matrix is also limited. |
| • The applied electric field contributes to the transportation and transformation of species through the soil toward the electrode wells. | • In full-scale studies, excessive heat is generated in the electrode vicinities, thereby causing desiccation or cracking in the soil. |
| • Metal species, polar organic molecules, ionic micelles, and colloidal electrolytes can be transported using this treatment method. | • Generates some undesirable products, such as chlorine gas, as a result of electrolytic decomposition (redox reactions) of water. |
| • Do not require robust equipment, excavation, or installation of large plants, thereby reducing its implementation cost. | • Cannot apply high voltages because its efficiency is decreased due to the increase in temperature. |

## 2.4 Electrokinetic Removal of Heavy Metals in Field-Scale Tests

Several bench-scale studies have demonstrated the enhancement of removal efficiency for heavy metals using various electrolytes, pH control, complex agents, pulsating direct and alternating currents in EKR [11, 12]. Environmental disasters, such as the breakdown of a dam in Brazil that discharged 43 million cubic meters of mineral sludge, have occurred due to the soil contamination. However, only a few studies have evaluated the EKR efficiency of soils contaminated with heavy metals under a full-scale environment [34–41] and the results are listed in Table 2.

Kim et al. [42] investigated the influence of four different electrode configurations as illustrated in Fig. 1 toward the removal of Cu, Pb, and As in real time. The four-electrode configurations (A1, A2, A3, and A4) were built by varying the spacing between the anode and cathode as well as the spacing between the same polarity

**Table 2** Electrokinetic remediation of heavy metals from soils

| Pollutant/type of soil/ concentration | Process fluid: Anode | Process fluid: Cathode | Reactor volume | Observations and best results | Ref. |
|---|---|---|---|---|---|
| As/real soil/219 mg $kg^{-1}$ | Tap water | 0.1 M oxalic acid | 1 $m^3$ | DC-powered system removed 32% and solar-powered removed 27% | [35] |
| As, Cu and Pb/real soil/184; 220, 505 mg $kg^{-1}$ | 0.5 M NaOH | 0.5 M EDTA | 12 $m^3$ | As (40% removal)<br>Cu (17% removal)<br>Pb (19% removal) during 14 weeks | [37] |
| As, Cu and Pb/real soil/55, 86 and 170 mg $kg^{-1}$ | 0.5 M NaOH | 0.5 M EDTA | 8 $m^3$ | As (61% removal)<br>Cu (11% removal)<br>Pb (1% removal) | [34] |
| Cu/real soil/173 mg $kg^{-1}$ | 50 mM citric acid | Water | 20 $m^3$ | Cu (85% removal) | [36] |
| As, Cu, and Pb/real soil/11, 17 and 39 mg $kg^{-1}$ | 0.01 M NaOH | Tap water | 26 $m^3$ | As (43% removal)<br>Cu (18% removal)<br>Pb (81% removal) | [39] |
| Cr/real soil/180–1100 mg $kg^{-1}$ | Tap water | Citric acid | 64 $m^3$ | Cr (78% removal) | [38] |
| Cd/real soil/5–20 mg $kg^{-1}$ | | | | Cd (70% removal) | |
| As, Cu, and Pb/real soil/107, 191 and 325 mg $kg^{-1}$ | 0.1 M EDTA | Tap water | 268 $m^3$ | As (48.7% removal)<br>Cu (48.9% removal)<br>Pb (54.5% removal) | [41] |
| As, Cu and Pb/real soil/25, 150 and 300 mg $kg^{-1}$ | Tap water | 0.1 M EDTA | 331 $m^3$ | As (49% removal)<br>Cu (49% removal)<br>Pb (54% removal) during 24 weeks | [40] |

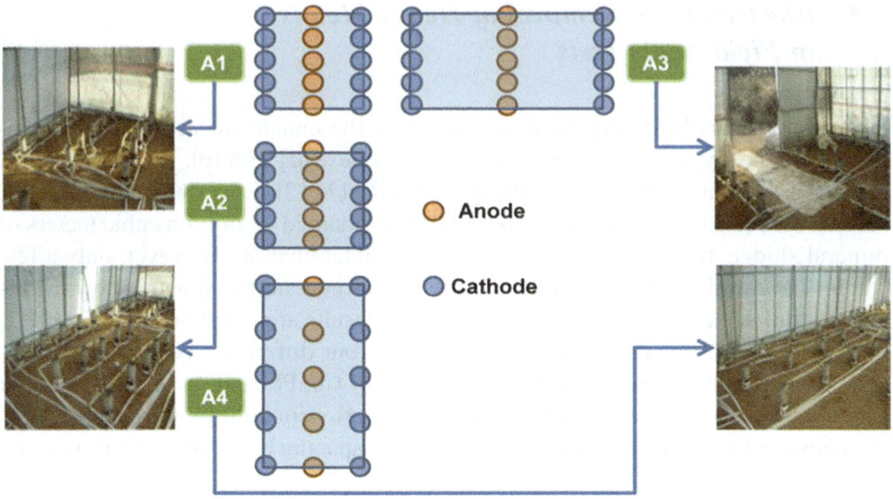


**Fig. 1** Schematic diagram of electrokinetic remediation on based different electrode configurations. (Reprinted with permission from [42])

electrodes. A constant voltage of 100 V was applied across all electrode configurations. Subsequently, the voltage gradient across A1, A2, and A4 were 1 V $cm^{-1}$, whereas it was 0.5 V $cm^{-1}$ for A3. The area covered by A1 and A2 were 4 $m^2$, and that covered by A3 and A4 were 8 $m^2$. In addition, an anolyte (0.5 M NaOH) and catholyte (0.5 M EDTA) were used. The lower removal efficiency of As, Cu, and Pb was observed across A1–A4 because of the soil heterogeneity. Among them, the best removal efficiency of 40%, 17%, and 19%, for As, Cu, and Pb, respectively, was observed in the A4 configuration after 4 weeks. Furthermore, the observed efficiencies for As, Cu, and Pb removal after 14 weeks of treatment were 21%, 12%, and 24%, respectively.

Chung et al. [36], investigated the removal of Cu (173 mg $kg^{-1}$) from spiked real soil by coupling electrokinetics with a permeable reactive barrier (PRB). They examined the performance of iron powder, zeolite, slag powder, and tire chips as reactive materials and observed the Cu sorption reactive rates of 68%, 93%, 75%, and 88%, respectively. In addition, they observed that the Cu migrated from the anode toward the cathode due to electroosmotic flow and electromigration.

EKR has shown improved soil remediation through electromigration and electroosmosis as a result of the mobilization of heavy metals. Furthermore, several researchers have shown that employing a suitable extraction agent contributes to the elimination of soil contaminants from soil by ion exchange, the redissolution of precipitates, or complexation reactions [1].

Jeon et al. [35] investigated the removal of As from the polluted soil (volume of 1 $m^3$) using a normal- and solar-powered systems. They demonstrated the As removal efficiency of 32% and 27% for DC- and solar-powered systems, respectively, after 5 weeks of treatment. In addition, the energy expenditure of the solar-powered system was 50% less than that of the DC-powered system. Therefore, the solar-powered system owing to its renewable energy source and environmental compatibility has shown high potential for remediation of soils contaminated with heavy metals.

### 2.5 *Electrokinetic Remediation of Organic Pollutants from Contaminated Soils*

A few review articles on the EKR of soils contaminated with organic compounds, such as pesticides [1], polycyclic aromatic hydrocarbons (PAHs) [41], organochlorinated hydrocarbons [18], and total petroleum hydrocarbon (TPH) [3] have been published. The volatile and/or soluble organic pollutants were relatively easier to remove using EKR when compared to other simpler organic compounds. However, it is imperative to understand the phenomena that affect the EKR treatment efficiencies. Table 3 lists the EKR performance for soils contaminated with different pollutants obtained from various pre-pilot and pilot-scale studies.

**Table 3** Electrokinetic remediation of organic pollutants from soils

| Pollutant/type of soil/ concentration | Process fluid | | Reactor volume | Observations and best results | Ref. |
|---|---|---|---|---|---|
| | Anode | Cathode | | | |
| PHE/ kaolinite/600 mg $kg^{-1}$ | Tap water | 0.07 M SDS[a] Concentrated | 0.175 $m^3$ | PHE (80% removal) | [43] |
| OXY/ kaolinite/30 mg $kg^{-1}$ | Tap water | Tap water | 0.175 $m^3$ | OXY (94% removal) | [26] |
| PNP[b]/real soil/370 mg $kg^{-1}$ | Tap water | Tap water | 0.453 $m^3$ | PNP (98% removal) | [17] |
| Atrazine/ kaolinite/2 mg $kg^{-1}$ | Tap water | Tap water | 0.551 $m^3$ | Atrazine (60% removal) | [44] |
| Gasoline/real Vertisol/1126 mg $kg^{-1}$ | 0.7 μM NaOH | 0.7 μM NaOH | 3.3 $m^3$ | Gasoline (80% removal) | [45] |
| TCE[c]/real soil/1500 mg $kg^{-1}$ | Tap water | Tap water | 4.73 $m^3$ | TCE (99% removal)<br>The removal efficiency influenced by electric heating of soil | [46] |
| OXY and 2,4-D/ kaolinite/<br>5 and 6 mg $kg^{-1}$ | Tap water | Tap water | 32 $m^3$ | OXY (92.9% removal)<br>2,4-D (90.78% removal) | [47] |
| TCE/real soil/507 mg $kg^{-1}$ | Tap water | Tap water | 159 $m^3$ | TCE (98% removal) | [18] |
| TCE/real soil/507 mg $kg^{-1}$ | Tap water | Tap water | 806 $m^3$ | TCE (99% removal) | [32] |

[a]Sodium dodecyl sulfate
[b]PNP
[c]TCE

Surfactants have shown to enhance soil remediation and produce a stable contaminant in water, thereby favoring the emulsion generation. Lopez-Vizcaíno et al. [43] investigated the applicability of the electrokinetic soil flushing processes toward the removal of phenanthrene (PHE) from the polluted soil using an anionic surfactant (sodium dodecyl sulfate). The primary transport mechanisms of PHE removal were desorption, dragging, and electrophoresis occurring due to the electrical heating of soil, electroosmotic flow in cathodic wells, and surfactant interactions in anodic wells, respectively. They demonstrated 25% PHE removal and approximately 350 mg $m^{-2}$ $day^{-1}$ of PHE volatilization. However, after remediation, 600 mg $kg^{-1}$ of surfactant was observed in the soil. This can be related to the high surfactant concentration used for soil flushing.

Meanwhile, electro-bioremediation of soils contaminated with pesticides was investigated by Barba et al. [47]. They demonstrated the remediation of nonpolar (oxyfluorfen) and polar (2,4-dichlorophenoxyacetic acid) pesticides with a soil-treatment capacity of 32 $m^3$. However, no significant difference in the removal of these pesticides was observed using both bioremediation and electro-bioremediation. About 9% and 7% of oxyfluorfen and 2,4-D, respectively, remained in the soil after 30 days of EKR treatment. These values were similar to those obtained using a nonbiological EKR fence process (13% of oxyfluorfen and 10% of 2,4-D). The mechanism responsible for removing both pesticides was found to be the volatilization.

Maini et al. [41] investigated the removal of PAHs, benzene, toluene, ethylbenzene, xylenes (BTEX), Pb, Zn, Mn, Cu, and As using EKR by performing bench and pre-pilot tests for 23 days and 112 days, respectively. The study compared the scale and distribution of electrodes using six electrode feeding systems, which were eventually replaced with a platinized titanium rod inserted directly into the carbon felt. In the bench tests performed using planar electrodes, Zn and Pb moved toward the soil at the cathode, whereas PAH next to the cathode was decreased by 94% after 23 days of treatment. A larger scale test was developed using a hexagonal array of tubular anodes surrounded by a central tubular cathode. The movement of organic compounds and heavy metals due to electroosmotic flow in the direction of the cathode was observed. After 22 days, PAH and BTEX compounds were reduced from 720 to 5 mg $kg^{-1}$. Moreover, granular activated carbon (GAC) columns were used to recover all the organic compounds.

## 3 Integrated Remediation Technologies

Electrokinetic technology has been effectively used in the remediation of organic and inorganic compounds, particularly in the removal of polar organic contaminants, such as diesel, petroleum, TCE, pesticides, and phenol [18, 43, 45]. Nevertheless, the removal of hydrophobic organic contaminants from subsurface environments with conventional EKR is difficult owing to their low solubility and slow desorption rates. Soil remediation in complex sites can be achieved using innovative and hybrid remediation techniques.

Though the soil remediation has mainly been investigated using EKR technology, EKR integrated with physical, chemical, or biological technologies has also been developed [8, 9, 19, 24, 25]. Recently, Popescu et al. reported the combination of soil flushing and simultaneous degradation of organic compounds present in the soil using EK–Fenton treatment [48]. They investigated the influence of hydrogen peroxide dosage, Fe concentration, and porosity of the soil contaminated with Rhodamine B and found that the higher the hydrogen peroxide dosage (7–10%), the faster the color removal rate. Moreover, the study confirmed that the addition of citric acid contributes to the Fenton catalyst availability and electroosmotic transport of active species. Furthermore, they showed that the removal efficiency of diesel fuel was approximately 78% using EKR, as opposed to 87% obtained using the EK–Fenton method.

Paixão et al. [9] studied the elimination of petroleum from kaolinite using Fe electrodes and different electrolytes (tap water, $H_2O_2$, and citric acid) based on the EK–Fenton technology. They observed that $Fe^{2+}$ ions generated by magnetite ($Fe_3O_4$) electrodes reacted with $H_2O_2$ to produce hydroxyl radicals, thereby resulting in the TPH removal of 89% in the presence of citric acid. However, when the pH was not controlled, only 27% of the hydrocarbons were removed after 15 days of treatment. This can be attributed to the precipitation of Fe ions into $Fe(OH)_2$ and $Fe(OH)_3$.

The EKR approach has shown to improve the soil washing results through the electrically induced mobilization of metal species present in the soil. Several research groups have observed that suitable extraction agents can help eliminate pollutants from the soil by promoting ion exchange, redissolution of precipitates, or dragging pollutants through complexation reactions [48]. As a result, the contamination is transferred from the soil to the pore water, thereby creating a wastewater treatment problem.

Another soil remediation process is in situ chemical oxidation (ISCO), wherein chemical oxidants, such as hydrogen peroxide, permanganate, and ozone are widely used. However, their practical application is limited because chemical oxidants have low stability (hydrogen peroxide and ozone) and high affinity for natural soil organics (permanganate) [49]. On the other hand, though persulfate has received increased interest in soil remediation over the last decade, its transportation in low-permeability soil is limited. This results in the lower degradation efficiency of soil pollutants. Therefore, EKR could promote the transportation of persulfate in low-permeability soil, and different activation methods using zerovalent Fe, citric-chelated $Fe^{2+}$, Fe electrode, alkaline pH, and peroxide have been evaluated. The activity of persulfate delivered during EKR is described in Eqs. (1)–(4) [49].

$$S_2O_8^{2-} \xrightarrow{\text{Heat}} 2SO_4^{\bullet -} \tag{1}$$

$$Mn^{+} + S_2O_8^{2-} \rightarrow Mn^{+1} + SO_4^{\bullet -} + SO_4^{\bullet 2-} \tag{2}$$

$$S_2O_8^{2-} + \bullet OH \rightarrow SO_4^{\bullet -} + SO_4^{\bullet 2} + 1/2O_2 + H^+ \quad (3)$$

$$SO_4^{\bullet -} + OH^- \rightarrow SO_4^{\bullet 2} + \bullet OH \quad (4)$$

Fan et al. [49] explored an electrokinetic approach coupled with persulfate to remediate polychlorinated biphenyls (PCBs) using the above-mentioned activation methods. They showed that the removal efficiency of PCBs from the soil followed the activation order of alkaline (40%) > peroxide (35%) > citric acid chelated $Fe^{2+}$ (34%) > zerovalent Fe (32%) > without activation (31%) > Fe electrode (30%). Furthermore, the activation was highly dependent on the ratio of activating agent to persulfate concentrations.

Electrokinetics coupled with PRB (EK–PRB) could represent the most sustainable approach for removal of pesticides from contaminated soils [50]. Advantages of this coupling method over conventional pump-and-treat methods include (1) degradation or immobilization of pollutants can occur in situ without bringing them up to the surface, thereby avoiding potential cross-media contamination, (2) no need for transportation, storage, or disposal; (3) does not require continuous input energy, (4) degradation of contaminants is achieved along with a change in phase, (5) does not require the discharge of effluent, which eliminates the related technical and regulatory problems, and (6) relatively low cost. However, due to the lack of long-term data, its cost-effectiveness has not yet been proven [64].

Barba et al. [51] reported an 85% removal efficiency of 2,4-D after 10 days of EK–PRB treatment using activated sludge (biological barrier) and synthetic groundwater as electrolytes. Authors demonstrated the electro-bioremediation process in two stages. First, a fixed-bed biofilm reactor was used to biodegrade 2,4-D in wastewater. Second, the biofilm was added to the contaminated soil and used as a biological barrier. They obtained a 25% removal efficiency, which proved that the EK transport mechanism increases the contact between contaminants and microorganisms during bioremediation.

The elimination method of organic and/or inorganic compounds from contaminated soil using the combination of plants and electrokinetics has been developed. This integrated technology is appealing because of its low cost and energy requirements, ease of field implementation, and enhancement of soil properties. Acosta-Santoyo et al. [3] demonstrated the removal of hydrocarbons from the spiked real soil by coupling electrokinetic and phytoremediation technologies under laboratory, pilot, and field environments. The seed germination and growth of maize (Zea mays L.) were measured using a 2D circular arrangement of electrodes ($IrO_2$-$Ta_2O_5$|Ti||Ti) Fig. 2.

**Laboratory scale** They observed that at 0.2 V $cm^{-1}$ the seed germination reached 83% and 100% on Days 5 and 6, respectively. However, seeds did not germinate after the fifth day of treatment at 0.8 V $cm^{-1}$.

**Pilot-scale** The seed germination rate was measured at 90% and 32% in clean and hydrocarbon-contaminated soils, respectively. When electro-phytoremediation technology was employed with the treatment occurring at 0.2 V $cm^{-1}$ for 4 h a day, the seed germination rate was 63%.

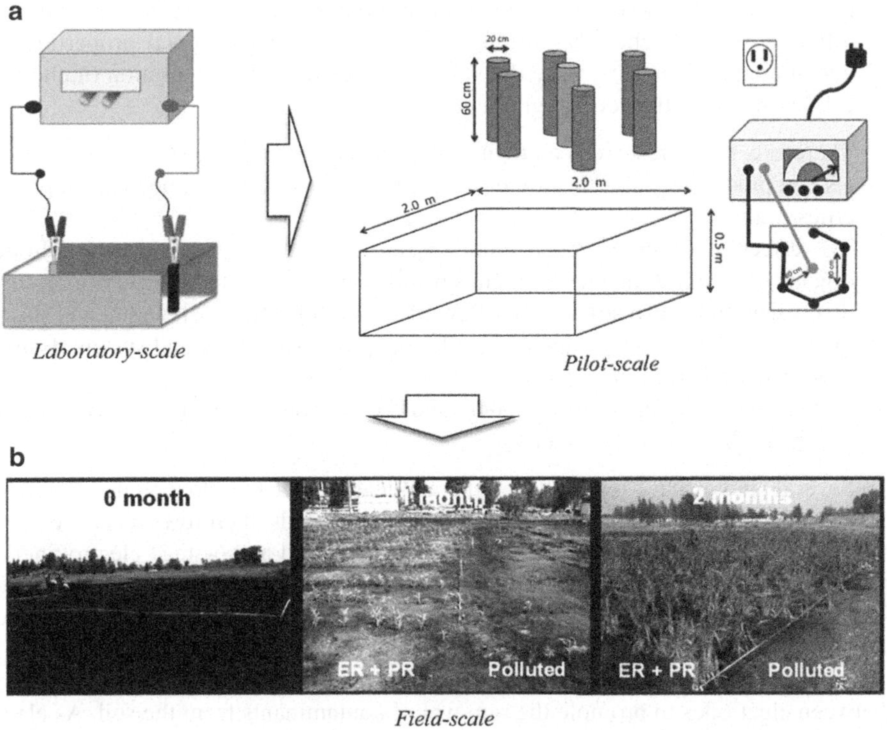


**Fig. 2** Schematic systems of laboratory (**a**) and pilot-scale (**b**) for electro-culture of maize seeds. (Reprinted with permission from [3])

**Field-scale** Electrokinetic treatment was performed at an oil refinery in Guanajuato, Mexico for 30 days. The hydrocarbon removal rate after the EKR treatment was 60%, whereas 90% of hydrocarbons were removed after the application of integrated electro-phytoremediation technology.

# 4 Research and Directions for Future Research on Electrokinetic Remediation

In the last few decades, the performance of EKR for removing inorganic and organic compounds has been successfully demonstrated in bench-, field-, and pilot-scale tests. From this chapter, it is clear that some of the discussed laboratory and pilot applications have potential technical effectiveness toward real-time environmental remediation applications [1]. The key points are summarized below.

- Most of the studies focused on strengthening and improving the mechanical and engineering characteristics of low conductivity soil. From a soil remediation perspective, it is important to study how the complex microstructural behavior of different soil and chemical species influence the environmental properties of contaminants. This is because the soil properties limit the integration of chemical, hydraulic, and electrical gradients.

- Therefore, future research can focus on the development and testing of new models that predict the modification of the chemical, physical, and mechanical properties of the soil.
- The effect of different enhancement agents toward the removal of insoluble organic compounds and heavy metals needs to be investigated. Furthermore, further research that investigates the effect of soil composition, pH, nature and spacing of electrodes, voltage and current levels, processing time, and system design needs to be carried out.
- In addition, it is important to understand the transport mechanisms involved in removing the soil contaminants.

Several studies have reported that the application of EKR alone is not effective for the removal of some hydrophobic organic compounds; therefore, it is necessary to use integrated technologies. Silva et al. [7] combined a two-stage electrokinetic process and demonstrated enhanced electrolyte (to favor the transportation of the dye from the soil) and electrochemical oxidation of the wastewater without the addition of external oxidants.

Another question that often arises is when should the electric field be applied between electrodes to promote the removal of contaminants from the soil. As electricity is required for the transportation of pollutants, the use of renewable energy as an alternative source can reduce the cost of EKR treatment and make it environment friendly. Furthermore, the use of renewable energies is more sustainable, as it would involve minimum water and carbon footprints. Rodrigo's group [52, 53] investigated the application of EKR powered by standalone/off-grid PV modules for the treatment of pesticide-contaminated soils in a laboratory- and pre-pilot settings Fig. 3.

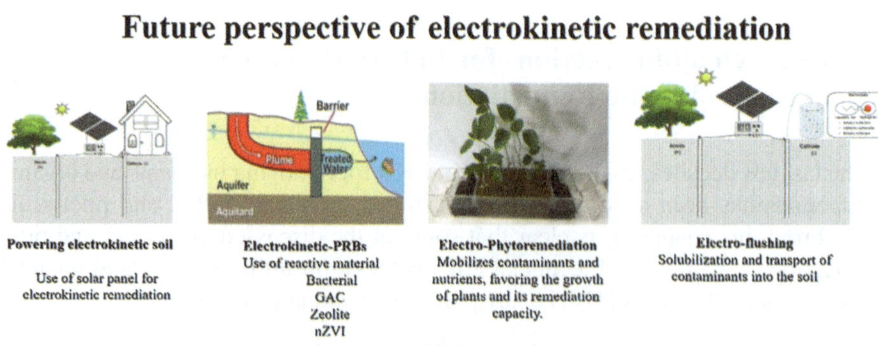


**Fig. 3** Electrokinetic remediation coupled with other technologies

EK–PRB technology has a wide range of applications and is more beneficial because of its high economic benefits and less secondary contamination. Furthermore, compared to conventional technologies, the major advantages of EK–PRB are its potential application toward remediation of low-permeability soil and in situ remediation. Though effective at a laboratory scale, its applicability in full scale has to be evaluated. Furthermore, it is necessary to understand the position of reactive barriers using a mathematical model.

Electro-phytoremediation has been attractive because of its effective remediation of both heavy metals and organic compounds [8, 44]. The advantage includes the mobilization of contaminants and nutrients, which favors the growth of plants and remediation efficiency. However, it is important to investigate the influence of the electric field intensity and chemical nature of electrodes in scaled-up testing environments.

## 5 Conclusion

EKR technology has a wide range of applications toward the remediation of different contaminants present in soils. As discussed in this chapter, it is important to consider the possibility of coupling EKR with other processes. Furthermore, for the selection of suitable soil remediation technology, understanding the influence of soil pollutants, pH, water content, electrode materials, electric field, types of organic contaminants, and their solubility is important. In the case of highly contaminated soil, it is not recommended to use alone biodegradation technology. However, EKR coupled with soil flushing can be employed. Finally, additional studies are necessary to understand the importance of the involved transport mechanisms and to develop specific methods for contaminants present at the remediation site.

**Acknowledgments** The authors gratefully acknowledge the financial support from the National Council for Scientific and Technological Development (CNPq—430121/2016-4 and CNPq—306323/2018-4) and, L'ORÉAL-ABC-UNESCO for Women in Science. Fundação de Amparo à Pesquisa do Estado de São Paulo—FAPESP (Proc. FAPESP 2014/50945-4).

## References

1. M.A. Rodrigo, N. Oturan, M.A. Oturan, Electrochemically assisted remediation of pesticides in soils and water: a review. Chem. Rev. **114**, 8720–8745 (2014)
2. Y. Song, G. Fang, C. Zhu, F. Zhu, S. Wu, N. Chen, T. Wu, Y. Wang, J. Gao, D. Zhou, Zero-valent iron activated persulfate remediation of polycyclic aromatic hydrocarbon-contaminated soils: an in situ pilot-scale study. Chem. Eng. J. **355**, 65–75 (2019)
3. G. Acosta-Santoyo, S. Solís, G. Hernández-Silva, J. Cárdenas, Z. Plank, E. Bustos, Analysis of the biological recovery of soils contaminated with hydrocarbons using an electrokinetic treatment. J. Hazard. Mater. **371**, 625–633 (2019)

4. Y.B. Acar, A.N. Alshawabkeh, Principles of electrokinetic remediation. Environ. Sci. Technol. **27**, 2638–2647 (1993)
5. E.V. Dos Santos, C. Sáez, P. Cañizares, C.A. Martínez-Huitle, M.A. Rodrigo, Reversible electrokinetic adsorption barriers for the removal of atrazine and oxyfluorfen from spiked soils. J. Hazard. Mater. **322**, 413–420 (2017)
6. E.V. Dos Santos, S. Ferro, M. Vocciante, Electrokinetic remediation, in *The Handbook of Environmental Remediation: Classic and Modern Techniques*, (The Royal Society of Chemistry, London, 2020) 600 p
7. S.S.M. Paiva, I.B. Silva, E.C.M.M. Santos, I.M.V. Rocha, C.A. Martínez-Huitle, E.V. Santos, Coupled electrochemical processes for removing dye from soil and water. J. Electrochem. Soc. **165**(9), E318–E324 (2018)
8. I.M.V. Rocha, K.N.O. Silva, D.R. Silva, C.A. Martínez-Huitle, E.V. Santos, Coupling electrokinetic remediation with phytoremediation for depolluting soil with petroleum and the use of electrochemical technologies for treating the effluent generated. Sep. Purif. Technol. **208**, 194–200 (2019)
9. I.C. Paixão, R. López-Vizcaíno, A.M.S. Solano, C.A. Martínez-Huitle, V. Navarro, M.A. Rodrigo, E.V. dos Santos, Electrokinetic-Fenton for the remediation low hydraulic conductivity soil contaminated with petroleum. Chemosphere **248**, 126029 (2020)
10. A. Alshawabkeh, R.J. Gale, E. Ozsu-Acar, R.M. Bricka, Optimization of 2-D electrode configuration for electrokinetic remediation. J. Soil Contam. **8**, 617–635 (1999)
11. S.W. Park, J.Y. Lee, J.S. Yang, K.J. Kim, K. Baek, Electrokinetic remediation of contaminated soil with waste-lubricant oils and zinc. J. Hazard. Mater. **169**, 1168–1172 (2009)
12. M.T. Ricart, M. Pazos, S. Gouveia, C. Cameselle, M.A. Sanromán, Removal of organic pollutants and heavy metals in soils by electrokinetic remediation. J. Environ. Sci. Health A Tox. Hazard. Subst. Environ. Eng. **43**, 871–875 (2008)
13. Y.B. Acar, A.N. Alshawabkeh, Electrokinetic remediation. I: Pilot-scale tests with lead-spiked kaolinite. J. Geotech. Eng. **122**, 173–185 (1996)
14. S. Pamukcu, C.P. Huang, In-situ remediation of contaminated soils by electrokinetic processes, in *Hazardous and Radioactive Waste Treatment Technologies Handbook*, (CRC Press, Boca Raton, FL, 2001)
15. D. Rosestolato, R. Bagatin, S. Ferro, Electrokinetic remediation of soils polluted by heavy metals (mercury in particular). Chem. Eng. J. **264**, 16–23 (2015)
16. A.N. Alshawabkeh, R.M. Bricka, D.B. Gent, Pilot-scale electrokinetic cleanup of lead-contaminated soils. J. Geotech. Geoenviron. Eng. **131**, 283–291 (2005)
17. S. Ho, P.W. Athmer, A.P. Shapiro, Scale-up aspects of the Lasagna super (TM) process for in situ soil decontamination. J. Hazard. Mater. **55**, 39–60 (1997)
18. S.V. Ho, C. Athmer, P.W. Sheridan, B.M. Hughes, R. Orth, D. McKenzie, P.H. Brodsky, A. Shapiro, R. Thornton, J. Salvo, D. Schultz, R. Landis, R. Griffith, S. Shoemaker, The lasagna technology for in situ soil remediation. 1. Small field test. Environ. Sci. Technol. **337**, 1086–1091 (1999)
19. R. López-Vizcaíno, V. Navarro, M.J. León, C. Risco, M.A. Rodrigo, C. Sáez, P. Cañizares, Scale-up on electrokinetic remediation: engineering and technological parameters. J. Hazard. Mater. **315**, 135–143 (2016)
20. R. Lopez-Vizcaíno, C. Risco, J. Isidro, S. Rodrigo, C. Saez, P. Cañizares, V. Navarro, M.A. Rodrigo, Scale-up of the electrokinetic fence technology for the removal of pesticides. Part II: Does size matter for removal of herbicides? Chemosphere **166**, 549–555 (2017)
21. E. Méndez, M. Pérez, O. Romero, E.D. Beltrán, S. Castro, J.L. Corona, A. Corona, M.C. Cuevas, E. Bustos, Effects of electrode material on the efficiency of hydrocarbon removal by an electrokinetic remediation process. Electrochim. Acta **86**, 148–156 (2012)
22. J. Virkutyte, M. Sillanpää, P. Latostenmaa, Electrokinetic soil remediation - Critical overview. Sci. Total Environ. **289**, 97–121 (2002)

23. C. Risco, S. Rodrigo, R. López-Vizcaíno, A. Yustres, C. Sáez, P. Cañizares, V. Navarro, M.A. Rodrigo, Electrochemically assisted fences for the electroremediation of soils polluted with 2,4-D: a case study in a pilot plant. Sep. Purif. Technol. **156**, 234–241 (2015)
24. A.T. Yeung, Y.-Y. Gu, A review on techniques to enhance electrochemical remediation of contaminated soils. J. Hazard. Mater. **195**, 11–29 (2011)
25. R. Lopez Vizcaíno, A. Yustres, L. Asensio, C. Saez, P. Canizares, M.A. Rodrigo, V. Navarro, Enhanced electrokinetic remediation of polluted soils by anolyte pH conditioning. Chemosphere **199**, 477–485 (2018)
26. C. Risco, S. Rodrigo, R. López Vizcaíno, A. Yustres, C. Saez, P. Cañizares, V. Navarro, M.A. Rodrigo, Removal of oxyfluorfen from spiked soils using electrokinetic soil flushing with linear rows of electrodes. Chem. Eng. J. **294**, 65–72 (2016)
27. K.N.O. Silva, S.S.M. Paiva, F.L. Souza, D.R. Silva, C.A. Martínez-Huitle, E.V. Santos, Applicability of electrochemical technologies for removing and monitoring Pb2+ from soil and water. J. Electroanal. Chem. **816**, 171–178 (2018)
28. R. López-Vizcaíno, C. Sáez, P. Cañizares, V. Navarro, M.A. Rodrigo, Influence of the type of surfactant on the mobility of flushing fluids for electro-remediation processes. J. Environ. Sci. Health A Tox. Hazard. Subst. Environ. Eng. **46**, 2148–2156 (2011)
29. C. Yuan, C.H. Hung, W.L. Huang, Enhancement with carbon nanotube barrier on 1,2-Dichlorobenzene removal from soil by surfactant-assisted electrokinetic (SAEK) process - the effect of processing fluid. Sep. Sci. Technol. **44**, 2284–2303 (2009)
30. E.V. Dos Santos, F. Souza, C. Saez, P. Cañizares, M.R.V. Lanza, C.A. Martinez-Huitle, M.A. Rodrigo, Application of electrokinetic soil flushing to four herbicides: a comparison. Chemosphere **153**, 205–211 (2016)
31. K.R. Reddy, Electrokinetic remediation of soils at complex contaminated sites: technology status, challenges, and opportunities, in *Coupled Phenomena in Environmental Geotechnics*, (Taylor & Francis Group, London, 2013)
32. S.V. Ho, C. Athmer, P.W. Sheridan, B.M. Hughes, R. Orth, D. Mckenzie, P.H. Brodsky, A.M. Shapiro, T.M. Sivavec, J. Salvo, D. Schultz, R. Landis, R. Griffith, S. Shoemaker, The lasagna technology for in situ soil remediation. 2. Large field test. Environ. Sci. Technol. **33**, 1092–1099 (1999)
33. O. Cuevas, R.A. Herrada, J.L. Corona, M.G. Olvera, S. Sepúlveda-Guzmán, I. Sirésd, E. Bustos, Assessment of IrO2-Ta2O5|Ti electrodes for the electrokinetic treatment of hydrocarbon-contaminated soil using different electrode arrays. Electrochim. Acta **208**, 282–287 (2016)
34. W.S. Kim, S.O. Kim, K.W. Kim, Enhanced electrokinetic extraction of heavy metals from soils assisted by ion exchange membranes. J. Hazard. Mater. **118**, 93–102 (2005)
35. E.K. Jeon, S.R. Ryu, K. Baek, Application of solar-cells in the electrokinetic remediation of As-contaminated soil. Electrochim. Acta **181**, 160–166 (2015)
36. H.I. Chung, Field applications on electrokinetic reactive pile technology for removal of cu from in-situ and excavated soils. Sep. Sci. Technol. **44**, 2341–2353 (2009)
37. W.S. Kim, E.K. Jeon, J.M. Jung, H.B. Jung, S.H. Ko, C.I. Seo, K. Baek, Field application of electrokinetic remediation for multi-metal contaminated paddy soil using two-dimensional electrode configuration. Environ. Sci. Pollut. Res. **21**, 4482–4491 (2014)
38. D.B. Gent, R.M. Bricka, A.N. Alshawabkeh, S.L. Larson, G. Fabian, S. Granade, Bench- and field-scale evaluation of chromium and cadmium extraction by electrokinetics. J. Hazard. Mater. **110**, 53–62 (2004)
39. B.K. Kim, G.Y. Park, E.K. Jeon, J.M. Jung, H.B. Jung, S.H. Ko, K. Baek, Field application of in situ electrokinetic remediation for As-, Cu-, and Pb-contaminated paddy soil topical collection on remediation of site contamination. Water Air Soil Pollut. **224**, 1698.1–1698.10 (2013)
40. E.K. Jeon, J.M. Jung, S.R. Ryu, K. Baek, In situ field application of electrokinetic remediation for an As-, Cu-, and Pb-contaminated rice paddy site using parallel electrode configuration. Environ. Sci. Pollut. Res. **22**, 15763–15771 (2015)

41. G. Maini, A.K. Sharman, C.J. Knowles, G. Sunderland, S.A. Jackman, Electrokinetic remediation of metals and organics from historically contaminated soil. J. Chem. Technol. Biotechnol. **75**, 657–664 (2000)
42. W.-S. Kim, G.-Y. Park, D.-H. Kim, H.-B. Jung, S.-H. Ko, K. Baek, In situ field scale electrokinetic remediation of multi-metals contaminated paddy soil: influence of electrode configuration. Electrochim. Acta **86**, 89–95 (2012)
43. R. López-Vizcaíno, J. Alonso, P. Cañizares, M.J. León, V. Navarro, M.A. Rodrigo, C. Sáez, Electroremediation of a natural soil polluted with phenanthrene in a pilot plant. J. Hazard. Mater. **265**, 142–150 (2014)
44. V. Sánchez, F.J. López-Bellido, M.A. Rodrigo, F.J. Fernández, L. Rodríguez, A mesocosm study of electrokinetic-assisted phytoremediation of atrazine-polluted soils. Sep. Purif. Technol. **233**, 116044 (2020)
45. B. Ochoa, L. Ramos, A. Garibay, M. Perez-Corona, M.C. Cuevas, J. Cardenas, M. Teutli, E. Bustos, Electrokinetic treatment of polluted soil at pilot level coupled to an advanced oxidation process of its wastewater. Phys. Chem. Earth **91**, 68e76 (2016)
46. C. Athmer, In-situ remediation of TCE in clayey soils. Soil Sedim. Contam. Int. J. **13**, 381–390 (2004)
47. S. Barba, R. López-Vizcaíno, C. Sáez, J. Villaseñor, P. Cañizares, V. Navarro, M. Rodrigo, Electro-bioremediation at the prototype scale: what it should be learned for the scale-up. Chem. Eng. J. **334**, 2030–2038 (2017)
48. M. Popescu, E. Rosales, C. Sandu, J. Meijide, M. Pazos, G. Lazar, M.A. Sanromán, Soil flushing and simultaneous degradation of organic pollutants in soils by electrokinetic-Fenton treatment. Process. Saf. Environ. Prot. **108**, 99–107 (2017)
49. G.-P. Fan, L. Cang, G.-D. Fang, W.-X. Qin, L.-Q. Ge, D.-M. Zhou, Electrokinetic delivery of persulfate to remediate PCBs polluted soils: effect of injection spot. Chemosphere **117**, 410–418 (2014)
50. D.C. Andrade, E.V. Dos Santos, Combination of electrokinetic remediation with permeable reactive barriers to remove organic compounds from soils. Curr. Opin. Electrochem. **22**, 136–144 (2020)
51. S. Barba, M. Carvela, J. Villaseñor, M.A. Rodrigo, P. Cañizares, Fixed-bed biological barrier coupled with electrokinetics for the in situ electrobioremediation of 2,4-dichlorophenoxyacetic acid polluted soil. J. Chem. Technol. Biotechnol. **94**, 2684–2692 (2019)
52. F.L. Souza, M.R.V. Lanza, J. Llanos, C. Sáez, M.A. Rodrigo, P. Cañizares, A windpowered BDD electrochemical oxidation process for the removal of herbicides. J. Environ. Manag. **158**, 36–39 (2015)
53. S.O. Ganiyu, C.A. Martínez-Huitle, M.A. Rodrigo, Renewable energies driven electrochemical wastewater/soil decontamination technologies: a critical review of fundamental concepts and applications. Appl. Catal. B Environ. **270**, 118857 (2020)
54. C. Yuan, C.H. Weng, Electrokinetic enhancement removal of heavy metals from industrial wastewater sludge. Chemosphere **65**, 88–96 (2006)

# Index





M. A. Rodrigo, E. V. Dos Santos (eds.), *Electrochemically Assisted Remediation of Contaminated Soils*, Environmental Pollution 30,
https://doi.org/10.1007/978-3-030-68140-1

**R**

## U



## V



## W

The manufacturer's authorised representative in the EU is Springer Nature Customer Service Centre GmbH, Europaplatz 3, 69115 Heidelberg, Germany. If you have any concerns regarding our products, please contact ProductSafety@springernature.com

Printed and bound by CPI Group (UK) Ltd, Croydon, CR0 4YY
30/09/2026
02225495-0015